WILLIAM F. MAAG LIBRARY
YOUNGSTOWN STATE UNIVERSITY

ANNUAL REVIEW OF PHYSICAL CHEMISTRY

EDITORIAL COMMITTEE (1988)

C. AUSTEN ANGELL
GERALD T. BABCOCK
JOSEPH E. DEMUTH
JAMES L. KINSEY
DONALD H. LEVY
C. BRADLEY MOORE
ROBERT J. SILBEY
HERBERT L. STRAUSS

Responsible for the organization of Volume 39
(Editorial Committee, 1986)

HANS C. ANDERSEN
C. AUSTEN ANGELL
GERALD T. BABCOCK
JAMES L. KINSEY
DONALD H. LEVY
C. BRADLEY MOORE
HERBERT L. STRAUSS
JOHN T. YATES
CHARLES W. BAUSCHLICHER, JR. (Guest)
I. D. KUNTZ (Guest)

Production Editor	SUZANNE PRIES COPENHAGEN
Indexing Coordinator	MARY A. GLASS
Subject Indexer	STEVEN SORENSEN

ANNUAL REVIEW OF PHYSICAL CHEMISTRY

VOLUME 39, 1988

HERBERT L. STRAUSS, *Editor*
University of California, Berkeley

GERALD T. BABCOCK, *Associate Editor*
Michigan State University

C. BRADLEY MOORE, *Associate Editor*
University of California, Berkeley

ANNUAL REVIEWS INC   4139 EL CAMINO WAY   P.O. BOX 10139   PALO ALTO, CALIFORNIA 94303-0897

ANNUAL REVIEWS INC.
Palo Alto, California, USA

COPYRIGHT © 1988 BY ANNUAL REVIEWS INC., PALO ALTO, CALIFORNIA, USA. ALL RIGHTS RESERVED. The appearance of the code at the bottom of the first page of an article in this serial indicates the copyright owner's consent that copies of the article may be made for personal or internal use, or for the personal or internal use of specific clients. This consent is given on the condition, however, that the copier pay the stated per-copy fee of $2.00 per article through the Copyright Clearance Center, Inc. (21 Congress Street, Salem, MA 01970) for copying beyond that permitted by Sections 107 or 108 of the US Copyright Law. The per-copy fee of $2.00 per article also applies to the copying, under the stated conditions, of articles published in any *Annual Review* serial before January 1, 1978. Individual readers, and nonprofit libraries acting for them, are permitted to make a single copy of an article without charge for use in research or teaching. The consent does not extend to other kinds of copying, such as copying for general distribution, for advertising or promotional purposes, for creating new collective works, or for resale. For such uses, written permission is required. Write to Permissions Dept., Annual Reviews Inc., 4139 El Camino Way, P.O. Box 10139, Palo Alto, CA 94303-0897 USA.

International Standard Serial Number: 0066-426X
International Standard Book Number: 0-8243-1039-X
Library of Congress Catalog Card Number: A-51-1658

Annual Review and publication titles are registered trademarks of Annual Reviews Inc.

Annual Reviews Inc. and the Editors of its publications assume no responsibility for the statements expressed by the contributors to this *Review*.

TYPESET BY AUP TYPESETTERS (GLASGOW) LTD., SCOTLAND
PRINTED AND BOUND IN THE UNITED STATES OF AMERICA

PREFACE

This volume presents the customary wide range of topics of interest to physical chemists. The Committee attempts to achieve coverage representative of the field over the span of several volumes, and the current volume advances that aim. The range of states of matter covered has remained remarkably stable, though somewhat cyclic, over the nearly 40 years that this series has been published. For example, "colloids" are mentioned in the Table of Contents after an absence of nearly 30 years. In contrast, techniques have become more sophisticated, a trend illustrated in this volume by articles on demonstrating the uses of molecular beam techniques. Another trend is the ever more precise and detailed study of more and more complicated molecules and molecular systems, as, for example, in the study of biomolecules.

As part of the continuing debate on the subject matter appropriate to this series, it has been suggested that the words "Chemical Physics" be added to the title. It is the opinion of the Editorial Committee that the distinction between the practice of "Physical Chemistry" and "Chemical Physics" has become small indeed. There was a time starting with the founding of the *Journal of Chemical Physics* [J. W. Stout, *Annual Review of Physical Chemistry*, 1985] that the two names represented distinguishable fields. We feel that this is no longer the case, and so the addition of "Chemical Physics" to the title would not add anything substantial. Indeed, the situation may well be what it was in 1860 when "Josiah Parsons Cooke, at Harvard, published a book entitled *Chemical Physics*, by which he undoubtedly meant what we now call physical chemistry..." [E. B. Wilson and J. Ross, *Annual Review of Physical Chemistry*, 1973]. We are pleased, as always, to have both chemists and physicists, representing a wide range of subdisciplines, contribute to this volume.

The Committee gratefully acknowledges the contributions of the authors. Despite the advances in information retrieval technology, writing a review does not seem to have become easier.

Our Production Editor, Suzanne Pries Copenhagen, has done an outstanding job of guiding manuscripts toward production deadlines, among authors, editors, and compositors. Without her efforts and those of the Annual Reviews staff, these volumes would not be possible.

The Editorial Committee

ANNUAL REVIEWS INC. is a nonprofit scientific publisher established to promote the advancement of the sciences. Beginning in 1932 with the *Annual Review of Biochemistry*, the Company has pursued as its principal function the publication of high quality, reasonably priced *Annual Review* volumes. The volumes are organized by Editors and Editorial Committees who invite qualified authors to contribute critical articles reviewing significant developments within each major discipline. The Editor-in-Chief invites those interested in serving as future Editorial Committee members to communicate directly with him. Annual Reviews Inc. is administered by a Board of Directors, whose members serve without compensation.

1988 Board of Directors, Annual Reviews Inc.

Dr. J. Murray Luck, Founder and Director Emeritus of Annual Reviews Inc.
 Professor Emeritus of Chemistry, Stanford University
Dr. Joshua Lederberg, President of Annual Reviews Inc.
 President, The Rockefeller University
Dr. James E. Howell, Vice President of Annual Reviews Inc.
 Professor of Economics, Stanford University
Dr. Winslow R. Briggs, *Director, Carnegie Institution of Washington, Stanford*
Dr. Sidney D. Drell, *Deputy Director, Stanford Linear Accelerator Center*
Dr. Eugene Garfield, *President, Institute for Scientific Information*
Mr. William Kaufmann, *President, William Kaufmann, Inc.*
Dr. D. E. Koshland, Jr., *Professor of Biochemistry, University of California, Berkeley*
Dr. Gardner Lindzey, *Director, Center for Advanced Study in the Behavioral Sciences, Stanford*
Dr. William F. Miller, *President, SRI International*
Dr. Harriet A. Zuckerman, *Professor of Sociology, Columbia University*

Management of Annual Reviews Inc.

John S. McNeil, Publisher and Secretary-Treasurer
William Kaufmann, Editor-in-Chief
Mickey G. Hamilton, Promotion Manager
Donald S. Svedeman, Business Manager

ANNUAL REVIEWS OF
Anthropology
Astronomy and Astrophysics
Biochemistry
Biophysics and Biophysical Chemistry
Cell Biology
Computer Science
Earth and Planetary Sciences
Ecology and Systematics
Energy
Entomology
Fluid Mechanics
Genetics
Immunology
Materials Science
Medicine
Microbiology
Neuroscience
Nuclear and Particle Science
Nutrition
Pharmacology and Toxicology
Physical Chemistry
Physiology
Phytopathology
Plant Physiology
 & Plant Molecular Biology
Psychology
Public Health
Sociology

SPECIAL PUBLICATIONS
Excitement and Fascination of Science, Vols. 1, 2 and 3A, B

Intelligence and Affectivity, by Jean Piaget

A detachable order form/envelope is bound into the back of this volume.

Annual Review of Physical Chemistry
Volume 39, 1988

CONTENTS

50 Years of Physical Chemistry, a Personal Account, *Sidney W. Benson*	1
Rotational Distributions in Direct Molecular Photodissociation, *R. Schinke*	39
Vibrational Raman Spectra of Simple Fluids, *M. J. Clouter*	69
Lattice Vibrations and Heat Transport in Crystals and Glasses, *David G. Cahill and R. O. Pohl*	93
Laser Spectroscopy of Large Polyatomic Molecules in Supersonic Jets, *Mitsuo Ito, Takayuki Ebata, and Naohiko Mikami*	123
Recent Developments in Dynamical Theories of the Liquid-Glass Transition, *Glenn H. Fredrickson*	149
Ab Initio Studies of Transition Metal Systems, *Stephen R. Langhoff and Charles W. Bauschlicher, Jr.*	181
Synchronicity in Multibond Reactions, *Weston Thatcher Borden, Richard J. Loncharich, and K. N. Houk*	213
Models for Colloidal Aggregation, *Paul Meakin*	237
Fractal Time in Condensed Matter, *Michael F. Shlesinger*	269
A Photochemical Investigation of the Dynamics of Oligonucleotide Hybridization, *John E. Hearst*	291
Quantum Effects in Gas Phase Bimolecular Chemical Reactions, *George C. Schatz*	317
The Nature of Simple Photodissociation Reactions in Liquids on Ultrafast Time Scales, *A. L. Harris, J. K. Brown, and C. B. Harris*	341
Kinetics of Radical Reactions in the Atmospheric Oxidation of CH_4, *A. R. Ravishankara*	367
The Semiclassical Way to Molecular Dynamics at Surfaces, *J. W. Gadzuk*	395
Chain Molecules at High Densities at Interfaces, *Ken A. Dill, J. Naghizadeh, and J. A. Marqusee*	425
(*Continued*)	vii

CONTENTS (*Continued*)

THEORY OF PURE DEPHASING IN CRYSTALS, *J. L. Skinner*	463
DISSOCIATIVE CHEMISORPTION: DYNAMICS AND MECHANISMS, *Sylvia T. Ceyer*	
HIGH-RESOLUTION SOLID-STATE NMR OF PROTEINS, *Steven O. Smith and Robert G. Griffin*	511
UV RESONANCE RAMAN STUDIES OF MOLECULAR STRUCTURE AND DYNAMICS: APPLICATIONS IN PHYSICAL AND BIOPHYSICAL CHEMISTRY, *Sanford A. Asher*	537
VIBRATIONAL SPECTROSCOPIC STUDIES OF THE STRUCTURE OF SPECIES DERIVED FROM THE CHEMISORPTION OF HYDROCARBONS ON METAL SINGLE-CRYSTAL SURFACES, *Norman Sheppard*	589
INDEXES	
Author Index	645
Subject Index	663
Cumulative Index of Contributing Authors, Volumes 35–39	673
Cumulative Index of Chapter Titles, Volumes 35–39	675

SOME RELATED ARTICLES IN OTHER *ANNUAL REVIEWS*

From the *Annual Review of Biophysics and Biophysical Chemistry*, Volume 17 (1988):

Water: An Integral Part of Nucleic Acid Structure, E. Westhof

Secondary Structure of Proteins Through Circular Dichroism Spectroscopy, W. C. Johnson, Jr.

Conformational Substates in Proteins, H. Frauenfelder, F. Parak, and R. D. Young

Fourier Transform Infrared Techniques for Probing Membrane Protein Structure, M. S. Braiman and K. J. Rothschild

From the *Annual Review of Earth and Planetary Sciences*, Volume 16 (1988):

Phase Relations of Peraluminous Granitic Rocks and Their Petrogenetic Implications, E-an Zen

Ore Deposits as Guides to Geologic History of the Earth, C. Meyer

From the *Annual Review of Materials Science*, Volume 18 (1988):

Laser Spectroscopy of Ultrafast Solid-State Phenomena, D. von der Linde

Recent Advances in Solid-State NMR of Zeolites, J. Klinowski

Precision X-Ray Techniques for Semiconductors, A. T. Macrander

Semimagnetic and Magnetic Semiconductor Superlattices, R. L. Gunshor, L. A. Kolodziejski, A. V. Nurmikko, and N. Otsuka

50 YEARS OF PHYSICAL CHEMISTRY, A PERSONAL ACCOUNT

Sidney W. Benson

Donald P. and Katherine B. Loker Hydrocarbon Research Institute, Department of Chemistry, University of Southern California, University Park, Los Angeles, California 90089-1661

I have always been interested in history, ancient history, modern history, chemical history, and even just plain gossip. However, I have never had either the time or motivation to learn any of these very thoroughly. What knowledge I may have in these areas is a somewhat random collection of memories, not necessarily accurate, integrated over a lifetime of episodes. Thus it was with some pleasure and numerous misgivings that I considered the invitation from the Editors to write this chapter. I have been actively involved with chemistry for over half a century, and my activities have always seemed to me very personal and possibly idiosyncratic. It was not very clear to me that I had any profound messages for my colleagues or the current generation of chemists. So I shall not attempt to invent any here. What may be of interest to the reader are my personal recollections of what has turned out in retrospect to have been a very exciting period both in human history and in chemistry. They are intimately related.

There are five billion of us on this planet, of whom about 250 million are located in the United States. Of that latter number, about 200,000 are currently involved in the diverse activity we call chemistry. Taken together with their families, chemists today represent a no-longer trivial fraction of the domestic American scene. Since 1900, their numbers have increased about 20-fold. These numbers are testimony to the profound and rapid changes that have occurred in the western world during the twentieth century and that will probably point the direction for what we call today the Third World.

George B. Kistiakowsky was fond of saying to his graduate students,

"The days of the gentleman chemist are over." As a 20-year-old member of that group, reared on the streets of New York City, much of the meaning of that remark escaped me. Few, if any, of the motley crew of graduate students in chemistry entering Harvard in the Fall of 1938 looked much like the Hollywood version of a gentleman, and few of us seemed to cherish that image as an immediate or distant goal. All of us had been reared in the background recollection of the bloodiest period in human history, World War I, and we were keenly aware, struggling as the United States still was with the effects of the Great Depression, that we were on the threshold of still another, possibly bloodier period—World War II.

Curiously enough, this social setting did little to change the interest or intensity with which we entered into graduate classes and research activities. There is nothing more immediate than the present, and the world of science and chemistry was witnessing its own rapid developments. Quantum mechanics was a relatively new discovery just beginning to make impacts on chemistry. Special and General Relativity were still exciting philosophers and physicists. Having been myself reared in the classical, rational world of Newtonian mechanics and thermodynamics, these were not particularly welcome changes. The Uncertainty Principle was not of great comfort to a student raised in a causal universe.

Many of my classmates were looking forward to careers in chemical industry. My own horizon was somewhat cloudy but, even more than I was then aware, I had already been infected by the bug of research. It was to be a chronic disease from which I have still not recovered.

At 20, I believed quite fervently that life was goal-oriented. One did research to learn something. The research was only a means. Was there life after finishing a PhD degree? Well it was clearly training for some more ambitious goal. But what? The future was, indeed, cloudy. If I have learned anything since then it is that while the ends may justify some means, life itself consists mainly of means, the things one does from day-to-day. If these are not on the whole satisfying, the end is not terribly important. For those who like it, research can be a very satisfying way.

My earliest interest in chemistry dates back to when I was 11 years old and had acquired my first chemistry set. I was fascinated by turning red water into colorless water and then back-again with a "magic wand" that had a capsule of sodium carbonate embedded in one end and crystals of tartaric acid in the other. Even more fascinating were the recipes for making gunpowder and colored gunpowders. Carbon, sulfur and potassium nitrate were the basic ingredients. The arcane world of strontium, calcium, barium, and sodium salts to achieve a color spectrum was able to provide fascinating hours of uninterrupted experiments.

However, one particular set one summer evening in the Bronx was interrupted by my frantic mother together with some neighbors, who burst into the kitchen, moving through a fog of fumes to find me, quite oblivious, concocting another potion to ignite. I had earlier noticed a very strong smell of burning sulfur and nitric oxides, and I had opened the doors and windows to create some ventilation, but neither the gases nor the smoke seemed to disturb me. I was quite surprised by their agitation. But that was the end of fireworks in the kitchen.

In the Fall of 1938, the world of chemistry was quite small. In fact the entire world of science was quite small. A few months after I had entered Columbia College as a freshman in the Fall of 1934, Harold Urey gave a lecture to which interested chemistry students were invited. He had just won the Nobel Prize for his separation of heavy water, and he demonstrated the properties of this exotic substance to the audience of about 250 students and faculty who had assembled in the then Victorian Havemeyer Hall. Most of the chemistry faculty of about eight were there, as were such notables as Professors G. Pegram and I. I. Rabi from the physics department. This distinguished event was still being eclipsed in the local press by the victory of the Columbia College football team over Stanford University in the Rose Bowl earlier that year. An ardent sports fan, I equally enjoyed the excitement of both occasions. Looking back at myself now, I am amused by the blasé casualness with which I found myself participating in these events. Part of it certainly must be attributed to the usual arrogance of being a native New Yorker, which was clearly identified with being in the center of the known universe. Provincialism may exist in the provinces, but I suspect it was exported from the big city centers.

My first serious interest in chemistry started with High School. I had always been interested in doing things with my hands and learning how things worked. Robinson Crusoe was one of my childhood heroes. When I finished public school, I knew that I wanted to do something in science or engineering. Through an older friend who had already graduated, I was encouraged to apply to the New York City Science High School of the time, Stuyvesant. My parents took a very dim view of it. We lived 20 miles away in the then, still rural West Bronx. Stuyvesant was located in a somewhat decrepit brick building in a slum area of the lower East side of New York, sandwiched in between an emergency hospital which forever reeked of ether and some equally foreboding dirty brown-stone, cold water, tenement buildings. It took a 30–40 minute ride in the subway followed by about a three quarter mile walk to reach it.

My three and a half years in this most unengaging environment were to turn out to be among the most exciting, exhilarating, and memorable of my early life. This unprepossessing exterior contained a group of the most dedicated teachers I have met anywhere, together with the most extraordinary facilities I could imagine at this time. Following a fairly uneventful first year in what was termed "the afternoon session" (due to a lack of space), I was promoted in my second year to the "morning session." We started classes at 8:00 A.M. and finished by 12:30 P.M., with only a six minute break between five solid classes. A quick lunch followed in the school auditorium, which doubled as a makeshift, fast food lunchroom. But then, for this privileged "morning session," the school resources were made available. Shops and facilities were open to interested students. There were clubs for every taste from Latin to math and chemistry, with enthusiastic faculty in charge. For me it was like entering an enchanted world. I seldom arrived home before 6:30 P.M.!

For the first time I met many boys (no girl students then) my age with similar interests in science. Some had labs at home—many very sophisticated, with chemicals I had never heard of and some with then newly discovered "electronic" gear. There was a wonderful young math teacher, Hyman Marcus, who first got me interested in mathematics and then later in fencing. He had just graduated from Columbia and, although I didn't realize it at the time, that pointed me in my future direction. Another wonderful chemistry teacher, Abraham Kerner, with whom I maintained contact until his death some ten years ago, opened up the chemistry labs to me. Perhaps most important was a dimunitive English teacher, Hyman Mostow, seemingly possessed of infinite sources of energy. He assumed planning my career as well as introducing me in his extraordinary course in English to Shakespere and English poetry. He had about eight other students in his menage who he also guided with a gentle but iron hand. In his spare time he coached the somewhat nebulous basketball team.

Leading this most unexpected coterie of teachers and mentors was the even more unlikely Principal of the High School, Dr. Earnest R. von Nardoff. Each semester he filled the 2000 seat auditorium for a special lecture, marvellously demonstrated, on some topic in science. They were extremely well-done, effective and for me a constant example of how science should be presented. I can still remember most of them, so vivid were they! Most dramatic was a lecture titled "The Spheroidal State." The large lecture stage was lined with a sand trough and half-way through the discourse on the cooling effects of evaporation and the pressure produced by gases, several students in asbestos aprons and boots, wearing asbestos gloves appeared, carrying a large, white hot crucible containing molten iron. They proceeded to the podium where they tilted the crucible into the trough, producing a small stream of white hot molten iron. Dr. von Nardoff, having earlier explained his protocol, carefully washed his hands and then passed his open palm through the stream, scattering a fountain of incandescent, sputtering globules all over the stage. The audience gasped the loudest gasp of astonishment I've ever heard, while three assistants with pails of sand extinguished flames from potential fires started where globules had penetrated the sand to the wooden floor below. Thunderous applause then elicited a repetition of the demonstration!

Stuyvesant provided more than a fertile soil for future scientists. It opened a view onto the intellectual world of ideas of which I had never been aware. My classmates came from many diverse social, ethnic, and political backgrounds, and passionate discussions could be heard and attended on politics, economics, sex, as well as on science. The once large family and neighborhood world of the Bronx and Brooklyn in which I had been raised steadily diminished in size and importance during those exciting years. I also discovered the incredible used book stores of lower Manhattan where one could browse for hours through books on subjects I had never heard of. Equally exotic was the chemical supply house of Eimer and Amend on Third Avenue, where a trip through their apparatus stockroom could intoxicate the imagination for weeks trying to guess the purpose of the various metal and glass constructions. Their chemical catalog was no less evocative. Only my mother expressed some dismay at the increasing arsenal of chemicals residing in her once splendid, now slowly invaded glass china cabinet. Weekend experiments had become less visible, more sophisticated, but no less odorous. When I began to explore qualitative analysis, the schemes for separation of metals based on solubilities became an adventure in devising new routes and new combinations of metallic groups. I found hydrogen sulfide a socially deplorable reagent and hunted for more acceptable means of separation, some of which would turn out later in college to save me many hours in the laboratory.

No experiences at Stuyvesant, however, were to prove as tense, anxious, as well as exciting as those with the Math Team. While I might have preferred a role in the somewhat pathetic but nevertheless, to me, glamorous football team, nature had not designed me in size, mass, or speed for a role in American football. The Math Team turned out to be a more than adequate substitute. Invited at first by classmates, and having to overcome an initial unfamiliarity and just plain ignorance, I discovered after a few months an unexpected facility in problem solving. Then encouraged, despite numerous failures, by an enthusiastic Mr. Marcus, I quickly began to develop a real interest in algebra, geometry, and mathematical puzzles. Most remarkable, I discovered books in the school library on mathematics that I could read and, to my amazement, from which I could learn. It was a heady experience. By my fourth semester, I was a regular on the school math team, and in my last year I was elected captain.

The New York City Official Math Meets were something else. I don't remember ever being able to sleep the night before a Meet. And during the actual 40 minutes of the two- or three-school competition my teeth clattered so hard I had to put a pencil between them to keep from attracting attention. Problems were written on a blackboard. We had one minute to read them and then, depending on their complexity, from one to a maximum of ten minutes to solve them. Only the answer counted. The team would spend hours practicing on all the available examples from texts, puzzle books, and examples from previous meets. It was altogether nerve-wrecking event softened only by the more relaxed post-mortems that followed.

If Stuyvesant was an introduction to the arena of science and ideas, Columbia was the big time. Unlike Stuyvesant it did not emphasize science. Rather its main thrust was the humanities. However, it did have very active and respectable chemistry, physics, and mathematics departments. In 1934, I would guess that perhaps 20 American universities carried on serious research efforts in the natural sciences. Perhaps another 20 supported token research efforts. There were then only three American journals of importance in physical chemistry, one of which, *The Journal of Chemical Physics*, had just begun publication in 1933, while another, *The Journal of Physical and Colloid Chemistry*, was considered an eccentric enterprise started by Professor Bancroft of Cornell. If we add to these the three or four major foreign journals, one could easily read all new significant articles in any given area of physical chemistry and still have time for the Sunday comics. Deciding to become a research chemist was very much like joining a rather small, select club.

The once miniature American chemical industry was just beginning to grow. With a huge domestic market and a very rich material base in petroleum, gas, coal, and iron, the stage had been set for a huge expansion. DuPont advertised "Better Living Through Chemistry." Vitamins had been recently discovered and plastics such as nylon were about to come into the market place. Despite the use of poison gas in World War I, chemicals did not have the ominous significance they were later to gain starting about 1950. For the neophyte chemist, chemistry was clearly a potentially hazardous profession, but that, if anything, gave it a romantic aura. The reality of chemistry classes, however, was something different.

After a somewhat neutral and repetitious first year my second year nearly ended my interest in chemistry. I have never found anything more excruciatingly dull than quantitative analysis as it was then practiced. A full two semesters were required. The intellectual content was provided in about half a semester. The remainder was a grinding drill, which even the instructor found boring.

Quantitative analysis is the bread and butter of chemistry, the necessary foundation for about any chemical procedure, but it is not the most stimulating way to spend a year. Only chemists who have experienced the tedium of counting swings on a beam balance and have spent two weeks calibrating weights can truly savor the sense of delivery brought about by automation and instrumental methods of analysis. It took great self-discipline to survive two semesters of "wet" analysis, and it was to be several years before I again had any pleasure in walking into a chemistry laboratory.

The interesting subjects during this period were in physics and mathematics. Newtonian mechanics, electricity, magnetism, probability theory, and theory of functions were challenging and exciting.

My introduction to research came unexpectedly enough in the second semester of quantitative analysis. Our laboratory assistant, James "Spike" Coles, suggested doing an original research problem, the determination of the bicarbonate content of sodium carbonate using a titration technique. It was easy enough to calculate that titration with a strong acid would produce a weak end-point at the pK of sodium bicarbonate, about 8.4, corresponding to the carbonate content, while further titration would yield a sharp end-point at pH 6, corresponding to the sum of both. After much experimenting with pH-sensitive dyes, I finally found one that gave reasonably sharp color changes at both pH's. Then began my first contact with the real world with the discovery that so-called solid sodium carbonate would if exposed to the laboratory atmosphere pick up CO_2 and moisture and perversely change its composition from day to day! A fairly elementary finding—but it took about six weeks of very frustrating efforts.

Louis P. Hammett was my advisor. I was but dimly aware of his growing reputation. He was an extremely shy person who was at first meeting rather taciturn. His colleagues many years later told me that it took two martinis to really make him relax. I found him kind and helpful if not loquacious. Columbia had a wonderful policy of waiving course requirements in any area upon consent of the faculty and the passing of a written examination in the course. He allowed me to take a special exam in physical chemistry, for which I studied in the Summer and passed in the Fall. That opened to me a huge menu of electives in my junior year which renewed my waning interest in chemistry. Very memorable was an encyclopedic course in Chemical Kinetics given by Victor K. La Mer. The idea of connecting time and chemistry in a quantitative form was new and stimulating and this course evoked an interest which has never ceased. Another course by Ray Crist on the new field of photochemistry gave further fuel to this interest in kinetics. It also for the first time made me aware of bond strengths, and the significance of energy in elementary reactions.

By my final year, I was entirely relaxed and enjoying the "intellectual" life. A $400 math prize in my junior year had made it possible for me to purchase my first automobile, an incredible luxury. I had finished all the required courses for my degree and could take any elective courses I wished. I took a few, notably Professor Harold Urey's courses in Quantum Chemistry and Thermodynamics and Professor Biot's course in Vector and Tensor Analysis. These and a very demanding Physical Chemistry laboratory were not enough to keep me from weekend and holiday drives outside of New York with my friends. I also had a job with the newly founded "Young Students Administration," organized by then-President Roosevelt to assist needy college students. It paid the magnificent wage of 50 cents an hour for 30 hours a month. That was enough to pay for most of my meals in the Columbia dormitory cafeteria. Originally assigned to dusting animal and human skeletons in the Schermerhorn Museum of Anthropology, I succeeded in getting Professor Urey to request my transfer to his laboratory, where I assisted in maintaining the one and only mass spectrometer.

My major job was to maintain the distilled water level in some 200 storage batteries used as a constant voltage source for his home-made magnetic sector instrument. Fortunately, the current drain was so small that the water level needed only occasional attention. When Urey found out I had taken courses in mechanics and electricity, he asked me to investigate the focusing quality of the magnetic field for different speed ions coming in at a variety of angles. I worked on this for awhile but soon decided it was not soluble in closed form. I didn't have then the background in the real essence of mathematical physics, namely, the fine art of approximation.

During most of my four years at Columbia, I had only a very vague idea of what I would do upon graduation. Teaching seemed the most attractive, but it clearly required a PhD at the University level. Jobs in the sciences were very few and, in the absence of any concrete ideas of how professional chemists spent their time, not very real. Reality started to intrude one Fall day in my senior year when Professor Hammett suggested that if I was interested in a PhD I should start writing to graduate schools. I asked about continuing at Columbia and he replied that it was best for me to go elsewhere; also, it was also against their policy to retain undergraduates. This produced a real shock! It had never seriously occurred to me that I would live anywhere else but New York City and the thought of a three or four year period away seemed like permanent exile. A follow-up audience with Professor Urey elicited the same advice with the additional suggestions that I write to Cal Tech, Princeton, and Harvard. I was fortunate enough to receive offers at each of them. Much soul searching and conversations with anyone who would listen finally led to my choosing Harvard, based on their fellowship offer that would give me the freedom to use my own time as I saw fit. Later that Spring, when I celebrated my graduation by driving with my mother, sister, and brother for a vacation and visit to some friends in California, I visited Cal Tech and got to know California. It was love at first sight. Although committed to Harvard, I made the decision there that when I finished my studies, I would try to come back to California. About six years later, in the most unexpected circumstances, I did.

Graduate school is a psychological quantum step removed from college. In college, I was busy learning, living, and observing. In graduate school, I was suddenly aware of doing things—research, and also preparing for a professional career of some kind. Columbia represented an existence, exciting and stimulating, since at that age so many experiences are new. It carried with it the feeling that it would go on forever. Graduate school and Cambridge, while no less interesting, were clearly transient states. Life had suddenly become more serious.

My introduction to Harvard started dramatically. My cousin, who was entering law school, and I set out on a sunny day for the five hour drive. It proved to take two days. The day before we started New England had its first hurricane of the century and we spent a night stranded at the Connecticut-Massachusetts border.

Although he doesn't truly realize it at the time, the most significant decision a graduate student will make is to choose his research director. In my own case it was very simple. None of the research being done by the physical chemists on the faculty then seemed terribly interesting to me. George B. Kistiakowsky had the practice of giving parties about once a month during the year for his graduate students and post-doctorates. They were interesting parties with good food to eat, a variety of interesting guests, and real liquor. My choice was made very easy. Choosing a research problem was much more difficult.

Kisty, as we called him, was a great lecturer, a very easygoing and gregarious person and an extraordinarily gifted experimentalist. He was a skilled machinist and a talented glass blower able to work with quartz, pyrex, soft glass or any grade in between. He had learned from Max Bodenstein, in Berlin, all the techniques for doing gas phase kinetics. I learned to construct the famous greaseless quartz helical manometers and the glass bellows–activated, greaseless, Bodenstein glass stopcocks. Kisty required his students to construct all their own equipment, except for these last two. After I completed my thesis, my last act in the laboratory was to destroy my vacuum system, first salvaging the mercury diffusion pump, the ubiquitous McCloud Gauge, and the several stopcocks. The next student came to a bare wooden frame!

Although I had a general interest in kinetics from my experiences at Columbia, I had little knowledge of the important or interesting problems. During the first semester, a course by Paul Bartlett, then a young Assistant Professor in physical organic chemistry, inspired my interest in mechanism with examples of how one could draw conclusions about mechanism from kinetic studies. Free radicals were coming into prominence at about that time, and I was intrigued by the Paneth experiments on pyrolysis using metallic mirrors as radical scavengers. F. O. Rice and K. Herzfeld had just published their classic paper on chain pyrolysis of organic molecules, and F. O. Rice and K. K. Rice had just written a very readable monograph, *The Aliphatic Free Radicals*, which summarized the known data. Although many chemists questioned the existence of free radicals in any given pyrolysis or photochemical study, the evidence to me seemed overwhelming for their existence. By this stage, I had become an avid reader, both of monographs and research articles. An article by Pearson and Purcell on biradicals had a special appeal. How to prepare these exotic species, then demonstrate their existence, and finally measure some of their properties seemed a real challenge.

The photochemical decomposition of acetone and higher ketones had already been studied. This was described in great detail by another just-published classic book by W. A. Noyes, Jr. and P. Leighton on *Photochemistry*. Absorption of a photon more energetic than 90 kcal appeared sufficient to cleave the carbon–carbon bond in acetone or ketones generally and produced two free radicals. What if the ketone were a cyclic molecule like cyclopentanone? Would photochemical excitation at 2900 Å produce a biradical with free valences at the two ends of the chain? W. G. R. Norrish and a student had performed such studies in 1938 and found good evidence for CO production from cyclopentanone, cyclohexanone, cycloheptanone, and cyclobutanone. This seemed to me unequivocal evidence for ring cleavage and biradical production. Thus, in the case of cyclopentanone a possible reaction sequence might be:

$$\text{(cyclopentanone)} \xrightarrow{h\nu} \cdot\overset{\overset{\displaystyle O}{\|}}{C}-CH_2-CH_2-CH_2\dot{C}H_2$$
$$\downarrow$$
$$CO + \dot{C}H_2-CH_2-CH_2-\dot{C}H_2$$
$$\downarrow$$
$$CH_2=CH-CH_2-CH_3.$$

It would appear as though the products, butene-1 and CO, must arise from a biradical. Similar products were found with all cyclic ketones. The sophisticated reader needs to understand that many chemists at the time believed that the many steps outlined above could all take place in one concerted act.

Not wanting to repeat someone's experiment, I spent a semester trying to synthesize 1,2-diazacyclopentane, better known as pyrazoline. The ring cleavage should produce N_2 and a trimethylene bi-radical.

$$\underset{N=N}{\overset{CH_2}{CH_2\diagup\diagdown CH_2}} \xrightarrow{h\nu} \underset{\cdot N=N}{\overset{CH_2}{CH_2\diagup\diagdown CH_2}} \longrightarrow \overset{CH_2}{CH_2\diagup\diagdown \dot{C}H_2} \qquad \overset{CH_2}{\dot{C}H_2\diagup\diagdown \dot{C}H_2} + N_2$$

Extrapolating from Norrish's results, one might suppose that the biradical might end up as propylene or dimerize to cyclohexane. In any case, the reaction should be very clean and free of side products, one hopes.

It proved to be a very frustrating experience, smelly and hazardous as well. Acrolein (CH_2=CHCHO), an extremely potent lachrymator, and hydrazine (N_2H_4), a very corrosive and metastable substance, were the starting reagents. There was no trick to following the published recipe, but several months' work went into futile purifications before I realized that pyrazoline was in tuatomeric equilibrium with its ene-imine:

$$\underset{\underset{H\ H}{|\ |}}{\underset{N=N}{\bigcirc}} \rightleftarrows \underset{\underset{H}{|}}{\underset{N-N}{\bigcirc\!\!\!/}}$$

The photochemistry of the tautomer was totally unrelated to that of pyrazoline. So back to the cyclic ketones! Kisty was kind enough to hire an organic student during the Summer to prepare cyclobutanone. It was a nearly disastrous experience. The starting materials were the very toxic and explosively unstable diazomethane (CH_2N_2) and the very labile ketene. One major explosion nearly lost us our organic student. He fortunately suffered only superficial injuries and was finally able to provide the magic ketone.

By the end of my first year, I was beginning to enjoy graduate life. I finished all my formal course work, was well begun in research, and had become a reasonably skillful glassblower, mostly by virtue of having broken several pieces of equipment in the lab and then having had to repair them. It had been necessary for me to do tutoring to supplement my fellowship, and Kisty generously offered to hire me to do some research for him over the Summer while he made a trip to Europe. It was my first real job and paid the unbelievable salary of $40 per week. That came to about 70 cents an hour, but it added up to nearly $180 a month (no withholding!) More importantly, three months would guarantee me full-time eating the following year.

If the New England hurricane of 1938 had been dramatic, it was minor compared to what was happening in Europe. The Chamberlain–Hitler pact in 1938 and the subsequent invasion of Czechoslovakia was a grim prelude to World War II. We had already witnessed the Italian invasion of Ethiopia in 1935, and the prolonged and bloody civil war in Spain, which only ended in the Spring of 1939. Many of my friends had been active in the loyalist cause and two had joined the famous Lincoln Brigade. Kisty, when he left, had felt that this might be the last chance to visit Europe. He proved to be correct. He managed to return on the last boat to sail from England just after the war was declared in August 1939. To those of us who, like myself, had been sympathetic to the communists and the Soviet Union, the Stalin–Hitler Pact was both a political bombshell and a profound disillusionment. The rapid dismemberment of Poland, followed only months later by the invasion of Finland by the Soviets, was a last cynical disenchantment.

My generation had been brought up on the profound disillusionment that followed World War I. The First World War had appeared a totally pointless war, and I had never understood, until quite recently from reading Barbara Tuchman's *The Zimmerman Telegram*, just how the United States had become involved.

The aftershock from the Stalin–Hitler pact was not long in coming. The invasion by Germany of Holland and Belgium and then, in May 1940, the Fall of France provided a watershed in American thinking. Suddenly, Hitler and his prophesised 1000-year Reich started to look very real and the European scene very grim. I had been very anti-war and active in the large anti-war student group on campus during 1939–1940. The events in Europe began to look like an inescapable tragedy approaching.

Roosevelt had immediately cast his lot with England and the now fallen France and many Harvard faculty had already responded to the Washington initiative to become active in what might follow. From the day after his return until the Fall of France in 1940, Kisty had begun to take part in some government-sponsored research projects that were just being organized. He spent about half his time in Washington during that year and much more after Pearl Harbor. He was kind and thoughtful enough, even though we did not quite agree politically, to arrange for me to be hired for the Summer of 1940 at the famous General Electric labs at Schenectady. As he put it, "They owe me one!" It was to be a marvellous experience.

The General Electric labs were predominantly a "physicist's" laboratory, presided over scientifically by Irving Langmuir, who was perhaps America's first chemical physicist. He was then still pursuing with Katherine Blodgett an active program on surface films, for which in 1932 he had been awarded the Nobel Prize in chemistry. He had earlier in his career endeared himself to General Electric by his work on tungsten filaments, atomic hydrogen (and welding), and vacuum tube getters. Chemistry was distinctly a minor actor in the General Electric script, but it had a very prestigious staff. Abe (Lincoln) Marshall headed the chemistry lab. Ray Fuoss, Lewis Tonks, and Saul Dushman were senior staff members along with Gene Liebhafsky. Eugene Rochow was just at the beginning of his famous researches on silicon chemistry, and the silicone polymers were then in the experimental stages. Marshall assigned me to a laboratory and a project related to work on alkyl polymer resins. A few weeks after I started, a summer visitor, John Kirkwood, agreed with some misgivings to share the office space located in my small laboratory. He never quite appreciated the noise of the vacuum pumps I used or the way in which my equipment progressively cluttered the available space.

Weekly seminars in any area of interest to the laboratory were presided over by Langmuir, who seemed to have an uncanny insight into whatever topic was being presented. My own seminar on photochemistry was explored by him in enervating detail. Although I succeeded in answering most of his questions, I breathed a sigh of relief when he had finished, and I had no difficulty falling asleep early that night, despite my usual custom of reading or writing until about 1:00 A.M.

The experience at General Electric was an important and interesting one, giving me my first insight into how industrial laboratories operated and how they selected their problems. Coolidge (he of the x-ray tube) was nominal head of the lab and an extremely gracious man. Although in principle all work was project-related, Fuoss was working on electrolyte theory, Tonks was exploring the statistical mechanics of two-dimensional gases, and Dushman was experimenting with the semiconductor properties of lead sulfide crystals. What I found fascinating in talking to them was their ready ability to relate what seemed like abstract research to the practical problems of the company. My experience at General Electric was also to be a practical introduction to big business and bureaucracy.

Once my problem had been selected and mutually agreed upon, I outlined my approach and the equipment to be used. I set about building the equipment myself. It took me about three weeks to finish it, whereupon a visit from the head electrician produced the incredible (to me) instruction that it didn't meet code specifications and would have to be redone by their department. No pleading or cajoling could alter that decision, and for the next two weeks I watched a perfectly functioning apparatus disassembled and reconstructed in ponderous B-X cable and conduit. It took two weeks of precious time because the labs opened at 8:00 A.M. and closed at 4:00 P.M.! Executive orders were required to stay late and weekend work was unheard of. It took me some time to reconcile myself to this totally absurd schedule. Research is not something like bread that is cut in so many slices and then consumed. Few things are more frustrating than to begin an experiment you know will take at least four hours only to realize that at 2:00 P.M. it is two hours too late. How the famous laboratories of Edison and Steinmetz reached this state of organization was never made clear to me.

During my last month, a photochemical project, the first in the laboratory, was begun. It was a project on the photochemistry of uranium salts. I didn't know it, but the nuclear age had begun.

Both chemistry and physics were usually represented in my high school and college courses in a very abstract and formal manner. Industrial processes were dragged in as an example of the application of chemistry or physics to some practical goals, but the real motivations were never explored. Outside of the alchemists' dreams of changing lead into gold—which always seemed to me a total waste of time—there appeared very little connection between science and everyday life. Scientists appeared to be a brand of medieval monks exploring some self-appointed tasks in natural phenomena. The lively give and take between hypothesis and laboratory, the scaffolding of science, was hygienically removed, leaving what seemed like a most improbable structure—the final result. The infrastructure of scientific research, how research was paid for or motivated, was also totally missing. At Columbia and Harvard one was aware of the Rockefeller Foundation and a few benevolent companies or individuals who made gifts. At Harvard, the unexplained benevolence of the Mallenkrodt Chemical Company made available a fund from which every graduate student was allowed $50 a semester for chemicals or equipment—a handsome sum actually, which was seldom exceeded by physical chemists. The university also made a research allowance (about $2000 a year) to each professor doing research—a very different system from our current federally supported research.

Years later when I started on the faculty of the University of Southern California, it was understood that the school would make funds available to me for my research, at a modest level of course. Grants didn't exist then, but I was asked to write a short summary of the work I wanted to do and include some statement as to why it was of importance. I never learned who read these summaries or what response they triggered.

When I returned to Harvard in September for my third year I had decided to devote myself to making it my final year. General Electric glass and machine shops had been very helpful. In exchange for teaching their glassblowers how to make the Bodenstein helical quartz manometers they built me a beautiful Blacet–Heldman apparatus for doing micro-analysis of gases. It was a technique of the time for quantitatively analyzing as little as 0.1 cc STP of a gas (about 4 micromole) and was to speed my research enormously. The military draft approved in the Summer of 1940 began in October. My best friend at the time, Milton Soffer, an organic chemist, had the first number picked by lot. Ironically he had been President of the very large Harvard Anti-War Society. He became very upset at his "luck" in the drawing; his state of mind was not helped by the group of us listening to the radio, laughing uproariously at his being the first chosen. An hour later the second number was announced. It was mine! I had been secretary of the same society. I have always had a somewhat fatalistic approach to such events. In any case nothing imminent was about to occur in October 1940. We were not at war, and students were given automatic deferments. However, the sudden turn of events reinforced my decision to graduate quickly, inaugurating a prolonged period of what was to be the hardest, most sustained working period in which I have ever been involved.

For the next seven months, I worked in the laboratory six days a week, usually for ten to 12 hours. Each days' work was outlined the night before and the end of the day was determined by the experiment at hand. By this time, the work had become routine, punctuated only by accidents and breakage of equipment. Accidents, I quickly observed, came at night generally, when I was tired and trying to hurry. It took me a long time to realize that at some point in a working day the probability of a destructive event began to exceed the probability of doing something useful. Fortunately, I survived the learning, but there were more than a few scares along the way.

I used large quantities of liquid refrigerants in my gas handling and analyses, particularly dry ice/acetone slurries, which I kept in a large Dewar container. A similar Dewar was used for shaved ice. One day I reached into what I thought was the ice Dewar only to find I had immersed my hand in dry ice/acetone at $-80°C$. I withdrew my numbed hand from the Dewar, staring at it with horror. I had no sensation in my fingers, which were still curled about some dry ice. The skin had turned yellow, and I was sure I had lost all my fingers. Instinctively I plunged my hand into my mouth to warm it and then as sensation slowly returned, put it under the faucet. Within about ten minutes everything was back to normal but my fright lasted for some time. Thereafter, I was very cautious about where I put my hands.

On another occasion, in the late evening at about 1:00 A.M. after an unusually long day, I managed to break part of my vacuum line. I stopped the experiment and began the laborious reconstruction using the lab oxygen torch for the glassblowing. In the middle of the repair I hung it on the vacuum rack, the flame still going strong, not realizing that it was pointing at my lab notebook on the other side of the metal frame. Suddenly, I smelled smoke and glanced up to observe, in awed disbelief, large flames coming from my lab notebook, containing all my results!—six months of arduously won information. I didn't think. I just picked it up and smothered the flames with my arms and lab coat. Again, I was lucky. No serious damage was done to my precious data. The book cover and edges were badly charred but the contents had been spared. I only had minor burns on my fingers, but I can still recall the anxious moment after as I scanned through the precious pages.

I succeeded in completing my studies on cyclohexanone and cyclopentanone photolysis with very gratifying results. Contrary to the earlier published reports, the major products were cyclopentane+CO from the former and cyclobutane, ethylene+CO from the latter. The earlier workers had misidentified the hydrocarbons. It looked like solid evidence for biradicals with a sequence:

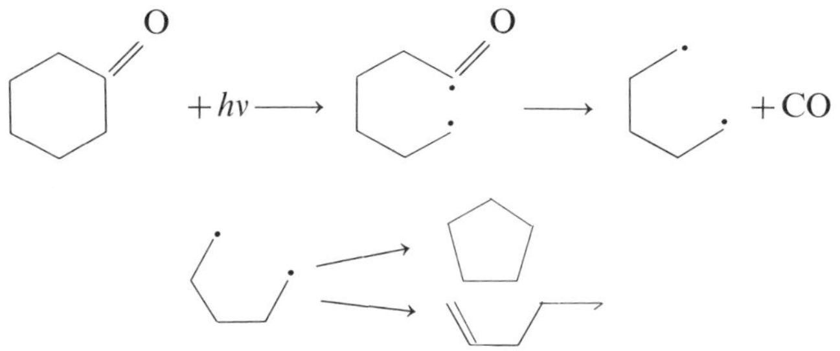

In the case of cyclopentanone, the tetramethylene biradical had two competing paths that appeared to be temperature independent and hence with little or no activation energy:

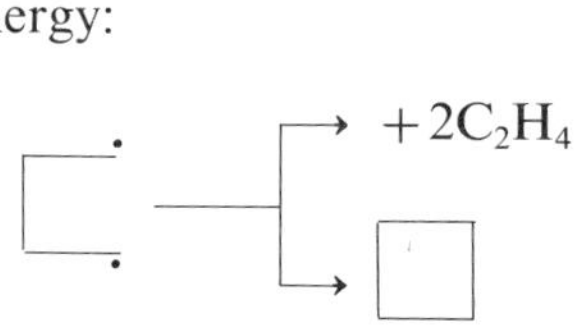

Kisty induced the Cabot Corporation to patent the method as a cheap praparation of the very rare cyclobutane. He thought it might be a useful anesthetic. It did become a standard research synthesis but found no greater use.

While doing cyclobutanone I seemed to have found a similar result with a gas product whose molecular weight was 42, as expected for cyclopropane or propylene, and whose boiling point was also about right, $-48°C$. However, attempts at analysis were totally disconcerting. Gas volumes were measured by isolating a sample in a 2 mm capillary between mercury threads. This gas burette had worked well with everything else I had examined, but with the presumed cyclopropane/propylene the gas volume kept shrinking in the capillary. After three days I gave up and went to the movies, convinced I was going to spend another year at Harvard. I didn't come back to the laboratory over the weekend. Monday I returned to consider my plans.

It was my custom to save all condensable gas samples by freezing the samples in small capillaries and sealing them off under vacuum. This morning, I looked at my collected samples, gathered in a large beaker, and saw something totally unexpected; the usual colorless hydrocarbons in three of them were a brilliant, burgundy red. Eureka! What I thought was cyclopropane was something else, ketene as it turned out. Ketene is notoriously reactive, and it didn't take long in the library to confirm a red dimer. Cyclobutanone, like cyclopentanone under photolysis, gave two sets of products:

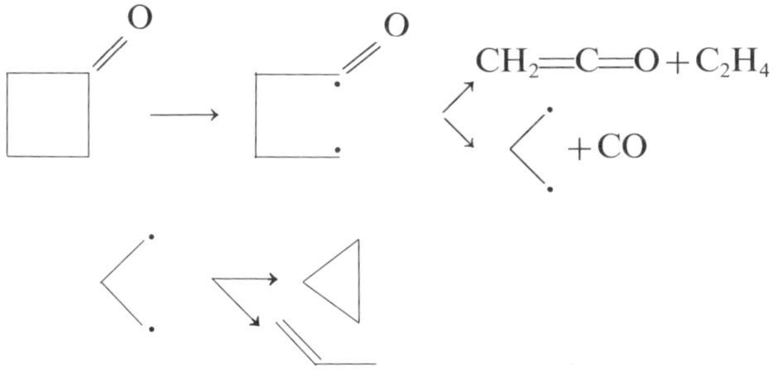

I completed my German language exam a week before the December deadline required for June graduation. My thesis was accepted three days before its April deadline and my final oral exam two days before the May deadline. I didn't wait for June. In May 1941, I left Harvard, I thought, forever. Four months later I returned.

During my race to complete the PhD requirements I gave little thought to a job. In April 1941, I had become qualified, as a result of my employment at General Electric the summer before, for unemployment insurance, one of the reforms introduced under Roosevelt's New Deal. The $15.00 per week I was able to receive made the last month at Harvard and the subsequent transition to civil life much easier once I returned to my beloved New York City.

I set about in earnest to write to various schools, to consult the want ads, and to visit the few employment agencies in downtown Manhattan that specialized in scientists. Jobs were still very scarce. The "Great Depression" had not yet abated and a PhD in chemistry was of questionable significance in the New York job market. Finally, just as my last thirteenth weekly unemployment check was spent I received a good offer doing classified, war-related work at the Rockefeller Institute in midtown Manhattan under the direction of a very distinguished physical chemist, Duncan MacInnes. My acceptance was made with very mixed feelings. Although I was not a pacifist, the words "War" and "War Work" evoked very negative reactions in me. The idea of spending my efforts in creating or improving destructive instruments was not part of my self-image. Even less appealing, our assignment was to improve the technique of dispersion of poison gases, nothing to proudly discuss with friends or family.

Rockefeller Institute was already a famous place. Wendell Stanley had done his classical work on viruses there, and the scientific staff were very interesting. After about six more weeks of a relatively pleasant readjustment to New York life my equilibrium was suddenly shattered by an early morning telegram from MacInnes saying that my services were no longer required. A special delivery letter followed, containing my paycheck and unexpectedly an equal amount from MacInnes. No reasons for the sudden decision were given, and MacInnes refused to talk to me on the telephone. Reflecting as calmly as I could on the circumstances, I concluded that my "political" activities were responsible. All war work required security clearance and evidently I had not been cleared.

Let me explain. After the Stalin–Hitler pact I had become disillusioned, critical of my communist-leaning friends, but also more interested in politics. I met and became friendly with some of the Socialist and Trotskyite students on campus. My anti-war activities brought me in even closer contact with them, and an off-campus incident in the Spring of 1940 in which I had been active in organizing a Cambridge protest against a particularly corrupt, local politician with the picturesque name of "Ratty" Hamilton brought me to the attention of the local police. Boston politics in 1940 were only a bit more covert than that of the notorious "Boss Tweed" of old Manhattan. Once returned to New York, I expanded my circle of Trotskyite friends. All of this I decided must not have passed FBI muster. Swallowing much of my by now battered pride I wrote to Kistiakowsky, asking if I could do some post-doctoral work for a while. He was extremely kind. He evidently had learned of my situation and offered a job at Harvard on the condition that I not engage in any political activity. He was by now spending about two thirds of his time in Washington DC on war work and he felt somewhat ambiguous about my political sentiments.

It was a moment of great soul-searching from which emerged the realization that my scientific training and underriding skepticism were more important to me than transient commitments to political opinions. I accepted his offer. I have never regretted the choice.

The post-doctoral year was a tranquil contrast to the preceding graduate student existence. I enjoyed the great luxury of a regular salary, time to read and attend seminars, and good laboratory facilities. Although Kisty had made me the offer, Professor George Shannon Forbes was my sponsor. Forbes was the very epitome of the stereotypical, Victorian, university Professor. From his, even by 1940 standards, somewhat archaic, celluloid collar and bow tie to his trimmed moustache, polished shoes, and very formal lecture style, he was the very soul of decorum. He had been one of the American pioneers in physical photochemistry and an extremely careful and meticulous experimenter. He was exceptionally shy in conversation, but one of the kindest men I have ever met. His hobby was photography, and he would spend each summer out west in either Colorado or California taking black and white pictures of the glorious views, mountains, and forests. One of these would be carefully selected, enlarged, mounted on a white mat, a small calendar attached and a copy sent to each of his students for the New Year. I received one faithfully every year thereafter until a year before his death at the age of 93. The last few years, he accompanied the calendars with apologies for the fact that the pictures were old since his health prevented him from taking hikes.

I succeeded in completing what I considered to be a thorough study of the photolysis of acetone at 2560 Å and proved that every photon absorbed produced CO + two methyl radicals. However, the effort invested in these modest conclusions was so great that I decided then and there that if I pursued research it would not be in photochemistry.

After the attack on Pearl Harbor in December 1941, everything suddenly took on a different significance. I received a letter from one of the schools I had written to the year earlier. New York City College offered me a job as instructor in chemistry at the magnificent salary of $2300 for nine months. It was a dream come true!

The reality was to be less than the dream. CCNY, as it was known, had the pick of the brightest students in the New York area. I enjoyed teaching very much. However, the students proved much less sanguine. Despite their intelligence and grades they were not as eager to learn all the chemistry I wanted to teach them. I suffered the usual disillusion of the beginning teacher, but it was quickly absorbed. Less attractive were the very poorly equipped and crowded laboratories and the one half desk that I shared in shifts with another instructor. Most disillusioning was the rapid realization that I could not live in New York on my salary. During Christmas and Easter I was reduced to taking a job in a store selling shoes. So much for academic "free" time. Before the second semester had begun, I was back interviewing for jobs. I was offered one by the Kellex Corporation of New Jersey. It was a subsidiary of the H. W. Kellog Company, a well known chemical engineering firm. They had been newly formed to perform special war work for the government. The salary was almost triple what I was making, and the people I spoke to, Manson Benedict and Charles Johnson, were scientifically impressive. I explained my history with war work. They felt it wouldn't matter and a month later when I next heard from them I was told that I had passed security clearance. I accepted the job with great enthusiasm, and as soon as my last spring lecture was over in early May I began my new work on what was to become better known as the Manhattan Project.

My announcement to my colleagues at CCNY that I would leave at the end of the semester was greeted first with stark disbelief and then with incredulity at my naive behavior. "No one had ever resigned from the department. Some people had waited 14 years at the Assistant Professor rank in just the hope of gaining tenure! Why would I throw away so promising a career?" All of this helpful advice gave further support to the wisdom of leaving.

The discovery of atomic fission had appeared in the scientific literature without inspiring much publicity, and I was but dimly aware of it. When I learned what we were engaged in and the magnitude of the energy available in an atomic bomb I was stunned. It was a difficult time. It took several days to absorb the facts and the potential implications. For many months I softened the impact by trying to see how it might work and decided that it was an impossible effort. I had no knowledge of neutron cross-sections and no idea of how, eventually, implosions would be used to achieve critical mass, so my skepticism seemed well founded. It was to be reinforced in the months ahead as our research proceeded.

The Kellex Corporation was responsible for the construction of the Gas Diffusion Plant to be built at Oak Ridge. U^{235} was to be separated from U^{238} by diffusion of UF_6 (a gas at room temperature) through a thin, porous nickel membrane. The lighter molecules of $U^{235}F_6$ had a higher average velocity by about 0.5% than $U^{238}F_6$, hence at each stage its concentration would be enriched by this amount. To go from a natural abundance of U^{235} of about 0.7% in uranium to, let us say, 25% would require about 7000 stages of diffusion if each one worked perfectly. Even small leaks in the barriers in the early stages would lower the efficiency to the point where the system would not operate. Of great concern was the fact that UF_6 was potentially reactive with nickel. Such a reaction would plug the pores with non-volatile UF_4 and NiF_2 and also close the system down. Since the system acted in a cascade fashion, no stage could begin until the earlier stages had begun to operate and had reached steady state. A failure in a few of the many units could shut the entire system down.

My research problem was to study the stability of the barriers to corrosion and plugging. At the time I began work, in early 1943, the barriers had not been developed, so only experimental samples were available to us. When Kellex had begun work on the plant design only 18 months earlier, Benedict and Johnson had been assigned to do whatever research might be needed! Within six months they concluded that a large effort was needed, and within the incredibly short time of six months, a swamp in New Jersey had been drained and a 20,000 square foot laboratory was built. This was at the height of the war effort when materials were tightly rationed and trained manpower had become scarce. Within a month after I was hired, I was promoted to group leader and our group was expanded from six to 24 people working three shifts around the clock.

For three months my two other group leaders and I took turns on the different shifts. I found the graveyard shift, 12:00 midnight to 8:00 A.M., most difficult to adjust to but it was an educational experience in a different style of living. It was also a period of learning for me—learning an engineering approach to scientific problems, learning to work with metal apparatus instead of my accustomed glass, and finally learning to direct research.

Because of the urgency of the project we had the highest AAA-1 priority in buying scarce materials, and every effort was made to facilitate our work. When technicians were unavailable, drafted personnel from the Army with scientific training were reassigned to us. However, the Army bureaucracy still exercised visible control. All capital equipment in excess of $100 had to be purchased by competitive bidding. Our solution to this was to relabel all items "supplies" and purchase them with petty cash by distributing the cost among as many vouchers as were needed to keep the total under $100 per item. When our petty cash account went suddenly from about $200 a month to about $10,000 we had a visit from the Army auditors and our purchasing scheme was terminated. We replaced it by one in which we purchased capital items in pieces, each a part of the whole, for less than $100.

At one point General Groves and the scientific directors, including Professor Urey, were to visit the labs. Incredibly, all work was brought to a halt while we spent two days cleaning up the lab for the visit. After a half-day show our visitors left and we plunged back into our usual disorderly routine.

It was an exciting period because of the urgency of our work and the understanding that we were also possibly in a race with the Nazis. It also carried with it many of the frustrations of mission-oriented research. A number of very intriguing observations we made seemed to cry out scientifically for further research and understanding. Because they were peripheral to our goals they had to be ignored.

There were the usual mishaps, accidents, and one very dramatic fluorine fire involving a large tank containing F_2 gas at 600 pounds pressure. Within seconds the lab was crossed by an intense 20-foot blue flame issuing from the tank nozzle. It was hard to understand how such a flame of F_2 in air could be maintained with no evident fuel but the moisture in the air. The entire lab was evacuated in less than a minute until we realized that the ventilation system was making the air outside heavy with HF and much more dangerous than the labs and offices. We quickly returned despite warnings from the firemen.

We made one major contribution, as usual in my experience, quite by accident. One day, opening up one of the copper lines used in our experiments for handling UF_6 and F_2 at temperatures up to 500°C, we discovered it to be bright and uncorroded by the usual green CuF_2. Following up on its history led to a method for passivating metals like Ni and Cu against corrosion by F_2 and UF_6. It is one of the few findings made on the Manhattan Project that has still, inexplicably, not been declassified. The discovery made possible a huge cost savings in the diffusion plant.

Near the end of the year, we had successfully completed our assigned work, and it was announced that the labs were to be moved to Oak Ridge. A few months earlier, in a response to an advertisement in *Chemical and Engineering News*, I had received a letter from Anton Burg at the University of Southern California (USC) in Los Angeles. He offered me a job as Assistant Professor at the salary of $2300 per year. I turned it down quickly, explaining my difficulties with this princely sum in New York. Three months later a return letter brought an offer of $2600, "In Los Angeles," I was assured, "an excellent salary." Discussions with Kellex elicited the counter-proposal to go either to Oak Ridge—I couldn't find it on the map and would go only if carried—or to Site X in New Mexico. Site X was somewhere in the mountains near Albuquerque. It sounded even less appealing. But I didn't appear to have much choice. Fate intervened most unexpectedly. I have never done well in New York winters and the Winter of 1943 had been most severe. I had come down with severe sinus infections and a chronic cold. Visits to the doctor brought the unhelpful suggestion that I move to a warmer, drier climate. Now armed with a letter from the doctor, I was able to obtain a release from the Manhattan Project. So near the end of 1943, I piled all my clothing, books, and furniture in the back of my once new 1939 Plymouth and set out again to California.

Los Angeles was beautiful, gleaming white in the brilliant Fall sun, composed of mainly single-story homes and low office buildings, small in population and vast in area. In strong contrast to the East and Midwest, it was spanking clean and surrounded by high mountains just beginning to be covered with snow. Smog hadn't yet made its appearance, and from the top of any building it seemed you could see forever.

USC was a private and almost rural University then, small in size and known mainly for its football team and professional schools. A former classmate from Harvard, Ron Brown, had just started as assistant professor in organic chemistry. UCLA, its neighbor, was the recently formed southern branch of the state university and was still struggling to establish itself. Two UCLA faculty members I had known while at Harvard, Tom Jacobs and Saul Winstein, were just starting their research programs. Best known to me by reputation was Francis Blacet, well recognized for his researches on photochemistry. Cal Tech was much better known. Linus Pauling had already become one of the dominant figures in American physical chemistry, and Don Yost and R. Badger were figures on the national chemical scene.

Although I had been hired (at the prompting of four younger, ambitious members of the six-man chemistry department) for my research promise, a 12-hour assignment in teaching relegated that to a hope more than a reality. In the winter semester, still inspired by experiences on the Manhattan Project, I organized for the evening program what I believed may have been the first course in Instrumental Analysis. In attendance were scientists from the local industry ranging in age from 40 to 78. They were a great class, motivated, knowledgeable, and attentive. My college program was largely freshman chemistry, two sessions each of about 200–250 students. They also were well-motivated and attentive, but not nearly so knowledgeable. Within six months, I was moved into war work once again. A government program was directed by Anton Burg, chairman of the chemistry department and on his way to becoming world famous for his research on boron chemistry. It was supported by Division 10 of the National Defense Research Council and was engaged in an intense effort to find a chemical stabilizer for cyanogen chloride (CNCl), a newly explored poison gas, discovered by the Germans, but not known to the Japanese. It had "great" lethal properties and could pass through the Japanese poison gas cannisters. Enormous quantities of 500 and 1000 pound bombs of the liquid were stockpiled in the Pacific Islands and in the West in preparation for the invasion of Japan to begin in 1945. It had the unfortunate property of polymerizing very exothermically and explosively to a dense plastic (melamine) or to an aromatic ring trimer.

This was for me a somewhat unwelcome digression and not exactly in my strongest areas of research. But it was again a useful learning period. I picked up some very ingenious techniques from Burg on the handling of sensitive, volatile materials in greaseless systems. One of them, using liquid propane as a refrigerant in gas separations, was nearly disastrous. Liquid propane was obtained from a large tank by allowing the gas to bubble into a Dewar of dry ice. As the gas liquified the dry ice evaporated. In about 15 minutes one could obtain 100 cc of liquid propane. While waiting for this to happen one day I stopped to make some calculations on a lab table in the middle of the room. Suddenly I became aware of a hissing sound and looking up saw a sheet of thin yellow flame approaching me. It stretched from the floor to the ceiling and came from near the lab door. I had just time to see it approach. It bifurcated in front of me, one sheet crossing the sheet of yellow paper on which I had been writing and burning a neat line across it. The other sheet went around my back. they joined on the other side and continued toward the casement window where the tank of propane was standing. Finally there was a "VROOM" and a roar. The door was blown shut, the window was blown open and several large bushes outside were blown absolutely flat on the ground. The dry ice Dewar was converted to tiny fragments, but strangely enough no other damage occurred. Some of my colleagues and students came in to see what had happened but I, standing stunned but unharmed, was relatively noncommittal. A lady passing by in the street outside complained to the university authorities and Anton Burg was questioned about the damaged foliage. I never told him of the explosion. I was lucky and relieved to come out in one piece.

Shortly after April 1945, and the end of the War in Europe, work on our CNCl project was intensified. Francis Blacet at UCLA was engaged in a parallel program. Two students in Army uniform were helping him at the time, Jim Pitts and Jack Calvert. After the war they completed their PhDs with Blacet. We had pleasant but intermittent contact with each other. With the shut-down of the European theatre of operations, Professor Morris Kharasch of the University of Chicago was assigned to our program, and within three months the problem was dramatically solved.

The Army had imposed some interesting boundary solutions on our problem. Because of the scale of the problem they decided that the CNCl bombs had to be stabilized in situ by means of some additive. Because an Army sergeant and privates would effect the remedy no solid additives could be used. The reason was that in opening the containers, a large screw closure had to be removed and then tightly resealed. It was unlikely, the Army decided, that available personnel could be relied upon not to get solid granules or powder into the exposed threads, thus jeopardizing the tightness of the seal when the screw was replaced. A loose seal was as hazardous as the potential explosions, ergo no solid additives.

Kharash in typical fashion did some quick and dirty experiments, sufficient to convince him he was dealing with a general acid, metal ion–catalyzed reaction. Chelating agents such as phosphates were shown to suppress this, so a good dose of solid sodium metaphosphate was the answer. There was no scientific problem remaining—only a personnel problem. The case was summarily closed and the ensuing discussion with Army brass led to a change in guidelines! It was frustrating to Burg, the thorough scientist, but illuminating to me.

World War II had catalysed transformations in American life that were little appreciated at the time. The sudden release to civilian status of millions of enlisted men, together with the GI Bill giving them educational benefits, sent a crashing wave of mature, motivated, hard-working students into the universities. It was a golden age in higher education. The GIs not only influenced the perceptions of the faculty as to how well students might perform, they also set a standard for their younger peers. This was to continue with the Korean War in 1949–1952, until almost 1960. For research scientists in physics and chemistry it was a golden age in a more material sense. For the first time, under the stimulus of the Office of Naval Research, government funds were made available to university professors. Summer employment could be paid and stipends made available for graduate students and post-doctoral research helpers. A new world had started! Within 20 years the US had changed from a minor to the major center of scientific research on the world scene. For me, 1945 started a process of examining for the first time what I wanted to do and how I might do it.

An assistant professor starting at a research university has five years in which to establish a research program and publish significantly. There is a very strong compulsion to continue work along the lines of a just completed post-doctorate research or the earlier PhD dissertation. The cost-benefit analysis of continuing research on photochemistry in 1946 was for me very negative. Too much input, very little output. Theoretical work, particularly in statistical mechanics, had a strong appeal for me. On the other hand, if I expected to work with students I needed an experimental program. Biochemistry seemed a relatively new field where physical chemistry might provide useful insights. A number of students expressed an interest in understanding catalysis and surface phenomena. Thus, I began a period of reading, learning, and reorientation. Kinetics was abandoned and I began some work on the theory of liquids and surfaces.

This kind of change in interest was not uncommon at the time. I had a number of current examples to inspire me. Linus Pauling at neighboring Cal Tech, after establishing a brilliant reputation in quantum chemistry and in x-ray and electron diffraction studies of solids, was now wholly involved in studying protein structure. John Kirkwood, just newly at Cal Tech, who had done such elegant work in statistical mechanics, had shifted to studying the theory of electrophoresis of protein molecules.

This was the prelude to what was to turn out to be for me a ten-year program on the structure of proteins explored by examination of their surface properties and their adsorption of polar and nonpolar gases. It was to be a very interesting program for me. In a few years, I was able to attract generous support from the National Institute of Health. Ironically, I was eventually forced to abandon it when it proved increasingly difficult for me to interest physical chemistry students in biochemical problems.

Another new research program began quite accidentally. During one of my lectures in freshman chemistry I had explained to the class how small polar water molecules could interact strongly enough with the ions in solid ionic salts to overcome the strong lattice forces completely and dissolve them. After the lecture one of the students approached me with the very innocent question, "Why doesn't steam dissolve salts?" I wrestled with that for some time after he had left, apparently satisfied with my facile answer, "There aren't enough H_2O molecules in steam." Finally, I decided that at high enough densities steam must be capable of dissolving salts. Some work in the library showed that no real information was available. A few engineering studies on steam turbines showed that, indeed, salt fouling of turbine blades could be a significant problem. In discussing the question further with a colleague, Charles Copeland, we conceived a simple experiment to resolve the question. We built a heavy-wall, stainless cell fitted with an auto spark plug. We filled it about one third full of a 0.1 Molar NaCl solution, tightened the spark plug, and breathlessly watched for a current flow as we heated it in an oven to the critical temperature of water, 374°C. At about 340°C we saw a small current and by 374°C it was much larger! A grant from the Research Corporation on the basis of these observations launched a more controlled study of supercritical salt solutions. Subsequent support by the Office of Naval Research made possible a fascinating study of supercritical solutions, which we extended to alcohols, SO_2 and NH_3. It was a nice field to work in, since we were pioneers and for many years the sole workers.

What was to be my last excursion into politics, in this case, very local, school politics, came just after the war ended in 1945. USC was then a very small school with about 6000 students and a faculty of about 350. Within a few months after arriving I had met most of my fellow faculty. The faculty lunch room had a very large round table where up to 16 at a time could be served. It was a real public forum. Very quickly I became aware that my salary was one of the highest in the school. Salaries hadn't changed since 1939 although the cost of living had slowly climbed during the war and then very rapidly after. Dissatisfaction with salaries was widespread and vocal. A Fall meeting called by President Rufus von Kleinschmidt to greet incoming faculty and boost morale turned into a fiasco when one of the older professors, Fred Weirsing, disappointed by what had turned into an empty pep talk, stood up and asked why, in the face of mounting income from all the new students, no changes had been made in salaries. Spontaneous applause from all sides suddenly made the mood very tense. V.K., as he was popularly known, a tall, stately, fatherly figure with a flowing white mane of hair, tried first to dissipate the mounting feeling with some cursory statements and then finally adjourned the meeting.

As a relative newcomer to the faculty I had a somewhat dispassionate reaction to the proceedings, but months of lunchtime discussions had made me keenly aware of widespread injustices and a very high-handed and dictatorial management. I decided that the moment was ripe for stronger action. I spent the next few hours in conversations with some friends who I believed had the support and the credibility of their colleagues and agreed to convene a rump meeting of the faculty. The response was overwhelming. Two days later about 100 faculty met to elect an executive committee to draw up a program for action. My experiences at Harvard stood me in good stead, and my facility at mental arithmetic established me as one of the active members. Not much in the way of arithmetic was required to calculate from the number of new students and the tuition the additional revenue and, even further, to estimate the salary budget. There could be no financial reason for not adjusting salaries! In a short time what became the largest single branch of the Association of University Professors had organized and become a virtual trade union. It was a real faculty revolution. Within six months over 95% of the faculty had deposited signed pledges saying that they would not sign their next year's contracts before holding a joint meeting in the Fall. I was given charge of these pledges, which also unprecedentedly contained the first revelation of the faculty salaries. Now I was able to do a very precise estimate of the faculty budget. Salaries turned out to be much lower than anyone had imagined!

Some very stormy meetings of our executive committee with the President and finally with the Board of Trustees led to meeting all of our demands. The revolution had succeeded! It was an exhilarating experience, but a few disagreeable events along the way convinced me that my future efforts would be better spent doing science. In 1949, I was elected Vice-President of the newly organized Faculty Senate, and six months later I resigned.

Before the war a serious American scientist could expect to spend one to two years as a post-doctoral in England, Germany, or France. Europe was clearly the world center for all major sciences. Although the war had completely altered that situation, my psychological outlook had not changed. A sabbatical year in Europe was still a fond goal. I had studied French and German in high school and had continued French in college. France had become a romantic dream. Guggenheim and Fulbright Fellowships in 1950 turned the dream into reality. In September, I arrived in Paris, where I had been offered space at the Laboratoire de Chimie Physique. Professor E. Bauer, a distinguished scientist of the old school, presided over physical chemistry, and Professor Michel Magat was actively in charge of the programs, principally in kinetics and polymer chemistry. European science was a relatively compact, tight community. Everyone knew everyone else, not only in their own fields, but in neighboring fields. My visit was propitious. Victor Weiskopf, a physicist from MIT, was doing a year's sabbatical in the next building, and Sam Eilenberg from Columbia was a guest of the local mathematicians. Paris was a central point in Europe, and it was a rare week that didn't bring from three to five distinguished visitors to the laboratory.

It was a stimulating year. For the first time in many years I had no fixed schedule to meet and no obligations. I could read, discuss and think. My project was to write a book on chemical kinetics. I had taught it for five years at USC at the graduate level, my notes were fairly extensive, and I thought I could finish it in one year. As it turned out I wrote about half in France and then spent the next eight years, mainly summers, completing it. Had I known how long it would take I would probably never have started. My decision to write the book stemmed from some recent advances in kinetics that seemed to have profound implications.

In 1950 Noel Slater, a mathematician at Manchester, used a theorem discovered by Mark Kac on the roots of trigometric equations to solve a very important problem in statistical mechanics. Treating a molecule as made up of coupled harmonic oscillators, he had derived the result of unimolecular rate theory on the frequency of passage through a given energy distribution. It yielded an A-factor in the range 10^{13}–10^{14} sec^{-1} for all unimolecular reactions and related it to the actual molcular frequencies. This had rekindled my interest in kinetics. In addition, Rudy Marcus, a post-doctorate student working with Oscar Rice, had been able to unify Eyring's Transition State Theory with the Rice–Ramsperger–Kassel theory of unimolecular processes to provide a quantitative model for all such reactions. I felt the time was ripe to review the relatively small amount of experimental data and reestablish chemical kinetics on a firmer base. The historical background needs some elaboration.

Chemical kinetics was a relatively recent science. Gas phase kinetics had started with Bodenstein in 1890 and the theoretical basis for it rested on the Arrhenius formulation from his famous 1888 paper plus the prior background in statistical mechanics. Chain reactions were discovered in 1916 by Nernst, and branching chains were discovered by Semenov in 1923. In 1923 Lindemann proposed the collisional mechanism for energy transfer, and within three years Rice and Ramsperger and then Kassel showed how this could account for the pressure dependence of the apparent rate constant for unimolecular reactions at low pressures. This was followed by a deluge of activity in the 1920s and 1930s led by Hinshelwood in England and Steacie in Canada to characterize all pyrolysis reactions as having rates determined by unimolecular fission processes. The development in 1934 of "Absolute Rate Theory" by Eyring and his colleagues seemed to complete the molecular framework of chemical kinetics.

The picture slowly unravelled after the war. By 1948 every reaction considered unimolecular before had been shown to be a complex chain process except for two, the isomerization of cyclopropane to propylene and the decomposition of nitrogen pentoxide:

$$N_2O_5 \rightarrow N_2O_4 + \tfrac{1}{2}O_2.$$

However, the latter fit none of the criteria of unimolecular rate theory (RRK) whereas the former had an abnormally large Arrhenius A-factor. Richard A. Ogg was able to resolve the N_2O_5 dilemma by showing that it must be a complex radical process. Thus by 1950 kineticists had an elegant theory and no reliable data. The stage seemed ready for a second effort to do some selected experiments in gas kinetics.

The year in Europe was an unparalleled opportunity for me to visit all the active laboratories. I went to London, Manchester, Birmingham, Oxford, and Cambridge. M. C. Evans and M. Polanyi were the leading figures in British kinetics at this time. Hinshelwood had just turned to biochemical kinetics of cell growth and, with the recent third edition of his famous text, seemed to have lost all interest in gas phase reactions. He was still directing a few graduate students when I visited him but not with much real interest. Evans was at the peak of his career at Manchester, while Polanyi had just resigned from chemistry to go into economics. Evans's premature death of cancer a few years later and Polanyi's retirement were an enormous loss to physical chemistry. However, Porter and Norrish were just completing their classical studies on flash photolysis at Cambridge, while a young Fred Dainton was establishing a reputation at Oxford. Christopher Ingold and colleagues at University College, continuing their elegant work on solution kinetics, were at their peak. H.

Melville was completing a dazzling career in kinetics and polymer chemistry at Birmingham and would soon become an elder statesman.

Eric Rideal, already elder statesman, was at the Royal Institution but still active in studying catalysis and surfaces, while Dennis Riley was doing his so-important x-ray studies of DNA. The Faraday Society Symposia that I was able to attend were a thrilling introduction to how scientific meetings could be run. Authors' manuscripts had been pre-distributed, they were given five minutes to summarize their findings, and then the fun began. The Discussions were just that—real discussions! Guggenheim, then at Reading, was his famous caustic self and woe to the unwary scientist who might draw his scathing attention!

In France the leading kinetics group was that of Letort at Nancy. Sadron was just starting polymers at Strasbourg. Lefort at Lyons had an active program in catalysis. Best known on the European mainland was the Brussels' group, with Prigogine in theoretical work and Goldfinger in kinetics. In Germany, Jost and Bonhoeffer were just beginning to reconstruct their laboratories and Eigen was becoming known for his work on fast reactions. I was not prepared for the warmth with which I was received on my visits. Even more surprising was the respect and even deference shown. European science was organized on very authoritarian lines. The professor of a department usually had lifetime tenure and total command of a state-provided budget. He could literally hire and fire at will. I was a recently promoted associate professor at USC. The title didn't exist in Europe. My hosts everywhere treated me like a European professor, even those who knew the plethora of assistant associate and full professor titles in the States. They couldn't divorce the title Professor from the European image.

We had quite a number of American visitors to Paris. These visits gave me an opportunity to get better acquainted with some of my compatriots. These included Jack Aston of Pennsylvania State University, then on sabbatical in Leyden, Ken Pitzer of Berkeley, then Commissioner of the Atomic Energy Commission, and Gerald Oster, a post-doctoral at the lab. Stan Ulam, a former colleague in mathematics briefly at USC and a once-active participant in the Manhattan Project, most recently from Los Alamos, made a surprise visit in the Spring of 1951. He cheerfully informed me at lunch one afternoon that the US had detonated an H-bomb and made a Pacific Island vanish. It was still secret and he didn't tell me about the role he had played in designing it. This was in the heyday of the McCarthy Committee and during the Korean War, a time punctuated by scandalous accounts on all sides of American and British spies serving the Soviet interests. I asked how he managed to come to Europe with his information. He directed my attention to the next table (at the Cafe de la

Paix where we were lunching). "They go everywhere with me, those two men. I'm sure if I talked to the wrong people I might have an accident." It didn't seem to perturb his usual good cheer.

I returned from Europe with plans for beginning a new program in kinetics. It had become clear to me during my sabbatical that one of the basic problems of kinetics was mechanism. It was pointless to speculate on the transition state of a chemical reaction until one was very certain as to what reaction one was observing. I searched long and hard for a simple system as free from chemical complexity as possible. The decomposition of N_2O to give $N_2 + O_2$, which was well studied, seemed too complex. Small amounts of NO were produced and it was not clear what role they might play, since NO was a known catalyst for the decomposition. Among the simply systems only ozone, O_3, fit the requirements. Only three species existed in the system, O, O_2, and O_3. Only one of these was reactive, O atoms. Arthur Axworthy, a new student, was interested and we began what became my most hazardous research. Our first experiment with pure O_3 at 100°C exploded, destroying our beautifully constructed equipment and leaving only fine particles of pyrex. Axworthy, miraculously, came out of it with only minor cuts. Over the next three years about one out of three experiments exploded. But we were well prepared with explosion barriers, so only a day was required to rebuild. Axworthy became a superb glassblower.

I describe the ozone work because it provides an excellent example of the connectedness of science and technology. We began it with no thought of any practical applications or idea that ozone might be of more general interest. I had no idea of the stratospheric ozone layer in 1952, let alone its significance. By 1957, when we published our findings, we had become "world experts" on a material that had become important enough to have an International Symposium devoted to it. We were able to verify the original two-step Chapman mechanism (1906) for its decomposition, prove that the 1934 16-step mechanism was incorrect, and provide the first detailed rate constants for the two steps. We also showed the significance of excited $O(^1D)$ in the photolysis and explained the surprising catalytic effect of H_2O on the photolysis.

This experience was typical of the connections that have occurred over and over in my own work and in the work of many of my colleagues. When photochemical smog was discovered in Los Angeles in 1946 and later ozone was shown to be one of its active agents, I was pulled into work on air pollution that has continued over 30 years. This work on air pollution excited an interest in oxidation and peroxides and this in turn has involved me in combustion research since 1964.

The decision to go back into chemical kinetics was a fortunate one. It was a propitious moment in the development of the field, and some of my most important discoveries were to be in kinetics and its closely coupled field, thermochemistry. Perhaps one of the best known, Group Additivity, began very innocently in 1957 with an analysis of elementary radical-molecule reactions. One of my very dedicated graduate students at this time was Jerry Buss. He had the most important gift of learning— thoroughness. We had many protracted discussions during the work on his degree. I never felt certain about an idea until I had succeeded in convincing Jerry of its validity. When he came up with a new idea it was almost certain to be important and insightful. His self-critical ability was unusual.

One bright winter day, just months before his thesis was completed, Jerry approached me with some intriguing observations. We were interested in the C–H and C–C bond strengths in some alkanes, ketones, and aldehydes. In almost every case the available data were insufficient for a comparison. However, closely related thermochemical data on heats of formation ($\Delta_f H°$) were available, and Jerry had noticed that $\Delta_f H°$ for CH_3CHO was equal to the average of CH_3COCH_3 and CH_2O. Then followed an exciting two weeks of scouring the thermochemical literature and finding similar results for $ClCH_2CH_3$ and the related symmetrical pair $ClCH_2CH_2Cl$ and CH_3CH_3. The obvious generalization was that in disproportionation reactions of the type

$$RAR + SAS \rightleftharpoons 2RAS$$

$\Delta H \simeq 0$.

We quickly discovered that it worked for molar heat capacities, for intrinsic entropies, and for virtually all molecular properties. Two further months of euphoric discoveries were finally distilled into a series of five short notes to the Editor of the *Journal of the American Chemical Society* (*JACS*). Jerry graduated with honors and went on to his first well-paid job, with A. D. Little in Boston. I started a senior post-doctorate year at Cal Tech, ostensibly to complete my exhausting *Foundations of Chemical Kinetics*. I've never had a busier year. My son Nick was born, my wife started law school, and I commuted to Cal Tech.

In August a deflating letter from the Editor informed me that the notes could not be published. They were not that new, they should be consolidated, and they should be related to earlier reported and similar observations. Thinking wicked thoughts about the long-since departed but sorely missed Jerry, I put aside my writing and spent a month in reflection and rewriting. The result was the discovery of what I like to think of as Raoult's law for molecules, or a generalization of all laws of additivity of molecular properties. Group Additivity was the second order level and the most accurate. It has since received widespread acceptance in physical chemistry and engineering.

Jerry Buss returned to Los Angeles, made an outstanding record in Aerospace Research, first with Douglas Aircraft and later with STL (now Aerospace). He soon decided that the challenges in aerospace were neither sufficiently interesting nor rewarding. He went instead into real estate and within ten years became an American success story. Today, he is a popularly recognized sports figure, and current owner of the world-famous Los Angeles Lakers basketball team. Most sports audiences don't realize that Dr. Buss, as he is popularly referred to, has his degree in physical chemistry.

In 1964, just a year after I had begun research in the newly installed department of Chemical Kinetics and Thermochemistry at the Stanford Research Institute (SRI), I had a similar experience. I had received a short paper to review from *JACS* on the B–B bond strength in diborane (B_2H_6). Since I had just published a paper on the subject that I felt fixed the bond at 35 ± 1.5 kcal/mole, I was somewhat outraged to read an experiment that raised it to over 54 kcal. My first impulse at such moments was simply to stop reading—the result was certainly in error! However, I had a very persistent, bright young post-doctorate working for me, one David Golden. David was to be one of my closest friends and colleagues during my 13-year stay at SRI. He proved a wonderful counterpoise for my impatience (impetuosity?). He was and is a meticulous and careful research worker, bright, energetic and indefatigable. When David felt some piece of research was done, it was really done; sometimes, in my opinion, overly done. I never felt comfortable with a result until it had Dave's blessing.

In this instance of diborane he kept nagging me. Why was the experiment wrong? Finally convinced that it deserved some thought, I pondered over it, eventually realizing that the experiment had been done at such low pressures (~ 1 mtorr) that the unimolecular rate had fallen off many powers of ten from its high pressure limit. The very low rate constant could only be rationalized by an apparently high activation energy, hence bond strength. This reflection was to be very rewarding. No one who has worked in close proximity to Anton Burg is unaware of the fact that boron hydrides are extremely sensitive to surfaces and impurities. In this case, I was astonished at the apparent stability of B_2H_6 (up to 700°C) in the presence of a very catalytic stainless steel reactor. Under normal conditions (760 torr) it decomposes rapidly in glass at 100°C. Surface-catalyzed reactions usually slow down at high temperatures and low pressures, since adsorption equilibrium favors the gas phase under these conditions. This suggested an experiment, and the technique of Very Low Pressure Pyrolysis (VLPP) was born. Another colleague, Neil Spokes, and I designed it.

In VLPP a reactant was allowed to flow into a Knudsen Cell (300–

1100°C) at pressures less than 1 mtorr. Under these conditions all unimolecular reactions tend to be at their low pressure limits, and all collisions are with the walls. A molecular beam exiting the reactor orifice (1–5 mm) is analyzed for reactants and products by a mass spectrometer. High pressure limiting rate constants can be estimated using RRKM Theory with the wall as a strong collider. The technique has given valuable insight into unusual fission reactions and has been adapted to studying both bimolecular and surface reactions. A variant technique developed in the past ten years at USC, the Very Low Pressure Reactor, has been used to get very accurate rate and equilibrium constants in systems such as:

$$RH + X \rightleftharpoons R + HX.$$

Under VLPR conditions, concentrations of radicals and atoms can be so low that secondary rections do not occur.

I was very grateful to Dave for his nagging. An important incentive for his persistence was the fact that Fred Stafford, the author of the B_2H_6 note, was an old friend. It later turned out that Fred was an alumnus of Camp Rising Sun, a small, unusual, summer scholarship camp in New York. A generation earlier, in 1934, I had attended the camp. It had had a strong influence on my life and attitudes, which I still recollect with fondness. Science is a small world!

World War II sponsored a vast mobilization of scientific and technical manpower, mainly on the home front. The Cold War that followed led to a revolution in scientific research both in its conduct and its funding. Starting in the physical sciences with the Office of Naval Research, then Air Force, the Army, and the Atomic Energy Commission (AEC), a parallel explosion in the life sciences was supported by the Public Health Services (PHS) (now the National Institute of Health). This culminated in 1952 with the installation of the National Science Foundation (NSF). By 1957 the scientific population had more than quadrupled. With Sputnik in 1957 and the development of the National Aeronautical and Space Administration (NASA) from the early NAACA, scientific funding in American universities had become predominantly federal.

In this same period, the Chemistry Department at USC grew from a 7 to a 12 to a 17-member department. Its members were young, almost uniform in age distribution, enthusiastic, and devoted to research. Although living in the shadow of Cal Tech in Pasadena and the more distant shadow of UC Berkeley, physical chemistry was surprisingly strong. A strong surface science group was possibly unique in the US, consisting of Robert and Marjorie Vold, Todd Dosher, Karol Mysels, and Arthur Adamson. Physical chemistry had, in addition to Charles Copeland (ionic solutions) and myself, Harold Friedman, Harry Frisch, Jerry

Donohue (X-rays), and Robert Simha. These together with Anton Burg, Wayne Wilmarth, and Jim Warf in inorganic chemistry and Jerry Berson and Norman Kharasch in organic chemistry, managed to attract enough research support to be counted among the top ten departments in research productivity (measured in papers/faculty). We were also fortunate in attracting a good, talented, hard-working graduate student enrollment. In retrospect I would be inclined to say anomalously good.

For me, the postwar period was a very happy and productive one. I could indulge my scientific tastes, collaborate with many of my colleagues, and enter the world of consulting. With support from Goodyear, we did work on polymers. NSF and AEC supported free radical studies. Funding from ARO allowed us to do ozone and chlorine kinetics. PHS continued supporting our protein work while the Office of Naval Research (ONR) supported our work on supercritical salt solutions. Consulting work at Douglas Aircraft allowed us to initiate some of the pioneering studies of collisional energy transfer by computer simulation.

When in 1963 my wife's illness forced us to move from Los Angeles to the Stanford Research Institute (now SRI International), very little of this changed, including the consulting, which extended by then to the Jet Propulsion Lab (JPL) and the Aerospace Corporation. This is probably a result of the federal nature of research funding.

SRI was an interesting introduction into the world of sponsored research, both federal and industrial, and provided a wonderful opportunity for me to see how the chemical industry operated. Most importantly, it gave me immediate and broad contact with a large and extraordinarily talented group of scientific and engineering professionals. When it finally severed its close ties to Stanford after 1970, it became more income-driven, and I decided after a few years with many mixed feelings to return to academia. In 1976, I returned to USC.

Fifty years have now passed since I started my first graduate studies at Harvard—a period of extraordinary development in physical chemistry. The explosive growth in electronics, optical methods of analysis, gas chromatography, mass spectrometry, and most recently computers has given us the facility to ask questions of nature we would never have dreamed of asking even 20 years ago. As a veteran of this period, I am not sure that I want even to know the answer to some of the questions my junior colleagues are now asking, but I cannot help but recognize, admire, and salute their expertise and enthusiasm. In the 60 years since Dirac added the last piece to the conceptual background of atomic science, chemistry has exploded. This explosion might be described as being application-driven. In this sense it is amusing to me to hear the endless discussions that go on among scientists concerning the various virtues of applied or basic research. No research is funded today without the author describing succintly the possible results of that research and their potential importance. As beauty is in the eye of the beholder so does importance lie in the ear of the importuned.

More clearly than ever before scientific research is a societal endeavour. It draws its support and its ambiance from society. If it is not responsive to the needs of that society it will not long endure. This is not to say that all research must be practical. Society has an endless need to be amused or entertained, to indulge in fantasy, and this is only slightly less to be addressed than more pressing problems such as hunger, disease, and war. A large amount of research in astrophysics and particle physics I would place in the entertainment category, but that should not diminish the consideration it is given.

A different problem affects research today, namely, the resources devoted to its performance. Big and small science are no longer negligible items in the national (or international) budgets. Science has been so successful that it has grown huge over the last 50 years. How much support can it expect? I do not know of any simple answers to that question. How many symphony orchestras should there be? How many hospitals?

As I approach my fiftieth year of research and teaching, I rediscover a typical phenomenon that touches teachers. Everyone around me seems to be the same age as when I started teaching but my reflection in the morning mirror has unaccountably aged. My feelings of excitement about science and discovery are just as strong as they always were. Periodically, just when I feel that no new findings can be as important as the ones already made, I find myself looking at phenomena from a totally new perspective and getting as agitated as I did at 20.

Things are easier for me now. I have worked in almost every area of physical chemistry so that most problems seem familiar. Pondering the more difficult, still unsolved problems is fun. Nonbonded interactions in large molecules, the structure of H-bonded strongly polar liquids, intrinsic activation energies, metallic bonding, the energetics of bonds on catalyst surfaces, correlation of electron motions in atoms and molecules, how big is a photon—these are some of the more interesting of the unsolved problems. Every now and then we stumble on some totally unexpected discovery. Within the last three years, one of my more enthusiastic and hard-working post-doctorates, Mohammed Hisham, succeeded in finding a set of simple empirical relations that permit us to estimate the heats of formation of any salt to within about 2 kcal/mole. We had started by trying to understand the thermochemistry and kinetics of some common catalytic process and unexpectedly ended predicting the thermochemistry of solids. We also feel we have discovered some new insights in catalysis and metal bonding. Also unexpectedly, out of this has come a new and economical process for the chemical conversion of HCl into Cl_2.

In the last year, I have been amusing myself by going back to quantum mechanics and reexamining the way the wave equation is solved. My intuitive feeling is that electrons are localized relative to each other in both atoms and molecules—not only in radial shells but angularly. I do not believe, for example, that the two electrons in the ground state He atom ever find themselves on the same side of the nucleus. Our best wave functions say they do. I think they are wrong. They have too much electron-electron repulsion and compensate by decreasing the electron-nucleus separation to give a good energy—but possibly a bad wave function. Can I do better? Should I even try? I do not know, but it is fun to think about.

ROTATIONAL DISTRIBUTIONS IN DIRECT MOLECULAR PHOTODISSOCIATION

R. Schinke

Max-Planck-Institut für Strömungsforschung, D-3400 Göttingen, Federal Republic of Germany

INTRODUCTION

Photodissociation of polyatomic molecules through the absorption of UV photons is an important and interesting topic of physical chemistry. Although it has been amply studied in the past, photodissociation is still an active field of current research. It combines aspects of absorption, molecular motion in electronically excited states, dissociation, and unimolecular reactions. A photodissociation process is usually envisioned to proceed in two, more or less separate steps according to

$$\text{ABC}(\tilde{X}, E_i) + h\nu \xrightarrow{(1)} \text{ABC}^*(\tilde{A}, \tilde{B}, \ldots) \xrightarrow{(2)} \text{A} + \text{BC}(e, n, j) \qquad 1.$$

where we assumed, for simplicity, a triatomic molecule ABC with one dissociation channel A+BC. Here E_i specifies a particular rotational-vibrational level within the ground electronic state $\tilde{X}$, ABC* denotes the activated complex within an excited electronic state $\tilde{A}, \tilde{B}, \ldots$, and (e, n, j) specifies the internal (electronic-vibrational-rotational) state of the product molecule. Step 1 in Eq. 1 represents the absorption process whereas Step 2 describes the fragmentation of the excited complex into products A and BC.

The observables are the *absorption spectrum* (total dissociation cross section), $\sigma_{\text{abs}}^{(i)}(\nu)$ as a function of the photon energy and the *product state distributions* $P^{(i)}(e, n, j | \nu)$ defined as

$$P^{(i)}(e, n, j | \nu) = \sigma^{(i)}(e, n, j | \nu) / \sigma_{\text{abs}}^{(i)}(\nu) \qquad 2a.$$

where $\sigma^{(i)}(e,n,j|v)$ are the state resolved (partial) dissociation cross sections and

$$\sigma^{(i)}_{abs}(v) = \sum_{e,n,j} \sigma^{(i)}(e,n,j|v). \qquad 2b.$$

The superscript (i) identifies the initial state of the parent molecule.

The total dissociation cross section is the probability for absorbing a photon with frequency v irrespective of the particular product state populated after the fragmentation. The partial dissociation cross sections are the probability for absorbing a photon hv and for populating a specific fragment state. Total absorption spectra were measured for most stable molecules a long time ago (1, 2). With the advent of powerful laser sources and efficient detection methods, modern experiments focus primarily on the resolution of internal state distributions (3–8), which undoubtedly contain richer information on the dissociation mechanism. The energy dependence of the spectrum "reflects" mainly the absorption Step 1 whereas the product distributions (under certain conditions) "reflect" primarily the nuclear motion in the excited state, i.e. the fragmentation Step 2.

If the potential energy surface within the excited electronic state, $V_{ex}(\tilde{A}, \tilde{B}, \ldots)$ is purely repulsive along the dissociation coordinate R_{A-BC} (the bond that is broken), the decay of the complex is fast (~ 1 fs) and *direct*. In this case the absorption spectrum is broad and in general structureless. If V_{ex} has a barrier along the dissociation coordinate, the activated complex might live for some picoseconds (or longer) before it breaks apart. The absorption spectrum for *indirect* (or pre-) dissociation processes usually exhibits "vibrational" structures, and the width of such "diffuse bands" is more or less directly related to the lifetime of the complex.

The second step in Eq. 1 can be considered as the second half of a scattering process evolving on the excited state potential energy surface. Indeed, all the features known from collision dynamics, such as rainbows and resonances, for example, can be also observed in dissociation. Moreover, all approximations and numerical methods normally utilized in scattering theory can be adapted for photodissociation with only slight modifications. Differences between scattering and dissociation arise through the "initial conditions": The initial conditions in scattering are defined when the "reactants" are infinitely far apart, whereas in dissociation, they are determined through the motion of the parent molecule within the ground electronic state. The dissociation process starts at close distances where the intermolecular forces are strongest. Secondly, in collisions many terms have to be included in the partial wave expansion of the scattering amplitude. This "averaging" not only complicates the calculations but it also might obscure some distinct dynamical features, such as resonances, for example. Due to the dipole selection rule for optical transitions, normally very few partial waves have to be included in the dissociation amplitude, especially if only low rotational states of the parent molecule are initially populated, as in a molecular beam, for example. This can enormously facilitate theoretical studies and the interpretation of measured data.

The final product state distributions are determined by both steps in Eq. 1: According to the Franck-Condon (FC) principle, the absorption step defines an initial probability $P_o(e, n, j)$ at $t = 0$. As the molecule dissociates P_o will be changed through the intermolecular forces induced by the excited state potential V_{ex}. This is known in photodissociation as *final state interaction*, or similarly in collisions as translational-vibrational-rotational energy redistributions. The extent to which P_o is modified depends exclusively on the strength of the final state interaction, i.e. it depends ultimately on the multidimensional potential energy surface V_{ex}. Therefore, model studies that completely neglect intermolecular coupling in the exit channel are in principle unrealistic.

Because of lack of space I discuss in this review only rotational state distributions that are nowadays routinely recorded for many diatomics, including OH, NO, CO, SH, etc. Most of the conclusions are readily applicable to vibrational distributions as well. Moreover, I confine the discussion to direct photodissociation for which particularly clear interpretations have been established in recent years. It is not our goal to review the extensive list of experimental results that has been compiled elsewhere (3, 6). I discuss some (but not all) general aspects and illustrate them by a few selected examples. In each case the excited state potential energy surface, which plays the central role in the dynamics of photodissociation, is known from ab initio calculations, and therefore the experimental distributions can be directly compared with theoretical predictions. According to the strength of the final state interaction I distinguish among *weak, strong, and extreme coupling* and organize the article following this scheme. In each class the fragmentation dynamics is quite different. In the beginning I outline very briefly the standard quantum mechanical and classical methods.

THEORIES OF MOLECULAR PHOTODISSOCIATION

Theoretical methods to calculate detailed photodissociation cross-sections have been recently reviewed by several authors (9–11). Therefore I describe here only the basic aspects necessary to follow the subsequent discussion of the various examples. For simplicity I consider the triatomic ABC system and include only the rotational (i.e. bending) degree of freedom. Extensions to incorporate also product vibration are formally trivial. In addition we assume that the total angular momentum is zero within the ground as well as the excited electronic state. This simplifies the notation remarkably and makes the equations more transparent.

The best choice of coordinates is particularly important for the theoretical treatment of photodissociation processes. Although the motion of the parent molecule is usually described in *normal coordinates*, the fragmentation process within the excited state must be described in *Jacobi-coordinates*, which are defined in Figure 2(a). Since the calculation of the dissociation (i.e. scattering) wavefunction is the most difficult part, it is advisable to use Jacobi-coordinates in both electronic states (9). Although it is not essential to use the same set of coordinates, it will significantly simplify all quantum mechanical as well as classical equations without complicating the actual calculations.

Time-Independent Quantum Mechanical Theory

Assuming weak light intensities, photodissociation cross-sections are quantum mechanically calculated from the well-known Golden Rule expression (9)

$$\sigma^{(i)}(j|v) = v |\langle \Psi_{gr}^{(i)}(R,\gamma|E_i) | \mu(R,\gamma) | \Psi_{ex}^{(-j)}(R,\gamma|E) \rangle|^2 \qquad 3.$$

where $\Psi_{gr}^{(i)}$ is the nuclear wavefunction of the parent molecule in state (i), μ is the transition dipole function for the electronic transition, and $\Psi_{ex}^{(-j)}$ is the dissociative nuclear wavefunction within the excited electronic state. The superscript $(-j)$ indicates outgoing boundary conditions in rotational state j. $\Psi_{gr}^{(i)}$ is determined by the ground state potential $V_{gr}(R,\gamma)$ and $\Psi_{ex}^{(-j)}$ is determined by the excited state potential $V_{ex}(R,\gamma)$. In Eq. 3 we omit all prefactors that do not affect the frequency- or the rotational state-dependence of the dissociation cross section. Both wavefunctions are exact solutions of the corresponding Schrödinger equation. They can be calculated, for example, by the close-coupling (CC) method, which is nowadays routinely used in collision studies.

The rigid-rotor Hamiltonian for total angular momentum $J = 0$ is given by (α = gr or ex)

$$\mathscr{H}_\alpha(R,\gamma) = -\frac{1}{2m}\frac{\partial^2}{\partial R^2} + \left(B_{rot} + \frac{1}{2mR^2}\right)\mathbf{j}^2 + V_\alpha(R,\gamma) \qquad 4.$$

where B_{rot} is the rotational constant of BC and m is the A–BC reduced mass. Expanding both wavefunctions in terms of eigenfunctions of $\mathbf{j}^2$ according to

$$\Psi(R,\gamma) = \sum_j \chi_j(R) Y_{j0}(\gamma,0) \qquad 5.$$

yields the CC equations

$$\left[-\frac{d^2}{dR^2} - k_j^2 + \frac{j(j+1)}{R^2}\right]\chi_j(R) = 2m \sum_{j'} V_{jj'}^{(\alpha)}(R)\chi_{j'}(R). \qquad 6.$$

The wavenumbers are defined by

$$k_j^2 = 2m[E - B_{\text{rot}} j(j+1)] \qquad 7.$$

with $E = E_i$ for the ground state and $E = E_i + h\nu$ for the excited state. The coupling elements that contain all the translational-rotational inelasticity ($T \to R$ energy redistribution) are given by

$$V_{jj'}^{(\alpha)}(R) = 2\pi \int_0^\pi d\gamma \sin\gamma\, Y_{j0}(\gamma,0) V_\alpha(R,\gamma) Y_{j'0}(\gamma,0). \qquad 8.$$

The coupled equations for the ground state wavefunction have to be solved subject to the boundary conditions $\chi_j(R) \sim 0$ for $R \to 0$ and $R \to \infty$. The coupled equations for the continuum wavefunction have to be solved subject to the boundary conditions $\chi_{j'} \sim 0$ for $R \to 0$ and

$$\chi_{j'}^{(-j)}(R) \underset{R \to \infty}{\sim} k_j^{-1/2} \exp(+ik_j R)\delta_{jj'}$$
$$+ k_{j'}^{-1/2} \exp(-ik_{j'} R) S_{jj'} \qquad 9.$$

for open channels ($k_{j'}^2 > 0$) and $\chi_{j'}(R) \sim 0$ as $R \to \infty$ for closed channels ($k_{j'}^2 < 0$).

Inserting Eq. 5 for the two nuclear wavefunctions into the Golden Rule expression yields for the rotationally resolved dissociation cross sections ($\mu = 1$)

$$\sigma^{(i)}(j|\nu) = \nu |t^{(i)}(j|\nu)|^2 \qquad 10a.$$

$$t^{(i)}(j|\nu) = \sum_{j'} \int_0^\infty dR\, \chi_{j',\text{gr}}^{(i)}(R) \chi_{j',\text{ex}}^{(-j)*}(R) \qquad 10b.$$

where $t^{(i)}(j|\nu)$ is the dissociation amplitude. The final cross-section expression would be much more involved if different coordinates and different expansion functions were used to describe the ground state wavefunction.

If the ground state wavefunction is separable in R and γ, i.e.

$$\Psi_{\text{gr}}^{(i)}(R, \gamma) = \phi_{\text{stretch}}^{(i)}(R)\phi_{\text{bend}}^{(i)}(\gamma), \qquad 11.$$

the dissociation cross section can be written in the more compact form

$$\sigma^{(i)}(j|v) = v \left| \sum_{j'} a_{j'}^{(i)} S_{jj'}^{hc(i)} \right|^2 \qquad 12a.$$

$$S_{jj'}^{hc(i)} = \int_0^\infty dR\, \phi_{\text{stretch}}^{(i)}(R) \chi_{j',\text{ex}}^{(-j)*}(R) \qquad 12b.$$

$$a_{j'}^{(i)} = 2\pi \int_0^\pi d\gamma \sin\gamma \phi_{\text{bend}}^{(i)}(\gamma) Y_{j'0}(\gamma, 0) \qquad 12c.$$

where $S_{jj'}^{hc}$ can be interpreted as a *half collision S-matrix*. The numbers $a_{j'}^{(i)}$ are the coefficients of the expansion of ϕ_{bend} in terms of free rotor states. Equation 12 is exact, provided that the factorization in Eq. 11 is correct—such is usually the case for low vibrational states. Formal expressions similar to Eq. 12a have been given before, using, however, more severe approximations (see, for example, the discussion of Shapiro & Bersohn (9)). Equation 12 is a convenient starting point to discuss quantum mechanically the influence of the final state interaction in the exit channel. As pointed out by Shapiro & Bersohn (9), the coefficients $a_{j'}$ can be regarded as the relative probability amplitude to populate state j' of the free rotor by photon absorption, and the half collision S-matrix $S_{jj'}^{hc}$ can be viewed as the amplitude for making a transition from state j' to state j within the exit channel.

Several methods have been suggested and applied in the literature to calculate dissociation cross sections exactly (9–13). Also, time-dependent propagation methods to calculate the matrix element in Eq. 3 have been developed (14–16). The CC approach is in principle exact and applicable to any degree of freedom. However, it has two major drawbacks: Applications are limited by the number of coupled channels, which, unfortunately, increases rapidly with the complexity of the molecule. Secondly, it can be quite cumbersome to interpret the final results in order to unravel the dissociation dynamics. Any approximation established in collision theory can be readily applied to dissociation. Most notable is the so-called *energy sudden approximation* (ESA), which is based on the fact that all wavenumbers k_j^2 in Eq. 6 can be replaced by an effective wavenumber k_{eff}^2 independent of j. This requires that the total energy E is much larger than the rotational energy $B_{\text{rot}} j(j+1)$ for all states populated during the fragmentation. This is equivalent to the statement that the collision time is short compared to the rotational period of the rotor, i.e. BC does not rotate significantly during dissociation. The ESA has been applied to photodissociation by several groups (17–22). Particularly interesting for interpretation purposes is the semiclassical limit of the sudden amplitude as $\hbar \to 0$ (23, 24).

Classical Theory

The classical theory of photodissociation follows the scheme in Eq. 1. The first step is a vertical Franck-Condon (FC) transition from the ground to the excited electronic state. It defines the initial conditions for the classical coordinates (R_o, γ_o) and momenta (P_o, j_o) at $t = 0$ for the nuclear motion in the excited state. P is the linear momentum corresponding to R and j is the molecular angular momentum corresponding to the angular coordinate γ, respectively. The second step describes the dissociation of the ABC* complex in the excited state according to Hamilton's equations of motion, i.e.

$$\frac{dR}{dt} = \frac{P}{m} \qquad \frac{dP}{dt} = -\frac{\partial V_{ex}}{\partial R} + \frac{j^2}{mR^3}$$

$$\frac{d\gamma}{dt} = 2j\left(B + \frac{1}{2mR^2}\right) \qquad \frac{dj}{dt} = -\frac{\partial V_{ex}}{\partial \gamma}. \qquad 13.$$

Here, $j(t)$ is the classical analogue to the (discrete) quantum number j. For zero total angular momentum $j(t)$ and $\gamma(t)$ are conjugate action-angle variables and therefore the variation of $j(t)$ along the trajectory is directly related to the anisotropy of V_{ex}.

The classical state resolved photodissociation cross section is defined as (25)

$$\sigma_{cl}^{(i)}(j|v) = v \int d\tau_o |\mu(\tau_o)|^2 W_{gr}^{(i)}(\tau_o) \delta[H_{ex}(\tau_o) - E] \delta[J_{ex}(\tau_o) - j] \qquad 14.$$

where integration is over the four-dimensional phase space at $t = 0$, i.e., $d\tau_o = \sin \gamma_o \, dR_o \, d\gamma_o \, dP_o \, dj_o$. The first delta-function selects those phase space points (i.e. "trajectories") which have the specified energy $E = E_i + hv$ whereas the second delta-function selects those trajectories which lead asymptotically ($t \to \infty$) to the specified rotational state $j = 0, 1, 2, \ldots$. $J_{ex}(\tau_o) = j(t \to \infty | \tau_o)$ is the classical excitation function, which is nothing other than the final angular momentum as a function of the initial coordinates and momenta. It is solely determined by the interaction in the exit channel indicated by the index ex. The excitation function is a quantity very helpful for analyzing the final data (26, 27) and is used in the same way as the *deflection function* in elastic scattering. If it is known from the trajectory calculations, we can evaluate the various cross-sections by numerical integration acording to Eq. 14. Examples of $J_{ex}(\tau_o)$ are given below.

In Eq. 14, $W_{gr}^{(i)}(\tau_o)$ is the weighting function for the nuclear motion of the parent molecule within the ground electronic state. It is most plausible to choose a quantum mechanical rather than a classical distribution function. Commonly used is the so-called *Wigner distribution* (28–33), although other choices, e.g.

$$W_{gr}^{(i)}(R_o, \gamma_o, P_o, j_o) = \Psi_{gr}^{(i)^2}(R_o, \gamma_o)\Phi_{gr}^{(i)^2}(P_o, j_o), \qquad 15.$$

are also possible (25) where $\Phi_{gr}^{(i)}$ is the nuclear wavefunction of the parent molecule in momentum space. For two uncoupled harmonic oscillators both in their ground vibrational state, the Wigner distribution and the distribution of Eq. 15 are identical and simply given by a product of four Gaussians, i.e.

$$W_{gr}(R_o, \gamma_o, P_o, j_o) = \exp[-2\alpha_R(R_o - R_e)^2]\exp[-2\alpha_\gamma(\gamma_o - \gamma_e)^2]$$
$$\times \exp[-P_o^2/2\alpha_R]\exp[-j_o^2/2\alpha_\gamma]. \qquad 16.$$

The classical calculations are straightforward and easily extended to several degrees of freedom. Classical mechanics is expected to be reliable whenever the wavenumber $k = (2mE)^{1/2}$ is large and when many quantum levels are populated. Such conditions are very often met in photodissociation experiments. Recently we compared classical versus exact quantum mechanical calculations for several models (11, 25, 34) as well as real systems (35–37) and found in all cases very good agreement. Another asset of the classical theory is the possibility of interpreting the cross sections in terms of the dynamics within the exit channel.

WEAK COUPLING: THE FRANCK-CONDON LIMIT

The product rotational state distributions are determined by the initial motion of the parent molecule (primarily the bending vibration) and by the final state interaction in the dissociation channel, i.e. the *anisotropy of the excited state potential energy surface* $\partial V_{ex}(R, \gamma)/\partial \gamma$. This is best seen in Eq. 13, where the variation of $j(t)$ is directly proportional to the anisotropy. If the anisotropy is zero or at least very weak, then the initially prepared distribution of angular momentum states j_o will not be changed during the fragmentation. In that case the final product distribution will primarily reflect the bound state wavefunction: The motion of the parent molecule is directly transferred to the product side without any reshuffling in the exit channel. This is the Franck-Condon (FC) limit, which has been amply discussed in the literature (38–40). Since the bound state wavefunction is the central quantity, it is advisable and in most cases rigorously possible to treat the FC limit purely quantum mechanically.

Franck-Condon Mapping

If the exit channel potential is purely elastic, i.e. independent of the orientation angle γ the potential coupling matrix $V_{jj'}(R)$ and therefore the matrix of radial functions $\chi_{jj'}^{(-j)}(R)$ are diagonal. The rotational channels of the continuum wavefunction are uncoupled. As a consequence the half collision S-matrix is also diagonal (elastic scattering) and the cross section reduces to the single term

$$\sigma^{(i)}(j|v) = v|a_j^{(i)}|^2|S_{jj}^{hc(i)}|^2. \qquad 17.$$

The radial integrals S_{jj}^{hc} depend on the rotational state j through the wavenumbers k_j that enter the one-dimensional Schrödinger equation for the radial function $\chi_j^{(-j)}$. They are primarily determined by the overlap $\phi_{\text{stretch}}(R)$ and the first (Airy-) maximum of the continuum wavefunction beyond the classical turning point defined by $V_{\text{ex}}(R_t) = E - B_{\text{rot}}j(j+1)$. The available translational energy decreases with increasing j and therefore the turning point $R_t(j)$ shifts outward. This implicit j-dependence is usually very small for rotational channels, and therefore the final rotational product distribution within the FC-limit is approximately given by

$$P^{(i)}(j) = |a_j^{(i)}|^2 = |2\pi \int_0^\pi d\gamma \sin\gamma \phi_{\text{bend}}^{(i)}(\gamma) Y_{j0}(r,0)|^2 \qquad 18.$$

where the a_j are the expansion coefficients of the bending wavefunction in terms of free rotor states.

This is the FC limit in its simplest form: The final distribution is solely determined by the bending part of the bound state wavefunction. Equation 18 is derived for $J = 0$. The expression becomes obviously more involved for non-zero total angular momenta but the generic structure remains the same (9, 40). The FC rotational state distribution is to a good approximation independent of the photon energy. A slight energy-dependence is contained in the more exact Expression 17 through the energy-dependence of S_{jj}^{hc}. It depends, however, significantly on the initial total angular momentum state as well as the initial bending state as indicated by the superscript (i) (11, 38-40). As a consequence, the FC-distribution is temperature-dependent according to

$$P(j|T) = \sum_i g(i|T)|a_j^{(i)}|^2 \qquad 19.$$

where $g(i|T)$ is the distribution function for the initial states.

In conclusion, the rotational state distribution within the FC-limit is a perfect map of the initial parent wavefunction (41, 42). The distribution prepared through the photon absorption is not destroyed by final state interaction. Within the FC-limit we expect (*a*) that only low rotational states ($j \lesssim 10$ or so) are populated, (*b*) that the distribution is independent or at most only weakly dependent on the photon energy, and (*c*) that the distribution is significantly temperature-dependent.

Incidentally, note that the situation is somewhat different for vibrational distributions within the FC-limit. Because the energy spacing between vibrational levels n is usually much larger than for rotational levels, the elastic half-collision S-matrix $S_{nn}^{hc}(E)$ can be significantly state-dependent. This in turn can induce a strong energy dependence of the final distribution (11, 38), although the translational-vibrational coupling in the exit channel is negligible. A nice example for this effect is the collinear model for the photodissociation of CH_3-I (15, 43–45). Although these model calculations formally include final state interaction, it is so weak that the CH_3 vibrational distribution can be well described within the vibrational FC-limit.

Photodissociation of Water in the First Absorption Band

A beautiful example for the rotational FC-limit is the photodissociation of water in the first absorption band around 165 nm, i.e.

$$H_2O(\tilde{X}) + h\nu \rightarrow H_2O(\tilde{A}) \rightarrow H(^2S) + OH(^2\Pi, j). \qquad 20.$$

Andresen et al (46a,b) and Grunewald et al (47) have measured the OH rotational state distributions resolved in the two spin components, $^2\Pi_{1/2}$ and $^2\Pi_{3/2}$, and the two Λ-doublet components, $\Lambda_{\pm}$, following the dissociation of room temperature and jet-cooled water at 157 nm (46a,b) and 193 nm (47), respectively. The general behavior of the measured distributions follows very closely the FC predictions discussed above. Only very low rotational states ($j \lesssim 6$) are populated or, expressed differently, only 1–3% of the total available energy goes into OH rotation. The distributions are temperature-dependent (beam vs bulk) but they depend only very weakly on the photon energy (193 nm vs 157 nm).

The $\tilde{A}$-state potential energy surface of H_2O has been calculated by Staemmler & Palma (48) with ab initio quantum chemical methods. $V_{ex}(\tilde{A})$ is *globally rather anisotropic* and a scattering experiment at energies comparable to photodissociation would yield strong rotational excitation (19). However, $V_{ex}(\tilde{A})$ is *locally isotropic* near the transition region around the ground state equilibrium angle, i.e. $\partial V_{ex}/\partial \gamma \approx 0$ near 104°, and therefore no torque is exerted on OH during the final fragmentation process. The initially prepared distribution through the electronic excitation is not changed in the exit channel. This has been confirmed in dynamical calculations on different levels of approximation (19).

In an elegant experiment Andresen and co-workers (49, 50) resolved, for the first time in direct photodissociation, true state-to-state cross sections. Recorded were OH rotational state distributions in both spin and in both Λ-doublet mainfolds for dissociation of *single* rotational states ($J K_a K_c$) of water. Water is an asymmetric top molecule and each rotational level is split into ($2J+1$) nondegenerate sublevels distinguished by the two projection quantum numbers K_a and K_c. The main experimental findings can be summarized as follows: The rotational distributions for dissociation of a single state of the parent molecule are structured and cannot be described by a temperature. In addition, they depend significantly on the specific initial rotational state. Three examples, all for the same J but for different projection quantum numbers, are given in Figure 1.

In order to compare with these detailed experimental data, the FC-theory described above must be modified in several respects. First, the ground state wavefunction must be expanded in terms of total angular momentum eigenstates for arbitrary J as described by Balint-Kurti & Shapiro (51), for example. Secondly, the electronic structure of OH($^2\Pi$) must be included in the final analysis. This is usually not done in photodissociation theories for polyatomic molecules but has been worked out for the special case of water by Balint-Kurti (52). The resulting FC expression is quite lengthy. Besides the molecular constants of OH and several "geometrical" terms (Clebsch-Gordon factors), it contains only the expansion coefficients of the ground state wavefunction, which must be determined numerically for each individual angular momentum state ($J K_a K_c$) separately.

The agreement of theoretical FC distributions with the truely state resolved experimental results in Figure 1 is excellent (53). All details and trends are nicely reproduced. This is a comparison on the level of six quantum numbers: Three to specify the initial H$_2$O state, J, K_a, K_c, and three to specify the final OH state, j, $^2\Pi_{1\pm1/2}$, $\Lambda_{\pm}$. Since the FC-theory works so well, it is possible to calculate rotational distributions for many initial states and to generate distributions, in principle, for any temperature. This has been done (50), and the beam as well as the bulk data could be satisfactorily reproduced. In the course of this exercise we found that averaging over a few final states (the two Λ-doubled components, for example) or over a few initial states immediately smears out all the structures seen in Figure 1 and Boltzmann distributions are obtained.

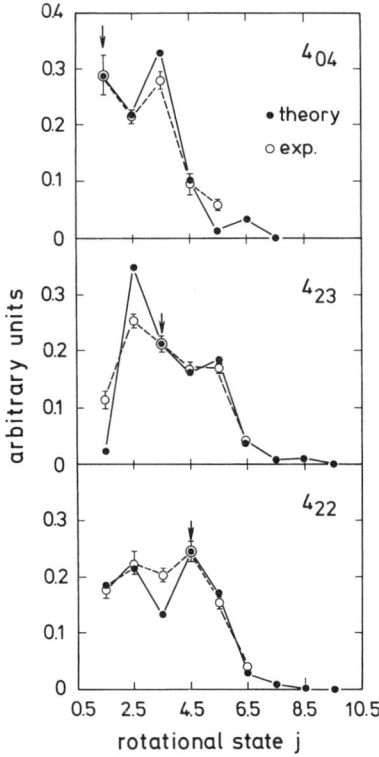

Figure 1 Rotational state distributions of OH($^2\Pi_{3/2}$) within the lower Λ-doublet state following the dissociation of single rotational states (JK_aK_c) of water, as indicated. The *arrows* indicate the rotational state where theory and experiment are normalized. Because of the electronic spin the rotational quantum numbers are half-integer. From Ref. (53).

In conclusion, the rotational distributions of OH in the photodissociation of H_2O in the first band are mainly determined by the motion of the parent molecule, and final state interaction for the rotational degree of freedom is very weak, although not negligible. Incidentally, the vibrational distribution cannot be described in the FC-limit. The translational-vibrational coupling induced by $V(\tilde{A})$ is apparently strong and leads to broad, energy-dependent distributions (54). There are many more interesting aspects, especially the population of the two Λ-doublet states, which cannot be discussed here (55). Similar results (although not fully state-resolved) have been found for H_2S (56), and a FC analysis identical to that for H_2O indeed gives good agreement with the experimental data (R. Schinke, G. G. Balint-Kurti, to be published).

STRONG COUPLING: THE ROTATIONAL REFLECTION PRINCIPLE

The FC-limit is a very special case. In general, however, the final state interaction, i.e. the anisotropy $\partial V_{\text{ex}}/\partial \gamma$, cannot be ignored. Nobody would try to study rotational excitation in collisions with an elastic potential. The final state interaction will change the distribution of rotational states initially prepared by the photon absorption. Quantum mechanically inelastic effects are manifested by nonzero off-diagonal elements of the half collision S-matrix $S_{jj'}^{hc}$ in Eq. 12. Classically redistribution processes are directly linked to the torque $\partial V_{\text{ex}}/\partial \gamma$, as it follows from Hamilton's equations in Eq. 13. If the final state interaction is strong enough, the initial distribution $P(j_o)$ will be completely destroyed during the fragmentation process, and the final rotational state distribution will be primarily determined by the second rather than the first step in Eq. 1. In this case the final distributions can be explained by the *rotational reflection principle* (57, 58), which is somehow similar to the *rotational rainbow* effect in full collisions (27, 59). If final state interaction is strong, the dissociation process is best described by classical mechanics.

Dynamical Mapping

Assuming that both initial momenta, P_o and j_o, are zero, the classical cross-section reduces to the one-dimensional integral ($\mu = 1$)

$$\sigma_{\text{cl}}^{(i)}(j|v) = v \int_0^\pi d\gamma_o \sin \gamma_o W_{\text{gr}}^{(i)}(\gamma_o, R_t) \delta[J(\gamma_o) - j] \qquad 21.$$

where R_t is the γ_o-dependent classical turning point and $J(\gamma_o)$ is the classical excitation function defined as $j(t \to \infty | \gamma_o)$. Assuming separability for the ground state wavefunction acording to Eq. 11, the weighting function due to the motion of the parent molecule is given by

$$W_{\text{gr}}^{(i)}(\gamma_o) = \phi_{\text{stretch}}^{(i)2}(R_t(\gamma_o)) \phi_{\text{bend}}^{(i)2}(\gamma_o). \qquad 22.$$

Note that now γ_o is the only independent initial variable. The weighting function is primarily determined by the parent molecule although the excited potential also enters through the turning point. The excitation function is exclusively determined by the excited state potential. Inclusion of nonzero initial momenta normally broadens the final distribution without changing it significantly (25).

Equation 21 can be rewritten to give

$$\sigma_{\text{cl}}^{(i)}(j|v) = v \sum_\alpha \sin \gamma_{\text{o},\alpha} W_{\text{gr}}^{(i)}(\gamma_{\text{o},\alpha}) \left| \frac{dJ}{d\gamma_\text{o}} \right|_{\gamma_{\text{o},\alpha}}^{-1} \qquad 23.$$

where we sum over all trajectories determined by $\gamma_{\text{o},\alpha}$, which lead to the specified rotational state $j = 0, 1, 2, \ldots$, i.e., which are solutions of

$$J(\gamma_{\text{o},\alpha}) = j. \qquad 24.$$

Equation 24 establishes a direct relation between the initial angle variable γ_o and the final "action" variable j.

As in full collisions (27), the rotational excitation function is the central quantity. It reflects directly the anisotropy of the excited state potential. Under sudden conditions ($\Delta E_{\text{rot}} \ll E$), Eq. 13 can be directly integrated to approximately give

$$J(\gamma_\text{o}) = -\int_0^\infty dt \frac{\partial V_{\text{ex}}}{\partial \gamma} = \int_0^\infty dt \sin \gamma \frac{\partial V_{\text{ex}}(\cos \gamma)}{\partial \cos \gamma} \qquad 25.$$

where the last equality follows from the fact that the potential is usually a function of $\cos \gamma$ rather than γ. If the excited state potential is very steep along the dissociation coordinate R it is reasonable to approximate it by a hard potential with angle-dependent contour $\hat{R}(\gamma)$ (60a,b). Then the excitation function is approximately given by (23)

$$J(\gamma_\text{o}) = (2mE)^{1/2} \frac{d\hat{R}(\gamma_\text{o})}{d\gamma_\text{o}} \qquad 26.$$

where $d\hat{R}/d\gamma$ is the anisotropy of the hard potential, and again the sudden condition $\Delta E_{\text{rot}} \ll E$ has been assumed.

A typical excitation function is shown (schematically) in Figure 2(c). Figure 2(b) depicts the underlying potential energy surface $V_{\text{ex}}(R, \gamma)$ as a contour plot. The *arrows* represent three inelastic trajectories for the same energy but starting at different γ_o. The degree of rotational excitation is roughly proportional to the variation of γ with respect to R. Trajectory 1 leads to a low rotational state whereas Trajectory 3 leads to a high final angular momentum. The torque starts to rotate the diatomic molecule immediately after the photon absorption. In view of Eq. 25, $J(\gamma_\text{o})$ is zero at $\gamma_\text{o} = 0$ and $90°$ (for a homonuclear molecule) with a maximum at intermediate angles. According to Eq. 26 the maximum stems (approximately) from the points of inflection of the potential when $d^2\hat{R}(\gamma_\text{o})/d\gamma_\text{o}^2 = 0$.

Equation 24 has usually two (homonuclear molecule) or four (heteronuclear molecule) independent solutions for each rotational state j as indicated in Figure 2(c). Because of the weighting function $W_{\text{gr}}^{(i)}(\gamma_\text{o})$, however, only one of them contributes significantly to the cross-section,

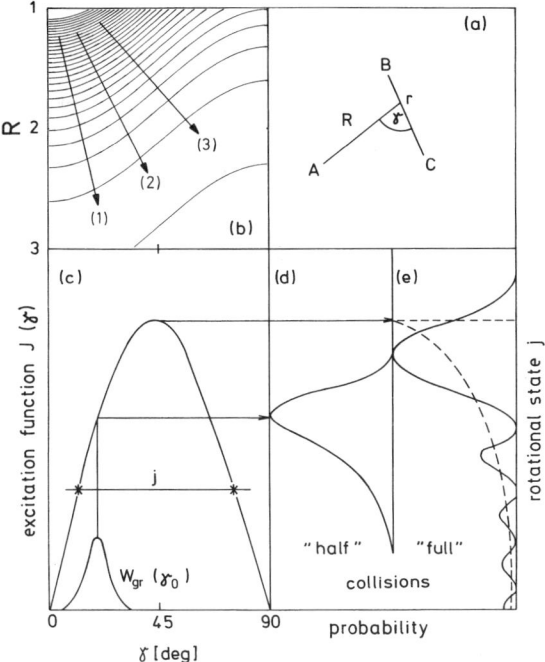

Figure 2 (a) Definition of Jacobi-coordinates R, r, and γ for the A+BC system. R measures the A to BC center-of-mass distance. (b) Contour plot of a typical excited state potential energy surface $V_{ex}(R, \gamma)$. The abscissa is the orientation angle γ and the ordinate is the distance R. The *arrows* schematically represent three dissociative trajectories. (c) Rotational excitation function $J(\gamma_0)$ and weighting function $W_{gr}(\gamma_0)$. The *horizontal line* indicates the solution of Eq. 24 for a particular rotational state j. (d) Rotational distribution for the photodissociation process ("half" collision). (e) Rotational distribution for a scattering process ("full" collision). The *solid line* is the quantum mechanical distribution with a broad rainbow maximum and supernumerary oscillations due to the interference of two trajectories. The *dashed line* is the classical distribution with a rainbow singularity at $j_R = J_{max}$.

and then Eq. 24 establishes a unique relation between the initial angular variable γ_0 and the final momentum variable j: Each final rotational state is primarily determined by one initial angle, and its probability is roughly given by the modulus square of the bending part of the ground state wavefunction. This is the *rotational reflection principle*, as illustrated in Figure 2(d) for $P(j)$. The final state distribution is roughly a reflection of the ground state bending wavefunction mediated by the excitation function above the ground state equilibrium. It reflects features of both the ground state as well as the excited state potential. W_{gr} would be bimodal for the first excited bending state, and the same bimodality would obviously occur in the probability. The maximum of $P(j)$ would shift to higher (lower) states if the anisotropy of V_{ex} were increased (decreased). Likewise, it would shift up or down if the ground state equilibrium angle were changed (57, 58). The final state distributions are more or less direct maps of the anisotropy of the excited state potential around the ground state equilibrium.

The rotational reflection principle is completely analogous to the well-known one-dimensional reflection principle (28), which explains the energy dependence of the absorption spectrum as a reflection of $\phi_{\text{stretch}}(R)$ via the excited state potential, i.e.

$$\sigma_{\text{abs}}(E) = v\phi_{\text{stretch}}^2[R_t(E)] \left|\frac{dV_{ex}}{dR}\right|_{R_t}^{-1} \qquad 27.$$

where $R_t(E)$ is the classical turning point defined by

$$V_{ex}(R_t) = E. \qquad 28.$$

The modifications to treat the vibrational degree of freedom are quite obvious and lead naturally to the *vibrational reflection principle* (11, 34, 36).

The final rotational state distributions in both the weak and the strong coupling cases are reflections of the ground state wavefunction. This is not surprising because the photodissociation cross-section is given as a matrix element that explicitly contains the wavefunction of the parent molecule. However, the mapping mechanism is quite different in the two cases. In order to distinguish the weak and the strong coupling limit, one might use the terms *Franck-Condon mapping* and *dynamical mapping*, respectively. FC-mapping is of course the limit of dynamical mapping as the final state interaction vanishes.

At this point it is worthwhile to discuss the differences between direct photodissociation and direct scattering. First of all it is easy to show that, under the same conditions as assumed above, the excitation function in scattering is approximately twice the excitation function in dissociation, i.e.

$$J_{\text{scatt}} \sim 2J_{\text{diss}}. \qquad 29.$$

This follows immediately from Eq. 25 if the range of integration is extended to $t = -\infty$ or from the hard shell model (23). Equation 29 manifests intriguingly the difference between "full" and "half" collisions.

The classical scattering cross-section is formally identical to Eq. 23 except for the weighting function $W_{\text{gr}}^{(i)}$, which is obviously not present in collisions ($W_{\text{scatt}} = 1$) because the initial conditions are prepared for the isolated fragments. Therefore all solutions of Eq. 24 contribute more or less equally to the scattering cross section. The superposition of several contributions to the scattering amplitude, including their phases, quantum mechanically gives rise to the pronounced oscillations in Figure 2(e) (26, 27, 61a,b), which indeed have been resolved experimentally (62a,b). The maximum of $J(\gamma_o)$ leads to the *rotational rainbow* singularity in the classical cross section, which appears as a broad maximum in the quantum mechanical cross section. Rotational rainbows are prominent features of atom-molecule collisions and have been resolved for many systems (59). They have been also observed in nuclear scattering (63a–c), electron-molecule scattering (64a,b), and molecule-surface scattering (65a,b). Rotational rainbows also exist in photodissociation, provided the weighting function $W_{gr}^{(i)}$ samples the maximum of the excitation function (66, 67).

In the strong coupling case the initially prepared distribution $P(j_o)$ is more or less completely destroyed, and the final state distribution is primarily determined by the fragmentation step. We therefore expect (a) that many rotational states ($j > 10$) are populated, (b) that the final distributions are inverted and confined to a relatively narrow interval (rotational reflection principle), (c) a clear energy- and mass-dependence as predicted by Eq. 26, and (d) no dramatic temperature-dependence. The last point follows because the final distribution is, to a large extent, independent of the particular initial rotational state of the parent molecule. The expectations in the strong coupling limit are therefore contrary to what one expects in the FC-limit. The situation is certainly more complicated for dissociation at the steep onset of the absorption spectrum when mainly higher excited states of the parent molecule are photolyzed.

Many measured rotational distributions originating from direct photodissociation show these predictions. Examples are the photodissociation of Cl–CN (68), CH_3O–NO (69, 70), H_2–CO (71), OH–OH (72–74), I–CN (75–77), S–CO (78a,b), Br–CL (79), and HN–CO (80). There are several other systems that might be explained by the rotational reflection principle. In addition, similar effects have been observed in classical (31–33) as well as exact quantum mechanical (57, 58) calculations. In cases in which quantum mechanical and classical calculations could be directly compared for the same (model) systems, usually very satisfactory agreement was found (25, 57, 58). We believe that the rotational reflection principle in photodissociation is rather universal, similar to rotational rainbows in scattering.

The analysis of real systems can be more complicated (a) if more than one degree of freedom leads to rotational excitation (H_2O_2, for example), (b) if highly excited rotational states of the parent molecule are dissociated, a condition that can lead to inverted distributions without any final state interaction (39, 40), (c) if the final state interaction is weak but not negligible such that FC-mapping as well as dynamical mapping overlap, or finally (*d*) if more than one excited state are involved (OCS and ICN, for example). The central quantity is the excited state potential energy surface; without knowing it, at least qualitatively, it is almost impossible to understand a real system. For a few systems it has been calculated with quantum chemical methods, and in these cases rigorous dynamical studies are possible: Cl–CN (81), $H_2O(\tilde{A})$ (48), $H_2O(\tilde{B})$ (82), H_2O_2 (83), H_2–CO (84), and O_3 (33). In the following I discuss two examples for which detailed comparison with experimental distributions has been made.

Photodissociation of Formaldehyde

The photodissociation of formaldehyde has been extensively studied in the last decade by Moore and his group (85). It is, in principle, not a direct dissociation process by proceeds according to

$$H_2CO(S_0) + h\nu \xrightarrow{(1)} H_2CO(S_1) \xrightarrow{(2)} H_2CO^*(S_0)$$

$$\xrightarrow{(3)} H_2(j_{HH}) + CO(j_{CO}) \quad 30.$$

where Step 2 is an internal conversion (fast electronically nonadiabatic transition) from S_1 to S_0. In this way ground state formaldehyde is prepared in a very high vibrational state with a total energy just above the barrier for dissociation into H_2 and CO (71). All dissociative trajectories must pass through a narrow transition state (ts) in the multidimensional coordinate space. According to the discussion of Bamford et al (71), the H_2 and CO final state distributions are therefore determined in the H_2–CO exit channel (Step 3) rather than through the motion of the highly excited parent molecule. In this sense the fragmentation of formaldehyde can be understood as *direct dissociation starting at the transition state.*

The H_2–CO potential energy surface $V(R, \gamma_{CO}, \gamma_{HH})$ in the exit channel is known from quantum chemical calculations originally performed to study H_2–CO collisions at low energies (84). R is the H_2–CO separation and γ_{CO} and γ_{HH} are the orientation angles of CO and H_2, respectively, with respect to the intermolecular vector **R**. The forces $\partial V/\partial \gamma_{CO}$ and $\partial V/\partial \gamma_{HH}$ are responsible for rotational excitation in the exit channel. The geometry of the transition state, especially the two angles γ_{CO}^{ts} and γ_{HH}^{ts} that determine the starting points of the dissociative trajectories, are known from the work of Goddard & Schaefer (86). Using these ab initio data we performed simple quantum mechanical calculations in the energy sudden approximation. They do not include any adjustable parameter (66, 67).

Figure 3 depicts the experimental (71) and the theoretical (66) CO rotational distributions for dissociation of H_2CO, HDCO, and D_2CO. All

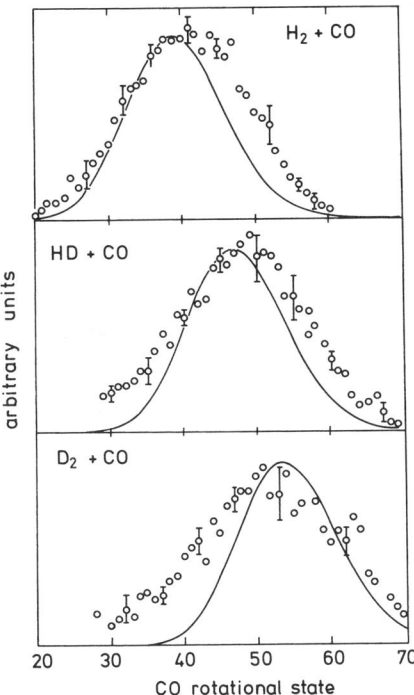

Figure 3 Theoretical (—) and experimental (◯) rotational state distributions of CO following the photolysis of H_2CO, HDCO, and D_2CO, respectively. Theory and experiment are normalized at the maxima. From Ref. (66).

distributions are highly inverted, smooth, and relatively symmetric around the peak center. Despite the simplicity of the model, the calculated distributions agree remarkably well with the experiment. The analysis of the classical excitation function $J_{CO}(\gamma_{CO})$ reveals unambiguously that the distributions are determined by the rotational reflection principle. Since the maximum of J_{CO} is probed, the results could, however, also be interpreted in terms of rotational rainbows (66). As demonstrated in Ref. (66), the maximum of the distributions is directly related to the anisotropy $\partial V/\partial \gamma_{CO}$ near the transition state geometry γ_{CO}^{ts}. Therefore, the good agreement with the experimental data underlines the accuracy of both quantum chemical calculations (84, 86). The shift of the maximum as H_2 is substituted by HD and D_2 scales with the square root of the reduced mass and is thus a nice manifestation of the general prediction in Eq. 26.

The rotational distribution of *ortho*-H_2 has been measured by Debarre et al (87) and calculated within the same model as for CO by Schinke (67).

The agreement between theory and experiment is also satisfactory. In this case, however, the anisotropy of the overall potential with respect to the H_2 orientation angle, i.e. $\partial V/\partial \gamma_{HH}$, is extremely weak and therefore the final H_2 distribution is solely determined by the Franck-Condon limit: The H_2 rotational distribution reflects the wavefunction at the transition state expanded in terms of free rotor states. There is no redistribution within the exit channel. This observation is interesting (although not surprising) because it shows that in the photodissociation of a polyatomic molecule, one fragment can be determined by FC-mapping while the other fragment is determined by dynamical mapping. The central quantity is the potential energy surface, or to be more precise, the forces $\partial V/\partial \gamma_{HH}$ and $\partial V/\partial V_{CO}$, and they of course can be quite different.

Photodissociation of Hydrogen Peroxide

The photodissociation of hydrogen peroxide in the wavelength region $\lambda \gtrsim 193$ nm has been studied extensively by several groups in recent years (72–74). Earlier experiments by Ondrey et al (88) were not performed under reported free conditions and therefore the collision OH rotational distributions were not the nascent distributions following from photodissociation. The absorption spectrum within the first band is smooth (89), a condition that indicates a fast and direct dissociation process. It is ascribed to transitions to the two lowest excited states $\tilde{A}$ and $\tilde{B}$ (83) according to

$$H_2O_2(\tilde{X}) + h\nu \rightarrow H_2O_2(\tilde{A}, \tilde{B}) \rightarrow 2\,OH(^2\Pi, j). \qquad 31.$$

The dissociation of H_2O_2 is particularly interesting because rotational excitation of the two OH products can result from torsional motion around the O–O axis (associated with the tetrahydral angle φ) as well as bending motion (associated with the two HOO angles γ_1 and γ_2). The coordinates are defined in Figure 4(a). The potential energy surfaces $V(R_{OO}, \varphi, \gamma_1, \gamma_2)$ for both excited states $\tilde{A}$ and $\tilde{B}$ have been calculated by Meier et al (83).

We performed purely classical calculations following the general recipe as outlined before by using Jacobi-coordinates $(R, \varphi, \theta_1, \theta_2)$, which are defined with respect to the centers-of-mass of the two OH fragments. A full description of these calculations will be given in a forthcoming paper (89a). Here we discuss only those aspects which unambiguously prove that the OH rotational distributions are primarily induced by the torsional motion and that they can be explained by the rotational reflection principle. In preliminary calculations neglecting completely the bending degree of freedom we tested the applicability of classical mechanics for this system by comparison with exact quantum mechanical calculations (11, 36).

Figure 4(b) shows the φ-dependence of the $\tilde{A}$-state potential energy

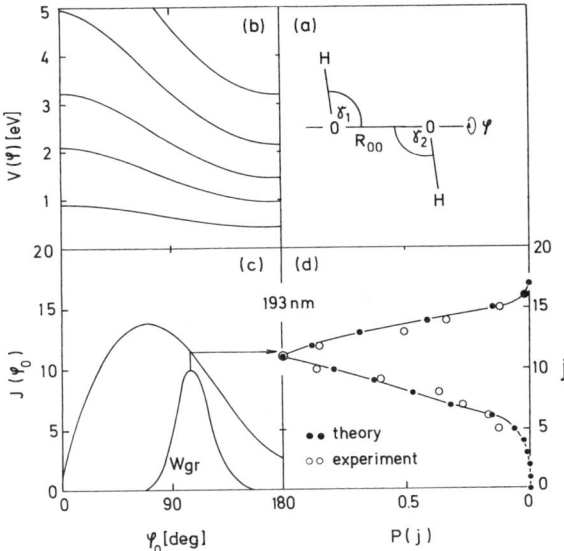

Figure 4 (a) Coordinates for the photodissociation of H_2O_2. (b) The $\tilde{A}$-state potential energy surface as a function of the torsional angle based on the calculations of Meier et al (83). The two bending angles γ_1 and γ_2 are fixed at 94.8° and the R_{OO} separation is $2.75a_o$, $3a_o$, $3.25a_o$, $3.5a_o$, and $4a_o$, respectively. (c) Rotational excitation function $J(\varphi_o)$ and weighting function $W_{gr}(\varphi_o)$ calculated as described in the text. (d) Comparison of experimental (91) and theoretical rotational state distributions for dissociation at 193 nm.

surface as calculated by Meier et al (83) for various O–O separations. The two bending angles γ_1 and γ_2 are fixed at the equilibrium value of the ground state, i.e. $\gamma_e = 94.8°$. The corresponding equilibrium value of torsional angle is $\varphi_e = 111.5°$ (90). The first excited state ($\tilde{A}$) has a minimum at $\varphi = 180°$, and therefore the two OH radicals are immediately excited after excitation near 111° with $\mathbf{j}_{OH(1)} \approx -\mathbf{j}_{OH(2)}$. At 193 nm, rotation is primarily about the O–O axis. The angle between $\mathbf{j}_{OH}$ and $\mathbf{R}$ is, roughly speaking, a measure of the influence of the bending degree of freedom.

Figure 4(c) depicts the rotational excitation function $J(\varphi_o)$ vs initial torsional angle φ_o for an excitation wavelength of 193 nm. The initial bending angles $\gamma_{1,o}$ and $\gamma_{2,o}$ are 94.8° and all initial momenta are set to zero. The excitation function has the general form as illustrated in Figure 2. It is small at $\varphi_o = 0°$ and $\varphi_o = 180°$ with a maximum at around 70°. The amount of rotational excitation at 0° and 180° stems exclusively from the bending degree of freedom. Also shown in Figure 4(c) is the weighting function $W_{gr}(\varphi_o)$ as defined through the motion of the parent molecule. The torsional part of the ground state wavefunction is calculated quantum mechanically using the hindering potential of Hunt et al (90).

In Figure 4(d) we compare the calculated and the measured rotational state distributions of OH for $\lambda = 193$ nm. The classical calculations are performed by sampling an eight-dimensional phase space to select the initial coordinates and corresponding momenta. The experiment has been performed in a molecular beam (91). Compared to the earlier results (72, 73) obtained in the bulk ($T = 300$ K), the peak position is unchanged and only the FWHM is reduced by about two quanta. This relatively weak temperature dependence by itself suggests that the product state distribution is mainly determined by strong final state interaction. The agreement between the classical and the experimental distributions is excellent; this underlines the accuracy of the ab initio potential as well as the classical approximation. In order to be precise we must note that at 193 nm both surfaces, $\tilde{A}$ and $\tilde{B}$, contribute with a ratio of about 2:1 (91). The calculation includes only the $\tilde{A}$-state potential. However, the rotational distribution following from absorption into the $\tilde{B}$-continuum is very similar to that for the $\tilde{A}$-state, although the φ-dependence in both cases is drastically different. Taking both surfaces into account even improves the agreement with experiment (89a). Figures 4(c) and 4(d) clearly manifest the rotational reflection principle for the photodissociation of H_2O_2: The peak position is determined by the anisotropy (φ-dependence) of the excited state potential within the FC-region. There are many more interesting results for this system, which cannot be discussed in this review. H_2O_2 is a prototype system for the photodissociation of a tetratomic molecule and neither the experimental nor the theoretical studies are completely finished yet.

EXTREME COUPLING: MULTIPLE COLLISION EFFECTS

The rotational coupling for the examples discussed in the foregoing sections is either very weak ($H_2O(\tilde{A})$) or only moderate (H_2CO, H_2O_2). The energy associated with the rotational transition at the peak centers for D_2CO and H_2O_2 is only 20% and 15% of the total available energy, respectively. The individual trajectories $j(t)$ rise fast and monotonically to its final value, which defines the excitation function $J(\gamma_0)$. Typical examples are schematically shown in Figure 5(b). At each instant the rotational energy is much smaller than the total available energy, or expressed differently, the collision time is much shorter than the rotational period such that the molecule does not rotate very much during the fragmentation process. Under such conditions the rotational excitation function is more or less a direct map of the potential anisotropy, as is nicely shown by Eq. 26. The rotational reflection principle is a simple way to analyze and

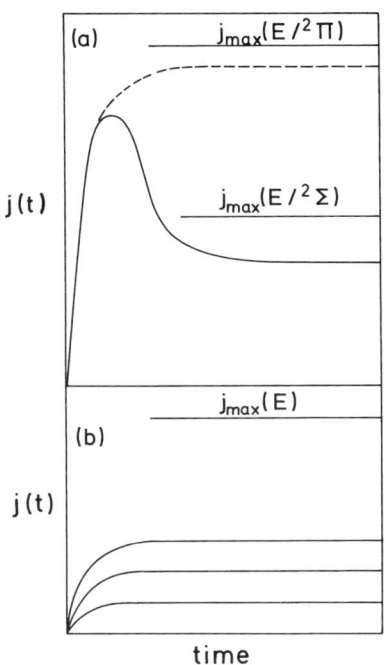

Figure 5 (*a*) Schematic illustration of a trajectory $j(t)$ for dissociation in the $\tilde{B}$-state of water. The *dashed line* represents a trajectory $j(t)$ that starts in the $\tilde{B}$-state and that is continued in the $\tilde{X}$-state after passing the line $\gamma = 180°$ (linearity). $j_{max}(E/^2\Sigma)$ and $j_{max}(E/^2\Pi)$ are the highest rotational levels that are accessible within the $^2\Sigma$- and $^2\Pi$-states of OH, respectively. (*b*) Schematic illustration of monotonical trajectories $j(t)$ in the case of moderate coupling strength.

interpret the final momentum distributions as simple reflections of the distribution of initial coordinates.

If the coupling is extremely strong or if it acts over a longer time period, the individual trajectories and consequently the rotational excitation function can be much more complicated. Under such conditions multiple collision effects can occur: The rotor might absorb most of the available energy in the first impact but subsequently a large portion of it may flow back to translation in a secondary collision. In this case $j(t)$ is not monotonical but might instead behave as schematically shown in Figure 5(*a*). In the presence of multiple collision effects the excitation function $J(\gamma_o)$ is no longer a "simple" map of the potential anisotropy. Multiple collision effects indeed have been observed in scattering (92) and they are of course also possible in dissociation. Their appearance depends sensitively on the strength of the potential anisotropy and the masses of the collision partners (92).

Photodissociation of Water in the Second Absorption Band

A nice example of multiple collision effects as qualitatively discussed above is the photodissociation of water in the second absorption band, i.e.

$$H_2O(\tilde{X}) + h\nu \longrightarrow H_2O(\tilde{B}) \xrightarrow{\lesssim 10\%} H + OH(^2\Sigma, j)$$

$$\xrightarrow{\gtrsim 90\%} H + OH(^2\Pi, j) \qquad 32.$$

with a maximum around $\lambda = 128$ nm. The $\tilde{B}$-state potential energy surface has a very unusual and interesting shape (82, 93): Due to an avoided crossing with the ground state near linearity, it has a deep well at $\gamma = 180°$ and an H–OH distance of ~ 1.6 Å. The well depth is more than 3 eV with respect to the $OH(^2\Sigma) + H$ asymptote (82). Because the potential well is confined to a relatively narrow region around 180°, the $\tilde{B}$-state potential energy surface is extremely anisotropic.

The dissociation of H_2O in the $\tilde{B}$-state proceeds in the following way: Following excitation from the ground state near $\gamma \sim 104°$ and $R_{H-OH} \sim 1$ Å, the water molecule immediately starts to bend because of the strong torque $\partial V(\tilde{B})/\partial \gamma$ (37, 93–95). This is equivalent to very strong rotational excitation in the first step of the dissociative trajectory as illustrated in Figure 5(a), which shows a typical trajectory $j(t)$ vs time. Because more than 3 eV of potential energy is released, rotational excitation far above the asymptotical value $j_{max}(E) \sim (E/B_{rot})^{1/2}$ is *temporarily* possible within the well. Although it is not measurable one can define a rotational probability $P(j|180°)$ at linearity $\gamma = 180°$ (37), which is shown in Figure 6(b). This is the probability that the rotor is in state j when the system becomes linear. It is highly inverted with a peak near $j \sim 30$, although at this energy [$E = 1$ eV above the $H + OH(^2\Sigma)$ limit] only rotational states up to $j \sim 21$ are asymptotically allowed. The strong anisotropy acts like a lens that focuses the majority of trajectories into a narrow interval of rotational states at linearity.

As the dissociation continues in the $\tilde{B}$-state, the rotor is significantly deexcited because the torque $\partial V(\tilde{B})/\partial \gamma$ changes its sign at linearity. The amount of potential energy that was locally available for rotation has to be repaid as the system dissociates. The final rotational distribution of $OH(^2\Sigma)$ molecules is shown in Figure 6(c) compared to the measurement of Simons et al (96). Similar distributions have been reported long before by Carrington (97) and Yamashita (98). This and similar distributions for different wavelengths have often been termed *unusual* or *abnormal* because they peak very close to the highest state that is asymptotically accessible. Most of the available energy goes into rotation rather than translation, a result that is very different from the other cases discussed above. The results can be understood in terms of the extreme anisotropy in the first part of the dissociation process. A relatively wide range of the initial phase space (i.e. initial conditions) is mapped onto a narrow range of final

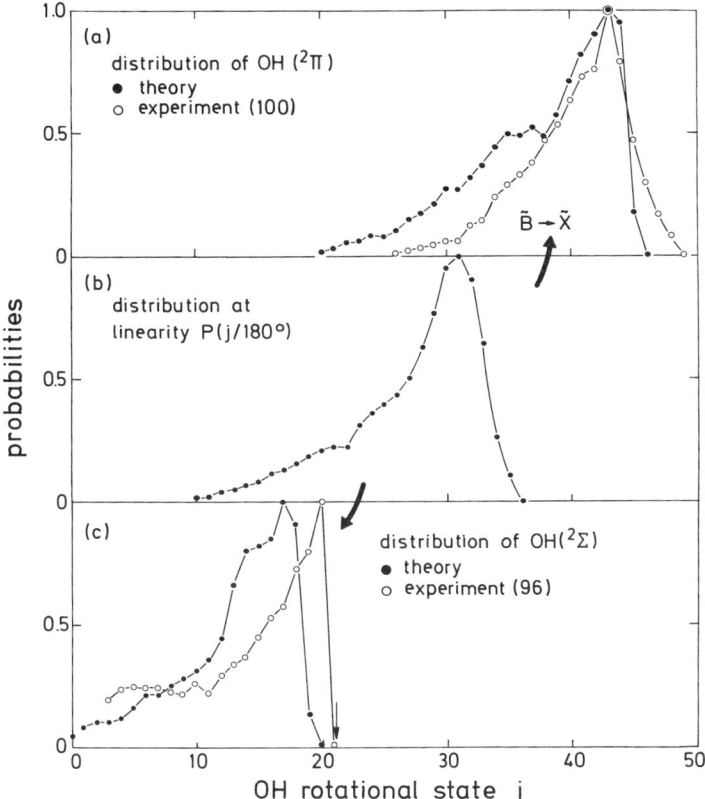

Figure 6 (a) Comparison of theoretical (37) and experimental (100) rotational state distributions for OH($^2\Pi$) molecules. (b) Rotational state distribution $P(j/180°)$ at linearity as described in the text. (c) Comparison of theoretical (37) and experimental (96) rotational state distributions for OH($^2\Sigma$) molecules. The *arrow* at $j \sim 21$ indicates the highest available state at this energy. In all cases the energy is 1 eV with respect to the H+OH($^2\Sigma$) asymptote.

rotational states. In this sense the highly inverted distribution can be interpreted as a rainbow catastrophe.

It is well known (99) that the quantum yield for OH($^2\Sigma$) is less than 10% whereas more than 90% of the OH molecules are produced in the $^2\Pi$ ground state as indicated in Eq. 32. Electronically nonadiabatic transitions can be either to the $\tilde{A}$-state or to the $\tilde{X}$-state, both of which correlate asymptotically with OH($^2\Pi$). According to some primitive surface hopping calculations (37), we believe that the $\tilde{B} \rightarrow \tilde{X}$ channel is more realistic. The calculations are performed in the following way: First, the trajectories are followed in the $\tilde{B}$-state from the Franck-Condon region up to linearity, where the probability of making a nonadiabatic transition is largest because of the avoided crossing. Then each trajectory is continued on the ground state potential energy surface until the molecule is dissociated. Since $\partial V(\tilde{B})/\partial \gamma$ for $\gamma < 180°$ and $\partial V(\tilde{X})/\partial \gamma$ for $\gamma > 180°$ have the same sign, the excitation process started in the $\tilde{B}$-state is continued in the $\tilde{X}$-state, as indicated in Figure 5(a). The final distribution for OH($^2\Pi$) is shown in Figure 6(a) in comparison with recent measurements of Krautwald et al (100). Since ca. 4 eV of electronic energy are additionally available, rotational states up to $j \sim 50$ can be excited for OH($^2\Pi$) molecules. The reasonable agreement between theory and experiment suggests that $\tilde{B} \to \tilde{X}$ is indeed the dominant dissociation channel. In conclusion, the rotational distributions of both OH($^2\Sigma$) and OH($^2\Pi$) are primarily determined by the extremely strong forces in the first part of the fragmentation process when the HOH bending angle instantaneously opens.

Whether multiple collision effects exist for a real system cannot be decided a priori without knowing the excited state potential energy surface. If a large portion of the total available energy is transferred to rotation (50% or more), it is quite likely that such effects are indeed operative. This is the case, for example, in the photodissociation of H_2O_2 at 157 nm (101, 102).

FINAL REMARKS

This review is admittedly very restricted. Only direct dissociation processes that evolve on a short time scale are considered. The reason for the restriction is that only for direct processes was a reasonably detailed understanding in terms of dynamical calculations achieved in recent years. This became possible because of three interrelated developments: the application of sophisticated, new experimental methods, progress in quantum chemical methods to calculate excited state potential energy surfaces, and progress in dynamical theories to treat the nuclear motion of complicated systems.

The beauty of direct photodissociation processes is based on the observation that the distribution of final angular momenta is more or less a reflection of the distribution of initial coordinates. The specific type of "mapping" depends very much on the strength of the coupling forces in the exit channel. We distinguished three basic cases: weak coupling (FC-mapping), strong coupling (dynamical mapping), and extreme coupling. In the second case, which is probably most common, final rotational state distributions can be elegantly interpreted in terms of the rotational reflection principle, which is the inelastic analogue to the well-known elastic (one-dimensional) reflection principle. Each coupling case is illustrated by real examples for which dynamical calculations based on ab initio potential energy surfaces are compared with new experimental data. The agreement between theory and experiment is very good in all cases, and this intriguingly manifests the power of rigorous dynamical theories if combined with accurate ab initio potential calculations.

The mapping of initial coordinates onto the final momentum space is a characteristic feature of direct processes. The situation is certainly more complicated for indirect processes when the dissociation proceeds via a long-lived intermediate complex (predissociation). In such cases the distribution of initial coordinates is destroyed within the complex and the final state distribution might be determined solely in the exit channel after the break-up of the complex. On the other hand, rotational distributions following the vibrational predissociation of rare gas–ICl (103) and rare gas–NO (104) van der Waals molecules are qualitatively similar to distributions following from direct dissociation. The distributions are inverted, and the variation of the peak positions with the energy or the reduced mass approximately follows Eq. 26. It seems that features prominent in direct processes also exist in indirect dissociation. The study of rotational state distributions following vibrational predissociation will be an interesting topic of future research.

The central quantity for a detailed understanding of photodissociation processes is, of course, the excited state potential energy surface. Because of significant progress in quantum chemical methods and the availability of efficient computing facilities, more and more systems are tackled today. The time when the potential was either completely ignored (FC-limit) or approximated by very rough models (impulsive forces, for example) is certainly over. It is not so important to know the potential very well on a quantitative level. In many cases it is sufficient to know the qualitative features and the variation of the potential with respect to the molecular degrees of freedom, i.e. the coupling forces that are ultimately responsible for energy redistribution.

Gelbart (105) stated in his 1977 review of the photodissociation dynamics of polyatomic molecules: "It is remarkable that, after the considerable theoretical and experimental work reviewed above, so little is understood about the photodissociation of polyatomic molecules. There is not a single example, in fact, where the actual 'mechanism' (dynamics) of a polyatomic photodissociation has been unambiguously established." This statement is not valid anymore! Due to the close collaboration of experiment, quantum chemistry, and dynamical theory today we understand at least a few systems in most details. Some of them have been discussed in this review.

Literature Cited

1. Robin, M. B. 1974. *Higher Excited States*. New York: Academic
2. Okabe, H. 1978. *Photochemistry of Small Molecules*. New York: Wiley
3. Leone, S. R. 1982. *Adv. Chem. Phys.* 50: 255
4. Simons, J. P. 1984. *J. Phys. Chem.* 88: 1287
5. Bersohn, R. 1984. *J. Phys. Chem.* 88: 5145
6. Jackson, W. M., Okabe, H. 1986. *Advances in Photochemistry*, Vol. 13, ed. D. H. Volman, K. Gollnick, G. S. Hammond. New York: Wiley
7. Buelow, S., Noble, M., Radhakrishnan, G., Reisler, H., Wittic, C., Hancock, G. 1986. *J. Phys. Chem.* 90: 1015
8. Ashford, M. N. R., Baggott, J. E. 1987. *Molecular Photodissociation Dynamics*. London: Royal Soc. Chem.
9. Shapiro, M., Bersohn, R. 1982. *Ann. Rev. Phys. Chem.* 33: 409
10. Shapiro, M., Balint-Kurti, G. G. 1985. In *Photodissociation and Photoionization*, ed. K. P. Lawley. New York: Wiley
11. Schinke, R. 1988. In *Collision Theory for Atoms and Molecules*, ed. F. A. Gianturco. New York: Plenum
12. Shapiro, M. 1972. *J. Chem. Phys.* 56: 2582
13. Kulander, K. C., Light, J. C. 1980. *J. Chem. Phys.* 73: 4337
14. Kulander, K. C., Heller, E. J. 1978. *J. Chem. Phys.* 69: 2439
15. Lee, S. Y., Heller, E. J. 1982. *J. Chem. Phys.* 76: 3035
16. Heather, R., Metiu, H. 1987. *J. Chem. Phys.* 86: 5009
17. Segev, E., Shapiro, M. 1983. *J. Chem. Phys.* 78: 4969
18. Atabek, O., Beswick, J. A., Delgado-Barrio, G. 1985. *J. Chem. Phys.* 83: 2954
19. Schinke, R., Engel, V., Staemmler, V. 1985. *J. Chem. Phys.* 83: 4522
20. Henshaw, J. P., Clary, D. C. 1987. *J. Phys. Chem.* 91: 1580
21. Kulander, K. C., Light, J. C. 1986. *J. Chem. Phys.* 85: 1938
22. Grinberg, H., Freed, K. F., Williams, C. J. 1987. *J. Chem. Phys.* 86: 5456
23. Schinke, R., Engel, V. 1985. *J. Chem. Phys.* 83: 5068
24. Schinke, R. 1986. *J. Phys. Chem.* 90: 1742
25. Schinke, R. 1987. *J. Phys. Chem.* In press
26. Miller, W. H. 1974. *Adv. Chem. Phys.* 25: 69; 1975. *Adv. Chem. Phys.* 30: 77
27. Schinke, R., Bowman, J. M. 1983. In *Molecular Collision Dynamics*, ed. J. M. Bowman. Heidelberg: Springer
28. Heller, E. J. 1978. *J. Chem. Phys.* 68: 2066
29. Goursaud, S., Sizun, M., Fiquet-Fayard, F. 1976. *J. Chem. Phys.* 65: 5453
30. Pattengill, M. D. 1982. *Chem. Phys.* 68: 73; 1983. *Chem. Phys.* 78: 229; 1984. *Chem. Phys. Lett.* 104: 462; 1984. *Chem. Phys.* 87: 419
31. Goldfield, E. M., Houston, P. L., Ezra, G. S. 1986. *J. Chem. Phys.* 84: 3120
32. Bersohn, R., Shapiro, M. 1986. *J. Chem. Phys.* 85: 1396
33. Hay, P. J., Pack, R. T., Walker, R. B., Heller, E. J. 1982. *J. Phys. Chem.* 86: 862; Sheppard, M. G., Walker, R. B. 1983. *J. Chem. Phys.* 78: 7191
34. Untch, A., Hennig, S., Schinke, R. 1988. *Chem. Phys.* In press
35. Engel, V., Schinke, R. 1988. *J. Chem. Phys.* In press
36. Schinke, R., Engel, V., Hennig, S., Weide, K., Untch, A. 1988. *Ber. Bunsenges. Phys. Chem.* In press
37. Weide, K., Schinke, R. 1987. *J. Chem. Phys.* 87: 4627
38. Freed, K. F., Band, Y. B. 1978. In *Excited States*, ed. E. C. Lim. New York: Academic
39. Morse, M. D., Freed, K. F., Band, Y. B. 1979. *J. Chem. Phys.* 70: 3604; Morse, M. D., Freed, K. F. 1981. *J. Chem. Phys.* 74: 4395; Morse, M. D., Freed, K. F. 1983. *J. Chem. Phys.* 78: 6045; Morse, M. D., Freed, K. F. 1980. *Chem. Phys. Lett.* 74: 49
40. Beswick, J. A., Gelbart, W. M. 1980. *J. Phys. Chem.* 84: 3148
41. Shapiro, M. 1981. *Chem. Phys. Lett.* 81: 521; 1986. *J. Phys. Chem.* 90: 3644
42. Child, M. S., Shapiro, M. 1983. *Mol. Phys.* 48: 111
43. Shapiro, M., Bersohn, R. 1980. *J. Chem. Phys.* 73: 3810
44. Henrikson, N. E. 1985. *Chem. Phys. Lett.* 121: 139
45. Gray, S. K., Child, M. S. 1984. *Mol. Phys.* 51: 189
46a. Andresen, P., Ondrey, G. S., Titze, B. 1983. *Phys. Rev. Lett.* 50: 486
46b. Andresen, P., Ondrey, G. S., Titze, B., Rothe, E. W. 1984. *J. Chem. Phys.* 80: 2548
47. Grunewald, A. U., Gericke, K.-H., Comes, F. J. 1986. *Chem. Phys. Lett.* 133: 501
48. Staemmler, V., Palma, A. 1985. *Chem. Phys.* 93: 63

49. Andresen, P., Beushausen, V., Häusler, D., Lülf, H. 1985. *J. Chem. Phys.* 83: 1429
50. Häusler, D., Andresen, P., Schinke, R. 1987. *J. Chem. Phys.* 87: 3949
51. Balint-Kurti, G. G., Shapiro, M. 1981. *Chem. Phys.* 61: 137; 1982. *Chem. Phys.* 72: 456
52. Balint-Kurti, G. G. 1986. *J. Chem. Phys.* 84: 4443
53. Schinke, R., Engel, V., Andresen, P., Häusler, D., Balint-Kurti, G. G. 1985. *Phys. Rev. Lett.* 55: 1180
54. Engel, V., Schinke, R., Staemmler, V. 1988. *Chem. Phys. Lett.* 130: 413; 1987. *J. Chem. Phys.* 88: 129
55. Andresen, P., Schinke, R. 1987. See Ref. 8, pp. 61–113
56. Hawkins, W. G., Houston, P. L. 1982. *J. Chem. Phys.* 76: 729
57. Schinke, R. 1986. *J. Chem. Phys.* 85: 5049
58. Schinke, R., Engel, V. 1986. *Faraday Discuss. Chem. Soc.* 82: 111
59. Buck, U. 1986. *Comments At. Mol. Phys.* 17: 143
60a. Schepper, W., Ross, U., Beck, D. 1979. *Z. Phys. A* 290: 131; Bosanac, S. 1980. *Phys. Rev. A* 22: 2617
60b. Korsch, H. J., Schinke, R. 1981. *J. Chem. Phys.* 75: 3850
61a. Schinke, R. 1978. *Chem. Phys.* 34: 65
61b. Korsch, H. J., Schinke, R. 1980. *J. Chem. Phys.* 73: 1222
62a. Hefter, U., Jones, P. L., Mattheus, A., Witt, J., Bergmann, K., Schinke, R. 1981. *Phys. Rev. Lett.* 46: 915
62b. Gottwald, E., Bergmann, K., Schinke, R. 1987. *J. Chem. Phys.* 86: 2685
63a. Landowne, S., Vitturi, A. 1986. In *Treatise of Heavy-Ion Science*, Vol. 1, ed. D. A. Bromley. New York: Plenum
63b. Massmann, H., Rasmussen, J. O. 1975. *Nuclear Phys. A*. 243: 155
63c. Levit, S., Smilansky, U., Pelte, D. 1974. *Phys. Lett.* 53B: 39
64a. Ziegler, G., Rädle, M., Pütz, O., Jung, K., Ehrhardt, H., Bergmann, K. 1987. *Phys. Rev. Lett.* 58: 2642
64b. Korsch, H. J., Kutz, H., Meyer, H. D. 1987. *J. Phys. B* 20: L433
65a. Kleyn, A. W., Luntz, A. C., Auerbach, D. J. 1981. *Phys. Rev. Lett.* 47: 1169
65b. Voges, H., Schinke, R. 1983. *Chem. Phys. Lett.* 100: 245
66. Schinke, R. 1985. *Chem. Phys. Lett.* 120: 129
67. Schinke, R. 1986. *J. Chem. Phys.* 84: 1487
68. Halpern, J. B., Jackson, W. M. 1982. *J. Phys. Chem.* 86: 3528
69. Benoist d'Azy, O., Lahmani, F., Lardeux, C., Solgadi, D. 1985. *Chem. Phys.* 94: 297
70. Brühlmann, U., Dubs, M., Huber, J. R. 1987. *J. Chem. Phys.* 86: 1249
71. Bamford, D. J., Filseth, S. V., Foltz, M. F., Hepburn, J. W., Moore, C. B. 1985. *J. Chem. Phys.* 82: 3032
72a. Grunewald, A. U., Gericke, K.-H., Comes, F. J. 1986. *Chem. Phys. Lett.* 132: 121
72b. Klee, S., Gericke, K.-H., Comes, F. J. 1986. *J. Chem. Phys.* 85: 40
72c. Gericke, K.-H., Klee, S., Comes, F. J., Dixon, R. N. 1986. *J. Chem. Phys.* 85: 4463
73. Jacobs, A., Wahl, M., Weller, R., Wolfrum, J. 1987. *Appl. Phys. B* 42: 173
74. Docker, M. P., Hodgson, A., Simons, J. P. 1986. *Faraday Discuss. Chem. Soc.* 82: 25
75. Fisher, W. H., Carrington, T., Filseth, S. V., Sadowski, C. M., Dugan, C. H. 1983. *Chem. Phys.* 82: 443
76. Marinelli, W. J., Sivakumar, N., Houston, P. L. 1984. *J. Phys. Chem.* 88: 6685
77. Nadler, I., Mahgerefteh, D., Reisler, H., Wittig, C. 1985. *J. Chem. Phys.* 82: 3885
78a. Sivakumar, N., Burak, I., Cheung, W.-Y., Houston, J. P., Hepburn, J. W. 1985. *J. Phys. Chem.* 89: 3609
78b. Sivakumar, N., Hall, G. E., Houston, P. L., Hepburn, J. W., Burak, I. 1988. In press
79. Russell, J. A., McLaren, I. A., Jackson, W. M., Halpern, J. B. 1987. *J. Phys. Chem.* In press
80. Spiglanin, T. A., Perry, R. A., Chandler, D. W. 1987. *J. Chem. Phys.* 87: 1568
81. Waite, B. A., Dunlap, B. I. 1986. *J. Chem. Phys.* 84: 1391
82. Theodorakopoulos, G., Petsalakis, I. D., Buenker, R. J. 1985. *Chem. Phys.* 96: 217
83. Meier, U., Staemmler, V., Wasilewski, J. 1988. In press
84. Schinke, R., Meyer, H., Buck, U., Diercksen, G. H. F. 1986. *J. Chem. Phys.* 80: 5518
85. Moore, C. B., Weisshaar, J. C. 1983. *Ann. Rev. Phys. Chem.* 34: 31
86. Goddard, J. D., Schaefer, H. F. 1979. *J. Chem. Phys.* 70: 5117
87. Debarre, D., Lefebvre, M., Pealat, M., Taran, J.-P. E., Bamford, D. J., Moore, C. B. 1985. *J. Chem. Phys.* 83: 4476
88. Ondrey, G., van Veen, N., Bersohn, R. 1983. *J. Chem. Phys.* 78: 3732
89. Suto, M., Lee, L. C. 1983. *Chem. Phys. Lett.* 98: 152

89a. Schinke, R., Staemmler, V. 1988. *Chem. Phys. Lett.* In press
90. Hunt, R. H., Leacock, R. A., Peters, C. W., Hecht, K. T. 1965. *J. Chem. Phys.* 42: 1931
91. Grunewald, A. U., Gericke, K.-H., Comes, F. J. 1988. *J. Chem. Phys.* In press
92. Buck, U., Otten, D., Schinke, R., Poppe, D. 1985. *J. Chem. Phys.* 82: 202
93. Flouquet, F., Horsley, J. A. 1974. *J. Chem. Phys.* 60: 3767
94. Akamatsu, R., O-Ohta, K. 1978. *J. Phys. Soc. Jpn.* 44: 589
95. Dunne, L. J., Guo, H., Murrell, J. N. 1987. *Mol. Phys.* In press
96. Simons, J. P., Smith, A. J., Dixon, R. N. 1984. *J. Chem. Soc. Faraday Trans. 2* 80: 1489
97. Carrington, T. 1964. *J. Chem. Phys.* 41: 2012
98. Yamashita, I. 1975. *J. Phys. Soc. Jpn.* 39: 205
99. Vinogradov, J. P., Vilesov, F. I. 1976. *Opt. Spectrosc.* 40: 32
100. Krautwald, H. J., Schnieder, L., Welge, K. H., Ashfold, M. N. R. 1986. *Faraday Discuss. Chem. Soc.* 82: 99
101. Becker, K. H., Groth, W., Kley, D. 1965. *Z. Naturforsch. Teil A* 20: 748
102. Gölzenleuchter, H., Gericke, K.-H., Comes, F. J. 1984. *Chem. Phys.* 89: 93
103. Skene, J. M., Drobits, J. C., Lester, M. I. 1986. *J. Chem. Phys.* 85: 2329
104. Sato, K., Achiba, Y., Nakamura, H., Kimura, K. 1986. *J. Chem. Phys.* 85: 1418
105. Gelbart, W. M. 1977. *Ann. Rev. Phys. Chem.* 28: 323

VIBRATIONAL RAMAN SPECTRA OF SIMPLE FLUIDS

M. J. Clouter

Department of Physics, Memorial University of Newfoundland, St. John's, Newfoundland, Canada A1B 3X7

INTRODUCTION

Vibrational Raman spectroscopy has been, and continues to be, an invaluable probe of molecular structure, particularly when employed under conditions of low density where intermolecular interactions can be neglected. In recent years, however, there has been a remarkable expansion of literature pertaining to the vibrational Raman spectra of dense fluids, and liquids in particular, with emphasis on the influence of intermolecular interactions and dynamics. The earliest experiments of this kind were concerned with measuring the effects of temperature and density on the peak frequency of the Raman profile, and thereby gaining insight into the nature of the intermolecular forces. At a more detailed level, similar studies have subsequently focused on the width and shape of the Raman spectrum as a source of information regarding the processes of vibrational relaxation and dephasing. When viewed in conjunction with related theoretical work, this has led to the development of vibrational Raman spectroscopy as an informative probe of the molecular environment that can contribute significantly to our understanding of dense fluids.

The reader may consult a number of review articles and books (1–5) that cover the material published prior to about 1983. The scope of the present review is determined by the observation that much of the work on the simplest molecular liquids is of more recent vintage, as also is most of the work pertaining to the critical region. The essential information pertaining to the molecular systems to be discussed is given in Table 1. Attention is focused on the systems that have been studied most extensively, namely, H_2 (with its isotopic variants) and N_2. In the case of the diatomics the discussion is limited to the polarized components of the

Table 1 Fixed-point data and vibrational Raman shifts for some simple molecular liquids

	Triple point		Critical point		Raman shift[a]
Liquid	T_t (K)	$\tilde{\rho}_t{}^d$	T_c (K)	$\tilde{\rho}_c{}^d$	ν_0 (THz)
n–H_2	14.0	859	33.2	335	124.8[b]
HD	16.6	910	35.9	357	109.0[b]
n–D_2	18.7	968	38.4	389	89.8[b]
N_2	63.2	695	126.2	251	69.9
CO	68.1	674	132.9	240	64.2
O_2	54.4	915	154.6	306	46.7
CH_4	90.7	631	190.6	224	87.4[c]
CF_4	89.2	490	227.7	164	27.1[c]

[a] Approximate (free-molecule) values.
[b] $Q(0)$ component.
[c] Totally symmetric mode.
[d] Densities relative to STP (see text).

fundamental ($\Delta v = 1$) vibrational bands, and for the polyatomics is further restricted (with minor exceptions) to the totally symmetric vibrational modes. These spectra arise from transitions for which the rotational quantum number is conserved (i.e. $\Delta J = 0$), and in spectroscopic terminology are referred to as Q branches, with the individual components identified as $Q(J)$ (see Figure 1). The term *pure vibrational* spectrum is also frequently

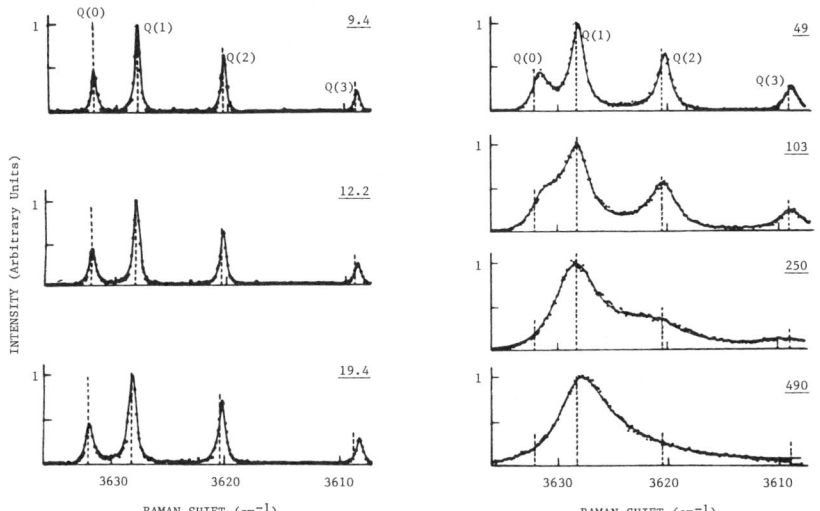

Figure 1 The Raman Q branch of gaseous HD at room temperature (courtesy of A. D. May). The relative densities, $\tilde{\rho}$ (see text), are shown at the *upper right* of each spectrum. The *vertical dashed lines* are positioned at the free-molecule transition frequencies.

used in this connection although, strictly speaking, the effect of intramolecular vibration-rotation interaction does lead to perturbation of the rotational energy.

It is not simply a matter of coincidence that the liquid spectra of such simple molecules as H_2 and N_2 were among the last to receive attention in the context of lineshape and vibrational relaxation studies. From the experimental point of view, these systems (and others in Table 1) are rather special cases in at least two respects. First, the critical points, and the cryogenic temperature ranges corresponding to liquid-vapor coexistence, have traditionally presented problems in that they are not readily achievable via the standard liquid coolants. It is only as a result of the relatively recent availability of economical closed-cycle helium refrigerators that these difficulties have been eliminated. Second, the widths of the vibrational Raman spectra can be less than 1 GHz, and require considerably higher resolution than is afforded by the grating spectrometers that are commonly employed. Most of the spectra discussed here were in fact obtained using the technique of Fabry-Perot interferometry whose development (6) to a high level of sophistication and reliability happens to have proceeded in parallel with the increased interest in liquid Raman spectra. As a high resolution technique it still provides distinct advantages over more recently developed methods, such as Raman-gain spectroscopy, when it is necessary to limit the size of the sample either to accommodate high pressures or to ensure thermal homogeneity (e.g. near a critical point).

It is first appropriate to review the gas-phase work as a vehicle for introducing the basic ideas, and much of the terminology, which are also applicable in the liquid regime. With respect to low densities in particular, no attempt is made to be comprehensive in this treatment because the material is not of primary interest in the present context. A key to the literature is provided by the recent publication of Smyth et al (7).

For the purposes of the discussion it is convenient to introduce a dimensionless *relative density*, $\tilde{\rho}$, which is defined as the density of a given material relative to that (0.0446 mol m^{-3}) of an ideal gas at standard temperature and pressure (0°C, 101 kPa).[1] The experimental work in question spans the range from $\tilde{\rho} \simeq 1$ to values near 1000, which are typical of the liquid phase. The hydrogens occupy positions of prominence in the discussion because the individual components of their Raman Q branches are easily resolvable over a wide range of densities. Although HD is later treated as a special case because of its heteronuclear character, its behavior in the range $\tilde{\rho} < 200$ is qualitatively similar to H_2 and D_2, and it is for both these reasons that it is shown in Figure 1.

---

[1] The traditional unit for $\tilde{\rho}$ is the *amagat*, i.e. $\tilde{\rho} = 1$ corresponds to 1 amagat.

Throughout the discussion, and in connection with spectral widths in particular, it is necessary to refer frequently to a number of relaxation mechanisms that involve the exchange of energy between the translational (T), vibrational (V), and rotational (R) degrees of freedom. For convenience I employ the abbreviation scheme that identifies the type of energy transfer by an appropriate combination of the letters T, V, and R: for example, VT transfer refers to the transfer of molecular vibrational energy to the center-of-mass translation motion.

GAS-PHASE RAMAN SHIFTS

The Hydrogens

A basic understanding of the density and temperature dependence of the $Q(J)$ Raman shifts is best conveyed with reference to H_2, HD, and D_2. The first studies of the H_2 spectrum were conducted by May and co-workers (8, 9) and Allin et al (10) using prelaser techniques with densities up to $\tilde{\rho} \simeq 800$ at room temperature and $\tilde{\rho} \simeq 400$ at 85K. Looi et al (11) subsequently examined the region $10 < \tilde{\rho} < 100$ for both H_2 and D_2 at temperatures from 85K to 315K. These earlier experiments, as well as related theoretical work (12–16), have been treated in some detail in previous reviews (1, 2). Recent experimental work on the homonuclear hydrogens (7, 17) has concentrated on the low density region $\tilde{\rho} < 20$, and the most recent theoretical calculations appear to be those of Kelley & Bragg (18). Spectra of HD at room temperature and 85K have been reported by Dion & May (19) and Witkowicz & May (20; see Figure 1) for the density range $9 < \tilde{\rho} < 500$. The following qualitative comments encompass only the essential features of this work.

The vibrational spectrum of a free molecule such as H_2 is determined entirely by its mass and the details of the internuclear potential function. This potential is, however, always anharmonic to some degree so that perturbations which affect the average internuclear separation cause small changes in the energies of the (v, J) quantum states. These quasi-adiabatic effects are smaller by a factor of ~ 10 for the (more nearly harmonic) v = 0 vibration as compared to v = 1 so that it is sufficient for the moment to consider the latter state only. In fact, it is the centrifugal stretching of the v = 1 internuclear bond with increasing rotational energy (i.e. J) that is primarily responsible for the observed separations of the $Q(J)$ components in the low density spectrum. Consequently, with reference to Figure 1, one can immediately correlate the increased bond length with decreasing Raman shift. A similar process can, of course, also result from external influences, arising in particular from interactions with other molecules. Since the time scale of these fluctuating interactions is typically in the picosecond range, it is clear that any experimental determination of the Raman shift (i.e. the peak position) must represent a time average of such events. One is thus led to think in terms of a mean-field effect in which the field in question is determined as an average over the configurational distribution function of the system. Although the difficulties involved in constructing a general theory of this effect are obvious, simplification is afforded by assuming pair-wise additivity of the interactions so that the process of configurational averaging involves only the pair distribution function.

The result thus obtained (9, 16) for an intermolecular pair potential of the Lennard-Jones type is that the difference (Δv) between the observed shift (v) and the free-molecule value (v_0) is, for a given $Q(J)$ component, expressible as a power series in the density, i.e. $\Delta v = v - v_0 = a_J \tilde{\rho} + b_J \tilde{\rho}^2 + \cdots$. The linear term in this expansion is associated with interactions that involve only isolated pairs of molecules (i.e. binary collisions), whereas higher order terms are associated with encounters involving three and four molecules, etc. At low densities ($\tilde{\rho} < 100$) the coefficient a_J is sufficient to describe the observed highly linear behavior, and a comparison of the data (11) for H_2 and D_2 reveals that the a_J's are considerably larger in magnitude for the lighter molecule. Given that the internuclear potential is identical for both molecules, because it is determined by the electronic configuration, this is a direct consequence of the larger amplitude of the $v = 1$ vibration in H_2, which consequently probes a more anharmonic region of the (intramolecular) potential.

The dependence of the coefficients a_J and b_J on temperature (T) is primarily via the pair distribution function. For fixed $\tilde{\rho}$ one expects that as T is decreased there will be fewer encounters that involve the strong repulsive core of the intermolecular (pair) potential. In other words, as liquid conditions are approached the molecules tend to congregate near the minimum of the potential well, and spend most of their time under the influence of attractive forces. The effect of the mean field is consequently to increase the average internuclear separation for given J and $v = 1$, and to cause a decrease in the value of v by way of the correlation noted above. This is reflected in the behavior of the a_J's, which were found (11) to be negative at room temperature and increased in magnitude as the temperature was lowered. Conversely, the more energetic collisions that are associated with increasing T eventually lead to the dominance of the repulsive potential core and a change in the sign of a_J. Bischel & Dyer (17) have observed $a_J > 0$ for H_2 at 474K, and the crossover point, i.e. $a_J = 0$, is estimated to occur at about 420K.

As already noted, the density dependence of the shift for $\tilde{\rho} < 100$ at constant T is determined by the leading term in the series expansion. The fact that $a_J < 0$ at room temperature implies average pair separations that are beyond the potential minimum, so that increasing density results in an enhancement of the attractive forces and a consistent decrease in the Raman shift. This is demonstrated in Figure 1. Eventually, of course, the repulsive potential core must come into play at high densities and the Δv values will become less negative. This behavior is reflected in the observed positive values of the b_J coefficients, and the fact that the quadratic term becomes more important as the density is increased. The room temperature data (8) for $\Delta v(\tilde{\rho})$ actually show minima at $\tilde{\rho} \simeq 300$ for each of the $Q(J)$ lines. That these minima are shifted to higher densities as the temperature is lowered is again consistent with the idea that the importance of repulsive interactions is correspondingly reduced. Indeed, the HD data (20) at 85K exhibit a linear behavior (i.e. $b_J = 0$) for $\tilde{\rho}$ values as high as 500.

Predictably, perhaps, quantitative agreement between experiment and theory is obtained only for the leading term in the expansion (i.e. a_J). The relatively poor agreement obtained for the b_J values has been shown (16) to arise at least in part from the additivity assumption. The theory nevertheless provides a valuable conceptual basis that is qualitatively consistent with the observed shift behavior for all the molecular systems listed in Table 1. In particular, it has been confirmed (21) that $b_J = 0$ for all these systems at temperatures characteristic of their respective liquid phases. The experimentally determined a_J coefficients (21) are therefore sufficient to describe the behavior under these conditions, and are an essential consideration in later discussions of vibrational dephasing effects.

GAS-PHASE LINE WIDTHS

H_2 and D_2

For $\tilde{\rho} < 5$ the width of the $Q(J)$ profiles is determined primarily by the Doppler effect, and in this range it has been found that the width decreases with increasing density. This phenomenon is known as *Dicke narrowing* (1, 2, 7). Since it is of little interest here, note only that it is analogous to other effects to be discussed below under the general heading of *motional narrowing*.

For $\tilde{\rho} > 5$ the $Q(J)$ profiles are observed to broaden linearly with increasing density at room temperature (1, 7). In attempting to account for this behavior there are at least four mechanisms that should be considered. First, there is the possible effect of collisions in limiting the lifetime of the upper (v = 1) vibrational state, i.e. *VT* transfer, or vibrational energy relaxation. There is no doubt that this is a negligible effect for the hydrogens because the measured (22) relaxation time for H_2 is of order 10 ms whereas the inverse of the observed (10–50 GHz) linewidths is of order 100 ps. It is this circumstance that in fact makes it possible to observe the (~ 100 MHz) low-density Doppler broadening of the profiles as mentioned above. Second, it is reasonable to expect that the line-shifting mechanism itself should play a role, i.e. that the width of a given $Q(J)$ profile should reflect the (quasi-adiabatic) effect of the time-varying local field on the vibrational transition frequencies. This particular mechanism is included under the heading of *vibrational dephasing*, which is sometimes taken further to include the broadening effect due to resonant VV transfer between molecules. Here, however, the latter is recognized as a separate, and third, broadening mechanism. Both of these are discussed in more detail below: with respect to gaseous H_2 and D_2 there are no a priori grounds for neglecting either. The fourth mechanism, as elucidated by Van Kranendonk and co-workers (12–15), involves the effect of anisotropic intermolecular forces in promoting rotationally inelastic collisions that limit the lifetime of states corresponding to different J in the Raman-active, or probe, molecule. For example, if the excited probe molecule is initially in the state (v = 1, J = 2), then collision with a (v = 0, J = 0) molecule can result in a simple exchange of rotational energy between them, leaving the probe molecule in the state (v = 1, J = 0). In the approximation that intramolecular vibration-rotation interaction is neglected, this particular exchange corresponds to exact resonance. Several other non-resonant exchanges involving the translational energy can also occur at room temperature, but always subject to the strict prohibition of odd ΔJ values, which is the selection rule responsible for the existence of *ortho* and *para* species in H_2 and D_2 (23). In keeping with the adopted scheme, this process is identified here as RR-exchange broadening. It is distinguishable from the vibrational dephasing effect because of its strong dependence on J. The calculation of this dependence (14) for the anisotropic quadrupole interaction, and comparison with experimental data (8), leaves no doubt that RR exchange is the dominant broadening mechanism in the pure gases at room temperature, even if additional possibilities such as RT transfer and rotational dephasing are also considered.

Mixtures of H_2 and D_2 with the Rare Gases

Linewidth studies of low concentrations of either H_2 or D_2 in the rare gases (24, 25, 7) are of interest because the RR-exchange and VV-exchange mechanisms are rendered inactive. Although RT transfer and rotational dephasing can still contribute to the linewidth, it has been concluded (24, 25) that in such cases that vibrational dephasing is predominant. It is consequently now appropriate to consider this phenomenon in greater detail.

The broadening associated with vibrational dephasing in principle reflects two essential characteristics of the line-shifting local field within the medium; namely, the variation in the strength of the field, and the time scale on which the variations occur. Any broadening contribution that arises from spatial variations in the strength of the local field is, for obvious reasons, described as *inhomogeneous*. In the extreme case one imagines each small volume element within the medium as producing a local Raman spectrum whose width is negligible compared to the rms variation (Γ) in the peak frequency that occurs from one volume element to another. In other words, the phase correlation time, τ_c, as determined by the inverse width of each local spectrum, is considered to be sufficiently long that the effects of translational dynamics can be completely neglected. This obviates the need to consider the consequences of local variations in τ_c. Various terms, such as *static limit, slow modulation*, or *rigid-lattice limit*, have been used to describe this regime, which is characterized by a Gaussian profile and the condition $\Gamma\tau_c \gg 1$. If τ_c is very much less than Γ (i.e. $\Gamma\tau_c \ll 1$), then the opposite extreme of *fast modulation* applies; all local spectra (and the observed spectrum) are characterized by identical Lorentzian profiles, and the broadening is described as *homogeneous*. The width of a given $Q(J)$ profile is not, however, inversely proportional to τ_c as might initially be expected by analogy with energy relaxation, for example. Rather, a short phase correlation time implies a very rapid switching through the distribution of $v = 1$ vibrational frequencies and has the effect of an averaging operation that becomes more and more efficient as τ_c decreases. The result is that the width actually decreases in inverse proportion to the density. This is the principal characteristic of the *motional narrowing* process, which applies with equal validity to other degrees of freedom besides vibration. Reference to the translational analog (Dicke narrowing) has already been made, and the spectrum of HD (discussed below) involves yet another example.

The intermediate case of $\Gamma\tau_c \sim 1$, which is the most commonly observed, is the least tractable theoretically. The departures of the lineshape from either the Gaussian or Lorentzian form may be quite subtle and difficult to detect, and the width may either decrease or increase with density. The theoretical treatment of this case has concentrated on symmetrical spectral profiles and has usually involved modeling of the phase correlation function under the assumption that τ_c itself is not subject to inhomogeneous variation at the local level. Further discussion of this point is presented below in the context of critical fluctuations.

It is not a trivial matter to predetermine which of the above modulation conditions applies in a given system, because of the interplay of factors that affect Γ and τ_c. Γ, in particular, is governed not only by the details of the intramolecular potential (e.g. the anharmonicity), but also by the details of the intermolecular potential, and by thermodynamic conditions. If vibrational dephasing is the only active mechanism, it is intuitively obvious that $\tau_c \to \infty$ for $\tilde{\rho} \to 0$ because the molecules do not interact at all for most of the time, and the phase of a given v = 1 vibration is only occasionally perturbed by binary collisions. At the same time, $\Gamma \to 0$ for essentially the same reason, namely, that a sampling of a frozen configuration of the gas will reveal only occasional instances in which the vibrational frequency is different from the free-molecule value (in a binary collision). Thus, not surprisingly, both the homogeneous and inhomogeneous contributions to the dephasing width vanish at zero density. As the density is increased isothermally, τ_c will decrease with increasing collision frequency, but Γ will increase because more molecules come within the range of the pair potential and a wider distribution of vibrational frequencies is sampled. Consequently, the increase in dephasing width which must occur at sufficiently low densities arises from homogeneous and inhomogeneous contributions that are both increasing in importance, and the modulation regime is a priori undetermined. From a practical point of view an indication of which, if either, is more important can be obtained with reference to the shift data, since Δv as defined above can be regarded as an upper limit on Γ. Thus, if the observed width is comparable to Δv it is reasonable to expect that the homogeneous contribution is secondary: this is consistent with the data for H_2–Ne, H_2–Ar, and H_2–Kr mixtures given by Robert et al (25), for example. If Δv is much greater than the width, then motional narrowing is indicated and observations should confirm that the width decreases with increasing density.

It must be added, of course, that only occasionally is it justifiable to neglect all but one line-broadening mechanism, especially over a wide range of densities. Consequently, the distinguishing characteristics of the dephasing contribution to the width may not be readily observable. For example, the low-density limit of the $Q(J)$ linewidths is in fact determined either by the natural (free-molecule) energy relaxation time or by the Doppler effect. At sufficiently high densities complications also arise when different $Q(J)$ components begin to overlap. Such an effect is evident in the HD spectra of Figure 1.

HD and HD–Ar Mixtures

The HD molecule occupies a special place among the hydrogens because it is the only stable isotopic variant that is heteronuclear. The most relevant consequences of this are that odd ΔJ values are no longer forbidden in collisional transitions, and that the non-coincidence of the centers of mass and charge gives rise to additional angle-dependent terms in the pair potential. Both of these combine to considerably enhance the effect of *RR*-exchange in pure HD. For example, in the experiments of May and co-workers (19, 20) the widths were found to be greater than the corresponding H_2 values by an order of magnitude. In the case of HD–Ar mixtures (26) these circumstances are also responsible for dominance of broadening due to *RT* transfer rather than vibrational dephasing (cf H_2–Ar). In both cases the rapid broadening of adjacent $Q(J)$ profiles causes them to overlap so that the individual components are no longer identifiable at densities greater than $\tilde{\rho} \simeq 300$ (see Figure 1). The shape of the resulting profile is, however, narrower and generally quite different from that expected on the assumption of independently broadened $Q(J)$ lines (19).

The characteristics of the high-density profile are the result of rather special circumstances that have been treated theoretically in some detail (27–33). The essential point is that, as a result of $\Delta J = \pm 1$ collisional transitions occurring in both the ground and excited states, the Raman *transitions* that comprise the Q branch are rendered degenerate with respect to the quantum number J. J is still a good quantum number in the primary sense that the components of the pure rotational spectrum are completely resolved, but as far as the Q branch is concerned there are no contributions to the Raman spectrum that can be unambiguously identified with specific values of J. It is only meaningful to think in terms of an average value of J that becomes more closely defined as the collisional effects are enhanced with increasing density. These ideas are consistent with the observed narrowing of the spectrum and the shift of its peak frequency toward a position corresponding to the center of gravity of the unperturbed band (after appropriate adjustments for the normal density dependence of the shift: see Figure 1).

The behavior is also analogous to that associated with the fast modulation condition for vibrational dephasing as described above, and is also a motional-narrowing effect. In fact, the collapse of the HD Q branch can be regarded as due to a kind of *indirect* vibrational dephasing that is mediated by the intramolecular vibration-rotation coupling. The process effectively quenches the broadening effects due to *RR* and *RT* transfer so that at sufficiently high densities the width is expected to be limited by vibrational dephasing. This may not be apparent, however, if the fast modulation condition holds for the dephasing, because under such conditions the spectral width will simply continue to decrease. The pure HD data were not carried to sufficiently high density to test this point, but the spectra of HD–Ar at 300K (26) clearly show a dephasing broadening for $\tilde{\rho} > 500$.

It is noted in passing that the collapse of the Q branch in H_2 and D_2 could only occur at much higher densities because components corresponding to $Q(J)$ and $Q(J\pm 2)$ would be required to overlap in response to weaker broadening mechanisms. Adjacent components could significantly overlap at densities comparable to the HD case, but the requirement for even ΔJ in collisional transitions preserves the identity of the $Q(J)$ contributions to the spectrum that, consequently, can be represented as a simple sum of these contributions. To date the only indications of Q-branch narrowing in H_2 are for cases (34) where it is dissolved in liquids such as CCl_4, CS_2, and SF_6.

N_2 and CO

The energy of molecular rotation for given J is inversely proportional to the moment of inertia of the molecule under consideration (23). For a heavier diatomic such as N_2 this means that the effect of centrifugal stretching in perturbing the vibrational frequency is greatly reduced, and the $Q(J)$ components are much more closely spaced. Only a slight broadening effect will therefore cause a merging of the components into a single-peaked profile. For both molecules considered here this occurs for $\tilde{\rho} < 1$ so that such systems offer the opportunity of examining the advanced stages of the motional narrowing process in more detail than is feasible for pure HD. It was with this objective that May et al (35) conducted a study of the Q branch spectra of N_2 and CO over the density ranges $10 < \tilde{\rho} < 359$ and $1 < \tilde{\rho} < 596$, respectively. Their results display a behavior similar to that for HD–Ar (26) in that the narrowing is limited by dephasing at high densities. As pointed out by Gray & Welsh (1), the width data of May et al (35) are consistent with a density dependence of the form $a\tilde{\rho}^{-1} + b\tilde{\rho}$, where the linear term attempts to account for the effect of direct dephasing.

LIQUIDS AND THE CRITICAL REGION

It is clear from the foregoing that any interpretation of liquid linewidth data must be based on a knowledge of the several possible broadening mechanisms that may be active in any given case. In addition, there is always the important question as to whether the different contributions are additive. As noted above, studies involving the dilution of the active species by the rare gases or by an isotopic variant are advantageous in this respect because resonant, or near-resonant, exchange effects are eliminated. For the most part, it is a distinguishing characteristic of the liquid systems considered here that VT transfer is known to be a minor effect, while vibrational dephasing is determined to be the most important of the remaining effects. This is significant not only because of the simplification it affords, but also because it renders the study of linewidths a more effective probe of the molecular environment. For the diatomics in particular, this follows from the fact that VT transfer is promoted primarily by environmental fluctuations on a time scale comparable to the period of the molecular vibration (see Table 1) and is consequently a probe of high frequency motions only. Vibrational dephasing, on the other hand, is sensitive to environmental fluctuations over the complete range of frequencies, including the hydrodynamic modes that are of importance in the critical region. Broadly speaking, the different molecular systems are discussed here according to the chronological order in which they were investigated experimentally.

N_2

The first high-resolution measurements of the Q branch in liquid N_2 were reported by Clements & Stoicheff (36), and shortly thereafter by Scotto (37). The conditions corresponded to the normal boiling point ($T = 77K$, $\tilde{\rho} = 646$), and the profile was found to be Lorentzian with a full-width at half-maximum (FWHM) of 2.0 GHz, which was considerably less than the separation of most of the individual $Q(J)$ components in the free-molecule spectrum. It was also more than an order of magnitude smaller than the minimum room temperature value subsequently reported by May et al (35) for $\tilde{\rho} = 359$. In attempting to explain this result it was clear that motional narrowing must be responsible for the dramatic reduction in width, but it was not clear whether this process had reached a limit determined by some other broadening mechanism, such as VV or VT transfer. The latter was eliminated as a possible candidate by Renner & Maier (38), who measured an energy relaxation time of 50 ms. Their suspicion that the relaxation time was limited by impurities was later confirmed by Calaway & Ewing (39), who obtained a value of 1.5 s, and by Brueck & Osgood (40), who measured remarkably long relaxation times of up to 56 s. As reviewed by Oxtoby (3), varying degrees of success were attained in several subsequent attempts to account for the observed linewidth on the basis of dephasing theory. A molecular dynamics study (41) of liquid N_2 near its normal point provided valuable insight in that the effect of VV transfer was found to be small: it actually caused a slight decrease in the linewidth, which emphasized the imprudence of assuming the different contributions to be additive.

The first comprehensive measurements of the liquid linewidth, which covered the complete range of saturated liquid conditions from the triple point to the critical point, were reported by Clouter & Kiefte (42, 21; see Figure 2). These results also included a linewidth measurement for 2% N_2 in Ar at 90K, which confirmed that the contribution of VV transfer to the linewidth was of order 10%. The data for the pure liquid showed the ~ 2 GHz linewidth to be essentially constant throughout the lower part of the liquid range (from 63.2K to about 100K), but that it increased quite rapidly by nearly a factor of 10 as the critical point was approached. Although the latter behavior was qualitatively expected on the basis of the normal density dependence of the dephasing mechanism, it was pointed out (21) that at densities comparable to $\tilde{\rho}_c$ one should also expect an important contribution from the intramolecular vibration-rotation coupling that is responsible for the free-molecule band shape. This possibility was further explored by Brueck (43), who calculated the T dependence of the saturated liquid linewidth assuming the only active mechanism to be the RR-exchange effect (as discussed above), which is responsible for the collapse of the branch at room temperature (35). His results revealed the correct qualitative behavior and were in good quantitative agreement over the

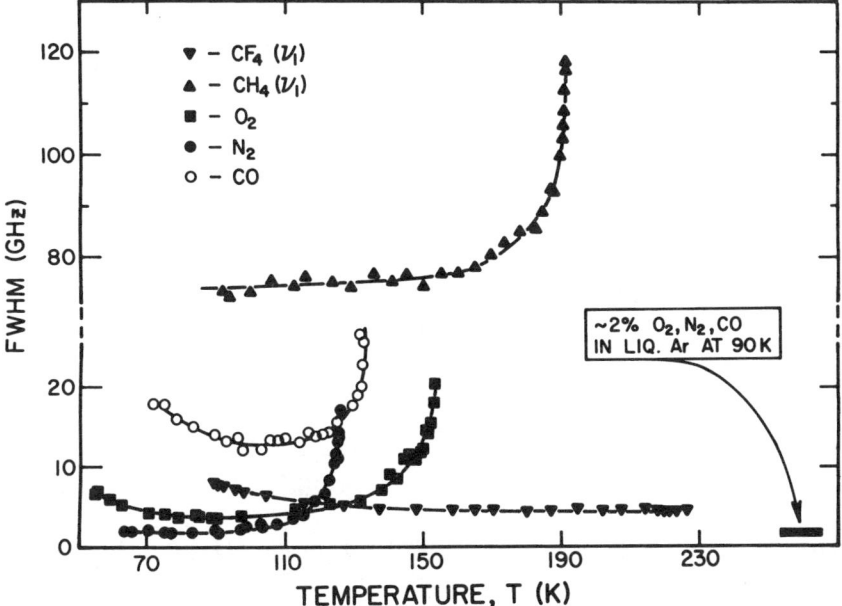

Figure 2 Vibrational Raman linewidth data for some simple fluids (from Ref. 21).

intermediate range from about 90K to 115K. At the triple point the calculated value was less than the observed by a factor of ~ 2 and suggested that RR exchange and vibrational dephasing were of comparable importance in this region, whereas the former became dominant at higher temperatures. A further discrepancy that was encountered near the critical point can, in retrospect, be associated with an additional critical contribution to the broadening.

The possibility of the latter contribution to the linewidth in the liquid phase was considered by Hills & Madden (44), who, however, included only the effect of vibrational dephasing in their theoretical calculations. The basis of their proposal was that the rapid motion of the molecules, which takes place on the picosecond timescale and is associated with the main peak in the dynamic structure factor, was presumed to be unaffected by the approach to the critical point. It was consequently taken to be responsible only for its normal (motionally narrowed) contribution to the linewidth. An additive critical contribution was associated with the hydrodynamic modes that are known to be responsible for the observed characteristics of the critical phenomenon (e.g. opalescence) and that describe collective molecular motion on the microsecond timescale near the critical point. The claimed agreement with experimental data (42) was, however, argued (21) to be a fortuitous consequence of neglecting the RR-exchange mechanism while, at the same time, using the simplest classical form for the dynamic structure factor as a description of the hydrodynamic fluctuations.

Experimental investigation of the critical broadening in the liquid phase was complicated by the normal increase in width associated with the rapid decrease in density that occurs just below T_c. A subsequent investigation (45) avoided this problem by recording the spectrum along the critical isochore in the temperature range $0.05 < (T-T_c) < 10$, where the linewidth was found to increase by a factor of about two as the critical point was approached. Most of this change occurred within about 2K of T_c and was clearly not associated with any normal temperature dependence because it was not observed on the isochore at twice the critical density. Two important characteristics of the near-critical spectrum were noted. First, the profile was distinctly asymmetric in a sense opposite to that associated with the normal shape of the low density spectrum. Second, the cosine transform of the profile was found to be accurately Gaussian in form and indicated that the broadening was largely inhomogeneous. The latter was consistent with the Hills & Madden idea (44) that long-lived spatial fluctuations in local density (associated with the "slowing-down" of the hydrodynamic modes) were contributing a broadening in excess of the normal motionally narrowed width as observed at $(T-T_c) \simeq 10K$. It was suggested (45) that information pertaining to the fluctuations in local density could perhaps be extracted from the near-critical spectrum.

In this connection, one possible approach would have been first to obtain a vibrational phase correlation function as the Fourier transform of the intensity profile (3, 5). However, the asymmetry of the spectrum would then lead to a complex correlation function of questionable physical significance. The presence of the spectral asymmetry therefore posed a fundamental question that needed to be addressed in an alternative analysis of the data. The approach taken by Clouter et al (46) utilized a modeling procedure based in part on the separability of time scales assumed by Hills & Madden (44) and on ideas presented by Strauss & Mukamel (47). It was emphasized that the local field experienced by a given probe molecule was essentially determined by the range of the pair potential and involved only the first two or three shells of neighboring molecules. As was apparent from the relatively wide temperature range over which broadening was significant, the Raman experiment must therefore be sensitive to variations in local density (molecular clusters) on a much smaller scale, and which persist much farther from the critical point, than those responsible for critical opalescence. Since the time scale for the latter fluctuations is known to be of order 1 μs near the critical point, it was reasonable to assume that the smaller scale fluctuations would persist for a comparable time, and probably much longer. It was consequently assumed that the hydrodynamic fluctuations made no significant homogeneous contribution to the spectral width in the immediate vicinity of the critical point. Rather, it was proposed (46) that the observed spectrum could be treated as an inhomogeneous superposition of local spectra of finite width. The hydrodynamic fluctuations were regarded as determining the metastable local densities, the peak frequencies of the local spectra, and consequently the inhomogeneous contribution to the spectral width. The high frequency motions determined the correlation time, τ_c, appropriate to each local (Lorentzian) spectrum, which should, in general, depend on the local temperature as well as the local density. It was pointed out, however, that measurements (46) of the normal spectrum along a noncritical isochore revealed only a weak temperature dependence, which could be neglected. The width of each local spectrum was therefore assumed to be identical to that actually observed for the equivalent noncritical (bulk) density and roughly comparable temperature.

The process of modeling the observed, $I(v)$, intensity profile was expressed (46) symbolically as

$$I(v) = \int_0^{\tilde{\rho}_m} \tilde{\rho} L(v, \tilde{\rho}) P(\tilde{\rho}) \, d\tilde{\rho}, \qquad 1.$$

where v is the Raman shift relative to the peak of the profile, $\tilde{\rho}$ is now the *local density* (as distinct from the bulk/average density, $\bar{\rho}$) with $\tilde{\rho}_m$ corresponding to some reasonable upper limit such as the triple point value $\tilde{\rho}_t$ (see Table 1), and $L(v, \tilde{\rho})$ is the Lorentzian local spectrum of unit area (intensity) with peak frequency, v_p, and FWHM = Δ determined by $\tilde{\rho}$. The intensity of the local spectrum is scaled according to the weighting factor $P(\tilde{\rho})$, which represents the probability distrubition for local density whose zeroth order approximation is the Gaussian function. An additional $\tilde{\rho}$ factor also appears in the integrand because, as in the normal fluid, the probability of a Raman event is proportional to the density. The linear dependence $v_p(\tilde{\rho}) = v_0 - 0.09\tilde{\rho}$ was taken to be identical to the previously determined (21) density dependence of the Raman shift under noncritical conditions, i.e. $v_p(\bar{\rho})$. $\Delta(\tilde{\rho})$ was also modeled in a number of different ways designed to mimic the known $\Delta(\bar{\rho})$ dependence. The simplest of these was $\Delta(\tilde{\rho}) = a/(1 + b\tilde{\rho})$, which incorporates the $\tilde{\rho}^{-1}$ effect of motional narrowing while avoiding an unphysical divergence for $\tilde{\rho} \to 0$. The a,b parameters were chosen to be consistent with the high-density (liquid) Δ values while approximately reproducing the known value of Δ for $\tilde{\rho} \to 0$. The peak of the Gaussian $P(\tilde{\rho})$ function was necessarily located at $\tilde{\rho} = \tilde{\rho}_c$, and its FWHM was the only adjustable parameter in the computerized fitting procedure corresponding to Eq. 1. For the near-critical condition $\varepsilon \equiv (T - T_c)/T_c = 8 \times 10^{-5}$ an accurate fit to the observed $I(v)$ profile was obtained with a value of $0.55\tilde{\rho}_c$ as the FWHM of $P(\tilde{\rho})$.

The asymmetry of the profile was satisfactorily reproduced, and was a direct consequence of the $\Delta(\tilde{\rho})$ dependence. This served to emphasize that the dephasing could only be properly described via a distribution of correlation times as opposed to a single, or effective, value. The latter idea was apparently the basis of the approach taken by Chesnoy (48), who ignored the asymmetry of the spectrum and defined a correlation function determined only by the cosine transform of the profile. This would seem to be an unnecessarily severe compromise by comparison with the assumed separability of timescales leading to Eq. 1.

The significance of this result (46) was that it provided experimental access to the $P(\tilde{\rho})$ distribution that had previously been studied only by computer simulations (49, 50). In particular, the extension of the data analysis to higher temperatures on the critical isochore showed that the width of the $P(\tilde{\rho})$ distribution approached its maximum critical value through an exponential dependence on ε. This behavior, which is different from that predicted by conventional theories of the critical phenomenon, was subsequently attributed (51, 52) to the finite character of the probed volume.

CO and O_2

The energy relaxation time in pure liquid CO has been determined (40, 53, 54) to be of order 10 ms and consequently makes no significant contribution to the observed linewidths in the GHz range. With respect to other broadening effects, a comparison of CO and N_2 is somewhat analogous to HD and H_2. The essential differences are that odd ΔJ transitions are allowed for CO but forbidden for N_2, and that the electric dipole moment resulting from the heteronuclear character of CO gives rise to additional terms in the intermolecular interaction potential. The first measurements were those of Brueck (55), who reported linewidth data for liquid CO as well as liquid solutions of CO with O_2, N_2, and Ar. The data covered the temperature range from 72K to 87K and revealed a pure-liquid linewidth of 17 GHz at 77K, which was reduced to approximately 3 GHz in each of the dilute solutions. The latter was comparable to the corresponding values for both pure N_2 and 2% N_2 in Ar (42, 21), and suggested that VV transfer was responsible for a large additional contribution to the linewidth in pure CO. This was supported by theoretical calculations that identified the additional terms in the pair potential as being responsible for the effect. The absence of any restrictions on ΔJ was presumably of little consequence (cf N_2), because in the lower temperature regions for both liquids the narrowing of the spectrum due to RR exchange is effectively complete.

Later linewidth data for liquid CO (21; Figure 2) were in substantial agreement with Brueck's results (55) and, in addition, covered the complete range of the saturated liquid (see Table 1). As for N_2 the width was found to increase sharply as the critical point was approached, but the fractional increase was reduced because of the large underlying contribution from VV transfer. It is for this reason that CO is less effective than N_2 as a probe of critical fluctuations. To date no explanation has been offered for the $\sim 50\%$ increase in the liquid CO linewidth that occurs on approaching the triple point.

The energy relaxation time for liquid O_2 (38, 56) is comparable to that for CO and can likewise be neglected as a line broadening mechanism. The available data (36, 37, 42, 21; Figure 2) cover the complete range of the saturated liquid and also include an Ar-dilution measurement (21) that indicates that the contribution from VV transfer is $\sim 25\%$ at 90K. For temperatures above their normal boiling points the linewidth behavior is similar for both O_2 and N_2. The principal difference between the two is that the O_2 linewidths are higher by an additive 1.5 to 2.0 GHz. In this range the line-broadening mechanisms are presumably the same as those for N_2. At temperatures below the normal boiling, the O_2 data are qualitatively similar to CO in displaying a definite, and unexplained, increase as the triple point is approached. The critical contribution to the O_2 linewidth has not been investigated in detail.

CH_4 and CF_4

Low density ($\tilde{\rho} > 0.3$) data (57, 58) for the totally symmetric (v_1) mode of CH_4 reveal an interesting behavior in that there is no evidence of motional narrowing with increasing density at room temperature. Rather, the spectral width continues to increase with density after the individual Q branch components have completely overlapped to produce a single-peaked profile. Motional narrowing is, however, apparent in the saturated liquid (59, 21; Figure 2), because the width decreases by $\sim 50\%$ in the region just below the critical point. The width is constant in the lower part of the liquid range and is greater than the zero-density value (60) by a factor of ~ 350, so the effects of intramolecular vibration-rotation coupling can clearly be neglected. Since liquid-phase dilution measurements (61) also show that resonant VV transfer is a minor effect, it appears that vibrational dephasing and VT transfer must account for the observed behavior. By comparison with the diatomics discussed above, an enhancement of the VT transfer mechanism can be expected because of the mediating effect of nonresonant (intramolecular) VV transfer between modes. However, there do not appear to be any reported measurements of the energy relaxation time in the liquid phase, so its possible contribution to the linewidth is unknown. On the other hand, the liquid-phase narrowing, noted above, and the presence of a significant ($\sim 20\%$) critical broadening as reported by Echargui & Marsault-Herail (59) are not consistent with energy relaxation being the dominant effect (21). It seems likely that the different behavior in the low density gas as compared to the liquid is associated with a transition from slow-modulation dephasing in the former case to fast modulation in the latter. In this connection it should be noted that the observed (21, 59) $\Delta v(\tilde{\rho})$ dependence for the v_1 mode of CH_4 shows a (negative) slope that is steeper by a factor of ~ 6 compared to N_2 (21), and is even steeper than that for H_2 (11) under comparable conditions. The large Γ values that can be expected as a result are therefore consistent with the occurrence of slow, or intermediate, modulation conditions at low density.

When the latter argument is applied to CF_4, it is concluded (21) that Γ must be less by a factor of ~ 20 as compared to CH_4. As a result the dephasing contribution to the width can be expected to be small, and perhaps even negligible, compared to energy relaxation. Such circumstances would indeed account for the complete absence (21; Figure 2) of critical broadening in CF_4. Further work, particularly in the area of detailed lineshape analysis, is prerequisite to a satisfactory understanding of the spectra of both these molecular systems.

The Hydrogens

At liquid-phase temperatures only the $J = 0$ and $J = 1$ rotational states of the hydrogens are significantly populated. For the homonuclear variants, H_2 and D_2, these populations do not in general correspond to thermal equilibrium because (*ortho* ↔ *para*) transitions for which $\Delta J = \pm 1$ are forbidden in the absence of magnetic interactions. In practice, however, a slow *ortho* ↔ *para conversion* rate with a characteristic time of several days is promoted by magnetic impurities in the walls of the containment cell and by the rotational magnetic moments of the $J = 1$ molecules themselves. This complicates the investigation of line-broadening effects because the width of either the $Q(0)$ or $Q(1)$ component can be expected to depend on the *ortho/para* ratio, which changes as equilibration proceeds during the course of an experiment. It is partly for this reason that the first comprehensive data to appear in the literature (62) were for HD, which is not subject to such restrictions on the rotational-state populations.

For HD at temperatures above about 40K the population ($> 5\%$) of the $J = 1$ state is sufficient to render the $Q(1)$ component detectable under low density conditions, where the overlapping of the two components is negligible. Because the critical density of HD is in the range where the effect of Q-branch collapse is operative, the weak $Q(1)$ component does not appear as a separate feature in spectra recorded on the critical isochore. It is, however, responsible for a spectral asymmetry that is observed for $(T - T_c) > 10K$ on this isochore, and which is opposite in sense to that which develops as the critical point is approached. Critical broadening accounts for a two-fold increase in width on the critical isochore, and the inhomogeneous character of this broadening was demonstrated as in the N_2 case. However, the application of the lineshape modeling procedure described above for N_2 was unsuccessful because of a significant $Q(1)$ intensity and complications associated with the effect of Q-branch collapse.

The linewidth data for HD are summarized in Figure 3. It should be noted that the increase in width that occurs for $T > 45K$ on the critical isochore is due to the emergence of the $Q(1)$ component with increasing intensity, and that the widths measured on the isochore at $0.5\tilde{\rho}_c$ are for the $Q(0)$ component only. Since at temperatures below T_c there were no detectable effects associated with the $Q(1)$ component, it is apparent that *RR* exchange is inoperative because it is dependent on an appreciable population of states for which $J > 0$. A similar comment also applies to the effects of *RT* transfer and rotational dephasing. Given that *VV* transfer was determined to be a secondary effect (62), it was consequently concluded

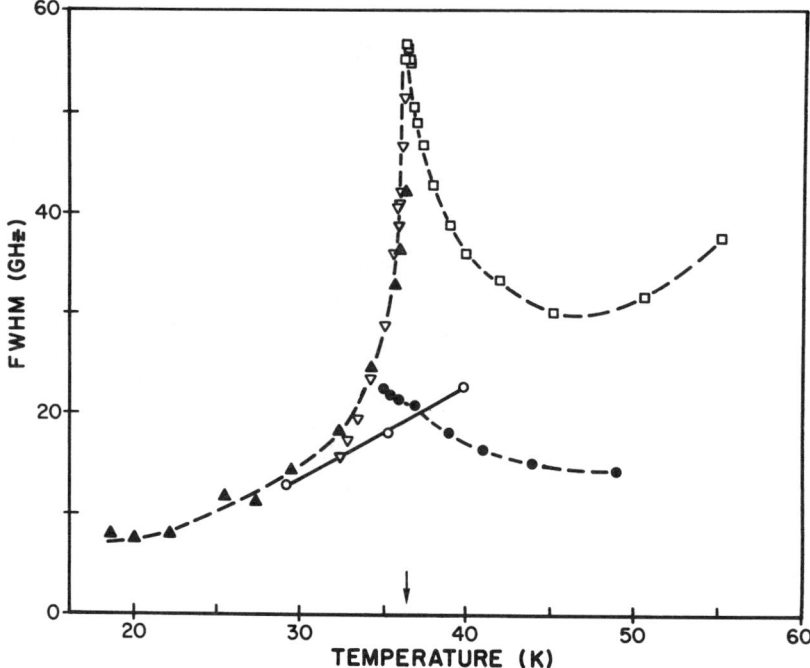

Figure 3 Vibrational Raman linewidth data for HD (from Ref. 62). ▲—saturated liquid; ▽—saturated vapor; □—critical isochore; ●—$0.50\tilde{\rho}_c$ isochore; ○—$2.3\tilde{\rho}_c$ isochore; ↓—critical temperature.

that the linewidth behavior in both the liquid and vapor phases of HD must be associated primarily with vibrational dephasing. As outlined above, dephasing is generally expected to cause a broadening (with increasing density) at low densities followed by a motional narrowing at high densities. This is consistent not only with the data (Figure 3) for the saturated liquid and vapor, but also with the data for the two noncritical isochores, where the densities differ by a factor of ~ 5 yet the widths are comparable. It is thus apparent that the maximum dephasing width occurs at densities near $\tilde{\rho}_c$, but the details of this (normal) process are obscured by the superimposed critical broadening.

The behavior of the linewidth for either the $Q(0)$ or $Q(1)$ component of H_2 is qualitatively similar to HD (M. J. Clouter, unpublished data). The most important difference is that the collapse of the Q branch does not occur for reasons that have already been discussed. For comparable *ortho* and *para* fractions there is significant overlap of the wings of the two components near the critical point, but for the special case of nearly pure

$J = 0$ (*para*) H_2, this problem is eliminated so that the critical broadening of the $Q(0)$ component is amenable to detailed treatment. An additional distinguishing feature of the spectrum was that the weak $Q(1)$ component of nearly pure *para* H_2 exhibited an inhomogeneous broadening, and sideband structure, due to the clustering of $J = 1$ (*ortho*) molecules. This effect is analogous to that previously observed in the solid phase (63), but was found to persist throughout the liquid phase to temperatures well above T_c.

The Q-branch spectrum of D_2, while otherwise similar to H_2, is unique in that it exhibits (64) a fine structure over a wide range of conditions around the critical point. The structure takes the form of a series of up to seven features with widths of ~ 2 GHz and a regular separation of 6 GHz that are superimposed upon the $Q(0)$ and $Q(1)$ profiles, whose overall widths are approximately 20 GHz near the critical point. It has been suggested (64) that the structure is indicative of local ordering in the fluid. Independent investigation of this possibility by neutron diffraction would be highly desirable in providing a direction for future work.

Others

There are at least two other molecular systems that should be mentioned in connection with the phenomenon of critical broadening. The v_1 and $2v_2$ vibrational spectra of CO_2 ($T_C = 31.1°C$) were first investigated by Garrabos et al (65). Their spectral resolution was instrument-limited, however, and they were consequently unable to detect any critical broadening effects. Subsequent work (66) quoted values of 60 GHz and 48 GHz, respectively, for the noncritical FWHM of each of these Q branches on the critical isochore. A critical broadening of 25% was reported (66) for the v_1 mode whereas the corresponding effect for the $2v_2$ mode was only marginally detectable. The v_3 band of C_2H_6 (ethane) was also investigated by Wood & Strauss (67), whose resolved spectra did not exhibit any critical broadening. Their result is similar to that for CF_4 and is probably a consequence of vibrational energy relaxation being the dominant broadening mechanism.

SUMMARY

The spectroscopic information for most of the diatomic liquids treated in this review is reasonably complete and understood. The hydrogens are an exception, since they are the most recently studied, and much remains to be done in both the experimental and theoretical areas. With respect to the polyatomics (CH_4, CF_4), additional information, particularly with respect to energy relaxation, is required before an unambiguous interpretation of the observed line broadening is possible. In all cases there is considerable scope for future experimental work in the area of isotopic and rare-gas dilution of the active species. Not only does this provide a means of assessing the importance of resonant-transfer mechanisms in the pure fluids, but in some cases (e.g. N_2) it permits the detailed study of vibrational dephasing as the only active broadening mechanism. While the paucity of work in this area is understandable because of the difficulties associated with the low signal levels involved, it can perhaps be anticipated that the wider use of nonlinear techniques such as Raman-gain spectroscopy will provide easier access to this information, at least in the noncritical fluids. Studies of the critical-broadening phenomenon, and its exploitation as a potentially valuable probe of the critical phenomenon, have yet to benefit from a definitive theoretical treatment. A similar comment also applies to the broadening effects that have been observed (e.g. in O_2 and CO) in the neighborhood of the triple point.

One of the most promising areas for future work involves the regime of ultra-high pressures ($\sim$ 10 GPa), which are now attainable by diamond-anvil techniques. Conventional Raman spectroscopy is well suited to studies of this kind because of the very small sample volumes involved. Although Raman-shift data have already been employed to advantage in exploring the behavior of a number of simple molecular systems (68) over a greatly expanded range of thermodynamic conditions, corresponding high-resolution measurements of the width and shape of the vibrational spectra have not been reported to date.

Literature Cited

1. Gray, C. G., Welsh, H. L. 1971. *Essays in Structural Chemistry*, pp. 163–88. London: Macmillan
2. Srivastava, R. P., Zaidi, H. R. 1979. *Raman Spectroscopy of Gases and Liquids*, pp. 167–201. Berlin: Springer-Verlag
3. Oxtoby, D. W. 1979. *Adv. Chem. Phys.* 40: 1
4. Oxtoby, D. W. 1981. *Photoselective Chemistry, Part 2*, ed. J. Jortner, R. D. Levine, S. A. Rice, pp. 487–519. New York: Wiley
5. Rothschild, W. G. 1984. *Dynamics of Molecular Liquids*. New York: Wiley
6. May, W., Kiefte, H., Clouter, M. J., Stegeman, G. I. 1978. *Appl. Opt.* 17: 1603
7. Smyth, K. C., Rosasco, G. J., Hurst, W. S. 1987. *J. Chem. Phys.* 87: 1001
8. May, A. D., Degen, V., Stryland, J. C., Welsh, H. L. 1961. *Can. J. Phys.* 39: 1769
9. May, A. D., Varghese, G., Stryland, J. C., Welsh, H. L. 1964. *Can J. Phys.* 42: 1058
10. Allin, E. J., May, A. D., Stoicheff, B. P., Stryland, J. C., Welsh, H. L. 1967. *Appl. Opt.* 6: 1597
11. Looi, E. C., Stryland, J. C., Welsh, H. L. 1978. *Can. J. Phys.* 56: 1102
12. Fiutak, J., Van Kranendonk, J. 1962. *Can. J. Phys.* 40: 1085
13. Fiutak, J., Van Kranendonk, J. 1963. *Can. J. Phys.* 41: 21
14. Van Kranendonk, J. 1963. *Can. J. Phys.* 41: 433
15. Gray, C. G., Van Kranendonk, J. 1966. *Can. J. Phys.* 44: 2411
16. May, A. D., Poll, J. D. 1965. *Can. J. Phys.* 43: 1836
17. Bischel, W. K., Dyer, M. J. 1986. *Phys. Rev. A* 33: 3113

18. Kelley, J. D., Bragg, S. L. 1986. *Phys. Rev. A* 34: 3003
19. Dion, P., May, A. D. 1973. *Can. J. Phys.* 51: 36
20. Witkowicz, T., May, A. D. 1976. *Can. J. Phys.* 54: 575
21. Clouter, M. J., Kiefte, H., Jain, R. K. 1980. *J. Chem. Phys.* 73: 673
22. Audibert, M. M., Joffrin, C., Ducuing, J. 1974. *Chem. Phys. Lett.* 25: 158
23. Herzberg, G. 1959. *Spectra of Diatomic Molecules.* New York: Van Nostrand
24. Robert, D., Bonamy, J., Marsault-Herail, F., Levi, G., Marsault, J. P. 1980. *Chem. Phys. Lett.* 74: 467
25. Robert, D., Bonamy, J., Sala, J. P., Levi, G., Marsault-Herail, F. 1985. *Chem. Phys.* 99: 303
26. Marsault-Herail, F., Echargui, M., Levi, G., Marsault, J. P., Bonamy, J. 1982. *J. Chem. Phys.* 77: 2715
27. Alekseyev, V. A., Grasiuk, A., Ragulsky, V., Sobelman, I., Faizulov, F. 1968. *IEEE J. Quantum Electron.* 4: 654
28. Alekseyev, V. A., Sobelman, I. 1969. *Sov. Phys. JETP* 28: 991
29. Temkin, S. I., Burshtein, A. I. 1976. *JETP Lett.* 24: 86
30. Temkin, S. I., Burshtein, A. I. 1979. *Chem. Phys. Lett.* 66: 52
31. Temkin, S. I., Burshtein, A. I. 1979. *Chem. Phys. Lett.* 66: 57
32. Temkin, S. I., Burshtein, A. I. 1979. *Chem. Phys. Lett.* 66: 62
33. Bonamy, J., Bonamy, L., Robert, D. 1977. *J. Chem. Phys.* 67: 4441
34. Altmann, K., Holzer, W., Leduff, Y. 1975. *Chem. Phys. Lett.* 36: 259
35. May, A. D., Stryland, J. C., Varghese, G. 1970. *Can. J. Phys.* 48: 2331
36. Clements, W. R. L., Stoicheff, B. P. 1968. *Appl. Phys. Lett.* 12: 246
37. Scotto, M. 1968. *J. Chem. Phys.* 49: 5362
38. Renner, G., Maier, M. 1974. *Chem. Phys. Lett.* 28: 614
39. Calaway, W. F., Ewing, G. E. 1975. *Chem. Phys. Lett.* 30: 485
40. Brueck, S. R. J., Osgood, R. M. Jr. 1976. *Chem. Phys. Lett.* 39: 568
41. Oxtoby, D. W., Levesque, L., Weis, J. J. 1978. *J. Chem. Phys.* 63: 5528
42. Clouter, M. J., Kiefte, H. 1977. *J. Chem. Phys.* 66: 1736
43. Brueck, S. R. J. 1977. *Chem. Phys. Lett.* 50: 516
44. Hills, B. P., Madden, P. A. 1979. *Mol. Phys.* 37: 937
45. Clouter, M. J., Kiefte, H. 1984. *Phys. Rev. Lett.* 52: 763
46. Clouter, M. J., Kiefte, H., Deacon, C. G. 1986. *Phys. Rev. A* 33: 2749
47. Strauss, H. L., Mukamel, S. 1984. *J. Chem. Phys.* 80: 6328
48. Chesnoy, J. 1986. *Chem. Phys. Lett.* 125: 267
49. Kaski, K., Binder, K., Gunton, J. D. 1984. *Phys. Rev. B* 29: 3996
50. Binder, K., Landau, D. P. 1984. *Phys. Rev. B* 30: 1477
51. Tuszyński, J. A., Clouter, M. J., Kiefte, H. 1985. *Phys. Lett.* 108A: 272
52. Tuszyński, J. A., Clouter, M. J., Kiefte, H. 1986. *Phys. Rev. B* 33: 3423
53. Calaway, W. F., Ewing, G. E. 1975. *J. Chem. Phys.* 63: 2842
54. Legay-Sommaire, N., Legay, F. 1977. *Chem. Phys. Lett.* 52: 213
55. Brueck, S. R. J. 1978. *Chem. Phys. Lett.* 53: 273
56. Protz, R., Maier, M. 1979. *Chem. Phys. Lett.* 64: 27
57. Clements, W. R. L., Stoicheff, B. P. 1970. *J. Mol. Spectrosc.* 33: 183
58. Clements, W. R. L. 1972. PhD thesis. Univ. Toronto
59. Echargui, M. A., Marsault-Herail, F. 1987. *Mol. Phys.* 60: 605
60. Kozlov, D. N., Prokhorov, A. M., Smirnov, V. V. 1979. *J. Mol. Spectrosc.* 77: 21
61. Orlova, N. D., Pozdnyakova, L. A. 1978. *Opt. Spectrosc. USSR* 44: 544
62. Staniaszek, P., Clouter, M. J., Kiefte, H. 1988. *Can. J. Phys.* 66: In press
63. Prior, W. R. C., Allin, E. J. 1972. *Can. J. Phys.* 50: 1471
64. Clouter, M. J., Deacon, C. G., Kiefte, H. 1987. *Phys. Rev. Lett.* 58: 1116
65. Garrabos, Y., Tufeu, R., LeNeindre, B., Zalczer, G., Beysens, D. 1980. *Chem. Phys.* 72: 4637
66. Deacon, C. G., Clouter, M. J., Kiefte, H. 1985. *J. Chem. Phys.* 83: 446
67. Wood, K. A., Strauss, H. L. 1983. *J. Chem. Phys.* 78: 3455
68. Schiferl, D., Kinkead, S., Hanson, R. C., Pinnick, D. A. 1987. *J. Chem. Phys.* 87: 3016.

LATTICE VIBRATIONS AND HEAT TRANSPORT IN CRYSTALS AND GLASSES

David G. Cahill and R. O. Pohl

Laboratory of Atomic and Solid State Physics, Cornell University, Ithaca, New York 14853-2501

Introduction

In dielectric solids, most of the thermal energy is contained in the form of lattice vibrations, which, therefore, play an essential role in all processes that involve temperature. The purpose of this paper is to review and illustrate the models developed for the description of lattice vibrations and their interactions. Because of our personal involvement we discuss measurements of specific heat, thermal conductivity, and the propagation of heat pulses, the subjects of our own work. We begin our historical review with the earliest model, as proposed by Einstein, and then show why for crystalline solids it had to be replaced by a model employing elastic travelling waves. We review scattering experiments that demonstrate the wave-like character of these excitations, and other experiments that are more readily understood with the particle picture of quantized vibrations. Finally, we show that in amorphous and highly disordered solids this picture of elastic waves has serious deficiencies at high frequencies, and that the original Einstein model is superior.

Einstein's Model of Lattice Vibrations

In the classical model based on the work of Petit & Dulong (1), every atom contains the vibrational energy $3k_BT$, where k_B is Boltzmann's constant and T the absolute temperature. This leads to a temperature-independent specific heat $3k_Bn$, where n is the number density of atoms (number per volume).

The limits of this picture were demonstrated by H. F. Weber (2) in 1875, who showed that the specific heat approached this so-called Dulong-Petit value only at high temperatures. By extending the measuring temperatures below room temperature, he observed a decrease in the specific heat of diamond by as much as a factor of 10. Because of the immense importance of Weber's work (his data could have led to the concept of energy quanta 25 years prior to Planck's work), we reproduce the original data in Figure 1, obtained on crystalline boron, silicon, graphite, and diamond.

It took 32 years until Einstein (3) recognized that the atomic vibrations were quantized. These Einstein oscillators, as they are now called, have a specific heat that approaches the Dulong-Petit value at high temperatures and that decreases exponentially at low temperatures.

Einstein (4) also discovered, in 1911, the most serious difficulty with his model of isolated atomic vibrations in analyzing Eucken's low temperature measurements (5) of the thermal conductivity, which are reproduced in Figure 2. Einstein noticed that in dielectric crystals the magnitude found by Eucken greatly exceeded the value derived by assuming a random walk of the elastic energy among Einstein oscillators, even if the oscillators are so heavily damped that they pass on their energy within half a period of oscillation. For KCl, for example, Einstein calculated a thermal conductivity at room temperature $\Lambda(300K) = 2.9 \times 10^{-3}$ W cm^{-1} K^{-1}, 22 times smaller than the value measured by Eucken $\Lambda(300K) = 6.3 \times 10^{-2}$ W cm^{-1} K^{-1}, see Figure 2. Furthermore, in Einstein's model the thermal conductivity should decrease with decreasing temperature, while Eucken observed it to increase. (We mention that Eucken also reported measurements on glasses, which Einstein, however, ignored. We return to them below.)

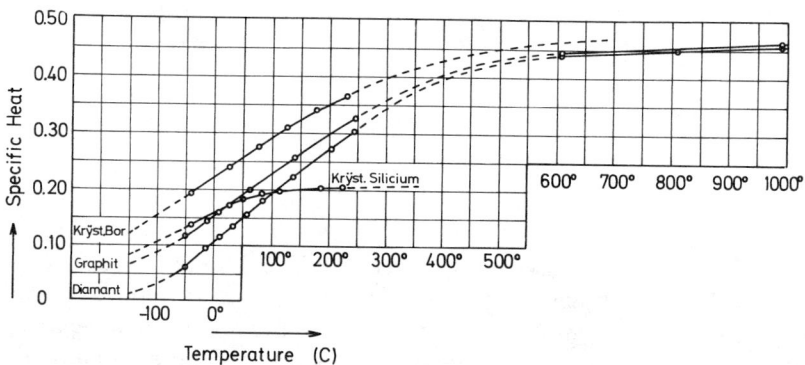

Figure 1 Temperature dependence of the specific heat of crystalline boron (*Kryst. Boron*), silicon (*Kryst. Silizium*), graphite, and diamond, after Weber (2). The Dulong-Petit value is reached only at high temperatures. The units for the vertical axis are not given in the original, but are (cal g^{-1} °C^{-1}).

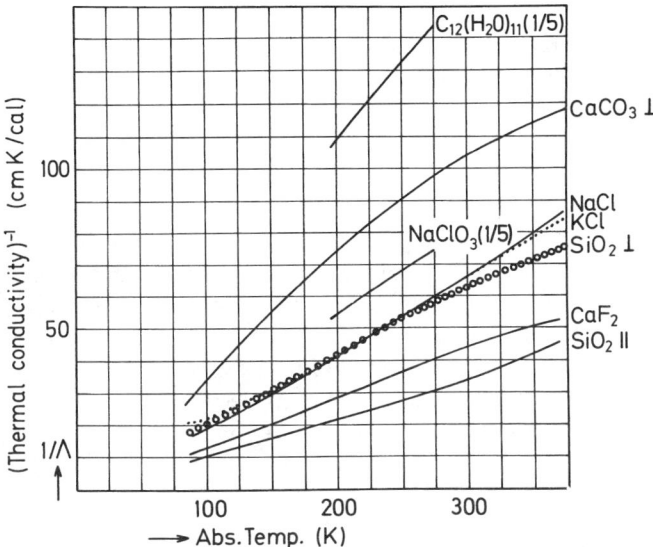

Figure 2 The inverse of the thermal conductivity Λ of several crystalline dielectrics between 83 and 373K, after Eucken (5). The values plotted for $C_{12}(H_2O)_{11}$, sucrose, and for $NaClO_3$ have been reduced by a factor of five, as indicated by the factor in parentheses.

The Debye and Born–von Karman Model

The way out of the fundamental dilemma inherent in the Einstein model was suggested by Debye (6) and by Born & von Karman (7). These authors argued that the atoms in a solid do not oscillate as isolated entities, but collectively as propagating waves. We do not review these theories, which are well covered in textbooks, and mention only the following essential points: Debye treated these elastic waves as dispersionless in a way identical to the electromagnetic waves in an empty cavity; the only difference was that he included longitudinal waves, and that he limited the number of normal modes to $3n$, where n is the number density of atoms in a three-dimensional solid. With these assumptions, he was able to predict, with no free parameters, the specific heat of solids, and found agreement with the measurements, whereas Einstein's theory employed the Einstein frequency as a free parameter. Debye also explained why at low temperatures the specific heat decreased less rapidly than predicted by Einstein (Debye's T^3 dependence of the specific heat). While Debye used a continuum approach, Born & von Karman started their calculation from the individual atoms in the crystal lattice and the interatomic force constants. This refinement led to a dispersion of the lattice waves and to their exact description even in complicated crystal structures.

A very important success of the elastic wave theory was that it opened the way to a qualitative understanding of the observed high thermal conductivity of dielectric crystals, and also of its temperature dependence. In analogy to the kinetic theory of gases, Debye (8) wrote the thermal conductivity Λ as

$$\Lambda = \tfrac{1}{3} C_V v l, \qquad 1.$$

where C_v is the specific heat (per volume), v the wave velocity, and l the mean free path between collisions with lattice defects and other waves. A mean free path of the order of 100 Å, or a few tens of wavelengths, was required to explain Eucken's findings for crystalline solids. A rapidly decreasing scattering probability could be expected to more than compensate for the decreasing specific heat at decreasing temperatures, thus leading to an increase of the thermal conductivity. Some of the important scattering mechanisms for elastic waves and their study through measurements of heat transport are reviewed below.

Phonon-Phonon Scattering

In 1929, Peierls (9) demonstrated how elastic waves could be scattered by each other, as a result of the anharmonicity of the interatomic potential. In the quantum picture, such scattering processes could be described through the destruction of some quanta of elastic energy, and the creation of new ones. The most likely process is three-quantum scattering in which two quanta are destroyed and one created or vice versa. Peierls showed that two conservation laws had to be obeyed:

$$\hbar\omega_1 + \hbar\omega_2 = \hbar\omega_3, \qquad 2.$$

where ω_i is the (angular) frequency of the quanta and $\hbar$ is Planck's constant divided by 2π; this equation express the conservation of energy. The other conservation law is written as

$$\hbar\mathbf{k}_1 + \hbar\mathbf{k}_2 = \hbar\mathbf{k}_3 + j\hbar\mathbf{G} \qquad 3.$$

where $k_i = 2\pi/\lambda_i$ is the wavevector, for a wave of wavelength λ_i, and $\mathbf{G}$ equals a reciprocal lattice vector; j can be either 0 or ± 1. If $j = 0$, the scattering process is called a Normal or N-process, and Eq. 3 is analogous to the conservation of momentum of the three quanta. This collision is illustrated in Figure 3, for a one-dimensional crystal lattice with lattice constant a, where a transverse (T) quanta with wave vector $\mathbf{k}_1$ and a longitudinal (L) quanta with wave vector $\mathbf{k}_2$ collide to form a longitudinal (L) one, with wave vector $\mathbf{k}_3$. For an N-process, it is essential that k_3 is smaller than $2\pi/a$. In this case the resulting wave continues to propagate in the same direction, and the total wavevector is conserved. On the

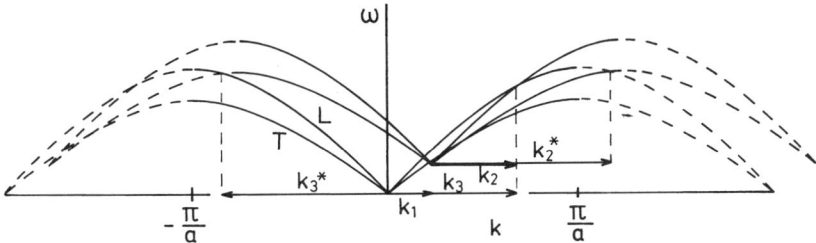

Figure 3 Normal and Umklapp collision in a one-dimensional crystal lattice of interatomic spacing a, after Peierls (9). A transverse (T) phonon of wavevector $\mathbf{k}_1$, collides with a longitudinal (L) phonon $\mathbf{k}_2$ to form a phonon of wavevector $\mathbf{k}_3$ in a Normal process ($j = 0$ in Eq. 3). The same phonon $\mathbf{k}_1$, colliding with a transverse phonon $\mathbf{k}_2^*$ leads in an Umklapp process to a reflected phonon $\mathbf{k}_3^*$ ($j = -1$ in Eq. 3).

average, such collisions by themselves do not inhibit the flow of energy. In conjunction with other, momentum-destroying collisions, however, they may have an important influence, either enhancing or reducing, or even qualitatively altering the flow of energy. In the section on Poiseuille flow of heat and second sound, below, we describe some dramatic consequences of N-processes for energy flow in carefully prepared crystals and show for a particularly transparent example how the N-process scattering rate can be determined.

Figure 3 also illustrates a so-called Umklapp or U-process, for which j in Eq. 3 is not zero. It occurs as follows. The addition of two quanta with the wavevectors $\mathbf{k}_1$ and $\mathbf{k}_2^*$ can lead to a quantum with wavevector $\mathbf{k}_3$ greater than $2\pi/a$. This wavevector corresponds to a wavelength shorter than twice the lattice spacing—which is physically meaningless—or to a wave that travels in the opposite direction with the wave vector $\mathbf{k}_3^* = \mathbf{k}_3 - \mathbf{G}$; this wave is physically meaningful. The scattering process will lead to thermal resistance. Since a U-process involves phonons with energies of the order of the zone boundary phonons, the U-process rate will depend on the thermal population of these phonons. Consequently, the probability for U-processes to occur will increase exponentially with temperature and can overwhelm the increase of the specific heat, thus leading to a decrease of the thermal conductivity with increasing temperature, in qualitative agreement with Eucken's findings.

Figure 4 shows the rapid increase of the thermal conductivity, and hence of the phonon mean free path l even more clearly than in Figure 2 on four rather perfect single crystals of Al_2O_3 (10), Si (11, 11a), CsI (12), and ^{4}He (13). The data extend to lower temperatures than those covered by Eucken. Figure 4 also illustrates boundary scattering, a scattering process that begins to dominate at the lowest temperatures (e.g. below 5K in CsI).

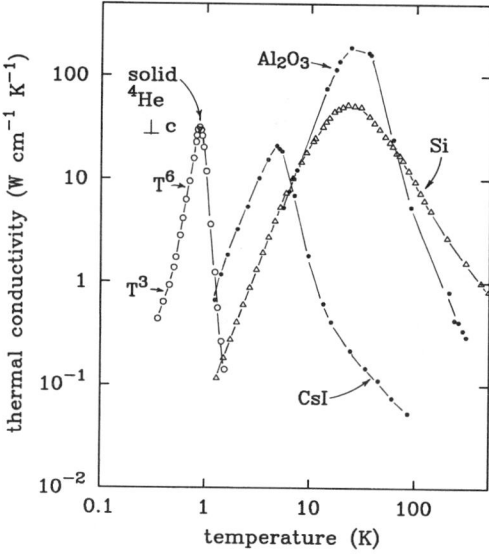

Figure 4 Thermal conductivity of high purity single crystals of sapphire (Al$_2$O$_3$) (Ref. 10 and this work), Si (11, 11a), CsI (12), and ^{4}He (13). The data for Si and Al$_2$O$_3$ overlap below 10K.

Boundary scattering was first observed by deHaas & Biermass (14, 14a) and was explained by Casimir (15). As the Umklapp mean free path begins to exceed the diameter of the (typically pencil-shaped) sample, scattering by the surfaces becomes the dominant scattering processes, leading to a mean free path that is independent of the phonon wavelength. For a sample of circular cross section, this mean free path is equal to the sample diameter d. Since in this temperature range the specific heat of dielectric crystals varies as the temperature cubed, a thermal conductivity proportional to dT^3 is predicted, in excellent agreement with the experiment. Figure 4 shows one peculiarity. In ^{4}He, just below the maximum, the conductivity decreases more rapidly than with the third power of T until it approaches Casimir's T^3 law near the end of the temperature range of measurement. This is evidence for Poiseuille flow of heat, a phenomenon that is discussed in detail below.

Although Umklapp combined with boundary scattering provides a qualitatively correct description of the thermal conductivity of rather perfect dielectric crystals shown in Figure 4 (ignoring the Poiseuille flow in ^{4}He crystals for the moment), a detailed understanding of imperfect crystals requires the additional knowledge of phonon scattering by lattice defects, and the role played by N-processes. In all crystals except those of extreme perfection, N-processes play only a secondary role in determining the thermal conductivity. For example, two long wavelength phonons can combine through an N-process to form a short wavelength phonon, which is then scattered by a lattice defect. These mode-conversion effects are not discussed here.

Phonon Scattering by Lattice Defects

Any perturbation of the crystalline order will lead to phonon scattering. The study of phonon scattering through measurements of the thermal conductivity of dielectric crystals was pioneered by Berman and his co-workers (16, 17) and by Klemens (18). In the present review, we consider only mass-mismatch scattering, the so-called isotope effect (19, 19a), and scattering by impurity modes (19b,c), i.e. by local variations of the lattice vibrations in disordered crystals (20, 20a).

In the isotope effect, elastic scattering of phonons results from local variations of the isotopic masses. The scattering rates increase, as in Rayleigh scattering, as the fourth power of the phonon angular frequency, leading to a mean free path

$$l_{\text{isot}} \propto \omega^{-4}. \qquad 4.$$

A study of the isotope effect has been performed in isotopic mixtures of ^{6}LiF and ^{7}LiF (21, 22); see Figure 5. The top curve was obtained on almost isotopically pure ^{7}LiF, the bottom curve on a nearly 50–50 mixture of the two isotopes. The effect is most pronounced near the conductivity maximum. The reason is, briefly, the following. At low temperatures, where most of the heat is carried by low frequency phonons, the Rayleigh scattering is weak. At high temperatures, although increasing rapidly, see Eq. 4, the scattering becomes relatively less important because of the exponentially increasing Umklapp scattering rate. This leaves the conductivity near the maximum as the one most sensitive to defect scattering.

It is useful for the subsequent discussion to illustrate the spectrum of the phonons that are carrying the heat at different temperatures. For a Debye solid, the spectral distribution dC_v/dv of the specific heat is plotted in Figure 6 for several temperatures. The maximum of this distribution shifts to higher frequency in linear proportion to the temperature T, in analogy to Wien's displacement law for black-body radiation. At 1K, it peaks at 80 GHz, at 10K, at 800 GHz, and so on. In referring to Eq. 1, we conclude from Figure 6 that at any given temperature, the heat will be carried predominantly by a spectral range of phonons that centers around a certain frequency. It must be kept in mind, however, that if the scattering rate varies rapidly with frequency, as in the isotope effect, the main contribution to the heat flow may actually come from phonons with rather

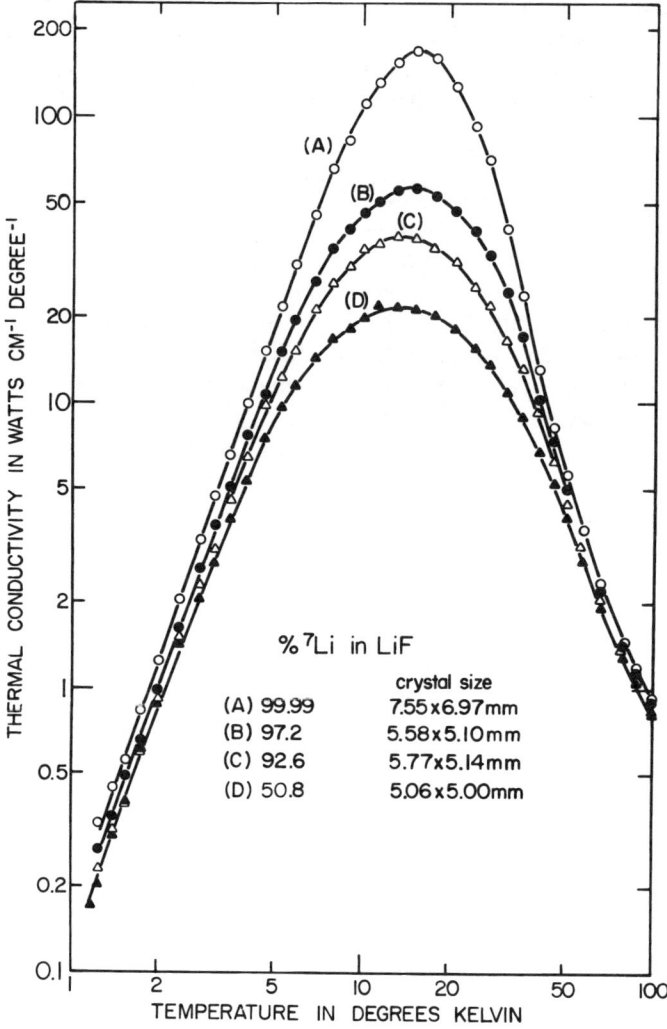

Figure 5 Phonon scattering through isotopic disorder in LiF (22).

different frequencies (lower ones in the case of the isotope effect). Nevertheless, the concept of the dominant phonon frequency, according to which the heat at a certain temperature is carried by phonons of a certain frequency, has provided such a transparent and convenient way of extracting average phonon scattering rates from thermal conductivity measurements, that we also use it in the following discussion. For reasons that are unimportant here (23) we use as the dominant phonon frequency v_{dom}

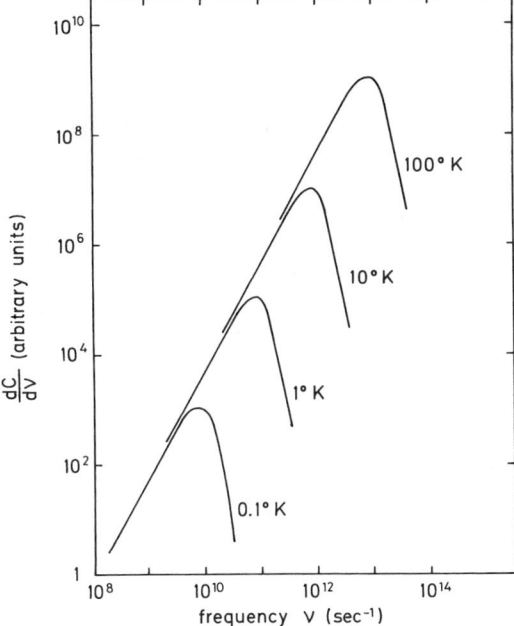

Figure 6 Spectral distribution of the specific heat C_v in a dispersionless solid, Debye (6). The limiting phonon frequency that would be shown as a cut-off in the distribution has been omitted.

$$v_{\text{dom}} = 90 \text{ GHz } K^{-1}T = 4.25\frac{k_B}{h}T. \qquad 5.$$

Thus, 1K corresponds to 3 cm^{-1} in the wave number measure.

Phonon scattering by impurity modes is the second example of defect scattering that is reviewed here. Substitution of atoms of the host lattice with heavier impurity atoms leads to local changes of the vibrational spectrum of the host lattice (24) as first suggested by Kagan & Iosilevskii (25) and by Brout & Visscher (26). The inset of Figure 7 illustrates how such an impurity mode arises, and the scattering it causes, for a one-dimensional crystal lattice. The heavy impurity has the mass $m + \Delta m$. In thermal equilibrium, the amplitude of vibration of the atoms will be different near the impurity (imagine Δm to be very large; the impurity will not move at all, and will slow down the motion of the neighboring atoms). If a plane wave is incident on the defect, it will be only partly transmitted. Part of the amplitude will be reflected (scattered). Resonant scattering will be observed at $\omega_{0,\text{theory}}$ given by

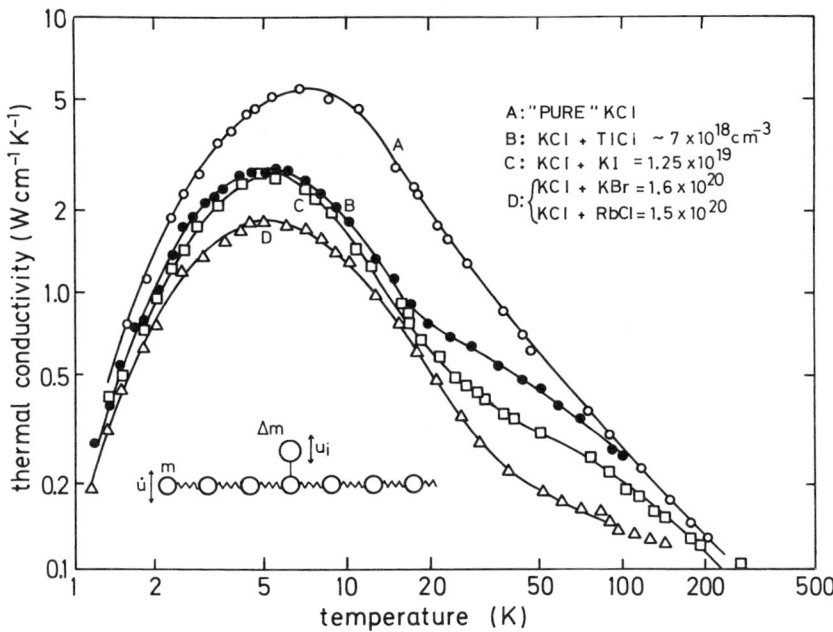

Figure 7 Thermal conductivity of KCl crystals containing substitutional monatomic impurities (27) as an example of phonon resonance scattering by impurity modes. The *inset* is a one-dimensional illustration of a crystal lattice containing a heavy impurity.

$$\omega_{0,\text{theory}}^2 = \frac{1}{3}\omega_D^2 \frac{m}{\Delta m} \qquad 6.$$

for a Debye solid with the characteristic frequency ω_D.

The phonon scattering by heavy substitutional impurities is revealed in the thermal conductivity, as illustrated in Figure 7 for KCl containing between 0.1 and 1.0 mole % TlCl, KI, KBr, and RbCl in solid solution (27). The resonant scattering leads to a dip in the thermal conductivity, which shifts from about 40K for the relatively lighter impurities Br^- and Rb^+ to about 20K for the heaviest ion, Tl^+. The location of the dips, defined as the temperature at which the inflection occurs in the conductivity curves, indicates the temperature of the strongest phonon scattering. The resonance frequencies determined with the aid of Eq. 5 in the dominant phonon approximation were found to agree well with the theoretically predicted ones, Eq. 6. At low temperatures, the defect scattering is again of the Rayleigh type. Its strength is close to that expected for the mass mismatch alone, as in the isotope effect.

Another class of phonon resonance scatterers is derived from quasi-rotational excitations of molecular impurities. Consider a molecular impurity, say a CN^- ion, substituting for a halogen ion in an alkali halide lattice. Its rotational degrees of freedom are constrained by potential barriers as sketched in the inset of Figure 8. These barriers alter the lower rotational states to become librational ones, which in turn may be split by tunneling through the barriers (28, 29). These tunneling states have shown resonant scattering of phonons in thermal conductivity experiments with exceptional clarity. An example is shown in Figure 8 (12, 30). The two upper curves, which differ from each other only around the conductivity maximum, were obtained on crystals grown with extreme efforts to avoid chemical contamination of the (naturally isotopically pure) NaF single

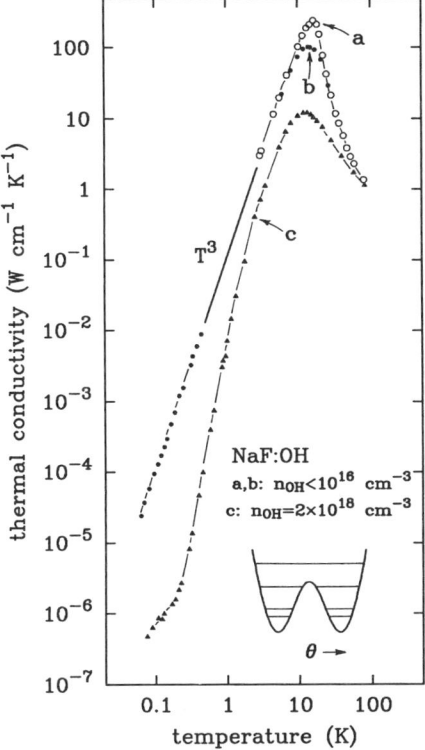

Figure 8 Thermal conductivity of pure (*top two curves*) and of OH^--doped NaF single crystals (*bottom curve*). Curve (*a*), sample 607167 J (30). Curve (*b*), sample 7208142 W, and curve (*c*), sample 910166 W, after McNelly (12). Note the boundary scattering in the pure crystals, leading to a T^3 variation over five orders of magnitude in thermal conductivity. The *inset* shows the influence of a crystal field on the rotational states of a molecular ion. The resonance scattering in the OH^--doped samples is caused by the tunnel-splitting of the librational ground state.

crystals. The lower of the two curves belongs to a sample with less than 1 ppm OH⁻ in solid solution. The lowest curve in Figure 8 was measured on a NaOH doped sample containing 50 ppm OH⁻. Its conductivity around 0.2K is 500 times lower than that of the undoped samples. The resonant scattering is shown even more clearly in Figure 9, where the phonon scattering rate τ^{-1}, which is proportional to the reciprocal phonon mean free path l^{-1}, determined with the help of Eq. 1, is plotted against temperature. The scattering rate peaks at 0.22K. Using the dominant phonon approximation, with the conversion given in Eq. 5, a resonance frequency of 0.66 cm^{-1} is determined. This agrees closely with the tunnel splitting derived from a specific heat anomaly in NaF crystals believed to be contaminated with OH⁻ (31). (For a more detailed discussion of tunneling defects, see Ref. 29.)

Due to the broad-band nature of the technique, the observation of resonant scattering in thermal conductivity requires that the width of the resonance be quite large, large enough to scatter a significant fraction of the heat-carrying phonons. This large width has been observed for other

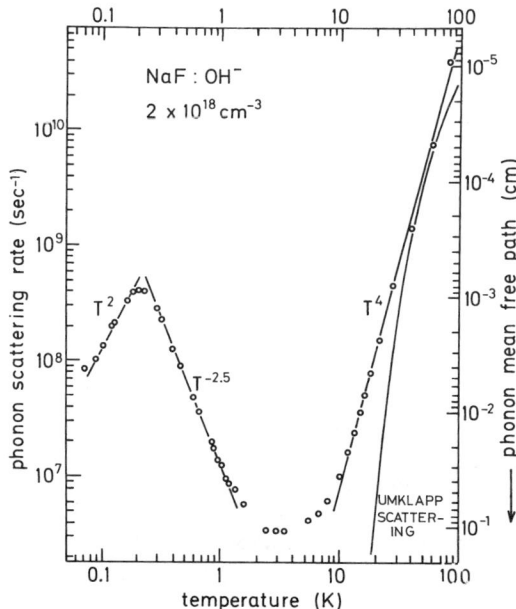

Figure 9 Phonon scattering rate (*left scale*) and phonon mean free path (defined in Eq. 1) for the OH⁻-doped NaF crystal of Figure 8. Umklapp scattering rate as determined in the pure samples (12). For NaF: Debye temperature $\Theta = 466K$; specific heat $C_v = 9.16$ erg g^{-1} K^{-4} T^3; mass density $\rho = 2.851$ g cm^{-3}; number density of ions $n = 81.797 \times 10^{21}$ cm^{-3}; Debye sound velocity $v = 3.608 \times 10^5$ cm sec^{-1}. Taken from McNelly (12).

tunneling defects (29). In the OH doped, NaF crystal of Figure 9, the width of the resonance probably results from random stresses in the crystal leading to local variations in the tunnel splitting of the OH^{-1} ground state (31a,b,c). The T^2 dependence of the phonon scattering rate reflects the distribution of tunnel splittings in the sample.

Above 10K, the scattering rate increases again, as T^4, and dominates the Umklapp scattering up to at least 30K. The reason for this is probably point defect, Rayleigh type scattering. The mismatch in force constants, rather than the mass mismatch, could be the cause of this scattering.

In summary, the similarity between electromagnetic and elastic waves has been further demonstrated through these experiments. Elastic waves, just as electromagnetic waves, can be scattered through Rayleigh as well as through resonance processes. In both cases, either the classical wave or the quantized excitation picture can be used to describe the phenomena.

Poiseuille Flow of Heat and Second Sound

In this section, we describe two experiments in which the behavior of the phonons is remarkably different from that of their electromagnetic counterparts. These experiments, however, emphasize particularly clearly the particle nature of the quanta of the elastic waves and bring out a striking similarity with a gas of particles that have energy and momentum. These experiments require that Normal processes are the dominant scattering events. This requirement can only be satisfied in crystals of exceptional perfection in a rather narrow temperature window. Therefore, although Poiseuille of heat and second sound are scientifically very important to our understanding of the lattice vibrations of crystals, these effects do not influence the flow of energy in materials of normal purity.

In a Normal process, not only is the energy $\hbar\omega$ of the colliding phonons conserved (see Eq. 2) but also their wave vector **k**, with $j = 0$ in Eq. 3. An analogy with colliding gas particles suggests itself, in particular if we realize that on the average the number of phonons is also conserved. The energy and momentum of the gas particle correspond to $\hbar\omega$ and $\hbar k$. The occurrence of Poiseuille flow of heat and second sound can be readily understood if we use this analogy. We begin with Poiseuille flow.

Consider the laminar flow of a gas through a hollow tube of radius r. The collision length l_{coll} for interatomic collisions shall be very small relative to r. Without interaction with the wall, the gas would move forward along the length of the tube, with a velocity equal to the average velocity of its atoms, called the *drift velocity* (note that they are injected into the tube with this velocity). In a real gas, however, atoms that hit the wall will temporarily stick to it. When they are released, their velocity will be randomized and their average drift velocity will be zero. The net momentum connected with the forward motion has been taken up by the wall. Thus the near-surface gas layer does not move. Atoms that move a distance from the wall will share their forward momentum with atoms in the near-surface layer through atomic collisions, and thus forward momentum will diffuse out of the gas. The larger the collision mean free path, l_{coll}, the faster this process will proceed, provided that $l_{\text{coll}} \ll r$. The resulting profile of the local drift velocity of the Poiseuille flow is sketched in Figure 10. The average mass flow rate is given by the Hagen-Poiseuille law

$$(\dot{m}/\pi r^2)/\nabla p = \frac{\rho}{8} \frac{r^2}{\eta}, \qquad 7.$$

where the ∇p is the pressure gradient, ρ the mass density of the gas, and η its viscosity. Note that η is proportional to the collision length l_{coll}, i.e. the larger the distance between interatomic collisions, the larger the viscosity of the gas.

In the analogous experimental situation for heat flow, phonons enter at the hot end, and leave at the cold one. Collisions with the rough wall remove momentum from the near-surface phonon gas; momentum-conserving N-processes transport momentum from the interior of the crystal toward the wall. Instead of a mass flow rate, a heat flow rate is observed, and instead of a pressure gradient, a temperature gradient. The Poiseuille flow of heat is given by (32)

$$(\dot{Q}/\pi r^2)/\nabla T = \Lambda = \frac{1}{3}\frac{5}{8}C_v v \frac{r^2}{l_N}, \qquad 8.$$

where C_v and v are the specific heat (per volume) and average sound velocity, and l_N the N-process mean free path. Note again that a large collision length leads to a small heat flow, because it "makes more phonons see the wall." However, since phonon momentum can also be destroyed in the bulk, through U-processes or through defect scattering, the observation of Poiseuille heat flow requires not only that $l_N \ll r$, but also that

Poiseuille Flow

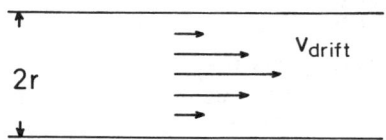

Figure 10 Profile of the local drift velocity for Poiseuille flow of a gas through a tube of diameter $2r$.

during the time it takes for a phonon (or rather its momentum) to diffuse to the wall, no momentum-destroying process occurs in the bulk. The diffusion follows a random walk. Thus it takes n steps to reach the wall, with

$$\sqrt{n}l_N = r. \qquad 9.$$

The requirement that no bulk scattering take place during those n steps means

$$nl_N < l_R, \qquad 10.$$

where l_R contains both Umklapp and defect scattering. Combining Eqs. 9 and 10 yields the second requirement for the occurrence of Poiseuille heat flow:

$$l_N l_R > r^2, \qquad 11.$$

in addition to the first condition,

$$l_N \ll r. \qquad 12.$$

Thus, Poiseuille heat flow requires not only that l_N is small, Eq. 12, but simultaneously that the momentum-destroying processes are very rare, Eq. 11. Because of these very stringent conditions, this form of heat flow has been seen only in solid helium ^{4}He (32) and in ^{3}He (33) through an enhancement of the thermal conductivity near and below the maximum, as shown for ^{4}He in Figure 4. The very rapid decrease of the conductivity as the temperature decreases below the conductivity maximum is an indication that the phonon scattering rate decreases rapidly at lower temperatures (i.e. the mean free path l_N increases). The open circles in Figure 11 are the experimental phonon mean free paths l_{exp} determined from similar data (32). Above 1K l_{exp} decreases exponentially, as expected for U-processes. Below 0.5K, l_{exp} approaches its predicted Casimir value, $2r$. The increase of l_{exp} between 0.5 and 1.0K indicates Poiseuille heat flow. The N-process mean free path determined from these data is also shown in Figure 11. It varies as T^{-3}. The mean free paths, l_N and l_U, satisfy Eqs. 11 and 12 in the temperature window where Poiseuille heat flow is observed.

The second example of the particle-like nature of phonons is the so-called second sound. Again, the analogy with the atoms of a gas can be used to explain this phenomenon, see Figure 12. An individual atom, or an atom in a noninteracting gas, moves with its thermal velocity. In an interacting (ideal) gas, a particle density pulse, produced for example by tapping on the left wall of the container, will move to the right with

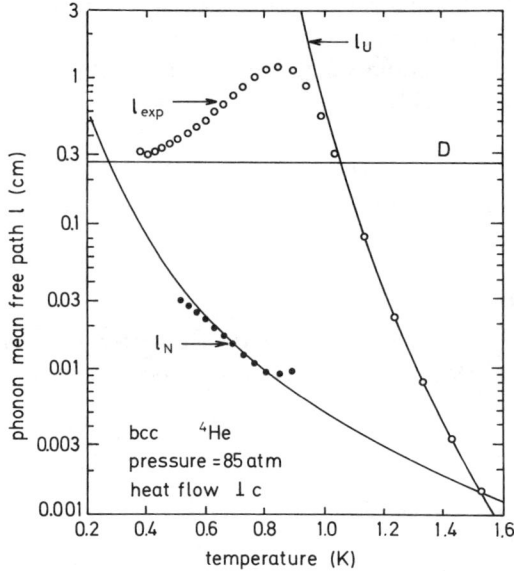

Figure 11 Average phonon mean free path determined from the thermal conductivity of bcc ^{4}He crystals grown under 85 atm; heat flow perpendicular to the c-axis (32). l_{exp} determined with the help of Eq. 1. Above 1K, the mean free path is determined solely by Umklapp scattering (l_U). The Normal process mean free path l_N is determined from the thermal conductivity in the region of Poiseuille heat flow (0.5K < T < 0.9K), using Eq. 8.

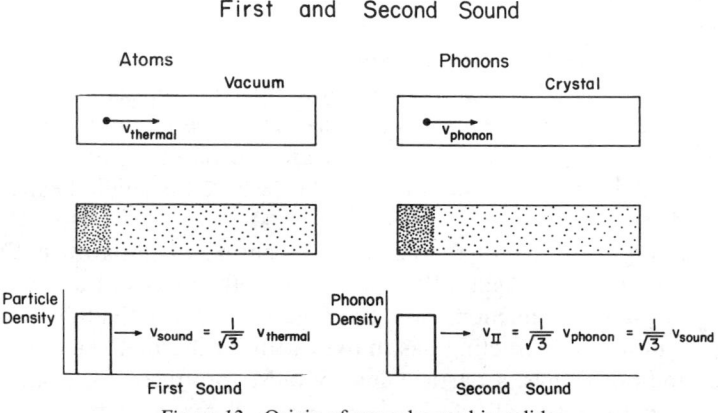

Figure 12 Origin of second sound in solids.

LATTICE VIBRATIONS 109

the speed of sound (we ignore any effects of the container wall in this experiment). Interatomic collisions will randomize the thermal velocity vectors, and hence the pulse will travel with only one average spatial component of v_{thermal}, say, the x-component, which, by Pythagoras' theorem, yields

$$v_{\text{sound}} = \frac{1}{\sqrt{3}} v_{\text{thermal}}. \qquad 13.$$

Note that the condition for the pulse to travel without spreading in the direction of propagation is that its width w is large relative to the collision length l_{coll}.

In a gas of phonons scattered by N-processes only, a phonon density pulse should also be expected to propagate through the crystal with a velocity that is equal to $1/\sqrt{3}$ of the individual phonon velocity. A phonon density pulse can be generated by passing a current pulse through a metal heater that has been evaporated onto one face of a crystal. On the opposite face, the phonon density pulse will be detected as a temperature change of a bolometer. Since its constitutents are quanta of elastic energy, or sound, the term *second sound* has been coined for this phonon density pulse. Essential for its occurrence is the interaction of the phonons within the pulse through N-processes, through which collisions their velocity vectors are randomized in their direction. Similarly, waves of phonon density, i.e. temperature waves, can be generated by periodically heating one end of the crystal, and detecting the temperature oscillation at the opposite end. The conditions for the occurrence of second sound are less stringent than for Poiseuille flow. They are (34)

$$\tau_N \ll \tau \ll \tau_R, \qquad 14.$$

where τ is approximately the period of oscillation (or the duration of the heat pulse). Thus, second sound can be observed whenever $\tau_R \gg \tau_N$, through the proper adjustment of the experimental conditions (τ and sample length). It should be mentioned in passing that second sound in superfluid helium is fundamentally the same process as in solids, but the phenomenon is more complicated in the liquid because of its more complex phonon dispersion (35).

Second sound has been observed in solid ^{4}He (36) and ^{3}He (37), in NaF (30, 38), and in Bi (39). Both of the latter substances are naturally isotopically pure, and very great efforts had to be invested in their chemical purification and physical perfection in order to detect this phenomena. Figure 13 shows the temperature recorded on a plate of NaF that carried on one face a thin metal film as thermometer, and on its opposite face a

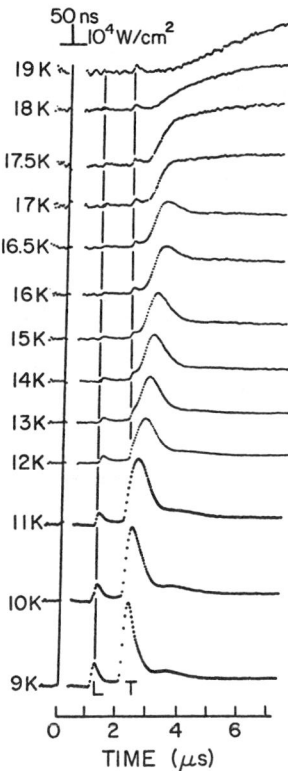

Figure 13 Heat pulses at various temperatures in NaF. Sample had a maximum thermal conductivity of 176 W cm^{-1} K^{-1} (intermediate to that of the pure samples in Figure 8). Sample length, 0.74 cm. Heater pulse of 50 ns duration, 10^4 W cm^{-2}, at $t = 0$. Ballistic heat pulses (longitudinal, L, and transverse, T) are clearly resolved at the lowest temperatures. Second sound is observed between approximately 13 and 16K. Ref. (12).

thin metal film as heater. The quality of the sample was intermediate between that of the two undoped, highly purified NaF samples measured in Figure 9.

At a sample temperature of 9K (bottom trace in Figure 13), the ballistic arrival of the longitudinal and the transverse phonon pulses, after a travel time determined by their respective speeds of sound, demonstrates that no phonon interactions occur (the delayed arrival of phonons that were reflected at the edges of the sample had been eliminated through the proper choice of sample geometry, a plate). At around 12K, the longitudinal pulse has significantly diminished, and instead of the transverse pulse a new, delayed pulse appears. At even higher temperatures (ca 16K), this pulse arrives so late that it is clearly resolved from the remnants of the transverse ballistic pulse. The new, delayed pulse, is the second sound pulse. Its arrival time approaches that predicted for the second sound in NaF according to Eq. 13; as first sound velocity the Debye average of the longitudinal and transverse speeds of sound has been used in this calculation. As the sample temperature is increased to 17K, the second sound pulse diminishes again, and merges in the diffusively arriving heat pulse; this gives evidence for the rapidly increasing strength of the Umklapp processes (a comparison with the thermal conductivity curves in Figure 8 is useful).

The evolution and the shape of the arriving second sound pulse can be analyzed to obtain N-process mean free paths. Thus, both Poiseuille flow and second sound yield information on N-processes, which is difficult to obtain otherwise from thermal conductivity measurements. Details would go too far at this point. The important lesson that has been learned from these experiments is that the quantized elastic waves or phonons are remarkably similar in their properties to atoms in a gas. This strengthens our picture of lattice vibrations as traveling waves as well as wave packets or particles. We see in the following section, however, that this picture runs into serious limitations when we consider highly disordered solids.

Lattice Vibrations of Glasses and of Certain Disordered Crystals—Return to Einstein's Model

In his first investigation of the thermal conductivity of nonmetals, Eucken (5) had noticed a fundamental difference between crystalline and amorphous solids. In Figure 14, we reproduce his data on SiO_2 as extended to lower temperatures in more recent work (40, 41, 41a, 42). The relatively low conductivity of the amorphous SiO_2, as well as its temperature dependence, has since been found to be an inherent property of amorphous solids, as shown in Figure 15, independent of chemical composition or bonding (42, 42a, 43). The data shown span the range of the thermal conductivity observed to date on all amorphous solids studied.

The temperature region below 1K, in which the thermal conductivity varies as T^n, with n ranging between 1.8 and 2.0 depending on the chemical composition of the glass, has attracted a great deal of attention. It is believed to be connected with low energy excitations that have been observed through a specific heat anomaly that varies close to linearly with temperature. It has been suggested that these excitations are caused by tunneling of atoms or groups of atoms between nearly identical sites in the amorphous lattice, with an almost uniform density of states (44, 45). The thermal conductivity results from phonons resonantly scattering off these states (we refer to the discussion of tunneling states in crystals presented above). The properties of the tunneling states in glasses have been reviewed extensively (e.g. 43, 46–50) and are not reviewed here. We refer to these papers for details. We wish to emphasize, however, that the

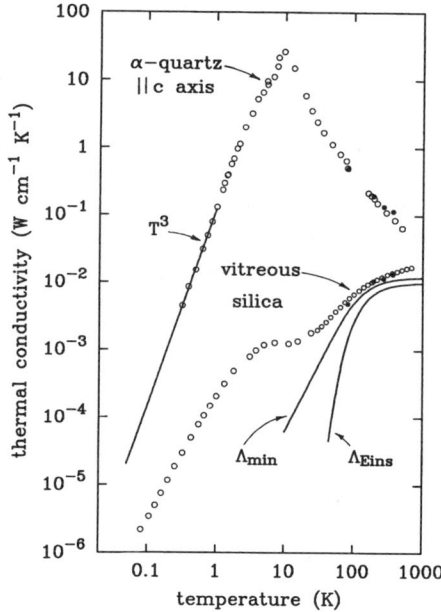

Figure 14 Extension of Eucken's (5) data (*solid circles*) on crystalline and amorphous SiO$_2$ to lower and higher temperatures (40, 41, 41a), *open circles*. Recent high temperature measurements on a-SiO$_2$ (Ref. 42 and this work) shown as *small open circles*. The *curves* marked Λ_{Eins} and Λ_{min} are the calculated minimum thermal conductivities for SiO$_2$ using Eqs. 15 and 18, respectively.

major question, namely the cause of the universality of the phenomenon, is not only unanswered, it has to date hardly been addressed at all (51). It is fair to say that a basic understanding of these excitations that dominate most of the thermal and elastic properties of this class of solids at low temperatures is still lacking.

In the present review, we concentrate our attention on the high temperature properties, and in keeping with the approach chosen so far, focus on the high temperature thermal conductivity, approximately above 30K.

In Figure 16, we analyze the thermal conductivity of several amorphous solids in terms of a phonon mean free path, Eq. 1. Just as a reminder, we have also included l for crystalline SiO$_2$ (data in Figure 14), and discern the Umklapp (plus defect) scattering, and the boundary scattering region (the sample diameter was 0.5 cm). In glasses, the phonon mean free path so calculated above 30K is less than 10^{-7} cm, i.e. it approaches the interatomic spacing as first noticed by Birch & Clark (54). Since the shortest possible wavelength in a solid is of that magnitude, the concept of waves carrying

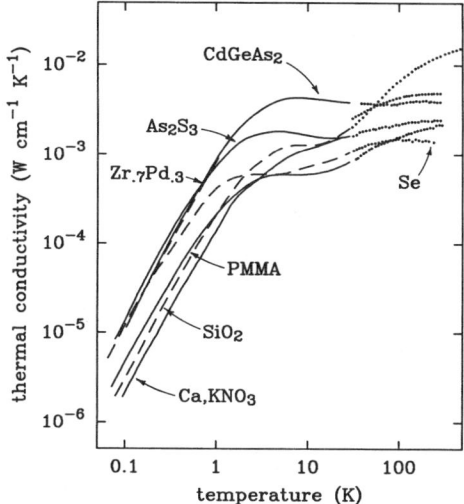

Figure 15 Thermal conductivity of seven different glasses that are characteristic for different bonding types, as reviewed in (43), extended to room temperature in recent work (42, 42a). The conductivity of all glasses measured to date falls into the range spanned by the data shown here. The $Zr_{0.7}Pd_{0.3}$ data are almost indistinguishable from those of $CdGeAs_2$ below 1K.

heat loses its significance above this temperature. Einstein's original concept of a random walk of localized (Einstein) oscillations suggests itself as a more appropriate description. Following Einstein (4), consider their oscillations to be so strongly damped that they pass on their energy within half a period of oscillation. In the kinetic expression Eq. 1, C_v is the Einstein specific heat of $3n$ oscillators (n is the number density of the atoms in the solid), v is the thermal velocity of the atom, $v = n^{-1/3}/\tau_E$, where τ_E is one half their period of oscillation, and the interatomic spacing l equals $n^{-1/3}$. Inserting these quantities into Eq. 1, we obtain

$$\Lambda_{\text{Eins}} = \frac{k_B^2}{\hbar} \frac{n^{1/3}}{\pi} \Theta_E \frac{x^2 e^x}{(e^x - 1)^2}, \qquad 15.$$

where Θ_E is the Einstein temperature, and $x = (\Theta_E/T)$. The Einstein temperature is an adjustable parameter. We choose $\Theta_E = 435$K, which is suggested by fitting the specific heat of a-SiO_2. The thermal conductivity given by Eq. 15 is shown in Figure 14. Above 100K the magnitude is reasonably close to the experimental value, and so is the temperature dependence. Before testing this model on other glasses, however, we apply one modification. Instead of considering solely individual atoms as oscillators, we also consider, following Slack (55), larger entities with larger

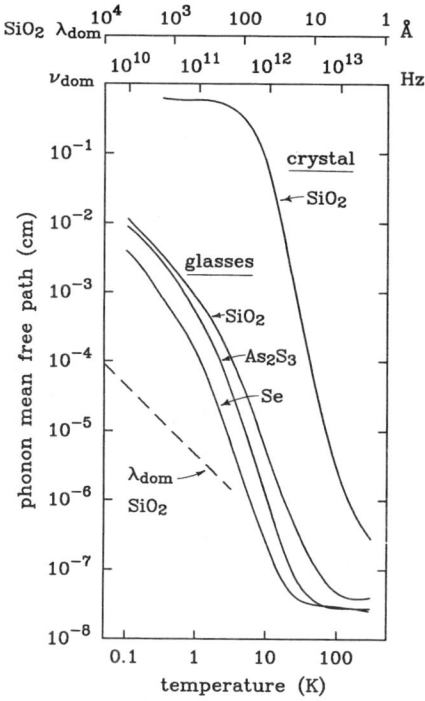

Figure 16 Phonon mean free path *l* calculated with the help of Eq. 1. The specific heat used for the calculation is the Debye specific heat based on low frequency speeds of sound. At the *top* is shown a scale of the dominant phonon frequency (see text) and dominant phonon wavelength (for a-SiO$_2$, using the Debye speed of sound $v = 4.1 \times 10^5$ cm sec^{-1}). The *dashed line* marked SiO$_2$, λ_{dom}, shows how much shorter this quantity is than the mean free path measured in the glasses (40).

masses and smaller eigenfrequencies. We imagine subdividing the sample volume into subgroups of atoms, with the linear dimension d, and have to determine their density of states and their eigenfrequencies, ω_d. The thermal conductivity is determined by integrating over all angular frequencies ω;

$$\Lambda = \frac{1}{3} \int_0^\infty \frac{dC}{d\omega} v(\omega) l(\omega) \, d\omega, \qquad 16.$$

where $dC/d\omega$ is the specific heat per frequency element, the mean free path is the separation of the subgroups, $l = d$, and the velocity v is given by the linear dimension of the subgroup divided by one half its period of oscillation $\tau(d)$;

$$v = \frac{d}{\tau(d)}. \qquad 17.$$

The problem is then to determine the density of states and the eigenfrequencies, ω_d, of these oscillators. A similar problem has been solved by Debye (6), who considered traveling waves, which effectively subdivided the sample into subgroups of half the dimension of the wavelength $\lambda/2$. We adopt his calculation here, by substituting $\lambda/2$ for d, and by assuming that the frequency of oscillation of the subgroup is given by the frequency of the corresponding wave. Or, to express it another way, we calculate the thermal conductivity within the Debye model, with the assumption that the scattering length is one half of the wavelength. This is what Slack called the *model of the minimum thermal conductivity* $\Lambda_{\min}$ (55). We repeat, however, that our physical picture is that of a random walk between Einstein oscillators, of varying sizes.

Applying the Debye formalism, Eq. 16 becomes

$$\Lambda_{\min} = \frac{1}{2.48} k_B n^{2/3} v_i 2\left(\frac{T}{\Theta_c}\right)^2 \int_0^{\Theta_c/T} \frac{x^3 e^x}{(e^x - 1)^2} \, dx, \qquad 18.$$

for one polarization with speed of sound v_i, where Θ_c is the cutoff frequency for this polarization (expressed in degrees K) $\Theta_c = \hbar/k_B v_i (6\pi^2 n)^{1/3}$. To get the total conductivity, we add the contributions from two transverse and one longitudinal mode. In the limit of high temperatures, $T \gg \Theta_c$, Eq. 18 becomes

$$\Lambda_{\min} = \frac{1}{2.48} k_B n^{2/3} (2v_t + v_l), \qquad 19.$$

where v_t and v_l are the transverse and longitudinal speeds of sound, respectively. This result is just 20% larger than the simple procedure of writing in Eq. 1 with $C_v = 3k_B$ per atomic volume, $v = 1/3(2v_t + v_l)$, and l = interatomic spacing.

The arbitrariness of choosing Θ_E in Eq. 15 has been replaced with the similar arbitrariness of assuming no dispersion in the model underlying Eqs. 18 and 19. The advantage of the latter model is that it transforms into the standard Debye model of thermal conductivity as the picture of elastic waves becomes appropriate for the heat flow, i.e. as the mean free path becomes large relative to the wavelength.

The minimum thermal conductivity $\Lambda_{\min}$ for SiO_2 has also been plotted in Figure 14. It fits the experimental data better than the single frequency Einstein model, Eq. 16, since it varies less rapidly (proportional to T^2) at low temperatures.

As a test of this model of the minimum thermal conductivity, we have measured the thermal conductivity of several glasses (42). Figure 17 shows a few examples selected to cover as wide a range of conductivity as possible. The agreement between measured and predicted thermal conductivity is encouraging, considering that the model contains no adjustable parameters. Thus we conclude that in amorphous solids, i.e. solids lacking translational symmetry, the lattice vibrations are more appropriately described through localized vibrations like Einstein modes rather than through wave-like motions, at least in temperature range above ca. 30K. It is, however, an open question why these modes are equally strongly damped in all glasses, so that they pass on their vibrational energy within one half of a period of oscillation. Equation 15 is an upper limit of the thermal conductivity by localized oscillations, as also noted by Einstein (4). The physical origin of the strong damping is not understood at this time.

We return briefly to the low temperature regime. Below ca. 30K, the model of Λ_{min}, obviously, quickly ceases to be appropriate; see Figures 14 and 17. the mean free path l rapidly exceeds the dominant phonon wavelength, λ_{dom}, which is also sketched in Figure 16. In all glasses below 1K, l exceeds λ_{dom} by two orders of magnitude. This observation further strengthens the picture that elastic waves do exist in glasses at these temperatures and frequencies, a picture confirmed by many other experiments (e.g. ultrasonic or light scattering). The rapid transition to localized oscillations sets in at a few degrees Kelvin. The transition indicates a Rayleigh scattering process (40), the origin of which is also still a puzzle (56).

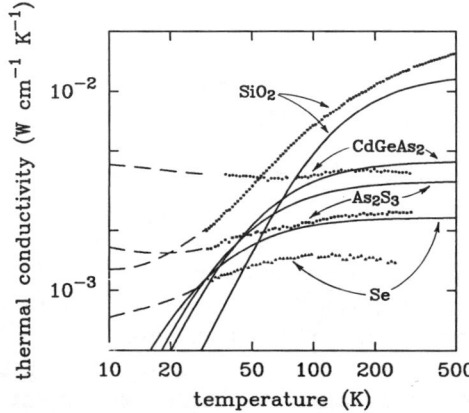

Figure 17 Thermal conductivity of several glasses above 10K compared with the minimum thermal conductivity (42, 42a, 55).

LATTICE VIBRATIONS

Considering the significant differences between the lattice vibrations of crystalline and of amorphous solids, it would be of interest to know what kind of disorder it takes for a crystalline solid to acquire glass-like lattice vibrations. One would also like to know whether thermal conductivities smaller than the minimum thermal conductivity, see Eq. 18, can ever be achieved. To date, only some partial success has been accomplished. We refer to some reviews of this effort (49, 57, 58) and discuss briefly two examples that illustrate these achievements.

The molecular ion CN^- substituted in certain alkali halide crystals for the halogen ion can retain its quasi-rotational degrees of freedom, as reviewed above, and can act as a strong phonon scatterer. At very high concentrations, the thermal conductivity approaches that of amorphous solids, as shown in Figure 18 for $(KBr)_{1-x}(KCN)_x$ for $x = 0.25$ and 0.50 (31c). At high temperatures, it reaches the minimum thermal conductivity, while at low temperature it exceeds this theoretical lower limit in exactly the same way that is observed in amorphous solids. This is particularly remarkable, since for x near 10^{-2}, the thermal conductivity below 1K is in fact smaller than that of glasses (though always larger than Λ_{min}).

The low-energy glass-like excitations discovered in these solids have

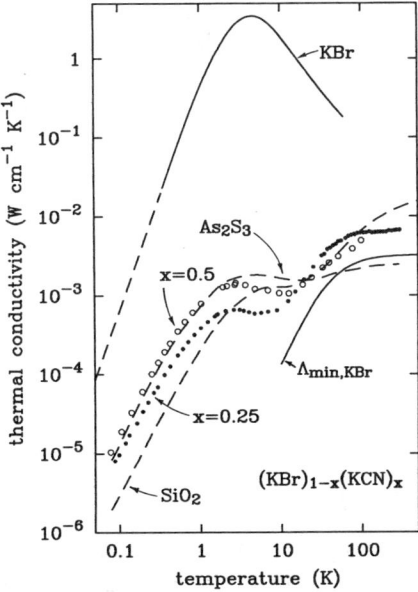

Figure 18 The thermal conductivity of the mixed single crystals $(KBr)_{1-x}(KCN)_x$ (Ref. 31c and this work) is identical to that of amorphous solids (*dashed lines*). At high temperatures it is close to the predicted minimum thermal conductivity for KBr.

been explained (60, 60a) through tunneling states of a relatively small fraction of the CN^- ions that retain their freedom in the random stress field that is set up by the surrounding CN^- ions, and thus distort the cubic lattice. In the temperature region of the plateau and at higher temperature, it has been suggested that the phonons are scattered by the librational states of all CN^- ions in this solid (61). Conceivably, these librational excitations could play the same role as the Einstein oscillators in amorphous solids.

The second example is crystalline YB_{68}. Its lattice consists of a simple cubic array of clusters of 156 boron atoms separated by 11.72 Å. In the center of the cube sits another cluster of 48 boron atoms. Halfway between the latter and the cube faces are six equivalent positions over which the three yttrium atoms are randomly distributed, conceivably causing stress dipoles in the lattice. The Y atoms are expected to be mobile at high temperatures (62). In several, slightly nonstoichiometric crystals of YB_{68},

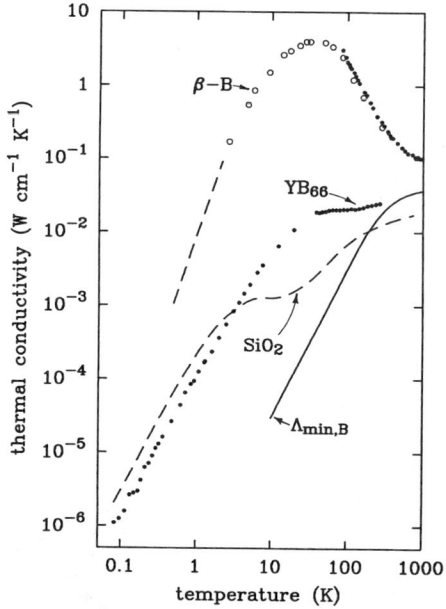

Figure 19 The thermal conductivity of slightly nonstoichiometric single crystal YB_{68} shows glassy behavior, and at high temperatures approaches the predicted minimum thermal conductivity (63–65) for B.

glass-like thermal properties have been observed (63–65). Thermal conductivity measurements are shown in Figure 19. At high temperatures, the theoretical Λ_{min} is approached. If the picture of a distortion associated with the yttrium atoms is valid, it follows that a relatively small number of elastic dipoles can lead to Λ_{min}.

These examples show how the study of crystalline solids can be used to reach an understanding of the lattice vibrations of glasses. Obviously, more disordered crystals will have to be found and studied, if we want to isolate the physical nature of these excitations.

Summary and Conclusions

The purpose of this review is to compare our level of understanding of lattice vibrations in crystals and glasses, relying primarily on studies of the heat transport. We have seen that in crystals elastic waves, or phonons in the quantum picture, can lead to a very satisfactory description of the observed phenomena. By contrast, in amorphous solids this same picture is incorrect except at low temperatures or for long wavelengths. Lattice vibrations as they are thermally excited above ca. 30K appear to be more appropriately described as localized Einstein oscillators; the heat being carried through the lattice by a random walk, rather than by wave-like motion. It is not understood, however, why these oscillators are so heavily damped in all glasses.

At temperatures below 30K, thermally excited elastic waves occur also in glasses; in fact, their scattering mean free path always exceeds 100 wavelengths below a few Kelvin, corresponding to phonon frequencies less than terahertz. However, in this energy range, some additional excitations have been found in all glasses. They are localized, and are most likely tunneling states. Their physical origin, however, is also not understood. We therefore conclude that many aspects of the vibrational spectrum of glasses over the entire frequency range are still poorly understood and require further study.

ACKNOWLEDGMENTS

Many fruitful discussions with Drs. J. Jaeckle, Tom Klitsner, M. Randeria, J. P. Sethna, and E. T. Swartz are gratefully acknowledged. Much of our work reviewed here has been supported through the National Science Foundation, Grant DMR-84-17557. One of us (R.O.P.) acknowledges the hospitality extended to him by Drs. K. Dransfeld and J. Jaeckle at the University at Konstanz, and the financial support of the Alexander von Humboldt Foundation while completing the manuscript.

Literature Cited

1. Petit, A. T., Dulong, P. L. 1819. *Ann. Chim. Phys.* (2nd Ser.) 10: 395
2. Weber, H. F. 1875. *Ann. Phys.* 154: 367, 553; Engl. transl. in *Philos. Mag.* (4th Ser.) 1875. 49: 161, 276
3. Einstein, A. 1907. *Ann. Phys.* 22: 180
4. Einstein, A. 1911. *Ann. Phys.* 35: 679
5. Eucken, A. 1911. *Ann. Phys.* 34: 185
6. Debye, P. 1912. *Ann. Phys.* 39: 789
7. Born, M., von Karman, Th. 1912. *Phys. Z.* 13: 297
8. Debye, P. 1914. *Vortraege ueber die Kinetische Theorie der Materie und der Elektrizitaet*, pp. 17–60. Berlin: Teubner
9. Peierls, R. E. 1929. *Ann. Phys.* 3: 1055
10. Berman, R., Foster, E. L., Schneidmesser, B., Tirmizi, S. M. A. 1960. *J. Appl. Phys.* 31: 2156–59
11. Kumar, G. S., Vandersande, J. W., Klitsner, T., Pohl, R. O., Slack, G. A. 1985. *Phys. Rev. B* 31: 2157
11a. Klitsner, T., Van Cleve, J. E., Fischer, H. E., Pohl, R. O. 1988. *Phys. Rev. B*. In press
12. McNelly, T. F. 1974. PhD thesis. Cornell Univ., Ithaca, NY
13. Lawson, D. T., Fairbank, H. A. 1973. *J. Low Temp. Phys.* 11: 363
14. de Haas, W. J., Biermasz, T. 1937. *Physica* 4: 752
14a. de Haas, W. J., Biermasz, T. 1938. *Physica* 5: 47
15. Casimir, H. B. G. 1938. *Physica* 5: 495
16. Berman, R. 1953. *Adv. Phys.* 2: 103
17. Berman, R. 1976. *Thermal Conduction in Solids*. Oxford: Clarendon
18. Klemens, P. G. 1958. In *Solid State Physics*, ed. F. Seitz, D. Turnbull, 7: 1. New York: Academic
19. Pomeranchuk, I. 1942. *J. Phys. USSR* 6: 237
19a. Klemens, P. G. 1955. *Proc. Phys. Soc. London* 68: 1113
19b. Wagner, M. 1963. *Phys. Rev.* 131: 1443
19c. Walker, C. T., Pohl, R. O. 1963. *Phys. Rev.* 131: 1433
20. Lifshitz, I. M. 1948. *Zh. Eksper. Teor. Fiz.* 18: 293
20a. Lifshitz, I. M. 1956. *Nuovo Cimento Suppl.* 3: 716
21. Berman, R., Brock, J. C. F. 1965. *Proc. R. Soc. London Ser. A* 289: 46
22. Thacher, P. D. 1967. *Phys. Rev.* 156: 975
23. Klitsner, T., Pohl, R. O. 1987. *Phys. Rev. B* 36: 6551
24. Wallis, R. F., ed. 1968. *Localized Excitations in Solids*. New York: Plenum
25. Kagan, Yu., Iosilevskii, Ya. A. 1962. *Zh. Eksperim. Teor. Fiz.* 42: 259; Engl. transl. in 1962. *Soviet Phys. JETP* 15: 182
26. Brout, R., Visscher, W. M. 1962. *Phys. Rev. Lett.* 9: 54
27. Baumann, F. C., Pohl, R. O. 1967. *Phys. Rev.* 163: 843; also Pohl, R. O. 1968. In Ref. 24
28. Narayanamurti, V. 1964. *Phys. Rev. Lett.* 13: 693
29. Narayanamurti, V., Pohl, R. O. 1970. *Rev. Mod. Phys.* 42: 201
30. Jackson, H. E., Walker, C. T. 1971. *Phys. Rev. B* 3: 1428
31. Harrison, J. P., Lombardo, G., Peressini, P. P. 1968. *J. Phys. Chem. Solids* 29: 557
31a. Pompi, R. L., Narayanamurti, V. 1968. *Solid State Commun.* 6: 645
31b. Rollefson, R. J. 1972. *Phys. Rev. B* 5: 3235
31c. De Yoreo, J. J., Knaak, W., Meissner, M., Pohl, R. O. 1986. *Phys. Rev. B* 34: 8828
32. Hogan, E. M., Guyer, R. A., Fairbank, H. A. 1969. *Phys. Rev.* 185: 356; also an excellent review of theory and experiment of Poiseuille heat flow
33. Thomlinson, W. C. 1969. *Phys. Rev. Lett.* 23: 1330
34. Guyer, R. A., Krumhansl, J. A. 1966. *Phys. Rev.* 148: 778
35. Dynes, R. C., Narayanamurti, V., Andres, K. 1973. *Phys. Rev. Lett.* 30: 1129
36. Ackerman, C. C., Guyer, R. A. 1968. *Ann. Phys.* 50: 128
37. Ackerman, C. C., Overton, W. C. 1969. *Phys. Rev. Lett.* 22: 764
38. McNelly, T. F., Rogers, S. J., Channin, D. J., Rollefson, R. J., Goubau, W. M., et al. 1970. *Phys. Rev. Lett.* 24: 100
39. Narayanamurti, V., Dynes, R. C. 1972. *Phys. Rev. Lett.* 28: 1461
40. Zeller, R. C., Pohl, R. O. 1971. *Phys. Rev. B* 4: 2029
41. Raychaudhuri, A. K., Pohl, R. O. 1982. *Solid State Commun.* 44: 711
41a. Vandersande, J. W., Pohl, R. O. 1980. *Rev. Sci. Instrum.* 51: 1694
42. Cahill, D. G., Pohl, R. O. 1987. *Phys. Rev. B* 35: 4067
42a. Cahill, D. G., Pohl, R. O. 1988. *Phys. Rev. B*. In press
43. Pohl, R. O. 1985. *Phase Trans.* 5: 239
44. Anderson, P. W., Halperin, B. I., Varma, C. M. 1972. *Philos. Mag.* 25: 1
45. Phillips, W. A. 1972. *J. Low Temp. Phys.* 7: 351
46. Hunklinger, S., Arnold, W. 1976. In *Physical Acoustics*, ed. W. P. Mason, R. N. Thurston, 12: 155. New York: Academic
47. Phillips, W. A., ed. 1981. *Topics in Current Physics, 24, Amorphous Solids: Low Temperature Properties*. Berlin: Springer

48. Hunklinger, S., Raychaudhuri, A. K. 1986. *Progr. Low Temp. Phys.* 9: 265
49. Pohl, R. O., De Yoreo, J. J., Meissner, M., Knaak, W. 1985. In *Physics of Disordered Materials*, ed. D. Alder, H. Fritzsche, S. R. Ovshinsky, pp. 529. New York: Plenum
50. Phillips, W. A. 1988. *Rep. Progr. Phys.* In press
51. Pohl, R. O. 1987. *Am. J. Phys.* 55: 240
52. Deleted in proof
53. Deleted in proof
54. Birch, F., Clark, H. 1940. *Am. J. Sci.* 238: 529, 613
55. Slack, G. A. 1979. *Solid State Phys.* 34: 1
56. Raychaudhuri, A. K. 1988. *Phys. Rev. B.* In press
57. Anderson, A. C. 1985. *Phase Trans.* 5: 301
58. Pohl, R. O. 1986. In *Transport and Relaxation in Random Materials*, ed. J. Klafter, R. J. Rubin, M. F. Shlesinger, p. 1. Philadelphia: World
59. Deleted in proof
60. Sethna, J. P., Chow, K. S. 1985. *Phase Trans.* 5: 317
60a. Meissner, M., Knaak, W., Sethna, J. P., Chow, K. S., De Yoreo, J. J., Pohl, R. O. 1985. *Phys. Rev. B* 32: 6091
61. Randeria, M. 1987. PhD thesis. Cornell Univ., Ithaca, NY
62. Slack, G. A., Oliver, D. W., Brower, G. D., Young, J. D. 1977. *J. Phys. Chem. Solids* 38: 45
63. Cahill, D. G., Fischer, H. E., Watson, S. K., Pohl, R. O., Slack, G. A. 1988. *J. Less Common Metals.* In press
64. Slack, G. A., Oliver, D. W., Horn, F. H. 1971. *Phys. Rev. B* 4: 1714
65. Tuerkes, P. R. H., Swartz, E. T., Pohl, R. O. 1986. In *Boron-Rich Solids, AIP Conf. Proc.*, ed. D. Emin, T. Aselage, C. L. Beckel, I. A. Howard, C. Wood, 140: 346. New York: Am. Inst. Physics

LASER SPECTROSCOPY OF LARGE POLYATOMIC MOLECULES IN SUPERSONIC JETS

Mitsuo Ito, Takayuki Ebata, and Naohiko Mikami

Department of Chemistry, Faculty of Science, Tohoku University, Sendai 980, Japan

INTRODUCTION

Spectroscopic study of large polyatomic molecules had long been prevented by spectral complexity. The complexity arises from the fact that a large molecule has a large number of vibrational degrees of freedom. Another complication comes from the thermal distribution of molecules over many ground-state vibrational levels, which causes many hot bands to appear in the spectrum. Molecular collisions in gas and condensed phases also contribute to broadening of spectral bands. All these complicating factors become increasingly significant as a molecule becomes larger. Eventually, the spectrum becomes a continuous feature, even in the gas phase, owing to heavy spectral band congestion. A typical example of this spectral congestion is seen in the electronic absorption spectrum of biphenyl in the gas phase. The $S_1 \leftarrow S_0$ absorption spectrum of benzene is well known to exhibit prominent vibrational structure. However, in biphenyl, which is composed of two benzene rings, the corresponding spectrum is completely structureless (1). We often encounter such a broad and structureless absorption spectrum when a molecule is larger than benzene. Clearly, detailed information on the energy levels cannot be obtained from such a broad spectrum.

The recent development of the supersonic jet technique (2–5) has greatly improved the above situation. Supersonic expansion of sample molecules, seeded in high-pressure rare gas, into vacuum through a small nozzle orifice produces a supersonic jet in which the molecule is cooled by the collision-induced transfer of thermal energy of the molecule to translational energy of the rare gas atoms. The molecules in a supersonic jet are in a nonequilibrium state, and the temperature of the molecules is generally different for the translational, rotational, and vibrational degrees of freedom. In most cases, the translational and rotational temperatures are less than a few degrees Kelvin and 10K, respectively. The vibrational temperature, however, generally differs from molecule to molecule. In general, the large polyatomic molecules having many low frequency vibrational modes are efficiently cooled and the vibrational temperature becomes very low. Therefore, the molecules are nearly all in the zero-point level of the ground state, and they are also in a collision-free condition. As a result, the two factors leading to the spectral congestion of large molecules—the appearance of hot bands and the collisional broadening—are almost completely removed, and sharp spectral features show up even for a large molecule that ordinarily gives a continuous spectrum. Thus, the range of large molecules that are subject to detailed spectroscopic study is now expanded to molecules as large as porphyrin (6).

Although a jet-cooled molecule is collision-free and is in a well-defined vibrational state, the concentration of the molecules is so low that normal spectroscopic methods such as absorption spectroscopy are difficult to apply. Therefore, we are forced to use indirect methods for spectral measurements. The major indirect methods that are widely used are fluorescence excitation spectroscopy and multiphoton ionization (MPI) spectroscopy. In the former, the electronic transition of a jet-cooled molecule is probed by the fluorescence emitted from the excited state reached by the transition. In the latter, the transition is probed by the ions generated by resonance-enhanced multiphoton ionization of a molecule. Since both fluorescence and ions can be detected with a high sensitivity, these spectroscopies are very useful for jet-cooled molecules. In the present paper, we do not discuss these spectroscopies in detail because they are well established (3, 4). Rather, we emphasize the new supersonic jet spectroscopies that have recently been developed. The principles, advantages, and disadvantages of these new spectroscopies are described in the following section. The remaining parts of the paper are devoted to the applications of supersonic jet spectroscopies to molecules that are nearly all larger than benzene. We then describe low-lying electronic states (S_0 and S_1) of large polyatomic molecules in supersonic jets, restricting our discussion to the rapidly growing fields of large amplitude motion and rotational isomerism.

At present, highly excited states of molecules are being explored by two-color optical-optical double resonance spectroscopy applied to jet-cooled molecules. Therefore, a section of this review is devoted to recent developments in studies of highly excited states of large molecules by the double resonance technique. We also include studies of molecular ions and radicals.

The supersonic jet technique created an attractive new field of van der Waals (vdW) molecules formed by weak intermolecular interactions. A vdW molecule may be regarded as a minisolution or minicrystal. In this regard, it provides us with a new approach to the fundamental problems of condensed phases. The vdW molecules containing large polyatomic molecules are briefly reviewed in the final section. Hydrogen-bonded complexes are also discussed.

Since the subjects we deal with are very broad, it is almost impossible to cover all the literature. We restricted ourselves to the references appearing during the last five years and also to those closely related to our own studies.

SUPERSONIC JET SPECTROSCOPIES

Fluorescence excitation and multiphoton ionization are the most popular spectroscopic means for studying jet-cooled molecules (3, 4). They are very useful techniques for highly fluorescent molecules and for molecules with high multiphoton ionization yields. However, many molecules are nonfluorescent but still have low ionization yields, for example, aromatic ketones and aldehydes such as benzaldehyde and benzophenone. Although these molecules are nonfluorescent, they are known to be phosphorescent. A spectrum corresponding to the absorption spectrum can be obtained by monitoring the total phosphorescence emitted from the molecule. This is called a *phosphorescence excitation spectrum* and is a popular method for studying molecules in the condensed phase. However, when applying the method to a molecule in a jet, one encounters great difficulty in detecting the phosphorescence signal with a space-fixed detector because of the long lifetime of the phosphorescence and the high velocity of the phosphorescent (triplet-state) molecule in the jet.

The difficulty was partly solved by Spangler & Pratt (7). Near a jet nozzle they installed an ellipsoidal reflector whose first focus coincides with the laser focusing point in the jet. The phosphorescence emitted from this point hits the reflector and is efficiently collected at the second focus of the ellipsoid, where a photomultiplier is placed. This collection system improves sensitivity by a factor of ~ 25 over the usual lens system for collecting emission. By using this method, Spangler & Pratt succeeded in measuring the $T_1 \leftarrow S_0$ transitions of jet-cooled glyoxal and its derivatives (7). The method was also applied to the $S_1 \leftarrow S_0$ transition of jet-cooled benzophenone (8). When a molecule is both fluorescent and phosphorescent, the excitation spectrum due to the $S_1 \leftarrow S_0$ transition obtained by this method represents a mixture of the fluorescence excitation spectrum and the phosphorescence excitation spectrum. The separation of these two spectra can be achieved by the sensitized phosphorescence excitation method described below.

Abe et al (9) placed an appropriate solid phosphor downstream from a jet. The triplet-state molecules produced directly by laser excitation or by intersystem crossing from a laser-excited singlet state travel downstream in the jet at a high speed and hit the solid phosphor. Collision-induced energy transfer to the phosphor then results in sensitized phosphorescence. When the sensitized phosphorescence is detected by a detector placed near the solid phosphor as the laser frequency is scanned, the *sensitized phosphorescence excitation spectrum* of the jet-cooled molecule is obtained. Abe et al found that for many molecules the sample solid prepared on a cooled copper surface by deposition of jet-cooled molecules serves as a good phosphor (10). Since the triplet-state molecules must live until the time of the collision, the triplet state lifetime of the molecule should be larger than the flight time (typically 40 μs) of the molecule from the laser focusing point to the solid phosphor. When a jet-cooled molecule is fluorescent as well as phosphorescent, both the fluorescence excitation spectrum and the sensitized phosphorescence excitation spectrum can be simultaneously measured with two detectors, one for the fluorescence and another for the phosphorescence, the former being placed near the crossing point of the laser beam and the jet, and the latter near the solid phosphor. Because of a long time delay corresponding to the flight time, the phosphorescence signal can be completely distinguished from the fluorescence signal. The intensity distribution of the vibronic bands in the $S_1 \leftarrow S_0$ transition of a jet-cooled molecule is in general different between the simultaneously measured fluorescence and phosphorescence excitation spectra. This difference provides us with information on intersystem crossing rates of the different vibronic levels of the excited singlet-state molecule.

The sensitized phosphorescence excitation spectroscopy was applied mainly by Ito's group to the observations of the $S_1 \leftarrow S_0$ transitions of glyoxal and its methyl derivatives (11, 12), benzaldehyde, acetophenone, benzophenone (9, 13), benzoic acid and its derivatives (14), naphthalene and its derivatives (15), and vdW complexes involving aromatic molecules (16) in jets. The $T \leftarrow S_0$ spectra were also measured for jet-cooled benzaldehyde, acetophenone, and benzophenone (10).

Spectroscopists have wanted direct measurements of the absorption spectra of jet-cooled molecules for a long time. However, in the usual conditions of supersonic expansion, the concentration of molecules in the jet is extremely low and the optical pathlength of the jet available for the absorption measurement is too short. To improve the situation, Jortner and co-workers (17) developed the *planar supersonic expansion technique*, in which they used a pulsed nozzle slit of 90 mm in length and 0.27 mm in width. The jet parallel to the slit produced by the supersonic expansion has sufficient optical pathlength for the detection of strong absorption. The absorption spectrum is obtained with an ordinary xenon flash lamp and a monochromator. The absorption spectra of large molecules in jets such as anthracene derivatives (17), phenanthrene (18), azulene (19), stilbene (20), and fluorene (21) have been measured by this planar supersonic jet technique. The direct absorption spectra of jet-cooled molecules were also measured by Vaida and her co-workers, who used an ordinary jet nozzle. The large molecules studied include hexadine (22) and octadine (23). The direct absorption measurements of jet-cooled molecules are indispensable for all excited-state problems. However, the direct absorption spectra so far obtained are still far from satisfactory in quality, and they are limited to the transitions having large cross-sections.

Recently, studies of highly excited electronic states of molecules have been greatly advanced by the use of two laser light sources. A jet-cooled molecule is excited to a low-lying excited state by the absorption of one-photon of laser frequency v_1; then the excited molecule is further promoted to a higher excited state by the one-photon absorption of another laser light of frequency v_2. In this double resonance spectroscopy, the highly excited states reached from the well-defined intermediate state by the v_2 absorption are severely restricted by selection rules depending upon the nature of the selected intermediate state. As a result, the spectrum becomes simple, making the assignment of the highly excited state easy. By selecting the intermediate state with the laser light of a fixed frequency v_1, one can sort out the congested highly excited states by the v_2 absorption. In double resonance spectroscopy, the preparation of molecules in a selected intermediate state is essential. Since selective excitation is easily achieved under jet conditions, the combination of double-resonance spectroscopy and supersonic jet is most useful for the study of highly excited states, especially of large molecules. Various methods for the detection of highly excited states by double resonance excitation have recently been developed (24).

Two-color multiphoton ionization spectroscopy is probably the most popular means for observing highly excited states of a molecule. Its principle is schematically shown in Figure 1*a*. The molecule in a particular vibronic level A pumped by v_1 absorption is further excited to a highly excited state R by the second tunable laser light of v_2. The highly excited molecules absorb another v_1 or v_2 photon to reach the ionization continuum of the molecule, thereby generating the molecular ion. Therefore, by detecting the ions while scanning the laser frequency v_2, one obtains the excitation spectrum corresponding to the $R \leftarrow A$ absorption spectrum.

Two-color multiphoton ionization spectroscopy was first applied to I_2 vapor by Williamson & Compton (25). Ebata et al (26, 27) measured the two-color double resonance enhanced four-photon ionization spectra of NO vapor in which the $A^2\Sigma^+$ state of the molecule was populated by the absorption of two photons of v_1 and another photon of v_2 was used to reach the higher excited Rydberg state from the $A^2\Sigma^+$ state. The technique was applied to a jet-cooled large molecule by Fujii et al (28) for p-difluorobenzene. They found four Rydberg series converging to the adiabatic ionization limit at 73871 cm^{-1} from the spectrum obtained via the first excited singlet state $S_1(^1B_{2u})$.

As shown in Figure 1a, when the one-photon transition from the A state with v_2 is resonant to the adiabatic ionization potential (IP_0) of a molecule, direct ionization from the A state to the ionization continuum occurs and ions are generated. Therefore, by detecting ions while scanning the frequency v_2, a threshold appears when v_2 reaches IP_0. The threshold energy represents the zero-point level in the electronic ground state of the positive ion. When the geometrical structure of the neutral molecule in the A state is similar to that of the ground state ion, the Franck-Condon factor is large exclusively for the transition with $\Delta v = 0$. Therefore, a strong IP_0 threshold can be observed when the zero-point vibrational level in the A state is selected as an intermediate state. Similarly, direct ionization preferentially occurs from a vibrational level in the A state to the ionization continuum belonging to the same vibrational level of the ion as that in the A state. The ionization threshold that occurs thus represents the vibrational level of the ground state ion. Therefore, so-called *two-color ionization threshold spectra* provide us with IP_0 of a neutral molecule and the vibrational structure of its ion. Two-color ionization threshold spectra are reported for many aromatic molecules in jets (29–40). For all the molecules, the S_1 state serves as an intermediate A state.

When the potential and geometrical structures differ greatly between the molecule in its A state and the ion, direct ionization gives a broad rather than a sharp threshold. In such a case, direct ionization is possible only from the A state to very high vibrational levels in the ion where the levels are heavily congested, thus resulting in a broad threshold. Such a broad threshold was found for the vdW complexes of fluorobenzene (30, 31) and pyrimidine (36), the hydrogen-bonded complexes of phenol (31, 33) and 7-azaindole (33), and 2-aminopyridine (37), indole (39), and p-xylene (41).

Many discrete bound states of a neutral molecule lie above the adiabatic

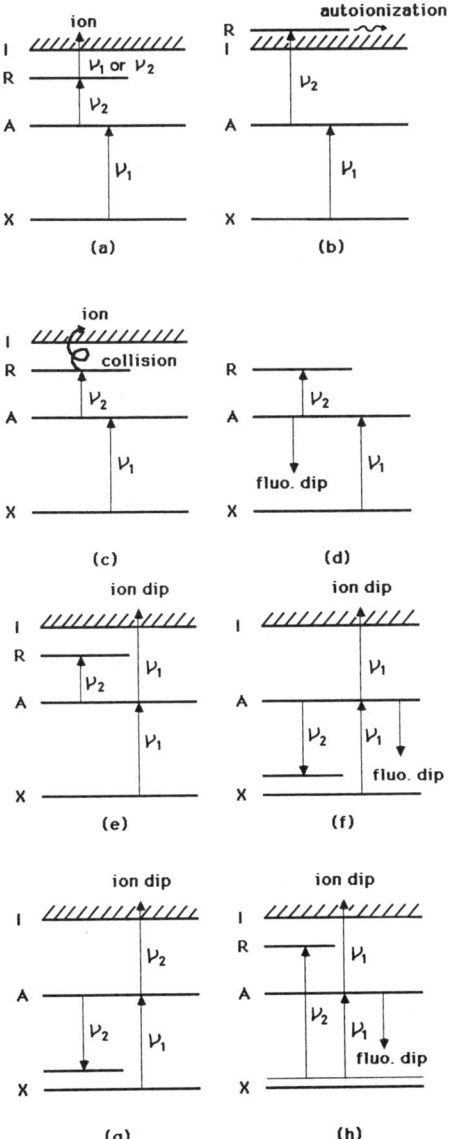

Figure 1 Various two-color double resonance spectroscopies.

ionization potential IP_0 of the molecule. These states can couple with the isoenergetic ionization continuum by a breakdown of the Born-Oppenheimer approximation and can then autoionize. Therefore, these states can be observed by *two-color MPI assisted by autoionization*, a technique that is schematically shown in Figure 1b. Of course the ionization peaks caused by autoionization appear only above IP_0. The autoionizing excited states of large molecules so far observed by two-color MPI spectroscopy are all Rydberg states of the molecules (28, 35, 42–46). Since the geometrical structure of a molecule in a Rydberg state is generally very similar to that of the ground state ion, vibrational or rotational autoionization is mainly responsible for the appearance of the Rydberg states.

For an isolated molecule, the highly excited states lying below IP_0 cannot be detected by the two-color MPI method if the cross-section for the ionization from the high excited state to the ionization continuum by the absorption of one more v_1 or v_2 photon is very small. In such a case, the two-color MPI spectrum gives a sharp threshold at IP_0 but no signal appears in the region below IP_0. However, when a molecule is subject to collisions, the high excited state molecule acquires collisional energy and can be promoted to the ionization continuum. This situation is schematically shown in Figure 1c. The observations of high excited Rydberg states by *two-color MPI assisted by collision* are reported for DABCO vapor at room temperature (46, 47), jet-cooled *trans*-stilbene (48), and jet-cooled *p*-difluorobenzene (28). The appearance of the Rydberg states of jet-cooled molecules indicates that collisional effects cannot be neglected even under supersonic jet conditions for highly excited Rydberg states having a large Rydberg electron orbit.

Another two-color double resonance spectroscopy recently developed is the so-called "*Dip spectroscopy*." The principle of this spectroscopy is shown in Figure 1d. A molecule in a jet is first excited by laser light of v_1 to a selected level in the fluorescent A state, and then tunable laser light v_2 is introduced to excite the molecule from state A to higher excited states, as is usually done in two-color experiments. In the presence of v_1, the molecule in state A emits fluorescence of a constant intensity. When the frequency v_2 is resonant to a higher excited state R, the fluorescence intensity decreases because of the depopulation of the A state caused by the $R \leftarrow A$ absorption. Therefore, observing the total fluorescence while scanning the frequency of v_2, one obtains the so-called *two-color fluorescence dip spectrum*. Two-color fluorescence dip spectroscopy is free from the ionization process of a molecule. Therefore, the method can be applied to higher excited states in a wide spectral region irrespective of the location of the ionization potential. Another advantage of this spectroscopy is that the dip intensity directly reflects the cross-section of the transition from the selected A level to a higher excited state by v_2.

The two-color fluorescence dip technique was successfully applied for the first time by Ebata et al (27) for observing the high Rydberg states of NO. It was also applied to the high Rydberg states of jet-cooled DABCO (44), ABCO (46), and benzene (49) by using the fluorescent S_1 states as intermediate states. Goto et al observed the two-color fluorescence dip spectra of jet-cooled glyoxal (50) and pyrazine (51) via the vibronic levels in the $S_1(n, \pi^*)$ states of these molecules. Since the transition from the $S_1(n, \pi^*)$ state to the higher excited $S_n(n, \pi^*)$ state is orbital-allowed, the latter state is selectively observed in the fluorescence dip spectra. Similar to fluorescence dip spectroscopy, *two-color phosphorescence dip spectroscopy* is employed in detecting the highly excited triplet states of a molecule by probing the phosphorescence emitted from a low-lying triplet state excited directly by v_1. The highly excited n, π^* triplet states of glyoxal were found by this method using the $T_1(n, \pi^*)$ state as an intermediate state (50).

Another kind of dip spectroscopy is *two-color ionization dip spectroscopy*, and its principle is shown in Figure 1e. With the first laser light of v_1, one-photon resonant two-photon ionization occurs by resonance to a particular vibronic level in the A state. We probed the ion signal generated by this two-photon ionization. When the second laser light of v_2 is in resonance with a higher excited state R, the ion signal exhibits a dip at resonance. In general, the two-color ionization dip spectrum gives the same result as that of the two-color fluorescence dip spectrum, but the former can be applied to nonfluorescent molecules. Two-color ionization dip spectroscopy was applied for the first time by Cooper et al (52, 53) for I_2 and by Murakami et al (54) for toluene and aniline. The high Rydberg states of jet-cooled DABCO (44), ABCO (46), and benzene (49) were studied by Fujii et al by this spectroscopy.

Two-color dip spectroscopy is useful not only for the study of highly excited states of a molecule but also for the study of vibrational states of a molecule in the ground electronic state. In the latter study, stimulated emission from an intermediate state is utilized. As shown in Figure 1f, the second laser light of v_2 induces the stimulated emission from the intermediate A state to a vibrational level in the ground state molecule. The stimulated emission appears as a dip in the fluorescence from the A state or as a dip in the ion signal generated by one-photon resonant two-photon ionization with v_1. The methods are often called *stimulated emission pumping*. Stimulated emission spectroscopy utilizing fluorescence dips was first developed by the MIT group (55, 56), and it was applied for the detailed rovibronic level structures of formaldehyde (57), acetylene (58), and glyoxal (59) in their ground electronic states. This spectroscopy is also a good tool for the study of relaxation processes in ground-state vibrational levels. Such studies were carried out by Knight and co-workers (60, 61) for p-difluorobenzene, by Suzuki et al (62) for *trans*-stilbene, and by Yamanouchi et al (59) and Kim et al (63) for glyoxal.

A different version of stimulated emission spectroscopy utilizing the ion dip technique was recently proposed by Suzuki et al (48, 62). Its principle is shown in Figure 1g. With this method, the photoionization is induced by $v_1 + v_2$. To meet this condition, the laser power of v_1 is suppressed as low as possible, so that the photoionization does not occur with v_1 alone. Only when v_1 and v_2 coexist can we observe the ion signal. The laser light of v_2 is used not only for the ionization but also for the stimulated emission from the A state to a vibrational level in the ground state. The stimulated emission is detected as a dip in the signal from ionization by $v_1 + v_2$. In this method, since there is no ion signal with v_1 or v_2 alone but only with the two laser beams present, we need not worry seriously about temporal and spatial mismatching of the two laser beams, which is always a great problem in double resonance experiments. However, the vibrational region studied by this method is restricted by $0 < E_{vib} < 2hv_1 - IP_0$, where IP_0 is the adiabatic ionization potential of a molecule. This method was applied by Suzuki et al for their studies of the ground-state vibrational level structures of *trans*-stilbene (62), aniline, and p-fluorotoluene (64) in jets and their intramolecular vibrational redistributions (IVR).

A final jet spectroscopy worth mentioning is *population labeling spectroscopy*; its principle is shown in Figure 1h. The first laser light of a fixed frequency v_1 is tuned to the well-defined $A \leftarrow X$ transition of a molecule which is associated with a particular rovibronic level in the ground state, and the fluorescence from the A state is monitored. Then, the second laser light of tunable frequency v_2 is scanned through the $R \leftarrow X$ absorption region. When the frequency v_2 coincides with that of the $R \leftarrow X$ transition involving the same rovibronic level in the ground state as that for the $A \leftarrow X$ transition, the population of this particular rovibronic level in the ground state decreases, and the fluorescence from the A state or the ion signal generated by the one-photon resonant two-photon ionization with v_1 exhibits a dip. In this way, definite assignment of the $R \leftarrow X$ transition can be made by labeling the common rovibronic state in X. This spectroscopy was first applied by Kaminsky et al (65) for Na_2 molecules and is now widely used for the assignment of hot bands in the vapor absorption spectra of diatomic molecules. Recently, the $S_1 \leftarrow S_0$ absorption spectrum of jet-cooled azulene was obtained by this population labeling technique by Suzuki & Ito (66), who demonstrated the advantage of this method for the study of excited states with very fast relaxation processes, like the S_1 state of azulene.

LARGE AMPLITUDE MOTIONS AND ROTATIONAL ISOMERISM

Large Amplitude Motion

The large amplitude motions of flexible molecules are an interesting subject in molecular spectroscopy and are also important in various intramolecular relaxation processes. These motions have been traditionally studied by vibrational and rotational spectroscopies such as Raman, infrared, and microwave. However, it was recently recognized that supersonic jet spectroscopy utilizing an electronic transition is often more powerful than vibrational or rotational spectroscopy, even for ground state molecules (5, 67). The internal rotation of a CH_3 group is a typical large amplitude motion in large molecules. Okuyama et al (68) observed the fluorescence excitation and dispersed fluorescence spectra ($S_1 \leftarrow S_0$) of o-, m- and p-fluorotoluene in jets. They analyzed the vibronic bands associated with the internal rotation of the CH_3 group and obtained accurate potentials of the internal rotation for both the S_0 and S_1 states. It was found that the barrier to the internal rotation dramatically changes in going from S_0 to S_1 for each isomer. The internal rotational levels of the cationic ions of these molecules were also obtained by two-color ionization threshold spectroscopy (67, 69), and the barrier to the internal rotation was found to greatly increase in the ionic state. Similar studies on internal rotation are also reported for o- and m-toluidine (70), m-cresol (69) toluene, xylenes (71), and propyltoluenes (72). Parmenter & Stone (73) carried out a comparative study of the fluorescence excitation spectra of the $S_1 \leftarrow S_0$ transitions of fluorobenzene and p-fluorotoluene and found a great acceleration of IVR by the internal rotation of the CH_3 group in the S_1 state of p-fluorotoluene. Moss et al considered theoretically the origin of the acceleration (74).

The torsional motions of *trans*-stilbene and diphenylacetylene (tolan) are large amplitude motions in the ground state that could not be found by vibrational spectroscopy but were recently revealed by supersonic jet spectroscopy. Suzuki et al (62) observed the stimulated emission spectrum of jet-cooled *trans*-stilbene by using the two-color ionization dip method. The stimulated emission from the S_1 state revealed the detailed vibrational level structure of this molecule in the ground state, and this in turn led them to the discovery of a large amplitude torsional motion. The motion is the out-of-phase torsion of the C–C bonds connecting the phenyl groups and the ethylene group. The potential for the torsion was found to be very anharmonic with a flat bottom. The maximum oscillational angle between the two phenyl groups is as large as 50° even in the zero-point level. This is probably the largest amplitude motion ever found and may play an essential role in the *cis-trans* isomerization of this molecule. Actually, it was found that the torsional motion greatly accelerates IVR (62). Very recently, Zewail and co-workers (75, 76) showed with picosecond spectroscopy that this torsion also accelerates IVR in the S_1 state. The large amplitude torsion of the C≡C bond of diphenylacetylene in the ground state was also found for the first time from the dispersed fluorescence spectrum of the jet-cooled molecule (77).

Some large molecules are rigid in the ground state but become flexible in the excited state. This was shown recently for 1,2,4,5-tetrafluorobenzene. Okuyama at al (78) observed the fluorescence excitation spectrum ($S_1 \leftarrow S_0$) of the jet-cooled molecule and analyzed the irregular vibronic structure in the spectrum. The structure analysis revealed that the excited-state molecule has a double minimum potential along the coordinate of the butterfly motion of the C–F bonds and that the butterfly tunneling takes place between the two minima. The appearance of the double minimum potential in the S_1 state was discussed by Orlandi & Zerbetto (79) in terms of vibronic coupling.

Other large amplitude motions revealed from supersonic jet spectroscopy are the C–C torsion of biphenyl (80), benzophenone (12), 9,9'-bianthryl (81) in S_1 and the cage torsion of DABCO (82).

Rotational Isomerism

A new field that has recently become active through the application of supersonic jet spectroscopy is rotational isomerism of large molecules. The subject has traditionally been studied by rotational and vibrational spectroscopies. The application of these traditional spectroscopies becomes increasingly difficult as a molecule becomes larger, because a small geometrical difference in the rotational isomers of a large molecule is reflected only by a subtle difference in rotational or vibrational levels. However, the electronic structure of a molecule may be more sensitive to geometrical differences than the rotational or vibrational structure (5). This was found to be true for some aromatic molecules. Oikawa et al (83, 84) observed the fluorescence excitation spectra ($S_1 \leftarrow S_0$) of m-substituted phenols and β-naphthol in jets and found that the *cis*- and *trans*-isomers of each molecule arising from the orientation of the OH group with respect to the m-substituent or to the naphthalene ring gave a difference in the electronic excitation energy of as large as $100 \sim 300$ cm^{-1}, large enough for clear discrimination. The dispersed fluorescence spectrum of each isomer can be obtained by exciting a vibronic band belonging to this isomer. The results show that the vibrational level structures of the rotational isomers in their ground states are almost the same, thus indicating the difficulty in discrimination by vibrational spectroscopy. So far, for any aromatic molecules anticipated to have rotational isomers, the isomers have been clearly identified by the electronic spectra of the jet-cooled molecules (69, 85, 86).

The ionization potential of each isomer was also measured by the two-color ionization threshold spectrum of the jet-cooled molecule by using the S_1 level of this isomer as an intermediate state. It was shown that the difference in the ionization potential between the different isomers of a molecule is also very large and amounts to 50–500 cm^{-1} for many phenol derivatives (85). The large difference might lead to a possible separation of the rotational isomers by ionization. The fluorescence excitation and dispersed fluorescence spectra of the complexes of p-dimethoxybenzene with various nonpolar and polar solvent molecules prepared by supersonic expansion were observed by Yamamoto et al (87). They found that only the *cis*-isomer of the molecule selectively forms the complexes with polar solvent molecules, whereas both the *cis*- and *trans*-isomers form complexes with nonpolar solvent molecules. This kind of selective complexation will be important in understanding the selective reactivities of rotational isomers.

HIGHLY EXCITED STATES, IONS, AND RADICALS

As described above, the advantage of using two-color double resonance spectroscopy for the observation of highly excited states and ionic states of molecules is to greatly simplify the spectrum. For diatomic molecules, this method has been extensively used as *optical-optical double resonance* spectroscopy for molecules such as halogens, alkali dimers, BaI, NO, CO, and HgAr. However, in the case of diatomic molecules it is not necessary to cool down the molecules by supersonic expansion, since the normal dye laser has enough spectral resolution for pumping to a specific rovibronic state at room temperature, static-gas conditions.

Several groups have extensively studied the highly excited states of large polyatomic molecules in supersonic free-jet conditions (29–43). Generally, however, for high energy valence states the spectra exhibit broad band structures even when the two-color technique is used. A typical example may be the observations reported by Kakinuma et al (49) of the $^1E_{2g}$ (~ 60000 cm^{-1} region) state of benzene, which was probed by dips in the fluorescence from the $S_1(^1B_{2u})$ state. Only the Rydberg states show sharp vibrational structure. The valence $^1E_{2g}$ state spectrum was obtained as a broad background overlapping the Rydberg spectrum. The result indicates that the high energy valence state has very fast relaxation such as internal conversion or dissociation.

Compared to the broad features of the high energy valence states, Rydberg states show surprisingly sharp structures. Fujii et al found long Rydberg series converging to the ionic states of DABCO (44), ABCO (46), and p-difluorobenzene (28) by two-color multiphoton ionization and fluorescence dip spectroscopy. Especially for the DABCO and ABCO molecules, they reported well-resolved Rydberg series ranging from $n = 5$ to $n = 39$. Their success was due to the choice of the 3s-Rydberg state as an intermediate state. Since all Rydberg states have parallel potentials, transitions to higher Rydberg states favor $\Delta v = 0$, thus resulting in the very simple spectrum. The Rydberg series of pyrazine have been observed by Goto et al (35) and those of aniline were reported by Hager et al (43).

An interesting phenomenon in these observations is the autoionization of the Rydberg states. Ion peaks were observed only in the energy region above the adiabatic ionization potential. It was concluded that the ionization occurs by the vibrational autoionization of the Rydberg states due to a breakdown of the Born-Oppenheimer approximation, as described above. It was also shown that the normally accepted $\Delta v = -1$ selection rule for the vibrational autoionization of diatomic molecules (88) also holds for polyatomic molecules.

It is interesting to inquire whether the vibrational autoionization probability of polyatomic molecules depends on the vibrational mode excited. From an analysis of aniline spectra Hager et al (43) reported an upper limit on the autoionization rate on the order of $10^{12} \sec^{-1}$. They found that the nontotally symmetric vibration induces autoionization more effectively than does the totally symmetric vibration. However, later Goto et al (35) explained the same phenomena in pyrazine in terms of the Fano model, in which the transitions to the discrete level and the continuum level interfere with each other. Therefore, mode dependence seems to remain an open question.

The structures of ions have been studied and the ionization potentials of the polyatomic molecules have been measured. Usually the S_1 state is chosen as the first intermediate state excited by v_1 and the second laser, v_2 is scanned in the ionization region. This method gives more accurate values, especially for the ionization potential, than the normal energy analysis of the photoelectron ejected by one-photon ionization using the HeI line. Ionization threshold spectra and vibrational frequencies of ions have been reported for large polyatomic molecules: DABCO (44, 45), ABCO (46), pyrazine (35), naphthalene (38), aniline (42, 43), phenol and its derivatives (31, 69, 85), and indole (39, 40). By combining two-color MPI with threshold photoelectron spectroscopy, Chewster et al (89) determined the precise ionization potential of benzene. The two-color MPI technique has been extended to vdW molecules and hydrogen-bonded complexes. A depletion of the adiabatic ionization potential in these systems is discussed below.

By monitoring the signal as a function of the delay time between the first laser and the second laser, we can measure the time evolution of the intermediate state. In the absence of collisions, molecules pumped to a specific vibronic level, for example, in the S_1 state, may fluoresce, redistribute energy to other vibrational levels (IVR), go to isoenergetic levels of the low-lying triplet state (ISC), or dissociate. Since the fluorescence lifetime of a large polyatomic molecule is typically on the order of a nanosecond and the time scale of these processes is less than that, picosecond time resolution is needed for the laser system. So-called two-color "pump and probe" experiments have been performed by Zewail's group (75, 76, 90). They first demonstrated this technique for the measurement of IVR in isolated t-stilbene molecules by changing the time delay and polarization of the pump and probe lasers. They extended the study to the predissociation of the phenol(cresol)-benzene vdW complex (90). The dissociation rate was explained by a modified RRKM calculation for the phenol-benzene complex.

Several studies have measured the collision-free decay rate of the triplet state as a function of excess energy on longer time scales (91–96). After the intersystem crossing from a specific vibronic level of the S_1 state pumped by v_1, the triplet state (usually T_1) decays radiatively and by intersystem crossing to high vibrational states of S_0. When the second laser has enough energy to ionize the triplet state molecule and the delay time between the first and second lasers is changed, we can measure the time evolution of the triplet state (T_1). Starting from benzene (91), Smalley's group has measured the triplet decay rates of toluene (92), pyrazine, and pyrimidine (93). Benzene (94) and aniline (95) were measured by Johnson's group. Recently, Lohmannsroben et al have studied alkylbenzenes (96). The lowest excess energy in this experiment is determined by the energy difference between the T_1 and S_1 states, and the excess energy range is from 2000 cm^{-1} to 10000 cm^{-1}. In every case the decay rate rises sharply within the first 2000 cm^{-1} of excess energy from the T_1 origin. For example, in pyrimidine the decay rate at an excess energy of 2000 cm^{-1} is more than five orders faster than that at the T_1 origin; above this energy the increase becomes very modest. Several theoretical treatments have been applied to give a quantitative explanation of the $T_1 \rightarrow S_0$ nonradiative process (97, 98).

Radicals and cations are usually produced by electron impact, discharge, or photolysis of parent molecules in the gas phase. The ions or radicals produced are therefore rotationally and vibrationally very hot. Vibrational analyses of the observed spectra have only been performed for small polyatomic radicals. If we can produce species in a supersonic expansion, the spectra become very simple and we can obtain even rotationally resolved spectra. Another purpose for producing these reactive species in cold jets is to study the dynamics of their reactions and ultimately to observe state-to-state chemistry.

A simple and popular method to obtain supersonically cooled radicals and ions is as follows: Parent molecules are dissociated or ionized by laser photolysis or electron impact just after expansion. The generated species are internally very hot but are cooled down by subsequent collisions during the expansion process until they reach the same velocity. Obi's group applied Hg-photosensitization to produce small polyatomic radicals (99, 100). These species were detected by fluorescence excitation or resonance enhanced MPI spectra.

Polyatomic radicals detected in the supersonic expansions are HNO (99), NH_2 (100), allyl (101), alkyl (102), alkoxy (103, 104), vinoxy (105), and benzyl radicals (105). Precise rovibrational analyses have been performed, and the improvement in the quality of the spectroscopic constants is remarkable. Some of these radicals, such as HNO and alkoxy radicals, show predissociation, and the onsets of the predissociation have also been determined.

Not as many polyatomic cations exhibit fluorescence. The ions studied by laser-excited fluorescence are halobenzenes (106, 107), halogenated acetylenes (108), diacetylenes (109), and furan (110). Miller's group studied the spectra of halobenzene ions extensively and discussed the Jahn-Teller effect in the ground state ion of hexafluorobenzene (107). Klapstein's group was mainly concerned with the spectroscopy of cations of acetylene derivatives (108, 109).

MOLECULAR COMPLEXES

Van der Waals Complexes

Since the pioneering work by Levy's group (111) for the s-tetrazine-rare gas complexes, a number of studies have dealt with the generation and characterization of polyatomic complexes, with the following interests: (a) spectral shifts of the electronic transitions, (b) low-frequency intermolecular modes, (c) dissociation energies of the vdW bonds, (d) geometrical structures, (e) degree of the aggregation, and (f) conformational isomers. The complexes are prototype systems for the studies of intermolecular interactions between solute and solvent molecules and also of physisorption of molecules on solid surfaces. In this respect, the above information provides the fundamental basis for solvent perturbations and molecule–surface interactions. Recent studies of polyatomic vdW complexes may be classified into two groups: one seeks to obtain precise structures by using high resolution spectroscopy and the other to find rather broad characteristics of the molecular interactions by using model calculations.

The rotationally resolved electronic spectra of the tetrazine-rare gas complexes were first observed by Haynam et al (111). The tetrazine dimers (112), aniline-rare gas (113), pyrimidine-Ar (114), tetrazine-HCl (115), tetrazine-H_2O (116), benzonitrile-Ar (117), and tetracene-rare gas (118) complexes were observed by using high resolution spectroscopy, and their precise geometrical structures were determined by analysis of their rotational fine structures. Detailed information about intermolecular distances, angles, and bonding sites contributes greatly to our understanding of weak molecular interactions.

The characterization of complexes through model calculations was initiated by Ondrechen et al (119) for the tetracene–rare gas systems. The observed spectral shifts, low-frequency vdW modes, and dissociation energies were elucidated in terms of pair-wise atom–atom interactions expressed by Lennard-Jones type potentials. The studies by research groups of Jortner and of Leutwyler have been recently summarized in their review article (120), which emphasizes the analogies between complexes and atoms on graphite surfaces. Similar studies were performed for the complexes between mono-ring aromatics (benzene, toluene and diazines) and small alkanes (CH_4, C_2H_6, C_3H_8) by Bernstein and co-workers (121). The complexes of pyrene and perylene with various hydrocarbon solvent molecules were also studied by Mangle & Topp (122).

The structures and molecular interactions of vdW dimers have been discussed in relation to structures in the crystalline phase. In benzene crystal, for example, the translationally inequivalent nearest neighbors are known to be in a T-shape geometry. The fluorescence excitation and MPI spectra of benzene dimer were observed for the first time by Hopkins et al (123) and by Langridge-Smith et al (124), respectively. The T structure of the gas phase dimer was suggested by Klemperer and co-workers (125). SCF calculations (126) at the ab initio restricted Hartree-Fock level were performed by using the GAUSSIAN-80 set and gave the T structure as the most stable one. However, arguments against this structure's stability were made on the basis of the site splitting of the dimer transitions, which is similar to exciton splitting in crystal. Law et al (127) concluded from the study of the site splitting that the parallel stacked and displaced configuration having C_{2h} symmetry, rather than the T-shape dimer, is the most stable conformation. Recently, this problem was elegantly solved by Börnsen et al (128), who, by using mass-separated MPI spectra of various isotopically substituted benzene dimers, showed the absence of the site splitting in the $(h_6^* - d_6)$ hetero-dimer. It was concluded from the result that the two benzene rings are in a V-type configuration (C_{2v} symmetry) with an angle of about 70° between the two molecular planes.

Various dynamical processes occur after the photo-excitation of vdW complexes: emission from the pumped state, vibrational energy redistribution (IVR), internal conversion, intersystem crossing, conversion to isomer, and dissociation to fragments. Among these processes, vibrational predissociation of an electronically excited vdW complex is of particular interest. Vibrational predissociation occurs when the complex is excited to the vibronic levels lying higher than the dissociation energy of the vdW bond of the electronically excited molecule. The predissociation of large polyatomic vdW complexes are first studied by Levy's group (129) for the complex of tetrazine with a rare gas atom and for the tetrazine dimer (130). Similar studies have been done for the complexes of glyoxal (131), benzene (132), isoquinoline (133), pyrimidine (134), difluorobenzene (135), and benzonitrile (136) with rare gas atoms or simple molecules such as N_2 and H_2O.

In the predissociation process the following interrelated aspects emerge: competition with other processes, mode selectivity or propensity, and time evolution. In order to demonstrate the processes competitive with predissociation, let us select the pyrimidine-Ar complex as an example. The fluorescence excitation spectrum due to the $S_1 \leftarrow S_0$ transition of the complex (134) showed anomalous intensities for some vibronic bands, particularly, the extremely weak 0, 0 band and the strong vibronic bands higher than $6a^1$ (614 cm^{-1}). The strong intensity was found to be due to the vibrational predissociation from these vibronic levels that produces the pyrimidine fragment in the S_1 state; the latter has a higher fluorescence quantum yield than the complex. The branching ratios to the fragment levels were also obtained and were found to be determined by an energy (or momentum) gap law (137). When the predissociative levels of $6a^1$ and $6b^2$ (669 cm^{-1}) were excited, no resonance fluorescence or fluorescence from the vibrationally relaxed levels were found, thus suggesting that IVR is slower than the predissociation. The competition between predissociation and IVR occurring in the pyrimidine-Ar complex was also investigated with two-color photoionization spectroscopy (36). Since photoionization is a process induced by the radiation field of the laser used, the photoionization rate is usually much larger than the spontaneous fluorescence decay rate. Therefore, a very fast predissociation process is not detected by the fluorescence but by the ionization. Based on this idea, the ionization threshold spectra of the pyrimidine-Ar complex was observed after pumping the predissociative levels of $6a^1$ and $6b^2$. It was shown that the threshold of the fragment pyrimidine and that of the complex simultaneously appear in the spectra. The result indicates that the predissociation to the pyrimidine fragment occurs prior to IVR.

Mode selectivity in the predissociation of electronically excited vdW complexes is a subject of current study. Systematic studies of selectivity were done by Rice and co-workers (132, 138) for benzene-rare gas complexes. They observed the redistribution of the vibrational energy after predissociation of an S_1 complex. Highly selective dissociation pathways were found to occur to a few particular levels of the fragment among the energetically accessible levels. Very recently, Rice et al (139) reported the branching ratios of variously deuterated benzene-He complexes after predissociation, and suggested different relaxation pathways for the different deuterated complexes. The mode-selective vibrational predissociation of p-difluorobenzene-Ar was shown by Butz et al (135). They found quite different branching ratios for the 6^1 vs the 0^0 levels of the fragment following the excitation of 5^1 and 6^2 levels of the complex that were separated by only 3 cm^{-1}. The mode dependence was qualitatively explained by the Morse oscillator model and by Fermi resonances between the initially prepared levels and the isoenergetic levels involving the vdW mode.

Exciplex formation in jet-cooled vdW complexes has been studied by several groups (140–142). The first observation of exciplex transformation from the vdW complex was reported by Saigusa & Itoh (140) for the complex between 1-cyanonaphthalene and triethylamine. The fluorescence excitation spectrum in the UV region exhibits sharp vibrational structure, which is ascribed to the vdW complex. On the other hand, the dispersed fluorescence spectra obtained by the excitation of various vibronic bands of the complex consist of two components: one is the UV fluorescence originating from the vdW complex and the other is the visible fluorescence from the exciplex. The visible exciplex fluorescence was found to be enhanced as higher vibronic bands were excited. Exciplex formation is induced efficiently with excess energy larger than about 400 cm^{-1}. Very recently Saigusa & Itoh reported the fluorescence decays of the UV and exciplex emissions as a function of excess vibrational energy (143).

Ions of vdW complex may be regarded as model systems of molecular ions in the condensed phase and thus spectroscopic study is of great interest. The precise ionization potential of the benzene-Ar vdW complex was first determined by Fung et al (144) using two-color ionization threshold spectroscopy. The observed reduction in the ionization potential (-171 cm^{-1}) of the complex from that of bare benzene was found to be reproduced by the model calculations (145), taking account of short-range repulsions, dispersive interactions, and charge-induced dipole interactions. Similar studies (146) for the vdW complexes of fluorobenzene with various solvent molecules such as Ar, CCl_4, H_2O, CH_3CN, and $CHCl_3$ showed a correlation between the ionization potential and the polarizability of the solvent molecule. The ionization energies of the p-xylene-$(Ar)_n$ complexes ($n = 1-6$) were also observed (41), and showed a smooth decrease of the ionization potential as n was increased.

Hydrogen-Bonded Complexes

Hydrogen bonding has long been a fascinating subject in chemistry, and has been extensively studied in vapor, solution, and solid phases. The electronic spectra of jet-cooled complexes of phenol with various solvent molecules, including water, alcohols, ethers, and benzene, were observed by Abe et al (147, 148). Large spectral red shifts of the $S_1 \leftarrow S_0$ transitions of these complexes were found to be in good correlation with the known heat of formation of the hydrogen bond in the S_0 state. Low-frequency vibrations of the complexes in the S_0 and S_1 states were assigned to the bending and stretching modes with respect to the hydrogen bond. Conformational isomers of the complexes and complexes involving more than one solvent molecule were also found. The growth of the (1:2) complexes was investigated (149) by studying the dependence of the spectra upon the vapor pressures of the solvents. Heavily solvated phenol with H_2O and clusters of phenol were also studied (150) using mass-selected MPI spectra.

Spectroscopy of jet-cooled hydrogen-bonded complexes is very useful in the study of proton transfer processes. Tomioka et al (151) observed the fluorescence excitation spectrum of jet-cooled tropolone and found doublet structures that they interpreted as due to the proton tunneling between the two equivalent oxygen atoms in the molecule. Complex formation with H_2O or CH_3OH destroys the symmetrical double minimum potential and leads to an asymmetrical potential by hydrogen bonding of the solvent molecule with one of the oxygen atoms. As a result, the tunneling splittings disappear in the complex.

Dimers having two equivalent intermolecular hydrogen bonds exhibit tautomerism due to double proton transfer. This was studied by Fuke et al (152, 153) for the 7-azaindole dimer and the 1-azacarbazole dimer in their excited states. The normal form of the dimer, which is stable in the ground state, exhibits UV fluorescence from its excited state. When vibronic levels involving the low-frequency hydrogen-bond mode (N–H ... N stretching) were excited, Fuke et al found visible fluorescence that originates from the tautomer in the excited state. The results were explained by the acceleration of the double proton transfer due to the excitation of the intermolecular stretching vibrations.

A large reduction in ionization energy was found by two-color ionization threshold spectroscopy (31) for the hydrogen-bonded complexes of phenol with benzene, dioxane, and methanol. For example, the ionization threshold of the (1:1) complex of phenol-dioxane is lower by 5140 cm^{-1} than that of phenol, thus suggesting a large geometrical change of the complex upon ionization. The ionization threshold of the (1:2) complex with two dioxane molecules, on the other hand, was found to be higher than that of the (1:1) complex. Such a nonadditive reduction of the ionization energy is seen commonly in the hydrogen-bonded complexes studied. The characteristic reduction was explained as being caused by the strong ion-dipole attraction in the ion complex resulting from the sudden positive charge upon ionization. A large reduction of the ionization energies was also reported for the complexes of 2-aminopyridine with H_2O, NH_3, dioxane, CH_3OH, and C_2H_5OH (154).

Dissociation of hydrogen-bonded complex ions is of particular interest because of proton transfer through the hydrogen bond. Using mass-selected two-color photoionization spectroscopy, dissociation processes occurring after the photoionization of the complexes of phenol with NR_3 (R = H and CH_3) were investigated by Mikami et al (155, 156). When the complexes are ionized with sufficient excess energy, two dissociation pathways are opened: dissociation into (*a*) phenol ion and neutral NR_3 and (*b*) phenoxyl radical and the protonated fragment ion of HNR_3^+. For the complex of phenol–NH_3, Mikami et al found that the former dissociation occurs in preference to the latter, even though the ionization energy exceeds both dissociation limits. For the complex of phenol-$N(CH_3)_3$, on the other hand, the protonated fragment, $HN(CH_3)_3^+$, was found to be produced efficiently from the excited state of the complex ion generated by further excitation following photoionization. The difference in dissociation processes of the two systems was discussed in terms of the proton affinity of NR_3.

CONCLUDING REMARKS

The application of laser spectroscopy to supersonic jets greatly contributes to deepening our understanding of a bare molecule and its interaction with other molecules. The advantage of supersonic jet laser spectroscopy is its high state selectivity. It is now quite feasible to prepare large polyatomic molecules in a selected level of a low-lying electronic excited state. By using this level as an intermediate state, we easily gain access to both lower and higher states through transitions with laser light of low frequency. The spectra due to the transition from this particular level also become very simple even for large polyatomic molecules. By using these advantages of optical-optical double resonance spectroscopy, the high energy region we can explore with low frequency photons is now extended to the deep vacuum ultraviolet region. Moreover, the quality of information obtained for the high excited states is often much better than that gained by using synchrotron radiation because of the high selectivity. Even for the electronical ground state of a molecule, double resonance spectroscopy provides us with more detailed vibrational structure of the molecule than ordinary vibrational spectroscopy. The use of two laser beams also enables us to study directly the dynamics occurring in the intermediate state by using a suitable time delay between the two laser pulses. Double resonance spectroscopy combined with the supersonic jet technique has become the most useful tool for studying the dynamics and energetics of a molecule in any particular state in a wide energy range. Many more variations are possible in triple resonance spectroscopy using three laser beams. Triple resonance spectroscopy has already been used in several laboratories and will be popular in the near future.

ACKNOWLEDGMENT

We thank T. Amano, T. Aota, A. Haijima, and T. Maeyama for their help in surveying the literature.

Literature Cited

1. Almasy, F., Laemmel, H. 1950. *Helv. Chim.* 33: 2092–2100
2. Levy, D. H., Wharton, L., Smalley, R. E. 1977. Laser spectroscopy in supersonic jets. In *Chemical and Biochemical Applications of Lasers*, ed. C. B. Moore, 2: 1–38. New York: Academic. 288 pp.
3. Levy, D. H. 1980. *Ann. Rev. Phys. Chem.* 31: 197–225
4. Mikami, N. 1980. *Appl. Phys.* 49: 302–12 (In Japanese)
5. Ito, M. 1986. Electronic spectra in a supersonic jet as a means of solving vibrational problems. In *Vibrational Spectra and Structure*, ed. J. R. Durig, 15: 1–55. Amsterdam: Elsevier. 507 pp.
6. Even, U., Jortner, J., Friedman, J. 1982. *J. Phys. Chem.* 86: 2273–76
7. Spangler, L. H., Pratt, D. W. 1986. *J. Chem. Phys.* 84: 4789–96
8. Holtzclaw, K. W., Pratt, D. W. 1986. *J. Chem. Phys.* 84: 4713–14
9. Abe, H., Kamei, S., Mikami, N., Ito, M. 1984. *Chem. Phys. Lett.* 109: 217–20
10. Ohmori, N., Suzuki, T., Ito, M. 1988. *J. Phys. Chem.* 92: 1086–93
11. Kamei, S., Okuyama, K., Abe, H., Mikami, N., Ito, M. 1986. *J. Phys. Chem.* 90: 93–100
12. Kamei, S., Mikami, N., Ito, M. 1986. *J. Phys. Chem.* 90: 2321–23
13. Kamei, S., Sato, T., Mikami, N., Ito, M. 1986. *J. Phys. Chem.* 90: 5615–19
14. Kamei, S., Abe, H., Mikami, N., Ito, M. 1985. *J. Phys. Chem.* 89: 3636–41
15. Suzuki, T., Sato, M., Mikami, N., Ito, M. 1986. *Chem. Phys. Lett.* 127: 292–96
16. Goto, A., Fujii, M., Mikami, N., Ito, M. 1986. *J. Phys. Chem.* 90: 2370–74
17. Amirav, A., Sonnenschein, M., Jortner, J. 1986. *Chem. Phys.* 102: 305–12
18. Amirav, A., Jortner, J. 1986. *J. Chem. Phys.* 84: 1500–7
19. Majors, T. J., Even, U., Jortner, J. 1984. *J. Chem. Phys.* 81: 2330–38
20. Bersohn, R., Even, U., Jortner, J. 1983. *J. Chem. Phys.* 79: 2163–67
21. Bersohn, R., Even, U., Jortner, J. 1984. *J. Chem. Phys.* 80: 1050–58
22. Leopold, D. G., Pendley, R. D., Roebber, J. L., Hemley, R. J., Vaida, V. 1984. *J. Chem. Phys.* 81: 4218–29
23. Leopold, D. G., Vaida, V., Granville, M. F. 1984. *J. Chem. Phys.* 81: 4210–17
24. Ito, M., Fujii, M. 1988. Two-color resonance spectroscopy for the study of

high excited states of molecules. In *Advances in Multiphoton Process and Spectroscopy*, ed. S. H. Lin, Vol. 4. Singapore: World Scientific. In press
25. Williamson, A. D., Compton, R. N. 1979. *Chem. Phys. Lett.* 62: 295–99
26. Ebata, T., Imajo, T., Mikami, N., Ito, M. 1982. *Chem. Phys. Lett.* 89: 45–47
27. Ebata, T., Mikami, N., Ito, M. 1983. *J. Chem. Phys.* 78: 1132–39
28. Fujii, M., Kakinuma, T., Mikami, N., Ito, M. 1986. *Chem. Phys. Lett.* 127: 297–302
29. Duncan, M. A., Dietz, T. G., Smalley, R. E. 1981. *J. Chem. Phys.* 75: 2118–25
30. Gonohe, N., Shimizu, A., Abe, H., Mikami, N., Ito, M. 1984. *Chem. Phys. Lett.* 107: 22–26
31. Gonohe, N., Abe, H., Mikami, N., Ito, M. 1985. *J. Phys. Chem.* 89: 3642–48
32. Smith, M. A., Hager, J. W., Wallace, S. C. 1984. *J. Chem. Phys.* 80: 3097–3105
33. Fuke, K., Yoshiuchi, H., Kaya, K., Achiba, Y., Sato, K., Kimura, K. 1984. *Chem. Phys. Lett.* 108: 179–84
34. Oikawa, A., Abe, H., Mikami, N., Ito, M. 1985. *Chem. Phys. Lett.* 116: 50–54
35. Goto, A., Fujii, M., Ito, M. 1987. *J. Phys. Chem.* 91: 2268–73
36. Mikami, N., Sugahara, Y., Ito, M. 1986. *J. Phys. Chem.* 90: 2080–85
37. Hager, J., Wallace, S. C. 1985. *J. Phys. Chem.* 89: 3833–41
38. Cooper, D. E., Frueholz, R. P., Klimcak, C. M., Wessel, J. E. 1982. *J. Phys. Chem.* 86: 4892–97
39. Hager, J., Ivanco, M., Smith, M. A., Wallace, S. C. 1985. *Chem. Phys. Lett.* 113: 503–7
40. Hager, J., Ivanco, M., Smith, M. A., Wallace, S. C. 1986. *Chem. Phys.* 105: 397–416
41. Dao, P. D., Morgan, S., Castleman, A. W. Jr. 1985. *Chem. Phys. Lett.* 113: 219–24
42. Hager, J., Smith, M. A., Wallace, S. C. 1985. *J. Chem. Phys.* 83: 4820–22
43. Hager, J., Smith, M. A., Wallace, S. C. 1986. *J. Chem. Phys.* 84: 6771–80
44. Fujii, M., Ebata, T., Mikami, N., Ito, M. 1984. *J. Phys. Chem.* 88: 4265–71
45. Fujii, M., Ebata, T., Mikami, N., Ito, M. 1983. *Chem. Phys. Lett.* 101: 578–81
46. Fujii, M., Mikami, N., Ito, M. 1985. *Chem. Phys.* 99: 193–206
47. Fisanick, G. J., Elchelberger, T. S., Robin, M. B., Kuebler, N. A. 1983. *J. Phys. Chem.* 87: 2240–46
48. Suzuki, T., Mikami, N., Ito, M. 1985. *Chem. Phys. Lett.* 120: 333–36
49. Kakinuma, T., Fujii, M., Ito, M. 1987. *Chem. Phys. Lett.* 140: 427–33
50. Goto, A., Fujii, M., Mikami, N., Ito, M. 1985. *Chem. Phys. Lett.* 119: 17–21
51. Goto, A., Fujii, M., Ito, M. 1987. *Chem. Phys. Lett.* 135: 407–12
52. Cooper, D. E., Klimcak, C. M., Wessel, J. E. 1981. *Phys. Rev. Lett.* 46: 324–28
53. Cooper, D. E., Wessel, J. E. 1982. *J. Chem. Phys.* 76: 2155–60
54. Murakami, J., Kaya, K., Ito, M. 1982. *Chem. Phys. Lett.* 91: 401–5
55. Kittrel, C., Abramson, E., Kinsey, J. L., McDonald, S. A., Reisner, D. E., Field, R. W., Katayama, D. H. 1981. *J. Chem. Phys.* 75: 2056–59
56. Vaccaro, P. H., Kinsey, J. L., Field, R. W., Dai, H. L. 1983. *J. Chem. Phys.* 78: 3659–64
57. Reisner, D. E., Field, R. W., Kinsey, J., Dai, H. L. 1984. *J. Chem. Phys.* 80: 5968–78
58. Abramson, E., Field, R. W., Imre, D., Innes, K. K., Kinsey, J. L., 1984. *J. Chem. Phys.* 80: 2298–2300
59. Yamanouchi, K., Yamada, H., Tsuchiya, S. 1986. *Chem. Phys. Lett.* 132: 361–64
60. Lawrance, W. D., Knight, A. E. W. 1982. *J. Chem. Phys.* 77: 570–71
61. Kable, S. H., Knight, A. E. W. 1987. *J. Chem. Phys.* 86: 4709–11
62. Suzuki, T., Mikami, N., Ito, M. 1986. *J. Phys. Chem.* 90: 6431–40
63. Kim, H. L., Reid, S., McDonald, J. D. 1987. *Chem. Phys. Lett.* 139: 525–27
64. Suzuki, T., Hiroi, M., Ito, M. 1988. *J. Phys. Chem.* In press
65. Kaminsky, M. E., Hawkins, R. T., Kowalski, F. V., Schawlow, A. L. 1976. *Phys. Rev. Lett.* 36: 671–73
66. Suzuki, T., Ito, M. 1987. *J. Phys. Chem.* 91: 3537–42
67. Ito, M. 1987. *J. Phys. Chem.* 91: 517–26
68. Okuyama, K., Mikami, N., Ito, M. 1985. *J. Phys. Chem.* 89: 5617–25
69. Mizuno, H., Okuyama, K., Ebata, T., Ito, M. 1987. *J. Phys. Chem.* 91: 5589–93
70. Okuyama, K., Mikami, N., Ito, M. 1987. *Laser Chem.* 7: 197–211
71. Breen, P. J., Warren, J. A., Bernstein, E. R., Selman, J. I. 1987. *J. Chem. Phys.* 87: 1917–26
72. Breen, P. J., Warren, J. A., Bernstein, E. R., Selman, J. I. 1987. *J. Chem. Phys.* 87: 1927–35
73. Parmenter, C. S., Stone, B. M. 1986. *J. Chem. Phys.* 84: 4710–11
74. Moss, D. B., Parmenter, C. S., Ewing, G. E. 1987. *J. Chem. Phys.* 86: 51–61
75. Semmes, D. H., Baskin, J. S., Zewail,

A. H. 1987. *J. Am. Chem. Soc.* 109: 4104–6
76. Scherer, N. F., Shepanski, J. F., Zewail, A. H. 1984. *J. Chem. Phys.* 81: 2181–82
77. Okuyama, K., Hasegawa, T., Ito, M., Mikami, N. 1984. *J. Phys. Chem.* 88: 1711–16
78. Okuyama, K., Kakinuma, T., Fujii, M., Mikami, N., Ito, M. 1986. *J. Phys. Chem.* 90: 3948–52
79. Orlandi, G., Zerbetto, F. 1987. *J. Phys. Chem.* 91: 4238–40
80. Murakami, J., Ito, M., Kaya, K. 1981. *J. Chem. Phys.* 74: 6505–6
81. Yamasaki, K., Arita, K., Kajimoto, O., Hara, K. 1986. *Chem. Phys. Lett.* 123: 277–81
82. Gonohe, N., Yatsuda, N., Mikami, N., Ito, M. 1982. *Bull. Chem. Soc. Jpn.* 55: 2796–2802
83. Oikawa, A., Abe, H., Mikami, N., Ito, M. 1984. *J. Phys. Chem.* 88: 5180–86
84. Ito, M., Oikawa, A. 1985. *J. Mol. Struct.* 126: 133–40
85. Oikawa, A., Abe, H., Mikami, N., Ito, M. 1985. *Chem. Phys. Lett.* 116: 50–54
86. Dunn, T. M., Trembreull, R., Lubman, D. M. 1985. *Chem. Phys. Lett.* 121: 453–57
87. Yamamoto, S., Okuyama, K., Mikami, N., Ito, M. 1986. *Chem. Phys. Lett.* 125: 1–4
88. Berry, R. S. 1966. *J. Chem. Phys.* 45: 1228–45
89. Chewster, L. A., Sander, M., Müller-Dethlefs, K., Schlag, E. W. 1987. *J. Chem. Phys.* 86: 4737–44
90. Knee, J. L., Khundkar, L. R., Zewail, A. H. 1987. *J. Chem. Phys.* 87: 115–27
91. Duncan, M. A., Dietz, T. G., Liverman, M. G., Smalley, R. E. 1981. *J. Phys. Chem.* 85: 7–9
92. Dietz, T. G., Duncan, M. A., Smalley, R. E. 1982. *J. Chem. Phys.* 76: 1227–32
93. Dietz, T. G., Duncan, M. A., Pulu, A. C., Smalley, R. E. 1982. *J. Phys. Chem.* 86: 4026–29
94. Otis, C. E., Knee, J. L., Johnson, P. M. 1983. *J. Chem. Phys.* 78: 2091–92
95. Knee, J. L., Johnson, P. M. 1984. *J. Chem. Phys.* 80: 13–17
96. Löhmannsröben, H. G., Luther, K., Stuke, M. 1987. *J. Phys. Chem.* 91: 3499–3503
97. Hornburger, H., Kono, H., Lin, S. H. 1984. *J. Chem. Phys.* 81: 3554–58
98. Heller, E., Brown, R. C. 1983. *J. Chem. Phys.* 79: 3336–51
99. Obi, K., Matsumi, Y., Takeda, Y., Mayama, S., Watanabe, H., Tsuchiya, S. 1983. *Chem. Phys. Lett.* 95: 520–24
100. Mayama, S., Hiraoka, S., Obi, K. 1984. *J. Chem. Phys.* 80: 7–12
101. Sappey, A. D., Welsshaar, J. C. 1987. *J. Phys. Chem.* 91: 3731–36
102. Chen, P., Colson, S. D., Chupka, W. A., Berson, J. A. 1986. *J. Phys. Chem.* 90: 2319–21
103. Fuke, K., Ozawa, K., Kaya, K. 1986. *Chem. Phys. Lett.* 126: 119–23
104. Foster, S. C., Hsu, Y. C., Damo, C. P., Liu, X., Kung, C. Y., Miller, T. A. 1986. *J. Phys. Chem.* 90: 6766–69
105. Heaven, M., DiMauro, L. F., Miller, T. A. 1983. *Chem. Phys. Lett.* 95: 347–49
106. Lester, M. I., Zegarski, B. R., Miller, T. A. 1983. *J. Phys. Chem.* 87: 5228–33
107. Kennedy, R. A., Miller, T. A., Scharf, B. 1986. *J. Chem. Phys.* 85: 1336–47
108. See, for example, Klapstein, D., Kuhn, R., Maier, J. P. 1984. *Chem. Phys.* 86: 285–93
109. See, for example, Klapstein, D., Kuhn, R., Maier, J. P., Oschsner, M., Wyttenbach, T. 1986. *Chem. Phys.* 101: 133–46
110. Smith, R. S., Anselment, M., DiMauro, L. F., Frye, J. M., Sears, T. J. 1987. *J. Chem. Phys.* 87: 4435–46
111. Haynam, C. A., Brumbaugh, D. V., Levy, D. H. 1984. *J. Chem. Phys.* 80: 2256–64
112. Haynam, C. A., Brumbaugh, D. V., Levy, D. H. 1983. *J. Chem. Phys.* 79: 1581–91
113. Yamanouchi, K., Watanabe, H., Koda, S., Tsuchiya, S., Kuchitsu, K. 1984. *Chem. Phys. Lett.* 107: 290–94
114. Sugahara, Y., Mikami, N., Ito, M. 1986. *J. Phys. Chem.* 90: 5619–22
115. Haynam, C. A., Morter, C., Young, L., Levy, D. H. 1987. *J. Phys. Chem.* 91: 2519–25
116. Haynam, C. A., Morter, C., Young, L., Levy, D. H. 1987. *J. Phys. Chem.* 91: 2526–29
117. Kobayashi, T., Honma, K., Kajimoto, O., Tsuchiya, S. 1987. *J. Chem. Phys.* 86: 1111–16
118. van Herpen, W. M., Meerts, W. L., Dymanus, A. 1987. *J. Chem. Phys.* 87: 182–90
119. Ondrechen, M. J., Berkovitch-Yellin, Z., Jortner, J. 1981. *J. Am. Chem. Soc.* 103: 6586–92
120. Leutwyler, S., Jortner, J. 1987. *J. Phys. Chem.* 91: 5558–68
121. Wanna, J., Menapace, J. A., Bernstein, E. R. 1986. *J. Chem. Phys.* 85: 1795–1805
122. Mangle, E. A., Topp, M. R. 1987. *Chem. Phys.* 112: 427–42

123. Hopkins, J. B., Powers, D. E., Smalley, R. E. 1981. *J. Phys. Chem.* 85: 3739–42
124. Langridge-Smith, P. R. R., Brumbaugh, D. V., Haynam, C. A., Levy, D. H. 1981. *J. Phys. Chem.* 85: 3742–46
125. Janda, K., Hemminger, J. C., Winn, J. S., Novic, S. E., Harris, S. J., Klemperer, W. 1975. *J. Chem. Phys.* 63: 1419–21
126. Pawliszyn, J., Szczęśniak, M. M., Scheiner, S. 1984. *J. Phys. Chem.* 88: 1726–30
127. Law, K. S., Schauer, M., Bernstein, E. R. 1984. *J. Chem. Phys.* 81: 4871–82
128. Börnsen, K. D., Seizle, H. L., Schlag, E. W. 1986. *J. Chem. Phys.* 85: 1726–32
129. Brumbaugh, D. V., Kenny, J. E., Levy, D. H. 1983. *J. Chem. Phys.* 78: 3415–34
130. Young, L., Haynam, C. A., Levy, D. H. 1983. *J. Chem. Phys.* 79: 1592–1604
131. Halberstadt, N., Soep, B. 1984. *J. Chem. Phys.* 80: 2340–51
132. Stephenson, T. A., Rice, S. A. 1984. *J. Chem. Phys.* 81: 1083–1101
133. Jameson, A. K., Forch, B. E., Chen, K. T., Okajima, S., Saigusa, H., Lim, E. C. 1984. *J. Phys. Chem.* 88: 4937–43
134. Abe, H., Ohyanagi, Y., Ichijo, M., Mikami, N., Ito, M. 1985. *J. Phys. Chem.* 89: 3512–21
135. Butz, K. W., Catlett, D. L., Ewing, G. E., Krajnovich, D., Parmenter, C. S. 1986. *J. Phys. Chem.* 90: 3533–41
136. Kobayashi, T., Kajimoto, O. 1987. *J. Chem. Phys.* 86: 1118–24
137. Ewing, G. E. 1979. *J. Chem. Phys.* 71: 3143–44
138. Rice, S. A. 1986. *J. Phys. Chem.* 90: 3063–72
139. Rosman, R. L., Rice, S. A. 1987. *J. Chem. Phys.* 86: 3292–3300
140. Saigusa, H., Itoh, M. 1984. *J. Chem. Phys.* 81: 5692–99
141. Castella, M., Tramer, A., Piuzzi, F. 1986. *Chem. Phys. Lett.* 129: 112–16
142. Anner, O., Haas, Y. 1986. *J. Phys. Chem.* 90: 4298–4302
143. Saigusa, H., Itoh, M., Baba, M., Hanazaki, I. 1987. *J. Chem. Phys.* 86: 2588–96
144. Fung, K. H., Seizle, H. L., Schlag, E. W. 1983. *J. Phys. Chem.* 87: 5113–16
145. Jortner, J., Even, U., Leutwyler, S., Berkovitch-Yellin, Z. 1983. *J. Chem. Phys.* 78: 309–11
146. Gonohe, N., Abe, H., Mikami, N., Ito, M. 1983. *J. Phys. Chem.* 87: 4406–11
147. Abe, H., Mikami, N., Ito, M. 1982. *J. Phys. Chem.* 86: 1766–71
148. Abe, H., Mikami, N., Ito, M., Udagawa, Y. 1982. *J. Phys. Chem.* 86: 2567–69
149. Oikawa, A., Abe, H., Mikami, N., Ito, M. 1983. *J. Phys. Chem.* 87: 5083–90
150. Fuke, K., Kaya, K. 1982. *Chem. Phys. Lett.* 91: 311–14
151. Tomioka, Y., Ito, M., Mikami, N. 1983. *J. Phys. Chem.* 87: 4401–5
152. Fuke, K., Yoshiuchi, H., Kaya, K. 1984. *J. Phys. Chem.* 88: 5840–44
153. Fuke, K., Yabe, T., Chiba, N., Kohida, T., Kaya, K. 1986. *J. Phys. Chem.* 90: 2309–11
154. Hager, J. W., Leach, G. W., Demmer, D. R., Wallace, S. C. 1987. *J. Phys. Chem.* 91: 3750–58
155. Mikami, N., Suzuki, I., Okabe, A. 1987. *J. Phys. Chem.* 91: 5242–47
156. Mikami, N., Okabe, A., Suzuki, I. 1988. *J. Phys. Chem.* 92: 1858–62

RECENT DEVELOPMENTS IN DYNAMICAL THEORIES OF THE LIQUID-GLASS TRANSITION

Glenn H. Fredrickson

AT&T Bell Laboratories, Murray Hill, New Jersey 07974

SCOPE OF THE REVIEW

Due to size constraints, the present review is necessarily limited in its coverage. We restrict consideration to recent theoretical developments in the kinetics of supercooled liquids and glasses and, in particular, to those theories that are in some sense microscopic. Hence, little attention is given to advances in phenomenological theories of glassy relaxation, which have proven to be quite valuable in the prediction of material behavior. Readers interested in such theories are referred to the excellent books by Brawer (1) and Scherer (2), and to the original papers (3–5).

We also provide limited discussion of the many structural or packing models that have been proposed for amorphous solids (6–9). Although short-ranged frustration arising from packing constraints undoubtedly contributes to the sluggish dynamics of glassy systems, these models do not directly address the kinetics of structural relaxation considered here. Some discussion of structural models is given in the recent review article of Jäckle (10) and by several authors in the three New York Academy of Sciences volumes on glasses (11–13).

In the present review we use the term *structural* glass (STG) to refer to conventional amorphous materials with configurational disorder, prepared for example by quenching a liquid. The glass transition phenomena in STGs have some similarities to, but are in many ways distinct from, corresponding phenomena observed in so-called *spin* glasses (25, 26). We focus exclusively on conventional STGs throughout the manuscript.

Finally, the present review gives little attention to computer simulation methods for the investigation of structural relaxation. The extent to which glass transitions observed in molecular dynamics simulations reflect the behavior of laboratory liquids is still unclear. Although there is evidence that the glass transition phenomena obtained under the extremely fast cooling rates in computer experiments are related to those seen in the laboratory (14, 15), more work needs to be done to clarify the relationship between quantities measured in the two types of experiments. Sources of information on this subject include Refs. (1, 14–20).

The organization of the present review is as follows. In the next section we discuss the relaxation phenomena associated with the laboratory glass transition. The discussion focuses on those aspects of structural relaxation that are relatively material-independent. We then describe some of the physical mechanisms that have been proposed to lead to the anomalous kinetics of glassy materials. We attempt to identify a consistent picture of structural relaxation from these mechanisms. A few recent theoretical "models" of STG dynamics are then presented and discussed. We describe their implementation of the physical mechanisms discussed in the preceding section and give a critical analysis of their predictions. Finally, we summarize what has been learned from studying these simple kinetic models and comment on future theoretical developments in STG dynamics.

THE GLASS TRANSITION PHENOMENA

Introduction

Reference to the glassy state usually implies that the material being described has the amorphous structure of a liquid, yet possesses mechanical properties that are more similar to those of a crystalline solid. It is not clear, however, whether glasses belong to a state of matter that is truly distinct from the liquid and crystalline states. Indeed, as we discuss further below, a glass can be viewed as an extremely viscous liquid, one that cannot undergo structural rearrangements or viscous flow on accessible timescales. Moreover, some materials frequently described as glasses are actually believed to be polycrystalline, with grain boundaries delimiting microscopic crystallites (21). Certain metallic glasses also possess limited short-ranged icosohedral order that appears to be related to the long-ranged icosohedral order that characterizes quasicrystals (22).

Glasses are typically prepared by rapid cooling of a molten liquid. The cooling rates required to prevent crystallization, however, vary widely among materials. For example, metallic glasses often require extremely high cooling rates, whereas atactic polymer melts can be cooled arbitrarily slowly with no danger of crystallization. Good glass formers are usually substances that, for some reasons of molecular asymmetry or chemical disorder, either possess no crystalline configurations or have great difficulty in accessing these configurations. While preparation from the liquid state is the norm, glasses can also be produced by sputtering or electrodeposition, or even from crystalline alloys by solid state chemical reactions (23). In the present review, we restrict attention to the liquid route to the glassy state.

The glass transition observed in the laboratory by supercooling a liquid is almost certainly a kinetic phenomenon. As depicted schematically in Figure 1, measurements of a property P (such as the enthalpy or specific volume) under continuous cooling in the glass transition region depend on the cooling rate imposed. Consider a hypothetical experiment in which a supercooled liquid has been prepared at a temperature T_A shown in Figure 1. The solid curve on which point A lies, the *liquidus*, describes the variation of the property P with respect to temperature for the metastable supercooled liquid. Although the supercooled liquid is higher in energy than the crystal (if one exists), the time required for the system to find the crystalline configuration can be very large for a good glass-former. Hence, it is reasonable to think of the supercooled liquid as being in thermo-

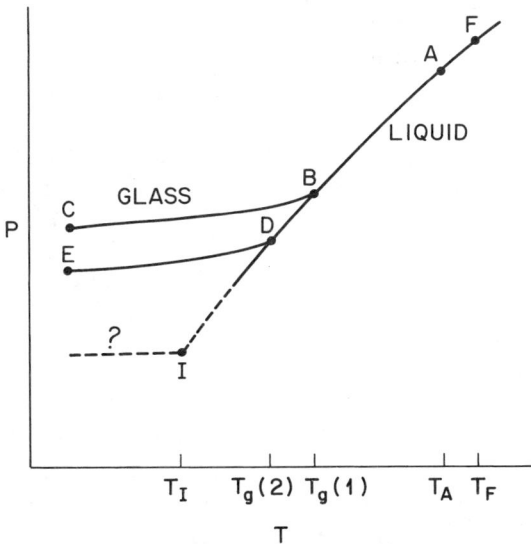

Figure 1 Variation of a thermodynamic property P for a typical supercooled liquid in the glass transition region. The temperature T_F and T_I correspond to the crystallization temperature and the hypothetical ideal glass transition (IGT) temperature, respectively. Laboratory glass transitions occur for different cooling rates at points B and D on the metastable equilibrium liquidus curve.

dynamic equilibrium at points along the liquidus. (Thermodynamic averages along this curve can be computed from a canonical ensemble in which the crystalline configuration has been removed.) Our hypothetical experiment proceeds by cooling the liquid, starting at point A, at a constant rate α_1. If the cooling rate is sufficiently slow, the supercooled liquid will remain in metastable equilibrium and track the liquidus curve along the segment AB. As the temperature of the liquid drops, however, its viscosity and various structural relaxation times increase dramatically (i.e. often faster than Arrhenius growth; this observation is of tantamount importance to glass formation and is discussed in detail in the next section). At some point, which corresponds to B in Figure 1, the fluid becomes so sluggish that it can no longer follow the imposed cooling program. On subsequent cooling the liquid falls out of metastable equilibrium and departs from the liquidus curve, following the path from B to C. The variation of property P along BC is similar to the temperature dependence of P for the crystalline phase and results primarily from local vibrational degrees of freedom, not from large-scale structural rearrangements. Along the path from B to C the fluid still has the amorphous character of a liquid, yet it does not respond with viscous flow on the timescale imposed by our experiment, namely $1/\alpha_1$. Hence, we refer to the substance at point C as a *glass* and the temperature $T_g(1)$, corresponding to the point B where the fluid fell out of equilibrium, as the *glass transition temperature*. Because of the change in slope of property P (for example, enthalpy) at the glass transition, experiments that probe temperature derivatives of P (such as the heat capacity) show rounded discontinuities on crossing $T_g(1)$. Thus, the laboratory glass transition, a nonequilibrium phenomenon, might be naively mistaken for an equilibrium phase transition that is second-order in the Ehrenfest scheme.

Now consider repeating the above experiment, but with a slower cooling rate $\alpha_2 < \alpha_1$. On starting from point A the fluid can follow the slower cooling program to a lower temperature than $T_g(1)$ before falling out of equilibrium. Hence, the liquidus curve is traced out to a new temperature $T_g(2) < T_g(1)$, where a glass transition occurs. Subsequent cooling leads to the glass at point E. The glass at E, while prepared from the same material as the glass at point C, has distinct mechanical and relaxational properties. Actually, each point along BC and along DE represents a distinct glass. However, the microscopic structure of the material changes very little in cooling along these curves, so we frequently refer to BC and DE as single glasses and specify their respective structures by the glass transition temperatures $T_g(1)$ and $T_g(2)$. In this context, when used to characterize the structure of a nonequilibrium glass, the glass transition temperature is referred to as the *fictive temperature*. The fictive temperature is the temperature at which the (metastable) equilibrium supercooled liquid has approximately the same structure as the nonequilibrium glass.

The lesson to be learned from these thought experiments is that the glass transition, as observed in the laboratory, is a kinetic phenomenon that reflects a falling-out of thermodynamic equilibrium. Although the physical mechanisms responsible for the sluggish kinetics of liquids near their glass transition are poorly understood and may involve an underlying phase transition, the restriction to finite experimental timescales mandates that the *laboratory glass transition* (LGT) is a nonequilibrium event. Furthermore, the phenomena described above and depicted in Figure 1 are not restricted to a particular class of materials. Amorphous polymers, network materials, metallic alloys, and certain organic liquids all exhibit qualitativley similar behavior under cooling rates attainable in the laboratory. In principle any material can be vitrified if sufficiently high quench rates can be achieved. Noble gases, for example, have been prepared in high fictive temperature glassy states by using molecular dynamics simulations with extremely rapid quenches (15). (A note to avoid confusion: In the remainder of the paper, when we refer to the temperature of the LGT, we are actually referring to the range of temperatures for which a LGT can be attained under the slowest of laboratory coolings.)

There is a second type of glass transition that cannot be realized in the laboratory, which we shall refer to as an *ideal glass transition* (IGT). An ideal glass transition is easily explained by returning to Figure 1. Suppose that we continue the series of experiments described above and in the nth experiment cool the metastable liquid, starting at point A, at a rate α_n, where $\alpha_n < \alpha_{n-1}$. Furthermore, suppose that in the nth experiment a LGT is observed at a temperature $T_g(n)$. The existence of an IGT can be posed as follows. If the sequence of cooling rates is such that $\lim_{n\to\infty} \alpha_n = 0$, is the corresponding limit $\lim_{n\to\infty} T_g(n) = T_I$ nonzero? If the answer is affirmative, then we say the system has an ideal glass transition at temperature T_I. It follows that one or more structural relaxation times diverge at such a transition. An IGT might correspond to a thermodynamic phase transition, for example a second-order phase transition where certain order parameter relaxation times are singular, or to some sort of purely dynamical phase transition. By a dynamical phase transition, we refer to a transition in which there is broken ergodicity, but there are no singularities in the various thermodynamic functions (24). The existence of an ideal glass transition has not been demonstrated in any real STG, although various theoretical models (one of which is described below) predict thermodynamic or dynamical phase transitions that correspond to an IGT.

Linear Response Properties of Supercooled Liquids

Supercooled liquids in the vicinity of their LGT exhibit a number of unusual kinetic properties. In the present section we focus on the linear dynamical response of STGs, i.e. their relaxation behavior following a very small perturbation from metastable equilibrium. In the linear regime the response of the liquid is independent of the magnitude and sign of the perturbation.

Returning to Figure 1, suppose that an infinitesimal field $H(t')$, conjugate to the property P, is applied to the system at time t'. If the fluid were at metastable equilibrium along AB prior to imposition of the field, then measurements of P at times $t > t'$ determine the linear response function $\phi(t-t')$

$$P(t) = \int_{-\infty}^{t} dt' \, \phi(t-t') H(t'). \qquad 1.$$

We will work in units of P and H such that $\phi(t-t')$ is dimensionless and $\phi(0) = 1$. At temperatures well above the glass transition region, for instance at point A in Figure 1, the linear response function often has simple Debye (single-exponential) behavior

$$\phi(t) = \exp(-t/\tau) \qquad 2.$$

where $\tau = \tau(T)$ is a characteristic structural relaxation time that in the present temperature regime usually follows the simple Arrhenius expression, $\tau \sim \exp(E/k_B T)$. Eq. 2 tends to be more closely obeyed by low molecular weight glass-forming liquids than by polymeric liquids, but this is not crucial to the present discussion.

If the temperature of the hypothetical fluid in Figure 1 is lowered from A along the liquidus curve toward point B, but so slowly as to always remain in metastable equilibrium, the linear response properties change dramatically. It has been established on the basis of a large number of experiments, such as dielectric relaxation, dynamic light scattering, and shallow temperature jump experiments, that the time-dependent structural relaxation of many supercooled liquids near (but above) their LGT is *nonexponential* (11–13). An empirical expression, referred to as the *stretched exponential* or Kohlrausch-Williams-Watts (KWW) function, has been found to describe the results of these linear response experiments remarkably well

$$\phi(t) = \exp[-(t/\tau)^\beta]. \qquad 3.$$

Arguments can be made that suggest Eq. 3 with $\beta < 1$ is not correct in either limit of asymptotically short or long times, but it is physically acceptable in the experimental window of intermediate times.

Fits of Eq. 3 to experimental data on a variety of supercooled liquids indicate that the temperature dependence and magnitude of the parameters τ and β can be used to categorize different materials according to their relaxation behavior. In this regard the classifications of *strong* and *fragile* liquids of Angell (27a–c) are useful. Strong liquids are typically network materials, such as SiO_2 and GeO_2, which possess a three-dimensional structure that is highly resistant to change on application of temperature or pressure. It follows that such materials are also characterized by a very small heat capacity anomaly at the LGT. The linear dynamical response of strong liquids, when described by Eq. 3, usually indicates an Arrhenius temperature dependence of the average relaxation time τ

$$\tau = \tau_0 \exp(E/k_B T) \qquad 4.$$

with a large, but temperature independent, activation energy E. Furthermore, the exponent β for strong liquids typically lies in the range 0.8–1.0 and is weakly temperature dependent. The exponent tends to lie close to unity for the majority of experimentally accessible temperatures and only begins to drop slightly as the LGT is very closely approached. Hence, strong undercooled liquids have linear response properties that are quite similar to simple liquids above the freezing temperature, namely (nearly) single-exponential structural relaxation with Arrhenius temperature dependence of the average relaxation time.

At the other extreme of Angell's classification scheme are the so-called *fragile* liquids (27a–c). These are materials whose short-range structure is susceptible to change in the glass transition region and that typically exhibit a large heat capacity anomaly at the LGT. Examples of substances belonging to the fragile class are toluene and orthoterphenyl. Fragile liquids are characterized by approximately Arrhenius temperature dependence of τ at high temperatures (near T_F in Figure 1), but with a much lower activation energy than that of a strong liquid, and complex *non-Arrhenius* behavior of τ as the temperature is lowered to approach the LGT. The effective activation energy in the latter regime is a monotonically increasing function of decreasing temperature. There is some evidence of a return to Arrhenius temperature dependence with a very high activation energy (i.e. even larger than that of strong liquids) at the lowest temperatures for which equilibrium data can be obtained, but this has not been conclusively demonstrated for a large number of fragile liquids.

Coincident with these changes in the temperature dependence of τ are changes in β. In the high temperature regime fragile liquids exhibit nearly single-exponential structural relaxation, and hence have $\beta \approx 1$. In the transition regime showing non-Arrhenius behavior, β is found to drop from unity to somewhere in the range 0.3–0.5. Finally, at the lowest temperatures, linear response data indicate that β is again only weakly temperature dependent. The observation that β changes with temperature is consistent with the lack of *thermorheological simplicity* for the fragile liquids, namely that the temperature dependence of $\phi(t)$ cannot be described by a simple scaling of the time by τ. (However, over certain narrow temperature ranges, fragile liquids can appear thermorheologically simple.) A decrease of the exponent β when the temperature is lowered can be interpreted as a broadening of the spectrum of relaxation times in the material.

Most glass-forming liquids have linear response properties that fall somewhere between the extremes of fragile and strong behavior, although typically closer to the fragile category. Moreover, the distinction between strong and fragile is often not clear-cut, as are the structural reasons for classification of a given material. Alcohols and amorphous polymers are examples of "intermediate" materials showing many aspects of fragile behavior, but with somewhat weaker temperature dependence of the effective activation energy, $E(T) = \partial \ln \tau / \partial (1/k_B T)$, than for a very fragile liquid such as ortho-terphenyl.

The temperature dependence of structural relaxation times for fragile and intermediate liquids is often described by the Vogel-Tamman-Fulcher (VTF) equation (11–13, 28)

$$\tau = \tau_0 \exp[E_0/(T - T_0)] \qquad 5.$$

where τ_0, E_0, and T_0 are temperature-independent parameters that vary among materials. Equation 5 is fundamentally empirical, but it also emerges from several approximate theories. It predicts an effective activation energy $E(T)$ that increases as the temperature is lowered and that diverges at T_0. Hence, if the VTF equation is a correct description of certain supercooled liquids, then such materials should possess an IGT at T_0. From fits to Eq. 5 the IGT temperature is predicted to lie 10–50 degrees Kelvin below the experimental LGT.

Another expression used to describe the non-Arrhenius temperature dependence of linear response data is the Adam-Gibbs (AG) equation (29)

$$\tau = \tau_1 \exp[E_1/(TS_c)] \qquad 6.$$

where τ_1 and E_1 are new temperature-independent parameters and $S_c = S_c(T)$ is the *configurational entropy* of the supercooled liquid, i.e. the configurational (nonvibrational) part of the excess entropy. The excess entropy is defined as the entropy of the metastable liquid relative to that of the crystal. Equation 6 was derived on the basis of some rather tenuous arguments (29), but has given good agreement with experiment for a number of materials (28).

The AG equation reflects the entropy catastrophy pointed out by Kauzmann (30), now referred to as the *Kauzmann Paradox*. Namely, Kauzmann noted that for a variety of materials, extrapolations of the configurational entropy to temperatures below the LGT predict that S_c vanishes at a positive temperature T_K (the Kauzmann temperature). Such extrapolations suggest that the configurational entropy is negative at temperatures below T_K, implying that the entropy of the amorphous supercooled liquid is less than that of the crystal. It has been suggested (30) that some sort of phase transition, such as an IGT or simply crystallization, intervenes at $T \geq T_K$ to prevent this violation of the Third Law. In the Gibbs-DiMarzio theory (72), for example, a second-order phase transition is predicted at T_K. As one might expect, the Kauzmann Paradox is most evident in those materials classified as fragile.

The Adam-Gibbs theory argues that the smallness of the configurational entropy for a supercooled liquid (at metastable equilibrium) governs its sluggish kinetic behavior and, according to Eq. 6, predicts an IGT at T_K. We note that the AG and VTF equations become identical (with $T_0 = T_K$) if a hyperbolic expression (31) for the temperature dependence of the liquid heat capacity is assumed in computing the configurational entropy. This expression for the heat capacity and the coincidence of the Kauzmann and VTF temperatures are consistent with experimental measurements on several materials (28, 31).

Nonlinear Response Properties of Supercooled Liquids

Supercooled liquids in the vicinity of the LGT also exhibit a number of unusual time-dependent properties in response to large perturbations from equilibrium (1–5). An important aspect of nonlinearity in glass-forming liquids is the sensitivity to both the *sign* and the *magnitude* of the perturbation. As the magnitude of a perturbation is gradually increased, starting in the linear regime, the dynamical response of the fluid begins to depend on the distance of departure from equilibrium. At this point, which is rapidly approached for supercooled liquids, the system can no longer be described by linear response theory. The importance of the sign of the imposed perturbation can be seen by considering the simple example of a temperature jump experiment depicted in Figure 2. Suppose that we prepare a supercooled liquid in metastable equilibrium at a temperature $T_1 = T + \Delta T$. Then at some time, which is chosen to be the origin $t = 0$, we instantaneously drop the temperature by an amount ΔT to T. As illustrated in Figure 2, the property P will relax in a monotonic fashion from its equilibrium value prior to the temperature jump to its new equi-

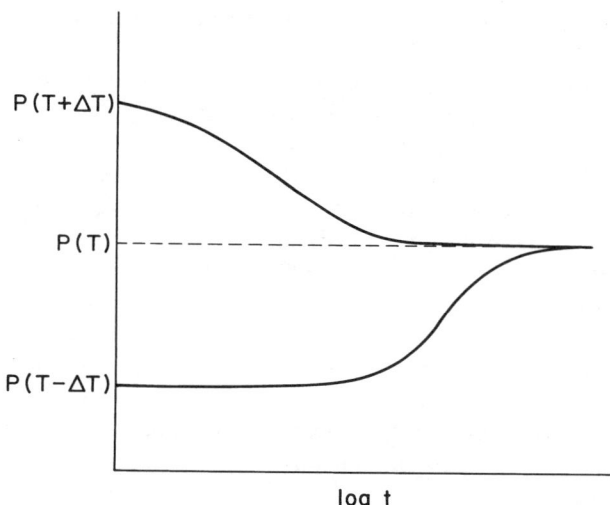

Figure 2 Nonlinear structural relaxation in a supercooled liquid. The *upper curve* describes the evolution of property P following a temperature jump from $T+\Delta T$ to T. The system is assumed to be in metastable equilibrium at $T+\Delta T$ prior to the jump. The *lower curve* describes the variation of P for a second temperature jump experiment also ending at temperature T, but where the temperature increment was opposite in sign. The asymmetry in the shape, as well as the timescale, of the structural relaxation curves is characteristic of nonlinear experiments on many glass-forming liquids.

librium value after the jump, $P(T)$. Now consider repeating the experiment, but starting at a lower temperature $T_2 = T - \Delta T$ and *raising* the temperature by ΔT at $t = 0$. The final temperatures and the magnitudes of the temperature steps in the two experiments are the same, but the jumps of ΔT are opposite in sign. If ΔT is sufficiently large, then the qualitative features of relaxation depicted in Figure 2 are observed. Namely, the relaxation back to equilibrium is highly asymmetric with regard to the sign of the temperature jump. Starting from a lower temperature T_2 in the second experiment, the fluid is much slower to loosen its structure toward configurations characteristic of the temperature T than in the higher temperature experiment starting at T_1.

There are many experimental manifestations of nonlinearity in glass-forming liquids other than the simple temperature jump experiment. For example, experiments in which the temperature is changed continuously at a constant rate sometimes exhibit marked hysteresis on heating and cooling in the glass transition region. Heat capacity measurements often illustrate such hysteresis when a large peak (or overshoot) that is absent on cooling appears near the LGT on heating. Another type of experiment that probes the nonlinear aspects of structural relaxation in supercooled liquids is the so-called "crossover" experiment. Here a fluid is subjected to a first temperature jump, usually negative in sign and to a new temperature below the LGT, but before relaxation is complete a second temperature jump is imposed with opposite sign. The magnitude of the second jump is such that the final temperature is intermediate between the first two. Due to the nonlinearity of the structural response, which in this instance gives rise to a memory effect, the fluid property P is often observed to overshoot its equilibrium value dramatically at the final temperature. Other examples of such experiments are discussed in Refs. (1–3).

Most of the theories that have been proposed to describe the intriguing (and technologically important) features of nonlinear relaxation are phenomenological and hence are not discussed in the present review. The book by Brawer (1) and the article by Moynihan et al (3) provide excellent introductions to this subject. However, the nonlinear response of two microscopic lattice models of supercooled liquids have recently been investigated by computer simulation and are discussed below.

CONCEPTS IN STRUCTURAL GLASS DYNAMICS

The objectives of a microscopic theory of structural relaxation in supercooled liquids are rather self-evident. Namely, one would like to understand how the anomalous relaxation properties described in the previous section emerge as a fluid is cooled or densified. In particular, what physical principles govern the nonexponential and non-Arrhenius behaviors observed in the laboratory? A successful theory should also explain why the qualitative features of these phenomena persist when classes of materials are changed and should address related questions of universality. Ultimately, such a theory should allow predictions to be made for the kinetic properties of new materials.

In principle, a microscopic kinetic theory for supercooled liquids can be constructed by selecting appropriate pair and/or three-body potentials, thus defining a Hamiltonian, and writing classical evolution equations for the molecules of the system. This approach, however, is best suited to the description of gaseous phases, for which Boltzmann-like equations can be derived. At liquid densities, accurate analytical results are much more difficult to obtain (32, 33). Significant progress has been made in developing approximate closure schemes, moment expansion methods, and renormalization techniques to treat the kinetics of dense gases and even high temperature liquids. Such methods, however, are increasingly difficult to justify as liquids are supercooled below their crystallization temperature.

An alternative approach to the kinetic theory description of fluids, and one that has proved invaluable in the theory of critical phenomena (34, 35, 44), is to project out of the full microscopic dynamics a few stochastic equations for a small set of "slow" variables. These are the variables that are expected to control the low-frequency, long-wavelength dynamics of the system. For example, in describing critical dynamics the set of variables usually contains one or more order parameter fields and any conserved variables, such as the energy, mass, or momentum density fields. In contrast to the original Hamiltonian dynamics, the stochastic equations are dissipative and take the form of nonlinear Langevin equations (35). The difficulty with this approach in treating the dynamics of supercooled liquids is that the choice of slow variables is far from obvious. One might retain just the mass and momentum density fields, as was done in one set of studies (36–38). Alternatively, one or more structural order parameter fields could be required (39). In general, the appropriate choice of structural order parameters and conserved fields to describe supercooled liquid dynamics is an important problem that is largely unsolved.

The uncertainty over how to appropriately coarse-grain the kinetic equations describing a supercooled liquid arises, of course, from a more fundamental lack of understanding about the physics involved in structural relaxation. In the first place, it is not known whether fluids possess an underlying IGT, either in the form of a thermodynamic phase transition or some purely dynamical transition. If such a transition could be identified, then by analogy with critical phenomena, it might prove fruitful to search for an appropriate order parameter. Another difficulty is that the role that molecular architecture plays in the glass transition phenomena is not understood. Although many of the features of structural relaxation in high molecular weight polymers, for example, are similar to those observed in simple organic liquids, topological entanglements undoubtedly play a more important role in the dynamics of the former systems. The question arises as to how much molecular detail is required in a theory to describe glassy relaxation without neglecting important physics.

Even though our understanding of the physics of structural relaxation is still quite poor, there are a number of common ingredients among the various theories of supercooled liquids. These ingredients are no more than reasonable hypotheses about the actual physics, but their implementation in simple models has given promising results. Palmer (26) has written a marvelous article in which he discusses physical mechanisms that can lead to sluggish or glassy dynamics. Many of the mechanisms have been identified in spin glasses, but there is evidence that they act in structural glasses as well. The overall picture that emerges is the following. Supercooled liquids, by virtue of their high densities, possess strong constraints on the dynamics of the individual atoms or molecules. A particular molecule is trapped by its neighbors in a "cage," which may persist for long periods of time. To destroy the cage (required for viscous flow), very cooperative dynamical events in the vicinity of the molecule are required. This follows because the neighboring molecules that constitute the cage are themselves caged. The spatial extent over which cooperative rearrangements must occur to relax a cage very likely increases as the fluid is densified. Long times are required for such cooperative rearrangements involving large numbers of molecules. Hence, when describing the events leading to glassy dynamics in Euclidean space (and using the terminology of Ref. 26), we say that *dynamical constraints* at molecular distances lead to a high degree of *cooperativity* in the dynamics at larger distances, and this in turn to anomalously slow structural relaxation.

Many models and theories of STGs, as well as spin glasses, suggest that the physical mechanisms for sluggish dynamics are more transparent in *configuration space*. Namely, one works in a space with dimensionality equal to the number of degrees of freedom in the liquid and studies dynamical evolution (essentially diffusion) of the system among points in the space. As discussed by various authors (26, 40), dynamical constraints among molecules in Euclidean space can lead to a rarification of configuration space. More precisely, dynamical constraints give rise to forbidden pathways between phase points. There is some reason to believe, at least for the case of long-ranged spin glasses, that the rarified configuration space has a hierarchical or ultrametric structure (41). For conventional glasses (or short-ranged spin glasses) it is not clear that such an organized structure is present or even necessary for sluggish dynamics. Because the network of connections is sparse, dynamical evolution proceeds by means of tortuous trajectories in configuration space. This diffusion on a sparsely connected network likely corresponds to the cooperative dynamics in real space described above.

Some authors are also of the opinion that some type of quenched disorder, such as that employed in spin glass models, is involved in the anomalous relaxation properties of supercooled liquids. At temperatures below the LGT one could certainly view the "frozen" slow dynamical modes as quenched and capable of exerting some influence on faster modes (e.g. modes associated with the beta-transition). However, above the LGT it is not apparent that such a separation is legitimate or desirable. Indeed, pure fluids describable by a Hamiltonian with no intrinsic disorder, such as a Lennard-Jones fluid (15), have shown glass-like behavior during molecular dynamics simulations. Furthermore, several of the models to be described in the next section exhibit virtually all of the anomalous kinetic properties characteristic of supercooled liquids just above their

LGT, but have no intrinsic disorder. It is my opinion that quenched randomness plays little or no role in structural glass dynamics.

SIMPLE MODELS

This section discusses a few of the theoretical models that have been recently proposed to describe the dynamics of supercooled liquids. A difficulty with many of the more established theories is that the underlying model is quite complicated and often poorly defined, allowing the investigators freedom to introduce approximations or assumptions that cannot be tested. The trend in recent years has been to invent simpler models, although frequently more abstract, that are precisely specified and that embody some of the physical mechanisms described in the previous section. These simpler models often permit exact analytical solutions, approximate solutions that can be systematically improved, or solutions that can be efficiently obtained by simulation methods. We restrict attention to such models in the present section. This restriction narrows the choice of models considerably, but to further reduce the number to a manageable level we consider only a few illustrative examples.

Generalized Hydrodynamic Models

Some of the most interesting developments in the dynamical theory of the liquid-glass transition have been associated with extensions of fluctuating hydrodynamics or kinetic theories of fluids to dense, supercooled liquids. Great interest in such approaches was stimulated by the paper of Leutheusser (42) and by the independent work of Bengtzelius, Götze & Sjölander (43). On the basis of detailed kinetic theory calculations for the hard sphere fluid, Leutheusser proposed a nonlinear feedback mechanism by which density fluctuations could lead to the ultimate structural arrest of a supercooled liquid. He wrote a simple nonlinear equation for a time correlation function, with a structure suggested by, but not rigorously derived from, self-consistent mode-coupling approximations (32, 34, 44) for dense fluids. The equation can be written in the following form

$$\ddot{C}(t) + \gamma \dot{C}(t) + \Omega^2 C(t) + \Omega^2 \int_0^t dt' M(t-t') C(t') = 0 \qquad 7.$$

where $C(t)$ is a time correlation function, such as the density correlation function at a particular wave number, and satisfies $C(0) = 1$, $\dot{C}(0) = 0$. The constants γ and Ω play the role of a damping coefficient and an oscillator frequency, respectively. Equations similar to Eq. 7 can be derived by projection operator methods (32, 33) for various correlation functions, with formally exact expressions for the "memory function" $M(t)$. Unfortunately, however, such expressions are so complicated as to defy analytical evaluation in all but a few exceptional cases. For the present (difficult) case of a dense fluid, Leutheusser made the self-consistent ansatz that $M(t) \approx 4\lambda C^2(t)$, a type of lowest-order mode-coupling approximation. Here λ is a coupling coefficient (which depends on static correlations in the fluid) that is expected to increase as the temperature is lowered or the fluid is densified. This approximation is suggested by renormalized diagrammatic expansions of coupled-mode theories (34, 44) and by detailed kinetic theory calculations on the hard sphere fluid (46, 47). While the approximation can be justified at asymptotically short times on the basis of high frequency expansions (32), there is no obvious means of establishing its validity at longer times. At present, no small parameter has been identified that would allow a systematic approximation scheme for calculating $M(t)$ in very dense supercooled liquids.

These reservations aside, one can use the above approximation for $M(t)$ in Eq. 7 and analyze the behavior of the resulting nonlinear equation. Leutheusser (42) carried out such an analysis and found the interesting result that the equation predicts a purely dynamical singularity, a type of IGT. This transition is not associated with any divergence in a static susceptibility, but corresponds to a kinetic singularity in which small density perturbations from metastable equilibrium cannot relax in a finite amount of time. At the transition, the long-time limit of $C(t)$ [a spin glass-like order parameter (24)] jumps from zero above the transition to a nonzero value below, the signature of a first-order transition. Furthermore, on approaching the IGT from above, Leutheusser's equation predicts that the average relaxation time of the fluid exhibits a power law singularity of the form $\tau \sim (\lambda_c - \lambda)^{-\mu}$, with $\lambda_c = 1$ the value of the coupling coefficient at the transition and $\mu \approx 1.765$. Because the dependence of the coupling coefficient on the physical parameters of the fluid is expected to be smooth (i.e. there is no thermodynamic phase transition), the model suggests a similar power law singularity for structural relaxation times as the density or temperature is increased or decreased, respectively. Equation 7 also predicts that the time decay of the correlation function is nonexponential over certain time windows, in agreement with the experimental observations for fragile liquids discussed above. If the predicted IGT actually occurs in real fluids, then, due to the divergence of relaxation times at the ideal transition temperature, a LGT would be expected at a slightly higher temperature.

The model proposed by Leutheusser, if one interprets it strictly as an equation for the density correlation function, is somewhat incomplete in the sense that it neglects couplings between density fluctuations with different wave vectors. The approaches by Bengtzelius et al (43) and Kirkpatrick (45) employed essentially the same lowest-order mode-coupling approximation as Leutheusser, but provided a better treatment of the wave vector dependence of the correlation function. The conclusions from both of these studies were that the IGT of Leutheusser persists, albeit with slightly modified exponents and a broader relaxation time spectrum, and that the density fluctuations with wave vectors close in magnitude to the main peak in the static structure factor participate to the largest extent in the feedback mechanism leading to glassy behavior. Various workers (48–50) have extended the theory of Bengtzelius et al (43) and performed detailed comparisons with computer simulations and a few experiments. The results seem to indicate that the simple mode-coupling approximation discussed above does a reasonably good job of describing the dynamics of laboratory and computer fluids up to viscosities of order $10P$. At higher viscosities, substantial discrepancies begin to appear and the analytical theory predicts an IGT at a temperature where the experimental viscosity is only of order $100P$. Although a fluid with a viscosity of $10P$ is far from being glassy [the LGT is sometimes defined (27a–c) as a viscosity of $10^{13}P$], such a fluid is two or three orders of magnitude more viscous than "normal" liquids for which kinetic theory is reasonably successful (32). Furthermore, the present mode-coupling theories seem to identify correctly the temperature range where experiments on fragile or intermediate liquids (51) begin to show a marked qualitative change in the temperature dependence of the viscosity. Hence, I view the developments in Refs. (42, 43, 45–50) as significant progress toward the goal of a kinetic theory for supercooled liquids.

Another interesting development was the suggestion by Das et al (36) that a similar mode-coupling treatment of a set of nonlinear fluctuating hydrodynamic equations could exhibit an IGT of the type found by Leutheusser. This hydrodynamic model is much simpler than the kinetic theory description of a fluid and hence, if capable of accurately describing supercooled liquid dynamics, could allow a much more detailed and systematic investigation of its kinetic properties. The model equations (36–38) consist of the usual continuity equation expressing conservation of mass

$$\frac{\partial \rho}{\partial t} = -\nabla \cdot \mathbf{g} \qquad 8.$$

where $\rho(\mathbf{x})$ is the mass density and $\mathbf{g}(\mathbf{x})$ is the momentum density, and the stochastic momentum conservation equation

$$\frac{\partial g_i}{\partial t} = -\rho \nabla_i \frac{\delta F_u}{\delta \rho} - \sum_j \nabla_j (g_i g_j/\rho) - \sum_j L_{ij}(g_j/\rho) + \theta_i. \qquad 9.$$

Here, $F_u = F_u[\rho]$ is the "potential energy" part of the effective Hamiltonian for the fluid. As discussed by Das & Mazenko (37), the first term on the rhs of Eq. 9 can be interpreted as a generalization of the pressure term in the Navier-Stokes equation to account for spatial fluctuations in mass density. The second term in Eq. 9 is the usual convective term in the Navier-Stokes equation and the third term with

$$L_{ij}(\mathbf{x}) = -\eta_0(\tfrac{1}{3}\nabla_i\nabla_j + \delta_{ij}\nabla^2) - \zeta_0 \nabla_i \nabla_j \qquad 10.$$

is the conventional dissipative term with (bare) shear viscosity η_0 and bulk viscosity ζ_0. The last term, $\theta_i(\mathbf{x}, t)$, is a Gaussian noise representing the collective effect of the faster dynamical variables not explicitly retained in the hydrodynamic description of the fluid. For internal consistency of the statics and dynamics of the model, the covariance of the noise is related to the damping matrix L_{ij} by the usual fluctuation-dissipation theorem (34, 35)

$$\langle \theta_i(\mathbf{x},t)\theta_j(\mathbf{x}',t')\rangle = 2k_B T L_{ij}(\mathbf{x})\delta(\mathbf{x}-\mathbf{x}')\delta(t-t'). \qquad 11.$$

Finally, we note that Eqs. 8 and 9 satisfy the appropriate Poisson bracket relations (35, 37) and include all terms allowed by the symmetries and conservation laws of a simple compressible fluid. However, no structural order parameters are retained as dynamical variables.

To complete the specification of the model it is necessary to have an expression for the potential energy functional $F_u[\rho]$. The original choice studied by the workers in Refs. (36, 37) was quadratic and local

$$F_u[\rho] = \frac{1}{2}\int d\mathbf{x}\, \chi^{-1}(\delta\rho)^2 \qquad 12.$$

where $\delta\rho(\mathbf{x}) = \rho(\mathbf{x}) - \rho_0$, with ρ_0 the average density, and χ is the static susceptibility. If Eq. 12 if rewritten in terms of the Fourier components of $\delta\rho(\mathbf{x})$

$$F_u[\rho] = \frac{1}{2}\int \frac{d\mathbf{q}}{(2\pi)^3}\chi^{-1}\delta\rho(\mathbf{q})\delta\rho(-\mathbf{q}) \qquad 13.$$

then we see that the locality of Eq. 12 implies a *structureless* fluid. This follows because the static structure factor obtained from Eq. 13 is

$$S(\mathbf{q}) = \langle \delta\rho(\mathbf{q})\delta\rho(-\mathbf{q})\rangle = \chi. \qquad 14.$$

Because there is no evidence for a divergent susceptibility in the glass transition region, it is believed that the retention of only quadratic terms in Eq. 12 is sufficient. The kinetic theory results of Bengtzelius et al (43) and Kirkpatrick (45) regarding the importance of density fluctuations with wave numbers near the peak in $S(\mathbf{q})$, however, suggests that the *locality* of Eq. 12 should be relaxed. In a very recent study, *Das* (38) has used a more realistic nonlocal quadratic functional in place of Eq. 12

$$F_u[\rho] = \frac{1}{2}\int d\mathbf{x}(A(\delta\rho)^2 + \kappa[\delta\rho(\mathbf{x})(\nabla^2 + q_0^2)^2\delta\rho(\mathbf{x})]) \qquad 15.$$

that implies a static structure factor of the form

$$S(\mathbf{q}) = \frac{1}{A + \kappa(q^2 - q_0^2)^2}. \qquad 16.$$

Even with the simple quadratic expressions for the effective Hamiltonian given above, substitution into Eq. 9 for $F_u[\rho]$ leads to nonlinear terms in the momentum conservation equation. In the original letter by Das et al (36), it was argued that treatment of these nonlinearities by low-order renormalized perturbation theory gives rise to the same density-driven feedback mechanism proposed by Leutheusser, and hence to an IGT. In the later studies of Das & Mazenko (37) and Das (38), a careful analysis of Eqs 8 and 9 led to the prediction that although the feedback mechanism of Leutheusser is present, other nonlinearities ultimately cut off the singularity and the fluid remains ergodic. Hence, the sharp IGT is lost, but its signature is still felt at intermediate times and a rounded transition is observed.

The perturbation analysis of Eqs. 8 and 9 performed in Refs. (37, 38) is very detailed, yet clearly presented, so I refer the reader to the original papers. However, some discussion of the method is appropriate. Note that the second and third terms on the rhs of Eq. 9 are complicated by factors of $\rho(\mathbf{x})$ (a fluctuating variable) in the denominator. These factors are very inconvenient when formulating a perturbation theory, but they are easily removed by introducing a velocity field $\mathbf{v}(\mathbf{x})$, related to the other fields by

$$\mathbf{g}(\mathbf{x}) = \rho(\mathbf{x})\mathbf{v}(\mathbf{x}). \qquad 17.$$

The dynamics of the fluid are now described by the three fields $\{\rho, \mathbf{g}, \mathbf{v}\}$, related by the nonlinear constraint Eq. 17. Das & Mazenko (37) showed how these dynamics can be described by a Martin-Siggia-Rose (MSR) (52) generating functional and how the nonlinear constraint is easily incorporated into this functional integral description. Having obtained the "action" in the MSR generating functional, the perturbative solution of Eqs. 8 and 9 is a standard problem in field theory, but is very tedious because the appropriate field for the present problem has 14 components! Das & Mazenko carried out the perturbation theory to one-loop order (53) (with all internal propagators renormalized) and found an extra nonhydrodynamic mode that cuts off the IGT singularity predicted in the earlier studies. This nonhydrodynamic mode results from consistent treatment of the nonlinear constraint, Eq. 17, the convective nonlinearity in Eq. 9, and the density nonlinearities arising from the assumed expressions for $F_u[\rho]$. Its physical interpretation, however, is not yet clear.

Because of the overly simplistic effective Hamiltonian chosen by Das & Mazenko, it was not appropriate to compare the predictions of the theory with experiment. Das (38), however, used the more realistic form given in Eq. 15 and reworked the theory. For a Lennard-Jones fluid he fit the structure factor obtained from standard liquid state theories (33) to Eq. 16 and identified A, κ, and q_0 as the best-fit parameters. He then used the approximate theory at one-loop order to calculate the shear viscosity of the fluid and found quite good agreement with the available molecular dynamics data for the viscosity of the Lennard-Jones fluid ($\eta \lesssim 10P$). Other predictions of the theory, such as the time decay of the density correlation function, were found to be in qualitative agreement with both simulations and experiments on fragile liquids.

It is worth noting that subsequent to the work by Das & Mazenko, Götze & Sjögren (54) extended the kinetic theory approach to include mode-coupling of the mass density field to momentum current fluctuations. They find that this modification also removes the IGT singularity, leaving a rounded transition with behavior similar to that calculated previously (48–50) and in reasonable agreement with experiment. The relationship of these calculations to those of Das (38) is not entirely clear, although the predictions of the two approaches are in qualitative agreement.

The fluctuating hydrodynamic model has a number of advantages over the kinetic theory approach. First, it is not restricted to any particular type of fluid, but merely expresses the appropriate conservation laws, symmetries, and Poisson bracket relations of a compressible liquid. Along these lines, one could imagine extending it to a polymeric melt (although other slow variables describing the effects of topological entanglements might be required), whereas the kinetic theory method could be exceedingly difficult for this case. Another advantage is that the equations defining the fluctuating hydrodynamic model are simple and precise. Furthermore, the MSR perturbation theory developed by Das & Mazenko is systematic and could in principle (but with great effort) be extended to higher order.

One drawback of the fluctuating hydrodynamics model is that it requires independent static liquid theory calculations to determine the phenomenological parameters entering the equations. Furthermore, the legitimacy of retaining liquid structure in $F_u[\rho]$ on the scale of q_0, while neglecting wave vector dependence of the bare viscosities on the same scale, is not obvious. In addition, the loop expansion, although systematic, is not a series in a small parameter. The robustness of the one-loop approximation to higher-order corrections is not known. This latter criticism, however, applies equally well to the kinetic theory method. Finally, in both approaches it is not clear whether one or more structural order parameter fields should be included to describe relaxation very close to the LGT.

An interesting calculation along these lines was performed by Sachdev (39) to describe structural relaxation in metallic glasses. Working on the assumption that frustration associated with the inability to achieve icosohedral order dominates the dynamics of these materials, he coupled an icosohedral order parameter field to the momentum density field. Sachdev analyzed the resulting nonlinear hydrodynamic equations by the MSR method and found that short distance sluggishness of the icosohedral order parameter could propagate out to large distance scales and lead to glassy behavior for the renormalized macroscopic viscosity.

Spin Models

Although extensively employed as simple models to study phase transitions and critical phenomena (35, 53), Ising-like spin models can also be constructed to mimic the sluggish dynamics of STGs. I emphasize that in the present section *spin models of structural glasses* are discussed, not *spin glasses*. The spin variables in such models can be given various interpretations, but in all cases they represent some degrees of freedom in the fluid system that are allowed to fluctuate. When used to describe critical dynamics, the low-frequency, long-wavelength dynamics of spin models are not affected by the details of the lattice and the discrete nature of the spin variables. However, the expectation that spatially short-ranged dynamical constraints are involved in the glass transition phenomena suggests that this insensitivity to the underlying lattice will not be obtained for similar models of STGs. This should not deter one from employing such models to describe supercooled liquids, however. Indeed, the dependence of the spin model dynamics on the lattice structure is merely another manifestation of the dependence of the hydrodynamic calculations described above on the details of the static structure factor. While spin models are more abstract than the corresponding hydrodynamic models, they lend themselves to very efficient computer simulation and hence can be investigated (essentially exactly) in great detail.

An interesting class of spin models to describe structural glasses are the so-called n-spin facilitated Ising models (nSFM) introduced by Fredrickson & Andersen (55, 56). These are conventional spin-1/2 Ising models on hypercubic lattices in d-dimensions with the Hamiltonian

$$H(\sigma) = h\sum_i \sigma_i - J\sum_{i<j} \sigma_i \sigma_j \qquad 18.$$

where the first sum goes over the N spins on the lattice and the second is over all nearest-neighbor pairs. The state of the ith spin is denoted σ_i and can be either $+1$ (spin-up) or -1 (spin-down). h is a magnetic field that favors spin-down configurations and J describes the strength of ferromagnetic interactions between spins. In the following I simplify the model further by taking $J = 0$, unless explicitly stated otherwise. One possible interpretation of the spin state up [down] is a "region" of supercooled liquid that has a larger [smaller] than average compressibility or flexibility. The characteristic size of such a region is microscopic and should be of order the static correlation length of the liquid. For positive h the equilibrium population of up-spins, and hence more compressible regions, is expected to decrease as the temperature is lowered.

Equation 18 precisely specifies the thermodynamic properties of the nSFM, but there remains the choice of stochastic dynamics. The conventional choice of dynamics if σ_i is a non-conserved variable is the master equation dynamics proposed by Glauber (57). For a conserved spin variable, the Kawasaki prescription (58) is conventional. Fredrickson & Andersen (55) chose nonconserved dynamics, but selected spin-flip transition probabilities very different from those employed by Glauber. (The use of nonconserved dynamics is reasonable for the interpretation of σ_i given above, but if the spin states are to be interpreted as states of high and low density, for example, then conserved dynamics might be more appropriate.) The master equation that governs the time dependence of the probability distribution function $P(\sigma, t)$ for the nSFM is

$$\frac{\partial}{\partial t} P(\sigma, t) = -\sum_i W_i(\sigma) P(\sigma, t) + \sum_i W_i(\sigma') P(\sigma', t) \qquad 19.$$

where $W_i(\sigma)$ is the probability per unit time of flipping the ith spin in configuration σ, and σ' is the N-spin configuration obtained from σ by flipping the ith spin. The choice of flip rates that define the nSFM are (for $J = 0$ and $n \geq 1$)

$$W_i(\sigma) = \frac{\alpha}{n!} \exp\left[h(\sigma_i - 1)/k_B T\right] \prod_{k=0}^{n-1} [m_i(\sigma) - k] \qquad 20.$$

where $m_i(\sigma)$ is the number of nearest neighbors of the ith spin in configuration σ that are in the spin-up state and α is a constant that sets the time scale of the dynamics. This choice of dynamics satisfies the detailed balance condition, the analog of Eq. 11 for the hydrodynamic model, and thus is consistent with the Ising Hamiltonian in Eq. 18.

Consideration of Eq. 20 indicates that the nSFM has the property that a particular spin has a *nonzero* flip rate only if it has *n or more up-spins* in its immediate environment. The neighboring up-spins "facilitate" the relaxation of the central spin—hence the name n-spin facilitated model. At high temperatures and for small n, the equilibrium population of up-spins is sufficient to allow most spins to flip in their immediate surroundings. Hence, the strong dynamical constraints implied by Eq. 20 are not felt. At low temperatures, however, the population of up-spins is very small (due to the magnetic field) and hence the vast majority of spins cannot flip in their immediate surroundings. Relaxation at such temperatures can only occur by cooperative spin flipping events, possibly involving large numbers of spins. Thus, the nSFMs embody the hypothesized mechanism discussed above in the section on Concepts in Structural Glass Dynamics in which short-ranged dynamical constraints lead to cooperative dynamics on longer distance scales and cause a slow-down in structural relaxation.

Although the flip rates defined in Eq. 20 satisfy detailed balance, it is possible that the dynamical constraints render the Markov process generated by Eq. 19 *reducible*. This would imply that the 2^N element configuration space is divided by the constraints into partitions (of finite measure), with no dynamical transitions connecting the partitions. Nonergodic behavior would necessarily ensue. To ensure that the Markov process is not reducible, I restrict consideration to facilitated models with $1 \leq n \leq d$ for hypercubic lattices in $d \geq 1$ dimensions (coordination number $z = 2d$). Strong arguments, although not proofs, have been made (56) that suggests this subset of the nSFMs are not reducible in the thermodynamic limit, $N \to \infty$. These models, however, are reducible for the finite-sized systems employed in computer simulations, but this has been shown to have trivial consequences on their static and dynamic properties (59).

The choice $n = 1$ in Eq. 20 leads to the one-spin facilitated model (1SFM). The low temperature relaxation properties of this model in any dimension $d \geq 1$ are easily understood in terms of a defect diffusion picture. At such temperatures, the concentration of up-spins is very small and hence a representative equilibrium configuration of the system will contain isolated up-spin "defects" embedded in a "sea" of down-spins. The defects can propagate, however, because any neighbor of a defect spin can flip up, following which the original defect spin can flip down. The net result is that the defect has effectively hopped to a neighboring lattice site. Hence, although the fundamental dynamics described by Eqs. 19–20 are not conserved, the dynamical constraints in the 1SFM lead to defect diffusion (a spin-up–conserving process) as a dominant low-temperature relaxation mechanism. Of course, two defects can collide, annihilating one, and additional daughter defects can be spontaneously created next to an existing defect. These mechanisms ultimately cut off the apparent conserved dynamics, but at times that can become very long as the temperature is lowered. Because isolated defects can diffuse in the 1SFM, it can be argued (55, 56) that the model exhibits relaxation times that obey the Arrhenius expression Eq. 4 at low temperatures, and this behavior is confirmed by Monte Carlo simulations (G. H. Fredrickson, unpublished). Furthermore, the decay of time correlation functions can be fitted to the KWW expression Eq. 3, but with exponents β that are rather large (≈ 0.8). Hence, the 1SFM seems to exhibit relaxation behavior that is qualitatively similar to the behavior of strong laboratory liquids. The 1SFM is obviously related to the defect diffusion models proposed to describe glassy relaxation (61), but in contrast to these models, its thermodynamic properties are well-defined.

Apparently a more cooperative model is required to give fragile behavior. One would anticipate that the dynamical constraints would be more strongly felt in the nSFM as the ratio n/d is increased, although not to exceed unity. Hence, we studied in detail the kinetic properties of one of the most cooperative models, the 2SFM in two dimensions. In the 2SFM, isolated spin-up defects or small clusters of such defects cannot propagate in a sea of down-spins. A much more cooperative relaxation mechanism is required. Although the precise mechanism of relaxation in the 2SFM is not understood, visual observations during Monte Carlo simulations of the model indicate that connected surfaces of up-spins, possibly fractal objects, can move in concert to relax surrounding regions of down-spins. [An argument that relaxation is indeed possible in the 2SFM was given in Ref. (56).] These highly cooperative dynamics were analyzed by a renormalized perturbation theory in Refs. (55, 56) and it was predicted that the 2SFM would exhibit a Leutheusser-like purely dynamical IGT at a nonzero temperature. This predicted IGT, as in the hydrodynamic models, is apparently an artifact of the low-order approximation because subsequent Monte Carlo simulations (59, 60, 63) exhibited no such singularity. The simulations, however, demonstrate that the 2SFM has the kinetic behavior of fragile liquids, namely highly nonexponential decay of time correlation functions and strongly non-Arrhenius temperature dependence of the average relaxation time. Fits of simulation data for the single-spin correlation function

$$\phi(t) = \frac{\langle \sigma_i(t)\sigma_i(0) \rangle - \langle \sigma_i \rangle^2}{\langle \sigma_i(0)^2 \rangle - \langle \sigma_i \rangle^2} \qquad 21.$$

to the KWW expression Eq. 3 led to values of the exponent β that decreased weakly with decreasing temperature and that could be described by the empirical correlation (60)

$$\beta^{-1} = B + Ch/k_B T \qquad 22.$$

with $B = 1.3$ and $C = 2.3$. This proportionality between $1/\beta$ and $1/T$ is in good agreement with one set of data on a polymeric liquid (62), although no effort has been made to compare it with experiments on other fragile systems.

The average relaxation time of the 2SFM in two dimensions, $\bar{\tau}$, defined as the integral of the simulated $\phi(t)$ over $t \in (0, \infty)$, could be fitted to the VTF expression in Eq. 5, but a more remarkable agreement with the Adam-Gibbs (AG) Eq. 6 was found. For $J = 0$ the exact equilibrium entropy that follows from the Ising Hamiltonian, Eq. 18, is

$$S(T) = k_B \{\ln[2\cosh(h/k_B T)] - (h/k_B T)\tanh(h/k_B T)\}. \qquad 23.$$

Using this expression for the configurational entropy that enters Eq. 6, an extremely good fit of the AG equation to the simulation data was obtained (59, 60). In Figure 3 this fit is illustrated for data spanning six decades in relaxation time ($\bar{\tau}$ is expressed in units of α^{-1}). It is important to note that

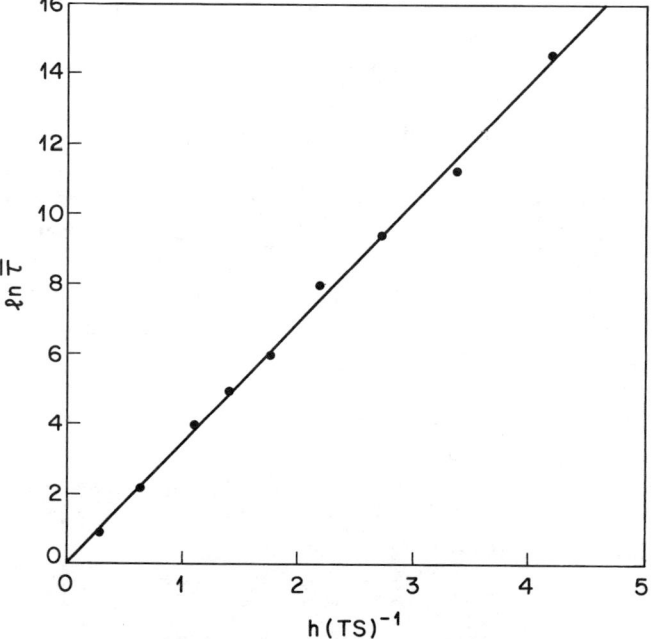

Figure 3 Adam-Gibbs correlation for the 2SFM in two dimensions. The *points* represent Monte Carlo data for the average relaxation time of the model, obtained by methods discussed in Refs. (59, 60). The *solid line* is a linear regression, yielding the relation $\bar{\tau} = 1.0\,\alpha^{-1}\exp[3.4h/TS]$, where $S = S(T)$ is the exact equilibrium entropy given in Eq. 23.

Eq. 23 is a smooth function of temperature and vanishes only at $T = 0$. Hence, the AG equation with this entropy expression is consistent with the absence of an IGT. Because of the curvature of this entropy function, however, linear extrapolations to temperatures below those for which the 2SFM can be conveniently equilibrated would lead to incorrect predictions of a Kauzmann temperature (74).

I have also performed Monte Carlo simulations (unpublished) to investigate the linear response properties of other facilitated Ising models with $J = 0$. The 2SFM on a hypercubic lattice in three dimensions (63; G. H. Fredrickson, unpublished) is less cooperative than in two dimensions and, although the average relaxation time still shows non-Arrhenius behavior describable by the VTF equation, the AG equation is violated for this model. Adam-Gibbs behavior is recovered, however, for the 3SFM in three dimensions. One might naively anticipate that the entire diagonal class of nSFMs in n dimensions, $n \geq 2$, will obey the AG relation, although this has certainly not been proven.

The motivation for choosing $J = 0$ in the Ising Hamiltonian was that it greatly simplifies the equilibrium properties of the nSFM and that, with the interpretation of a spin as a region of fluid with a size on the order of the static correlation length, the states of adjacent spins should be only weakly correlated at equilibrium. There has been one Monte Carlo study (65), however, of the 2SFM in two dimensions with $J \neq 0$ and $h = 0$. This simulation obviously has more to do with dynamic critical phenomena in the presence of strong kinetic constraints than with STG dynamics, but it yielded very interesting and complicated relaxation behavior.

Because of the ease with which the nSFMs can be simulated, it is also possible to investigate their nonlinear dynamical response. Fredrickson (60) performed Monte Carlo simulations of finite temperature jump experiments for the 2SFM in two dimensions. Asymmetry with respect to the sign of the temperature jump, characteristic of nonlinear relaxation in laboratory glasses, was observed for the spin model. Furthermore, the AG-based phenomenological theory that Scherer (5) and Hodge (66) have applied very successfully to nonlinear relaxation data in experimental systems was found to describe the 2SFM accurately (60). For the case of the spin model, however, all the parameters in the phenomenological theory were predetermined on the basis of simulation data for the linear response functions, i.e. the time correlation functions. This allowed a direct comparison between the temperature jump simulations of the 2SFM and the corresponding predictions of the Scherer-Hodge phenomenological theory. Excellent agreement was obtained throughout the temperature range investigated for jumps not exceeding 15% in absolute temperature (60). At the highest temperatures studied, good agreement was found even for jumps as large as 50%!

It is worth commenting that the nSFM has features in common with other spin models that have been proposed to describe STG dynamics. In particular, the nSFM on a Bethe lattice (26) appears to be related to the spin model of Palmer et al (67). The latter model has a hierarchical arrangement of dynamical constraints, and although it is not precisely defined in the sense of having consistent thermodynamics and kinetics, it shows KWW and VTF behavior characteristic of fragile liquids.

The Square Tiling Model

The square tiling model (STM), another lattice model having interesting relaxation properties, was introduced by Stillinger & Weber (68). In the STM, which is two-dimensional, an underlying unit (square) lattice is tiled by squares of various sizes. The tiling of the plane is complete, i.e. without gaps or overlap, and squares of all sizes $n \times n$ ($n = 1, 2, 3, \ldots, L$) are permitted, where $L^2 = N$ is the total area of the system. Periodic boundary conditions are imposed. (See Figure 4 for an illustrative tiling.) The various square domains are to be interpreted as regions of a supercooled liquid that contain only well-coordinated, well-packed molecules or atoms. The boundaries between contiguous square domains represent fluid regions of strained and weakened bonds, i.e. regions of defective packing. As such, there should be an energy penalty for the presence of boundary in the

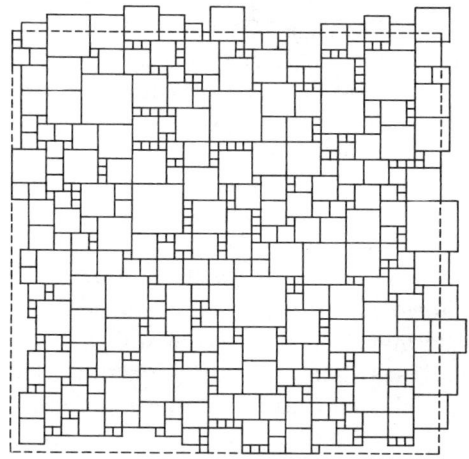

Figure 4 A "glassy" configuration of a 50 × 50 square tiling model, generated by Monte Carlo simulation with the minimal aggregation kinetic rules. The details of the simulation method and the cooling program employed to generate this configuration are discussed in Ref. (69).

interior of the system. Stillinger & Weber proposed the following Hamiltonian to describe the potential energy of a particular tiling of the system

$$H = 2\lambda \sum_{j \geq 1} jn_j \qquad 24.$$

where n_j is the number of $j \times j$ domains in the tiling and λ is a parameter describing the energy associated with a unit length of boundary. Because the total area of the system is fixed, the n_j must satisfy the following constraint

$$\sum_{j \geq 1} j^2 n_j = N. \qquad 25.$$

The Hamiltonian Eq. 24 is interesting in that it leads to a first-order thermodynamic phase transition. On the basis of Flory-like mean field approximations (68), combined high-temperature expansions and Monte Carlo simulations (69, 70), and transfer matrix calculations on semi-infinite strips (71), the phase transition is predicted to occur at $(\lambda/k_B T)_c \approx 0.270$. The transition corresponds to a temperature at which the STM becomes unstable with respect to expulsion of domain walls from the interior. Above the transition temperature, there is a finite concentration of squares of all sizes. At the transition, if equilibrium could be attained, the system would condense into a single domain of size $L \times L$. The system is expected to remain in this state at lower tempertures.

The kinetic properties of the STM were defined by a master equation similar to Eq. 19, again with transitions that satisfy detailed balance. Two kinetic rules were investigated by Monte Carlo simulation. In the first, defining the so-called *minimal aggregation model*, square domains of size $(pq) \times (pq)$ are permitted to fragment into a square cluster of p^2 $q \times q$ domains if and only if p is the smallest prime factor of pq. The inverse aggregation process, with a rate related to the fragmentation rate by a Boltzmann factor consistent with Eq. 24, is also permitted. This restrictive set of kinetic rules leads to sparse connections among configurational states of the STM, although all configurations can be accessed by the dynamics (i.e. the Markov process is irreducible). The relaxation properties of the minimal aggregation model were investigated in Ref. (69), yielding both KWW and non-Arrhenius behaviors. As in the case of the 2SFM, the linear response properties of this version of the STM are similar to those of fragile liquids. However, the AG equation is not satisfied for the minimal aggregation model. Nonlinear phenomena related to those described above in Glass Transition Phenomena were observed under continuous cooling and during temperature jump simulations (69). It was also found for the present choice of kinetic rules that the first-order condensation transition described above corresponds to an IGT. Namely, the model could not be equilibrated at temperatures below the transition temperature. This IGT, however, is somewhat unconventional in that the vast majority of dynamical modes have relaxation times that do not become singular at the transition. Only those associated with transitions to the single-square configuration break ergodicity. A "glassy" configuration of an $L = 50$ system, obtained by slow cooling through the IGT, is shown in Figure 4.

The second choice of kinetic rules, defining the so-called *boundary shift model*, was investigated in Ref. (70). In this model a square domain of size $(p+1) \times (p+1)$ can fragment into a $p \times p$ domain and $2p+1$ domains of unit squares. The inverse process in which an L-shaped arrangement of unit squares is combined with a $p \times p$ domain to form a $(p+1) \times (p+1)$ square is also permitted. Monte Carlo simulations of the boundary shift model indicate that it shows KWW and non-Arrhenius behaviors qualitatively similar to the minimal aggregation model, but in general the boundary shift model relaxes more quickly. The AG equation is again violated for this version of the STM and equilibration is apparently not possible below the condensation temperature.

DISCUSSION AND OUTLOOK

Much progress has been made toward the goal of a kinetic theory of supercooled liquids and the glass transition. We now possess simple models, with static and kinetic properties that are both consistent and precisely defined, that mimic the dynamics of real fluids near their LGT. As a consequence of their simplicity, various quantities can be obtained by analytical means or efficiently generated by numerical or simulation methods. For example, the equilibrium properties of the fluctuating hydrodynamic model and the nSFM (with $J = 0$) can be computed analytically, while those of the STM are provided with limited computational effort by Monte Carlo simulations. The kinetic properties of the latter two models are also easily accessed by simulation. In spite of their simplicity, the models show practically all the characteristic features of structural relaxation in fragile liquids. Namely, nonexponential and non-Arrhenius linear dynamical response, memory effects, and hysteresis. Simpler models, such as the 1SFM, can be constructed to describe relaxation for the (less interesting) case of strong liquids.

All three classes of models discussed in the present review embody some of the physical mechanisms delineated in Concepts in Structural Glass Dynamics, above. Most possess dynamical constraints that are spatially short-ranged. In the hydrodynamic model based on Eq. 15, the constraints are weak and arise from nonlinear interactions of density fluctuations on the distance scale $2\pi/q_0$, i.e. of order the molecular size. In the nSFM the dynamical constraints are explicitly incorporated in the choice of transition probabilities and reflect the ability (or inability) of a spin to change its state in the local environment of surrounding spins. The distance scale of the constraints is that of the lattice spacing, which in our previous interpretation of the spin model corresponds to a static correlation length. Hence, the dynamical constraints built into the nSFM are stronger than those of the hydrodynamic model, but they act over distances that are comparable. In contrast, the minimal aggregation version of the STM possesses arbitrarily long-ranged dynamical constraints due to restrictions on how very large, prime number-sized squares can form and dissociate. This artifact is not present in the boundary shift STM, where the constraints are again confined to a lattice spacing.

Consistent with the picture of structural relaxation embraced in the Concepts section above, the short-ranged constraints in the various models lead to dynamical cooperativity at larger wavelengths. In the 2SFM, for example, isolated clusters of up-spins that are present at low temperatures are, as a consequence of the constraints, trapped and unable to relax for long periods of time. Such clusters must await a very cooperative long-wavelength event involving passage of a connected "surface" of up-spins in order to relax. Similar cooperativity is present in both versions of the STM. As the condensation temperature is approached from above, several high-energy defect structures can be identified that require long times and cooperative rearrangements of large regions of the system to relax (69, 70). Presumably the hydrodynamic models also possess cooperativity in their long-wavelength dynamics, which is fed via nonlinear interactions among shorter-wavelength density modes. Unfortunately, no simulations are yet available for these models to aid in visualization of the relaxation mechanism. Such simulations, when coupled with graphics devices capable of delineating fluid regions of varying density or packing, would be useful in studying real-space cooperativity as well as for testing the analytical predictions of Refs. (37, 38).

It is of interest to compare the STM with the other two types of models considered here. The STM, by construction, has an IGT that coincides with the first-order condensation transition. Neither the 2SFM nor the hydrodynamic model based on Eq. 15 possess such an underlying thermodynamic phase transition. Moreover, the simulation results of Refs. (59, 60) for the 2SFM and the refined calculations of Das & Mazenko (37, 38) for the hydrodynamic model show no evidence for a purely kinetic IGT. Thus, an underlying IGT does not seem to be required for a model to describe laboratory glass transition phenomena. Indeed, Stillinger (73) has recently given strong arguments against the presence of an IGT for substances with moderate molecular weight and conventional intermolecular interactions. These arguments, however, do not apply to the (hypothetical) case of a melt of infinite molecular weight polymers. Such a system could conceivably exhibit an IGT of the type predicted by Gibbs & DiMarzio (72).

An important question that remains unanswered involves the relationships among the kinetic theory, hydrodynamic, spin, and square models. Clearly the hydrodynamic model considered by Das (38), for example, is less abstract and more straightforward in interpretation than the 2SFM. Unfortunately, our present (rather crude) analytical methods only permit accurate approximations for the kinetic properties of the hydrodynamic model to viscosities of order 10^2–$10^3 P$—orders of magnitude less than those that identify the laboratory glass transition region. Furthermore, numerical simulations of stochastic hydrodynamic equations are difficult and computationally expensive. In contrast, the 2SFM can be conveniently studied via Monte Carlo simulation over many decades in time. Hence, it would be advantageous if one could make a formal connection between the hydrodynamic model and a spin model, such as the 2SFM. Undoubtedly the 2SFM is not the proper spin model, as is does not possess the same symmetries and conservation laws as the fluid model described by Eqs. 8–11. However, one might expect that a spin model could be constructed, with some of the same features as the facilitated models, that has long-wavelength, low-frequency dynamics similar to those of the hydrodynamic model. That is, on appropriate coarse-graining, both models should assume the same field-theoretic form. How to identify such a spin model or carry out the coarse-graining procedure is not clear at present.

Now that we have simple models that exhibit glass transition phenomena, the next step is to try to understand the role that the chemical details of a particular substance play in distinguishing its relaxation behavior from that of another material. This is probably best carried out by starting with a more fundamental kinetic description in which the molecular details can be incorporated in a convincing manner. These models could then be coarse-grained by projection operator methods into simpler hydrodynamic models of the type described above and the material-dependence of the resulting phenomenological parameters investigated. Molecular dynamics simulations of the starting detailed models might be helpful in identifying slow, collective variables, such as structural order parameters, to be retained in the coarse-graining process. Because the existence of an underlying IGT appears unlikely, strict universality in the sense of critical phenomena is not to be expected. However, the experimental evidence that diverse classes of materials exhibit qualitatively similar glass transition behavior suggests that some sort of "weak" universality may be present. One might hope that such limited material-dependence could emerge merely as parametric variations in simple models like those of the previous section.

ACKNOWLEDGMENTS

I am indebted to my collaborators H. C. Andersen and S. A. Brawer for introducing me to the field of structural glass dynamics and for numerous invaluable discussions. In addition, I have benefited from discussions and/or collaborations with C. A. Angell, T. A. Weber, F. H. Stillinger, E. Helfand, S. Matsuoka, J. Weeks, and S. Yip.

Literature Cited

1. Brawer, S. 1985. *Relaxaton in Viscous Liquids and Glasses*. Columbus, Ohio: Am. Ceramic Soc.
2. Scherer, G. W. 1986. *Relaxation in Glass and Composites*. New York: Wiley
3. Moynihan, C. T., Macedo, P. B., Montrose, C. J., Gupta, P. K., DeBolt, M. A., et al. 1976. See Ref. 11, pp. 15–35
4. Kovacs, A. J. 1981. See Ref. 12, pp. 38–64
5. Scherer, G. W. 1984. *J. Am. Ceram. Soc.* 67: 504; 69: 374
6. Bernal, J. D. 1964. *Proc. R. Soc. London Ser. A* 280: 299
7. Frank, F. C. 1952. *Proc. R. Soc. London Ser. A* 215: 43
8a. Nelson, D. R. 1983. *Phys. Rev. B* 28: 5515
8b. Sachdev, S., Nelson, D. R. 1985. *Phys. Rev. B* 32: 1480
9. Phillips, J. C. 1979. *J. Non-Cryst. Solids* 34: 153
10. Jäckle, J. 1986. *Rep. Prog. Phys.* 49: 171
11. Goldstein, M., Simha, R., eds. 1976. *Ann. NY Acad. Sci.: The Glass Transition and the Nature of the Glassy State*, Vol. 279. New York: NY Acad. Sci.
12. O'Reilly, J. M., Goldstein, M., eds. 1981. *Ann. NY Acad. Sci.: Structure and Mobility in Molecular and Atomic Glasses*, Vol. 371 New York: NY Acad. Sci.
13. Angell, C. A., Goldstein, M., ed. 1986. *Ann. NY Acad. Sci.: Dynamic Aspects of Structural Change in Liquids and Glasses*, Vol. 484 New York: NY Acad. Sci.
14. Angell, C. A., Torell, L. M. 1983. *J. Chem. Phys.* 78: 937
15. Fox, J. R., Andersen, H. C. 1984. *J. Phys. Chem.* 88: 4019
16. Stillinger, F. H., Weber, T. A. 1979. *J. Chem. Phys.* 70: 4879
17. McTague, J. P., Mandell, M. J., Rahman, A. 1978. *J. Chem. Phys.* 68: 1876
18. Abraham, F. F. 1980. *J. Chem. Phys.* 72: 359
19. Woodcock, L. V. 1981. See Ref. 12, pp. 274–98
20. Brawer, S. 1980. *J. Chem. Phys.* 72: 4264
21. Phillips, J. C. 1986. See Ref. 13, pp. 271–86
22. Nelson, D. R. 1986. See Ref. 13, pp. 264–70
23. Johnson, W. L. 1986. See Ref. 13, pp. 13–25
24. Palmer, R. G. 1982. *Adv. Phys.* 31: 669
25. Chowdhury, D. 1986. *Spin Glasses and Other Frustrated Systems*. Princeton, NJ: Princeton Univ. Press
26. Palmer, R. G. 1986. *Lect. Notes Physics: Heidelberg Colloq. Glassy Dynam.* 275: 275–86
27a. Angell, C. A., Dworkin, A., Figuiere, P., Fuchs, A., Szwarc, H. 1985. *J. Chim. Phys.* 82: 773
27b. Angell, C. A. 1985. *J. Non-Cryst. Solids* 73: 1
27c. Martin, S. W., Angell, C. A. 1986. *J. Phys. Chem.* 90: 6736
28. Angell, C. A., Smith, D. L. 1982. *J. Phys. Chem.* 86: 3845
29. Adam, G., Gibbs, J. H. 1965. *J. Chem. Phys.* 43: 139
30. Kauzmann, W. 1948. *Chem. Rev.* 43: 219
31. Angell, C. A., Sichina, W. 1976. See Ref. 11, pp. 53–67
32. Boon, J. P., Yip, S. 1980. *Molecular*

33. Hansen, J. P., McDonald, I. R. 1976. *Theory of Simple Liquids*. New York: Academic
34. Kawasaki, K. 1970. *Ann. Phys.* 61: 1
35. Hohenberg, P. C., Halperin, B. I. 1977. *Rev. Mod. Phys.* 49: 435
36. Das, S. P., Mazenko, G. F., Ramaswamy, S., Toner, J. 1985. *Phys. Rev. Lett.* 54: 118
37. Das, S. P., Mazenko, G. F. 1986. *Phys. Rev. A* 34: 2265
38. Das, S. P. 1987. *Phys. Rev. A* 36: 211
39. Sachdev, S. 1986. *Phys. Rev. B* 33: 6395
40. Stillinger, F. H. 1984. *J. Phys. Chem.* 88: 6494
41. Rammal, R., Toulouse, G., Virasoro, M. A. 1986. *Rev. Mod. Phys.* 58: 765
42. Leutheusser, E. 1984. *Phys. Rev. A* 29: 2765
43. Bengtzelius, U., Götze, W., Sjölander, A. 1984. *J. Phys. C* 17: 5915
44. Keyes, T. 1977. *Statistical Mechanics Part B: Time-Dependent Processes*, ed. B. J. Berne, pp. 259–309. New York: Plenum
45. Kirkpatrick, T. R. 1985. *Phys. Rev. A* 31: 939
46. Leutheusser, E. 1982. *J. Phys. C* 15: 2801
47. Leutheusser, E. 1982. *J. Phys. C* 15: 2827
48. De Raedt, H., Götze, W. 1986. *J. Phys. C* 19: 2607
49. Götze, W., Sjögren, L. 1987. *J. Phys. C* 20: 879
50. Bengtzelius, U., Sjölander, A. 1986. See Ref. 13, pp. 229–40
51. Taborek, P., Kleiman, R. N., Bishop, D. J. 1986. *Phys. Rev. B* 34: 1835
52. Martin, P. C., Siggia, E. D., Rose, H. A. 1973. *Phys. Rev. A* 8: 423
53. Amit, D. J. 1984. *Field Theory, the Renormalization Group, and Critical Phenomena*. New York: World
54. Götze, W., Sjögren, L. 1987. *Z. Phys. B* 65: 415
55. Fredrickson, G. H., Andersen, H. C. 1984. *Phys. Rev. Lett.* 53: 1244
56. Fredrickson, G. H., Andersen, H. C. 1985. *J. Chem. Phys.* 83: 5822
57. Glauber, R. J. 1963. *J. Math. Phys.* 4: 294
58. Kawasaki, K. 1966. *Phys. Rev.* 145: 224
59. Fredrickson, G. H., Brawer, S. A. 1986. *J. Chem. Phys.* 84: 3351
60. Fredrickson, G. H. 1986. See Ref. 13, pp. 185–205
61. Shlesinger, M. F., Montroll, E. W. 1984. *Proc. Natl. Acad. Sci. USA* 81: 1280
62. Matsuoka, S. 1985. *Polym. J. Jpn.* 17: 321
63. Leutheusser, E., De Raedt, H. 1986. *Solid State Commun.* 57: 457
64. Deleted in proof
65. Nakanishi, H., Takano, H. 1986. *Phys. Lett. A* 118: 415
66. Hodge, I. M. 1986. *Macromolecules* 19: 936
67. Palmer, R. G., Stein, D. L., Abrahams, E., Anderson, P. W. 1984. *Phys. Rev. Lett.* 53: 958
68. Stillinger, F. H., Weber, T. A. 1986. See Ref. 13, pp. 1–12
69. Weber, T. A., Fredrickson, G. H., Stillinger, F. H. 1986. *Phys. Rev. B* 34: 7641
70. Weber, T. A., Stillinger, F. H. 1987. *Phys. Rev. B* 36: 7043
71. Bhattacharjee, S. M., Helfand, E. 1987. *Phys. Rev. A* 36: 3332
72. Gibbs, J. H., DiMarzio, E. A. 1958. *J. Chem. Phys.* 28: 373
73. Stillinger, F. H. 1987. Preprint
74. Matsuoka, S., Fredrickson, G. H., Johnson, G. E. 1985. *Lect. Notes Physics: Molec. Dynam. Relax. Phenomena in Glasses* 277: 188–202

AB INITIO STUDIES OF TRANSITION METAL SYSTEMS[1]

Stephen R. Langhoff and Charles W. Bauschlicher, Jr.

NASA Ames Research Center, Moffett Field, California 94035

INTRODUCTION

In the last 20 years ab initio electronic structure calculations on first-row molecules have progressed to a level of accuracy that is capable of providing new insights into both electronic structure and chemistry. In contrast, theoretical calculations on molecules containing transition metal atoms have proven to be much more difficult, although recent advances in methodology and the availability of fast vector processors have now made it possible to also obtain quantitative information for many molecular systems containing transition metals. Nonetheless, the quantitative description of multiple metal-metal bonding, such as the well-known problem associated with calculating the dissociation energy and bond length of the $^1\Sigma_g^+$ state of Cr_2, remain a challenge to theory. Because of the many important materials and catalytic applications of the transition metals, there is considerable incentive to develop methods capable not only of accurately describing small systems, but also for modeling larger systems. In this review we discuss many of the features of molecular systems containing transition metals that arise from the participation of the *d* electrons. We present applications to selected transition metal hydrides, halides, oxides, dimers and trimers. Finally, we discuss recent work on the $Ni+H_2$ system as a model for the dissociation of H_2 on a metal surface.

UNIQUE FEATURES OF THE TRANSITION METALS

In the transition metal atoms, the nd, $(n+1)s$, and $(n+1)p$ orbitals are similar in both spatial extent and energy. This gives rise to many low-lying

---

[1] The US Government has the right to retain a nonexclusive, royalty-free license in and to any copyright covering this paper.

atomic states (1), and, in turn, results in very complicated spectra in many transition-metal-containing molecules. From an experimental viewpoint this complexity is frequently exacerbated by states of high multiplicity and different isotopes with large nuclear spins and magnetic moments. Transition metal diatomics therefore also represent a challenge to experimentalists. Another consequence of many low-lying states with different occupations of the nd, $(n+1)s$, and $(n+1)p$ orbitals is the variety of bonding mechanisms that occur. Consider, for example, Ni atom where the lowest state, $^3F_4(3d^84s^2)$, is nearly degenerate with the lowest excited state, $^3D_3(3d^94s^1)$. The bonding in the $^2\Delta$ ground state of the NiH molecule includes a contribution from both asymptotes: the ^3F state contributes by first forming $4s4p$ hybrid orbitals, whereas the ^3D state forms directly a $4s$–$1s$ bond (2). The bonding in NiH$_2$ derives primarily from the ^3F state of Ni, which first forms $4s4p$ hybrid orbitals (3). Each hybrid then bonds to hydrogen, yielding a linear molecule with a triplet ground state. In NiCO, the bonding arises from a mixture of the $^1D(3d^94s^1)$ and $^1S(3d^{10})$ states of Ni (4, 5), whereas the Ni atom in Ni(CO)$_4$ has essentially the $3d^{10}$ configuration (6). The similar spatial extent of the valence orbitals is responsible for the fact that Ni can $4s$–$4p$ hybridize in NiH$_n$ and $3d$–$4s$ hybridize in NiCO.

To obtain an accurate description of molecular bonding that arises from a mixture of two atomic asymptotes, it is necessary to account quantitatively for the relative separation between the atomic states. The $3d^n4s^2$–$3d^{n+1}4s^1$ separation varies greatly across the first transition row (1). As the $3d$ orbital becomes more stable relative to the $4s$ orbital with increasing nuclear charge, there is a general decrease in the $3d^n4s^2 \rightarrow 3d^{n+1}4s^1$ and $3d^n4s^2 \rightarrow 3d^{n+2}$ excitation energies from left to right in the row. The large loss in exchange energy when the d orbitals are paired results in a discontinuity in the separation for half-filled shells. For example, in Cr atom the $^7S(3d^54s^1)$ state is 1.00 eV lower than the $^5D(3d^44s^2)$ state, while in Mn atom the $^6D(3d^64s^1)$ state is 2.14 eV higher than the $^6S(3d^54s^2)$ state (1). In the second transition row, the more diffuse $4d$ orbital, combined with the increasing stability of the $4d$ with respect to the $5s$ orbital, leads to smaller $4d^n5s^2 \rightarrow 4d^{n+1}5s^1$ and $4d^n5s^2 \rightarrow 4d^{n+2}$ excitation energies, and a smaller discontinuity when the d shell is half-filled. It is the extra stability of the $4d$ orbital that leads to a $^1S(4d^{10})$ ground state for Pd atom, whereas the corresponding state lies 1.74 eV above the $^3F_4(3d^84s^2)$ ground state of Ni atom. In the third transition row, relativistic effects stabilize the $6s$ orbital relative to $5d$, resulting in a $^3D(5d^96s^1)$ ground state for Pt atom (7). Thus the relative energy separations between the low-lying atomic states varies greatly down the column Ni, Pd, and Pt, resulting in a large variation in the chemical and material properties of these metals.

Since the nd orbital contracts more rapidly than the $(n+1)s$ orbital with increasing nuclear charge, the propensity for s–d hybridization decreases across the row. For the first transition row, the ratio of the radial extent of the $4s$ to $3d$ orbital varies from about 2.0 in Sc to 3.4 in Cu (8). The bonding in the $^1\Sigma^+$ ground state in ScH involves s–d hybridization (9, 10). The $4d$ and $5s$ orbitals are more similar in spatial extent, varying from about 1.6 in Y to 2.7 in Ag. As a consequence, the ground states of both YH and ZrH involve s–d hybridization of the metal (11).

The very large d–d exchange energy and the inherent weakness of metal-metal d–d bonds (owing to their small overlap), distinguish bonding involving transition metals from carbon chemistry. For example, the overlap of the $3d\pi$–$3d\pi$ orbitals in Cr_2 is only one third that of the C–C $2p\pi$–$2p\pi$ orbitals in C_2H_2, and the $3d\delta$–$3d\delta$ overlap is ten times smaller still (12). As bonding results in a loss of d–d exchange energy, the bond energy of Cr_2 is quite small (≈ 1.8 eV) (13), despite being nominally a hextuple bond. Even the $4s$ orbital contribution is small as there is little net $4s$–$4s$ bonding at the optimal $3d$–$3d$ bond length. Thus, a high level of theory is required to describe the delicate balance between weak, low-overlap bonding and large $3d$–$3d$ exchange. In contrast, low levels of theory often give a qualitative description of the bonding for first-row chemistry.

Another feature of transition metal atoms is their ability to form clusters ranging in size from two atoms to bulk. Recent experiments on small metal clusters have shown a large, non-monotonic variation in the rate constant for dissociation of H_2 with cluster size; in Fe clusters with up to 25 atoms the rate constant varies by four orders of magnitude (14). Thus the chemistry of transition metals can be greatly modified by changing the cluster size, adding ligands, or by alloying or doping the surface. Although some of these effects are chemical in nature, experiments on neutral and singly positively charged Nb clusters suggest that geometry is one of the critical features in determining the reaction rate (15). The properties of transition metals can also be markedly changed by adding ligands, as manifested by transition metals in enzymes where steric effects can give great specificity to chemical reactions.

Theory is contributing to an improved understanding by elucidating the spectroscopy and qualitative features of the bonding in transition-metal diatomics, thereby giving some insight into the varied chemistry of these systems. Also, the study of clusters and model studies of reactions on these clusters provides insight into such processes as chemisorption and catalysis.

METHODS

In this section we give an overview of the ab initio methods used to study transition metals, with emphasis on recent developments. In all cases we solve the nonrelativistic Schrödinger equation in the Born-Oppenheimer approximation. Our first approximation is then to assume that relativistic effects can either be neglected or included later as a perturbation within the L-S coupling scheme (16, 17). This limits the discussion to the first two transition metal rows where these approximations are valid. In the ab initio molecular orbital approach a double basis set expansion technique is used to solve (approximately) the Schrödinger equation. In the studies presented here the molecular orbitals are expanded in terms of a linear combination of Gaussian-type functions (the one-particle basis). Electron correlation is included with the configuration-interaction (CI) method (18, 19), where the wave function is expanded as a linear combination of antisymmetrized products (the n-particle basis) of molecular orbitals. With the availability of high-speed vector processors with large memories (such as the CRAY 2), it has recently become possible to carry out full configuration-interaction (FCI) calculations in realistic 1-particle basis sets (20–23). This has given new insight into approximate methods of including electron correlation as well as a means of calibrating these methods for specific problems. FCI calculations for a large number of properties have indicated that in most cases the largest remaining errors are in the 1-particle basis sets. It is thus appropriate to first discuss the 1-particle basis sets that have been developed for transition metals.

For the first transition row the most widely used primitive Gaussian basis is the $(14s\,9p\,5d)$ set optimized by Wachters (24) for the $3d^n4s^2$ occupation, supplemented with his two optimized p functions to describe the $4p$ orbital. In addition, to describe the $3d^{n+1}4s^1$ states accurately where the $3d$ orbital is more diffuse, it is necessary to add an additional $3d$ function; these functions have been optimized for the first transition row by Hay (25). The Wachters basis set supplemented with the Hay $3d$ yields a $3d^n4s^2$–$3d^{n+1}4s^1$ separation in good agreement with the numerical Hartree-Fock (NHF) result (26). Without the diffuse $3d$ function in the basis set, the $(^3D)3d^94s^1$–$(^3F)3d^84s^2$ separation for Ni is in error by more than 1 eV, and large basis set superposition errors may occur in molecular calculations (see discussion in Ref. 4). The Wachters basis set, supplemented with a diffuse $3d$ and two $4p$ functions, contains $14s$, $11p$, and $6d$ primitive functions. Such a basis set is most often contracted in a segmented fashion to give basis sets ranging in size from $[8s\,6p\,4d]$ to $[5s\,4p\,4d]$. These basis sets are of about equal quality to the $(9s\,5p)/[4s\,2p]$ treatment of first-row atoms.

Very recently larger primitive sets have been optimized for the first transition row (27). For example, H. Partridge (in preparation) has optimized $(20s\,12p\,9d)$ sets, which provides a triple-zeta description of the $4s$ orbital. When supplemented with three p functions to describe the $4p$ orbital and a diffuse d function to improve the description of the states arising from the $3d^{n+1}4s^1$ occupation, these new large primitive sets are approximately equivalent to the $(13s\,8p)$ sets for first-row atoms. To provide for polarization it is then necessary to add f and g functions to the transition metal basis sets, in analogy with adding d and f functions for the first-row atoms.

To avoid significant contraction errors when using very large primitive sets, it is necessary to use a general contraction (28) (each primitive occurs in each contracted function), as opposed to segmented contractions. A recent development for implementing the general contraction scheme is that of Almlöf & Taylor (29), where the contractions are based on the atomic natural orbitals (ANOs) of a single and double excitation configuration-interaction (SDCI) calculation for the atom. Often the average natural orbitals for several low-lying states are obtained to minimize any orbital bias, and as much as possible to reproduce the uncontracted results for the atomic separations. The outermost diffuse s and p basis functions are not included in the ANO contraction to provide greater flexibility in molecular calculations (30). These ANO basis sets are capable of yielding results as accurate as for the first row without making the subsequent CI expansion prohibitively long.

For the second transition row, the $(17s\,11p\,8d)$ primitive basis sets of Huzinaga (31), augmented with two functions to describe the $5p$ orbital and a diffuse d function to improve the description of the $4d^{n+1}5s^1$ occupation, are commonly used (32). As this basis set is rather large, the innermost functions are frequently replaced with an effective core potential (ECP) (33, 34). However, for accurate results it is necessary to include the $4s$, $4p$, $4d$ and $5s$ orbitals in the valence treatment (34, 35). In addition, the dominant relativistic effects (Darwin and mass-velocity) (17) are generally included—a so-called relativistic ECP or RECP. When an all-electron treatment is employed for the second row, the Darwin and mass velocity effects are usually included by using first-order perturbation theory (16).

Not only are there greater demands on the one-particle basis sets for transition metals, but the choice of the molecular orbital basis is also more critical. This is well illustrated by comparing the computational requirements for describing the ^3D–^3F splitting in Ni atom to the ^3P–^5S splitting in C atom. For the latter case, if the optimum orbitals determined for the ^3P state are used for the ^5S state or vice versa, the error in the splitting is only about 0.1–0.2 eV. With the addition of correlation, the error is reduced to 0.002–0.004 eV. The situation is dramatically different for Ni: the splitting at the single configuration level is in error by ≈ 4 eV in a common orbital basis. Although an SDCI treatment reduces the error to ≈ 1 eV, the error is much too large for quantitative studies (12, 36, 37).

The choice of the molecular orbital basis is intimately tied both to the method of including external correlation and to whether two-state properties such as transition moments are required. When the state is relatively well described by a single configuration, the spin-restricted Hartree-Fock equations are solved for that particular state with symmetry and equivalence restrictions. External correlation is generally included by allowing single and double excitations from only this one configuration (SDCI). The effect of higher than double excitations is then included by using either the Davidson correction (+Q) (38) or by using a size-consistent reformulation of SDCI, namely either the coupled-pair functional (CPF) method of Ahlrichs et al (39) or the modified CPF method of Chong & Langhoff (40), which is much more stable to near degeneracies. The MCPF method has even given excellent results for cases where the single-reference description is fairly poor—see, for example, discussion below on the dipole moment function of NiH. Even when the MCPF method is not entirely adequate owing to the multireference character of an electronic state, it is still useful for calibrating basis set extension and inner-shell correlation effects.

A consequence of nearly isoenergetic orbitals of varying occupation is the large near-degeneracy effects that occur in transition metal systems. This results in a large degree of multireference character in the wave functions. As an extreme example consider the $^1\Sigma_g^+$ state of Cr_2, where no single configuration has a coefficient greater than 0.4. Clearly, in these cases an MCSCF procedure must be used to optimize the orbitals. The MCSCF approach that we use is the complete-active-space-SCF (CASSCF) method (41a–d). If single and double excitations are generated from all configurations in the CASSCF, we refer to this as a second-order CI (SOCI). However, since the SOCI expansions can be quite lengthy, the reference configurations are generally selected based on their coefficients in the CASSCF wave function. More specifically, an occupation is included if any of its component spin couplings is above a given threshold. These calculations are denoted as multireference CI (MRCI). Higher excitations can be estimated by using the multireference analog of the Davidson correction.

The most difficult case for orbital optimization occurs when several states must be described with one molecular orbital basis set. This situation is encountered when transition probabilities are to be determined at the CASSCF/MRCI level (see discussion below on the spectroscopy of ScF and YCl). To provide the best molecular orbital basis for several states, we have used the state-averaged CASSCF method (42a–d), generally with equal weights for each state included in the averaging. It is essential to provide sufficient flexibility in the active space to describe each of the states equivalently, especially if they are of mixed valence and Rydberg character. The subsequent MRCI treatments then must include single and double replacements from any of the configurations that have coefficients above a designated threshold in any of the states.

RESULTS AND DISCUSSION

In this section we present examples of transition metal systems in order of increasing computational complexity. We first discuss the computational requirements for obtaining accurate state separations in the transition metal atoms, as quantitative results have demanded high levels of theory. These separations affect the bonding in molecular systems, especially the transition metal hydrides and their positive ions that are discussed next. Empirical rules have been developed to explain the ground states of the hydrides based on the relative mixing of the atomic asymptotes. The dipole moments are shown to be particularly sensitive to the relative mixings. We next consider the low-lying electronic states of ScF and YCl. The spectroscopy of these molecules is dominated by transitions between the valence s, p, and d orbitals localized primarily on the metal. Quite accurate calculations are possible for these systems, as the ionic bonding is relatively straightforward to describe. The transition metal oxides are considerably more challenging computationally, primarily owing to multiple bonding and the greater flexibility of the bonding with oxygen relative to the halogens. This is particularly true of the transition metal atoms in the middle of the row, where the presence of near degeneracies gives rise to a large number of low-lying states with a large degree of multireference character. However, the greatest challenge is the description of transition metal-metal bonding, where again the computational difficulty is greatest in the middle of the row. Studies for several diatomic and triatomic systems are presented. Finally, we discuss theoretical attempts to understand the activation of the H_2 molecule by Ni atom, which can be considered the initial stage of catalysis. These applications illustrate the diversity of the bonding in transition metal systems, and show that theory and experiment provide complementary information for these systems.

Transition Metal Atoms

The relative difficulty of accurately describing first-row and transition-metal systems is illustrated by calculations on the atoms. Consider again, for example, the computational requirements of determining the $^3P-^5S$ separation in carbon atom with $^3F(3d^84s^2)$ $^3D(3d^94s^1)$ separation in Ni atom. For the carbon atom, the SCF separation agrees with the NHF value to within 0.02 eV using a double zeta basis set. Augmenting the basis by a set of d polarization functions and including electron correlation with an SDCI(SDCI+Q) treatment gives an error in the separation of only 0.37 (0.29) eV with respect to experiment. Expanding the basis set to triple-zeta plus two sets of polarization functions further reduces this error to 0.18 (0.06) eV. In contrast, a triple-zeta description of the $3d$ orbital is required to approach the $^3F-^3D$ NHF separation in Ni atom. Electron correlation has a very large effect on the Ni atomic separations as a result of the very large correlation energy associated with the compact $3d$ orbitals. An SDCI treatment in a large ANO basis set reduces the error in the Ni atom $^3F-^3D$ separation from 1.3 to 0.3 eV, neglecting relativistic effects (43). To reduce the error much further, it is necessary to correlate the $3s$ and $3p$ electrons. The Ni atom $^3F-^3D$ and $^3D-^1S$ separations have also been recently studied (43) by using the CASSCF/MRCI method. Although little improvement over SDCI is found for the $^3F-^3D$ separation, the $^3D-^1S$ separation is improved by 0.6 eV, thereby reducing the error to only 0.1 eV. This reflects the increased importance of $3d \rightarrow 3d'$ correlation in the $^1S(3d^{10})$ configuration. As illustrated by the following discussion on Fe atom, much of the remaining error in the $^3F-^3D$ separation of Ni atom is probably due to neglect of $3s$ and $3p$ correlation.

Full CI calculations have been performed (44) for Fe atom to gain insight into the correlation requirements for accurately describing the $^5D-^5F$ separation. In these studies a compact one-particle basis set constructed by following the ANO procedure was used. This allowed the inclusion of f polarization functions in the one-particle basis set for the FCI calibration calculations that correlated the $3d$ and $4s$ electrons. The SDCI $^5D-^5F$ separation in Fe atom was found to be in excellent agreement with the FCI for each of the one-particle basis sets considered. In a very large one-particle basis set the SDCI separation (including an estimate for relativistic effects) still differed from experiment by 0.38 eV. Since the FCI calculations show that truncation of the n-particle space is not the source of the error, and since further basis set saturation is not expected to change the separation by more than 0.07 eV, the error must be due to the differential effect of $3s$ and $3p$ correlation. Although at the SDCI level the inclusion of $3s$ and $3p$ correlation accounts for only about half of the remaining error, it is expected that there is a differential effect of higher excitations when 16 electrons are correlated. This is consistent with the study of the electron affinity of O atom where the $2p$ valence correlation could be accurately included at the MRCI level, but this level underestimated the differential effect of adding the 2s correlation (45). Thus most of the remaining error in both the Fe and Ni separations probably arises from inner-shell correlation not accounted for with the SDCI procedure. This same error carries over into the separations between the $^6\Delta$ and $^4\Delta$ states of FeH (see discussion below).

An all-electron study of the ^3D, ^3F, and ^1S separtion in Pd showed that the differential correlation effects are smaller, and that the SDCI treatment yields a better description of Pd than the equivalent treatment does for Ni (46). Calculations for the entire second transition row using an RECP and correlating the 4d and 5s electrons show superior results to those for the first row (11). The mixing of atomic asymptotes in molecular systems can therefore be expected to be easier to describe than for the first row.

Transition Metal Hydrides

The nature of the bonding in the electronic states of the transition metal monohydrides (TMH) reflects to a large degree the relative mixing of the atomic states of the metal. Except for several cases on the left side of the row where the ground states involve s–d hybrid bonds, the ground states of both the first- and second-row systems can be predicted on the basis of simple empirical rules involving these mixings (2). Although the spectroscopic constants r_e and ω_e are rather insensitive to the details of the bonding, the dissociation energies (D_0) and particularly the dipole moments (μ) are a sensitive measure of the degree of mixings that also determine the d populations (10, 47). Unfortunately, experimental dipole moments are available only for the $^2\Delta$ ground state of NiH (48), and relatively accurate experimental dissociation energies are available only for the positive ions of the transition-metal hydrides (TMH$^+$) (49–52). We therefore discuss the theoretical determination of the D_0 values of the TMH$^+$ systems to help calibrate analogous calculations on the neutrals, where the experimental data are somewhat incomplete and contradictory. Although the ground states for most of the TMH are definitively known, some controversy exists as to whether the ground state of FeH is $^4\Delta$ or $^6\Delta$ (53, 54). From a theoretical standpoint these states are very difficult to treat equivalently, and, from an experimental point of view, although the photoelectron experiments (53) strongly support a $^4\Delta$ ground state, the failure to observe the $^4\Delta$–$^4\Delta$ infrared system in absorption in low-temperature matrix isolation experiments (54) does not corroborate the photoelectron results. We therefore discuss the latest theoretical attempts to compute the $^4\Delta$–$^6\Delta$ separation in FeH accurately (55). Several examples are presented that contrast the bonding between the first and second transition rows. As the dipole moments are a sensitive measure of the charge distribution, additional experimental dipole moment determinations would help to calibrate theory further.

As our first example of the diversity of the bonding that occurs in the TMH systems, we discuss the $X^4\Phi$ and $a^2\Delta$ states of TiH. If only the five valence electrons in the Ti $3d$ and $4s$ orbitals and in the H $1s$ orbital are correlated, FCI calculations can be carried out (56) in a large one-particle basis set to evaluate approximate methods for including the effects of electron correlation. In the $X^4\Phi$ ground state, the bonding arises from a mixture of both the $3d^34s^1$ and the $3d^24s^2$ occupations. The $3d^34s^1$ occupation contributes by forming a $4s$–$1s$ bond polarized toward the H atom. For the $3d^24s^2$ occupation, both $4s$–$4p$ and $4s$–$3d$ hybridization can occur, with one hybrid orbital bonding with the H, while the other is polarized away. The $^4\Phi$ state is of mixed-state character as manifested by the Mulliken $3d$ population of 2.40 electrons at the FCI level. In contrast, the bonding in the low-lying $a^2\Delta$ state involves mostly $3d$–$4s$ hybridization of the metal. There is very little contribution from the Ti $3d^34s^1$ occupation as manifested by the FCI $3d$ population of 2.07 electrons. The dipole moment of the $X^4\Phi$ state is considerably larger than the $a^2\Delta$ state, because the larger $3d^34s^1$ contribution enhances the dipole moment owing to the large polarization of the $4s$ orbital toward hydrogen.

In spite of the mixed-state character of the $X^4\Phi$ state, single-reference based methods such as CPF and MCPF give accurate spectroscopic constants and dipole moments. However, we find that the CASSCF procedure builds in a significant bias in the molecular orbital basis toward the $3d^24s^2$ occupation, which is not overcome in the MRCI treatment unless the active space is quite large (11 active orbitals are required for quantitative agreement with the FCI). However, if smaller active spaces are used, the orbital bias can be overcome at the MRCI level by using natural orbital (NO) iterations. Note that it is necessary to continue the NO iterations until the property is converged, even though the energy increases slightly. This procedure is generally not necessary for first-row chemistry, since the orbital bias is never this large.

The accuracy of the various computational approaches is quite different for the $a^2\Delta$ state. This state is not well described by a single configuration, even though it arises predominantly from the $3d^24s^2$ occupation. Relatively modest CASSCF/MRCI treatments give accurate results, but since the wave functions contain considerable multireference character, the CPF and MCPF results are not as accurate.

The bias introduced by the orbital optimization manifests itself in the dipole moments of other TMH systems as well. For FeH, CoH, and NiH, there is about a 1 D increase in μ with NO iterations (10). In each case the converged μ value is in good agreement with the MCPF value. For the

$^2\Delta$ state of NiH, both the converged MRCI value (2.59 D) and the MCPF value (2.56 D) (10) are in excellent agreement with experiment (2.4±0.1 D) (48). For each method, the magnitude of the dipole moment of NiH parallels the $3d$ population, and thus the relative contribution of the $3d^84s^2$ and $3d^94s^1$ atomic states, which contribute very differently to the dipole moment. The slow convergence of the Møller-Plesset perturbation series for this state has also been attributed to the mixing of these asymptotes (35).

All of the first- and second-row TMH systems have been studied at the MCPF level by using large one-particle basis sets (10, 11). Spectroscopic constants for the ground states of these systems are summarized in Table 1. The different relative mixings of the d^ns^2 and $d^{n+1}s^1$ asymptotes are apparent in the d populations. These indicate that the bonding in ScH, YH, and ZrH involve primarily the d^ns^2 state. This bonding mechanism involving s–d hybrids is somewhat more favorable for the second row owing to the more equivalent radial extent of the d and s orbitals and the smaller loss in d–d exchange energy. The $3d$ population of the $^7\Sigma^+$ state of MnH is very nearly five, as the $3d^54s^2$ state lies over 2 eV lower than the $3d^64s^1$ state. The greater stability of the d orbitals in the second transition

Table 1 Summary of theoretical spectroscopic constants for the ground states of the first- and second-row transition metal hydrides

Molecule	State	r_e (a_0)	ω_e (cm^{-1})	D_e (eV)	μ (D)	nd^a	atm sep^b
ScH	$^1\Sigma^+$	3.390	1587	2.27	1.37	0.84	1.44
TiH	$^4\Phi$	3.440	1548	2.06	2.19	2.30	0.81
VH	$^5\Delta$	3.249	1635	2.33	2.02	3.40	0.25
CrH	$^6\Sigma^+$	3.201	1647	2.13	3.81	4.83	−1.00
MnH	$^7\Sigma^+$	3.313	1530	1.67	1.24	5.05	2.14
FeH	$^4\Delta$	2.973	1915	1.67	2.90	6.52	0.88
CoH	$^3\Phi$	2.895	1842	1.94	2.74	7.60	0.42
NiH	$^2\Delta$	2.807	1987	2.69	2.56	8.65	−0.03
CuH	$^1\Sigma^+$	2.851	1852	2.63	2.95	9.80	−1.49
YH	$^1\Sigma^+$	3.706	1559	2.95	1.54	0.74	1.36
ZrH	$^2\Delta$	3.509	1483	2.45	1.23	1.90	0.59
NbH	$^5\Delta$	3.384	1583	2.60	2.45	3.69	−0.18
MoH	$^6\Sigma^+$	3.300	1642	2.19	3.03	4.89	−1.47
TcH	$^5\Sigma^+$	3.158	1797	1.95	2.15	5.35	0.41
RuH	$^4\Phi$	3.114	1801	2.34	2.70	7.00	−0.87
RhH	$^3\Delta$	2.976	2057	2.81	2.24	8.10	0.53
PdH	$^2\Sigma^+$	2.911	1958	2.22	2.03	9.24	−0.95
AgH	$^1\Sigma^+$	3.130	1703	2.22	2.95	9.80	−3.97

^{a} The Mulliken d populations obtained with the MCPF wave functions.
^{b} The d^ns^2–$d^{n+1}s^1$ atomic separation, except for Rh and Pd where it is the $4d^{n+1}5s^1$–$4d^{n+2}$ separation.

row, which results in considerably different atomic state separations (especially on the right-hand side of the row), also results in considerably different relative stabilities of the low-lying states of the first- and second-row TMH systems. Although the RuH population of 7.00 suggests that only the $4d^75s^1$ state is contributing, there is actually some contribution from the $4d^8$ state. However, the $4d$ involvement in the bonding orbital, which donates charge to H, reduces the total $4d$ population to 7.00. The contribution of the low-lying $4d^{n+2}$ asymptote results in a $^4\Phi$ ground state for RuH. In contrast, FeH has a $^4\Delta$ ground state, as the $3d^8$ state is too high in energy to contribute significantly to the bonding (10, 11, 57). The extra stability of the $4d^{n+2}$ relative to the $3d^{n+2}$ asymptote also yields different ground states for RhH and PdH relative to their first-row counterparts. The increased contribution of the d^{n+2} state in the second row is also apparent in the d populations.

There are significant trends in the spectroscopic constants and dipole moments of the first- and second-row TMH systems. In the first row, the r_e values decrease as the number of $3d$ electrons increase from Ti to Cr, and then from Mn to Cu (when a correction for relativistic effects is included). A discontinuity occurs at Mn where the $3d$ shell is half filled. The bond distance in the $^1\Sigma^+$ state of ScH is somewhat shorter because of the formation of the $3d$–$1s$ bond. Similarly in the second row the r_e values for the lowest state arising mostly from the $4d^{n+1}5s^1$ configuration (this is not always the ground state) decreases monotonically across the row, except for the $^1\Sigma^+$ ground state of AgH. The dipole moments of those states that involve s–d hybridization or bond primarily from the d^ns^2 atomic state are small, since the nonbonding hybrid orbital balances the charge donation in the bonding hybrid. In contrast, those states which involve primarily $d^{n+1}s^1$ character have much larger dipole moments due to the donation of charge.

The greater stability of d bonding in the second row generally leads to larger bond dissociation energies for these systems. Exceptions to this trend, such as NiH and PdH, are also made consistent if both molecules are referenced to the d^9s^1 atomic asymptote. This is reasonable considering that the bonding in NiH is predominantly d^9s^1 with some admixture of d^8s^2, whereas PdH is predominantly d^9s^1 with some admixture of d^{10}. The theoretical D_e values in Table 1 are generally in good agreement with the somewhat uncertain experimental values. The theoretical values are expected to increase slightly with further improvements in the one-particle basis sets.

To gain additional insight into the accuracy of our D_e values for the TMH systems, we have carried out analogous calculations on the TMH$^+$ systems (58), where more accurate experimental values are available for comparison. The bonding in the ions is much like that in the neutral systems, except that there is somewhat more d character to the bonding, as the d orbital is stabilized with respect to the s orbital in the ions. The positive ions, unlike the neutrals, are amenable to studies with guided ion beam mass spectroscopic techniques, since the reaction with H_2 can be made exothermic by accelerating the ions (49–52). It is still necessary to account for the remarkably different reaction rates of the different atomic states of the $d^n s^1$ and d^{n+1} ions: for Fe^+ the rate of reaction differs by a factor of 80, even though the states are separated by only 0.25 eV. Once the differences in electronic state reactivity are accounted for, quite accurate dissociation energies can be deduced from the threshold behavior of the reaction cross-sections, as a function of the relative kinetic energy. The ab initio results, with a correction (a maximum of less than 5 kcal/mole) for errors in the atomic separations and basis set incompleteness, are compared with the experimental values in Figures 1 and 2. Previous theoretical determinations for the first-row systems by Schilling et al (59) and Alvarado-Swaisgood et al (60, 61) are also in good agreement with the theoretical values plotted in Figure 1. Our theoretical values suggest that the experimental values for MoH^+ and AgH^+ are slightly too large. However,

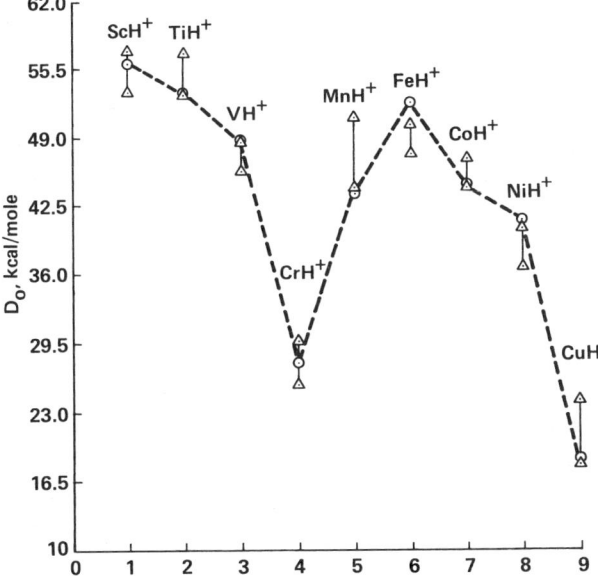

Figure 1 Comparison of theoretical dissociation energies (D_0) in kcal/mole for the first-row transition metal hydride positive ions (connected by *dashed line*) with experimental values (Refs. 49–52) given with error bars.

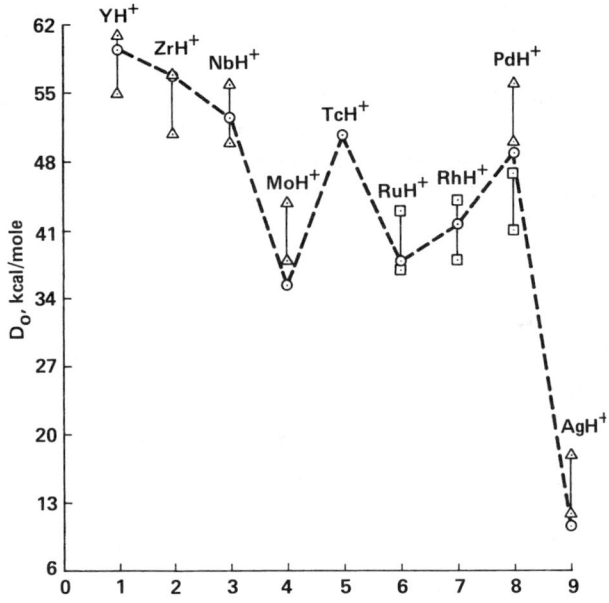

Figure 2 Comparison of theoretical dissociation energies (D_0) in kcal/mole for the second-row transition metal hydride positive ions (connected by *dashed line*) with experimental values (Refs. 49–52) given with error bars.

the overall good agreement for the positive ions gives us considerable confidence that our values for the neutrals are also very accurate.

The theoretical calculations are sufficiently accurate to delineate definitively the ground state of both the first- and second-row TMH systems in almost every case. However, the FeH molecule has presented a real challenge to both experiment and theory. The laser photoelectron spectroscopy studies of Stevens et al (53) have been interpreted in terms of a $^4\Delta$ ground state and low-lying (0.25 eV) $^6\Delta$ state. However, the $^4\Delta$–$^4\Delta$ infrared system, which has been observed in the absorption spectra of heated mixtures of iron metal and hydrogen, has not yet been observed at low temperatures in matrix studies (54). Although previous theoretical studies (62) supported the interpretation of Stevens et al, direct calculations (10, 62) placed the $^6\Delta$ state lower. However, more accurate calculations are now possible. For example, a CASSCF/MRCI calculation correlating the eight valence electrons in a large [$8s\ 7p\ 5d\ 3f\ 2g/4s\ 3p\ 2d$] ANO Gaussian basis set, predicts the $^4\Delta$ state to be 0.06 eV lower upon inclusion of a $+Q$ correction for higher excitations (55). This is further increased to 0.16 eV when a correction for $3s$ and $3p$ inner-shell correlation is included based on MCPF calculations. Most of the remaining discrepancy with the experimental estimate of 0.25 eV for the splitting is probably due to underestimating the differential effects of inner-shell correlation. Therefore, the recent theoretical calculations provide strong support for the interpretation of the photodetachment spectra of Stevens et al that places the $^6\Delta$ state 0.25 eV above the $^4\Delta$ ground state. Nevertheless, FeH represents the case most difficult for theory, namely the presence of two nearly degenerate states of different multiplicity. Very high levels of correlation treatment are required to account for the separation quantitatively, as the correlation energy is generally substantially larger for the lower spin state. The failure to observe the $^4\Delta-^4\Delta$ infrared system in absorption at low temperature is probably due to the broadening of this already weak band system by the matrix (53).

Although most of the theoretical studies on TMH systems have focused on the low-lying states, a few studies have been more concerned with elucidating the spectroscopy. Particularly noteworthy in this regard is the study of the optical spectrum of NiH by Blomberg et al (63). The spectrum of NiH is described to have a band-like structure, where the atomic lines of Ni atom are split only slightly by the weak ligand field from hydrogen. The bands in the 15,000–16,000 cm^{-1} region had previously been interpreted as involving two $^2\Delta-X^2\Delta$ transitions. However, using the state-averaged CASSCF/MRCI approach, they convincingly demonstrated that only one $^2\Delta$ state is possible in this energy region. Therefore, it is necessary to reinterpret some of the bands in this energy region in terms of transitions from the $X^2\Delta$ state to either a $^2\Pi$ or $^2\Phi$ state.

The differences between the first and second rows is well illustrated by comparing the ordering of the electronic states in NiH with PdH. In PdH the lowest three states are ordered as $^2\Sigma^+ < {}^2\Delta < {}^2\Pi$ as compared to the ordering in NiH $^2\Delta < {}^2\Pi \approx {}^2\Sigma^+$. The switch to a $^2\Sigma^+$ ground state for PdH can be explained in terms of a larger participation of the $4d$ electrons, as only the $^2\Sigma^+$ can mix in the ground state $^1S(d^{10})$ atomic configuration. Spin-orbit effects are also relatively important for PdH. Balasubramanian et al (64) have shown that the $^2\Delta$ and $^2\Pi$ states (in L-S coupling) mix quite strongly via the spin-orbit operator to produce two $J = 3/2$ states (in j-j coupling) that differ considerably from their L-S counterparts. For PtH it becomes essential to include both relativistic and correlation effects equivalently for quantitative results (65). Interestingly, the ground state of PtH is also $^2\Delta_{5/2}$, because the $^3D_3(d^9s^1)$ ground state of Pt atom is about 6140 cm^{-1} below the $^1S(d^{10})$ state, which is the ground state of Pd atom.

Transition Metal Halides

In this section we discuss the halides of scandium and yttrium, and contrast these with the corresponding alkaline-earth halides. As only single ionic bonds are formed, these systems are straightforward computationally compared with the transition-metal oxides and dimers where multiple bonds occur. Since these molecular systems are quite amenable to theoretical study (66, 67), the goal of the calculations was to provide not only spectroscopic constants for the low-lying states, but also to position the singlet and triplet manifolds accurately, to determine approximate radiative lifetimes, to help assign observed band systems, and to identify other band systems that should be observable spectroscopically. The discussion is focused primarily on the prototypical ScF and YCl systems. The ScF molecule is the best studied of these systems, both from an experimental and theoretical point of view.

The spectroscopic constants for the $X^1\Sigma^+$ and $a^3\Delta$ states of the fluorides, chlorides, and bromides of scandium and yttrium are compared in Table 2. These results are obtained at the CPF level correlating the eight valence electrons. A noteworthy observation is that there is better correlation between the scandium and yttrium monohalides when the $a^3\Delta$ state is compared. For example, the D_0 values of the $a^3\Delta$ states of the yttrium monohalides are only 0.07 eV, 0.13 eV, and 0.15 eV larger than scandium for the fluoride, chloride, and bromide, respectively. This is a consequence of the fact that the bonding in the $a^3\Delta$ state is very ionic, with little d involvement. There is also considerable similarity between the $a^3\Delta$ states of the scandium and yttrium monohalides and the $X^2\Sigma^+$ states of the alkaline-earth monohalides (68), because the principal difference is the addition of a predominantly nonbonding electron into the $d\delta$ orbital. The larger D_0 values and larger $a^3\Delta$–$X^1\Sigma^+$ separations for the yttrium systems probably arise from the larger degree of covalent character in the $X^1\Sigma^+$ state. This is manifested by the very similar dipole moments for the scandium and yttrium systems, even though the bond lengths are longer for the yttrium monohalides.

Table 2 Comparison of the 8-electron CPF results for the scandium and yttrium monohalides

System	$r_e(a_0)$		μ (D)		D_0 (eV)		T_e (cm^{-1})
	$X^1\Sigma^+$	$a^3\Delta$	$X^1\Sigma^+$	$a^3\Delta$	$X^1\Sigma^+$	$a^3\Delta$	
ScF	3.412	3.562	1.410	2.733	6.02	5.61	3320
ScCl	4.283	4.474	2.623	3.881	4.55	4.40	1226
ScBr	4.590	4.789	2.689	3.941	3.90	3.81	729
YF	3.715	3.799	1.681	3.227	6.72	5.68	8383
YCl	4.600	4.709	2.670	4.537	5.36	4.53	6699
YBr	4.922	5.034	2.692	4.650	4.74	3.96	6254

Spectroscopic constants for the low-lying singlet and triplet states of ScF are compared with experiment (69) in Table 3. Since this study included several states of the same symmetry, and the higher lying states were not well described by a single reference configuration, it was necessary to perform all calculations at the CASSCF/MRCI level. The state-averaged CASSCF method was used to provide an optimal common set of molecular orbitals for the calculation of electronic transition moments. The constants are reported including the effect of higher excitations, and the Darwin and mass velocity contributions to the relativistic energy (denoted MRCI+Q+Rel). As correlation and relativistic effects both preferentially lower the ground state, the T_e values are uniformly larger when these corrections are included. In Figure 3 we have drawn the singlet energy levels for ScF and labeled the dipole-allowed transitions by the magnitude of the electronic transition moment at 3.6 a_0. Throughout we use our recommended revised spectroscopic designations for both the singlet and triplet manifolds.

The ScF molecule has an $X^1\Sigma^+$ ground state and a rather low-lying excited $a^3\Delta$ state. The $A^1\Delta$ state, which has not been observed, is also

Table 3 Spectroscopic constants for the excited states of ScF[a]

State	MRCI+Q+Rel			Expt.[b]		
	r_e	ω_e	T_e	r_e	ω_e	T_e
$X^1\Sigma^+$	3.394	730	0	3.378	735.6	0
$A^1\Delta$	3.561	604	4482			
$B^1\Pi$	3.644	600	9578	3.625	586.3	≤10735.5
$C^1\Sigma^+$	3.643	582	16769	3.603	589.6	16165
$D^1\Pi$	3.539	643	21150	3.524	622.1	20384
$E^1\Delta$	3.589	584	23791			
$F^1\Phi$	3.544	642	24036			
$G^1\Delta$	3.611	613	26089			
$H^1\Sigma^+$	3.624	586	27446			
$I^1\Pi$	3.635	589	27755	3.609	565.3	26892
$a^3\Delta$	3.545	632	2362	3.507	649.0	
$b^3\Pi$	3.595	585	6434			
$c^3\Sigma^+$	3.634	574	9246			
$d^3\Phi$	3.579	577	18364		570.4	
$e^3\Pi$	3.589	552	19900			
$g^3\Delta^c$	3.637	572	20449			
$h^3\Pi$	3.473	624	21520			

[a] r_e is reported in a_0, and ω_e and T_e are given in cm^{-1}.
[b] Huber & Herzberg (69).
[c] The letter "f" is reserved for the lowest $^3\Sigma^-$ state, which is expected to lie in this region.

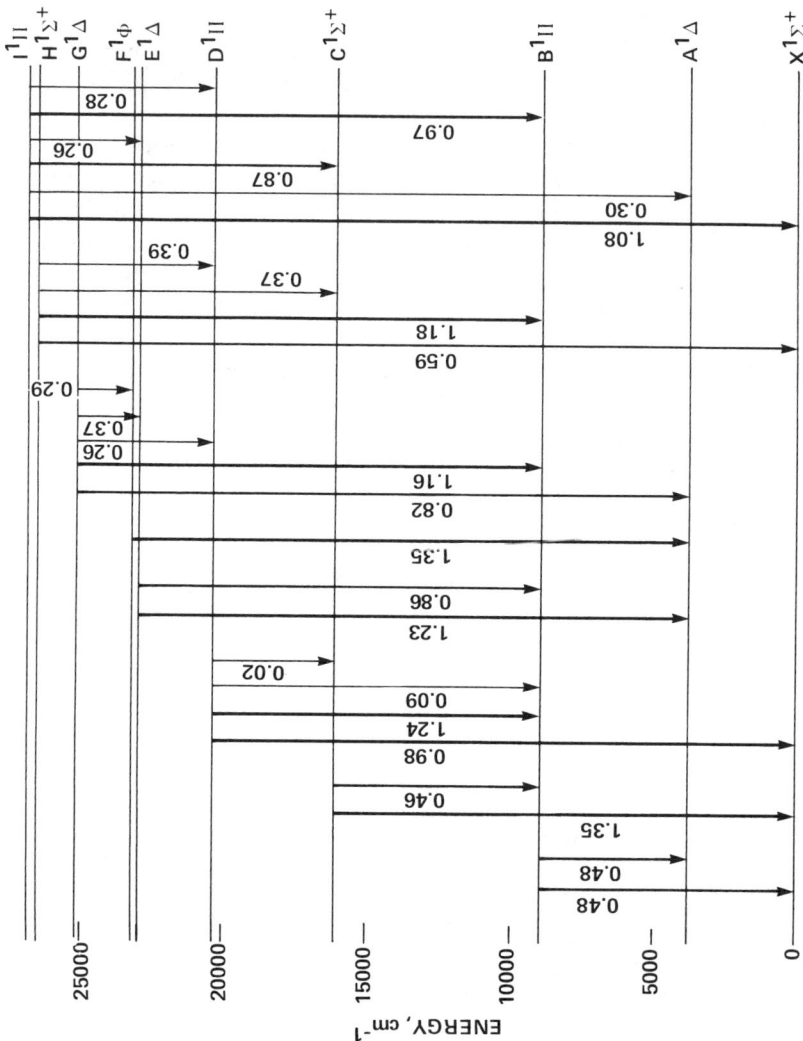

Figure 3 Energy level diagram for the singlet states of ScF denoting the magnitude of the dipole-allowed transitions. The stronger transitions are denoted by the bolder lines.

rather low-lying ($T_e \approx 4482$ cm^{-1}). It should be possible to observe this state in emission or in absorption at elevated temperatures from one of the following four band systems: $D^1\Pi-A^1\Delta$ at 6000 Å, $E^1\Delta-A^1\Delta$ at 5179 Å, $F^1\Phi-A^1\Delta$ at 5117 Å and $G^1\Delta-A^1\Delta$ at 4628 Å.

Transitions to seven vibrational levels of the $B^1\Pi$ state from the lowest vibrational level of the ground state have been observed in a neon matrix at 4K (70). The intensities of the first six peaks are in the approximate ratio of 11:8:6:3.5:1:0.5. To gain insight into the assignment of the absorption spectrum, we have computed Franck-Condon factors for the $B-X$ transition based on Morse functions fitted to the experimental spectroscopic constants. The agreement with the experimental peak heights is quite striking when the 9405 Å band is assigned to the 2 ← 0 transition of the $B-X$ system. When the 2-0 band is normalized to 11.0, the Franck-Condon factors for the 0-0 to 7-0 bands are in the ratio of 4.8:9.9:11.0:8.7:5.6:3.0:1.5:0.6. Therefore it is likely that the correct T_e value for the $B^1\Pi$ state is between 9500–9600 cm^{-1}, in good agreement with our MRCI+Q+Rel value. The rotational analysis of the $B-X$ system (71) shows that the Λ-doubling in the $B^1\Pi$ state is large and decreases rapidly from $v = 0$ to $v = 3$ (note that this may in fact be $v = 2$ to $v = 5$). This requires either a $^1\Sigma^-$ state or the 0^- component of a $^3\Sigma^+$ state to lie just slightly below the $B^1\Pi$ state. As can be seen from Table 3, the $c^3\Sigma^+$ state is perfectly positioned to cause the observed Λ-doubling in the $B^1\Pi$ state.

The low-lying states of the transition metal monohalides involve primarily excitations on the metal with relatively little involvement of the halide. The spectroscopy of ScF is not significantly different from that of ScCl or ScBr, and the strong transitions in these systems can be related to correspondingly strong transitions in Sc$^+$ and Y$^+$. For example, the strong radiative transitions in ScF can to a large extent be correlated with a large degree of atomic Sc$^+$ $4p \to 4s$ transition character. This is particularly true, for example, of the $C^1\Sigma^+-X^1\Sigma^+$ and $E^1\Delta-A^1\Delta$ transitions. However, there is some component of this strong atomic transition in many of the transitions, so that the radiative lifetimes of most of the excited singlet states are less than 50 ns. The radiative lifetime of only one state of ScF has been measured. This state ($I^1\Pi$ according to our designation) is computed to have a radiative lifetime of 15.2 ns, as compared with the measured value of 20±2 ns (72).

Comparable theoretical calculations have been performed for the excited states of YCl. Although the relative ordering of the lowest six singlet states is the same as for ScF, considerable differences exist for the higher states. For example, the nearly degenerate $F^1\Gamma$ and $G^1\Sigma^+$ states of YCl, which are described principally by the δ^2 valence configuration, are considerably lower in YCl than in ScF owing to the greater stability of the $d\delta$ orbitals in the second row. Two transitions were observed for YCl at 22,787 and 27,116 cm^{-1} by Fischell et al (72) using laser excitation fluorescence. The theoretical calculations provide convincing evidence for assigning these transitions as $D^1\Pi-X^1\Sigma^+$ and $J^1\Pi-X^1\Sigma^+$, respectively. Thus, theoretical calculations on the halides of Sc and Y are capable of yielding quantitative information that can be used to aid in the interpretation of experiment. Although calculations on the transition metal halides in the middle of the transition row may prove to be somewhat more difficult, it should be possible to characterize the low-lying states accurately for these systems as well.

Transition Metal Oxides

The transition metal oxides (TMO), especially those in the middle of the transition row, represent a challenge to both theorists and experimentalists. Although some systems, such as TiO, are reasonably well characterized [see for example Ref. (69)], even the ground states have not been definitively identified in other cases. The existence of several low-lying atomic states of oxygen atom gives rise to a large number of low-lying molecular states, especially when there is also a large degree of flexibility in occupying the d shell on the metal. Consider first CuO, where one expects the $X^2\Pi$ state to involve primarily a bond between the Cu(4s) and O(2$p\sigma$) orbitals. The bond contains both an ionic, Cu$^+$(3d^{10})+O$^-$(2$p\sigma^2 2p\pi^3$), and a covalent, Cu(3$d^{10}4s^1$)+O(2$p\sigma^1 2p\pi^3$), component. At the SDCI level this is precisely the picture that emerges (73). However, at the SDCI level in an extended Gaussian basis the bond length is 0.07 Å too long and the dissociation energy is about 0.7 eV too small, even when relativistic effects are included. This large discrepancy with experiment is removed at the CPF level, which enhances considerably the contribution from the low-lying Cu(3$d^9 4s^2$) and Cu$^+$(3$d^9 4s^1$) asymptotes, as manifested by a very large reduction in the 3d population on Cu. The importance of this contribution is increased by the occurrence of back donation into the Cu 4$p\pi$ orbital. The importance of the np configuration of the metal positive ion has recently been stressed (V. I. Srdanov and D. O. Harris, private communication). The dipole moment is particularly sensitive to this mixing, undergoing a significant reduction in response to the increased covalent character in the bond. Further support of this effect comes from the recent sub-Doppler optical Stark measurement (74) of the permanent electric dipole moment of the $v = 0$ level of the $X^2\Pi$ state of CuO. The observed value is in excellent agreement with the CPF results.

The dipole moment is a much more sensitive measure of the charge distribution in a molecule than are most other spectroscopic constants.

Unfortunately, apart from the recent experimental determination of the $X^2\Pi$ dipole moment of CuO, only one other experimental determination of the dipole moment of a first- or second-row transition metal oxide is available, namely for the $v = 1$ level of the $A^2\Pi$ state of ScO (75). This experimental value is also in quite good agreement with the CPF treatment of this state (76). Thus, for the oxides on either the far left or far right of the transition row, the CPF method is capable of providing a quantitative description of the states near their respective minima.

In the middle of the transition row there is much greater flexibility in occupying the d orbitals on the metal. These near-degeneracy effects give rise to many low-lying electronic states that are inherently of multireference character. Consider, for example, the $^4\Sigma^-$, $^4\Delta$, and $^4\Gamma$ states of CoO shown in the diabatic representation in Figure 4 (S. R. Langhoff, C. W. Bauschlicher, and L. G. M. Pettersson, unpublished). The lowest quartet state at the CASSCF level is $^4\Delta$, but this state is crossed on the outer limb by a myriad of other low-lying states (including many not shown in Figure 4). It is computationally very expensive to carry out MRCI calculations for these states, as the reference space is very large. Thus, it is necessary

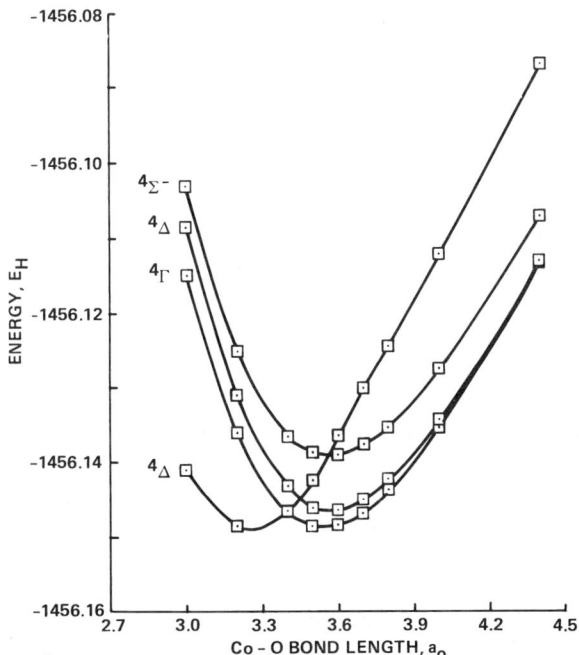

Figure 4 CASSCF potential energy curves for the lowest $^4\Sigma^-$, $^4\Delta$, and $^4\Gamma$ states of CoO.

to extract 5–10 roots of at least a several million configuration expansion to describe just the states below about 2 eV in the region of their minima. Our work has, therefore, been restricted to the more modest goal of determining the ground state. At the CASSCF level the lowest state is found to be $^6\Delta$. However, correlation preferentially stabilizes the $^4\Delta$ state, making this the ground state. Our study is therefore consistent with a recent laser-induced fluorescence study of CoO that has predicted the ground state to be a $^4\Delta$ state based on a rotational analysis of the red system (77). We also find that the lowest $^4\Delta$ state is of predominantly $\sigma^2\delta^3\pi^2$ character in agreement with the nuclear hyperfine splitting data. Thus our calculations fully support the recent assignment of the ground state symmetry of CoO. They also illustrate the extreme computational challenge that these systems pose for the theorist. A theoretical study (57) of FeO and RuO shows that there are substantial differences between the first- and second-row transition metal oxides, just as was found for the hydrides. Although the oxides are much more difficult than the corresponding hydrides and halides, the difficulties reach their culmination in the dimers discussed next.

Transition Metal Dimers

In this section we present some illustrative calculations on selected transition metal dimers. Experimental and theoretical studies of both first- and second-row transition metal dimers through about 1985 have been compared in an excellent review by Morse (78). Also noteworthy are the reviews of Weltner & Van Zee (79) and the review of theoretical studies by Walch & Bauschlicher (8). Therefore, we make no attempt to be complete here, but discuss only selected topics that illustrate the unique challenge of these systems to the theorist.

Probably the most studied of the transition metal dimers is dicopper, Cu_2. Accurate spectroscopic parameters have been determined experimentally for the $^1\Sigma_g^+$ ground state (69, 80, 81), and this system has therefore become a benchmark for testing theoretical methods. Although the bonding in Cu_2 involves primarily a single 4s–4s bond, the 3d electrons must be included in the correlation treatment for accurate results (36). A complicating factor for theory is that improvements in the one-particle basis set, inclusion of electron correlation (including the effects of higher excitations), and relativistic effects all contribute additively to reduce the bond length and increase the dissociation energy. High angular momentum functions in the basis set and high levels of correlation treatment are required to include the angular correlation terms. Inclusion of 3d–4s correlation effects also mixes in more $3d^94s^2$ character into the wave function. Excellent agreement with experiment is obtained only after the basis set is expanded to include at least f functions, higher excitations are accounted for by using (for example) the CPF or coupled-electron-pair-approximation methods, and relativistic effects, which reduce r_e by about 0.07 a_0, are included (82–84).

Theoretical calculations have recently contributed to understanding the photoelectron spectra of Cu clusters (85). Leopold et al (86) observed that the dissociation energy and vibrational frequency of Cu_2^- were 80% of the corresponding Cu_2 values. This was somewhat surprising considering that the extra electron occupies an antibonding orbital. Theoretical calculations show that, although the extra electron in Cu_2^- does occupy a Cu_2 σ_u antibonding orbital, the orbital is polarized to reduce the density in the internuclear region. Hence, although the bond order is formally reduced from one to one half, the D_e and ω_e for Cu_2^- are still 80% of the Cu_2 values. A plot of the antibonding orbital in Figure 5 shows that the $4s$–$4p$ near degeneracy allows the orbital to polarize out of the bonding region and thus its impact on the bonding is far less than would be expected.

Theoretical studies (87–90) of the ground state of Ag_2 show some striking differences from Cu_2. Although relativistic effects are larger as expected, the bond distance is much less sensitive to both improvements in the one-particle basis set and level of correlation treatment. This has been attributed to less admixture of $nd^9(n+1)s^2$ in the wave function, as this state lies 3.74 eV above the ground state in Ag, as compared to 1.49 eV in Cu. However, the best theoretical values for r_e are about 0.2 a_0 longer than that estimated (91a,b) by using the empirical Morse-Clark formula

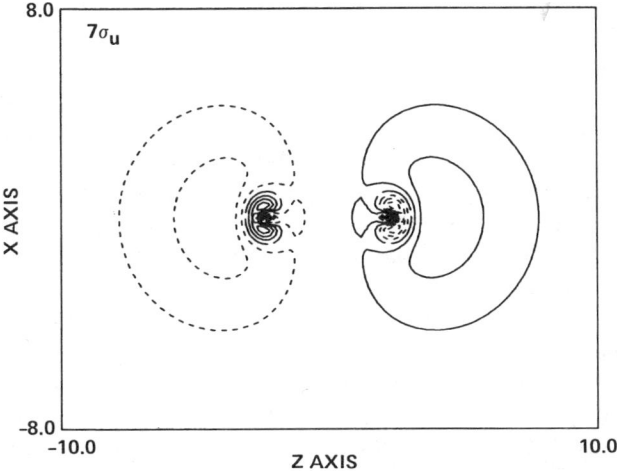

Figure 5 The SCF $7\sigma_u$ antibonding orbital in Cu_2^-.

($r_e^3\omega_e$ = constant) and the known B_e values for Cu_2 and Au_2. Since this difference is much larger than the estimated uncertainties in the calculated values, this suggests that Cu_2 and Au_2 are not that analogous to Ag_2, possibly as a result of the much larger d^9s^2–$d^{10}s^1$ separation in Ag. An accurate experimental determination of the bond length in the $^1\Sigma_g^+$ state would be very valuable for further calibration of theoretical calculations.

Theoretical calculations are capable of yielding accurate spectroscopic constants for the dimers of the coinage metals, even though high levels of theoretical treatment are required. Next we consider calculations on Cr_2 and Mo_2, which not only illustrate the difficulties in describing weak multiple d–d bonding, but also important differences between the first and second transition rows. The spectroscopic constants for the $X^1\Sigma_g^+$ ground state of Cr_2 are accurately known from laser-induced fluorescence studies (92). The dissociation energy has been estimated as 1.78 ± 0.35 eV using the second-law method (13). Although local spin density (LSD) functional methods (93, 94) and empirical approaches (95) have produced spectroscopic constants in reasonable agreement with experiment, to date we are unaware of any ab initio calculations that have obtained a realistic potential. One of the problems with Cr_2 is that a relatively short bond must be formed to maximize the inherently weak overlap of the d orbitals. At this bond length there is considerable repulsion from the $3p$ electrons that are in orbitals of comparable radial extent to the $3d$ orbitals and the s–s bonding is effectively eliminated, as this favors much longer bond lengths. CASSCF calculations yield only a weak outer well due to $4s$–$4s$ bonding. In large basis sets there is a small inflection in the CASSCF curve near the observed inner well minimum (96). However, analogous calculations for Mo_2 give a double well potential, the outer well corresponding to the $5s$–$5s$ bond and the much deeper inner well to $4d$–$4d$ bonding (97). The Mo_2 potential is easier to describe than the Cr_2 potential for several reasons. First, the d–d overlaps are larger in the second row, resulting in considerably stronger d–d bonds. Second, the $4p$ orbital is less spatially extended than the $4d$ orbital, and thus interferes to a lesser extent. Finally, the radial extent of the $5s$ and $4d$ orbitals are more comparable than their first-row counterparts, so that some s–s bonding can occur near the optimal $4d$–$4d$ bond length.

Let us consider then the expected computational requirements for obtaining a quantitative $^1\Sigma_g^+$ ground state potential for Cr_2 from ab initio calculations. From the work of McLean & Liu (98) we know that f functions contribute substantially to the strength of the Cr_2 bond. Therefore, we expect that the one-particle basis set will have to contain at least [$6s\,5p\,3d\,2f$] contracted functions, even if the ANO procedure is used. Since a CASSCF calculation yields no dominant coefficient in the wave function as a result of the large near-degeneracy effects, it will probably be necessary to include single and double excitations from the entire CASSCF reference space. This SOCI calculation would include about 141 million configurations. However, it is unlikely that quantitative results would be obtained without also correlating the 3s and 3p electrons. This would not only increase the one-particle basis set requirements, but would increase the CI expansion to over 10 billion configurations. It may be possible to circumvent some of these problems by placing constraints on the coefficients in the CI expansion as is done with the internally (99) or externally (100) contracted CI methods. Nevertheless, the reason that alternative approaches to describing these weak multiple d–d bonds have been developed is clear.

One such method that has led to the prediction of a double-well ground state potential for Cr_2 is the modified generalized valence-bond method (MGVB) of Goodgame & Goddard (95). This method incorporates a correction to the one-center self-Coulomb integrals to match the difference between the experimental ionization potential and electron affinity of Cr. This is to correct for the fact that the atomic correlation error is larger in the ionic part of the GVB wave function. This approach gave accurate bond distances and bond energies when applied to the ground states of H_2, N_2, and Mo_2. Although this approach is empirical, the potentials shown for Cr_2 in Figure 1c of Ref. (95) are probably the most quantitative available. Note that the local spin density approaches do not give an outer well for either Cr_2 or Mo_2, which is expected to arise from s–s bonding with the d shells coupled antiferromagnetically into a singlet state. Goodgame & Goddard suggest that the LSD formalism may fail in this regard owing to varying spin contamination in the unrestricted Hartree-Fock-type reference wave function.

Although not all of the transition metal dimers are as difficult as Cr_2, the theoretical determination of the ground states of other dimers, especially in the first row, has proven to be challenging. In several cases it has taken a combination of experimental and theoretical studies to produce a definitive determination. Consider, for example, the Sc_2 dimer, which the matrix isolation ESR spectrum (101) indicates has a $^5\Sigma$ ground state. Until recently theoretical calculations had failed to consider the $^5\Sigma_u^-$ state, as it correlates with one ground state atom $^2D(3d^14s^2)$ and one excited state atom $^4F(3d^24s^1)$. The bonding in this state is best described in terms of three one-electron bonds (102). One-electron bonds are much more common for the transition metal dimers, owing to the relatively small overlaps of the d orbitals and the benefit of preserving the large atomic exchange energy. The probability is also much larger that the ground state of the TM dimers will correlate to at least one excited atomic state. For example, in the case of Fe$_2$ (103), the interaction of two $3d^64s^2$ atoms is expected to lead to a repulsive interaction, whereas a bond can be formed if one of the atoms is first excited to $3d^74s^1$.

The Ti$_2$ molecule represents a case in which theory (8) and experiment (104) give complementary information. Theoretical calculations find that the $^1\Sigma_g^+$ state, which correlates with two excited $3d^34s^1$ atoms, lies about 0.4 eV above the $^7\Sigma_u^+$ state. However, the vibrational frequency observed in resonance Raman experiments (104) is in far better agreement with the calculated value for the $^1\Sigma_g^+$ state. Considering that additional improvements in the treatment of electron correlation is expected to preferentially stabilize the $^1\Sigma_g^+$ state, it is expected that the $^1\Sigma_g^+$ state is the true ground state for Ti$_2$.

Transition Metal Trimers

Considering the problems that the transition metal dimers have posed for theory, it is not surprising that very limited results exist for the transition metal trimers. An excellent review of existing theoretical and experimental information is given in the review by Morse (78). Theoretical studies of transition metal trimers have also been reviewed by Walch & Bauschlicher (105). Here we focus on recent results for Cu$_3$ and its negative ion, where theory has substantially helped in the interpretation of the experimental studies, and where some controversy exists for the value of the symmetric stretching frequency and in the assignment of the optical spectrum. The ESR spectrum of Cu$_3$ in an adamantane matrix (106) has been interpreted in terms of a $^2B_2$ ground state, as most of the unpaired electron density is located on the terminal atoms. This is consistent with ab initio calculations (107) that also indicate a $^2B_2$ ground state corresponding to a Jahn-Teller distortion of a $^2E'$ equilateral triangle geometry. The surface, however, is extremely flat and quite sensitive to the level of treatment. For example, the LSD method (108) incorrectly predicts a $^2A_1$ ground state (with the $^2B_2$ state 0.154 eV higher in energy). The Xα method (109) does not show the distortion away from the equilateral triangle, whereas calculations using the Anderson-Hückel approach (110) predict a linear molecule. SCF calculations have also favored the linear structure. Considering the sensitivity of the bond length, vibrational frequency, and dissociation energy of Cu$_2$ to the level of treatment (82–84), it is not surprising that high levels of theory are also required to determine the topology of the very flat ground-state surface of Cu$_3$ correctly.

Truhlar et al (111) have shown that the fluorescence spectrum of Rohlfing & Valentini (112) requires a pseudorotation barrier of 95 cm^{-1} and a Jahn-Teller stabilization energy of 221 cm^{-1} on the ground-state surface. Considering the large success of the CPF method in determining the spectroscopic constants of Cu_2, this approach was applied to the ground state surface of Cu_3. This produced a pseudorotation barrier and Jahn-Teller stabilization energy in excellent agreement with those deduced from experiment. Also, the calculated stretching frequency of 268 cm^{-1} gives definitive resolution of this controversy. Theoretical calculations (113) have helped in assigning the fluorescence spectrum, by showing that only states of $^2A_1'$ or $^2E''$ symmetry were in the correct energy region, and further that only the $^2A_1' \leftarrow X^2E'$ transition carried significant oscillator strength. The recent observation (112) of fluorescence originating from the 146 cm^{-1} level of the upper state now provides convincing evidence for this assignment (78).

Recent photodetachment experiments for Cu_3^- were interpreted (86) as showing an adiabatic electron affinity (EA) of 2.35–2.55 eV, which is significantly larger than that of Cu (1.235 ± 0.005 eV) or Cu_2 (0.842 ± 0.010). Some ambiguity remained in the interpretation owing to a weak unexplained feature in the spectrum at about 1.5 eV. Theoretical calculations (85) indicate that Cu_3^- is linear and has a $^1\Sigma_g^+$ ground state. The variation in the EA from Cu to Cu_3 was shown to arise from the changes in the degree of bonding character in the orbital to which the electron was added. Although it is difficult to compute EAs from first principles, a rather accurate value of the EA of Cu_3 can be deduced by scaling the computed value by the ratio of the experimental and theoretical values for Cu and Cu_2. The scaled value confirms the experimental EA of Cu_3 and further demonstrates that the weak feature at about 1.5 eV in the photodetachment spectrum arises from the excited $^3A_2'$ state of Cu_3^-, which is obtained by adding an electron to the non-bonding e orbital of the $^2E'$ ground state of Cu_3.

$Ni+H_2$ Reaction

The study of catalytic processes involving transition metals is an active area of research. In this section we discuss the reaction of Ni atom with H_2 as a model for the important reductive and oxidative addition reactions that are involved in metal-catalyzed processes. Although most of the theoretical work in this area has been of qualitative nature, recently Blomberg & Siegbahn (3) have presented CASSCF and contracted CI calculations for both the concerted and stepwise dissociation of NiH_2 into Ni and H_2. Their work clearly illustrates the involvement of the Ni $3d$ electrons in the reaction. In this section we summarize their work to illustrate the insights that ab initio quantum chemistry can give into reactions involving transition metals.

One of the important conclusions from the Blomberg & Siegbahn study of the singlet and triplet surfaces of NiH_2 was that the lowest singlet state of NiH_2, $^1A_1$, has a H–Ni–H bond angle of about 50° and is bound by 8 kcal/mole with respect to the singlet asymptote. This fact together with its ability to undergo symmetry-allowed (by Woodward-Hoffmann rules) elimination and addition reactions probably means that the $^1A_1$ is the chemically most interesting state. However, triplet reactions may also be important, as many nickel complexes have triplet ground states. Therefore, it is of interest to consider how the reaction proceeds on both the lowest singlet and triplet surfaces.

In Figure 6 we present a subset of the dissociation curves plotted in Figure 2 of Blomberg & Siegbahn. Consider first the $^3A_1$ and $^3B_1$ triplet states, which correlate to both the lowest state of linear $NiH_2(^3\Delta_g)$ and to ground state dissociated products, $Ni(^3F) + H_2(^1\Sigma_g^+)$. The reactions on the triplet surfaces are Woodward-Hoffmann forbidden, as they involve moving two electrons from the b_2 orbital of one of the NiH bonds to the a_1 symmetry $4s$ orbital of Ni. The importance of the $3d$ electrons in reducing the barrier for the concerted dissociation process is illustrated by the considerably different barrier heights on the $^3A_1$ and $^3B_1$ surfaces. On the $^3B_1$ surface the dissociation reaction can be divided into three steps. First, the $3d^84s^2$ configuration of NiH_2 is converted to $3d^94s^1$ NiH_2 by moving one of the bonding b_2 electrons to a b_2-type $3d$ orbital. Next, the second bonding b_2 electron is moved to the Ni $4s$ orbital of a_1 symmetry, and the

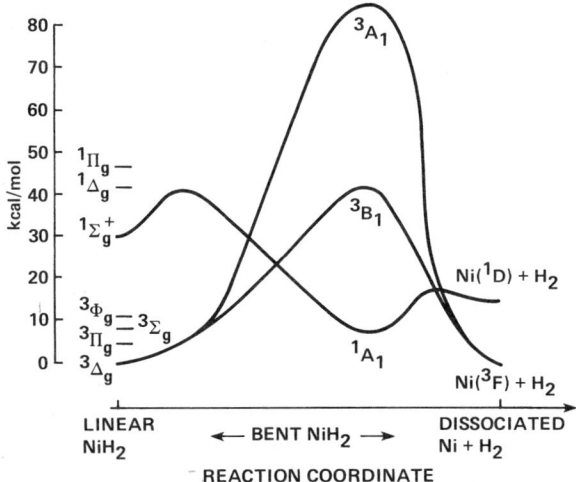

Figure 6 Dissociation curves for selected singlet and triplet states of NiH_2 shown relative to H_2 and the ^3F state of nickel atom. These curves are a subset of those given in Figure 2 of Ref. (3).

hole in the 3d-shell changes simultaneously from b_1 to a_2 to preserve overall symmetry. Finally, the $3d^94s^1$ Ni atom is converted to the nearly degenerate $3d^84s^2$ state of Ni. This three-step mechanism requires a specific initial occupation of the 3d shell in NiH$_2$, and therefore cannot occur in certain symmetries such as $^3A_1$, because the configuration cannot be changed to $3d^94s^1$ and still preserve the overall symmetry. Thus although the 3d orbitals are not explicitly involved in the bonding, their ability to recouple in different ways determines to a large extent the minimum energy pathway. In contrast, on the $^1A_1$ surface discussed next, the 3d orbitals are intimately involved in the bonding.

On the $^1A_1$ surface shown in Figure 6, the energy has minima at both linear and bent geometries. The $^1A_1$ state correlates with the excited linear $^1\Sigma_g^+$ state of NiH$_2$ and with the ^1D state of Ni atom. The bent configuration of NiH$_2$ is formed from the $3d^94s^1$ configuration of Ni and bonds by forming two s–d hybrids with hydrogen. This has been described as a type of ring bond with a weakened H$_2$ molecule bonding to Ni, as the H$_2$ bond distance is 2.3 a_0 compared to 1.4 a_0 in free H$_2$. The s–d hybrid Ni–H bonds are also about 0.2 a_0 shorter than the predominantly s–p hybrids formed for the linear states. The addition reaction goes readily since it is exothermic and there is only a 2.6 kcal/mole barrier to forming the bent NiH$_2$ molecule. Also, the elimination reaction should proceed easily considering that the barrier is less than 10.6 kcal/mole. This is all consistent with the fact that both elimination and addition reactions of H$_2$ are observed experimentally for Ni complexes. The bent $^1A_1$ NiH$_2$ structure is expected to be quite similar to an L$_2$NiH$_2$ complex (where L stands for a ligand such as a methyl radical). Addition of ligands is also expected to further stabilize the $^1A_1$ pathway.

Since the bent $^1A_1$ state corresponds to an almost dissociated H$_2$ molecule, this mechanism is probably also responsible for the dissociative adsorption of H$_2$ on a Ni(100) surface. Cluster studies (114) show that the only major difference is that the reaction on the Ni surface does not stop with a bent H–Ni–H system, but the dissociation proceeds to completion, forming two chemisorbed H atoms. Although the 3d electrons are intimately involved in breaking the H–H bond, there is little 3d involvement once the H atom is bound to the Ni surface.

CONCLUSIONS

The variation in the ratio of the spatial extent of the d and s orbitals and in the relative separation of the various atomic states leads to diverse bonding mechanisms in molecules containing transition metal atoms. High levels of theory are required to describe multiple d–d bonding, owing to the small overlap and large exchange energy associated with the d orbitals. Recent advances in methodology, such as the ANO procedure for contracting large primitive basis sets and the FCI method for calibrating approximate methods of including electron correlation, have led to an improved understanding of these systems. Also the development of reliable single-reference based methods such as MCPF has made it possible to consider the effect of basis set saturation and inner-shell correlation effects at a moderate computational cost.

In this review we have focused on examples in which theory has provided insight into both the spectroscopy and chemistry of transition metal systems. In many cases theory is competitive with experiment for determining spectroscopic constants and bond energies of these molecules. In turn, the recent experimental determinations of the electric dipole moments of $NiH(^2\Delta)$, $ScO(A^2\Pi)$, and $CuO(X^2\Pi)$ have been very valuable for calibrating theoretical methods. Theory and experiment thus often provide complementary information, which leads to a higher level of understanding than can be obtained by either approach alone.

ACKNOWLEDGMENTS

The authors would like to thank the *Journal of Chemical Physics* and P. E. M. Siegbahn for permission to use previously published figures.

Literature Cited

1. Moore, C. E. 1949. *Atomic energy levels*. Washington, DC: US Natl. Bur. Stand. (US) circ. no. 467
2. Walch, S. P., Bauschlicher, C. W. 1983. *J. Chem. Phys.* 78: 4597–4605
3. Blomberg, M. R. A., Siegbahn, P. E. M. 1983. *J. Chem. Phys.* 78: 5682–92
4. Bauschlicher, C. W., Bagus, P. S., Nelin, C. J., Roos, B. O. 1986. *J. Chem. Phys.* 85: 354–64
5. Blomberg, M. R. A., Brandemark, U. B., Siegbahn, P. E. M., Mathisen, K. B., Karlström, G. 1983. *J. Phys. Chem.* 89: 2171–80
6. Bauschlicher, C. W., Bagus, P. S. 1984. *J. Chem. Phys.* 81: 5889–98
7. Martin, R. L., Hay, P. J. 1981. *J. Chem. Phys.* 75: 4539–45
8. Walch, S. P., Bauschlicher, C. W. 1985. *Comparison of ab initio Quantum Chemistry with Experiment*, ed. R. Bartlett, pp. 17–51. Boston: Reidel
9. Bauschlicher, C. W., Walch, S. P. 1982. *J. Chem. Phys.* 76: 4560–63
10. Chong, D. P., Langhoff, S. R., Bauschlicher, C. W., Walch, S. P., Partridge, H. 1986. *J. Chem. Phys.* 85: 2850–60
11. Langhoff, S. R., Pettersson, L. G. M., Bauschlicher, C. W., Partridge, H. 1987. *J. Chem. Phys.* 86: 268–78
12. Bauschlicher, C. W., Walch, S. P., Langhoff, S. R. 1986. *Quantum Chemistry: the Challenge of Transition Metals and Coordination Chemistry*, ed. A. Veillard, pp. 15–35. Dordrecht: Reidel
13. Kant, A., Strauss, B. 1966. *J. Chem. Phys.* 45: 3161–62
14. Richtsmeier, S. C., Parks, E. K., Liu, K., Pobo, L. G., Riley, S. J. 1985. *J. Chem. Phys.* 82: 3659–65
15. Brucat, P. J., Pettiette, C. L., Yang, S., Zheng, L. S., Craycraft, M. J., Smalley, R. E. 1986. *J. Chem. Phys.* 85: 4747–48
16. Martin, R. L. 1983. *J. Phys. Chem.* 87: 750–54
17. Cowan, R. D., Griffin, D. C. 1976. *J. Opt. Soc. Am.* 66: 1010–14

18. Shavitt, I. 1977. *Methods of Electronic Structure Theory*, ed. H. F. Schaefer, pp. 189–275. New York: Plenum
19. Saunders, V. R., van Lenthe, J. H. 1983. *Mol. Phys.* 48: 923–54
20. Siegbahn, P. E. M. 1984. *Chem. Phys. Lett.* 109: 417–23
21. Knowles, P. J., Handy, N. C. 1984. *Chem. Phys. Lett.* 111: 315–21
22. Bauschlicher, C. W., Taylor, P. R. 1986. *J. Chem. Phys.* 85: 6510–12
23. Bauschlicher, C. W., Langhoff, S. R. 1987. *J. Chem. Phys.* 86: 5595–99
24. Wachters, A. J. H. 1970. *J. Chem. Phys.* 52: 1033–36
25. Hay, P. J. 1977. *J. Chem. Phys.* 66: 4377–84
26. Bauschlicher, C. W., Walch, S. P., Partridge, H. 1982. *J. Chem. Phys.* 76: 1033–39
27. Faegri, K., Speis, H. J. 1987. *J. Chem. Phys.* 86: 7035–40
28. Raffenetti, R. C. 1973. *J. Chem. Phys.* 58: 4452–58
29. Almlöf, J., Taylor, P. R. 1987. *J. Chem. Phys.* 86: 4070–77
30. Bauschlicher, C. W. 1987. *Chem. Phys. Lett.* 142: 71–75
31. Huzinaga, S. 1977. *J. Chem. Phys.* 66: 4245
32. Walch, S. P., Bauschlicher, C. W., Nelin, C. J. 1983. *J. Chem. Phys.* 79: 3600–2
33. Krauss, M., Stevens, W. J. 1984. *Ann. Rev. Phys. Chem.* 35: 357–85
34. Hay, P. J., Wadt, W. R. 1985. *J. Chem. Phys.* 82: 299–310
35. Rohlfing, C. M., Hay, P. J., Martin, R. L. 1986. *J. Chem. Phys.* 85: 1447–55
36. Bauschlicher, C. W., Walch, S. P., Siegbahn, P. E. M. 1982. *J. Chem. Phys.* 76: 6015–17
37. Pelissier, M., Davidson, E. R. 1984. *Int. J. Quantum Chem.* 25: 483–91
38. Langhoff, S. R., Davidson, E. R. 1974. *Int. J. Quantum Chem.* 8: 61–72
39. Ahlrichs, R., Scharf, P., Ehrhardt, C. 1985. *J. Chem. Phys.* 82: 890–98
40. Chong, D. P., Langhoff, S. R. 1986. *J. Chem. Phys.* 84: 5606–10
41a. Siegbahn, P. E. M., Heiberg, A., Roos, B. O., Levy, B. 1980. *Phys. Scripta* 21: 323–27
41b. Roos, B. O., Taylor, P. R., Siegbahn, P. E. M. 1980. *Chem. Phys.* 48: 157–73
41c. Roos, B. O. 1980. *Int. J. Quantum Chem.* S14: 175–89
41d. Siegbahn, P. E. M., Almlöf, J., Heiberg, A., Roos, B. O. 1981. *J. Chem. Phys.* 74: 2384–96
42a. Werner, H.-J., Meyer, W. 1981. *J. Chem. Phys.* 74: 5794–5801
42b. Cheung, L. M., Elbert, S. T., Ruedenberg, K. 1979. *Int. J. Quantum Chem.* 16: 1069–1101
42c. Docken, K. K., Hinze, J. 1972. *J. Chem. Phys.* 57: 4928–36
42d. Diffenderfer, R. N., Yarkony, D. R. 1982. *J. Phys. Chem.* 86: 5098–5105
43. Bauschlicher, C. W., Siegbahn, P., Pettersson, L. G. M. 1988. *Theoret. Chim. Acta.* In press
44. Bauschlicher, C. W. 1987. *J. Chem. Phys.* 86: 5591–94
45. Bauschlicher, C. W., Langhoff, S. R., Partridge, H., Taylor, P. R. 1986. *J. Chem. Phys.* 85: 3407–10
46. Bauschlicher, C. W. 1982. *Chem. Phys. Lett.* 91: 4–8
47. Walch, S. P., Bauschlicher, C. W., Langhoff, S. R. 1985. *J. Chem. Phys.* 83: 5351–52
48. Gray, J. A., Rice, S. F., Field, R. W. 1985. *J. Chem. Phys.* 82: 4717–18
49. Elkind, J. L., Armentrout, P. B. 1986. *Inorg. Chem.* 25: 1078–80
50. Elkind, J. L., Armentrout, P. B. 1986. *J. Phys. Chem.* 90: 5736–45
51. Elkind, J. L., Armentrout, P. B. 1986. *J. Phys. Chem.* 90: 6576–86
52. Mandich, M. L., Halle, L. F., Beauchamp, J. L. 1984. *J. Am. Chem. Soc.* 106: 4403–11
53. Stevens, A. E., Feigerle, C. S., Lineberger, W. C. 1983. *J. Chem. Phys.* 78: 5420–31
54. Balfour, W. J., Lindgren, B., O'Connor, S. 1983. *Phys. Scripta* 28: 551–60
55. Bauschlicher, C. W., Langhoff, S. R. 1988. *Chem. Phys. Lett.* 145: 205–10
56. Bauschlicher, C. W. 1988. *J. Phys. Chem.* In press
57. Krauss, M., Stevens, W. J. 1985. *J. Chem. Phys.* 82: 5584–96
58. Pettersson, L. G. M., Bauschlicher, C. W., Langhoff, S. R., Partridge, H. 1987. *J. Chem. Phys.* 87: 481–92
59. Schilling, J. B., Goddard, W. A., Beauchamp, J. L. 1986. *J. Am. Chem. Soc.* 108: 582–84
60. Alvarado-Swaisgood, A. E., Harrison, J. F. 1985. *J. Phys. Chem.* 89: 5198–5202
61. Alvarado-Swaisgood, A. E., Allison, J., Harrison, J. F. 1985. *J. Phys. Chem.* 89: 2517–25
62. Walch, S. P. 1984. *Chem. Phys. Lett.* 105: 54–57
63. Blomberg, M. R. A., Siegbahn, P. E. M., Roos, B. O. 1982. *Mol. Phys.* 47: 127–43
64. Balasubramanian, K., Feng, P. Y., Liao, M. Z. 1987. *J. Chem. Phys.* 87: 3981–85
65. Wang, S. W., Pitzer, K. S. 1983. *J. Chem. Phys.* 79: 3851–58

66. Langhoff, S. R., Bauschlicher, C. W., Partridge, H. 1988. *J. Chem. Phys.* In press
67. Harrison, J. F. 1983. *J. Phys. Chem.* 87: 1312–22
68. Langhoff, S. R., Bauschlicher, C. W., Partridge, H. 1986. *J. Chem. Phys.* 84: 1687–95
69. Huber, K. P., Herzberg, G. 1979. *Constants of Diatomic Molecules.* New York: Van Nostrand Reinhold
70. McLeod, D., Weltner, W. 1966. *J. Phys. Chem.* 70: 3293–3300
71. Barrow, R. F., Pedersen, L. 1971. *J. Phys. B* 4: L11–L13
72. Fischell, D. R., Brayman, H. C., Cool, T. A. 1980. *J. Chem. Phys.* 73: 4260–72
73. Langhoff, S. R., Bauschlicher, C. W. 1986. *Chem. Phys. Lett.* 124: 241–47
74. Steimle, T. C., Nachman, D. F., Fletcher, D. A. 1987. *J. Chem. Phys.* 87: 5670–73
75. Rice, S. F., Field, R. W. 1986. *J. Mol. Spectrosc.* 119: 331–36
76. Bauschlicher, C. W., Langhoff, S. R. 1986. *J. Chem. Phys.* 85: 5936–42
77. Adam, A. G., Azuma, Y., Barry, J. A., Huang, G., Lyne, M. P. J., Merer, A. J., Schröder, J. O. 1987. *J. Chem. Phys.* 86: 5231–38
78. Morse, M. D. 1986. *Chem. Rev.* 86: 1049–1109
79. Weltner, W. Jr., Van Zee, R. J. 1984. *Ann. Rev. Phys. Chem.* 35: 291–327
80. Rao, T. V. R., Lakshman, S. V. J. 1971. *J. Quant. Spectrosc. Radiat. Trans.* 11: 1157–61
81. Rohlfing, E. A., Valentini, J. J. 1986. *J. Chem. Phys.* 84: 6560–66
82. Werner, H.-J., Martin, R. L. 1985. *Chem. Phys. Lett.* 113: 451–56
83. Scharf, P., Brode, S., Ahlrichs, R. 1985. *Chem. Phys. Lett.* 113: 447–50
84. Langhoff, S. R., Bauschlicher, C. W. 1986. *J. Chem. Phys.* 84: 4485–88
85. Bauschlicher, C. W., Langhoff, S. R., Taylor, P. R. 1988. *J. Chem. Phys.* 88: 1041–45
86. Leopold, D. G., Ho, J., Lineberger, W. C. 1987. *J. Chem. Phys.* 86: 1715–26
87. Walch, S. P., Bauschlicher, C. W., Langhoff, S. R. 1986. *J. Chem. Phys.* 85: 5900–7
88. McLean, A. D. 1983. *J. Chem. Phys.* 79: 3392–3403
89. Basch, H. 1980. *Faraday Symp. Chem. Soc.* 14: 149–58; Basch, H. 1981. *J. Am. Chem. Soc.* 103: 4657–63
90. Shim, I., Gingerich, K. A. 1983. *J. Chem. Phys.* 79: 2903–12
91a. Brown, C. M., Ginter, M. L. 1978. *J. Mol. Spectrosc.* 69: 25–36
91b. Srdanov, V. I., Pesic, D. S. 1981. *J. Mol. Spectrosc.* 90: 27–32
92. Michalopoulos, D. L., Geusic, M. E., Hansen, S. G., Powers, D. E., Smalley, R. E. 1982. *J. Phys. Chem.* 86: 3914–16
93. Baykara, N. A., McMaster, B. N., Salahub, D. R. 1984. *Mol. Phys.* 52: 891–905
94. Delley, B., Freeman, A., Ellis, D. E. 1983. *Phys. Rev. Lett.* 50: 488–91
95. Goodgame, M. M., Goddard, W. A. 1985. *Phys. Rev. Lett.* 54: 661–64
96. Walch, S. P., Bauschlicher, C. W., Roos, B. O., Nelin, C. J. 1983. *Chem. Phys. Lett.* 103: 175–79
97. Goodgame, M. M., Goddard, W. A. 1982. *Phys. Rev. Lett.* 48: 135–38
98. McLean, A. D., Liu, B. 1983. *Chem. Phys. Lett.* 101: 144–48
99. Werner, H.-J., Reinsch, E-A. 1982. *J. Chem. Phys.* 76: 3144–56
100. Siegbahn, P. E. M. 1983. *Int. J. Quantum Chem.* 23: 1869–89
101. Knight, L. B., Van Zee, R. J., Weltner, W. 1983. *Chem. Phys. Lett.* 94: 296–99
102. Walch, S. P., Bauschlicher, C. W. 1983. *J. Chem. Phys.* 79: 3590–91
103. Leopold, D. G., Almlöf, J., Lineberger, W. C., Taylor, P. R. 1988. *J. Chem. Phys.* 88: 3780–83
104. Cosse, C., Fouassier, M., Mejean, T., Tranquille, M., DiLella, D. P., Moskovits, M. 1980. *J. Chem. Phys.* 73: 6076–85
105. Walch, S. P., Bauschlicher, C. W. 1986. See Ref. 12, pp. 119–34
106. Howard, J. A., Preston, K. F., Sutcliffe, R., Mile, B. 1983. *J. Phys. Chem.* 87: 536–37
107. Langhoff, S. R., Bauschlicher, C. W., Walch, S. P., Laskowski, B. C. 1986. *J. Chem. Phys.* 85: 7211–15
108. Wang, S-W. 1985. *J. Chem. Phys.* 82: 4633–40
109. Post, D., Baerends, E. J. 1982. *Chem. Phys. Lett.* 86: 176–80
110. Anderson, A. B. 1978. *J. Chem. Phys.* 68: 1744–51
111. Truhlar, D. G., Thompson, T. C., Mead, C. A. 1986. *Chem. Phys. Lett.* 127: 287–91; see also Thompson, T. C., Truhlar, D. G., Mead, C. A. 1985. *J. Chem. Phys.* 82: 2392–2407
112. Rohlfing, E. A., Valentini, J. J. 1986. *Chem. Phys. Lett.* 126: 113–18
113. Walch, S. P., Laskowski, B. C. 1986. *J. Chem. Phys.* 84: 2734–43
114. Siegbahn, P. E. M., Blomberg, M. R. A., Bauschlicher, C. W. 1984. *J. Chem. Phys.* 81: 2103–11

SYNCHRONICITY IN MULTIBOND REACTIONS

Weston Thatcher Borden

Department of Chemistry, University of Washington, Seattle, Washington 98195

Richard J. Loncharich and K. N. Houk

Department of Chemistry and Biochemistry, University of California, Los Angeles, California 90024-1569

INTRODUCTION

Pericyclic reactions have been defined by Woodward & Hoffmann (1) as "reactions in which all first-order changes in bonding relationships take place in concert on a closed curve." Among the familiar organic reactions that have been generally assumed to fall into this class are the Cope and Claisen sigmatropic rearrangements, the Diels-Alder and 1,3-dipolar cycloaddition reactions, 1,5 hydrogen shifts, and the electrocyclic ring opening of cyclobutene. In fact, underlying the highly successful Woodward-Hoffmann rules (1) for predicting which reactions are likely to occur via pericyclic transition states is this assumption—that in "allowed" reactions bond making and bond breaking occur synchronously.

This assumption has been vigorously challenged by Dewar (2). In a paper entitled "Multibond Reactions Cannot Ordinarily Be Synchronous" Dewar espouses a "new rule regulating the course of reactions; i.e. *synchronous multibond reactions are normally prohibited.*"

Dewar defines a *synchronous* reaction as "one where all the bond-making and bond-breaking processes take place in unison, having all proceeded to comparable extents in the transition state." Dewar is careful to distinguish between a synchronous reaction and a concerted one. A *concerted* reaction is defined by Dewar as "one that takes place in a single kinetic step without necessarily being synchronous."

A concerted reaction that is not synchronous is termed a *two-stage* reaction by Dewar. In such a reaction "some of the changes in bonding take place in the first part of the reaction, followed by the rest [of the changes in the second part of the reaction]." In contrast, "A *two-step* reaction takes place in two kinetically distinct steps, via a stable intermediate." We believe these definitions of Dewar's to be a valuable contribution, and they are used throughout this paper.

Dewar bases his contention that multibond reactions cannot ordinarily be synchronous on three types of evidence. First he presents a theoretical argument, based on Bell-Evans-Polanyi plots, in favor of his hypothesis. However, Dewar concludes this section of his paper with the statement, "These arguments show that no qualitative theory can on its own lead to reliable conclusions concerning the mechanism of any multibond pericyclic reaction." We agree with this conclusion.

The second type of evidence discussed by Dewar comes from the results of calculations, performed by Dewar and his co-workers, using semiempirical methods (MINDO, MNDO, and AM1) developed by Professor Dewar's group. These methods, when applied to specific multibond reactions, most often find transition states in which bond making and bond breaking are not synchronous.

One must inquire, however, whether this general finding is providing information about the reactions being investigated or about the computational methods being used to investigate them. In this review we discuss the results of ab initio calculations on the transition states for a number of important multibond organic reactions. These calculations generally lead to the opposite results from the semiempirical calculations performed by Dewar and his co-workers. Ab initio calculations on organic reactions for which a pericyclic transition state is "allowed" by the Woodward-Hoffmann rules usually find these reactions to involve synchronous bond making and bond breaking.

Despite the dramatic growth in this decade of the power of both the hardware and software available to chemists for performing ab initio electronic structure calculations, it is still the case that the Schrödinger equation is very far from being solved for any of the systems discussed in this review. Thus, it can be argued that the results of the ab initio calculations that have been performed for the reactions discussed here are, *a priori*, no more reliable than those obtained by Dewar's semiempirical techniques.

Comparison with experiment must be the ultimate recourse for testing which of two computational models is the more useful. Dewar interprets the available experimental evidence on multibond reactions in favor of the two-stage mechanisms obtained from his semiempirical calculations. This provides the third line of evidence that he adduces in favor of his hypothesis that multibond reactions cannot ordinarily be synchronous.

For each of the reactions discussed in this review we reexamine the available experimental evidence to see whether it is better accomodated by a synchronous or nonsynchronous mechanism. In contrast to the conclusion reached by Dewar, we find that the experimental evidence supports the existence of synchronicity in organic transition states.

COPE REARRANGEMENT

The Cope rearrangement of 1,5-hexadiene is depicted schematically in Figure 1. There is general agreement that the reaction passes through a species that has higher symmetry than the reactants or products. However,

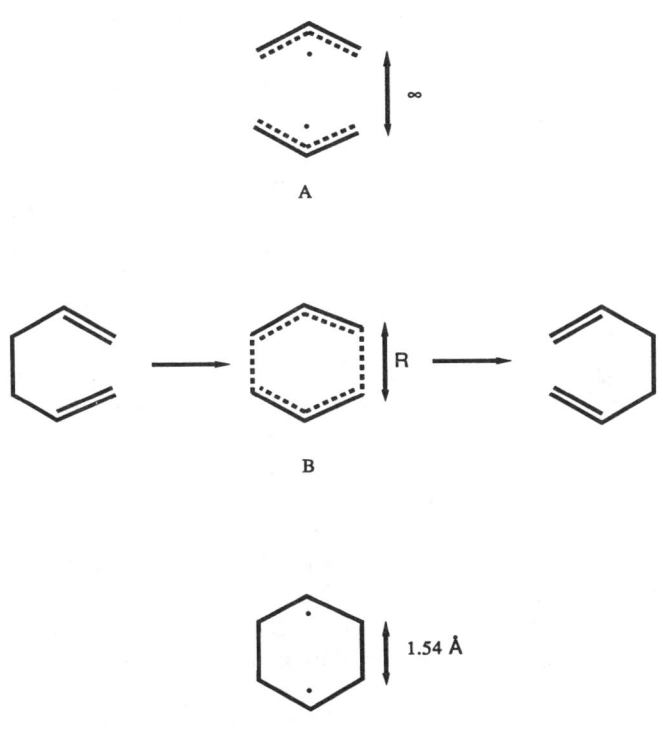

Figure 1 Schematic depiction of three possible high symmetry species (HSS) in the Cope rearrangement of 1,5-hexadiene. R, the length of the two equivalent bonds between the three-carbon fragments, would be infinite for two allyl radicals (A), about 1.54 Å in a completely localized cyclohexane-1,4-diyl (C), and somewhere between these extremes in the actual HSS.

as noted in a previous review of this reaction by Gajewski (3), there is considerable controversy surrounding the question of the nature of this high symmetry species (HSS).

As shown in Figure 1, if the Cope rearrangement is a two-stage reaction, there are two extreme structures for the HSS, which differ in the timing of the σ bond breaking and σ bond making steps. If σ bond breaking precedes σ bond making, the HSS would be essentially two allyl radicals (A). At the other extreme, if σ bond making is essentially complete before substantial σ bond breaking occurs, the HSS could be described as cyclohexane-1,4-diyl (C).

Of course, there is also the possibility that in the Cope rearrangement σ bond making and σ bond breaking occur synchronously, so that the HSS is really a hybrid of A and C. It can then be depicted as in B. Unlike A or C, which are diradicals, B is an "aromatic" transition state (4, 5) and, therefore, has a closed-shell electronic structure.

That the Cope rearrangement does not actually proceed by σ bond cleavage to afford two allyl radicals is indicated by labeling studies (3, 6) and by the fact that the bond dissociation energy of 56 kcal/mol for 1,5-hexadiene going to two allyl radicals (7) is substantially higher than the 33.5 kcal/mol required for its Cope rearrangement (8). On the other hand, Doering has pointed out that thermochemical estimates of the heat of formation of cyclohexane-1,4-diyl (C) are permissive of it being the HSS (8). Although a revised estimate of the heat of formation of C now places it about 7 kcal/mol above the transition state for the Cope rearrangement (9), this estimate is based on the assumption of no stabilizing interactions between the two radical sites.

There is, in fact, a stabilizing interaction between the nonbonding AOs on C-1 and C-4 of C, which occurs through the two σ bonds that are aligned with these orbitals (10, 11). The interaction involves mixing of the antisymmetric combination of the nonbonding AOs with the antibonding orbitals of the localized σ bonds. This mixing results in the lowering of the energy of the antisymmetric combination of nonbonding AOs. The symmetric combination of nonbonding AOs is actually destabilized by interaction with the σ bonding orbitals. Thus, the pair of ostensibly nonbonding electrons in C preferentially occupy the lower energy of the two resulting MOs, which involves the antisymmetric combination of AOs.

There are several consequences of the selective occupancy of this MO. One consequence is that the interaction between the AOs on C-1 and C-4 is antibonding, so that a geometry that maximizes the distance between these orbitals is favored. The experimental preference (3, 12–16) for a chair transition state, rather than a boat, in the Cope rearrangement has been attributed, at least in part, to this effect (1, 10, 12, 17).

Another consequence of the preferential occupancy of the MO that results from mixing of the antisymmetric combination of AOs on C-1 and C-4 with the antibonding orbitals of the two σ bonds is that these two bonds will be longer and the remaining four C–C bonds shorter than one would expect if the nonbonding electrons in C were really localized. Thus, the bond lengths of the HSS will be distorted in the direction of A away from that implied by the localized structure of C.

A key question in the Cope rearrangement, therefore, is not whether the HSS has a delocalized electronic structure but exactly where on the continuum between localized structures A and C it falls. The length, R, of the two equivalent C–C bonds in the HSS provides one useful indication of its electronic structure. At the extremes of A and C, R would be infinite for the former and about 1.54 Å for the latter.

Because the HSS has higher symmetry than the reactants or products, it must be a stationary point on the potential surface; this implies that its geometry can be optimized. A second important question regarding the Cope rearrangement concerns the nature of this stationary point. Is it a transition state, with one negative force constant, corresponding to motion along the reaction coordinate; or is it a stable intermediate, possessing only positive force constants?

Results of Semi-empirical Calculations

Komornicki & McIver (18) have performed MINDO/2 calculations on the potential surface for the Cope rearrangement, and Dewar and co-workers (19) carried out calculations using MINDO/3. Similar results were obtained by both groups. The HSS involved in both the chair and boat Cope rearrangement of 1,5-hexadiene was found to be a stable intermediate. The two equivalent C–C bond lengths, R, were calculated to be just 1.61 Å in the chair intermediate and only about 0.02 Å longer in the boat.

The transition states connecting 1,5-hexadiene to these higher symmetry intermediates were found to be approximately 2 kcal/mol higher in energy in both studies. The bond being formed between the terminal atoms of 1,5-hexadiene was calculated to be about 1.90 Å long in the chair transition state and slightly shorter in the boat. The bond between C-3 and C-4 in the transition states was computed to be barely longer than that in the reactant.

Both MINDO calculations thus found the Cope rearrangement to be a two-step reaction in which bond making and bond breaking are decidedly asynchronous. In the first step a σ bond is made, leading to formation of an intermediate in which the two equivalent C–C bond lengths, R, are only slightly longer than those of ordinary C–C bonds. In the second step of the reaction a σ bond is broken via a transition state that is the mirror image of the one that leads to σ bond formation.

The bond lengths, R, in the intermediate suggest rather little electron delocalization into them. However, Dewar (2, 19) has pointed out that through-bond interactions in the intermediate are sufficiently strong to keep it from being a true diradical, since 3×3 configuration interaction (CI) lowers the MINDO/3 energies of true diradicals by 40–60 kcal/mol; whereas, 3×3 CI lowers the energy of the intermediate in the Cope rearrangement by only 20 kcal/mol.

The MINDO/2 activation enthalpies for the chair and boat Cope rearrangements were computed to be 22.3 kcal/mol and 30.3 kcal/mol, respectively, values significantly lower than the experimental values of 33.5 kcal/mol (8) and 44.6 kcal/mol (13). The activation entropy of the chair rearrangement was computed by MINDO/2 to be -4.0 eu, which is considerably more positive than the experimental value of -13.8 eu (8). The boat activation entropy of -3.8 eu was in good agreement with the experimental value of -3 ± 3.6 eu (13).

The MINDO/3 activation enthalpies of 35.1 kcal/mol and 41.4 kcal/mol were in somewhat better agreement with experiment. The MINDO/3 activation entropies were -17.0 eu for the chair and -14.7 for the boat. The value for the chair is in reasonable agreement with experiment, but the value for the boat is now too negative. Neither MINDO/2 nor MINDO/3 reproduces the large difference in entropies between the boat and chair transition states that is found experimentally in both the parent 1,5-hexadiene (13) and in derivatives that are constrained to rearrange by the two different transition states (15). The experimental difference between the enthalpies of the two transition states is also somewhat underestimated by both sets of calculations.

Very recently Dewar (20) has published the results of AM1 calculations on the Cope rearrangement of 1,5-hexadiene and phenyl substituted derivatives. A major difference between the MINDO and AM1 results is that the latter calculations find both the structures and the energies of the transition states to be much closer to those of the HSS. In fact, Dewar states that the energy minima corresponding to the HSS are too shallow to have any definite significance, so that the HSS may actually be transition states. Thus, the AM1 results leave open the possibility that the Cope rearrangement is a concerted rather than a two-step reaction. However, the two equivalent bond lengths, R, in the HSS are still short enough (1.65 Å) that Dewar claims that they are inconsistent with a synchronous mechanism in which bond making and bond breaking occur simultaneously.

More recently still, Dewar (21) has reported the discovery of a second transition state on the AM1 potential surface for both the boat and chair Cope rearrangement. These boat and chair transition states, which were apparently overlooked in the previous AM1 study (20), both have values of R close to 2.0 Å and appear to be delocalized aromatic species, formed by bond breaking occurring synchronously with bond making. However, the aromatic transition states are each calculated to be higher in energy than their diradicaloid counterparts. The barrier height between the aromatic and diradicaloid transition states was, unfortunately, not reported for either the boat or chair.

Presumably because of their longer bond lengths and, hence, lower frequency vibrations, the aromatic transition states were each calculated to have a higher entropy than their diradical counterparts. The free energy differences between these two types of transition states are consequently smaller than the differences in their enthalpies. In the case of the boat the free energies of the two transition states were computed to be nearly identical at 523°K. The computed entropy of activation for passage through the aromatic boat transition state was computed to be close to that measured experimentally for the boat Cope rearrangement.

Results of Ab Initio Calculations

Ab initio calculations were first performed on the chair Cope rearrangement by Osamura and co-workers (22). Calculations were carried out with both the STO-3G minimal basis set and the 3-21G split-valence basis set. Several different types of wavefunctions were tested—SCF, two-configuration SCF (TCSCF) and multi-configuration SCF (MCSCF).

The MCSCF calculations included all the configurations that arise from distributing the six electrons that participate in the Cope rearrangement in six orbitals. With the 3-21G basis set, the MCSCF calculations proved so large that they had to be done in two separate steps. First, the orbitals were optimized by performing an MCSCF calculation with a smaller number of configurations. This was followed by a CI calculation with the optimized orbitals that included all the configurations.

The results obtained in this ab initio study depended upon the type of calculation that was performed. At the SCF level, with both basis sets, the HSS was found to be a pericyclic transition state. The bond length, R, was computed to be 1.79 Å with the STO-3G basis set and 2.02 Å with the more flexible 3-21G. The finding of a closed-shell, delocalized, aromatic transition state at the SCF level was expected, since an SCF wavefunction, consisting of a single configuration, gives a very poor description of a singlet diradical. Thus, because an SCF wavefunction is biased against both structures A and C in Figure 1, it was predictable that structure B would be obtained for the HSS.

With the STO-3G basis set, both TCSCF and MCSCF calculations gave a structure for the HSS with a value for R around 1.60 Å and a wavefunction with a very large coefficient for the second configuration that is necessary to describe cyclohexane-1,4-diyl. However, it is well known that the STO-3G basis set greatly overestimates the strength of σ, relative to π C–C bonds (23). Since structure C of Figure 1 has two more σ bonds than A, it is not surprising that STO-3G finds the HSS in the Cope rearrangement to resemble C more than A. In fact, not only was the HSS found to be an intermediate, but because it has one more σ bond than 1,5-hexa-diene, its energy was actually calculated to be lower than that of the reactant.

The 3-21G basis set provides a balanced description of σ and π C–C bond energies (23). However, when a TCSCF wavefunction was used, a cyclohexane-1,4-diyl type structure (C) with $R = 1.61$ Å was also found for the HSS with this basis set. Nevertheless, since the second configuration in the wavefunction was the one necessary to describe diradical C, but not diradical A, this finding is again suspect. Because A and C may be viewed as resonance contributors to the structure of the HSS in the Cope rearrangement, in order to obtain the correct structure of this species, it is essential that a calculation be unbiased and capable of describing both A and C equally well (22).

When an unbiased MCSCF wavefunction was used, the 3-21G MCSCF optimized geometry for the chair HSS was found to have $R = 2.06$ Å and to possess a negative force constant for asymmetric distortion of the two equivalent bond lengths. Thus, this species appears to be the transition state for a concerted pericyclic process in which bond breaking is synchronous with bond making.

The MCSCF energy of the transition state was 29.4 kcal/mol above that of 1,5-hexadiene, when the latter was optimized in C_{2h} symmetry. Hrovat and co-workers (D. A. Hrovat, K. Morokuma, W. T. Borden, unpublished results) have found that reoptimization of 1,5-hexadiene in either C_i or C_2 symmetry gives a 4.5 kcal/mol lower SCF energy. After correcting for the 1.7 kcal/mol larger vibrational energy of the reactant at 523 K, they obtained an enthalpy of activation for the chair Cope rearrangement of 32.2 kcal/mol, which is in excellent agreement with the experimental value of 33.5 kcal/mol (8).

The MCSCF calculations of Osamura et al (22) have been criticized by Dewar (20), who has objected to the use of a simulated MCSCF wavefunction, the lack of full geometry reoptimization in going from the

SCF to the MCSCF level, and the absence of a complete vibrational analysis to characterize the HSS as a transition state. Dewar's criticisms have been answered by Morokuma and co-workers (K. Morokuma, W. T. Borden, D. A. Hrovat, submitted for publication), who performed complete active space SCF (CASSCF) calculations, using an MCSCF wavefunction that contained all the configurations that arise from six electrons in six orbitals. Morokuma and co-workers optimized the chair and boat HSS and performed vibrational analyses on them with this MCSCF wavefunction, using the 3-21G basis set.

The vibrational analyses showed the HSS in both the chair and boat Cope rearrangement to be a transition state. The two equivalent bond lengths, R, were found to be 2.09 Å in the chair transition state and 2.32 Å in the boat. Obviously, considerable bond breaking accompanies bond making in both transition states.

The enthalpy of the boat transition state was computed to be 6.6 kcal/mol higher than that of the chair, and the entropy of the boat was found to be 2.8 eu greater. The difference in computed entropies for the two transition states is substantially less than the 11 eu found experimentally. The entropy of activation that was computed for the chair transition state is in good agreement with the experimental value; thus, it is the entropy computed for the boat that appears to be too small.

Ab initio calculations tend to overestimate vibrational frequencies. Since low frequency vibrations make the largest contributions to vibrational entropy, it is possible that the entropy of the boat transition state is underestimated by the ab initio calculations. Alternatively, as suggested by the results of Dewar's AM1 calculations (21), on the *free* energy surface for the boat Cope rearrangement the transition state may occur at a geometry that has both a higher enthalpy and entropy than the saddle point on the *potential* energy surface.

Discussion

Dewar's AM1 calculations, unlike the MINDO results, predict that the Cope rearrangement occurs in a concerted or nearly concerted fashion and does not involve formation of a chemically significant intermediate. The 3-21G MCSCF calculations also predict a concerted reaction. This prediction has been confirmed very recently by an elegant stereochemical experiment, performed by Owens & Berson (23a).

The semiempirical and ab initio calculations differ in the value of the bond length, R, that they predict for the HSS. AM1 finds the value of R to be 1.65 Å in the chair transition state; whereas, the 3-21G MCSCF value is 2.09 Å.

Despite the large difference in calculated bond lengths, the electronic structures of the transition state that are predicted by the two computational methods are not totally dissimilar. Dewar (2, 19, 20) has pointed out that the AM1 transition state geometry is not a diradical but a diradicaloid. Thus, it is not surprising that Dewar found experimentally (24), as well as computationally (20), that phenyl substituents placed at C-2 and C-5 of 1,5-hexadiene have a much smaller rate-accelerating effect than that which would be anticipated if the transition state were a true diradical like C. Moreover, Dewar found (20, 24) that a phenyl substituent at C-3 also has a rate-accelerating effect, which is hard to explain on the basis of localized diradical C as the transition state.

Dewar (25) originally interpreted his experiments on the rate accelerations caused by phenyl groups in terms of two types of transition states for the Cope rearrangement, one resembling A and the other C. However, the rate accelerations caused by the two phenyl groups in 2,4-diphenyl-1,5-hexadiene were found to be approximately multiplicative (24). This result is indicative of a single type of transition state for the Cope rearrangement, intermediate in structure between A and C, which can thus be stabilized in two ways.

The ab initio, MCSCF calculations are in full agreement with Dewar's assertion that the Cope transition state is not a diradical. The MCSCF wavefunction for the chair transition state is dominated by a single configuration (22). Thus, the wavefunction is best described as that for a closed-shell, delocalized, "aromatic," transition state.

Such a transition state should be stabilized by substituents placed at either C-2 and C-5 or at C-3 and C-4 of 1,5-hexadiene, as is found to be the case experimentally (24, 26). To test whether ab initio calculations would mirror the stabilizing effect of substituents at these positions, Hrovat and co-workers (D. A. Hrovat, W. T. Borden, R. L. Vance, N. G. Rondan and K. N. Houk, submitted for publication) have performed calculations on the chair Cope rearrangement of dicyano derivatives of 1,5-hexadiene.

With inclusion of electron correlation at the MP2 level, calculations with the 3-21G basis set found the following energy differences between the reactant and the SCF optimized chair Cope transition state: 1,5-hexadiene—28.8 kcal/mol; 1,4-dicyano-1,5-hexadiene—26.7 kcal/mol; and 2,5-dicyano-1,5-hexadiene—20.2 kcal/mol. Although the Cope rearrangement of 1,4-dicyano-1,5-hexadiene has not been studied, the predicted energy barrier for the Cope rearrangement of the 2,5-dicyano isomer is in good agreement with the activation enthalpy of 23.3 kcal/mol that has been measured (26).

An interesting finding of these ab initio calculations was that the optimized value of R in the transition state depended on the substitution pattern. At the SCF optimized transition states, the value of R was 2.09 Å for 1,4-dicyano, 1.93 Å for 2,5-dicyano, and 2.02 Å for the parent 1,5-hexadiene. Upon geometry reoptimization with inclusion of electron correlation, even larger differences in the optimized values of R were found between the three transition states.

It appears that 1,4-dicyano substitution, which would stabilize localized diradical A, increases the contribution of A to the electronic structure of the transition state and so moves the geometry of the transition state toward that of A. Similarly, 2,5-dicyano substitution, which would stabilize C, moves the structure of the transition state toward C. This computational result is in agreement with the interpretation of Gajewski & Conrad (27) of the secondary isotope effects measured by them.

Although the AM1 and ab initio calculations appear to be in general agreement with each other and with experiment on the response of the chair transition state in the Cope rearrangement to substituents, the very large difference (>0.4 Å) between the values predicted by the two methods for the bond length R in the transition state is striking. Unfortunately, since the experimental value of R at the chair Cope transition state is neither known nor likely to become known, the controversy regarding the geometry cannot now and probably will not in the future be settled by recourse to experiment.

Dewar has acknowledged that "state-of-the-art" ab initio procedures are more reliable than semiempirical methods like MINDO and AM1 (20). Ab initio calculations have the advantage over semiempirical techniques of being amenable to systematic improvement by expansion of the basis set size and by inclusion of more electron correlation. Very recently, Dewar (28) has published the results of ab initio calculations, designed to test whether an increase in basis set size and/or electron correlation might result in a considerably smaller value of R for the chair transition state than that predicted by the 3-21G MCSCF calculations.

Dewar performed calculations at the C_{2h} transition state geometry and at a C_{2h} geometry with R fixed at 1.61 Å. Both geometries were optimized at the MP2 level of theory with the 3-21G basis set. At the MP2 level of perturbation theory the transition state geometry was found to have $R = 1.92$ Å, which is a rather smaller interallylic separation than the 2.09 Å found with a strictly variational MCSCF wavefunction. Dewar then calculated the energy differences between the two optimized geometries at various levels of theory with the 6-31G* basis set, which contains polarization functions. The 6-31G* energy differences were compared with those obtained with the 3-21G basis set.

At the SCF level the $R = 1.61$ Å geometry was found to be higher in energy by 25.7 kcal/mol with 3-21G and 20.5 kcal/mol with 6-31G*.

Inclusion of electron correlation selectively stabilizes the $R = 1.61$ Å geometry, because it has the larger amount of diradical character. Thus, at the MP4 level of theory the energy difference between the two geometries was found to be substantially smaller, amounting to 10.4 kcal/mol with 3-21G and 5.1 kcal/mol with 6-31G*; but the $R = 1.61$ Å geometry still remained the higher in energy.

Dewar (28) argues that, because electron correlation plays a major role in determining the relative energies of the two geometries, the larger amount of correlation recovered with the larger basis set is responsible for the decrease in the MP4 energy difference between them. He goes on to state, "It therefore seems possible that a further comparable decrease in δE could be effected by further expansion of the basis set."

Although it is certainly true that electron correlation plays a major role in determining the relative energies of the two geometries, the 5.3 kcal/mol decrease in the MP4 energy difference between them on basis set expansion has little to do with electron correlation. In fact, 5.2 kcal/mol of the energy decrease occurs at the SCF level. Thus, although one cannot, *a priori*, rule out a further decrease in δE on further expansion of the basis set beyond 6-31G*, the reasoning that Dewar uses to make such a reduction seem plausible is specious.

From the results of Dewar's ab initio calculations, it is very difficult to understand how he could conclude (28), "There is therefore every reason to believe that the biradicaloid structure is in fact the lower in energy, as our [semiempirical] studies have indicated." On the contrary, the highest level calculations performed by Dewar (MP4/6-31G*) find an aromatic transition state to be 5.1 kcal/mol lower in energy than the biradicaloid structure. Thus, Dewar's MP4/6-31G* calculations actually provide additional computational evidence for a transition state in the Cope rearrangement in which σ bond breaking has occurred synchronously with σ bond making.

DIELS-ALDER REACTION

The Diels-Alder reaction of butadiene and ethylene is depicted schematically in Figure 2. The reaction can occur through a *synchronous* pathway with a C_s transition structure (D), through a concerted but asynchronous (*two-stage*) process with an unsymmetrical transition structure (E), or through a nonconcerted (*two-step*) reaction with a diradical intermediate (F). There is a long history of computations on this reaction (29–33). The most recent calculations, carried out with several of Dewar's recent semi-empirical methods (34, 35) or with ab initio techniques (36–39) are discussed here.

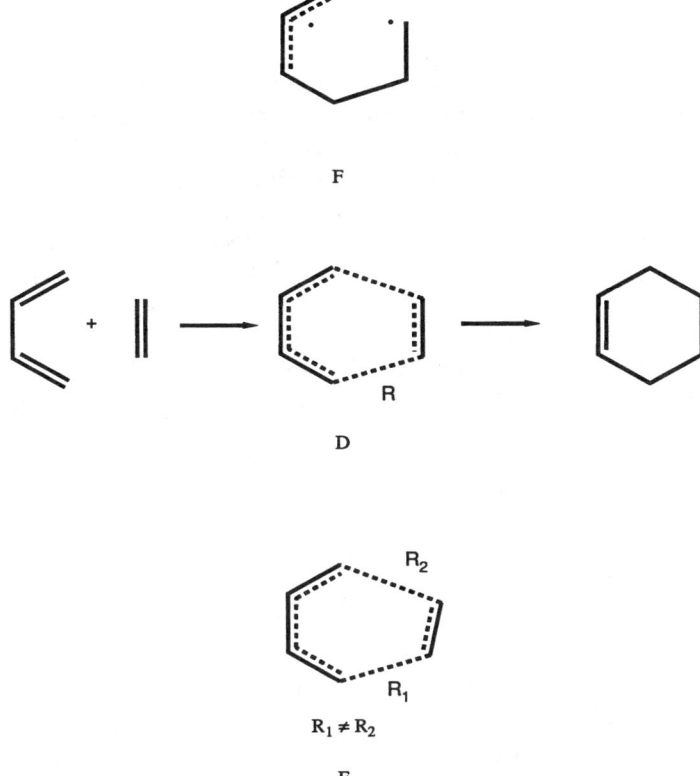

Figure 2 Schematic depiction of three possible species in the Diels-Alder reaction. D is the synchronous concerted transition structure, E depicts an asynchronous transition structure, and F represents a biradical intermediate, the extreme of an asynchronous biradicaloid transition structure.

In an early study of this reaction Townshend et al (37) used the STO-3G basis set with 3 × 3 CI for geometry optimizations, followed by single-point calculations on the geometries at the 4-31G+3 × 3 CI level. This work was only a partial optimization, constraining 17 of the 42 degrees of freedom to standard values, but was an enormous computation for its time. The computations done by Townshend et al (37) cannot be considered definitive, since they were done with a minimal basis set, and all variables were not optimized. Nevertheless, in spite of tremendous advances in hardware and software since the Townshend et al (37) work 12 years ago, and the enormous criticism heaped on that work by Dewar and others in the meantime, the conclusions of the Townshend calculations survive essentially unchanged today. This pioneering band of computational chemists explored the concerted and the two-step diradical pathways. The synchronous concerted transition state (D) was characterized by C–C forming bond distances of 2.21 Å. The activation energy for the two-step pathway involving F was only 4.1 kcal/mol above D.

The results of calculations using several semi-empirical techniques have been at odds with the ab initio results. Dewar et al (30, 31a,b) carried out an investigation of the prototype Diels-Alder reaction using a combination of the MINDO/3 (32), UMINDO/3, and MINDO/3-3 × 3 CI methods, and more recently employing MNDO and AM1 (34). The UMINDO/3 and MINDO/3-3 × 3 CI methods predict that the concerted pathway is very unfavorable, and that the two-step mechanism has a substantially lower activation energy. Forming C–C bond lengths of 2.06 Å and 5.29 Å are predicted with the UMINDO/3 method. Based on the UMINDO/3 and MINDO/3-3 × 3 CI calculations, Dewar states that there is "very definite evidence that Diels-Alder reactions involving unsymmetrical dienes, and/or unsymmetrical dienophiles, take place in a nonsynchronous manner, via very unsymmetrical transition states" (2).

However, Dewar (34) recently acknowledged that MINDO/3 and MNDO calculations inherently and incorrectly disfavor a symmetrical transition structure. Several reasons for this had been proposed in the literature by others (40, 41). Dewar suggests that this problem is due to an overestimation of repulsive interactions between atoms in the critical region where the C–C lengths are 1.5 to 2 times the length of a normal covalent bond. It is further claimed that "this overestimation of interatomic repulsions is corrected in AM1. AM1 consequently predicts the Diels-Alder Transition State (DA TS) to be symmetrical. ... AM1 should be used in all future semi-empirical studies of pericyclic reactions in preference to the MINDO/3 or MNDO (34)." Since Dewar has retracted the MINDO/3 and MNDO calculations on the Diels-Alder reaction, only the results of the AM1 semi-empirical method are discussed further here.

The RHF AM1 method predicts that the parent Diels-Alder reaction is concerted and synchronous, in agreement with ab initio calculations. The transition structure involves C–C bond formation of 2.119 Å with an activation energy of 23.8 kcal/mol. However, the UHF and RHF-CI versions of AM1 both predict a two-step mechanism.

The UHF version of AM1 predicts a first transition structure with two forming C–C bond lengths of 1.987 and 4.927 Å and an activation energy of 17.3 kcal/mol. The reaction pathway proceeds downhill in energy by 27.6 kcal/mol to a stable intermediate. The stable intermediate then rearranges via another transition structure of 4.0 kcal/mol activation energy to the product cyclohexene. This second transition structure has bond lengths of 1.530 and 2.409 Å (34).

The first transition structure found by the RHF-3 × 3 CI version of AM1 similarly possesses C–C forming bond lengths of 1.933 and 4.989 Å with an activation barrier of 34.3 kcal/mol. An intermediate is found at 15.7 kcal/mol above reactants. The second transition structure shows one fully formed C–C bond (1.532 Å) and one long C–C forming bond (2.903 Å). The activation energy from the intermediate to the second transition structure is 0.9 kcal/mol.

Dewar notes that the prediction by the UHF and 3 × 3 CI versions of AM1 of a two-step mechanism for the DA reaction is in conflict with the experimental evidence that the reaction is concerted. He claims that UHF and 3 × 3 CI AM1 both overestimate the stability of diradicals by 20 kcal/mol, and so he corrects the UHF and 3 × 3 CI energetics by this amount. When a correction factor of 20 kcal/mol is applied, the intermediate found by these two techniques is no longer an energy minimum. Passage over the second transition structure (now the only transition structure) becomes rate-determining (34).

Recent ab initio calculations of the reaction of 1,3-butadiene with ethylene have been carried out by Houk et al (39a,b). Full geometry optimizations with RHF STO-3G and 3-21G calculations gave C_s transition structures. The zero-point energy corrected activation energies are 38.6 and 38.3 kcal/mol, respectively. The C–C forming bond distances are 2.217 Å at the STO-3G level, whereas the transition structure is slightly later at the 3-21G level with the corresponding bonding distance of 2.210 Å. The experimental activation energy for the Diels-Alder reaction of butadiene with ethylene has been measured as 27.5 and 34.3 kcal/mol (42, 43a–c). At various levels of theory (39a,b), the calculated activation entropy is −42 eu, whereas the two experimental values are −30 eu and −41 eu (42, 43a–c).

Ab initio MCSCF calculations (36; F. Bernardi, A. Bottoni, M. J. Field, M. F. Guest, I. H. Hillier, M. A. Robb and A. Venturini, submitted for publication) contain enough electron correlation to find diradical transition states, but these calculations give a synchronous C_s transition structure as the one of lowest energy. CAS-STO-3G calculations were performed, involving six active electrons in six orbitals (CAS2) resulting in 175 configurations. CAS calculations with four active electrons in four orbitals (CAS1) with the STO-3G and 4-31G basis sets were performed. A more complete potential surface search was carried out at the CAS1-STO-3G level. The approach of *cis*-butadiene to ethylene gave a C_s symmetric transition structure, three diradical transition structures (*syn-gauche, trans, anti-gauche*), and three corresponding diradical minima.

The three minima are separated by two transition states for the required conformational changes.

The CAS-4-31G potential differs significantly from the CAS-STO-3G surface. Transition structures associated with the *gauche* and *trans* approaches disappear on the 4-31G surface; the C_s and *syn-gauche* diradical transition structures persist and were optimized. The forming C–C bond distances of the C_s transition structure were 2.244 Å. The fully formed bond of the *syn-gauche* transition structure was 1.656 Å. The C_s transition structure was 2.2 kcal/mol lower in energy than the *syn-gauche* diradical transition structure.

Experimental evidence is compatible only with a concerted mechanism (39a,b, 44). Reaction of tetradeuteriobutadiene with *cis*- and *trans*-1,2-dideuterioethylene showed no scrambling of stereochemistry in the product. It is concluded that the concerted stereospecific reaction has at least a 3.7 kcal/mol lower activation energy than the stepwise reaction that could scramble stereochemistry. Other estimates of the energy of concert of simple Diels-Alder reactions are in the range of 2–10 kcal/mol (45a,b, 46). All the combatants now agree that the butadiene-ethylene reaction has a synchronous transition structure! Sometimes the warriors lay down their arms . . .

. . . but not for long, since there is still adequate disagreement about substituted reactions to keep the epinephrine flowing. There have been few theoretical investigations of substituted Diels-Alder reactions, since the addition of one or more substituents on butadiene and ethylene drastically increases the computational time and disc space requirements for the calculations, often making such investigations prohibitively expensive.

Dewar, Olivella & Stewart have reported AM1 studies of the Diels-Alder reaction of butadiene with several substituted cyanoethylenes (34). RHF-AM1 calculations on reactions of substituted cyanoethylenes with ethylene predict transition structures that are symmetrical or nearly symmetrical. The RHF-AM1 activation energies for reaction of butadiene with dienophiles in the series acrylonitrile (24.3 kcal/mol), maleonitrile (25.5 kcal/mol), fumaronitrile (25.7 kcal/mol), and 1,1-dicyanoethylene (24.2 kcal/mol) are all greater than that calculated for the parent Diels-Alder reaction (23.8 kcal/mol). This is not the experimental order (44). RHF AM1+3 × 3 CI calculations predict a stepwise reaction. This level of theory also incorrectly predicts the experimental order of reactivity. The activation energy calculated for the parent Diels-Alder reaction is 16.6 kcal/mol (34). The activation energies of reactions of butadiene with maleonitrile (15.2 kcal/mol) and fumaronitrile (15.4 kcal/mol) are computed to be larger than that of butadiene with both acrylonitrile (12.6 kcal/mol) and 1,1-dicyanoethylene (11.5 kcal/mol). The numbers presented are for the second saddle point corresponding to ring closure of the biradicaloid intermediate. Again, this step was postulated to be rate-determining by the application of 20 kcal/mol correction factor (34).

The estimated relative activation enthalpies for the reaction of butadiene with the series ethylene, acrylonitrile, maleonitrile, fumaronitrile, 1,1-dicyanoethylene are thus computed to be 0.0, -4.0, -1.4, -1.2, -5.1 kcal/mol. The experimental values of the relative free energies of activation for the Diels-Alder reactions of cyclopentadiene with these dienophiles are 0.0, -6.3, -8.9, -8.9, -12.6 kcal/mol (34). Dewar et al proposed that the considerable disagreement results from solvation effects. The calculations represent gas phase reactions, whereas the experiments are done in solution. By applying a correction factor per cyano group, "calculated" $\Delta H^{\neq}$'s were obtained to follow the experimental trend. However, ample evidence for related reactions shows that there is little influence of solvation on rates (44). The so-called solvation corrections are actually more likely to be a correction of the AM1 method. Indeed, in Diels-Alder reactions of normal dienes, it is generally observed that electron-withdrawing substituents on the dienophile lower the activation energies (44). For example, the experimental activation energy for the Diels-Alder reaction of butadiene with ethylene has been measured as 27.5 and 34.3 kcal/mol (42, 43a–c), as compared to the measured activation energy of 19.7 kcal/mol for the reaction of butadiene with acrolein (47). The experimental value for the lowering of the activation value by the cyano substituent in acrolein is obviously somewhat uncertain, but a qualitative effect certainly exists, and is mirrored by ab initio RHF calculations. Using the minimal STO-3G basis set, the Diels-Alder activation energy is only lowered from 36.0 to 35.5 kcal/mol by replacing ethylene by acrolein. However, with the split-valence 3-21G basis set the activation energy drops from 35.9 to 30.5 kcal/mol; and with 6-31G*//3-21G calculations the activation energy drops from 45.9 to 41.9 kcal/mol. Experience with other reactions (62a,b) suggests that the inclusion of electron correlation would give a larger stabilization of the transition state by a cyano substituent. It is gratifying to see that even ab initio RHF calculations parallel the experimental substituent effect for reaction of butadiene with acrolein, when reasonable basis sets are used.

The question of substituent-induced asynchronicity is one of fundamental interest in substituted Diels-Alder reactions. For example, the transition structure for reaction of butadiene with s-cis acrolein is necessarily unsymmetrical, and so the concerted reaction must be asynchronous. However, the results of the RHF 3-21G calculations predict that this asynchronicity is small, with the forming C–C bond lengths of 2.353 and 2.088 Å differing by only 0.27 Å.

In an important experimental study of the secondary deuterium kinetic isotope effects on the Diels-Alder reactions of d_0, d_2, and d_4-isoprene with four dienophiles—acrylonitrile, fumaronitrile, vinylidene cyanide, and methyl β-cyanoacrylate—Gajewski et al (48a,b) have concluded that the reaction is not a two-step reaction in which the second step is rate-determining. Of particular interest to the discussion above is the reaction of isoprene with acrylonitrile. The magnitude of the inverse kinetic isotope effect (KIE) observed at the β site of acrylonitrile is larger than that at the α site, but is only half of the maximum value expected for a fully formed σ bond. The experimental results are indicative of an early, unsymmetrical transition state; but not one posessing one fully formed bond. This is precisely what is shown by the RHF calculations on the butadiene-acrolein reaction by both ab initio or AM-1 methods. However, as discussed above the AM1 calculations fail to predict the rate accelerations that are caused by adding cyano substituents.

ENE REACTION

The Alder ene reaction (3, 49) of propene with ethylene to form 1-pentene is depicted schematically in Figure 3. Few calculations (50, 51) have been done on the Alder ene reaction and its reverse, the retro-ene reaction. The ene reaction of propene with ethylene to form 1-pentene has been studied with ab initio molecular orbital calculations at the STO-3G and 3-21G basis levels (50). The calculations suggest a concerted reaction, characterized by an envelope conformation, that is necessarily asymmetrical. The forming C–C bond length of 2.110 Å is similar to that of the Diels-Alder reaction. The breaking C–H bond length of 1.349 Å is slightly less than the C–H forming bond length of 1.447 Å. The calculated activation energy of 51.7 kcal/mol at the RHF/3-21G basis level is lowered to 31.2 kcal/mol at the MP2/6-31G*//3-21G level. Although this calculated value is high as compared to an experimental value of 21.4 kcal/mol for the ene reaction of ethylene with propene (52), the calculated value is close to the experimental activation energies for the ene reactions of ethylene with *trans*-2-butene (37.0 kcal/mol) and *cis*-2-butene (36.5 kcal/mol) to form 3-methylpent-1-ene (53). It is likely that the experimental value of the activation energy for the reaction of propene with ethylene is in error and that the actual value is about 35 kcal/mol.

Dewar's MINDO/3 calculations of ene reactions predict a transition structure involving slightly more than 50% of hydrogen transfer between an oxygen atom and a carbon atom (51). Dewar suggests that the synchronicity of ene reactions should decrease in the order of a hydrogen atom transferring between (*a*) two heteroatoms, (*b*) a heteroatom and a

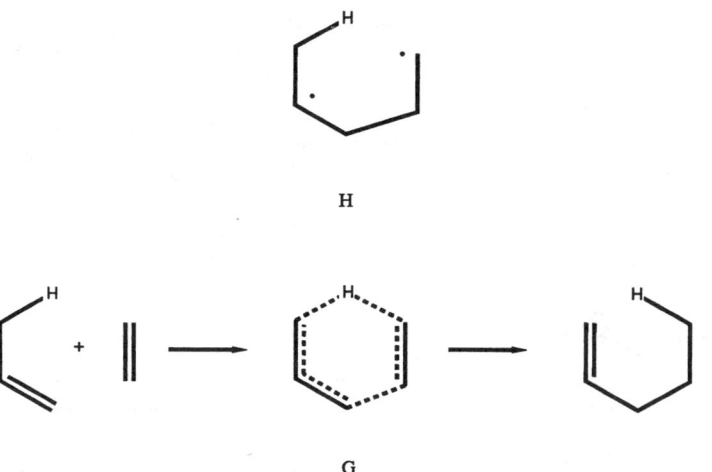

Figure 3 Schematic depiction of the concerted pericyclic transition structure (G) and a potential biradical intermediate (H) in the Ene reaction.

carbon atom, and (c) two carbon atoms (2). However, comparison of the RHF 3-21G transition structures for the ene reactions of propene with ethylene (breaking C–H = 1.349 Å, forming C–H = 1.447 Å, C–C = 2.110 Å) and propene with formaldehyde (C–H = 1.322 Å, O–H = 1.311 Å, C–C = 2.016 Å) find that the latter transition state as a whole is slightly more synchronous.

1,3-DIPOLAR CYCLOADDITION

The reaction of fulminic acid with acetylene depicted schematically in Figure 4 is the archetype 1,3-dipolar cycloaddition. Again, this reaction can occur through a concerted transition structure (I) or through a diradical transition structure or intermediate (J). Controversies have erupted as a result both of experimental data and of theoretical calculations.

Poppinger (54) was the first to investigate the reaction of fulminic acid and acetylene in 1976 with ab initio molecular orbital calculations. The STO-3G minimal basis set was employed to characterize the reactants, C_s concerted transition state, and product. Single-point calculations were subsequently performed with the 4-31G basis level. The forming C–C and C–O bond lengths in the transition state were 2.369 Å and 2.136 Å, respectively. The STO-3G activation energy of the concerted C_s process is 21.3 kcal/mol and increases to 30.4 kcal/mol at the 4-31G level.

Dewar (31b) has questioned the authenticity of the fulminic acid-accty-

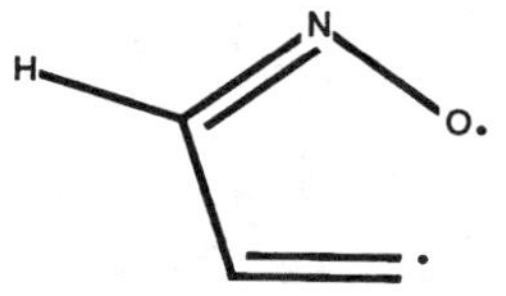

Figure 4 Schematic depiction of the concerted transition structure (I) and potential diradical intermediate (J) in the 1,3-dipolar cycloaddition of fulminic acid to acetylene.

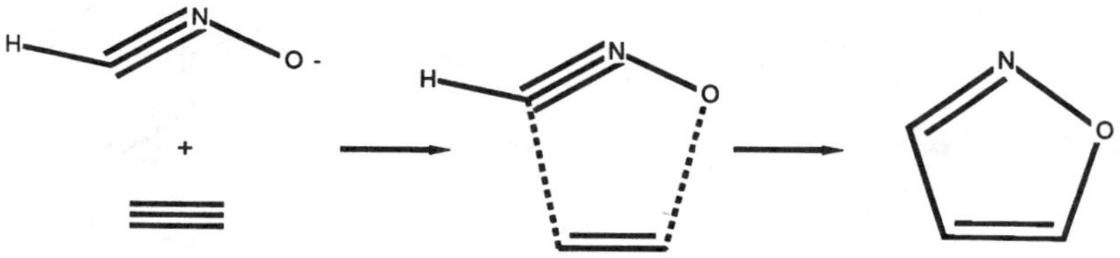

lene transition state, citing MNDO calculations that suggest that the Poppinger study has not dealt adequately with biradicaloid intermediates. The MNDO study, the results of which were cited only in a footnote, found a biradical transition state with an activation energy of 25.8 kcal/mol and forming C–C and O–C bond lengths of 1.85 Å and 3.71 Å, respectively. This transition state leads to a stable zwitterionic intermediate that is 2.9 kcal/mol above reactants. A second transition state barrier of only 0.8 kcal/mol separates this intermediate from the product isoxazole. This second transition structure has a forming C–C bond length of 1.40 Å and a forming O–C bond of 2.55 Å. Thus MNDO predicts stepwise formation of a stable diradical intermediate, paralleling the MNDO calculations on the Diels-Alder reaction.

However, Dewar (2) subsequently suggested that the MNDO calculations are in error due to the tendency of MNDO to overestimate repulsions between atoms separated by 1.5 to 2 times a normal bond length. As discussed above, Dewar (2) states that this problem vitiates the MNDO result on the Diels-Alder reaction. Nevertheless, he goes on to state his conviction that all these cycloaddition reactions proceed by a concerted mechanism but are nonsynchronous with the second stage being rate-determining (2).

Further ab initio refinement of the fulminic acid-acetylene transition state in the Poppinger study was carried out by Komornicki et al (56) with the 4-31G basis set and also the Huzinaga (9s5p/4s) double ζ basis set contracted to 4s2p/2s. This study, like Poppinger's, is consistent with a C_s concerted transition state. The RHF activation energies, using the 4-31G and double ζ basis set, were 28.4 and 30.8 kcal/mol, respectively. At the highest level of CI used on the double ζ geometries with 4a'+3a" molecular orbitals and 13,672 configurations, the activation energy was lowered to 19.5 kcal/mol, while the calculated exothermicity was 70.3 kcal/mol.

The vibrational analysis at the 4-31G SCF level gave only one imaginary vibrational frequency (684i cm^{-1}) corresponding to simultaneous stretching or contracting the forming C–C and C–O bonds, which is consistent with a concerted mechanism. Nevertheless, the force constant for the forming C–C bond is ten times that for the forming C–O bond. This suggests that although the reaction can be considered concerted, it is not synchronous. The concerted mechanism, whether the transition state is synchronous or asynchronous, is consistent with the experimental evidence (57, 58).

Hiberty et al (59), further investigated the fulminic acid-acetylene transition structure, exploring in detail the energy surface for an unsymmetrical transition structure and for an extended diradical intermediate. A diradical transition state was located at the UHF/4-31G level of theory. Since the UHF structure was heavily contaminated with triplet character, and hence the reliability of the structure was uncertain, the main parameters of the structure were reoptimized with the RHF/4-31G+3 × 3 CI level. Single-point calculations were then carried out on the synchronous transition state found by Komornicki et al (56) and on the asynchronous transition structure with inclusion of electron correlation. At the 3 × 3 CI level, the synchronous transition state was lower in energy than the asynchronous transition state by 6.4 kcal/mol. However, at the highest level of theory, the CIPSI (*c*onfiguration *i*nteraction by *p*erturbation with multi-configurational zeroth order wavefunction *s*elected by *i*terative process) method was applied to a full set of valence orbitals, and the asynchronous transition structure was favored over the synchronous transition structure by 3.6 kcal/mol. These calculations suggest that the diradical mechanism proposed by Firestone (60) should be considered seriously for 1,3-dipolar cycloadditions.

Recently McDouall et al (61) performed an MCSCF study of the fulminic acid addition to acetylene by using the 4-31G basis level with full geometry optimization. These calculations, which are the best to date, predict that in the gas phase the concerted pathway ($E_a = 26.0$ kcal/mol) is favored by 5 kcal/mol over the diradical pathway ($E_a = 30.7$ kcal/mol). The forming C–C and O–C bond lengths of 2.17 Å and 2.28 Å, respectively, are close to being equal in the cyclic C_s transition structure. This is very different from Dewar's MNDO-CI prediction of bond lengths of 1.40 and 2.55 Å, respectively.

McDouall et al (61) performed single-point calculations using multireference CI, with the 7270 configurations that result from a 20 configuration reference space. These calculations gave activation barriers for the concerted pathway and diradical pathways of 18.2 kcal/mol and 31.7 kcal/mol, respectively. The 13.4 kcal/mol activation energy difference between the two pathways is considerable. The semi-empirical and previous ab initio with CI calculations are in considerable error. The computed activation energy for the concerted, nearly synchronous pathways is in the range of the 8–18 kcal/mol estimated from experiments (56).

The MCSCF bond lengths and force constants in the transition structure indicate that both bonds are partially formed to similar extents. Thus, even the reaction of the very unsymmetrical reactant, fulminic acid, is predicted to follow a synchronous pathway. It is therefore no surprise that symmetrical reactants generally appear also to react via synchronous pathways.

CONCLUSION

Ab initio calculations and experimental evidence both indicate that multibond reactions not only can be synchronous but often are synchronous. Dewar's (2) conclusions to the contrary apparently were made largely on the basis of the results of semi-empirical calculations, which he now acknowledges (34) are unsuitable for the study of the transition structures that occur in pericyclic reactions.

There is agreement between various computational methods that some multibond reactions are synchronous. As discussed above, AM1 calculations now agree that the parent Diels-Alder reaction is synchronous. In addition, the electrocyclic reaction of cyclobutene and the 1,5-sigmatropic hydrogen shift of 1,3-pentadiene are two reactions in which very similar, synchronous transition structures are obtained by both semi-empirical calculations (2) and ab initio calculations, with or without correlation energy corrections (62a,b).

Finally, the authors would be remiss in not pointing out that the fascination with synchronicity is not limited to a small group of theoretical and experimental chemists. Evidence of a much wider interest in this subject comes from this quote from the popular literature (63):

*A connecting principle
linked to the invisible
almost imperceptible
Science insusceptible
logic so inflexible
causally connectible
but nothing is invincible—*

*Effect without a cause
subatomic laws, scientific pause
Synchronicity.*

Literature Cited

1. Woodward, R. B., Hoffmann, R. 1969. *Angew. Chem. Int. Ed. Engl.* 8: 781–853
2. Dewar, M. J. S. 1984. *J. Am. Chem. Soc.* 106: 209–19
3. Gajewski, J. J. 1981. *Hydrocarbon Thermal Isomerizations*, pp. 166–76. New York: Academic. 442 pp.
4. Zimmerman, H. E. 1966. *J. Am. Chem. Soc.* 88: 1564–65
5. Dewar, M. J. S. 1966. *Tetrahedron Suppl.* 8: 75–92
6. Humski, K., Malojcic, R., Borcic, S., Sunko, D. E. 1970. *J. Am. Chem. Soc*, 92: 6534–38
7. Rossi, M., King, K. D., Golden, D. W. 1979. *J. Am. Chem. Soc.* 101: 1223–30
8. Doering, W. von E., Toscano, V. G., Beasley, G. H. 1971. *Tetrahedron* 27: 5299–5306.
9. Doering, W. von E. 1981. *Proc. Natl. Acad. Sci. USA* 78: 5279–83
10. Borden, W. T. 1975. *Modern Molecular Orbital Theory for Organic Chemists*, pp. 129–31. Englewood Cliffs, NJ: Prentice-Hall. 305 pp.
11. Dewar, M. J. S., Krischner, S., Kollmar, H. W., Wade, L. E. 1974. *J. Am. Chem. Soc.* 96: 5242–44
12. Doering, W. von E., Roth, W. 1962. *Tetrahedron* 18: 67–74
13. Goldstein, M. J., Benzon, M. S. 1972. *J. Am. Chem. Soc.* 94: 7147–49
14. Hill, R. K., Gilman, N. W. 1967. *J. Chem. Soc. Chem. Commun.*, pp. 619–20
15. Shea, K. J., Phillips, R. B. 1980. *J. Am. Chem. Soc.* 102: 3156–62
16. Gajewski, J. J., Benner, C. W., Hawkins, C. M. 1987. *J. Org. Chem.* 52: 5198–5204
17. Brown, A., Dewar, M. J. S., Schoeller, W. 1970. *J. Am. Chem. Soc.* 92: 5516–17
18. Komornicki, A., McIver, J. W. 1976. *J. Am. Chem. Soc.* 98: 4553–61
19. Dewar, M. J. S., Ford, G. P., McKee, M. L., Rzepa, H. S., Wade, L. E. 1977. *J. Am. Chem. Soc.* 99: 5069–73
20. Dewar, M. J. S., Jie, C. 1987. *J. Am. Chem. Soc.* 109: 5893–5900
21. Dewar, M. J. S., Jie, C. 1987. *J. Chem. Soc. Chem. Commun.*, pp. 1451–53
22. Osamura, Y., Kato, S., Morokuma, K., Feller, D., Davidson, E. R., Borden, W. T. 1984. *J. Am. Chem. Soc.* 106: 3362–63
23. Hehre, W. J., Radom, L., Schleyer, P. von R., Pople, J. A. 1986. *Ab Initio Molecular Orbital Theory*, pp. 288–91. New York: Wiley-Interscience. 548 pp.
23a. Owens, K. A., Berson, J. A. 1988. *J. Am. Chem. Soc.* 110: 627–28
24. Dewar, M. J. S., Wade, L. E. 1977. *J. Am. Chem. Soc.* 99: 4417–24
25. Dewar, M. J. S., Wade, L. E. 1973. *J. Am. Chem. Soc.* 95: 290–91
26. Wehrli, R., Bellus, D., Hansen, H.-J., Schmid, H. 1976. *Chimia* 30: 416–22
27. Gajewski, J. J., Conrad, N. D. 1979. *J. Am. Chem. Soc.* 101: 6693–6704
28. Dewar, M. J. S., Healy, E. F. 1987. *Chem. Phys. Lett.* 141: 521–24
29a. Basilevsky, M. V., Shamov, A. G., Tikhomirov, V. A. 1977. *J. Am. Chem. Soc.* 99: 1369–72
29b. Basilevsky, M. V., Tikhomirov, V. A., Chlenov, I. E. 1971. *Theoret. Chim. Acta* 23: 75–92
29c. Kikuchi, O. 1971. *Tetrahedron* 27: 2791–2800
29d. Jug, K., Krüger, H.-W. 1979. *Theoret. Chim. Acta* 52: 19–26
29e. Pancir, J. 1982. *J. Am. Chem. Soc.* 104: 7424–30
30. Dewar, M. J. S., Griffin, A. C., Kirschner, S. 1974. *J. Am. Chem. Soc.* 96: 6225–26
31a. Dewar, M. J. S., Pierini, A. B. 1984. *J. Am. Chem. Soc.* 106: 203–8
31b. Dewar, M. J. S., Olivella, S., Rzepa, H. S. 1978. *J. Am. Chem. Soc.* 100: 5650–59

32. See also: MINDO/2 and MINDO/3 calculations in Oliva, A., Fernández-Alonso, J. I., Bertrán, J. 1978. *Tetrahedron* 34: 2029–33
33. McIver, J. W. Jr. 1974. *Acc. Chem. Res.* 7: 72–77
34. Dewar, M. J. S., Olivella, S., Stewart, J. J. P. 1986. *J. Am. Chem. Soc.* 108: 5771–79.
35. Dewar, M. J. S., Zoebisch, E. G., Healy, E. F., Stewart, J. J. P. 1985. *J. Am. Chem. Soc.* 107: 3902–9
36. Bernardi, F., Bottoni, A., Robb, M. A., Field, M. J., Hillier, I. H., Guest, M. F. 1985. *J. Chem. Soc. Chem. Commun.*, pp. 1051–52
37. Townshend, R. E., Ramunni, G., Segal, G., Hehre, W. J., Salem, L. 1976. *J. Am. Chem. Soc.* 98: 2190–98
38a. Burke, L. A., Leroy, G. 1977. *Theoret. Chim. Acta* 44: 219–21
38b. Burke, L. A., Leroy, G., Sana, M. 1975. *Theoret. Chim. Acta* 40: 313–21
38c. Ortega, M., Oliva, A., Lluch, J. M., Bertrán, J. 1983. *Chem. Phys. Lett.* 102: 317–20
39a. Houk, K. N., Lin, Y.-T., Brown, F. K. 1986. *J. Am. Chem. Soc.* 108: 554–56
39b. Brown, F. K., Houk, K. N. 1984. *Tetrahedron Lett.* 25: 4609–12
40. Caramella, P., Houk, K. N., Domelsmith, L. N. 1977. *J. Am. Chem. Soc.* 99: 4511–14
41. Gordon, M. D., Fukunaga, T., Simmons, H. E. 1976. *J. Am. Chem. Soc.* 98: 8401–7
42. Rowley, D., Steiner, H. 1951. *Discuss. Faraday Soc.* 10: 198–213. (The estimated E_a at 0K is 25.1 kcal/mol.)
43a. Calculated from E_a of cyclohexene cycloreversion and the experimental heat of reaction: Uchiyama, M., Tomioka, T., Amano, A. 1964. *J. Phys. Chem.* 68: 1878–81
43b. Tsang, W. 1965. *J. Chem. Phys.* 42: 1805–9
43c. Tardy, D. C., Ireton, R., Gordon, A. S. 1979. *J. Am. Chem. Soc.* 101: 1508–14.
44. Sauer, J., Sustmann, R. 1980. *Angew. Chem. Int. Ed. Engl.* 19: 779–807 (and references therein)
45a. Doering, W. v. E., Franck-Neumann, M., Hasselman, D., Kaye, R. L. 1972. *J. Am. Chem. Soc.* 94: 3833–44
45b. Frey, H. M., Pottinger, R. 1978. *J. Chem. Soc. Faraday Trans. 1* 74: 1827–33
46. Bartlett, P. D., Schueller, K. E. 1968. *J. Am. Chem. Soc.* 90: 6021–77
47. Kistiakowsky, G. B., Lacher, J. R. 1936. *J. Am. Chem. Soc.* 58: 123–33. (The activation entropy is determined from the A value of 1.46×10^6 using a temperature of 516.8K.)
48a. Gajewski, J. J., Peterson, K. B., Kagel, J. R. 1987. *J. Am. Chem. Soc.* 109: 5545–46
48b. Gajewski, J. J. 1986. 192nd ACS Meet. Anaheim, Sept. 7–12
49. Hoffmann, H. M. R. 1969. *Angew. Chem. Int. Ed. Engl.* 8: 556–77
50. Loncharich, R. J., Houk, K. N. 1987. *J. Am. Chem. Soc.* 109: 6947–52
51. Brown, S. B., Dewar, M. J. S., Ford, G. P., Nelson, D. J., Rzepa, H. S. 1978. *J. Am. Chem. Soc.* 100: 7832–36
52. Walsh, R., unpublished results cited by Egger, K. W., Vitius, P. 1974. *Int. J. Chem. Kinet.* 6: 429–35
53. Richard, C., Scacchi, G., Back, M. H. 1978. *Int. J. Chem. Kinet.* 10: 307–24
54. Poppinger, D. 1976. *Aust. J. Chem.* 29: 465–78
55. Deleted in proof
56. Komornicki, A., Goddard, J. D., Schaefer, H. F. III. 1980. *J. Am. Chem. Soc.* 102: 1763–69
57. Husigen, R. 1963. *Angew. Chem. Int. Ed. Engl.* 2: 565–98, 633–45
58. Houk, K. N., Firestone, R. A., Munchausen, L. L., Mueller, P. H., Arison, B. H., Garcia, L. A. 1985. *J. Am. Chem. Soc.* 107: 7227–28
59. Hiberty, P. C., Ohanessian, G., Schlegel, H. B. 1983. *J. Am. Chem. Soc.* 105: 719–23
60. Firestone, R. A. 1972. *J. Org. Chem.* 37: 2181–91
61. McDouall, J. J. W., Robb, M. A., Niazi, U., Bernardi, F., Schlegel, H. B. 1987. *J. Am. Chem. Soc.* 109: 4642–48
62a. Jensen, F., Houk, K. N. 1987. *J. Am. Chem. Soc.* 109: 3139–3140
62b. Spellmeyer, D. S., Houk, K. N. 1987. *J. Am. Chem. Soc.* In press
63. "Synchronicity," by Sting © 1983. Reggatta Music, Ltd. All rights reserved. Used by permission.

MODELS FOR COLLOIDAL AGGREGATION

Paul Meakin

Central Research and Development Department, E. I. du Pont de Nemours and Company, Wilmington, Delaware 19898

INTRODUCTION

Computer simulations have been used for more than a quarter of a century to develop a better understanding of nonequilibrium growth and aggregation processes. Important early examples include Vold's ballistic deposition model for aggregation and sedimentation (1, 2), Sutherland's ballistic cluster-cluster aggregation model for floc formation (3), and Eden's surface growth model for the generation of cell colonies (4). Because of limited computer resources these early simulations were carried out on a small scale using simple models. Since these early pioneering efforts computer speed, storage capabilities, and availability have increased enormously and the task of writing computer programs has become much easier and more reliable. These developments have allowed two almost divergent research directions to develop. On the one hand, models have been developed that incorporate as much as possible of what is known about the physics and chemistry of aggregation processes, and on the other hand, a variety of very simple models have been introduced that allow us to generate the very large structures needed to investigate asymptotic (large size limit) geometric scaling relationships. The exploration of very simple aggregation models has been stimulated by the realization (5–10) that real aggregation processes and simple aggregation models frequently lead to structures that can be described in terms of the concepts of fractal geometry (11). In particular, the recent interest in nonequilibrium growth and aggregation models was initiated by the discovery by Witten & Sander (6) that a simple diffusion-limited aggregation (DLA) model in which particles are added, one at a time, to a growing cluster or aggregate of particles leads to structures with a fractal geometry. This review is concerned mainly with simple aggregation models that generate fractal structures.

The more detailed models for colloidal aggregation are important in assessing our understanding of aggregation processes and indicating directions in which more work is needed. If we really understand a particular aggregation process we should (at least in principle) be able to express this understanding in terms of a computer algorithm and to reproduce the experimentally observed behavior.

Recently, these two research directions have begun to converge as those (mainly physicists) interested in the fractal aspects of aggregation have come to realize that many real aggregation processes cannot be understood in terms of the most simple aggregation models and those (mainly chemists and chemical engineers) interested in comprehensive models for colloidal aggregation have become interested in their fractal nature. One of the main advantages of simple models is that they are reliable and can easily be evaluated. In almost all cases these models have been checked by completely independent (often simultaneous) duplication. These well-tested models provide a firm foundation for the development of more realistic models, and this is now an active area of research.

AGGREGATION MODELS

During the past five to six years since the introduction of the diffusion-limited (particle-cluster) aggregation model (6), a wide variety of new nonequilibrium growth and aggregation models has been developed (see Refs. 12–25 for some recent reviews, books, and conference proceedings). Many of the most important recently developed and earlier models can be described in terms of a closely related family of aggregation models.

In all of the simulations belonging to this family we start with N_0 particles, which may be considered as a list of $N = N_0$ single particle clusters. At each stage during the simulation a pair of clusters is selected from the list, combined, and returned to the list, which then contains N-1 clusters. This procedure is continued until a single cluster containing all of the particles is formed or until a predetermined maximum cluster size (s_{max}) or mean cluster size (S) has been reached. These simulations differ in how the clusters are selected and how they are combined. In particle-cluster aggregation models, the largest cluster and one of the remaining single particles is always selected. In hierarchical models the simulation is started with $N_0 = 2^M$ particles, which are combined to form $N_0/2$ binary clusters, which in turn are combined to form 2^{M-2} clusters of four particles, etc. In other models (described below), pairs of clusters are selected with probabilities that are determined by a reaction kernel [$K(i,j)$ where $K(i,j)$

is the rate constant for combination of clusters of size i and j]. In these models a polydisperse cluster size distribution evolves.

The most important quantity associated with how the clusters are combined is the fractal dimensionality, D_t, of the trajectories that they follow. For diffusion-limited aggregation the clusters follow random walk (Brownian) trajectories and $D_t = 2$. For ballistic aggregation the clusters follow linear trajectories ($D_t = 1$), and for reaction-limited aggregation the fractal dimensionality of the trajectory may be considered to be zero. In some cases aggregation simulations have been carried out by using fractal [Levy flights and walks (11)] trajectories (26; P. Meakin, unpublished). Simulations have also been carried out (P. Meakin, unpublished; 28) in which the closest points in two clusters are found and brought into contact with each other without rotation. This corresponds to $D_t \to \infty$. For all of these models the simulations can be carried out on a lattice or in a continuous space. In most cases the asymptotic scaling relationships that characterize their fractal geometries are the same for both lattice and off-lattice models (in this sense the models are universal). However, in some cases [such as the diffusion-limited aggregation model of Witten & Sander (6)] lattice anisotropy may have important effects.

Most simulations have been carried out by using two- or three-dimensional spaces or lattices because of their relevance to important real systems. However, one-dimensional simulations and simulations in spaces or lattices with a (Euclidean) dimensionality higher than 3 have also been performed. This work is generally motivated by a desire to develop a better theoretical understanding of simple aggregation models. Despite their apparent simplicity, many of these models have so far proven to be theoretically intractable. The theoretical challenge presented by these simple (to define) models is one reason for their current high level of interest.

Figures 1–4 show typical structures generated by some of these models (they are described in more detail below). All of these figures were obtained by using off-lattice models. Off-lattice models generate more realistic structures, but the corresponding lattice models are usually easier to program and more efficient. They are used widely to investigate asymptotic scaling relationships as well as to explore the effects of various types of anisotropy. Figures 1 and 2 show typical results generated by two-dimensional models and Figures 3 and 4 show clusters obtained from three-dimensional models. For all of these models, pairs of clusters are joined rigidly and irreversibly at their positions of initial contact. For all four figures, typical (randomly selected) clusters generated using trajectory dimensionalities (D_t) of 0, 1, and 2 are shown. The results shown in Figures 1 and 3 were obtained from particle-cluster aggregation models, and those in Figures 2 and 4 were obtained from polydisperse cluster-cluster aggregation models.

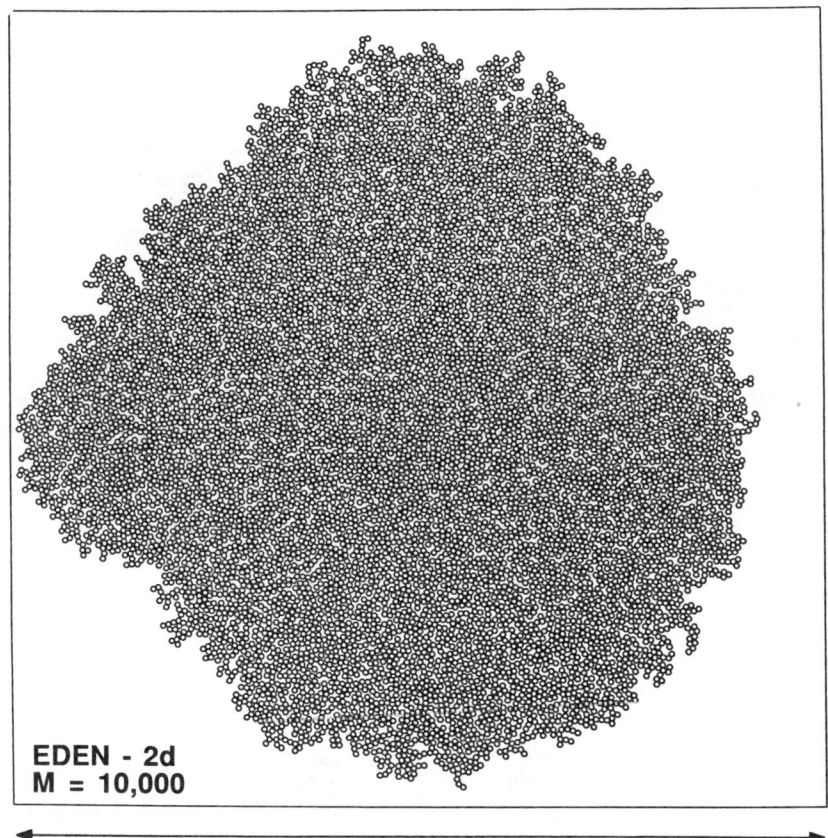

Figure 1 Clusters generated by two-dimensional off-lattice models for particle cluster aggregation. In Figure 1a the particle trajectory (D_t) is zero and we have an off-lattice version of the Eden model. In Figure 1b, $D_t = 1$ (ballistic aggregation) and in Figure 1c, $D_t = 2$ (diffusion limited aggregation or DLA).

Of these 12 models, four (1a, 1b, 3a, and 3b) have structures that are uniform on all but short length scales. All of the other clusters are fractals. A summary of their fractal dimensionalities is provided in Table 1. In addition, the surfaces of the clusters generated by models 1a, 1b, 3a, and 3b have recently been shown to be self-affine (11, 29, 30) fractals (31–34). Much of the work on fractal aggregates has been concerned with the use of theoretical approaches and computer models to investigate their structure and with an exploration of the implications of their fractal geometry for their growth kinetics, physical properties, etc. In addition, a

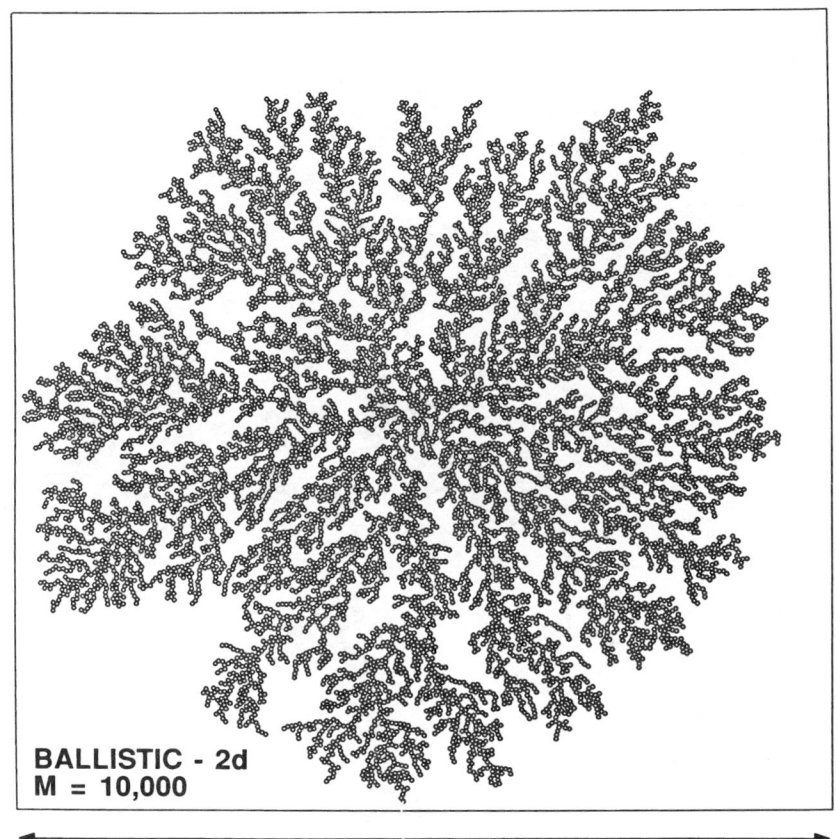

Figure 1b

variety of experimental studies have been carried out on real systems that seem to closely approximate the limiting behavior assumed in these models.

Fractal Geometry

The concepts of fractal geometry are based on mathematical ideas developed during the late nineteenth and early twentieth centuries. However, it is only during the past few decades, primarily as a result of the efforts of B. B. Mandelbrot, that these ideas have been synthesized and applied to the physical sciences. This work is summarized in a long series of publications (see Ref. 11 for a history of fractal geometry) and a series of beautifully illustrated books (11, 35, 36). For the purposes of this review,

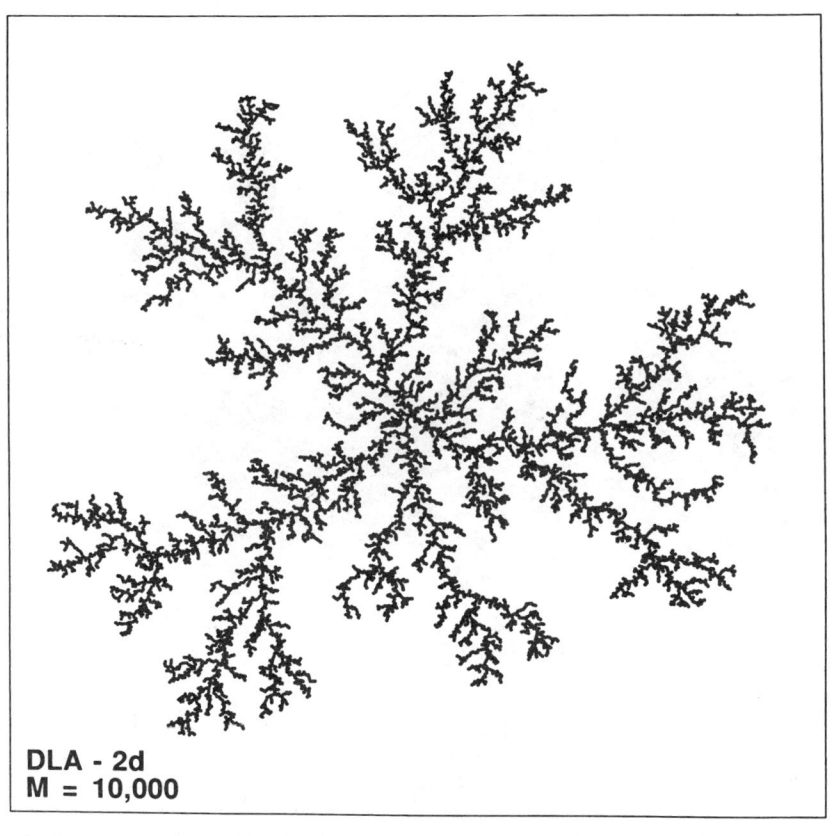

350 DIAMETERS

Figure 1c

a fractal can be regarded as a system in which the mass (M) is related to a characteristic length (L) by

$$M \sim L^D \qquad \qquad 1.$$

where the exponent in this mass-length scaling relationship is the fractal dimensionality (D). This is in accord with our intuitive understanding of dimensionality in ordinary (Euclidean) systems for which Eq. 1 is appropriate and the exponent D in the mass-length scaling relationship is the Euclidean dimensionality (d).

Figure 5 illustrates four stages in the construction of a simple, highly symmetric, fractal by a hierarchical process (37, 38). In Figure 5a five circles have been joined to form a cross. In Figure 5b five of the crosses

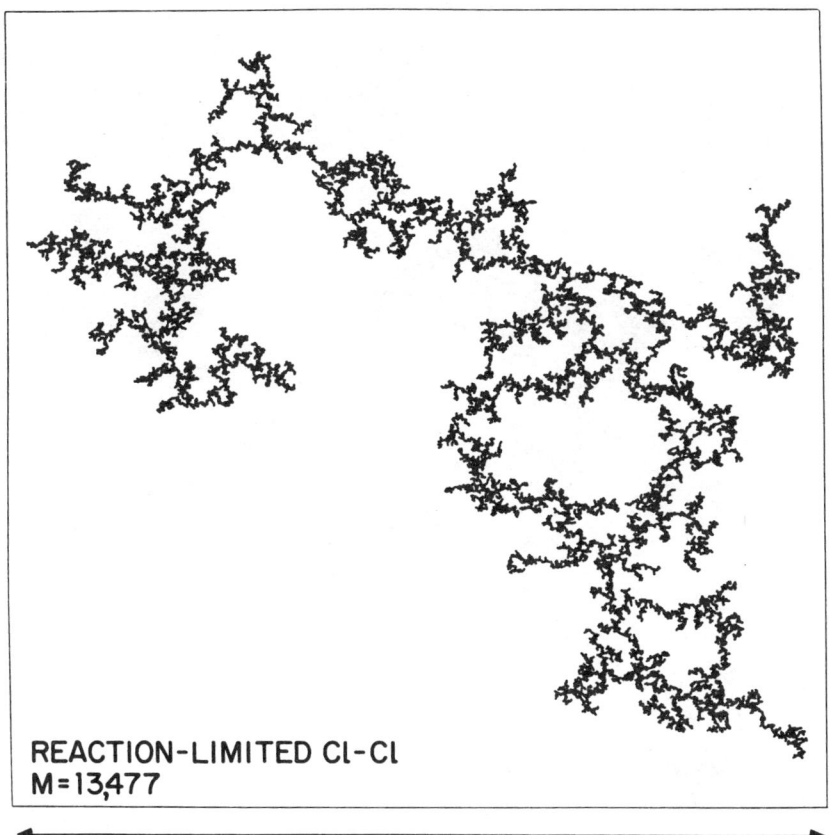

Figure 2 Typical clusters generated by two-dimensional cluster-cluster aggregation models. In Figures 2a–c, the dimensionality (D_t) of the trajectories was 0, 1, and 2, respectively.

shown in Figure 5a have been joined to form a still larger cross, and Figures 5c and 5d show the next two stages, containing 125 and 625 circles, respectively. Each time a characteristic length (the diameter of the cross, for example) is increased by a factor of 3, the mass is increased by a factor of 5 and the fractal dimensionality (from Eq. 1) is given by $D = \log(5)/\log(3)$ or 1.465. The hierarchical structure shown in Figure 5 can be extended down to smaller length scales by replacing each circle of radius r_0 in Figure 5a by a cross consisting of five circles, each of radius $r_0/3$, etc. If these construction processes are continued to infinitesimally small and infinitely long length scales, we will have generated a self-similar structure that can be mapped exactly onto itself after a change in length scales (by a factor

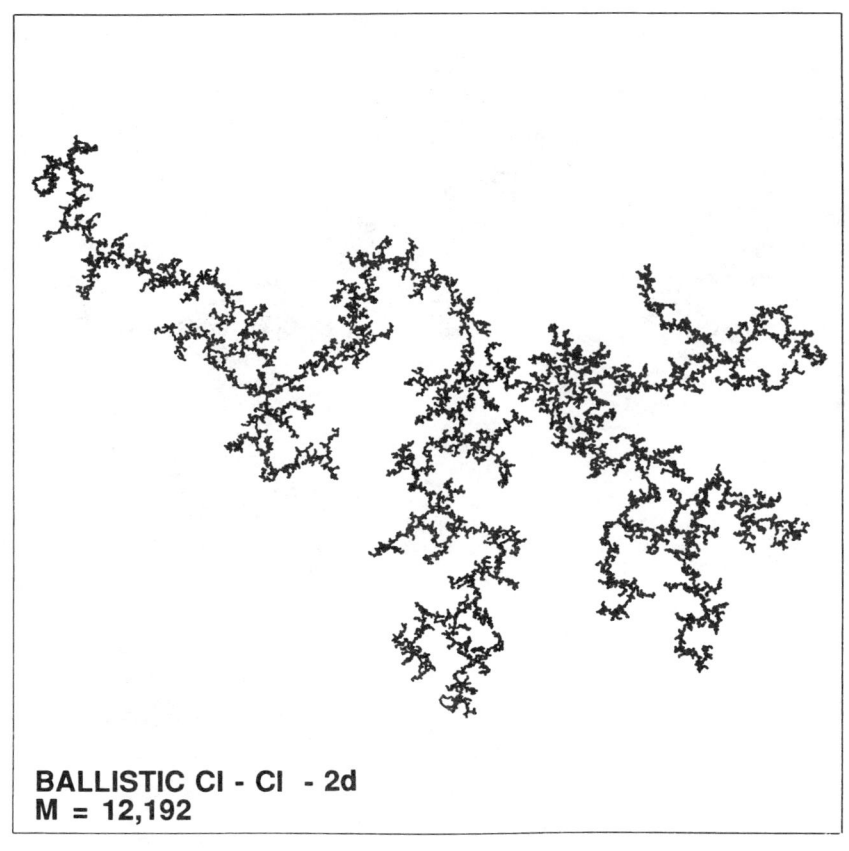

BALLISTIC Cl - Cl - 2d
M = 12,192

← 520 DIAMETERS →

Figure 2b

of 3^M where M is any integer). For the structure shown in Figure 5d this property of self-similarity extends over only a finite range of length scales limited by the size of a circle and by the overall size of cluster. Similarly, self-similarity (and the associated fractal scaling) extends over only a limited range of length scales in real aggregates between a lower cut-off length (generally associated with the size of a primary particle) and an upper cut-off length, which may be determined by the overall size of the aggregate or by other effects such as collapse due to external fields and/or thermal fluctuations (39, 40).

Fractal aggregates differ from the "aggregate" shown in Figure 5 in that they are not generally constructed by a precisely hierarchical process.

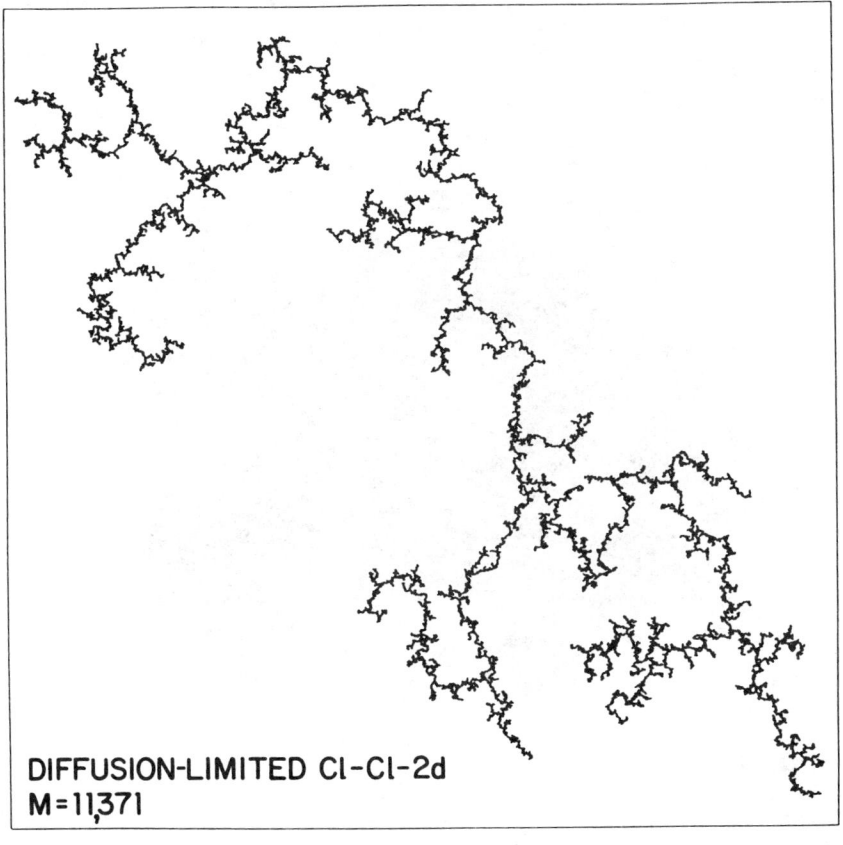

DIFFUSION-LIMITED Cl-Cl-2d
M=11,371

660 DIAMETERS

Figure 2c

Instead, they have a random structure that exhibits self-similarity only in a statistical sense. One of the important consequences of this statistical self-similarity is that the correlation functions that describe these structures [such as the two-point density-density correlation function $C(r)$] have a power law form (with no characteristic lengths other than the upper and lower cut-off lengths resulting from the limited range of fractal scaling). These power law correlations are crucial in understanding the physical properties of random fractals.

For a simple, statistically self-similar random fractal (generated by either an experiment or a simulation) the fractal dimensionality (D) can be determined by measuring a variety of mass-length scaling relationships.

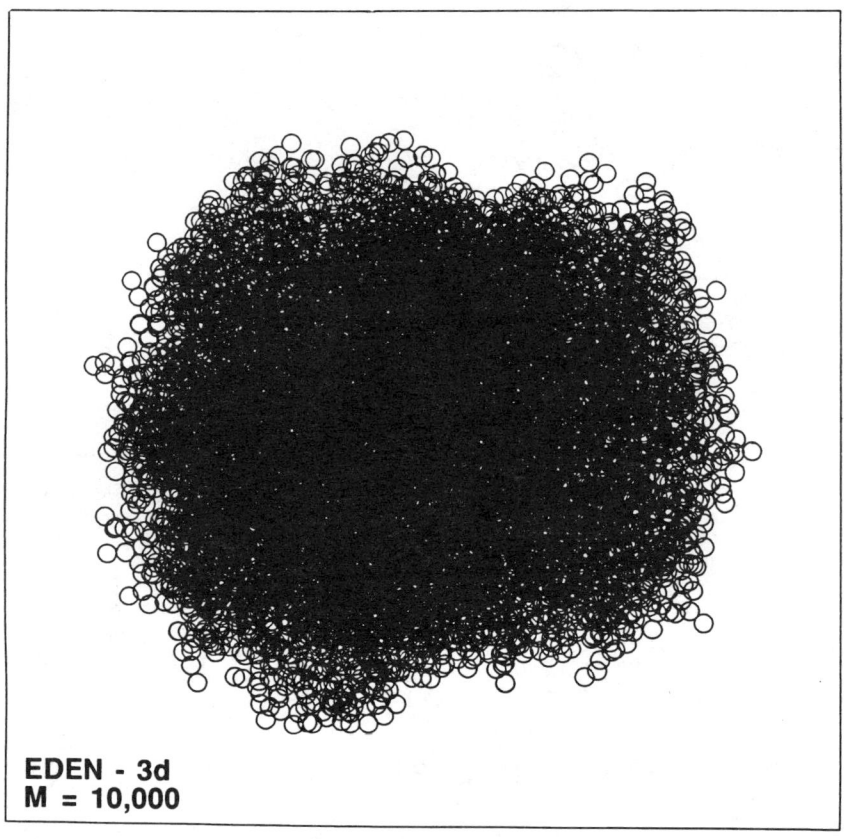

Figure 3 Projections of clusters generated by three-dimensional particle-cluster aggregation models. In Figures 3a–c the fractal dimensionality of the particle trajectories was 0, 1, and 2, respectively.

For example, if the cluster radius of gyration (R_g) is related to its mass by

$$R_g \sim M^\beta \qquad \qquad 2.$$

then the fractal dimensionality (D_β) is given by $D_\beta = 1/\beta$ (41). Similarly, if the two-point correlation function $C(r)$ is given by

$$C(r) \sim r^{-\alpha} \qquad \qquad 3.$$

then the corresponding fractal dimensionality is given by $D_\alpha = d - \alpha$ where

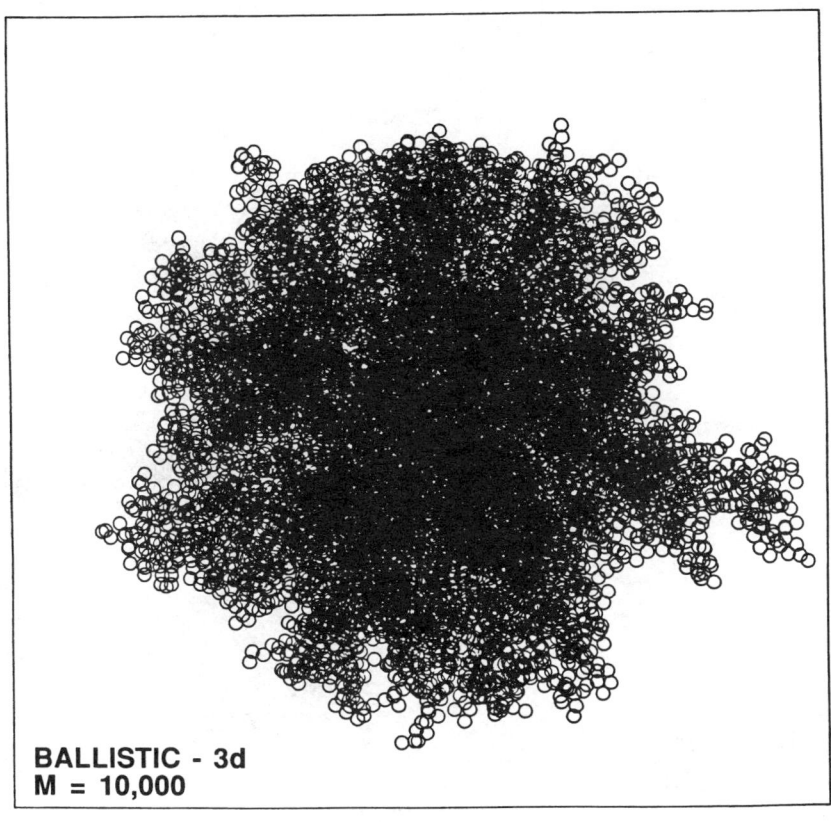

BALLISTIC - 3d
M = 10,000

◄──────── 60 DIAMETERS ────────►

Figure 3b

d is the Euclidean dimensionality of the embedding space or lattice. Alternatively, we could measure the amount of mass $M(L)$ contained within a distance L measured from an occupied point on the aggregate. In this case if

$$M(L) \sim L^{\gamma} \qquad 4.$$

the corresponding fractal dimensionality D_{γ} is equal to γ. It should be noted that equations such as Eq. 4 should be interpreted in terms of ensemble averages

$$\langle M(L) \rangle \sim L^{\gamma} \qquad 5.$$

since very large fluctuations in quantities such as $M(L)$ are characteristic

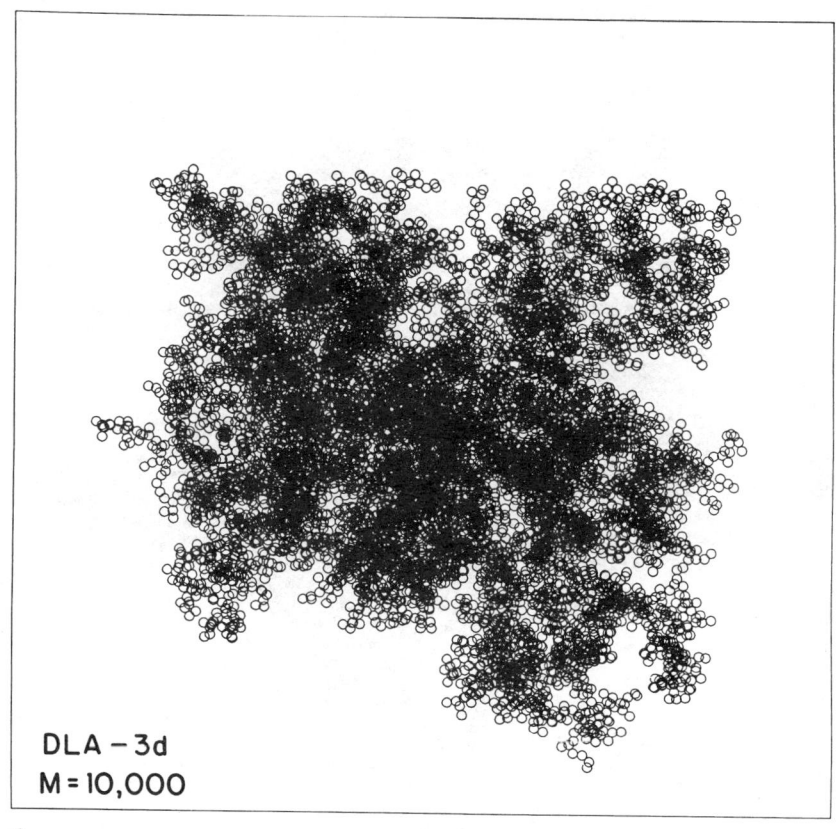

Figure 3c

of fractal structures. However, the ensemble-averaging symbols $\langle \rangle$ are omitted from all of our equations for the sake of simplicity. For self-similar fractals $D_\alpha = D_\beta = D_\gamma = D$ where D is the "all purpose" (42) fractal dimensionality. The field of fractal geometry is, of course, much broader and richer than these paragraphs indicate (11). However, this is all that is needed in this review.

For aggregates generated by experiments or computer simulations we do not expect to find pure power law relationships. Instead, corrections to the asymptotic scaling relationships are found as a result of finite size and other effects. Consequently, if, for example, β is found from a linear least squares fit to the coordinates [$\ln(R_g)$, $\ln(M)$] for a large number of clusters, then the fractal dimensionality D_β obtained from $1/\beta$ should be interpreted

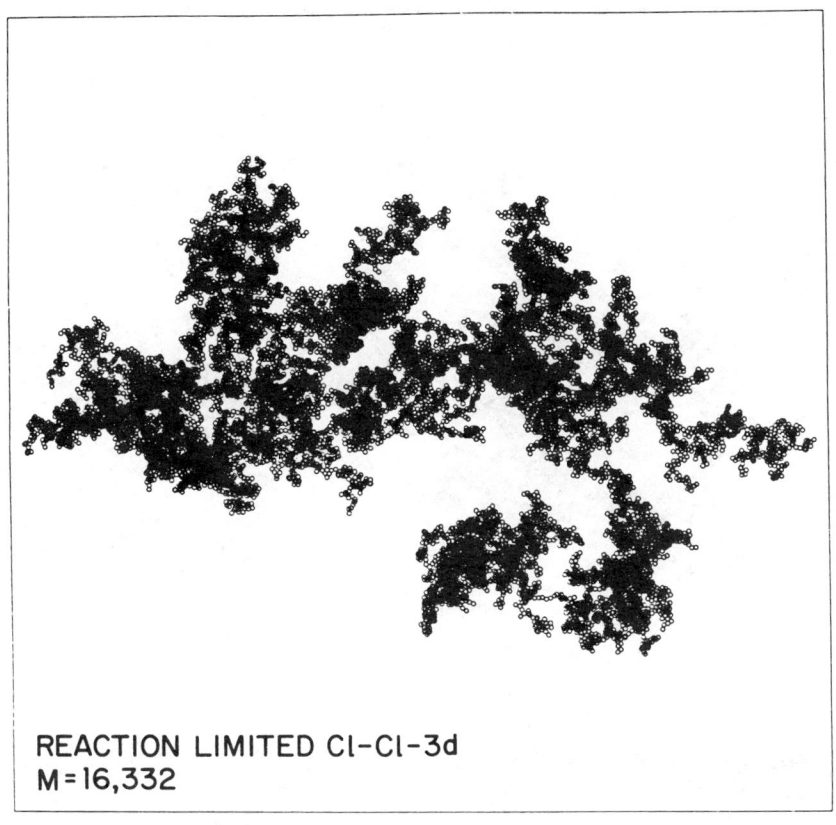

REACTION LIMITED Cl-Cl-3d
M = 16,332

◄─────────────────────────────►
180 DIAMETERS

Figure 4 Projections of three-dimensional clusters generated by off-lattice cluster-cluster aggregation models. Figures 4a–c were generated using polydisperse models in which the clusters follow trajectories with values of 0, 1, and 2, respectively, for D_t.

as an "effective" fractal dimensionality. In some cases the limiting or asymptotic fractal dimensionality can be estimated by measuring the effective fractal dimensionality over a range of length scales and extrapolating to infinite system size. However, in most cases the form of the corrections to the asymptotic scaling behavior is not known and the extrapolation to finite system size is arbitrary and uncertain. In many instances errors have been made in the interpretation of simulation results (even by experienced scientists using large scale simulations) as a result of not recognizing the importance of these "corrections to scaling."

The existence of these mass-length scaling relationships has important

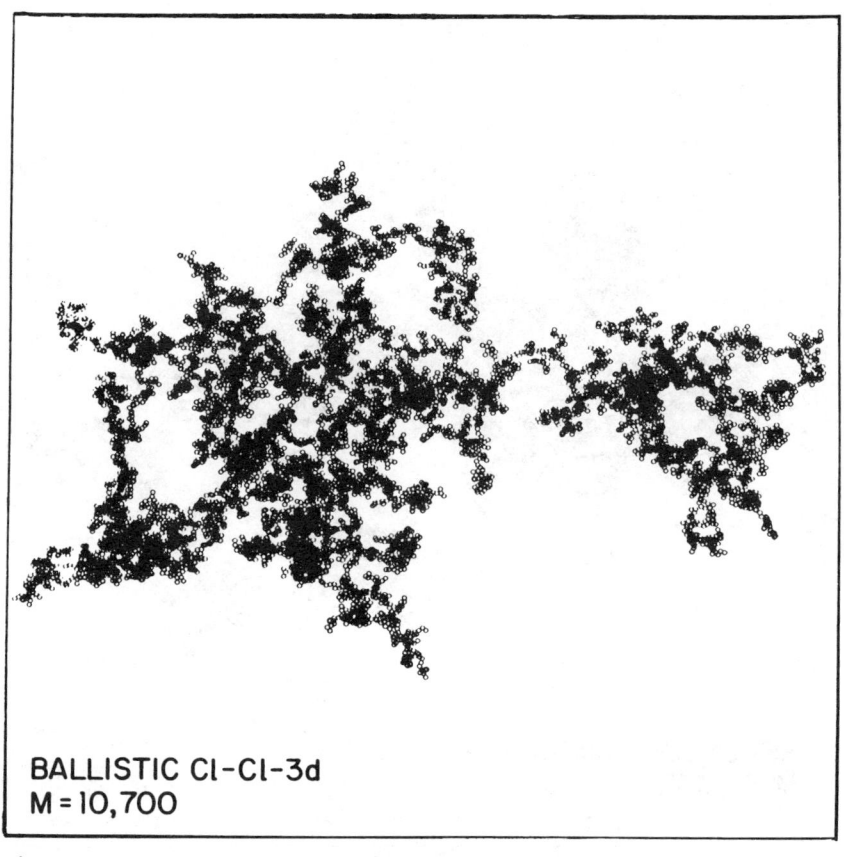

**BALLISTIC Cl-Cl-3d
M = 10,700**

◄─────────── 180 DIAMETERS ───────────►

Figure 4b

implications. For example, if we know how R_g depends (asymptotically) on M then other quantities (such as the hydrodynamic radius) will depend on the same power of M. In addition, we know how the aggregates will scatter light, X-rays, etc (via the density-density correlation function) and whether they will be opaque.

Diffusion-Limited Aggregation (DLA)

In the diffusion-limited aggregation model of Witten & Sander (6), particles are added, one at a time, to a growing aggregate. We imagine that the particles come from infinity but in practice they are launched from a randomly selected point on a circle that just encloses the cluster (Figure

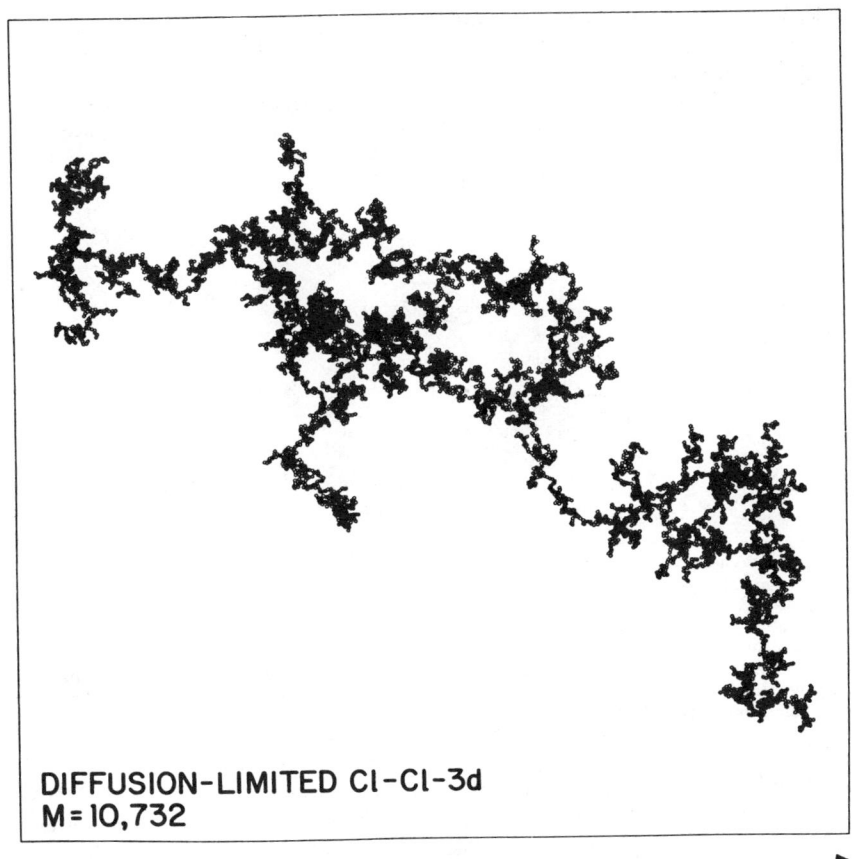

DIFFUSION-LIMITED Cl-Cl-3d
M = 10,732

260 DIAMETERS

Figure 4c

6). After launching, the particles follow a random walk trajectory ($D_t = 2$). Figure 6 shows a square lattice version of this model with two typical trajectories (t_1 and t_2). Trajectory t_1 eventually moves the particle a long distance from the cluster. In this event the trajectory is terminated and a new trajectory is started at a randomly selected point on the launching circle (typically the radius of this "killing circle" was $3R_{max}$, where R_{max} is the maximum cluster radius in early simulations, but $100\ R_{max}$ is more typical of recent simulations). Trajectory t_2 moves the "particle" onto an unoccupied site at the perimeter of the growing cluster. In this event the perimeter site is filled (growth has occurred), and a new random walk trajectory is started from the launching circle. This sequence of events is

Table 1 Fractal dimensionalities obtained from some simple aggregation models[‡]

Dimensionality of space or lattice	Particle-cluster	Polydisperse cluster-cluster	Hierarchical cluster-cluster
	Diffusion-limited ($D_t = 2$)		
2	$1.71^{o,(57)}$	$1.45^{o,(100)}$	$1.44^{l,(101)}$
3	$2.50^{o,(57)}$	$1.80^{o,(96)}$	$1.78^{o,(96)}$
4	$3.41^{o,(57)}$	$2.10^{o,(100)}$	$2.02^{l,(101)}$
	Ballistic ($D_t = 1$)		
2	$2.0^{o,l,(102-103)}$	$1.55^{o,(100),*}$	$1.51^{l,(106)}$
3	$3.0^{o,l,(102-103)}$	$1.95^{o,(83)}$	$1.89^{o,(95)}$
4	$4.0^{o,l,(103),+}$	$2.24^{o,(100),*}$	$2.22^{l,(106)}$
	Reaction limited $D_t = 0$		
2	$2.0^{l,(104),+}$	$1.61^{o,(105)}$	$1.54^{l,(92,93)}$
3	$3.0^{l,(104),+}$	$2.09^{o,(84)}$	$1.99^{l,(92,93)}$
4	$4.0^{l,(104),+}$	$2.48^{o,(105)}$	$2.32^{l,(92)}$

[‡] References are numbers in parentheses.
[*] Unrealistic (constant) reaction kernel.
[+] Theoretical result.
[o] Off-lattice model.
[l] Lattice model.

The results reported here are from the largest scale simulations reported up to the end of 1987. In many cases the statistical uncertainties are about 0.01 or smaller. The uncertainties resulting from finite size effects are, in most cases, larger than this.

repeated over and over again until a large cluster has grown. The cluster shown in Figure 1c was generated using an off-lattice version of this model.

It is evident from even a casual inspection of Figures 1c and 3c that this diffusion-limited aggregation model does not lead to structures that closely resemble those formed by most aggregating colloids. However, this simple model does provide a basis for understanding a surprisingly large variety of other phenomena, including electrodeposition (43, 44), dielectric breakdown (45), fluid-fluid displacement in Hele-Shaw cells (46, 47) and porous media (48), dissolution of porous materials (49, 50), random dendritic growth (51), and the morphology of some thin film systems (52–54).

In addition, the DLA model provided much of the motivation for the work discussed in this review. From the two-point density-density correlation function ($C(r)$) and the dependence of the radius of gyration on cluster mass (6, 55, 56), a fractal dimensionality of about 1.70 was obtained for $d = 2$. The largest scale three-dimensional simulations carried out so far give a fractal dimensionality (D_β) of about 2.50 (S. Tolman, P. Meakin, unpublished). Recent simulations indicate that lattice anisotropy has an important effect on the growth of DLA clusters. For example,

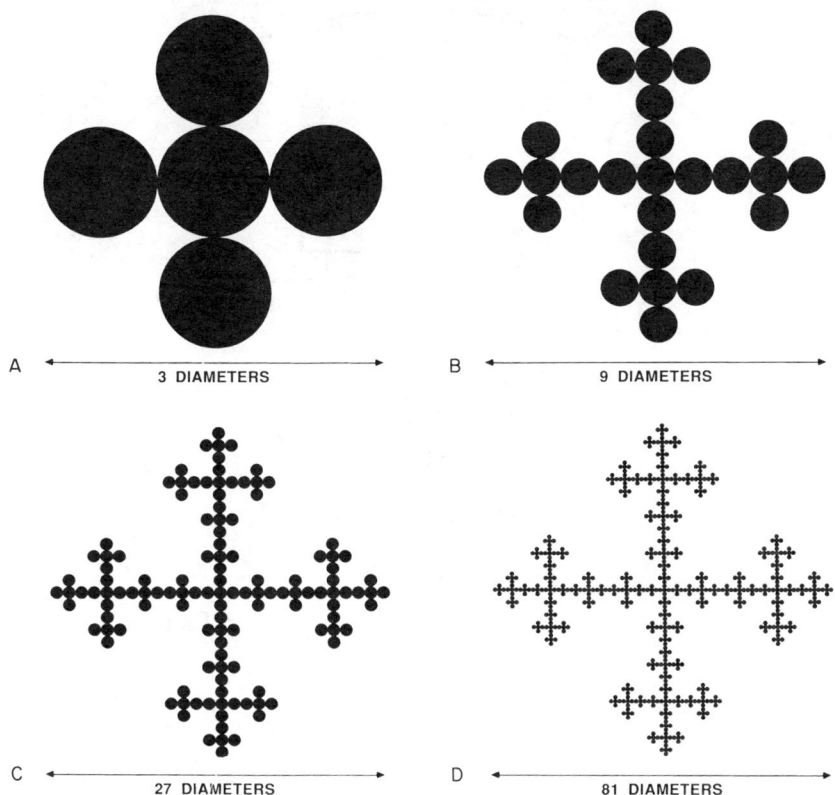

Figure 5 A fractal aggregate generated using a hierarchical deterministic procedure. In Figure 5a five circles have been joined to form a cross. In Figure 5b five of these crosses have been joined to form a still larger cross. Figures 5c and 5d show the next two stages in the generation process. As the aggregate grows, its average density becomes smaller and smaller.

Figure 7 shows a cluster of 3.8×10^6 sites grown on a square lattice using the algorithm described by Ball & Brady (58). It is evident from this figure that the shape of the cluster has evolved from a more or less circular shape characteristic of very small clusters (6, 55, 56) via a diamond-like shape for clusters containing about 10^5 sites (58, 59) into the cross-like shape shown in Figure 7 (60).

The effects of lattice anisotropy are very much smaller for hexagonal and triangular lattices than for the square lattice. Figure 8 shows the dependence of $\ln(R_g)$ on $\ln(M)$ obtained from DLA simulations carried out on the sites of a six-coordinate "honeycomb" lattice. For off-lattice DLA the radius of gyration exponent β has a value of about 0.6 (61),

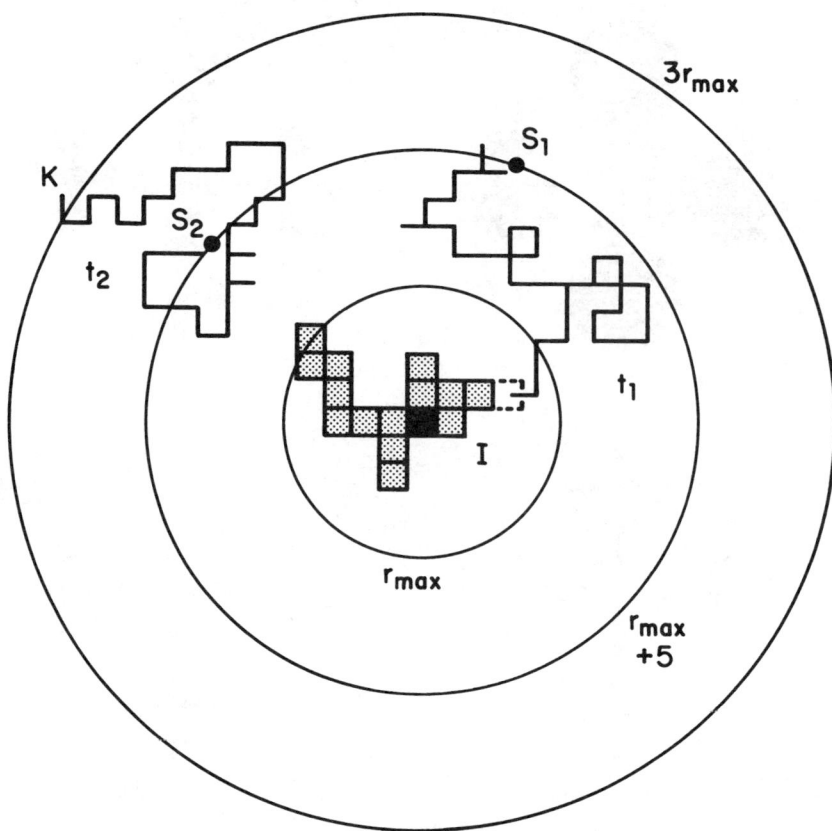

Figure 6 An early stage in a square lattice model simulation of diffusion-limited aggregation. The original seed or growth site is shown in *black* and the other sites that are occupied at this stage are *shaded*. Two typical trajectories starting at random positions on the launching circle are shown. Trajectory t_1 reaches an unoccupied surface site (growth site indicated by a *dashed border*) and this site is occupied. Trajectory t_2 reaches the termination circle, which in this case has a radius of $3R_{max}$ where R_{max} is the maximum radius of the cluster. This trajectory will be terminated and a new trajectory started at a random position on the launching circle.

corresponding to a fractal dimensionality of about 5/3. Consequently, in Figure 8a $\ln(R_g/M^{0.6})$ has been plotted against $\ln(M)$. Also shown in Figure 8a is the dependence of $\ln(R_g/M^{0.6})$ for honeycomb lattice model simulations carried out using sticking probabilities (σ) of (0.3, 0.1, 0.03, 0.01, and 0.003) and for simulations carried out using "dielectric breakdown model" boundary conditions at the surface of the growing cluster. Extensive off-lattice simulations (61) and simulations carried out on the

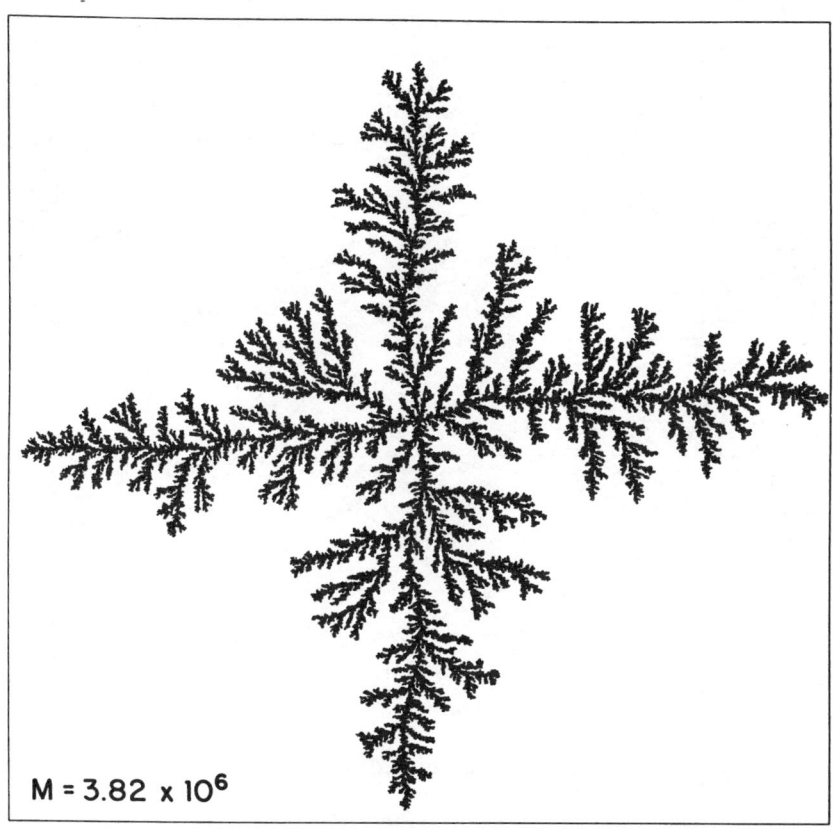

14,000 LATTICE UNITS

Figure 7 A 3.82 × 10⁶ site square lattice DLA cluster grown using the algorithm of Ball & Brady (58).

honeycomb lattice indicate that the exponent β in Eq. 2 has a value much closer to 0.585 than 0.6, corresponding to a fractal dimensionality of 1/0.585 or 1.71. In the models with sticking probabilities less than 1, a random number x uniformly distributed over the range $0 < x < 1$ was generated each time the random walker entered an unoccupied perimeter site. If $x < \sigma$ the site is filled, if $x > \sigma$ the random walk continues. Here σ is the sticking probability. If the random walker attempts to move onto an occupied site, it is returned to the perimeter site from which it attempted to move. In the model with dielectric breakdown boundary conditions (62), growth occurs in the last unoccupied perimeter site occupied by the random walker after it has stepped onto a site occupied by the cluster. The

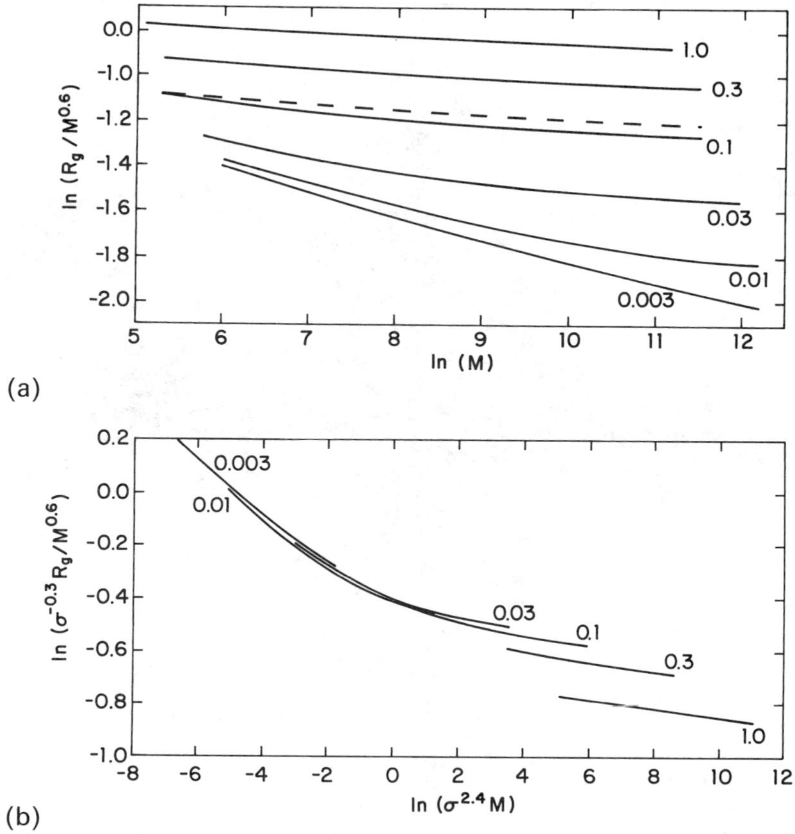

Figure 8 Dependence of $\ln(R_g/M^{0.6})$ on $\ln(M)$ obtained from simulations of diffusion-limited aggregation (DLA) on a six-coordinate honeycomb lattice. Results are shown for six values of the sticking probability ($\sigma = 1$, 0.3, 0.1, 0.03, 0.01, and 0.003) and for growth with dielectric breakdown boundary conditions and $\sigma = 1.0$ (*dashed curve*). Each curve was obtained from about 60 clusters containing 10^5 or more sites. Figure 8b shows the results of an attempt to scale the curve shown in Figure 8a (for ordinary DLA boundary conditions) onto a single curve.

results shown in Figure 8a indicate that for small clusters and small values of σ the effective fractal dimensionality is 2.0, whereas for large clusters and large values of σ the effective fractal dimensionality is 1.71. In Figure 8b an attempt has been made to scale the curves shown in Figure 8a by plotting $\ln(\sigma^{-0.3}R_g/M^{0.6})$ against $\ln(\sigma^{2.4}M)$. For the smaller values of the sticking probability ($\sigma \leq 0.1$), the data collapse seems reasonably good and indicates that the dependence of R_g on M and σ can be described by the scaling form

$$R_g \sim \sigma^\varepsilon f(\sigma^\phi M). \qquad 6.$$

The results shown in Figure 8b indicate that the exponent ϕ in Eq. 6 has a value of about 2.4 and ε has a value of about -1.14. For small values of $x(\sigma^\phi M)$, the scaling function $f(x)$ has the form $f(x) \sim x^{1/2}$ and $R_g \sim \sigma^{0.06} M^{0.5}$. In fact, it seems more probable that the radius of gyration should be independent of σ for small x. The value of 0.06 obtained from Figure 8b is quite close to this expected value. For large x, $f(x)$ has the form $f(x) \sim x^{0.585}$ and $R_g \sim \sigma^{0.3} M^{0.585}$. According to this picture, the asymptotic (large M) fractal dimensionality is 1.71 for all values of σ. In addition, the same fractal dimensionality is obtained using the dielectric breakdown boundary conditions. This "universality" associated with DLA models (6, 55, 56) is one of the reasons that they became the subject of extensive investigation.

Early simulations (6, 55, 56) also indicated that the fractal dimensionality was independent of lattice structure and other anisotropies. It now appears that this result was not correct (63–69).

It has recently been suggested that noise-reduced DLA algorithms (66–71) may allow us to approach the asymptotic (large size) limit of DLA without generating enormously large clusters. In these models random walkers are terminated when they reach a perimeter site, but growth does not occur until a perimeter site has been reached s times. Even for quite small values of s (2 or 3), clusters that resemble that shown in Figure 7 can be easily generated with less than 50,000 sites. For large values of s the clusters become distinctly cross-shaped, but it appears that both the length and width of the arms of the cluster grow as the same power of the mass in the large mass limit (66–68). This means that there is a specific nonzero angle associated with the tip of each arm. Both theoretical considerations (63, 64, 72) and computer simulations indicate that the fractal dimensionality in this limit is close to but larger than 1.5.

Cluster-Cluster Aggregation Models

The most obvious way in which the DLA model described in the previous section differs from colloidal aggregation in real systems is the existence of a "special" growth site or nucleation site. The approximate location of this growth site can be found by inspection, and quantitative analysis of clusters generated by DLA models (including off-lattice models) shows that the region surrounding the growth site is a region of anomalously high density (73). In addition, the density correlations measured at a distance r from the growth site have been found to be different in the radial and tangential directions (74, 75).

In cluster-cluster aggregation models there is no unique seed or growth site. Instead, aggregation is assumed to be occurring in a similar fashion at all locations in the aggregating system. Figures 2 and 4 show some of the clusters generated from models of this type. The cluster-cluster aggregation simulations used to obtain Figures 2 and 4 were carried out by using a list of particles and clusters. These models correspond to aggregation in the limit $\rho \to 0$ where ρ is the mean density. They are used to investigate the asymptotic scaling properties associated with fractal aggregates and can be used to explore some aspects of aggregation kinetics as well. However, finite concentration models (7, 8, 76–78) provide a more realistic representation of real aggregation processes. In these models a fixed number of particles (N_0) are initially placed at random in a hypercubic box or lattice (avoiding overlap between any two particles or sites). In the (more simple) lattice models a filled lattice site is used to represent each of the particles. As the simulation proceeds, clusters are selected randomly (with probabilities that depend on their sizes) and moved by one lattice unit in a randomly selected direction on the lattice. After a cluster has been moved, its perimeter is examined to determine whether any other clusters have been contacted via nearest neighbor occupancy. If contact does occur, all of the contacting clusters are combined to form a larger cluster that continues to move on the lattice. Figure 9 shows the results obtained using a small-scale two-dimensional version of this model. For the simulations illustrated in Figure 9 it was assumed that all of the clusters have the same diffusion coefficients irrespective of their masses (i.e. $\mathcal{D}(s) \sim s^0$ where $\mathcal{D}(s)$ is the diffusion coefficient for a cluster containing s sites.

These simple lattice model simulations can be made time-dependent in the following way (79, 80). After each cluster has been selected at random, the time is incremented by $1/(N\mathcal{D}_{max})$, where N is the number of remaining clusters and $\mathcal{D}_{max}$ is the maximum diffusion constant for any cluster in the system. After a cluster of size s has been selected (and the time incremented) a random number (x) uniformly distributed over the range $0 < x < 1$ is generated and the cluster is moved if $x < \mathcal{D}(s)/\mathcal{D}_{max}$. If $x > \mathcal{D}(s)/\mathcal{D}_{max}$ a new cluster is selected at random. This procedure can be used to study aggregation kinetics in time units that are equal to the average time required for a monomer (single lattice site) to move by one lattice unit. In these simulations no assumption need be made concerning the form of the reaction kernel. Instead, they have been used to determine the reaction kernel for diffusion-limited cluster-cluster aggregation (81). By measuring the number of times clusters of size i and j combine and their concentrations, it can be shown that for $d = 3$ the reaction kernel is given by

$$K(i,j) = (\mathcal{D}(i) + \mathcal{D}(j))(r_i + r_j) \qquad 7.$$

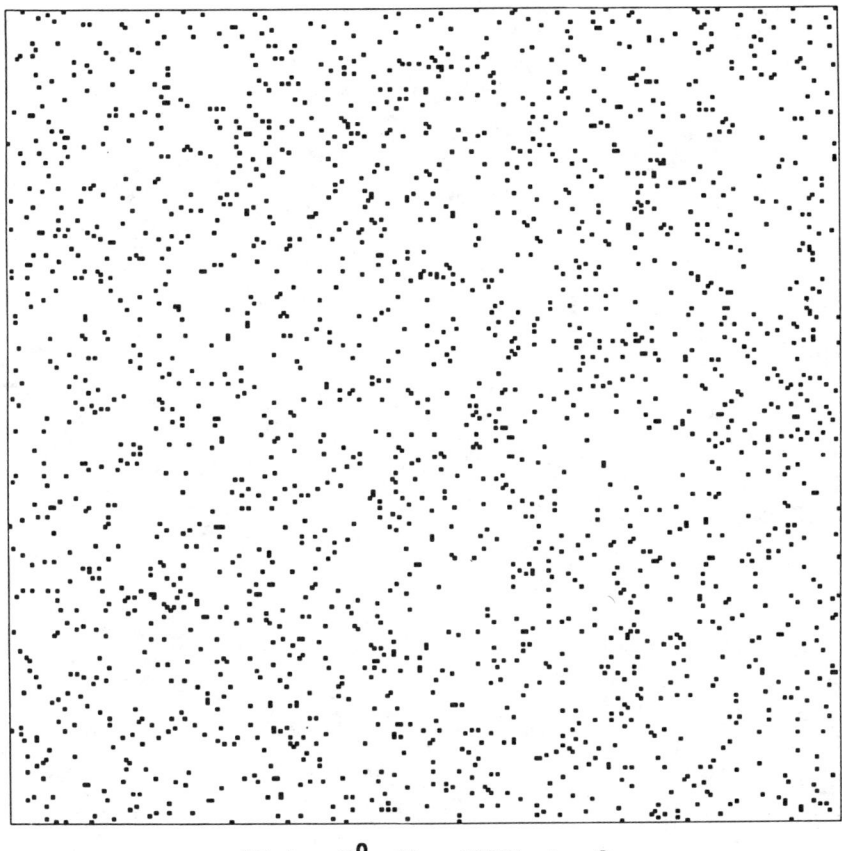

D(s) ~ S⁰, N = 1877, t = 0

Figure 9 Four stages in a small-scale simulation of diffusion-limited cluster-cluster aggregation carried out on a square lattice with a particle (occupied site) density of 0.0305.

where $\mathscr{D}(i)$ is the diffusion coefficient for a cluster containing i particles or sites and r_i is its radius. The radii r_i and r_j in Eq. 7 are defined by the cross section for the diffusion-limited collision between pairs of clusters. However, for a self-similar fractal we expect that all average radii (such as the radius of gyration and the hydrodynamic radius) and even quantities such as the maximum radius should all scale asymptotically as $i^{1/D}$ so that Eq. 7 can replaced by

$$K(i,j) \sim (i^{-1/D} + j^{-1/D})(i^{1/D} + j^{1/D}). \qquad 8.$$

The zero concentration limit models illustrated in Figures 1–4 can also

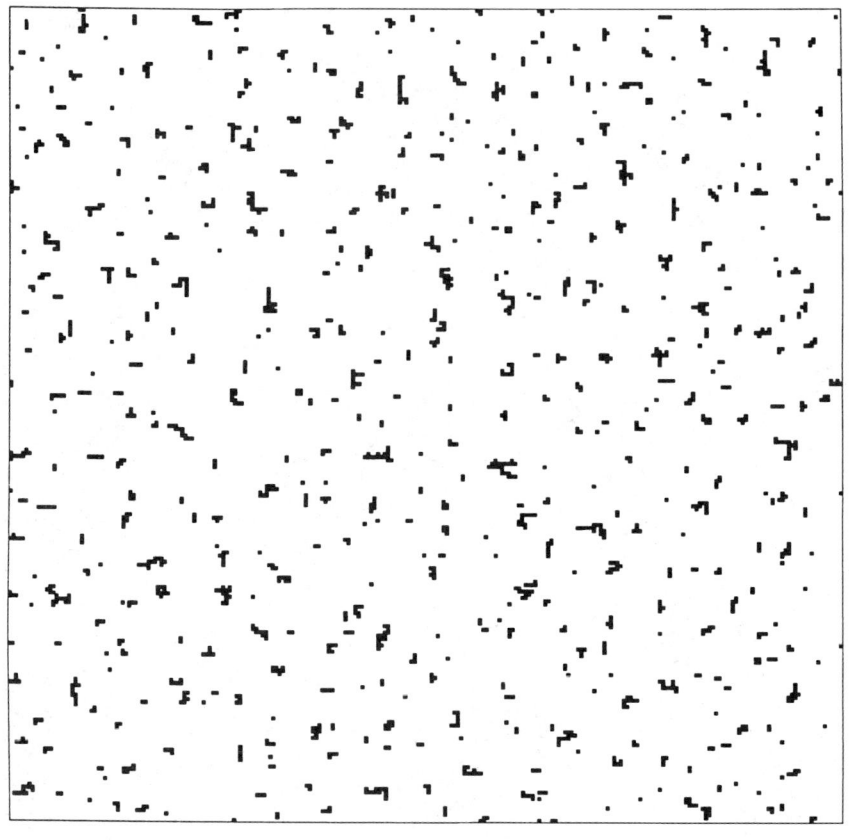

$D(s) \sim S^0$, N = 500, t = 70

Figure 9b

be used to obtain important information about aggregation kinetics. For example, the reaction kernel for ballistic cluster-cluster aggregation can be written as

$$K(i,j) = \sigma(i,j)((i+j)/ij)^{1/2} \qquad 9.$$

for the aggregation of small particles in a low density gas. The quantity $\sigma(i,j)$ in Eq. 9 is the collision cross-section for pairs of clusters of sizes i and j, and the second part of the right-hand side of Eq. 9 is obtained assuming a cluster velocity distribution given by the kinetic theory of gases (82). For $i \simeq j$ the collision cross-section is given by (83)

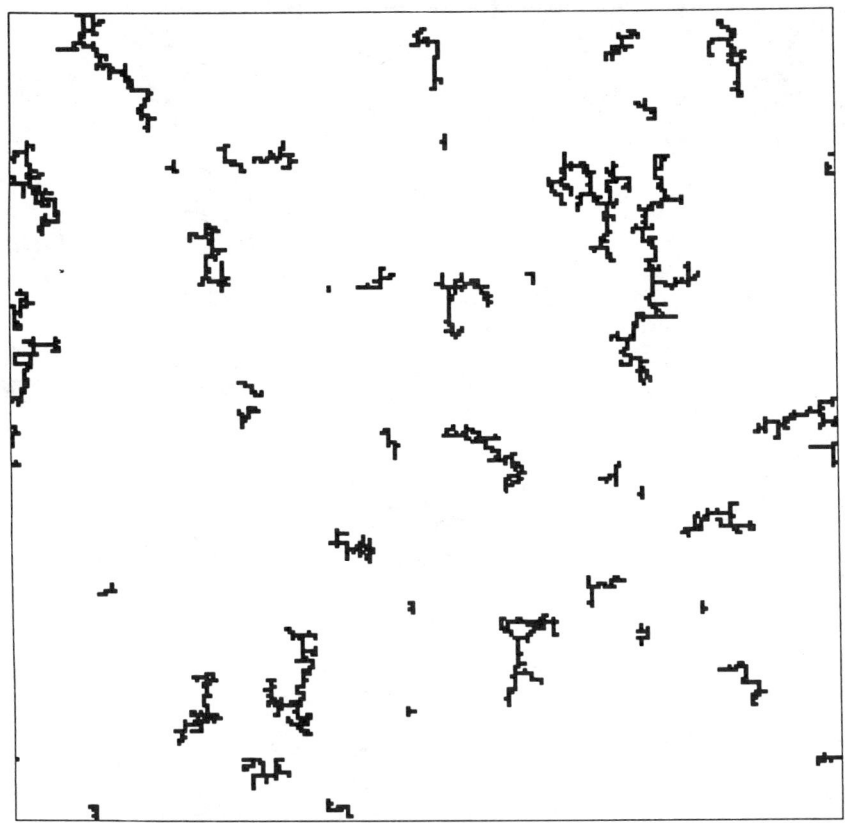

$D(s) \sim S^0$, N = 40, t = 980

Figure 9c

$$\sigma(i,j) \sim (i^{1/D} + j^{1/D})^2 \qquad 10.$$

and for $i \gg j$

$$\sigma(i,j) \sim j^{(2/D-1)}i \qquad 11.$$

for the cases $D < 2$ (the dimensionality appropriate for ballistic cluster-cluster aggregation in three-dimensional space). However, there is no general expression for $K(i,j)$ that is valid for all i and j, but $K(i,j)$ can be measured in ballistic cluster-cluster aggregation simulations. In practice, Eq. 11 describes quite well the dependence of $\sigma(i,j)$ on i and j for all values of i and j, and a reasonably accurate representation of the kinetics of

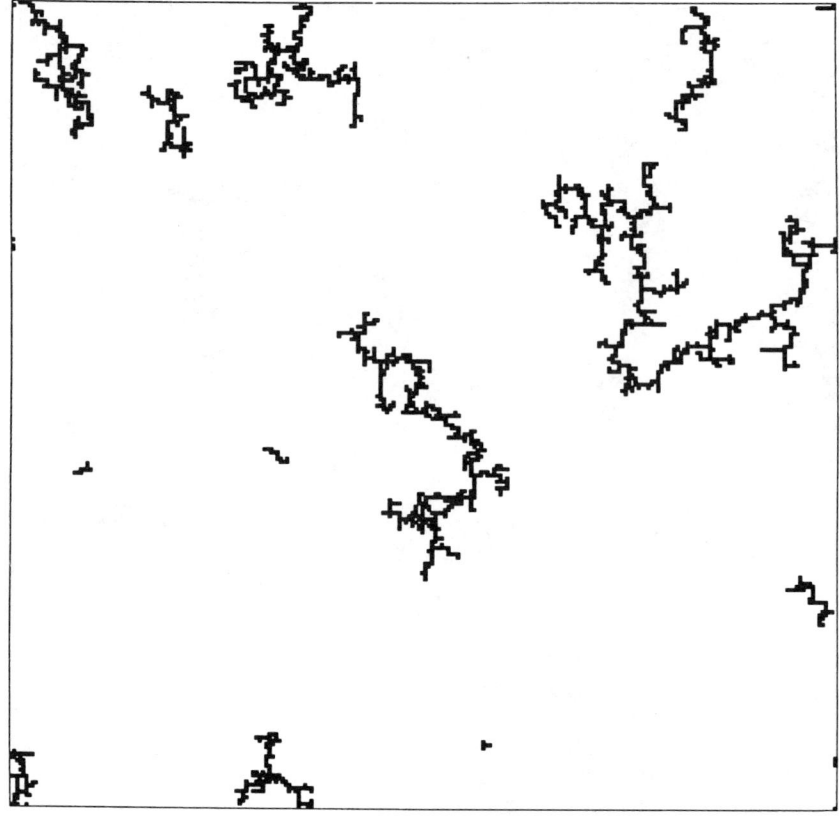

$D(s) \sim S^0$, N = 10, t = 2571

Figure 9d

ballistic cluster-cluster aggregation can be obtained by using Eqs. 9 and 11.

Efficient simulations of ballistic cluster-cluster aggregation kinetics (82) can be carried out using the quantity

$$k(12) = (R_1+R_2)^2((s_1+s_2)/s_1s_2)^{1/2}. \qquad 12.$$

Here R_1 and R_2 are the maximum radii for clusters 1 and 2 of sizes s_1 and s_2. After a pair of clusters (cluster 1 and cluster 2) has been selected from the list of clusters, the time is incremented by $1/(N^2 k_{max})$, where k is the maximum value of k (Eq. 12) for any pair of clusters in the system. A random number x ($0 < x < 1$) is then generated. If $x > k(12)/k_{max}$, the clusters are rotated to random orientations and cluster 1 is "fired" at cluster 2 with an impact parameter selected randomly from a circle of radius $R_1 + R_2$. If the two clusters contact each other, they are irreversibly combined at their position of first contact and returned to the list, which now contains $N-1$ clusters. If they miss, the two clusters are returned to the list, which still contains N clusters. In this model the time is incremented after each pair of clusters has been selected (irrespective of whether they are actually combined).

The reaction-limited cluster-cluster aggregation model illustrated in Figure 4c and described above can also be easily made time-dependent (84). Each time a pair of particles is randomly selected from the list of N_0 particles the time is incremented by a constant amount ($1/N_0^2$ is a convenient quantity), irrespective of whether the clusters containing the two particles do not overlap and the clusters are combined when the two particles are brought into contact with each other.

Figure 10a shows the cluster size distributions ($N_s(t)$ where $N_s(t)$ is the number of clusters of size s at time t) obtained from ballistic cluster-cluster aggregation model simulations. In Figure 10b these cluster size distributions have been scaled by plotting $\ln(s^2 N_s(t))$ against $\ln(s/S(t))$ where S is the mean cluster size defined by

$$S = \sum_{s=1}^{\infty} s^2 N_s(t) \bigg/ \sum_{s=1}^{\infty} s N_s(t).\qquad 13.$$

The success of this data collapse indicates that the cluster size distributions can be described by the scaling form

$$N_s(t) \sim s^{-2} f(s/S(t))\qquad 14.$$

or

$$N_s(t) \sim s^{-2} f(s/t^z)\qquad 15.$$

since $S(t)$ grows asymptotically as t^z where z has a value of about 2.06. This scaling procedure can be applied to a very wide range of aggregation processes (79, 80, 85–91). For systems in which only one characteristic cluster size [represented by $S(t)$] is found, it is natural to attempt to describe the cluster size distribution in terms of the scaling form

$$N_s(t) \sim S^{-\theta} f(s/S(t))\qquad 16.$$

and mass conservation requires the exponent θ to have a value of 2.0 (79).

SUMMARY

Interest in a broad range of nonequilibrium growth and aggregation models has developed rapidly in recent years. This is a natural consequence

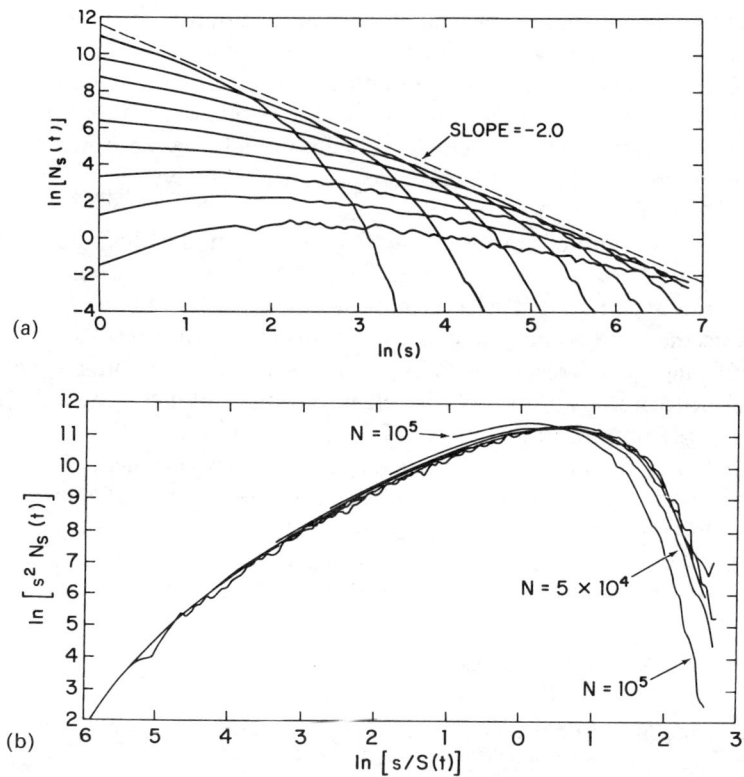

Figure 10 Time-dependent cluster size distributions obtained from a polydisperse off-lattice three-dimensional model for ballistic cluster-cluster aggregation. Figure 10a shows the cluster size distribution ($N_s(t)$) at nine stages during the simulation at which the number of clusters had been reduced from $N_0 = 2 \times 10^5$ to $N_0/2$, $M_0 4 \ldots N_0/2^9$. Figure 10b shows the scaling of the cluster size distributions in Figure 10a using the scaling form given in Eq. 13. The results from 32 simulations were averaged to obtain these results.

of the increased availability, capability, and ease of use of digital computers. At the present time much of our knowledge and understanding has come from computer simulations. Although our ultimate goal is to develop generally successful theoretical approaches to nonequilibrium processes such as those discussed in this review, additional progress will rely heavily on the continued enhancement of computer resources. As new theoretical ideas are developed, larger scale simulations will be needed to obtain results that are accurate and reliable enough to test these ideas and discriminate between the results of competing theories. Because of the scientific interest and practical importance of growth, aggregation, and related processes, research activity in this area should continue to increase. The use of simple computer models will continue to play an important role in helping us to understand the results of experiments and in evaluating and stimulating theoretical ideas.

A few years ago the idea that the concept of universality might be applied to nonequilibrium as well as equilibrium phenomena helped to stimulate interest in fractal aggregates. It now appears that the range of universality associated with most aggregation models is at best quite small. For example, the first investigations of reaction-limited cluster-cluster aggregation by Jullien & Kolb using a hierarchical (monodisperse) model indicated that $D \simeq 1.53$ for $d = 2$ and $D \simeq 1.98$ for $d = 3$ (92, 93). Brown & Ball (94) found larger values for the fractal dimensionality ($D \simeq 1.59$ for $d = 2$ and $D \simeq 2.11$ for $d = 3$) using a more realistic polydisperse model. These differences in the effective fractal dimensionalities (for the same Euclidean dimensionality) seem to reflect genuine differences in the asymptotic geometric scaling properties associated with these models. Similar, but smaller, differences have been found between the fractal dimensionalities associated with monodisperse and polydisperse ballistic cluster-cluster aggregation (83, 95, 96). For diffusion-limited cluster-cluster aggregation it also appears that polydisperse models lead to higher fractal dimensionalities than monodisperse models (96). However, in this case the differences are smaller than realistic statistical and systematic uncertainties. The idea that the fractal dimensionality should depend on the cluster size distributions is also supported by theoretical results (97).

In general, it appears that the fractal dimensionality associated with simple aggregation models often varies continuously with some parameter used to define the model (98, 99). This should be contrasted with the situation in critical phenomena, where it is rare to find models in which critical exponents vary continuously with model parameters. Unfortunately, it is often virtually impossible to distinguish, in practice, between continuously varying (asymptotic) exponents and crossover phenomena using finite size simulations and/or experiments. This is almost certainly the main source of uncertainty and ambiguity in the study of aggregation using simple computer models.

ACKNOWLEDGMENT

My interest in and understanding of models for colloidal aggregation has been enhanced by interactions with a large number of colleagues and collaborators. Without their direct and indirect contributions this work would not have been possible.

Literature Cited

1. Vold, M. J. 1963. *J. Colloid Sci.* 18: 684
2. Vold, M. J. 1959. *J. Colloid Sci.* 14: 168
3. Sutherland, D. N. 1967. *J. Colloid Interface Sci.* 25: 373
4. Eden, M. 1961. *Proc. 4th Berkeley Symp. on Math. Stat. Prob.*, ed. F. Neyman, 4: 223
5. Forrest, S. R., Witten, T. A. 1979. *J. Phys. A* 12: 109
6. Witten, T. A., Sander, L. M. 1981. *Phys. Rev. Lett.* 47: 1400
7. Meakin, P. 1983. *Phys. Rev. Lett.* 51: 1119
8. Kolb, M., Botet, R., Jullien, R. 1983. *Phys. Rev. Lett.* 51: 1123
9. Weitz, D. A., Oliveria, M. 1984. *Phys. Rev. Lett.* 52: 1433
10. Schaefer, D. W., Martin, J. E., Wiltzius, P., Cannell, D. S. 1984. *Phys. Rev. Lett.* 52: 2371
11. Mandelbrot, B. B. 1982. *The Fractal Geometry of Nature.* New York: Freeman
12. Family, F., Landau, D. P., eds. 1984. *Kinetics of Aggregation and Gelation.* Amsterdam: North-Holland
13. Stanley, H. E., Ostrowsky, N., eds. 1986. *On Growth and Form: Fractal and Non-Fractal Patterns in Physics*, NATO, ASI Ser. E100. Dordrecht: Martinus Nijhoff
14. Pietronero, L., Tosatti, E., cds. 1986. *Fractals in Physics. Proc. 6th Trieste Int. Symp. on Fractals in Physics.* Amsterdam: ICTP, North-Holland
15. Pynn, R., Skjeltorp, A. J., eds. 1986. *Scaling Phenomena in Disordered Systems.* NATO Ser. B133. New York: Plenum
16. Riste, T., Pynn, R. A., eds. 1987. *Time Dependent Effects in Disordered Materials.* Geilo, Norway: NATO ASI
17. Stanley, H. E., Ostrowsky, N., eds. 1984. *On Growth and Form: Fractal and Non-Fractal Patterns in Physics.* NATO ASI Ser. E100. Dordrecht: Martinus Nijhoff
18. Schlesinger, M., ed. 1984. *Proc. Symp. on Fractals in the Physical Sciences.* Gaithersburg: NBS. (*J. Stat. Phys.* 36(8/6))
19. Jullien, R., Botet, R. 1987. *Aggregation and Fractal Aggregates.* Singapore: World Scientific
20. Feder, J. 1988. *Fractals.* New York: Plenum
21. Herrmann, H. J. 1986. *Phys. Rep.* 136: 153
22. Jullien, R. 1986. *Ann. Telecommun.* 41: 343
23. Witten, T. A., Cates, M. E. 1986. *Science* 232: 1607
24. Sander, L. M. 1986. *Nature* 322: 789
25. Meakin, P. 1988. In *Phase Transitions and Critical Phenomena*, Vol. 12, ed. C. Domb, J. L. Lebowitz, p. 336. New York: Academic
26. Meakin, P. 1984. *Phys. Rev. B* 29: 3722
27. Deleted in proof
28. Jullien, R. 1985. *Phys. Rev. Lett.* 55: 1697
29. Mandelbrot, B. B. 1986. See Ref. 14, pp. 3, 17, 21
30. Voss, R. F. 1986. In *Scaling Phenomena in Disordered Systems*, ed. R. Pynn, A. J. Skjeltorp. NATO ASI Ser. B133. New York: Plenum
31. Family, F., Vicsek, T. 1985. *J. Phys. A* 18: L75
32. Kardar, M., Parisi, G., Zhang, Y.-C. 1986. *Phys. Rev. Lett.* 56: 889
33. Jullien, R., Botet, R. 1985. *Phys. Rev. Lett.* 54: 2055
34. Meakin, P., Ramanlal, P., Sander, L. M., Ball, R. C. 1986. *Phys. Rev. A* 34: 5091
35. Mandelbrot, B. B. 1975. *Les Objets Fractals: Form, Hasard et Dimension.* Paris: Flamarion
36. Mandelbrot, B. B. 1977. *Fractals: Form, Chance and Dimension.* San Francisco: Freeman
37. Fournier d'Albe, E. E. 1907. *Two New Worlds. I. The Infra World. II. The Supra World.* London: Longmans Green
38. Vicsek, T. 1983. *J. Phys. A* 16: L647
39. Kantor, Y., Witten, T. A. 1984. *J. Phys. Lett.* 45: L675
40. Lindsay, H. M., Lin, M. Y., Weitz, D. A., Sheng, P., Chen, Z., Klein, R., Meakin, P. 1987. *Faraday Discuss. Chem. Soc.* 83: 153
41. Stanley, H. E. 1977. *J. Phys. A* 20: L211
42. Mandelbrot, B. B. 1985. *Phys. Scr.* 32: 257
43. Brady, R. M., Ball, R. C. 1984. *Nature* 309: 225
44. Matsushita, M., Sano, M., Hayakawa, Y., Honjo, H., Sawada, Y. 1984. *Phys. Rev. Lett.* 53: 286
45. Niemeyer, L., Pietronero, L., Wiesmann, H. J. 1984. *Phys. Rev. Lett.* 52: 1033
46. Nittmann, J., Daccord, G., Stanley, H. E. 1985. *Nature* 314: 141
47. Daccord, G., Nittmann, J., Stanley, H. E. 1986. *Phys. Rev. Lett.* 56: 336
48. Måløy, K. J., Feder, J., Jøssang, T. 1985. *Phys. Rev. Lett.* 55: 2688

49. Daccord, G. 1987. *Phys. Rev. Lett.* 58: 479
50. Daccord, G., Lenormand, R. 1987. *Nature* 325: 41
51. Honjo, H., Ohta, S., Matsushita, M. 1986. *J. Phys. Soc. Jpn.* 55: 2487
52. Elam, W. T., Wolf, S. A., Sprague, J., Gubser, D. V., Van Vechten, D., Barz, G. L. Jr., Meakin, P. 1985. *Phys. Rev. Lett.* 54: 701
53. Radnoczi, G., Vicsek, T., Sander, L. M., Grier, D. 1987. *Phys. Rev. A* 35: 4012
54. Madeleine, D., Hurd, A. J. 1987. Preprint
55. Meakin, P. 1983. *Phys. Rev. A* 27: 604
56. Meakin, P. 1983. *Phys. Rev. A* 27: 1495
57. Deleted in proof
58. Ball, R. C., Brady, R. M. 1985. *J. Phys. A* 18: L809
59. Meakin, P. 1985. *J. Phys. A* 18: L661
60. Meakin, P., Ball, R. C., Ramanlal, P., Sander, L. M. 1987. *Phys. Rev. A* 35: 5233
61. Meakin, P., Sander, L. M. 1985. *Phys. Rev. Lett.* 54: 2053
62. Pietronero, L. 1985. Private communication
63. Ball, R. C., Brady, R. M., Rossi, G., Thompson, B. R. 1985. *Phys. Rev. Lett.* 55: 1406
64. Turkevich, L. A., Scher, H. 1985. *Phys. Rev. Lett.* 55: 1026
65. Meakin, P. 1986. *Phys. Rev. A* 33: 3371
66. Nittmann, J., Stanley, H. E. 1986. *Nature* 321: 663
67. Kertesz, J., Vicsek, T. 1986. *J. Phys. A* 19: L257
68. Thompson, B. R. 1986. Preprint
69. Meakin, P. 1987. *Phys. Rev. A* 36: 332
70. Kertesz, J., Vicsek, T., Meakin, P. 1986. *Phys. Rev. Lett.* 57: 3303
71. Tang, C. 1985. *Phys. Rev. A* 31: 1977
72. Ball, R. C. 1988. *Physica A* 104A: 62. (Invited lectures presented at 16th Int. Conf. Thermodynamics and Statistical Mechanics, Boston Univ., 1986, ed. H. E. Stanley)
73. Halsey, T. C., Meakin, P. 1985. *Phys. Rev. A* 32: 2546
74. Meakin, P., Vicsek, T. 1985. *Phys. Rev. A* 32: 685
75. Kolb, M. 1955. *J. Phys. Lett.* 46: L631
76. Finegold, L. X. 1976. *Biochem. Biophys. Acta* 448: 393
77. Donnell, J. T., Finegold, L. X. 1981. *Biophys. J.* 35: 783
78. Sunada, H., Otsuka, H., Yamada, Y., Kawashima, Y., Takenaka, H., Carstensen, J. T. 1984. *Powder Technol.* 38: 211
79. Family, F., Vicsek, T. 1984. *Phys. Rev. Lett.* 52: 1669
80. Meakin, P., Vicsek, T., Family, F. 1984. *Phys. Rev. B* 31: 564
81. Ziff, R. M., McGrady, E. D., Meakin, P. 1985. *J. Chem. Phys.* 82: 5269
82. Jeans, J. 1954. *The Dynamical Theory of Gases.* New York: Dover
83. Meakin, P., Donn, B. 1987. Preprint
84. Meakin, P., Family, F. 1987. *Phys. Rev.* 36: 5498
85. Kolb, M. 1984. *Phys. Rev. Lett.* 53: 1653
86. Botet, R., Jullien, R. 1984. *J. Phys. A* 17: 2517
87. Lushnikov, A. A. 1973. *J. Colloid Interface Sci.* 45: 549
88. Friedlander, S. K. 1977. *Smoke, Dust and Haze.* New York: Wiley
89. Drake, R. L. 1972. *Top. Current Aerosol Res.* 3(2)
90. van Dongen, P. G. J., Ernst, M. H. 1985. *Phys. Rev. Lett.* 54: 1396
91. Leyvraz, F. 1984. See Ref. 17, p. 136
92. Jullien, R., Kolb, M. 1984. *J. Phys. A* 17: L639
93. Kolb, M., Jullien, R. 1984. *J. Phys. Lett.* 45: L977
94. Brown, W. D., Ball, R. C. 1985. *J. Phys. A* 18: L517
95. Jullien, R., Meakin, P. 1988. *J. Colloid Interface Sci.* In press
96. Meakin, P., Jullien, R. 1988. *J. Chem. Phys.* In press
97. Botet, R. 1985. *J. Phys. A* 18: 847
98. Meakin, P. 1984. *J. Chem. Phys.* 81: 4637
99. Meakin, P. 1983. *Phys. Rev. B* 28: 6718
100. Meakin, P. 1985. *Phys. Lett.* 107A: 269
101. Jullien, R., Kolb, M., Botet, R. 1984. *J. Phys. Lett.* 45: L211
102. Meakin, P. 1985. *J. Colloid Interface Sci.* 105: 240
103. Ball, R. C., Witten, T. A. 1984. *Phys. Rev. A* 29: 2966
104. Richardson, D. 1973. *Proc. Cambridge Philos. Soc.* 74: 515
105. Meakin, P. Unpublished
106. Jullien, R. 1984. *J. Phys. A* 17: L771

FRACTAL TIME IN CONDENSED MATTER[1]

Michael F. Shlesinger

Physics Division, Office of Naval Research, Arlington, Virginia 22217-5000

INTRODUCTION

Temporal scaling laws involving noninteger exponents have appeared in the physics of many systems, including charge transport in xerographic films (1–4), electron-hole recombination in amorphous materials (5–7) and other heterogeneous kinetics (8), dielectric, magnetic, and mechanical relaxation in glassy materials (9–12), and in the time-dependent reactivity of radiation-induced chemistry in frozen liquids (13). While at first quite puzzling, all of these complex phenomena have been explained using a concept called *fractal time* to describe the transport of charges and defects in these systems. Fractals are usually considered to be self-similar geometric objects in space with features on an infinite number of scales. Fractal time describes highly intermittent self-similar temporal behavior that does not possess a characteristic time scale. If the average time of an event were finite, then this would provide a time scale. Thus for fractal time, the average time for an event must be infinity. The event can be the time a charge is trapped at a site in an amorphous solid, the time the phase in a Josephson junction rotates in the same direction, the time a fluid particle spends in a given vortex, etc.

In 1963, Berger & Mandelbrot (14) first used the concept of fractal time to characterize transmission errors in telephone networks, and Mandelbrot coined the term in his 1977 book (15) to describe the Scher-Montroll (1) model of transport in amorphous media. We are able to trace the scaling inherent in fractal time back to an idea introduced by Nicolas Bernoulli

---

[1] The US Government has the right to retain a nonexclusive, royalty-free license in and to any copyright covering this paper.

in 1713. We digress into the history of Bernoulli's scaling in the following section. Those who are impatient to learn more about fractal time, and its application to transport and relaxation in complex systems, should skip ahead to the section, fractal time.

BERNOULLI SCALING THROUGH HISTORY
The St. Petersburg Paradox

Montmort (16) in 1713, in the second edition of his book *Essai d'Analyse: Sur les Jeux des Chances (On Games of Chance)*, published, as an appendix, his lengthy correspondence with Nicolas Bernoulli. This device made clear Bernoulli's many contributions to the book. It might prove interesting to resurrect this tradition. In any event, in one letter, Nicolas, a nephew of Jacob and John, introduced an unusual game of chance in which the mean winnings were infinite. This game was next discussed in 1724, this time by Daniel Bernoulli, in the Commentarii of the St. Petersburg Academy. It has since become known as the *Petersburg Game* or as the *St. Petersburg Paradox*. We discuss this game because the scaling introduced by Nicolas lies at the heart of fractal time, a concept we have found to be a key to understanding transport, reaction, and relaxation in amorphous materials. What is the game?

Flip a coin. If it comes up heads, win one coin. If it comes up tails, flip again and again until a head appears. If it takes $N+1$ flips for the first head to appear, then win 2^N coins. This occurs with probability $(\frac{1}{2})^{N+1}$. The mean winnings are infinite, i.e.

$$1 \text{ coin} \times \tfrac{1}{2} + 2 \text{ coins} \times \tfrac{1}{4} + \cdots + 2^N \text{ coins} \times \tfrac{1}{2}^{N+1} + \cdots = \infty \qquad 1.$$

If a billion players try this game, on the average half of the players win one coin, a quarter two coins, and an eighth four coins, etc, even though the mean winning is infinite. When the average winning is infinite, winning occurs on a multitude of scales with no scale dominating. This is the paradigm of fractals. The infinity of Eq. 1 is the signature of scale invariance, and not as originally thought, the failure of probability theory.

What is the fair ante to play this game? The player favors a low ante because his *median* winnings (those that occur with probability up to $\tfrac{1}{2}$) are only one coin. The banker favors an infinite ante because these are his *expected* losses. The player counters that this is so, but that it would take an infinite number of coin tossings to accomplish this feat. The paradox arises because neither the player nor the banker can convince the other about his idea of a fair ante. The problem arises because one is trying to determine a characteristic size from a distribution that does not possess one!

In the physics we discuss here, such types of long tail distributions with infinite moments do arise. However, no paradox is encountered because the physics lies in the distributions, and experiments are able to measure the distribution function directly.

The basic scaling idea of Bernoulli—*an order of magnitude more* (winnings), but *an order of magnitude less often* (probability)—is the basis for the concept of fractal time. I first wish to point out, however, that Bernoulli scaling is a remarkable (if unacknowledged) precursor to many famous landmarks in Mathematics. As the Petersburg Game was considered paradoxical, its various progeny were treated as pathologies. In today's era of fractals, hierarchies, and renormalization groups, we find such scaling to be natural.

The Cantor Set

The Cantor construction (see Figure 1) gives a point set representation of Bernoulli scaling. In each stage, an order of magnitude (in base 3) larger spacings between bars occurs an order of magnitude less often (in base 2).

The Weierstrass Function

The Weierstrass function

$$W(k) = \sum_{n=0}^{\infty} a^{-n} \cos(b^n k), \quad b > a \qquad 2.$$

is essentially a functional form of a Cantor set. It is everywhere continuous, but nowhere differentiable. Each new term in the sum places an order of magnitude more wiggles (in base b) on a previous wiggle (see Figure 2). Each new wiggle is reduced in amplitude by an order of magnitude (in base a). The nondifferentiable Weierstrass curve has fractal dimension $\ln a / \ln b$.

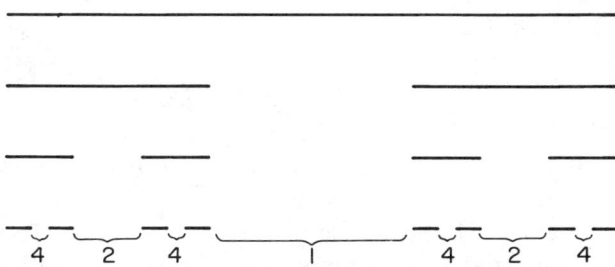

Figure 1 The Cantor Bar. The points remaining after an infinite number of iteration form a self-similar pattern with fractal dimension log 2/log 3.

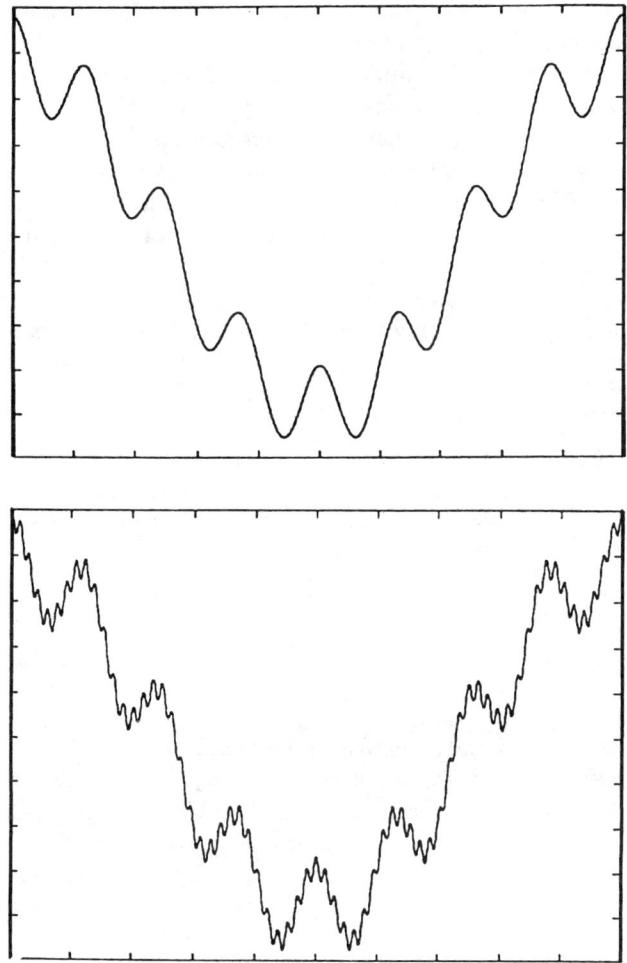

Figure 2 The Weierstrass function with $b = 8$ and $a = 4$. The first two terms (*top*) and the first three terms (*bottom*).

Levy Distributions

When the vogue in probability theory was to prove the Central Limit Theorem under ever weaker assumptions, Paul Levy in the 1920s sought exceptions to it (17). Basically, he proved that sums of identically distributed random variables (i.e. a random walk) need not eventually possess a Gaussian distribution if the variable's second moments are infinite. Consider the following example (18), called (as will become obvious) a

Weierstrass random walk. It is a special case of what is called a Levy flight (15). Let the probability for a jump from $(0,0)$ to (x,y) be

$$p(x,y) = \frac{a-1}{4a} \sum_{n=0}^{\infty} a^{-n}[\delta_{x,\pm b^n} + \delta_{y,\pm b^n}]. \qquad 3.$$

The Bernoulli scaling is apparent in that an order of magnitude larger jump has an order of magnitude less probability to occur, and in that the Fourier transform of $p(x,y)$ is a Weierstrass function, i.e.

$$p(k_x, k_y) = \frac{a-1}{2a} \sum_{n=0}^{\infty} a^{-n}[\cos(b^n k_x) + \cos(b^n k_y)] \qquad 4.$$

For this walk, a cluster of about "a" jumps of length unity occurs (on the average) before a jump of length b occurs. Then about "a" small jumps occur again before the next jump of length b. In this manner about "a" clusters each with "a" sites are formed before a jump of length b^2 occurs. Eventually a hierarchy of clusters is formed with point set dimension $\ln a/\ln b$ (18), which is also the dimension of the Weierstrass function (see Figure 6a). We encounter Levy flights below, in the discussion of TURBULENT DIFFUSION.

FRACTAL TIME

The Connection to Bernoulli Scaling

Consider a physical process in which the time between events (such as the jumping of a particle) can be considered to be a random variable. Let $\psi(t)\,dt$ be the probability that the time between events is between t and $t+dt$, and let $\langle t \rangle$ be the first moment of $\psi(t)$. If $\langle t \rangle$ is finite, then it provides a natural scale in which to measure time, i.e. events will occur at an average rate of $1/\langle t \rangle$. It is when $\langle t \rangle$ is infinite (actually much longer than the time of an experiment) that fractal time occurs. Montroll & Weiss (19) introduced $\psi(t)$ into the analysis of continuous-time random walks (see RELAXATION IN GLASSY MATERIALS, below). Scher & Lax (20) were the first to incorporate the randomness of a system into $\psi(t)$, a procedure that has been validated by Klafter & Silbey (21).

When $\langle t \rangle$ is infinite, the sequence of event times looks as dissimilar to a constant rate process as possible, with points appearing in self-similar clusters akin to points in a randomized Cantor Set.

The Laplace transform $\psi^*(s) \equiv \int_0^{\infty} \exp(-st)\psi(t)\,dt$ behaves for small s (the long time limit) as $1 - s\langle t \rangle$, as can be seen by expanding the exponential. To choose a fractal time $\psi(t)$ that explicitly incorporates Bernoulli scaling, i.e. an order of magnitude longer intermittency occurs, in base b, with an order of magnitude less probability (in base a), consider

$$\psi(t) = \frac{1-a}{a} \sum_{n=1}^{\infty} a^n b^n \exp(-b^n t), \quad b < a < 1. \qquad 5.$$

By Laplace transforming a scaling law becomes apparent:

$$\psi^*(s) = a\psi^*(s/b) + \frac{1-a}{a} \frac{ab}{b+s}. \qquad 6.$$

Any nonanalytic behavior in $\psi^*(s)$ must arise from the homogenous part of Eq. 6

$$\psi_h^*(s) = a\psi_h^*(s/b) \qquad 7.$$

where the subscript h denotes homogeneous, i.e. that the inhomogeneous analytic term is not included.

The solution (22) to Eq. 7 is

$$\psi_h^*(s) = s^\beta K(s) \qquad 8.$$

where $\beta = \ln a/\ln b$ and $K(s)$ is a function periodic in $\ln s$ with period $\ln b$, as can be seen by direct substitution. The full solution to Eq. 6 is (22)

$$\psi^*(s) = 1 - s^\beta K(s) + \frac{1-a}{a} \sum_{n=1}^{\infty} \frac{(-1)^n a s^n}{b^n - a}$$

with

$$K(s) = \frac{1-a}{a \ln b} \sum_{n=-\infty}^{\infty} \frac{\pi a b^x}{\sin(\pi x)} \exp(-2\pi i n \ln s/\ln b) \qquad 9.$$

and

$$x = -\ln a/\ln b + 2\pi i n/\ln b$$

The exponent β reflects the manner in which $\psi(t)$ decays asymptotically, i.e.

$$\psi(t) = 0(t^{-1-\beta}) \qquad 10.$$

and (when $\beta < 1$) the manner in which $\langle t \rangle = -\partial \psi^*(s=0)/\partial s$ diverges. In this case β is also the fractal dimension of the point set of event times. If $\beta > 1$ then $\langle t \rangle$ is finite and the s term dominates the s^β term for small s.

A Mechanism Generating Fractal Time

Many mechanisms can generate a fractal set of event times. A simple example is hopping of a particle over a distribution of activation barriers. Let $\psi(t)$ be a weighted distribution of Poisson processes, i.e.

$$\psi(t) = \int_0^\infty \lambda \exp(-\lambda t) \rho(\lambda) \, d\lambda$$

with

$$\int_0^\infty \rho(\lambda) \, d\lambda = 1. \qquad 11.$$

In regard to Eq. 5, if λ varies as b^n and $\rho(\lambda)$ like a^n, then fractal time will ensue for $b > a$.

This is accomplished if $\lambda = \nu \exp(-\Delta/kT)$ (activated hopping) and the distribution of barrier heights

$$f(\Delta) = (kT_0)^{-1} \exp(-\Delta/kT_0)$$

where kT_0 is the mean width of the barrier distribution. Since $f(\Delta) \, d\Delta = \rho(\lambda) \, d\lambda$,

$$\psi(t) = \int_0^\infty \lambda(\Delta) \exp(-\lambda(\Delta) t) f(\Delta) \, d\Delta, \qquad 12.$$

which for our choices has the asymptotic behavior $\psi(t) \sim t^{-1-\beta}$ with $\beta = T/T_0$ for $T < T_0$, which implies $\langle t \rangle$ is infinite. If $T > T_0$ then the first moment of $\psi(t)$, $\langle t \rangle$ is finite and fractal time does not occur. The form of Eq. 12 appears in the next section, in the discussion of relaxation.

RELAXATION IN GLASSY MATERIALS

The Stretched Exponential Law

In 1913 Debye treated the problem of how initially aligned small spherical molecules of radius R, with dipole moments $\mu(t)$, relax in a fluid with viscosity η when an external electric field is removed. The fluid particles collide with the spherical dipole molecules, causing them to undergo rotational Brownian motion. The relaxation function $\phi(t)$ defined by

$$\phi(t) = \langle \mu(t)\mu(0) \rangle / \langle \mu^2(0) \rangle \qquad 13.$$

measures the dipole-dipole correlation, and is related to the dielectric constant $\varepsilon(\omega)$ via

$$\frac{\varepsilon(\omega)-\varepsilon(\infty)}{\varepsilon(\infty)-\varepsilon(0)} \equiv \varepsilon'(\omega)+i\varepsilon''(\omega) = \int_0^\infty \exp(i\omega t)\,\frac{d\phi(t)}{dt}\,dt. \qquad 14.$$

Debye showed that this system's relaxation was governed by a characteristic time scale $\tau = (4\pi\eta R/kT)$ and $\phi(t)$ was given by

$$\phi(t) = \exp(-t/\tau) \qquad 15.$$

or equivalently

$$\varepsilon'(\omega) = \frac{1}{1+\omega^2\tau^2} \quad \text{and} \quad \varepsilon''(\omega) = \frac{\omega\tau}{1+\omega^2\tau^2}. \qquad 16.$$

A Cole-Cole plot of ε' vs ε'' yields the famous ideal semicircular result. Traditionally, dielectric relaxation analysis of more complex systems focused on $\varepsilon(\omega)$ and deviations from semicircular behavior. A change of focus from frequency to time [from $\varepsilon(\omega)$ to $\phi(t)$] occurred in 1970 when Williams & Watts (10) found empirically that the stretched exponential from

$$\phi(t) = \exp[-(t/\tau)^\beta], \quad 0 < \beta < 1 \qquad 17.$$

fit dielectric relaxation data exceedingly well. In fact, Kohlrausch in 1863 already used the stretched exponential to describe mechanical relaxation (creep) in glassy fibers. In the past few years the stretched exponential has been used to fit an ever-widening variety of experimental data for mechanical, NMR, dielectric, and magnetic relaxation phenomena.

Why Is The Stretched Exponential Universal?

A FROZEN DIPOLE IN A SEA OF MOBILE DEFECTS In Debye's model, fluid particles hitting dipolar molecules provided the mechanism for the relaxation of the system's dipole moment. Consider now an analog for the glassy state where mobile defects are able to move to a frozen-in dipole and bring about its reorientation—in a sense, by transporting a packet of "free volume" to the frozen configuration. When a defect and a dipole meet, the dipole is assumed to relax instantaneously. This model was first proposed in 1960, by Glarum (23), who solved it for one fixed dipole undergoing a one-dimensional random walk. The Williams-Watts form was not found. Philips, Barlow & Lamb (24) treated Glarum's 1D model but with two defects, and Bordewijk (25) considered an infinite system with a concentration of mobile defects. Bordewijk concluded that the defect diffusion model will not give a Williams-Watts result except in 1D, where $\beta = \frac{1}{2}$ is obtained. The value $\beta = 1$ is found in 3D. We now show that the missing ingredient is to treat the mobile defect motion as a fractal time random walk with $\psi(t) \sim t^{-1-\beta}$. This leads directly to the stretched exponential form. We do not specify the precise nature of the defects, although candidates such as dangling bonds, grain boundaries, vacancies, and local conformational fluctuations are possibilities.

Our model envisions a frozen dipole amidst a swarm of mobile defects. Let our dipole of interest be immobile and situated at the origin of a lattice of V sites. Consider that N independent mobile defects are also on this lattice, but not initially at the origin. The probability that a given defect is initially at site $\mathbf{l}_0$ is V^{-1}. Let $F(\mathbf{l}, t)$ be the probability density that a walker starting at $\mathbf{l}$ reaches the origin for the first time at time t. The survival probability of the dipole orientation $\phi(t)$ is the probability that no defect has reached the origin by time t, and it is given by

$$\phi(t) = \left[1 - \frac{1}{V} \sum_{\mathbf{l}_0 \neq 0} \int_0^t F(\mathbf{l}_0, \tau) \, d\tau \right]^N. \qquad 18.$$

The term in brackets is one minus the probability that a defect that starts at $\mathbf{l}_0$ reaches the origin (has its first passage to the origin) in the time interval $(0, t)$, i.e. the term in brackets is the probability that the defect does not reach the origin. Note we have averaged over all initial positions. The bracket is raised to the Nth power, as this is the probability that none of the N walkers reaches the origin. In the thermodynamic limit $N, V \to \infty$ but with $N/V = $ const. $= c$ (see Figure 3),

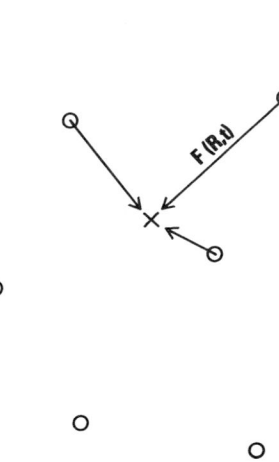

Figure 3 A frozen dipole (the target) is represented by an X at the origin. The *circles* represent mobile defects. Relaxation occurs when one reaches the origin for the first time.

$$\phi(t) = \exp\left[-c \sum_{\mathbf{l}_0 \neq 0} \int_0^t F(\mathbf{l}_0, \tau)\,d\tau\right]. \qquad 19.$$

THE MONTROLL-WEISS CONTINUOUS-TIME RANDOM WALK (19)
Generating functions for probabilities and first passage times To proceed further we need to calculate the first passage time in Eq. 19 in terms of the distributions governing the motion of a defect. Let us begin our discussion by considering a single random walker at the origin of a periodic lattice. The probability of a single jump having a displacement $\mathbf{l}$ is denoted by $p(\mathbf{l})$. The probability $P_{n+1}(\mathbf{l})$ of being at site $\mathbf{l}$ after $n+1$ steps is

$$P_{n+1}(\mathbf{l}) = \sum_{\mathbf{l}'} p(\mathbf{l}')P_n(\mathbf{l}-\mathbf{l}'), \qquad 20.$$

i.e. one can reach an intermediate site $\mathbf{l}-\mathbf{l}'$ after n steps and then cover the remaining distance $\mathbf{l}'$ in one jump to bring the walker to site $\mathbf{l}$. At this point it is useful to introduce the generating function

$$G(\mathbf{l}, z) \equiv \sum_{n=0}^{\infty} P_n(\mathbf{l}) z^n. \qquad 21.$$

Multiplying Eq. 20 by z^{n+1} and summing over n yields a Green's function equation for $G(\mathbf{l}, z)$, i.e.

$$G(\mathbf{l}, z) - z \sum_{\mathbf{l}'} p(\mathbf{l}') G(\mathbf{l}-\mathbf{l}', z) = \delta_{\mathbf{l}, 0}. \qquad 22.$$

We can also write Eq. 20 as

$$P_n(\mathbf{l}) = \sum_{m=0}^{n} F_{n-m}(\mathbf{l})P_m(0) + \delta_{n,0}\delta_{\mathbf{l},0} \qquad 23.$$

where $F_n(\mathbf{l})$ is the probability that on the nth step the walker reaches site $\mathbf{l}$ for the first time. Equation 23 counts all the ways a walker can get to site $\mathbf{l}$ after n steps. It allows for the walker reaching site $\mathbf{l}$ for the first time after $n-m$ steps and then returning in m steps (i.e. zero displacement in the last steps). Defining the generating function

$$F(\mathbf{l}, z) \equiv \sum_{n=0}^{\infty} F_n(\mathbf{l}) z^n \qquad 24.$$

it is easy to show, by using Eqs. 22–24, that

$$F(\mathbf{l}, z) = \frac{G(\mathbf{l}, z) - \delta_{\mathbf{l}, 0}}{G(\mathbf{l}=0, z)}. \qquad 25.$$

Number of distinct sites visited A quantity closely related to the first passage time probability is S_n, the mean number of distinct sites visited. This is an important quantity when calculating reaction rates, such as in our defect-diffusion model, because the probability of relaxation only increases when the defect goes to a new site, and not when revisiting an old site that does not hold the frozen-in dipole.

We write S_n as the sum of probabilities that steps 1 to n visited new sites (the origin counts as the first new site),

$$S_n = 1 + \sum_{\mathbf{l}} [F_1(\mathbf{l}) + \cdots + F_n(\mathbf{l})].$$

Next form the generating function,

$$S(z) = \sum_{n=0}^{\infty} S_n z^n = \frac{1}{1-z} + z \sum_{\mathbf{l}} F_1(\mathbf{l})$$

$$+ z^2 \sum_{\mathbf{l}} [F_1(\mathbf{l}) + F_2(\mathbf{l})] + \cdots + z^n \sum_{\mathbf{l}} [F_1(\mathbf{l}) + \cdots + F_n(\mathbf{l})] + \cdots$$

$$= \frac{1}{1-z} \sum_{\mathbf{l}} \delta_{\mathbf{l},0} + \frac{1}{1-z} \sum_{\mathbf{l}} [zF_1(\mathbf{l}) + \cdots + z^n F_n(\mathbf{l}) + \cdots]$$

$$= \frac{1}{1-z} \sum_{\mathbf{l}} F(\mathbf{l}, z) = \frac{z}{1-z} \frac{1}{G(\mathbf{l} = 0, z)} \qquad 26.$$

where Eq. 25 has been used together with $\sum G(\mathbf{l}, z) = 1/(1-z)$.

Continuous-time The probability density $\psi_n(t)$ that the nth jump of the random walker occurs at the time t is

$$\psi_n(t) = \int_0^t \psi_{n-1}(t-\tau)\psi(\tau)\,d\tau. \qquad 27.$$

The probability density $F(\mathbf{l}, t)$ for reaching site $\mathbf{l}$ for the first time at t is

$$F(\mathbf{l}, t) = \sum_{n=0}^{\infty} F_n(\mathbf{l})\psi_n(t) \qquad 28.$$

or in Laplace space

$$\int_0^{\infty} e^{-st} F(\mathbf{l}, t)\,dt \equiv F^*(\mathbf{l}, s) = \sum_{n=0}^{\infty} F_n(\mathbf{l})[\psi^*(s)]^n, \qquad 29.$$

which is precisely the generating function of Eq. 24 with $z = \psi^*(s)$. The number of distinct sites visited by time t, $S(t)$, is

$$S(t) = \sum_{n=0}^{\infty} S_n \int_0^t \psi_n(t-\tau) W(\tau) \, d\tau \qquad 30.$$

where

$$W(\tau) = 1 - \int_0^\tau \psi(x) \, dx \qquad 31.$$

takes account that the nth jump took place at time $t-\tau$, and no jump occurs in the remaining time τ. Laplace transforming, Eq. 30 becomes

$$S^*(s) = \sum_{n=0}^{\infty} S_n [\psi^*(s)]^n \frac{1-\psi^*(s)}{s} = S(z = \psi^*(s)) \frac{1-\psi^*(s)}{s}$$

$$= \psi^*(s) / [s(1-\psi^*(s)) G(\mathbf{l}=0, \psi^*(s))] \qquad 32.$$

where Eq. 26 for the generating function $S(z)$ has been used. Since (19)

$$G(0, z) = \begin{cases} 1.516 - \dfrac{3}{\pi} \left(\dfrac{3}{2}\right)^{1/2} (1-z) + \cdots & (3D) \\ (1-z^2)^{-1/2} & (1D) \end{cases}$$

asymptotically for large t

$$S(t) \sim \begin{cases} t/[\langle t \rangle G(0,1)] & (3D) \\ [8t/\pi \langle t \rangle]^{1/2} & (1D) \end{cases} \qquad 33.$$

assuming $\langle t \rangle$ is finite. For fractal time where $\psi^*(s) \sim 1 - s^\beta$ with $\beta < 1$ so $\langle t \rangle = \infty$, then asymptotically (2)

$$S(t) \sim \begin{cases} \text{const. } t^\beta & (3D) \\ \text{const. } t^{\beta/2}. & (1D) \end{cases} \qquad 34.$$

Finally, let us return to our relaxation law of Eq. 19 and consider the Laplace transform L of the argument of the exponential,

$$-Lc \sum_{\mathbf{l}} \int_0^t F(\mathbf{l}, \tau) \, d\tau = -\frac{c}{s} \sum_{\mathbf{l}} F^*(\mathbf{l}, s) = -\frac{c}{s} \sum_{\mathbf{l}} \frac{G(\mathbf{l}, \psi^*(s)) - \delta_{\mathbf{l},0}}{G(0, \psi^*(s))}$$

$$= -c \frac{\psi^*(s)}{s(1-\psi^*(s))} \frac{1}{G(0, \psi^*(s))}$$

$$= -cLS(t) \qquad 35.$$

where Eq. 32 has been used to make the identification with $S(t)$. Thus the grand result of the defect model is

$$\phi(t) = \exp(-cS(t)). \qquad 36.$$

For normal diffusion, using Eq. 33 one recovers the Bordewijk (25) results that $\phi(t) \sim \exp(-\text{const.} \sqrt{t})$ and $\phi(t) \sim \exp(-\text{const.} \, t)$ in 1D and 3D, respectively.

UNIVERSALITY From our analysis it is not now surprising that many 1D models (e.g. 26) give the $\beta = \tfrac{1}{2}$ result, since these models can probably be viewed having transfer matrices push fluctuations around a 1D system. As fluctuations move in a random fashion they relax the system.

For fractal time in 3D we have derived the Kohlrausch-Williams-Watts stretched exponential law using eqs. 34 and 36,

$$\phi(t) = \exp(-\text{const.} \, ct^\beta) = \exp(-(t/\tau)^\beta). \qquad 37.$$

In our theory there are essentially only two possible results in 3D, the Poisson law when $\langle t \rangle$ is finite, and the stretched exponential law when $\langle t \rangle$ is infinite. *Our answer as to why this law is so widespread is that it is a probability limit distribution.* Many similar models will all have the same final asymptotic behavior. For reasonable parameters, Bendler and I have shown (27) that in the defect-diffusion model, Eq. (37) can hold for 99% of the decay.

Our deviation of a stretched exponential law is very similar to one first given by Hamill & Funabashi (28) for calculating the luminescence decay of a system with electrons and electron acceptors in a frozen chemical solution. They considered the lifetime distribution for a mobile electron surrounded by many immobile acceptors. This is called the *trapping problem*. Considering relative coordinates, they transformed the problem into the target problem with one immobile electron and many mobile acceptors, and proceeded to give a derivation similar to the one included here. However, when fractal time is involved with many particles, this transformation cannot be used. The trapping problem has an algebraically decaying lifetime (7, 29), whereas the target problem has the stretched exponential law. Even though the mean time for a particular fractal time walker to move is infinite, the mean time that at least one from an infinite set moves, is finite.

This is essentially what is shown here, that even though each defect moves according to fractal time, the overall decay law, Eq. 37, has a finite mean time. In the trapping problem with only one walker, the decay law has an infinite first movement, i.e.

$$\phi(t) \sim t^{-\beta}.$$

Plonka (13) has suggested that perhaps defects, according to our model of relaxation, are the instrument responsible for the detrapping of electrons and this occurs with a stretched exponential distribution. The electrons then quickly combine with the acceptors, and the overall luminescence decay law would be stretched exponential.

Defect Diffusion and Mechanical Relaxation

NMR studies (30) have suggested that a likely candidate for a defect in polycarbonate (a polymeric glass) is a *cis-trans* conformational change ("kink") that can propagate along the polymer chain. Mechanical relaxation can occur when a kink leaves a chain. Relaxation studies (12) are well fit by a stretched exponential law with $\beta \sim 0.15$ to 0.2. Our defect diffusion will readily yield this behavior, but what can be said about the dependence of τ on mass (31)? An ideal polymer chain with M units has a radius of gyration varying as $M^{1/2}$. For a kink to leave the sphere of interaction of a given chain, and to lose contact with the original chain environment, it must move a distance varying as $M^{1/2}$. For unbiased fractal time (Eq. 10) motion it can be shown that $\langle R^2(t) \rangle \sim t^\beta$ (2). The time scaling t to move a distance $R \sim M^{1/2}$ varies as

$$t \sim M^{1/\beta}, \qquad 38.$$

which implies that in Eq. 17

$$\tau \sim M^{1/\beta} \sim M^5 \text{ to } M^7. \qquad 39.$$

This is in reasonable agreement with experiment.

Bendler and I (31) have also considered a phase transition in the number of single defects as a function of temperature in a theory of the glass transition. In Eq. 37 τ varies as $c^{-\beta}$. In our model (31), only the defects are mobile. When the temperature is lowered the defects must coalesce in order to lower entropy. We further assume that only single defects contribute to the relaxation. The probability of finding a single defect is $c_1 = c(1-c)^z$ where z is the number of sites within the correlation volume surrounding the single defect. In a mean field lattice gas model $z \sim |T-T_c|^{-3/2}$. Replacing c in Eq. 37 by $c_1 \tau$ then follows the law,

$$\tau \sim \exp\left(+\text{const.}/|T-T_c|^{3/2}\right) \qquad 40.$$

where T_c is the critical temperature at which single defects would no longer exist. This generalized Vogel law, Eq. 40, has been successful in fitting data, such as for B_2O_3, where the Vogel law $\tau \sim \exp(+\text{const.}/|T-T_c|)$ fails near the glass transition temperature. Note that single defects can exist below the glass transition temperature T_g so $T_c < T_g$.

OTHER MODELS (35)

A Different Approach

Consider again the defect diffusion, but now with the somewhat different theoretical approach of Redner & Kang (32). In one dimension consider a frozen dipole at the origin and let $f(R)$ be the probability of having no defects initially at a distance R from the origin. For randomly placed defects $f(R) = \exp(-cR)$. For Brownian motion of the defects, the probability that a defect at site R_i has not reached the origin by time t is given by $\exp(-Dt/4R_i^2)$ where D is the diffusion constant. For a given R value the relaxation law is

$$\phi(t|R) = \exp(-cR) \prod_{i=1}^{\infty} \exp(-Dt/4R_i^2)$$

$$\cong \exp(-cR) \exp\left(-cDt \int_R^{\infty} \frac{dl}{4l^2}\right)$$

$$= \exp(-cR + cDt/4R).$$

Integrating overall R, we arrive at

$$\phi(t) = \int_0^{\infty} \exp(-cR) \exp(-cDt/4R) \, dR$$

$$\sim \exp(-\text{const.} \, t^{1/2}). \qquad 41.$$

Hierarchical Constrained Dynamics

In general, in analogy to Eq. 41 one can consider integrals of the form

$$\phi(t) = \int_0^{\infty} \exp(-t/\tau(n)) f(n) \, dn \qquad 42.$$

as Palmer et al (33) have proposed. They treat systems that relax in a serial hierarchical fashion (see Figure 4). First a set of constraints must relax, say with a time scale $\tau(1)$, before the next level can relax with time scale $\tau(2)$, etc. This type of relaxation process is described in a continuum limit by Eq. 42 where $f(n)$ is the weight factor ascribed to the nth level. If one chooses $\tau(n)$ to be scale invariant and $f(n)$ to be exponential

$$\tau(n) = \text{const.} \, n^s$$
$$f(n) = \text{const.} \, e^{-cn} \qquad 43.$$

then the stretched exponential result is obtained with

$$\phi(t) \sim \exp(-\text{const.} \, t^{1/1+s}). \qquad 44.$$

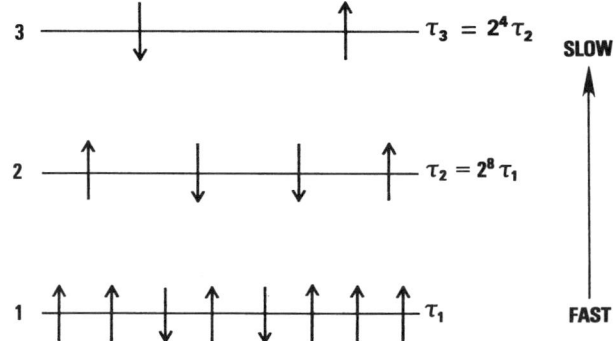

Figure 4 A portion of a realization of a hierarchically constrained dynamics model is shown. Here the first level of spins must attain one of its 2^8 configurations, before the second level of spins is free to change. The rate of change in level 2 is 2^8 times slower than that in level 1, and so on.

In a model of relaxation based on a liquid-like region percolating through a solid-like region, Cohen & Grest (34) introduce a relaxation time that depends on the surface area A of a liquid-like cluster $\tau = \tau(A)$ and a distribution of surface areas $f(A)$. This model has an integrated form for $\phi(t)$ similar to Eq. 42.

Direct Transfer

For a donor site at the origin and an acceptor site at $\mathbf{R}_i$ (see Figure 5), the relaxation law is

$$\phi_i(t) = \exp\left(-tW(\mathbf{R}_i)\right) \qquad 45.$$

where $W(\mathbf{R}_i) = 1/\tau(\mathbf{R}_i)$ is the rate of transfer. For a set of acceptors randomly placed with probability c at each site (29)

$$\phi(t) = \prod_i \{(1-c) + c \exp\left(-tW(\mathbf{R}_i)\right)\} \qquad 46.$$

where $\phi(t)$ is the probability that the donor is still at the origin at time t for $c \ll 1$

$$\phi(t) = \exp\left[-c \sum_i \{1 - \exp\left(-tW(\mathbf{R}_i)\right)\}\right]. \qquad 47.$$

Introducing the site density $\rho(\mathbf{R}) = \sum_i \delta(\mathbf{R} - \mathbf{R}_i)$ and going to the continuum limit in D dimensions

FRACTAL TIME 285

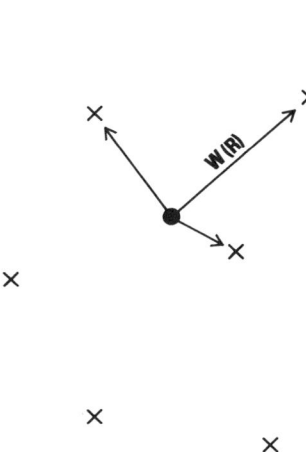

Figure 5a An initially prepared donor denoted by the *dark filled circle* can relax by a direct transfer to a configuration of randomly placed acceptor sites.

Figure 5b In one dimension the direct transfer is restricted to the nearest neighbor acceptor site.

$$\phi(t) = \exp\left[-c \int d\mathbf{R}\, \rho(\mathbf{R})\{1 - \exp[-tW(R)]\}\right] \qquad 48a.$$

$$\sim \exp(-\text{const.}\, t^{D/S}) \qquad 48b.$$

for a scale invariant $W(R) = \text{const.}\, R^{-s}$. Although Eq. 48a, the general result, is not in the form of Eq. 42, if the physics is restricted to 1D with only nearest neighbor transitions allowed (see Figure 5b) then

$$\phi(t) = \int_0^\infty f(R) \exp(-tW(R))\, dR, \qquad 49.$$

which is precisely Eq. 42 but with a different interpretation.

TURBULENT DIFFUSION (36, 37)

Levy Flights, Walks, and Fractal Space-Time

In the section on *Levy Distributions*, above, we discussed a Levy flight where a random walker visited a fractal set of sites. The mean square displacement per jump was $\langle l^2 \rangle$. Let us now consider a Levy walk (36, 37) that follows the path connecting the points of the Levy flight (see Figure 6). We call the sites visited by the Levy flight the turning points of the Levy walk. We define the joint space-time probability density for a single jump to be

$$\Psi(\mathbf{R}, t) \equiv \psi(t|\mathbf{R}) p(\mathbf{R}) \qquad 50.$$

where $\psi(t|\mathbf{R})$ is the conditional probability density that the time to traverse the distance between two turning points separated by a distance $\mathbf{R}$ is t. As before $p(\mathbf{R})$ is the probability that the distance between the turning points is $\mathbf{R}$. We choose a constant velocity between turning points, but this velocity $v(\mathbf{R})$ can depend on the length of the path. For example, in a turbulent flow larger vortices induce longer coherence lengths in a particle's motion. The larger the vortex, the more energy it has, and thus longer coherence lengths are traversed with larger velocities. We choose

$$p(\mathbf{R}) \sim |\mathbf{R}|^{-1-\beta}, \quad 0 < \beta < 1 \qquad 51.$$

so there is no space scale in the system.

We also choose

$$\psi(t|\mathbf{R}) = \delta\left(t - \frac{|\mathbf{R}|}{v(\mathbf{R})}\right) \qquad 52.$$

with

$$v(\mathbf{R}) = \text{const.} \, |\mathbf{R}|^{1/3}, \qquad 53.$$

which is the Kolmogorov (38) scaling law for fully developed turbulence. He hypothesized that the rate of energy transfer ε_R across a scale R is independent of R, i.e. $\varepsilon_R = \varepsilon = E_R/t_R = v_R^3/R$ where v_R is the relative velocity of two particles separated by a distance R, $E_R \sim v_R^2$ is the energy in the scale distance R, and $t_R = R/v(R)$ is the time to cross this region.

The probability density $Q(\mathbf{R}, t)$ for the Levy Walk to be at a turning point is given by the recursion Green's function relation

$$Q(\mathbf{R}, t) = \sum_{\mathbf{R}'} \int_0^t Q(\mathbf{R} - \mathbf{R}', t - \tau) \Psi(\mathbf{R}', \tau) \, d\tau + \delta(\mathbf{R}) \delta(t) \qquad 54.$$

The probability $P(\mathbf{R}, t)$ of being at any site $\mathbf{R}$ is given by

$$P(\mathbf{R}, t) = \sum_{\mathbf{R}'} \int_0^t Q(\mathbf{R} - \mathbf{R}', t - \tau) W(\mathbf{R}', \tau) \, d\tau \qquad 55.$$

where the Q terms brings the walker to $\mathbf{R} - \mathbf{R}'$ at time $t - \tau$ and the $W(\mathbf{R}', \tau)$ takes the walker the remaining distance $\mathbf{R}'$ in the remaining time τ.

FRACTAL TIME 287

Figure 6a The first-few points visited by a Levy flight. A fractal set of points (clusters within clusters within clusters, etc) is eventually visited.

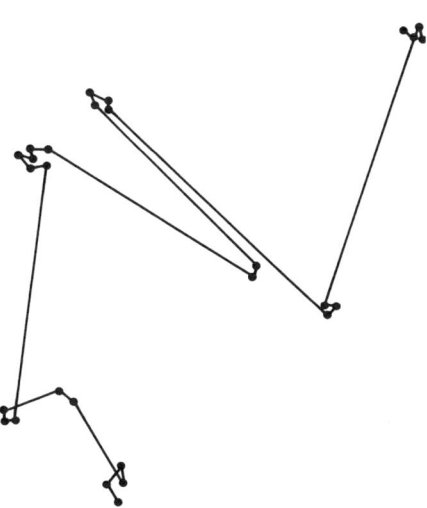

Figure 6b The set of points visited by the corresponding Levy walk. It includes the same set of points as the Levy flight, plus the trail connecting these points. The Levy flight points are turning points of the Levy walk. Different velocities can be associated with different path lengths.

The W term must take into account that the trajectory passes through R at time t, but does not necessarily end there.

We write W as[1]

$$W(\mathbf{R}', \tau) = \int_\tau^\infty f\left(v = \frac{|\mathbf{R}'|}{\tau}\bigg|\tau'\right)\psi(\tau')\, d\tau' \qquad 56.$$

where $f(v|\tau)$ is the conditional probability that the Levy walker has velocity v given that it traverses the path between its turning points in a time τ, and $\psi(\tau)$ is the probability density that the path traversal time is τ. Thus, if the walker has velocity $|\mathbf{R}'|/\tau$ for a time τ or greater, the walker will reach $\mathbf{R}$ at time t. If the walk takes a time τ' and has velocity R'/τ, then the path length is $(R'/\tau)\tau'$. If $V(R) \sim R^\beta$ then

$$R'/\tau = \left(\frac{R'}{\tau}\tau'\right)^\beta$$

implying that

$$f(v = R'/\tau|\tau') = \delta(R' - \tau'^{\beta/1-\beta}\tau^{(1+\beta/\beta-1)}).$$

Note that Eq. 55 differs from the usual hopping continuous-time random walk equation

$$P(\mathbf{R}, t) = \int_0^t Q(\mathbf{R}, t-\tau)W(\tau)\, d\tau \qquad 57.$$

where, as in Eq. 31, $W(\tau)$ is the probability that the walker has not jumped in a time τ, i.e. the walker reaches $\mathbf{R}$ at time $t-\tau$ and then does not move for at least a time τ. Our Eq. 55 is different because our random walker never stops moving.

Employing our Levy walk with Kolmogorov scaling we find for the mean square displacement of the random walker

$$\langle R^2(t)\rangle \sim \begin{cases} t^3 & \beta \leq \tfrac{1}{3} \\ t^{2+\tfrac{3}{2}(1-\beta)} & \tfrac{1}{3} \leq \beta \leq \tfrac{5}{3} \\ t & \beta \geq \tfrac{5}{3} \end{cases}. \qquad 58.$$

The t^3 result corresponds to walks with the mean time $\langle t\rangle$ spent in

---

[1] I thank Dr. A. Onuki from Kyoto University for pointing out an error in the derivation of the probability distribution in Ref. (37). The error is corrected by Eq. 56.

traversing a path being infinite. This is known as Richardson's law of turbulent diffusion (39). The second case corresponds to $\langle t \rangle$ finite, but $\langle t^2 \rangle$ infinite, and the Brownian motion result occurs when $\langle t^2 \rangle$ is finite. Note that the corresponding Levy flight has $\langle R_n^2 \rangle$ infinite. We have avoided this infinity by forcing the random walker to travel between the Levy turning points with a velocity $V(R)$. If $V(R) = $ constant, then in the $\langle t \rangle = \infty$ case $\langle R^2(t) \rangle \sim t^2$. This result has been used to analyze phase noise in Josephson junctions (40, 41) where R corresponds to the number of rotations the phase has made in a clockwise sense. The rate of rotation only depends on a fixed voltage so $V(R)$ does not depend on R.

SUMMARY

Fractal time is basically a very simple concept describing processes that do not have a characteristic time scale. This lack of an average time (through some intricate mathematics connected to Levy's theory of distributions) induces fractional exponents. Our prime example was the discontinuous change between the Debye exponential law and the Kohlrausch-Williams-Watts stretched exponential law. When Scher & Montroll first introduced fractal time in the analysis of charge propagation in amorphous films, it seemed like an obscure notion for a particular problem. Today it is recognized that fractal time is a generic attribute of the glassy state of matter and provides a useful paradigm along with such concepts as localization and percolation.

Acknowledgments

This work owes much to the late Elliott Montroll, who introduced me to Levy's theory of distributions, to Harvey Scher, who first applied the Montroll-Weiss random walk with fractal time to time dependent problems in condensed matter, John Bendler, who suggested an attack on the theory of the stretched exponential and the glass transition, Yossi Klafter, who explored the possible stochastic pathways for deriving the stretched exponential, Barry Hughes, who introduced Mellin transform techniques for the analysis of fractal stochastic processes, and Bruce West who melded Kolmogorov's scaling with Levy walks. All the above have been long-time colleagues and co-authors and share in whatever truths may have been uncovered in these fruitful investigations.

Literature Cited

1. Montroll, E. W., Scher, H. 1973. *J. Stat. Phys.* 9: 101
2. Shlesinger, M. F. 1974. *J. Stat. Phys.* 10: 421
3. Scher, H., Montroll, E. W. 1975. *Phys. Rev. B* 12: 2455
4. Pfister, G., Scher, H. 1978. *Adv. Phys.* 27: 747
5. Debye, P., Edwards, J. O. 1952. *J. Chem. Phys.* 20: 236
6. Vardeny, Z., O'Connor, P., Ray, S., Tave, J. 1980. *Phys. Rev. Lett.* 44: 1267
7. Shlesinger, M. F. 1979. *J. Chem. Phys.* 70: 4813.
8. Kopelman, R. 1986. In *Transport and Relaxation in Random Material*, ed. J. Klafter, R. J. Rubin, M. F. Shlesinger, pp. 177–208. Singapore: World Sci. Press
9. Kohlrausch, F. 1863. *Pogg. Ann. Physik* 119: 352
10. Williams, G., Watts, D. C. 1970. *Trans. Faraday Soc.* 66: 80
11. Chamberlin, R. V., Mozurkewich, G., Orbach, R. 1984. *Phys. Rev. Lett.* 52: 867.
12. Le Grand, D. G., Olszewski, W. V., Bendler, J. T. 1987. *J. Poly. Sci. B* 25: 1149.
13. Plonka, A. 1986. *Time Dependent Reactivity of Species in Condensed Media*, Springer-Verlag Lect. Notes Chem., Vol. 40
14. Berger, J. M., Mandelbrot, B. B. 1963. *IBM J. Res. Dev.* 7: 224
15. Mandelbrot, B. B. 1977. *Fractals: Form, Chance, and Dimension*. San Francisco: Freeman
16. Todhunter, I. 1865. *A History of the Mathematical Theory of Probability*. Cambridge: Cambridge Univ. Press. (A Chelsea Press Reprint, 1949).
17. Levy, P. 1937. *Theorie de l'addition des variable aleatoire*. Paris: Gauthier-Villars
18. Montroll, E. W., Shlesinger, M. F. 1984. In *Studies Stat. Mechanics* 11: 1–121
19. Montroll, E. W., Weiss, G. H. 1965. *J. Math. Phys.* 6: 167
20. Scher, H., Lax, M. 1973. *Phys. Rev. B* 7: 4491
21. Klafter, J., Silbey, R. 1980. *Phys. Rev.* 44: 55
22. Shlesinger, M. F., Hughes, B. D. 1981. *Physics* 109A: 597
23. Glarum, S. H. 1960. *J. Chem. Phys.* 33: 1371
24. Phillips, M. C., Barlow, A. J., Lamb, J. 1972. *Proc. R. Soc. London Ser. A* 329: 193
25. Bordewijk, P. 1975. *Chem. Phys. Lett.* 32: 592
26. Shore, J. E., Zwanzig, R. W. 1975. *J. Chem. Phys.* 63: 5445
27. Bendler, J. T., Shlesinger, M. F. 1985. *Macromolecules* 18: 591
28. Hamill, W. H., Funabashi, K. 1977. *Phys. Rev. B* 16: 5523
29. Blumen, A., Klafter, J., Zumofen, G. 1986. In *Optical Spectroscopy of Glasses*, ed. I. Zschokke-Granach. Dordrecht: Reidel
30. Jones, A. A., Inglefield, P. T., O'Gara, J. F., Roy, A. K. 1986. See Ref. 8, pp. 228–39
31. Bendler, J. T., Shlesinger, M. F. 1987. *J. Mol. Liq.* 36: 37 (also in *J. Stat. Phys.* In press)
32. Redner, S., Kang, K. 1984. *J. Phys. A* 17: L451
33. Palmer, R., Stein, D., Abrahams, E. S., Anderson, P. W. 1984. *Phys. Rev. Lett.* 53: 958
34. Cohen, M. H., Grest, G. S. 1981. In *Structure and Mobility in Molecular and Atomic Glasses*, ed. J. M. O'Reilly, M. Goldstein, p. 199. New York: Academic
35. Klafter, J., Shlesinger, M. F. 1986. *Proc. Natl. Acad. Sci. USA* 83: 848
36. Shlesinger, M. F., Klafter, J. 1985. In *On Growth and Form: A Modern View*, ed. H. E. Stanley, N. Ostrowski, pp. 279–83. Amsterdam: Martinus Nijhoff
37. Shlesinger, M. F., West, B. J., Klafter, J. 1987. *Phys. Rev. Lett.* 58: 1100.
38. Kolmogorov, A. N. 1941. *CR Dokl. Acad. Sci. URSS* 30: 301
39. Richardson, L. F. 1926. *Proc. R. Soc. London A Ser.* 110: 709
40. Geisel, T., Nierwetberg, J., Zacherl, A. 1985. *Phys. Rev. Lett.* 54: 616
41. Shlesinger, M. F., Klafter, J. 1985. *Phys. Rev. Lett.* 54: 2551

A PHOTOCHEMICAL INVESTIGATION OF THE DYNAMICS OF OLIGONUCLEOTIDE HYBRIDIZATION

John E. Hearst

Department of Chemistry, University of California and Lawrence Berkeley Laboratory, Berkeley, California 94720

INTRODUCTION

The nucleic acid double strand helix is now known to have several structural forms that are themselves rather variable or changeable in response to environmental conditions such as ionic strength, temperature, and the presence of a wide variety of solutes in the solutions in which the solvent is almost always water. A common chemistry among all of these helical forms is the reaction that separates the strands of the helices into single strands. The strands ideally may be thought of as random coils but, in reality, may fold back upon themselves and form local double stranded helical regions of secondary structure. The process of separation of the strands in the double helices is called melting by analogy to the melting of a crystal, because the transition of a long helix from double-stranded form to single-stranded form occurs in a narrow temperature interval with high temperature favoring the single strands and low temperature favoring the helix. Despite the descriptive language, it is clearly incorrect to think of the helix-coil transition as a phase transition, for there are major inhomogeneities in nucleic acid helices that invariably lead to stable intermediate states, each with its own melting temperature, in the transition from helix to coil. When a sample of melted single strands is allowed to reverse the melting process and reform the double helix, the process is called *renaturation*. This process follows second-order kinetics, and, because in very complex or heterogeneous DNA samples renaturation can take a long time, the process is often referred to as an *annealing* process.

Nucleic acid renaturation has become a remarkably important and specific analytical tool since it was first observed by Marmur & Lane (1) and Doty et al (2) that the complementary strands of DNA can be renatured by "annealing" solutions of denatured DNA at approximately 20°C below the melting temperature of the DNA for extended periods of time. *Hybridization* is a term used to describe the renaturation of a labeled single-stranded probe molecule with a denatured target sample. The formal distinction between a renaturation experiment and a hybridization experiment is that in true renaturation the desired goal is thermodynamically stable and is therefore the final product of all possible reactions. In the hybridization experiment the desired goal can be kinetically isolated, but if given enough time the hybridized probe, typically short relative to the target molecules, will ultimately be displaced when the target has completed the renaturation process if both the target and its complementary strand are initially present in the sample being probed. Initial procedures using hybridization analysis were performed in solution (3, 4).

Britten & Kohne (5) introduced Cot analysis in 1968. They discovered that when a DNA sample is fragmented into pieces of 400 base pairs, denatured, and renatured, the rate of renaturation was governed by the "complexity" of the original DNA. Thus, if one started with a sample of viral DNA with a total DNA content of 30,000 basepairs at a definite weight concentration, this sample will renature 100,000 times faster than the unique sequences in a sample of human DNA at the same weight concentration because the haploid genome of humans contains 3 billion basepairs. This result has been experimentally verified and provides proof that the renaturation process is described by a simple second-order process. The second-order rate constant derived from these data is

$k = 1.6 \times 10^6$ liter/mole second.

Hybridization has become an extraordinarily powerful tool because it provides an easy assay for base sequence. The probe will only hybridize to its complementary sequence or one with a small number of mismatches. Researchers have confronted many experimental difficulties with solution hybridization experiments, however, in which target renaturation competed with probe-target hybridization and with which a reliable separation method was needed to distinguish hybridized from unhybridized probe. The procedures involving DNA and RNA immobilization on nitrocellulose were developed to solve both of these problems and have been the mainstay of hybridization analysis for 12 years (6, 7). The Southern blot has been used to identify specific sequences in DNA fragments that have been separated by gel electrophoresis. Colony hybridization provides still another detection procedure in which sequence separations do not arise by gel electrophoresis but by cloning and plating of living cells (8–10).

The development and automation of the chemistry for the synthesis of oligodeoxyribonucleotides has rekindled an interest in the fundamental understanding of the hybridization of relatively short oligonucleotides (10 to 100 bases long) to denatured target nucleic acid samples of higher molecular weight in solution. Because such a hybridization is controlled both by the thermodynamic stabilities of the helical duplex strand interactions and by the kinetics of many competing processes, the ability to trap intermediates photochemically in these hybridization reactions has proven to be valuable. Experiments have revealed the important thermodynamic and kinetic parameters associated with such solution hybridizations and have established methods for increasing the sensitivity and discrimination and reducing the time required for the hybridization measurement.

SYNTHESIS OF A PHOTOCROSSLINKABLE OLIGONUCLEOTIDE PROBE

Psoralen (I) is a linear furocoumarin. Those psoralen derivatives which photoreact with nucleic acids do so in three distinct steps. First, a dark binding by intercalation between the DNA helix base pairs occurs. These intercalated psoralens are in equilibrium with psoralens that are free in solution. Thus, the efficiency and specificity of photoreaction is partially governed by the dissociation constants associated with this intercalative binding (11).

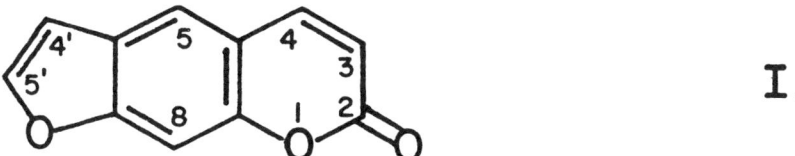

I.

Second, when an intercalated psoralen is excited by electromagnetic radiation with wavelength between 320 nm and 400 nm, addition products to the nucleic acid are formed. Cyclobutane addition occurs between the 5,6 double bond of a pyrimidine (primarily thymine) adjacent to the intercalated molecule and either the 4′,5′ double bond in the furan ring or the 3,4 double bond in the pyrone ring of the psoralen. In the case of 4,5′,8-trimethylpsoralen derivatives, the principal monoadducts are furan adducts (97%).

Third, since the furan monoadducts are substituted coumarins, which retain a strong absorbance between 320 and 380 nm, photon absorption by a monoadduct that is positioned in the DNA helix adjacent to a pyrimidine on the opposite strand may produce a second cyclobutane bridge between the 3,4 positions of the pyrone ring and the 5,6 positions of the pyrimidine. Such a reaction results in the formation of a covalent cross-link in the nucleic acid helix. Figure 1, summarizes these steps.

The description of this photochemical system requires noting two additional properties. First, psoralens free in solution dimerize and photo-

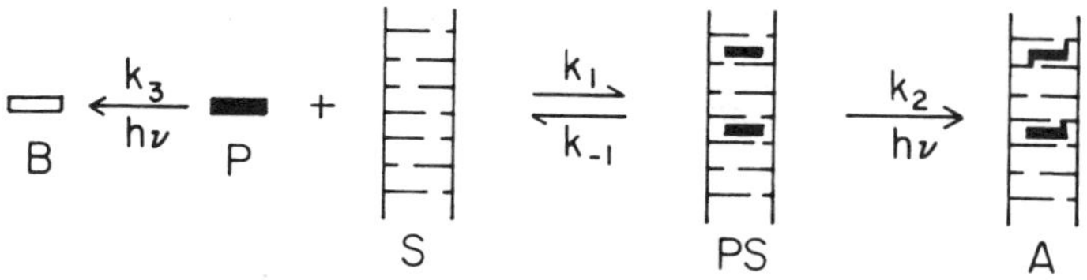

$$B \xleftarrow{k_3}{h\nu} P \; + \; S \; \underset{k_{-1}}{\overset{k_1}{\rightleftharpoons}} \; PS \; \xrightarrow{k_2}{h\nu} \; A$$

P = psoralen derivative in question
S = psoralen binding site in the DNA
PS = noncovalent intercalation complex between psoralen and the DNA
A = covalent adduct of the psoralen to the DNA
B = photobreakdown product of the psoralen

1. Noncovalent Binding Dissociation Constant K_D

$$K_D = \frac{k_{-1}}{k_1} = \frac{[P][S]}{[PS]} \qquad P + S \underset{k_{-1}}{\overset{k_1}{\rightleftharpoons}} PS$$

2. Photoaddition Rate Constant k_2

$$PS + h\nu \xrightarrow{k_2} A$$

3. Photobreakdown Rate Constant k_3

$$P + h\nu \xrightarrow{k_3} B$$

Figure 1 A schematic representation of the kinetic steps associated with the photoreaction of psoralen with DNA.

react with solvent to produce breakdown products that are no longer capable of photoreaction with nucleic acids. Second, as is the case with most cycloaddition photochemistry, the cyclobutane rings in both monoadducts and cross-links can be photoreversed by excitation at shorter wavelengths, typically at 260 nm.

The stereochemistry of the psoralen cycloadducts to DNA has been determined to be *cis-syn* by nuclear magnetic resonance of the thymidine nucleoside monoadducts and diadducts (12–14) and by X-ray crystallography of the thymine furan-side monoadducts (15, 16), which were isolated by degradation of DNA that had been photoreacted with 8-methoxypsoralen (8-MOP), 4,5′,8-trimethylpsoralen (TMP), or 4′-hydroxymethyl-4,5′,8-trimethylpsoralen. Although many other stereochemical products are found in the direct photochemical reaction of thymine and psoralen, the helix restricts the position of the intercalated psoralen so that only *cis-syn* adducts are formed. Although this places severe restrictions on the monoadducts that must be synthesized on oligonucleotide probes if these monoadducts are to serve as photocrosslinking sites in the nucleic acid helix, it also provides precise structural specificity for the existence of the nucleic acid helix formed between the probe and its target.

Gamper et al (17) showed that 5′-TpA-3′ sequences in DNA duplex helical regions were six times more likely to photoreact with 4′-hydroxymethyl-4,5′,8-trimethylpsoralen (HMT) than the average reactive site in DNA.[1] This observation provides the basis for a photochemical route to the formation of furan side monoadducted oligonucleotide probes. Two procedures are currently used. Both start with two oligonucleotides that are different in length, are complementary to each other, and have a single 5′-TpA-3′ on each strand in the complementary regions. The oligo sequences are often chosen so that no other internal thymines are part of the helical region formed between the two oligomers. Van Houten et al (18) used an 8-mer of sequence 5′-TCGTAGCT-3′ and a 12-mer of sequence 5′-GAAGCTACGAGC-3′. In a strategy first described by Cimino et al (19), these two oligomers are placed in approximately equimolar amounts at concentrations of 0.25 to 0.35 mg/ml in a buffer containing 0.1 M NaCl and 0.010 M $MgCl_2$, 0.03 mg/ml HMT, and 1% ethanol. The solution is irradiated for 3 min at 4°C with 320–380 nm ultraviolet light at 600 mW/cm^2. In Figure 2, crosslinked 8-mer–12-mer is readily purified by gel electrophoresis, yielding two resolved products, both crosslinked at the central 5′-TpA-3′ with 70% overall yield. One product has the furan adduct on the 8-mer, the other on the 12-mer, so the two differ only in the

---

[1] The DNA strand has polarity. The 5′-TpA-3′ refer to an internal thymine-adenine base sequence in such a DNA strand being read from the 5′ end to the 3′ end.

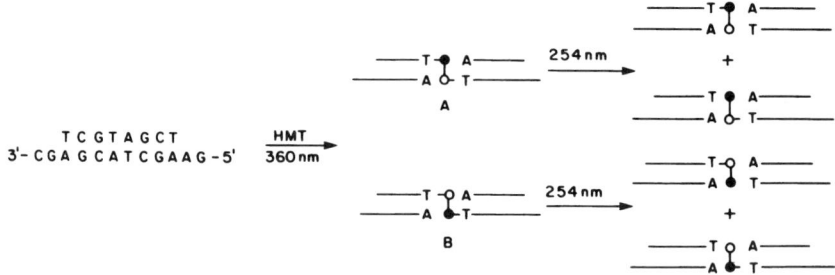

Figure 2 Preparation of HMT-monoadducted 8-mers and 12-mers. Photoreaction with HMT yields two orientational isomers of the crosslink. Partial photoreversal of these crosslinks generates four monoadducted oligonucleotides (18).

polarity of the crosslink. The partial photoreversal of each of these isolated orientational isomers of the crosslink with 254 nm light provides a route to each of four monoadduct products, the furan-side monoadducted 8-mer, the pyrone-side monoadducted 12-mer, the furan-side monoadducted 12-mer, and the pyrone-side monoadducted 8-mer. Several oligonucleotides are then synthesized having overlapping complementary ends so that the furan-side 8-mer or 12-mer can be ligated into a longer probe with desired length and sequence.

In recent months, it has been found that if this same system is irradiated at 390 nm, the photoreaction stops at monoadduct with 97% of the adduct of the desired furan-side stereoisomer. This procedure was first described by Chatterjee & Cantor (20), and it appears to be a worthwhile simplification relative to the above, if only furan-side monoadducts are desired (21).

EXPERIMENTAL METHODS

Most hybridizations performed today start with the target bound to a solid support, for example a nylon or nitrocellulose membrane. The target DNA is typically denatured and then attached to the membrane by drying and baking or by the use of short wave-length ultraviolet light. This attachment process yields a product that is poorly understood, both with respect to the number of membrane attachment sites per defined length of DNA, say per 1000 bases, and with respect to the stability of the physical attachment. In fact, studies have shown that the membrane attachment process is only 10 to 20% efficient following the standard hybridization and wash procedures (22). With such a small fraction of the initial target nucleic acid remaining on the membranes following these procedures, the use of such methods for analytical quality analysis must be questioned. Furthermore, while it is generally believed that the membrane attachment eliminates secondary structure from the target, making it more accessible to hybridize to the probe, quantitative analysis has shown that only 10 to 20% of the remaining targets of correct sequence hybridize to probe at saturation of probe, suggesting either target secondary structure or shielding of reactive target by the membrane. Thus, membrane hybridizations typically result in 1–3% target coverage overall (23). Membrane attachment apparently does eliminate the competing renaturation of complementary target strands, so the regulation of kinetic conditions is often not a significant consideration. A typical "dot blot or Southern blot" is allowed to hybridize "overnight." It is, in fact, the convenience of membrane hybridization rather than its precision and accuracy that have resulted in its nearly universal acceptance.

There are, however, two additional limitations to membrane hybridization. First, the probe solution has to cover the membrane, so solution volumes are relatively large. This is particularly a problem if ^{32}P radiolabel is used on the probe to identify successful attachment to the membrane. Far more radioactive isotope is used than required for the hybridization itself just to achieve sufficient hybridization solution volume. In such a hybridization, the probe concentration must be kept low for two reasons. It is experimentally observed that nonspecific background hybridization becomes severe at high probe concentrations (10^{-8} M or 10^{-9} M). In addition, the levels of radioactivity required for these high concentrations are prohibitively expensive and perhaps hazardous to human health.

The second limitation is the time required for hybridization. With probe excess, the half-life for the second order hybridization reaction is

$$\Gamma = 0.693/k[p]$$

where $[p]$ is the molarity of the probe, and k is the binary rate constant whose value is approximated by 1.6×10^6 for all the conditions considered in this review. Thus at 10^{-8} M, the half-life is 43 sec, but at 10^{-11} M, a concentration more typically utilized, the half-life is 12 hr. Probe concentrations of 10^{-8} M or higher are only feasible if solution hybridization is done in very small volumes (i.e. 10 μl). The crosslinking procedure makes such experiments feasible.

The major disadvantage of solution hybridization is that a subsequent separation is required between the high molecular weight target, with hybridized probe attached, and the remaining unhybridized probe. This separation is effected in the dot blot or Southern blot experiment by careful washing of the membrane at temperatures high or stringent enough to disrupt nonspecific interactions between probe and target sequences and yet not "melt" the probe from its true target. This is relatively easy to do with probes longer than 100 bases, since the differences in stabilities between the probe hybridized to its target and the probe hybridized to accidental short regions of sequence homology are substantial. With short probes, however, stability differences become smaller, and it is shown in the THEORY section that the best discrimination is achieved at the melting temperature of the probe from its true target. At such conditions, equilibrium can shift rapidly and it may not be possible to wash the hybrid without washing the probe away from the target. For this reason, it may be necessary to "fix" the equilibrium distribution of hybridized products rapidly relative to the half-life of the hybridization reaction. In this review, crosslinkable oligonucleotide probes and photochemistry are used for this "fixation" process.

The logic of this procedure is only valid if the monoadduct on the probe does not alter the stability and specificity of the probe-target hybrid relative to that observed with the unmodified probe. This question has been investigated in detail by Shi & Hearst (24), and they have concluded that HMT monoadducts at 5'-TpA-3' sites do not alter the melting temperatures or other thermodynamic parameters associated with the hybridization reaction. There is no reason for this to be true in general, but HMT has been chosen as the most appropriate psoralen adduct to create this situation. Adducts of 8-MOP would destabilize the hybridization relative to the unmodified interaction, and TMP would generate stronger hybrids, a conclusion suggested by the values of the intercalation binding constants of these derivatives (25).

METHODS OF SEPARATION

Once the hybrid between probe and target has been covalently fixed by the photochemical reaction, the unreacted labeled probe must be separated from covalently linked probe and target. This separation has been achieved in three ways. Crosslinked hybridization complexes between M13 circular DNA targets and HMT-modified 12-mer probes have been analyzed by gel filtration through a 1 cm × 7 cm agarose column (Bio-Rad; A-5m) (23). When hybridization samples containing single-stranded target were analyzed by gel filtration, the column was eluted with 0.03 M NaOH, 1 mM EDTA to avoid the chance formation of non-crosslinked probe-target complexes. Both crosslinked and free probe peaks were eluted, thus permitting a direct determination of the fraction of probe crosslinked to the target.

An alternative procedure for size separation involves centrifugation of the sample through a Centricon-30 cartridge (Amicon). These cartridges pass short oligonucleotide probes while the target DNA is concentrated into a small volume that is diluted and recentrifuged several times as a wash procedure. The crosslinked samples were diluted to 0.5 ml with 0.1 M phosphate (pH 4.0), 0.05% SDS, and 50% formamide. The sample was then heated at 65 to 70C for five min and centrifuged in the Centricon cartridge for 10 min at 5000 g. The retentates were diluted with 0.5 ml of the hot wash solution mentioned above, and centrifuged again. This procedure was repeated eight times. The final retentate was free of excess probe (23).

A very effective separation procedure involves the use of alkaline agarose gel electrophoresis. After hybridization and crosslinkage, each sample is supplemented with 6 µg of low molecular weight RNA [homochromatography mix; Hay et al (26)] and enough excess target DNA to bring it to 100 ng. After adding carrier nucleic acid, samples were dehydrated in a Speedvac. Each sample was then charged with dyes and diluted in volume to 10 µl with deionized formamide. Electrophoresis was conducted in horizontal 1% agarose gels immersed in 30 mM NaOH, 1 mM EDTA. Addition of the homochromatography mixture reduced background due to nonspecific interaction of the probe with agarose. This interaction was observed with both HMT-modified and unmodified oligonucleotides. In the presence of the indicated amount of RNA, up to 10^5 cts/min of radioactive probe could be loaded per well, with no background problems. Addition of carrier DNA permitted the visualization of labeled bands by ethidium bromide staining, making it possible to excise bands for direct counting. Results were analyzed either by autoradiography or scintillation counting. For autoradiography the gel was first neutralized and then transferred onto filter paper by using a gel dryer (23).

SOME EXPERIMENTAL RESULTS

Gamper et al (23) have studied the hybridization of three HMT-monoadducted oligonucleotides, a 12-mer, a 13-mer and a 25-mer; each complementary to a portion of the polylinker regions of the single stranded DNA of bacteriophages M13mp8' or M13mp19. Photofixation of the probe-target complex was studied under two conditions. By quick cooling the hybridization solution after the reaction had continued for the desired time and then irradiating the cold sample, the kinetics and equilibria for hybridization of these oligonucleotides to M13 were determined. By irradiating the hybridization solution at or near the melting temperature of the probe-target complex, rapid fixation of the reacting solution was possible, and the reaction between probe and unhybridized target could be driven further to obtain high yields of probe-target hybrid. This phenomenon has been called "photochemical pumping."

As an example of the important variables associated with such photochemically fixed hybridization reactions, the reaction of the 13-mer with M13mp19 DNA is now discussed (23). Figure 3 shows the photochemically fixed hybridization profile of this probe-target system. The control hybridization of this same probe to M13mp8 DNA showed no detectable hybridization, despite the fact that this target DNA contained a complementary sequence with only two mismatches to the 13-mer probe, providing clear evidence for the specificity of the photochemically fixed hybridization. In Figure 3a, the hybridization was fixed after 1 hr of reaction; in Figure 3b, the hybridization was fixed after 24 hr. The probe concentration of 4.8×10^{-9} M corresponds to a theoretical half-life for the probe-target hybridization of only 90 secs and yet it is clear from the two curves that far slower processes are occurring during the hybridization

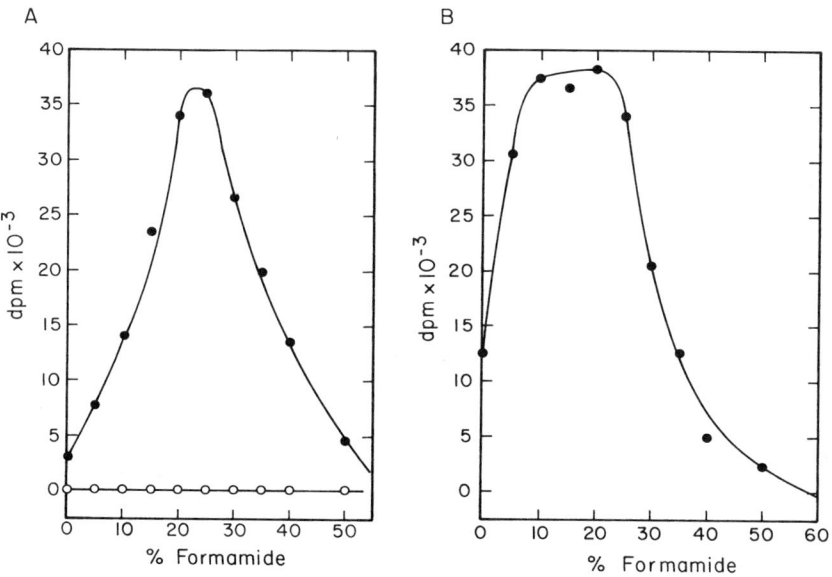

Figure 3 Hybridization profile as a function of percentage of formamide for the interaction of HMT-modified 13-mer with single-stranded M13 DNA. Photocrosslinkage was preceded by (*A*) 1 hr or (*B*) 24 hr incubation at 25°C. Each 10 μl sample was irradiated for 5 min at 25°C with 320 to 400 nm light. [M13DNA] = 4.16×10^{-9} M. [Probe] = 4.7 to 4.9×10^{-9} M. The peak in hybridization occurs with a target DNA containing sequence complementary to probe, —●—●—, single-stranded M13mp19 DNA; a control DNA not containing target sequence, —○—○—, single-stranded M13mp8 DNA. Each data point corresponds to the signal given by 100 ng M13 DNA (23).

process. These profiles should be interpreted as "melting" profiles, since the DNA helix is destabilized by high formamide concentration much as it is by high temperature.

How can we interpret these curves? The fall in hybridization at formamide concentrations above 25% in both curves is clearly due to the loss of stability of the helical base-pairing interaction and may be interpreted as a melt of the interaction. The similarity of the two profiles in this region suggests that the kinetics were rapid relative to one hour and that the signal was primarily determined by the probe-target equilibrium. Some photochemical pumping may have occurred in this region during the 5 min irradiation or fixation step. The region between 0 and 25% formamide was clearly not at equilibrium after 1 hr. Since the kinetics in this region were slow, the magnitude of the hybridization signal was determined by the degree of hybridization that had occurred at the time of fixation. As the percentage formamide became lower in this region, secondary and tertiary interactions in the M13 target DNA became more stable and accessibility of the target sequence, both in terms of the rate at which it became available through thermal rearrangements and in terms of the equilibrium fraction of target available, was reduced. In addition, it is possible that metastable interactions between probe and nonspecific target sequences might have reduced the free probe concentration, thus reducing the rate of the binary process still further.

The plateau in Figure 3b suggests that 24 hr was sufficient time for probe-target complex to achieve near equilibrium in the solutions containing 10 to 22% formamide. Notice that maximum target coverage (14%) was nearly the same for both time conditions, suggesting that under these equilibrium conditions only 14% of the target sequence was interacting with probe, because it was competed for by other secondary and tertiary interactions in the target M13 DNA molecule.

Thus, in summary, we must conclude that there are substantial advantages associated with performing hybridization at conditions that favor rapid equilibrium between probe and target and between the many single-stranded conformations the target is capable of forming. Such conditions involve either high temperatures or high formamide concentrations. At conditions near the melting temperature of the probe from its target, photopumping is possible and high coverage can be eventually achieved. Figure 4 (22) hows the time course of the photoreaction of an HMT-modified 25-mer with target M13mp19 DNA at or near melting conditions that satisfy the criteria for efficient photochemical pumping. Both hybridization and photofixation were carried out concurrently in 100 mM NaCl, 10 mM Tris-HCl, pH 7.0, 1 mM EDA (1X NTE) containing 35% formamide with 6.5×10^{-8} M probe. The probe-target melting tem-

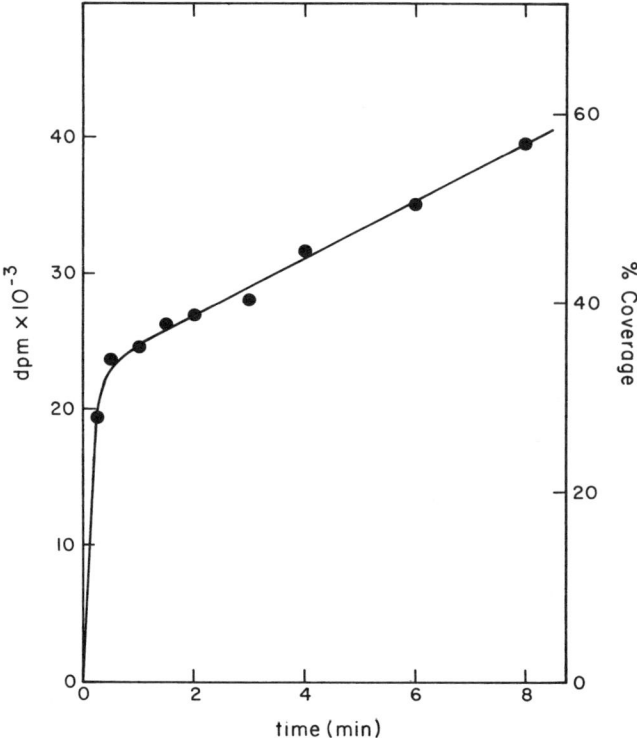

Figure 4 Continuous photochemical pumping of the hybridization equilibrium between HMT-modified 25-mer and single-stranded M13 mp19 DNA. Annealing and photofixation were carried out at 45°C in 1X NTE with 35% formamide. Aliquots of 5 μl were removed after the indicated times of irradiation for analysis by Centricon-30. [M13DNA] = 2.08×10^{-9}. [Probe] = 6.48×10^{-8} (22).

perature under these conditions is 48°C, and the half-life for the hybridization reaction is approximately 7 sec. Figure 4 shows a very rapid reaction of 30% of the target with probe, presumably reflecting the fact that 30% of the target under these conditions is readily accessible. The remaining sequences, which react more slowly, are masked by a secondary structure that becomes available with a much slower rate. After 8 min of photopumping, 57% of the target is reacted.

In addition to the advantages of time and degree of target coverage associated with photocrosslinkable probes, blotting and membrane hybridization can be avoided. The concept of a "Reverse Southern" was introduced by Gamper et al (22). They demonstrated that single stranded target can be covalently labeled with radioactive probe prior to gel electrophoresis, thus eliminating the need to blot and hybridize to the filter before autoradiography (Figure 5).

THE THEORY

The symbol τ is used to refer to target; the symbol p to probe; and the symbol τp to the helical duplex between target and probe. The equilibrium between these three forms is represented by the chemical reaction

$$\tau + p = \tau p.$$

The most elementary helix-coil transition theory for this system defines an initiation factor, β, and an equilibrium constant for the closure of a base pair once a chain has been initiated, s. The two-state helix-coil theory, in which each pair is viewed to be equally stable to all other base pairs in a helical run, leads to Eq. 1 for long chains (27–31)

$$K = \beta s^n \qquad \qquad 1.$$

where $n+1$ equals the number of base pairs in the helix (n equals the number of base-stacking interactions). The equilibrium constant, K, associated with this two-state model is defined by Eq. 2

$$K = \frac{[\tau p]}{[\tau][p]}. \qquad \qquad 2.$$

Upon introduction of the thermodynamic parameters, the enthalpy of base pair elongation, ΔH, and the entropy of base pair elongation, ΔS, and with an alternate representation of the initiation parameter, Eq. 2 for the equilibrium constant becomes

$$K = e^{-2.3\alpha} e^{-(n\Delta H/RT)} e^{(n\Delta S/R)} \qquad \qquad 3.$$

where T is the hybridization temperature and R is the gas constant.

The logarithmic form of Eq. 3 is written

$$\ln K = -2.3\alpha - n(\Delta H/RT) + n(\Delta S/R).$$

The temperature, T_∞, is defined by the equation, $T_\infty = \Delta H/\Delta S$. Furthermore,

$$\ln K = \ln \frac{[\tau p]}{[\tau]} - \ln [p].$$

An expression for the logarithm to the base 10 of the ratio of bound target to free target in terms of the thermodynamic parameters is obtained by rearranging the above equations, yielding Eq. 4:

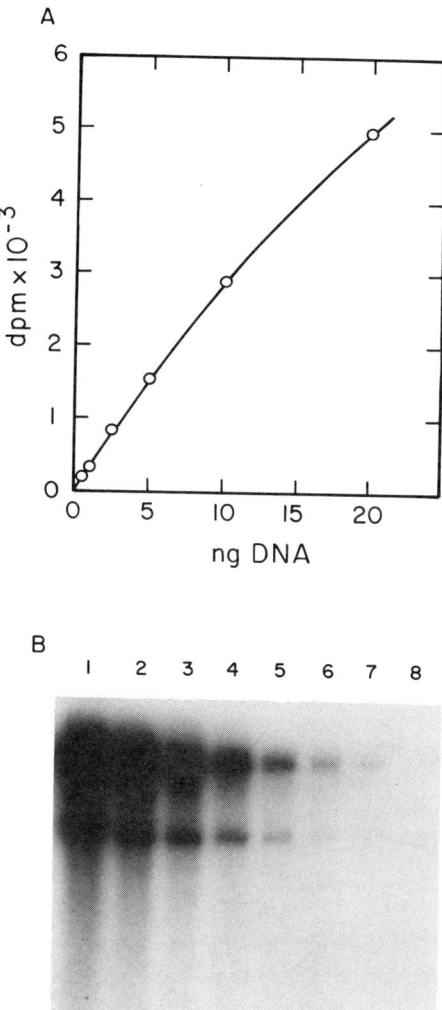

Figure 5 Detection of single-stranded M13 mp19 DNA with a high concentration of HMT-modified 25-mer probe. A concentration series of M13 DNA was hybridized and photofixed to HMT-modified 25-mer by irradiating each sample for 8 min at 45°C in 1X NTE with 35% formamide. The samples were then processed through Centricon-30. (*A*) Yield of photochemically fixed radiolabeled probe versus concentration of M13 DNA. (*B*) Same samples electrophoresed through a 1% alkaline agarose gel. In comparison to Fig. 5A, lane 1 is the 0.5 ng sample; lane 2, 2.5 ng; lane 3, 1.0 ng; lane 4, 0.50 ng; lane 5, 0.25 ng; lane 6, 0.10 ng, lane 7, 0.05 ng; lane 8, 0.025 ng each in 5 μl volume. [Probe] = 6.56×10^{-8} M. Target coverage was approximately 60% (22).

$$\log \frac{[\tau p]}{[\tau]} = \gamma n(\Delta T/T) - \{\alpha - \log [p]\} \qquad 4.$$

where $\gamma = -\Delta S/2.3R$ and $\Delta T = T_\infty - T$.

Discrimination Under Equilibrium Conditions (32)

A typical hybridization experiment involves the use of a single stranded probe of length $(n+1)$ to identify a specific sequence of the same length in a genome containing a total of N base pairs, which is assumed to be entirely composed of unique sequences of DNA. Background in such an experiment originates from two sources. Nonspecific background arises when probe molecules bind to material other than nucleic acid. "Specific" background arises when probe molecules bind specifically to nontarget regions of nucleic acids that are present during the hybridization assay. A hybridization of p to τ, which is devoid of a specific background signal, requires that the accidental homologous subsequences in the target or genomic nucleic acid that might be complementary to shorter stretches of adjacent bases in the probe are sufficiently infrequent and have insufficient strength of interaction with the probe to represent an appreciable background. At equilibrium, specific background is inherently present. The amount of specific background, however, is governed by the hybridization conditions under which equilibrium is established. To understand qualitatively the dependence of the specific background on the various factors that contribute to hybridization, a mathematical parameter, ∇, is defined as the number of decades of discrimination and is given by Eq. 5:

$$\nabla = \log \{[\tau p]/[\tau]_{\text{TOT}}\}_n - \log \{[\tau' p]/[\tau']_{\text{TOT}}\}_{n'} \qquad 5.$$

where the parameter $[\tau]_{\text{TOT}}$ represents the total target concentration whether covered or free, n equals the number of nucleotides in the probe, n' (which is less than n) equals the number of nucleotides in a subsequence of the probe to which complementarity may be found in nucleic acid sequences other than the true target sequence, $[\tau']_{\text{TOT}}$ is the total concentration of such complementary $(n'+1)$mers in the solution, whether covered or free, and $[\tau' p]$ is the concentration of probe hybridized to such $(n'+1)$mers.

Figure 6 shows plots of the log of the inverse fraction of occupied targets of lengths n and n', which have been drawn using Eq. 4. These plots should be interpreted as logarithmic melting curves for the short probes from the targets. These curves are plotted in terms of dimensionless parameters and are therefore independent of the details of the values of ΔH, ΔS, γ, and α. They deviate from Eq. 4 and develop curvature near $[\tau p] = [\tau]$ because the ratio $[\tau]_{\text{TOT}}/[\tau p]$, unlike $[\tau]/[\tau p]$, is unable to be less than 1. There are two sets of curves in Figure 6 that correspond to two values of $[p]$. Each set

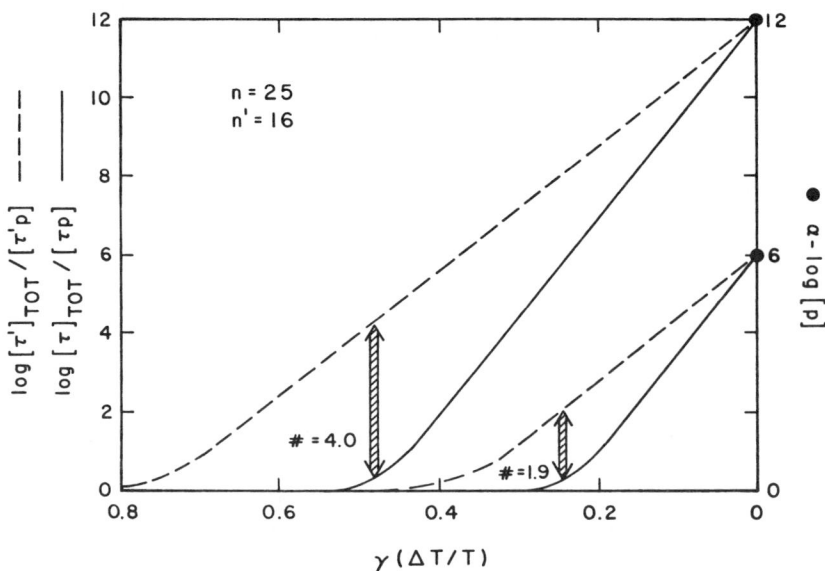

Figure 6 Plot of the logarithm to the base 10 of the ratio of total target to target hybridized with probe versus the dimensionless stability parameter $\gamma(\Delta T/T)$ for $n = 25$ and $\{\alpha\text{-log }[p]\}$ equal 12 and 6 (———). Plot of the logarithm to the base 10 of the ratio of total $(n'+1)$mer concentration with complementarity to probe sequences to the concentration of $(n'+1)$mer-probe hybrid for $n' = 16$ and $\{\alpha\text{-log }[p]\}$ equal 12 and 6 (-----). The *heavy vertical arrows* are equal in length to the number of decades of discrimination at the midpoint of the melting transition of probe with target for $\{\alpha\text{-log }[p]\}$ equal 12 and 6 (32).

contains a curve for which $n = 25$ and $n' = 16$. The $\log\{[\tau']_{TOT}/[\tau'p]\}$ is calculated by Eq. 4, where n' is substituted for n. The two curves of each set intersect at $\Delta T = 0$, near the melting temperature of the infinitely long helix. Note that at low $[p]$, the melting temperature of the probe from its target is low (large ΔT), and that at relatively high $[p]$, the probe melts from its target at much higher temperature. The maximum number of decades of discrimination occurs at the value of $\gamma(\Delta T/T)$ where there is a maximum difference between the n' curve and the n curve. This condition can be determined analytically, for it is the point where the tangents of the two curves are equal. However, for all useful conditions, an approximate value of the abscissa is the value for which the longer sequence, $n+1$ long, has $[\tau]/[\tau p] = 1$, or the midpoint of its melting curve. The points of maximum number of decades of discrimination of each of the two sets are indicated on Figure 6. Note that ∇ is much larger for the low value of $[p]$ than it is for the large $[p]$. An analytical expression for ∇ is shown in Eq. 6:

$$\nabla = \gamma(\Delta T/T)(n-n')+\delta \qquad 6.$$

where δ is the displacement from the straight line plot in Figure 6 associated with the curvature near $[\tau]/[\tau p] = 1$. The n' curve must be linear at the midpoint value of $\gamma n(\Delta T/T)$ for Eq. 6 to be valid. Now with the approximation that the evaluation of δ should be made at $[\tau]/[\tau p] = 1$,

$$\delta = \log 0.5 \quad \text{or} \quad \delta = -0.3.$$

At this midpoint of the transition of probe with target, Eq. 4 shows that

$$\gamma(\Delta T/T) = \{\alpha - \log [p]\}/n. \qquad 7.$$

The final expression for the number of decades of discrimination is shown in Eq. 8

$$\nabla = \{\alpha - \log [p]\}\{(n-n')/n\} - 0.3. \qquad 8.$$

Equation 8 is extremely interesting, for it shows that discrimination at equilibrium depends only on the initiation parameter of Eq. 1 and is completely independent of the enthalpy of helix formation/bp and the entropy of helix formation/bp. It depends logarithmically upon the concentration of probe and it depends on the relative lengths of base sequences being compared. The best available estimate of β for polynucleotide chains comes from Pohl (33) although several other references are consistent with this estimate. A typical range for values of β that has occurred in the literature is 10^{-3} to 10^{-5} (34–40). Using the value of $\beta = 1.6 \times 10^{-4}$ from Pohl, $\alpha = 3.8$. Thus our final Eq. 9 for the ∇ of decades of discrimination is

$$\nabla = \{3.8 - \log [p]\}\left(\frac{n-n'}{n}\right) - 0.3. \qquad 9.$$

It should be pointed out that Eq. 9 addresses only the discrimination between true target and partially homologous sequences that are n' in length and continuous in length. In a hybridization experiment it is likely that the probe will also associate with $(n'+1)$ bases through many noncontiguous base pairings. For these situations, Eq. 9 will yield an underestimate of the number of decades of discrimination, since the free energies associated with mismatches and bulges are not considered. Theoretically, Eq. 8 applies to the situation of $(n'+1)$ noncontinuous base-pairings if the parameter α is adjusted to include the free energies of the internal mismatches and bulges. Thus, the conclusions presented for the case of continuous base-pairings of $(n'+1)$ length are also valid for the general case of $(n'+1)$ total base pair interactions, although the thermodynamic parameters describing the more complex case will be altered and variable.

Kinetic Advantages of Crosslinkable Probes

All that has been derived up to this point applies to equilibrium distributions of probe and target. Is it really feasible to achieve equilibrium in these complex systems and, if so, how much time will this process take? The answers to these questions depend upon whether hybridization is achieved by a solution format or by immobilization techniques. Solution formats provide advantages because hybridization can be conducted in small volumes (~ 5–10 μM), which facilitates the use of higher probe concentrations and enhances the hybridization rates. Two central points must be recognized if solution hybridization with short probes is to be practical. First, the equilibrium distribution of base pairing interactions in solution will always favor "Watson (w) finding Crick (c)." The genomic DNA will always be in long enough sections so that a short oligonucleotide probe, even at 1 molar concentrations, cannot compete with the greater stability of wc over τp. The only way for a solution hybridization experiment to be successful is to entrap kinetically an intermediate state in which the probe-target interaction is much more frequent than the perfectly renatured natural interaction of complementary genomic strands. Second, since time is an essential variable, the ability to freeze the distribution of interactions in a manner that does not alter the distribution is very important. This review is restricted to a discussion of photochemical crosslinkage as the method of choice for this quenching process, for it can be accomplished in times several orders of magnitude shorter than any of the chemical reactions that alter the distribution. Nevertheless, quick cooling techniques as well as rapid mixing procedures are also effective under appropriate conditions.

If one is to trap a probe/target complex before genomic DNA renatures, the relative rates of hybridization of the two reactions are a central consideration. The hybridization reaction of probe with target, and the reaction between complementary genomic sequences, are second-order reactions. The rate at which they occur is determined by the rate of helix nucleation. The zippering up of the DNA after nucleation for pieces of DNA of interest to us (ie. fragments 5000 bases or shorter) is so rapid that it is not necessary to consider this process in the kinetic analysis. The rate of reaction of probe with target may be written in terms of a binary rate constant, k_p, where the subscript, p, refers to the probe

$$-\frac{d[\tau]}{dt} = k_p[p][\tau].$$

The reaction of any ith section of "Watson" with its appropriate ith section of "Crick" may also be written in terms of a binary rate constant, k_g, where the subscript, g, refers to the genome

$$\frac{d[w_i]}{dt} = k_g[w_i][c_i].$$

The times required for these two reactions to be half completed are

$$\Gamma_p = 0.693/k_p[p]; \quad \Gamma_g = 1/k_g[w].$$

For the first of these, the probe is considered to be in vast excess. For the second, $[w]$ equals $[c]$.

Now, how do the binary rate constants, k_p and k_g, compare? The binary rate constant can be calculated from the renaturation data or Cot curves of Britten & Kohne (5). These kinetics were done on fragments of DNA approximately 400 bases long, at a salt concentration of 0.18 M, and a temperature approximately 20°C below the melting temperature of the DNA. The resulting rate constant is $k_g = 1.6 \times 10^6$ l/mol sec. Riesner & Romer (41) present data on short oligonucleotides that are in excellent agreement with this number. Their numbers range from 10^5 to 10^7; the most frequent values are between 1.5×10^6 and 5.5×10^6. These values were obtained in 1 M salt at 17°C. This suggests the binary rate constant is length independent.

From a different point of view, the probability of nucleation should be proportional to the length of the fragments that are renaturing if there is an equal probability of initiating a nucleation at any site along the fragment. This would predict the binary rate constant to be proportional to the length of the fragments that are renaturing. Such a strong dependence of the binary rate constant on the length of renaturing fragments has never been observed experimentally. Wetmur & Davidson (42) observed a square root dependence of the rate constant on the number of nucleotides in the renaturing fragments, L, for high molecular weight DNAs. They presented the equation [modified in Bloomfield et al, p. 365, (29)] of the form

$$k_g = 3 \times 10^5 L^{1/2}.$$

The parameter, L, is equivalent to $(n+1)$ for the renaturation of short probes or, alternatively, L equals the average number of nucleotides in the fragments of genomic DNA with which the probe renaturation must compete. The Wetmur-Davidson expression also predicts a binary rate constant of 1.6×10^6 for a 25-mer probe, but it predicts a rate enhancement by a factor of 10 over this for restriction fragments of length 2500–5000 bases. The only issue is which of these two models is most accurate, an issue that will ultimately be settled by experimentation. Either possibility could be correct, for if there is a strong bias for nucleation at ends, the length dependence observed by Wetmur & Davidson might not manifest itself until lengths exceed several thousand nucleotides. The conclusions drawn here are based on the length-independent model or constant binary reaction rate constant.

If the probe is to compete kinetically with genomic renaturation, the probe concentration must exceed the target concentration, since the second-order rate constants are approximately equal for the two reactions. Clearly, photocrosslinkable probes can entrap a pseudo-equilibrium between probe and target. This kinetic advantage is optimized when the probe concentration exceeds the target concentration by at least a factor of 100. If this is the case, it will take the genome 100 times longer to renature than it will take the probe to find its target. The reaction should be stopped in four or five half-lives. If the probe concentration is 10^{-8} M, this corresponds to approximately 3–4 min of hybridization before the reaction is photochemically quenched. Solution hybridization of a crosslinkable probe at this concentration to a single copy sequence in the human genome satisfies this condition. A 1.5 mg/ml solution of human DNA is 10^{-12} mol/l in all unique human sequences. Thus, the probe is 100,000 times more concentrated than human unique sequences, so genomic renaturation will not compete significantly in the pseudo-equilibrium distributions achieved in our 3–4 min.

Probing an Entire Genome for a Single Unique Sequence

Technically the most challenging objective is the identification of a single unique sequence in a sample of eucaryotic DNA, for example in a sample of human DNA. Although examples of successful colony hybridization on cloned human libraries (43) and on fractionated human DNA on electrophoretic gels (10) have been reported, the goal of a direct hybridization with no significant background remains elusive. What are the necessary conditions for such a hybridization? Consider a photocrosslinkable probe of length $n+1$ with the crosslinkable modification at the center of the probe. If genomic sequences are complementary to an $n'+1$ long section of the probe containing the crosslinkable site, and if $n' < n/2$ then the number of crosslinkable $(n'+1)$mers in the $(n+1)$mer is n'. If $n' > n/2$, then the number of such crosslinkable $(n'+1)$mers is $n-n'$. If, on the other hand, the crosslinkable site is at one end of the probe, the number of such crosslinkable $(n'+1)$mers is always 1.

For each of these three cases discussed above, on a random sequence basis the number of complementary $(n'+1)$mers in a genome of size N is:

Case I—Conventional hybridization, all n'; site at the center, $n' > n/2$

$$\Phi = 2(n-n'+1)N(4)^{-(n'+1)}.$$

Case II—Site at the center, $n' < n/2$

$$\Phi = 2n'N(4)^{-(n'+1)}.$$

Case III—Site at one end

$$\Phi = 2N(4)^{-(n'+1)}.$$

Figure 7 contains a plot of these functions, log Φ, as a function of n'. For low background, the probability of a reaction at a complementary sequence other than the real target sequence should be less than some

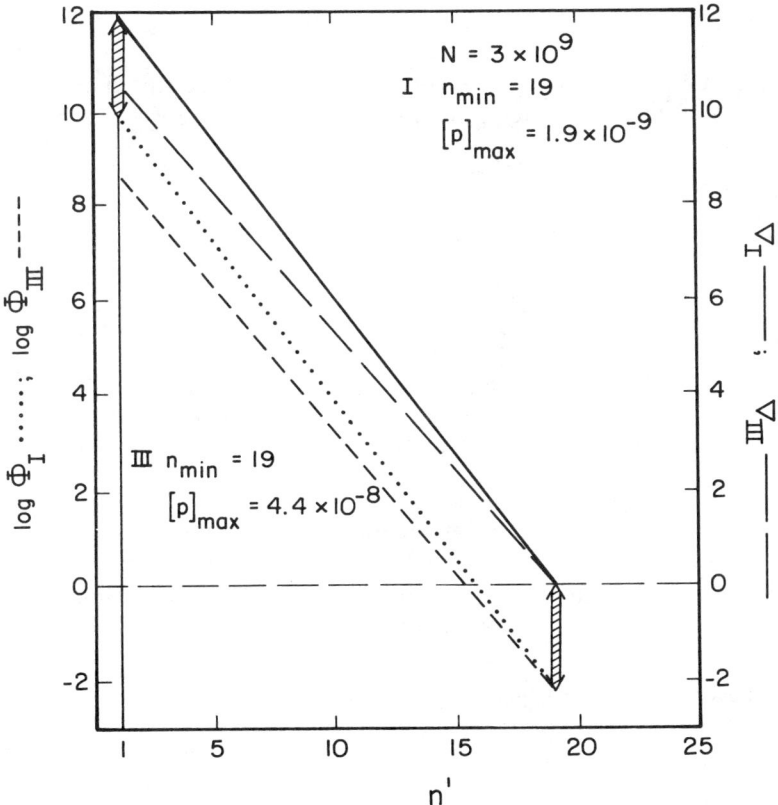

Figure 7 Minimal conditions required for low background. The conditions presented in Table 1 were used to generate the curves in this figure. The vertical distance between ∇ and log Φ on this plot determines the values of $10^{-\nabla} \Phi$ needed in the determination of the percentage specific background (32).

arbitrary small number, for example, 0.01. In order to achieve this condition, Eq. 10 must be satisfied for all values of n'

$$\nabla - \log \Phi > 2. \tag{10}$$

By observing Fig. 2, it is clear that the shortest possible probe that will satisfy this condition must do so at $n = n'$; this leads us to Eq. 11

$$4^n > 50N. \tag{11}$$

The minimum length probe that satisfies this equation for the human genome ($N = 3 \times 10^9$) is 19.

Background as a Percentage of Signal (32)

For any value of n', the background signal associated with hybridization to $(n'+1)$mers complementary to the probe relative to the true signal is $10^{-\nabla} \Phi$. The sum of all contributions from $n' = 1$ to $n' = n$ is $\Sigma\, 10^{-\nabla} \Phi$. Replacing this sum with an integration, the following closed expressions for each of the above three cases are obtained. The background signal, B, becomes:

Case III

$$B_{III} = (N/2\Psi)10^{-\Theta}[e^{\Psi n} - e^{\Psi}]$$

where $\Theta = \alpha - \log [p]$ and $\Psi = 2.303\Theta/n - 1.386$.

Case II $\hspace{10em} 12.$

$$B_{II} = (N/2\Psi^2)10^{-\Theta}[e^{\Psi n}(\Psi n - 1) - e^{\Psi}(\Psi - 1)].$$

Case I

$$B_I = (n+1)B_{III} - B_{II}.$$

Table 1 contains the data for the minimum length of probes based on Eq. 11 and the maximum probe concentrations for the same minimum probe lengths evaluated with Eq. 10 at $n' = 1$. Case I represents the conventional hybridization experiment with no use of a crosslinking agent.

Table 1 Minimal conditions of discrimination

Case	I	II	III
n_{min}	19	21	19
$[p]_{max}$, mol/l	1.9×10^{-9}	5×10^{-8}	4.4×10^{-8}
Half-life, Γ_p, sec	230	8.7	9.8
Background, B	50%	64%	27%

Case II represents the condition in which the probe contains a crosslinking agent at its center, and Case III represents the condition of a crosslinking agent at one of the ends of the probe. Note the advantage in reaction time of a factor of 30 provided by the crosslinkable probe over the non-crosslinkable probe.

Table 2 compares the same parameters at a constant, convenient probe length ($n = 25$) and a constant probe concentration of 5×10^{-8}. Note the enormous advantage in background (evaluated by using Eqs. 12) that the crosslinkable probe provides for these conditions.

SENSITIVITY LIMITED BY THE METHOD FOR DETECTION OF PROBE

An oligonucleotide probe that is labeled at one phosphate with 6000 Ci/mmol P32 can detect 100 attomoles of a unique gene at a level of radioactivity of 1200 dpm. At 1.5 mg/ml human DNA, 100 attomoles requires a volume of 0.10 ml, a volume somewhat inconvenient to load on a gel for separation of excess probe from the photocrosslinked target. Although the arguments presented here suggest that background should not create a significant problem in this experiment, the levels of radioactivity are more modest than what is desirable and the quantities of human DNA are high for typical clinical samples. We can hope that sensitive photon absorption or fluorescence techniques will provide an improvement upon this situation and avoid the need for radioactive probe.

CONCLUSIONS

A rapid hybridization assay for single copy unique sequences or low copy number pathogens is technically feasible. The only impediment to greater sensitivity is a more sensivitve detection system. Conditions capable of eliminating the contribution of binding to nontarget sequences have been defined. The necessary conditions for specific hybridization to the human

Table 2 Constant conditions for hybridization to single copy regions in the human genome

Case	I	II	III
n_{min}	25	25	25
$[p]$, mol/l	5×10^{-8}	5×10^{-8}	5×10^{-8}
Half-life, Γ_p, sec	8.7	8.7	8.7
Background, B	102%	9%	4.6%

genome include an oligonucleotide probe at least 20 bases long for which the maximum allowable probe concentration is 5×10^{-8}. Longer probes provide greater discrimination and allow for somewhat higher probe concentrations. Selection of the optimal condition for attainment of equilibrium requires that temperature and solvent conditions be found at the midpoint of the melting transition of the probe from its target. It is also essential that the probe be in molar excess of the target by a factor of 100 or more in order to avoid the complication of genomic renaturation, which competes with probe renaturation. The half-life of the approach to equilibrium of probe with target is simply related to free probe concentration. At $\log [p] = -6$, the half-life is 0.4 sec. At $\log [p] = -7$, the half-life is 4 sec. At $\log [p] = -8$, the half-life is 43 sec. Thus, if rapid approach to equilibrium is desired, probe concentrations lower than these should not be used. Finally, these procedures suggest the need for a method of freezing the equilibrium distribution in short times. Photocrosslinkage can be used for this purpose.

The ability of the hybridization procedures described in this review to distinguish single point mismatches between probe and target remains to be tested. In addition to providing new procedures for hybridization diagnostics, renaturation experiments under conditions of equilibrium with respect to probe and target coupled with photofixation of the equilibrium distribution will provide new insights into the thermodynamic and kinetic parameters that are important for a thorough understanding of this complex process.

ACKNOWLEDGMENTS

This review was supported by the Regents of the University of California.

Literature Cited

1. Marmur, J., Lane, D. 1960. *Proc. Natl. Acad. Sci. USA* 46: 453–61
2. Doty, P., Marmur, J., Eigner, J., Schildkraut, C. 1960. *Proc. Natl. Acad. Sci. USA* 46: 461–76
3. Hayashi, M., Hayashi, M. N., Spiegelman, S. 1963. *Proc. Natl. Acad. Sci. USA* 50: 664–72
4. Marmur, J., Greenspan, C. M. 1963. *Science* 142: 387–89
5. Britten, R. J., Kohne, D. E. 1968. *Science* 161: 529–40
6. Southern, E. M. 1975. *J. Mol. Biol.* 98: 503–17
7. Alwine, J. C., Kemp, D. J., Stark, G. R. 1977. *Proc. Natl. Acad. Sci. USA* 74: 5350–54
8. Grunstein, M., Hogness, D. S. 1975. *Proc. Natl. Acad. Sci. USA* 72: 3691–95
9. Benton, W. D., Davis, R. W. 1977. *Science* 196: 180–82
10. Berent, S. L., Mahmoudi, M., Torczynski, R. M., Bragg, P. W., Bollen, A. P. 1985. *BioTechniques* May/June: 208–20
11. Cimino, G. D., Gamper, H. B., Isaacs, S. T., Hearst, J. E. 1985. *Ann. Rev. Biochem.* 54: 1151–93
12. Straub, K., Kanne, D., Hearst, J. E., Rapoport, H. 1981. *J. Am. Chem. Soc.* 103: 2347–55
13. Kanne, D., Straub, K., Rapoport, H., Hearst, J. E. 1982. *Biochemistry* 21: 861–71
14. Kanne, D., Straub, K., Hearst, J. E.,

Rapoport, H. 1982. *J. Am. Chem. Soc.* 104: 6754–64
15. Peckler, S., Graves, B., Kanne, D., Rapoport, H., Hearst, J. E., Kim, S.-H. 1982. *J. Mol. Biol.* 162: 157–72
16. Kim, S.-H., Peckler, S., Graves, B., Kanne, D., Rapoport, H., Hearst, J. E. 1983. *Cold Spring Harbor Symp. Quant. Biol.* 47: 361–65
17. Gamper, H., Piette, J., Hearst, J. E. 1984. *Photochem. Photobiol.* 40: 29–34
18. Van Houten, B., Gamper, H., Hearst, J. E., Sancar, A. 1986. *J. Biol. Chem.* 261: 14135–41
19. Cimino, G. D., Shi, Y., Hearst, J. E. 1986. *Biochemistry* 25: 3013–20
20. Chatterjee, P. K., Cantor, C. R. 1978. *Nucleic Acids Res.* 5: 3619–33
21. Tessman, J. W., Isaacs, S. T., Hearst, J. E. 1985. *Biochemistry* 24: 1669–76
22. Gamper, H. B., Cimino, G. D., Isaacs, S. T., Ferguson, M., Hearst, J. E. 1986. *Nucleic Acids Res.* 14: 9943–54
23. Gamper, H. B., Cimino, G. D., Hearst, J. E. 1987. *J. Mol. Biol.* 197: 349–62
24. Shi, Y., Hearst, J. E. 1986. *Biochemistry* 25: 5895–5902
25. Isaacs, S. T., Shen, C. J., Hearst, J. E., Rapoport, H. 1977. *Biochemistry* 16: 1058–64
26. Jay, E., Bambara, R., Padmanabhan, R., Wu, R. 1974. *Nucleic Acids Res.* 1: 331–53
27. Crothers, D. M., Zimm, B. H. 1964. *J. Mol. Biol.* 9: 1–9
28. Crothers, D. M., Kallenbach, N. R., Zimm, B. H. 1965. *J. Mol. Biol.* 11: 802–20
29. Bloomfield, V. A., Crothers, D. M., Tinoco, I. 1974. *Physical Chemistry of Nucleic Acids.* New York: Harper & Row
30. Tinoco, I., Sauer, K., Wang, J. C. 1978. *Physical Chemistry—Principles and Applications in Biological Sciences*, pp. 514–21. Englewood Cliffs, NJ: Prentice-Hall
31. Cantor, C. R., Schimmel, P. R. 1980. *Biophysical Chemistry*, Part III: *The Behavior of Biological Macromolecules*, pp. 1183–1238. San Francicso: Freeman
32. Hearst, J. E. 1987. *Photobiochem. Photobiophys. Suppl.*, pp. 23–32
33. Pohl, F. M. 1974. *Eur. J. Biochem.* 42: 495–504
34. Borer, P. N., Dengler, B., Tinoco, I., Uhlenbeck, O. 1974. *J. Mol. Biol.* 86: 843–53
35. Breslauer, K. J., Frank, R., Blocker, H., Marky, L. A. 1986. *Proc. Natl. Acad. Sci. USA* 83: 3746–50
36. Levine, M. D. 1974. *The stability of ribonucleic acid in solution: Model calculations.* PhD thesis. Univ. Calif., Berkeley
37. Scheffler, I. E., Elson, E. L., Baldwin, R. L. 1968. *J. Mol. Biol.* 36: 291–304
38. Scheffler, I. E., Elson, E. L., Baldwin, R. L. 1970. *J. Mol. Biol.* 48: 145–71
39. Tinoco, I., Borer, P. N., Dengler, B., Levine, M. D., Uhlenbeck, O. C., Crothers, D. M., Gralla, J. 1973. *Nature New Biol.* 246: 40–41
40. Tinoco, I., Uhlenbeck, O. C., Levine, M. D. 1971. *Nature* 230: 362–67
41. Riesner, D., Romer, R. 1973. In *Physico-Chemical Properties of Nucleic Acids*, ed. J. Duchesne, 2: 237–318. London: Academic
42. Wetmur, J. G., Davidson, N. 1968. *J. Mol. Biol.* 31: 349–70
43. Studencki, A. B., Wallace, R. B. 1984. *DNA* 3: 7–15

QUANTUM EFFECTS IN GAS PHASE BIMOLECULAR CHEMICAL REACTIONS

George C. Schatz

Department of Chemistry, Northwestern University, Evanston, Illinois 60208

INTRODUCTION

Although the quantum mechanical underpinnings of chemical reactions have been understood on an abstract theoretical level since the discovery of quantum mechanics, and a variety of simple models of quantum effects such as tunneling have been available for a long time, the quantitative determination of how big quantum effects are and where they show up has only started to appear in the past ten years. Exciting progress of major significance in the experimental observation of several important types of quantum effects in gas-phase chemical reactions has been made during the last two years, so it seems appropriate to write a review that is specifically devoted to this topic. The first half of this review is a rather detailed discussion of what we mean by "quantum effects," with special attention given to the two most extensively studied quantum effects, tunneling and resonances. In the second half of the review, four reactions that show experimentally observable quantum effects are discussed in depth. These are: $H+H_2$, $O(^3P)+H_2$, $Cl+HCl$, and $H+CO$ (plus certain deuterated counterparts). The review is "results-oriented" in that I spend little time in discussing either theoretical or experimental methodology, but rather emphasize the results of calculations and measurements. No review with this scope and topic has been written previously, but a number of reviews have touched on parts of what is discussed here. Two recent reviews of tunneling in gas phase reactions are by Schatz (1) and Miller (2). Several reviews on variational transition state theory have been written by Truhlar and co-workers (3–6) and all of these include some discussion of tunneling.

Kuppermann (7) has written a review on resonances in reactive collisions, and Truhlar (8) has edited a monograph on resonances. Truhlar & Wyatt (9) have reviewed the H_3 kinetics, including some discussion of quantum effects, and Schatz has reviewed theoretical calculations on $H+H_2$ (10) and $O+H_2$ (11). There are also several reviews of theoretical dynamics methods that touch upon quantum effects in less detail (12–15). One reaction in which quantum effects are important but which I discuss only briefly is $F+H_2$. This has been reviewed before (16, 17), and the recently revised book by Levine & Bernstein (18) discusses it in detail.

WHAT ARE QUANTUM EFFECTS?

Zero Point Energies

Let us now discuss what we consider to be "quantum effects" in chemical reactions. Our starting point in this discussion is the Born-Oppenheimer separation of electronic and nuclear motion, which leads to the concept of a potential energy surface that governs nuclear motion. Since the electronic part of the problem is intrinsically quantum mechanical, it could be argued that potential surfaces are a kind of "quantum effect," but this review is restricted to nuclear motion quantum effects. Quantum effects associated with the "hopping" between Born-Oppenheimer potential surfaces during reaction are also not considered.

Once one has a potential surface, the motion of the nuclei can be described either classically or quantally, and one can consider that "quantum effects" refer to any difference between the results of these two descriptions. The completely classical description at this level is what one often uses in doing a *molecular dynamics* calculation (19), and one very large quantum effect that is left out in this is the *zero point energy* associated with vibrations of the nuclei in the reagents, in the products, and during the reaction. This zero point energy contributes significantly to the reaction dynamics for many reactions, especially hydrogen transfer reactions.

For reactions with a barrier between reagents and products, zero point effects cause the energetic threshold for reaction to be quite different from the classical barrier energy, often by several kilocalories per mole, and this leads to major differences between quantum and completely classical rate constants. These differences are in some sense uninteresting, however, as the zero point energy can to a certain degree of approximation be considered to represent a static effect that can be incorporated with the electronic potential surface to define an effective potential function upon which the nuclei move. There are, however, some ambiguities associated with what this effective potential means, since the concept of zero point energy is rigorously defined only for stationary states, and in the present context this means only for separated reagents and products. A standard way to circumvent this problem is to determine the local zero point energy associated with nuclear motions perpendicular to a reaction path that connects reagents and products. The technology associated with defining reaction paths and reaction path Hamiltonians has been discussed in detail by a number of researchers (2–6, 20–25), and if curvature along the reaction path is neglected and rotation is ignored, it leads to a well-defined one-dimensional effective potential function (electronic potential energy plus vibrational zero point energy) known as the *vibrationally adiabatic ground state* (VAG) *potential*.

The term "curvature" here is associated with a portion of the kinetic energy part of the reaction path Hamiltonian that couples along the path and those perpendicular to it (20). Curvature is usually not negligible for reactions involving hydrogen transfer, and, as a result, the zero curvature VAG potential is of limited usefulness for many properties. This has led to alternative partitionings of the Hamiltonian in which many dimensional adiabatic potentials are defined (26, 27). Another problem is that for many reactions, the modes perpendicular to the reaction path have vibrational periods that are short compared with the collision duration, so that the states for these modes are not even approximately stationary, and the adiabatic assumption is not valid (28, 29).

It turns out, however, that zero curvature VAG potentials are remarkably good indicators of the correct effective *threshold energies* for reaction (i.e. how much translational energy is needed to give an appreciable reaction probability), even when curvature and timescales are inappropriate. We know this as the result of a number of benchmark theoretical studies (10, 28, 30) in which the nuclear motion Schrödinger equation was solved accurately for motion on realistic potential surfaces in three dimensions, as well as from a variety of model calculations (1, 31, 32), including collinear exact quantum calculations in which the nuclear Schrödinger equation for motions in one dimension is solved exactly. It is also known that excited state vibrationally adiabatic (VA) potentials can be constructed that accurately determine reaction thresholds for reagents in excited states (33, 34a,b). Of course, the reaction dynamics away from threshold is not usually described very well by the VA potentials, but because threshold behavior is very important in determining rate constants, the comparison of quantum and classical dynamics relative to the VA potentials is a common way to define quantum effects, particularly tunneling.

Another approach to incorporating zero point energy effects into classical dynamics that is often used is the quasiclassical trajectory (QCT) method (35–37). In this method the initial coordinates and momenta in a molecular dynamics calculation are chosen to mimic quantum mechanics as determined by the WKB semiclassical approximation for one-dimensional problems (38), and by the EBK semiclassical approximation for nonseparable many-dimensional problems (39). Except for the initial conditions, the trajectory calculation is strictly classical, so the reagent zero point energy is not forced to be locked up in modes perpendicular to the reaction path as it is in the VAG potential. Because of this, quasiclassical trajectories can cross barriers with less than zero point energy in some modes, leading to a QCT reaction threshold that is often lower than the VAG threshold. The benchmark and model quantum dynamics studies referred to above (1, 10, 28, 30–34) all suggest that this behavior is qualitatively incorrect, and as a result the QCT method is not a reliable reference against which to measure quantum effects in the vicinity of the threshold for reaction.

The QCT method does, however, play a useful role in defining quantum effects at energies well above threshold, where curvature effects are more important [leading to internal centrifugal effects known as "bobsledding" (18)] and the zero curvature VAG description breaks down (40). It is in this region where resonances are significant, so much of our analysis of resonances here uses QCT results to provide a reference.

Tunneling

There is no unique definition of tunneling, but for one-dimensional problems, one usually associates tunneling with reaction that takes place at energies less than the barrier energy (41a,b). Since the VAG description given above reduces the multidimensional nuclear dynamics to motion on a one-dimensional potential curve, it is reasonable to define tunneling as reaction that occurs at energies less than the largest barrier on the VAG potential, which we denote V_a^G. It should be noted that the comparison that is implied in this definition is between exact quantum dynamics on the full potential surface and classical dynamics on the zero curvature VAG potential, and, as a result, this definition of tunneling implicitly incorporates the influence of curvature into tunneling. This is a reasonable association, for semiclassical calculations indicate (42–48a) that curvature enhances tunneling by shortening the path between the reagents and products [usually known as "corner cutting" (40)], thereby making the effective barrier narrower.

Other definitions of tunneling are possible (48b), but all suffer from the fact that they are based on a comparison of exact dynamics with a theoretical model of some type that cannot be directly related to experimental measurements. As a result, it is often difficult to decide from experimental data in the absence of detailed theory whether tunneling is playing an essential role in what is being measured, and this has led to tunneling being ignored in empirical modeling of chemical reaction rates in the gas phase

(49). Curiously enough, tunneling shows up in a more obvious way in many condensed phase rate measurements, where at very low temperatures, the temperature dependence of rate coefficients switches from being activated (i.e. obeying the Arrhenius formula) to being temperature independent. Examples where this is observed include photoexcitation of bacteriorhodopsin (50, 51), electron transfer in biological systems (52), hydrogen abstraction by methyl radicals in methanol glasses (53), and H atom diffusion on tungsten surfaces (54). For processes involving H atom motion, one often observes rate coefficients that are temperature independent at temperatures below 100K, so in gas phase reactions, which are typically done at much higher temperatures, this low temperature limit is never obtained. Instead, one usually sees modest curvature in Arrhenius plots (55). Recently, measurements on isotopomers of $H+H_2$ (56, 57) and $O(^3P)+H_2$ (58) have provided new information about how tunneling produces experimentally distinguishable results. I describe these results in detail below.

The quantitative influence of tunneling in bimolecular reactions is usually defined in terms of a transmission coefficient κ that is taken to be the ratio of the exact rate coefficient to one obtained from a variational transition state theory calculation in which tunneling has been ignored, but for which all the perpendicular modes are treated as quantum mechanical (same as the assumptions underlying the VAG potential). Thus we define

$$\kappa = k_{\text{exact}}/k_{\text{VTST}} \qquad 1.$$

where k_{VTST} refers to a transition state that is based on finding a free energy maximum for either a canonical or microcanonical ensemble. The energy at this maximum equals V_a^G only in the limit of zero temperature or energy, but for hydrogen transfer reactions, V_a^G is usually very close to this energy, up to quite high temperatures or energies. Although the definition in Eq. 1 is such that κ could be determined by factors other than tunneling (such as barrier recrossing), at low temperatures for H atom transfer reactions, κ is completely dominated by tunneling. Because of this I usually refer to κ as the "tunneling factor."

Recently, Schatz (1) has used Eq. 1 to estimate transmission coefficients for a variety of reactions for which accurate (to within 30%) quantum calculations have been done in three dimensions (the results of which were used to define k_{exact}). This approach to determining κ values is more precise than using experimental data in Eq. 1, since the same potential surface is used in determining both k_{exact} and k_{VTST}.

Table 1 presents a slightly updated version of this table, including results for $H+BrH(D)$ (59), $H+H_2$ and isotopic counterparts on both the older,

Table 1 Tunneling factors κ from accurate quantum scattering calculations in three dimensions

Reaction	β (deg)	κ (200K)	κ (300K)	$V^{\ddagger}$ (kcal/mol)
H+BrH	89.3	46.7	5.5	5.5
H+BrD	89.0	15	3.2	5.5
H+DH (PK2)	70.5	250	12	9.1
H+D$_2$ (LSTH)	65.9	19	3.2	9.8
H+H$_2$ (PK2)	60	556	23	9.1
H+H$_2$ (LSTH)	60	83	6.9	9.8
O+DH	57.0		7.1	12.5
D+H$_2$ (PK2)	54.7	303	17	9.1
D+H$_2$ (LSTH)	54.7	68	6.7	9.8
O+D$_2$	48.2		7.1	12.5
O+H$_2$	46.7		19	12.5
O+HD	37.6		12	12.6
Cl+HCl (LEPS)	13.6	16	7.8	8.5
Cl+HCl (sf-POLCI)	13.6	29	19	8.3

less accurate PK2 surface (11, 60) and the newer, more accurate LSTH potential surface (M. C. Colton and G. C. Schatz, unpublished results, which are slight revisions to results published in Ref. 61), O+H$_2$ and isotopic counterparts (62), and Cl+HCl on LEPS and ab initio (sf-POLCI) surfaces (from Refs. 45, 63 and unpublished results from B. C. Garrett). The parameter β in this Table is the "skew angle" between the reagent and product arrangement channels for the collinear (most favorable) geometry of the three atoms. β is related to the atom masses of the A+BC → AB+C system by

$$\beta = \tan^{-1}(m_B(m_A+m_B+m_C)/m_A m_C)^{1/2} \qquad 2.$$

and the magnitude of β is often inversely related to reaction path curvature. There is, however, no obvious correlation of κ with β in Table 1, and there is a similar lack of correlation with $V^{\ddagger}$, the classical barrier to reaction.

The most important result in Table 1 is in how big the κ values actually are. The smallest tunneling factor at 300K is 3.0, meaning that tunneling contributes 75% of the total rate constant, and most are larger than 10 in magnitude. At 200K the tunneling factors are even larger, but in all cases, reactions with heavier isotopes give rise to smaller tunneling. Notice that different potential surfaces can give rise to very different tunneling factors for the same reaction.

Since the accurate quantum calculations that were used to generate the tunneling factors in Table 1 are very time consuming, a great deal of effort has gone into developing methods for determining approximate tunneling factors. This point was reviewed in detail in (1), and the general conclusion was that although the simple tunneling factor expressions based on the VAG potential that one often finds in textbooks [Eckart, Wigner, inverted parabolic barrier (41b)] are usually very inaccurate, several recently developed semiclassical methods that are based on least action paths between reagents and products are quite accurate (42–48). However, one problem with using these more accurate methods is that global information about the potential surface, rather than properties of just a transition state, is needed.

Resonances

The term "resonance" in the context of a chemical reaction refers to a metastable state that is formed when the reagents collide and which subsequently decays to products (at least partially). Evidently, the total time needed for reaction to occur is substantially longer when resonances are populated than when reaction proceeds directly to products, and this is often used as a criterion for resonance formation. Resonances are quantum mechanical in character because they occur at discrete energies, much like bound states, except that resonance energies always have a finite width. In QCT calculations, one often observes trajectories that form long-lived complexes, which are in some sense the classical counterparts of resonances, but these complexes typically exist over a broad range of energies rather than over narrow energy intervals that are determined by the resonance lifetime. In addition, resonance scattering often gives rise to oscillating reaction probabilities which reflect quantum interference between direct and resonant scattering that is absent in classical mechanics.

Reactions in which a strongly bound intermediate exists often support a large number of resonance states, and in this case the reaction dynamics can be completely controlled by resonance formation. When this is the case, statistical theories such as RRKM theory are appropriate for describing the decay of the resonances. As the density of resonances decreases, we first encounter weakly bound molecules like HCO for which the number of resonances that can be populated in $H+CO$ collisions is small enough that discrete level structure is important in the dissociation dynamics, and then we pass to unbound molecules like H_3 or ClHCl for which resonances are still possible, but for which at most one resonance is populated at a given energy, and most of the time reaction occurs via direct scattering. Much attention has been directed in the past two years to the weakly bound and unbound cases, and further in this review I describe experiments that provide evidence for resonances in $H+H_2$, $Cl+HCl$, and $H+CO$. The celebrated but still controversial resonance in $F+H_2$ (64, 65, 16, 17) is also (apparently) a resonance that occurs in an unbound system.

Predissociating states of van der Waals clusters are also examples of resonances associated with scattering systems, but the systems that have been studied so far are ones for which no rearrangement is possible, so I do not consider them in this review. The use of van der Waals resonances as precursors in a chemical reaction [*precursor geometry limited reactions* (66)] is, of course, an interesting possibility that is being actively pursued by several groups.

Except for $F + H_2$, the study of resonances in unbound reactive systems has been largely confined to theoretical work until recently. Resonant features in reaction probabilities were noticed in very early model studies (67), and an important result of early collinear scattering calculations on realistic potentials (68–71) was the frequent appearance of resonances. Figure 1 illustrates this for collinear $H + H_2$ on the LSTH surface with a plot of the reaction probability [from unpublished work by G. C. Schatz that reproduces previous calculations (72, 73a,b)] versus total energy E (relative to separated $H + H_2$ with H_2 at equilibrium). The $v = 0$ to $v' = 0$ reaction probability P^R_{00} rises sharply at $E = 0.45$ eV, which corresponds to the classical threshold for reaction. Then for $E > 0.6$ eV, P^R_{00} gradually falls as a result of curvature induced recrossing of the saddle point, but interrupting this fall are a rapid oscillation at 0.89 eV and another at 1.29 eV that correspond to resonances. The $v = 0$ to $v' = 1$ probability P^R_{01} and the $v = 0$ to $v' = 2$ probability P^R_{02} also show oscillations at the same energies, reflecting the fact that the resonance decays into

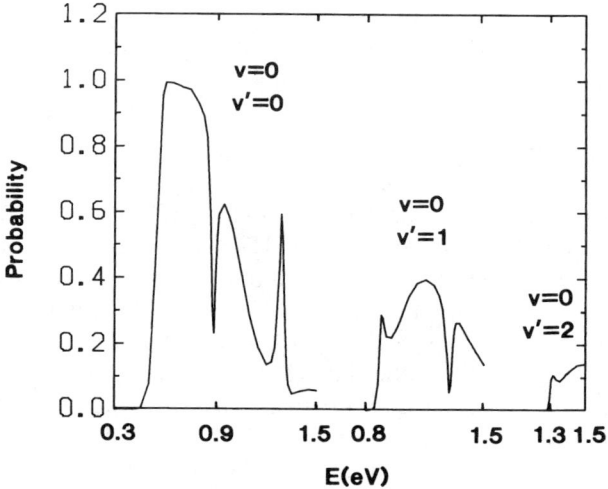

Figure 1 Reaction probability $P^R_{vv'}$ for $v = 0$ and $v' = 0, 1, 2$ versus total energy E for collinear $H + H_2$ on LSTH potential surface.

a different distribution of $v' = 0$–2 than is produced by direct scattering, plus there is interference between the resonant and direct scattering amplitudes.

The rather old results presented in Figure 1 have been greatly extended in recent years. Numerous studies of the properties of the collinear resonances have been done, both for $H+H_2$ and other reactions [reviews are in (7, 8, 12)]. A number of methods for approximately describing the resonances have been developed, including adiabatic approximations in both reaction path (74) and hyperspherical (75–78) coordinates [the latter is called the DIVAH model (77)]. Other approaches to the characterization of the resonances have been studied. Ref. (74) reviews progress through 1984; and see also Refs. (79, 80). Resonances have also been seen in three-dimensional accurate quantum calculations, first in $H+H_2$ (81) and later apparently in $F+H_2$ (17, 82). The $H+H_2$ reaction is discussed in the next section. $F+H_2$, which started out as the first experimentally observed resonance in an unbound system (83), was the subject of extensive theoretical study by several groups (17, 82, 84–89) using the Muckerman 5 surface (90a). The results of these studies appeared to correlate well with the measurements, but subsequent measurements by Neumark et al (65) revised the experimental picture in such a way that it is now incompatible with the older theory (90b). Revised potential surfaces have since been developed (91–94a) that correct the many errors in the Muckerman 5 potential. As of this writing, several groups are working on new quantum studies of $F+H_2$ (R. T Pack, private communication; D. G. Truhlar, private communication), and there is one report of converged calculations (94b). The question of whether the energy-dependent changes in angular distributions that Neumark et al observe is due to a resonance, or just to a strongly energy dependent direct scattering structure remains unresolved at this point (86). One report suggests the latter possibility (94c).

Other Interference and Symmetry Effects

A number of other quantum effects in chemical reactions have received less attention, either because they are hard to quantify or because they refer to reactions with special combinations of identical nuclei or special electronic state features. A list (not necessarily complete) is as follows:

1. It is well known that wave packets passing over a one-dimensional barrier with energies above the barrier have a finite probability of "non-classical reflection." The generalization of this to real systems is hard to quantify because curvature and other nonseparable effects lead to a finite probability of reflection classically at energies above the barrier. Non-classical reflection has been included by Truhlar and co-workers (45–47)

in many of their one-dimensional tunneling calculations, and in this context it has been simply lumped with tunneling in defining the transmission coefficient. In the few comparisons between QCT and accurate 3D quantum cross-sections that have been done for energies well above V_a^G (10, 62), the results are in good agreement, so one infers that nonclassical reflection is generally small for many-dimensional problems.

2. In many reactions one finds that a barrier is first crossed and an intermediate lives for one or two vibrational periods, and then that barrier may be recrossed. Sometimes this process is linked with the formation of resonances in which case quantum effects are expected, but sometimes the scattering is direct. Apparently, in the direct scattering case quantum effects are still possible. A good example of this occurs in the $F+H_2$ reaction, where at energies within 2 kcal/mol of threshold there are significant differences between the results of collinear QCT and quantum calculations for the Muckerman 5 surface (71); the quantum reaction probabilities are smaller, reflecting greater recrossing. One possible reason for this is that trajectories can be more chaotic in their vibrational motions than quantum wavepackets, with wavepackets sometimes seeing strong dynamical bottlenecks that trajectories easily penetrate (95–97). Although classical and quantum dynamics are the same at very short times (starting from equivalent initial conditions), noticeable differences due to bottleneck effects can occur after as little as one vibrational period (95). This intramolecular quantum effect could play an important role in addition reactions such as $H + CO \rightarrow HCO$, where it would lead to more recrossing quantum mechanically than classically. At this point, however, rather little is known about this potentially important issue.

3. The effect of identical particle symmetry is important in a few reactions, most notably $H+H_2$. This causes interference between reactive and nonreactive scattering amplitudes when both processes produce the same final state [as in $H+H_2(para) \rightarrow H+H_2(para)$] and it also leads to special numerical factors that multiply the scattering amplitudes for final states where interference is not possible [as in $H+H_2(para) \rightarrow H+H_2(ortho)$]. The basic theories involved have been worked out in detail by many groups (98–102). Most experiments, including the *para* → *ortho* conversion rate measurements in $H+H_2$ (103), are easily explained by simple statistical arguments (60, 104), but experiments at low temperatures or experiments with rotational and angular state resolution might produce interesting effects that require detailed dynamical theory to describe.

4. Reactions with several identical nuclei and a potential surface having a conical intersection may exhibit the Aharonov-Bohm (AB) effect (105–107). For $H+H_2$, this leads to a change in sign of the reactive scattering amplitude when it interferes with the nonreactive amplitude for *para* →

para and *ortho* → *ortho* scattering (106). Although the AB effect has been verified in other kinds of measurements, it has not yet been observed in gas phase reactions.

SPECIFIC SYSTEMS

$H + H_2$ and Isotopic Counterparts

TUNNELING   Tunneling in $H + H_2$ and isotopic counterparts was reviewed in 1986 (1), so I mainly discuss the few new results that have been published in the past two years.

During the past two years a burst of activity has occurred in theoretical studies of $H + H_2$ using accurate quantum methods (108–115), but all of these have been restricted to one or two partial waves and none have generated rate constants. What these new calculations do tell us is that the old results of Schatz & Kuppermann (60) on the PK2 surface are quite accurate, and those of Colton & Schatz (61) on LSTH are less accurate (103), though reasonably close near threshold. Thus the tunneling coefficients in Table 1 are close to being exact for the surfaces used. The accuracy of the LSTH surface has been confirmed by a variety of calculations (116–119), but a somewhat improved surface (DMBE) has also been developed (120).

New experimental results that show quantum effects have been obtained in three areas. First, Garner, Fleming and co-workers (121) have studied the reaction $Mu + H_2$ (where $Mu = \mu^+ e^-$ is an isotope of hydrogen with a mass 1/9 as large) at temperatures lower than in an earlier study (122) and have obtained rate coefficients in excellent agreement with accurate quantum calculations (123), and in somewhat less quantitative agreement with variational transition state theory, using a least action ground state (LAG) tunneling factor (72, 124). As might be expected, tunneling is quite significant in $Mu + H_2$ [κ for three-dimensional $Mu + H_2$ is 10.3 at 444K on the LSTH surface, as can be inferred by combining results from Refs. (123, 124)], but the light mass of the Mu causes the adiabatic barrier to be both high and flat, so the tunneling factor is only three to four times that for $H + H_2$ at the same temperature [see Refs. (61, 104)].

The second new experimental result refers to $H + H_2$ in solid matrices at extremely low temperatures ($T = 1-4K$), where Russian and Japanese (56, 57, 125–127) groups have measured decay rates for D in H_2, D in HD, H in H_2, and D in D_2 that can be interpreted as arising from bimolecular exchange reactions. A simple gas phase calculation (128) was able to reproduce the measured rates quite accurately, but this is partially fortuitous since solid state effects are expected to be significant. From the point of view of gas phase chemistry, these results are useful as they help define the low temperature limiting behavior of the thermal rate coefficient. This behavior is governed by the thermal rate coefficient versions of the Wigner threshold rules (128, 129), which predict temperature-independent rate coefficients for exothermic reactions, Arrhenius dependence on temperature for endothermic reactions, and a $T^{1/2}$ temperature dependence for thermoneutral reactions. For $H + H_2$ and isotopic counterparts, Wigner behavior holds for temperatures less than 10K (128).

The third new experimental result refers to the rate coefficient for $D + H_2(v = 1) \rightarrow HD + H$ (summed over product states), which has been remeasured by Wolfrum (130, 131). Table 2 shows the comparison of his new value, $(1.0 \pm 0.4) \times 10^{-13}$ cm^3/sec, with earlier theoretical estimates (34a,b, 132–145) based on QCT calculations (132–134, 140), various approximate quantum methods (135–137, 139, 141–144), and transition state theory (34a,b, 138, 145). The new value is in good agreement with many of the theoretical estimates, in particular with QCT and variational transition state theory. This situation is in stark contrast to that discussed in 1986 (10), when there were factor of 3 or more differences between theory and earlier experimental results (146, 147). The transition state theory results

Table 2 Rate coefficients (cm^3/sec × 10^{13}) for $H + H_2$ (1 → all) and $D + H_2$ (1 → all) at 300K

Method	$H + H_2$	$D + H_2$
A. PK2 surface		
QCT	4.0,[a] 6.1,[b] 8.4[c]	—
IOSA	5.6,[d] 4.0[e]	—
CEQB	2.6[f]	—
SCAD	1.8[g]	—
CS	1.7[h]	—
ICVT/LA	3.3[i]	2.9[i]
B. LSTH surface		
QCT	1.3[j]	1.7[j]
IOSA		2.9[k]
VADW		5.0[l]
BCRLM	4.8[m]	2.8[m]
SCAD	2.0[g]	2.1[g]
ICVT/LA	2.0[i]	2.1[i]
CEQB		0.9[n]
C. DMBE surface		
ICVT/LA	1.7[o]	1.6[o]
D. Experiment		1.0 ± 0.4[p]

References: [a] (132). [b] (133). [c] (134). [d] (135). [e] (136). [f] (137). [g] (138). [h] (139). [i] (34a,b). [j] (140). [k] (141). [l] (142). [m] (143). [n] (144). [o] (145). [p] (131).

describe tunneling using the VA potential, and they indicate that tunneling contributes a factor of 2 to the rate coefficient on the LSTH surface (34a,b). This result should be reliable, as the same VA reaction probability agrees with a recent accurate quantum calculation (148). The good agreement of the QCT results with experiment is due to a fortuitous cancellation of errors arising from tunneling and vibrationally nonadiabatic effects. This cancellation often but not always (123) occurs.

Two other recent experiments on the $D + H_2$ system include measurements of absolute and differential cross-sections for $D + H_2(v = 1)$ by Götting and co-workers (149, 150) and measurements of the angular and final vibrational state distributions for $D + H_2(v = 0)$ by Buntin and co-workers (151). Both measurements involve relatively high kinetic and/or internal energies, so QCT calculations should model the results accurately. This has been verified for the Götting experiment (149, 150).

RESONANCES Recently Nieh & Valentini (152) have presented the first experimental evidence for resonances in $H + H_2$. They measured the cross-section for $H + H_2(v = 0, para) \rightarrow H_2(v' = 1, j' = 1, 3) + H$ at total energies E between 0.95 eV and 1.37 eV, and found several distinct oscillations in the $j' = 1$ and 3 cross-sections and in the $v' = 1/v' = 0$ ratio. This behavior is quite different from earlier studies of $H + D_2$ by Valentini's and Zare's groups [reviewed in (153)], where the energy dependence of cross sections was smooth.

The significance of the Nieh & Valentini results is best appreciated by referring to Figure 2, which shows the predictions of a 3D quantum (coupled states) calculation by Colton & Schatz (154) for the LSTH surface in the 0.9–1.3 eV range. This figure presents the $v = 0, j = 0 \rightarrow v' = 1$ reaction probability for several partial waves J and body-fixed projections Ω. The $J = \Omega = 0$ probability shows a resonance at 0.98 eV that is the analogue of the 0.88 eV resonance in Figure 1, and can be labeled by the quantum numbers (10^00), corresponding to one quantum of symmetric stretch excitation, ground state bend, and ground state antisymmetric stretch. Other resonances seen in this figure include the excited bend states (01^10) at 1.10 eV in $J = \Omega = 1$, (02^00) at 1.20 eV in $J = \Omega = 0$, and (02^20) at 1.23 eV in $J = \Omega = 2$. Some or all of these results have been seen in other calculations (7, 8, 81, 108, 110, 155, 156), and it is known that the only resonances in H_3 over the energy range in Figure 2 are those in that figure.

The locations of the peaks found by Nieh & Valentini are indicated by the *arrows* in Figure 2, and we note that there is nearly perfect agreement with theory. This very exciting result needs to be viewed cautiously, however, for the experimental results refer to cross-sections whereas the

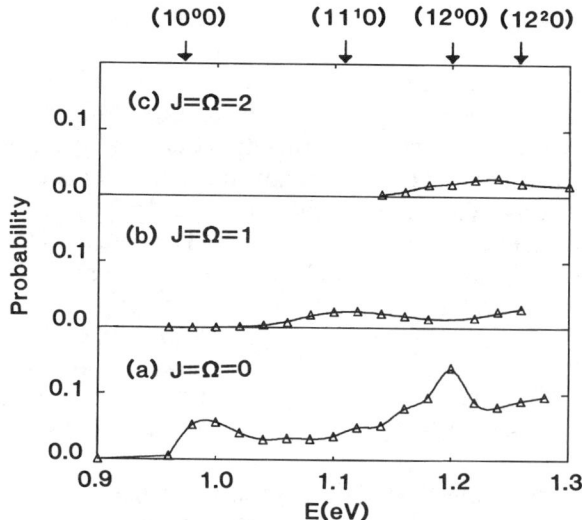

Figure 2 Reaction probability $P^R_{00 \to 1}$ ($v = j = 0$ to $v' = 1$, summed over j') versus E for 3D H+H$_2$ on LSTH potential surface. The total angular momentum quantum number J and its body fixed projection Ω (based on a coupled states calculation where Ω is conserved) have been selected to be 0 in the bottom panel, 1 in the middle, and 2 in the top (see Ref. 151 for details). Also plotted (*arrows* on top) are the positions and quantum number assignments of the observed peaks in the $v = 0 \to v' = 1$ cross-section as reported in Ref. (152).

theoretical results are for individual partial waves. Although only a few partial waves contribute to the $v' = 1$ cross-section, one theoretical calculation of $v' = 1$ cross-sections (139) found that the resonance peak was washed out in the partial wave sum. This calculation used the PK2 surface, so it will be important to repeat it on more accurate surfaces.

$O + H_2$ and Isotropic Counterparts

TUNNELING $O+H_2$ has been extensively studied for many years by both experimentalists and theoreticians, and much is known about thermal and vibrationally excited rate coefficients and isotope effects. The comparison of theory and experiment was last reviewed in 1981 (11), and since then a number of new calculations and measurements have been done that (*a*) place the experiment/theory comparison on a very solid footing, and (*b*) provide some of the clearest examples of experimentally distinguishable tunneling. These two points are discussed here. More general issues pertaining to the $O+H_2$ kinetics are reviewed elsewhere (157–160).

The improvements in the theory/experiment comparison have arisen from several studies. First, the thermal rate coefficient has been measured with higher accuracy at lower temperatures than in the past (160, 161) and for more isotopes (58). Second, very high quality abinitio calculations have been performed (162) that find a barrier (12.5 kcal/mol) that agrees closely with earlier empirically adjusted estimates (163, 164). Several new global surfaces have been developed (165, 166), and rate coefficients have been calculated with them using tunneling factors whose accuracy has been confirmed by comparison with distorted wave (30, 62) and accurate quantum calculations (167a,b, 168). The net effect of this is that on the best available surfaces, using accurate tunneling factors, the calculated rate coefficients and isotope effects are in excellent agreement with experiment (169, 170). Table 1 presents some representative tunneling factors, and we note that all are quite large.

Recently the OH/OD isotope ratio has been measured for the first time for O+HD (58). This, along with accurately measured $O+H_2/O+D_2$ isotope ratios (161), provides unambiguous indication as to the importance of tunneling. Both of these ratios increase with decreasing temperature, and at the lowest temperatures for which measurements have been done (339K) the values of these ratios are much larger than one can get either from variational transition state theory without a tunneling factor (58) or from trajectories (170). Simple theories of tunneling, such as the Wigner expression, also fail in this situation (171), unless the parameters in them are adjusted arbitrarily to fit the data (172).

$Cl+HCl$

RESONANCES The reaction Cl+HCl is an important prototype for (heavy)+(light)(heavy) reactions, and both its thermal (63, 64, 173) and state-resolved (174–176) kinetics have been studied in detail. Tunneling is of interest for this reaction (see Table 1), but much attention recently has been given to resonances because of an experiment by Metz et al (177), who appear to have observed them. In this experiment, the ClHCl$^-$ ion, which has a linear equilibrium geometry, is photodetached to give ClHCl. This neutral has no bound states (except for weak van der Waals states), so it dissociates to Cl+HCl, but the detached electron kinetic energy distribution shows several peaks, which Metz et al have interpreted as arising from ClHCl resonances. This interpretation is based on the similarity of the experimental resonance spectrum with one seen in collinear quantum calculations by Bondi et al (173).

Figure 3 plots the collinear reaction probability P_{00}^R, P_{11}^R and P_{22}^R for Cl+HCl as a function of energy (results from G. C. Schatz, unpublished, which reproduce those in Ref. 173). These probabilities show two types of oscillations: broad oscillations with peaks at 0.376, 0.502, and 1.17 eV, which are not due to resonances (instead they arise from systematic variation in the number of hops by the H atom between the two Cl's during

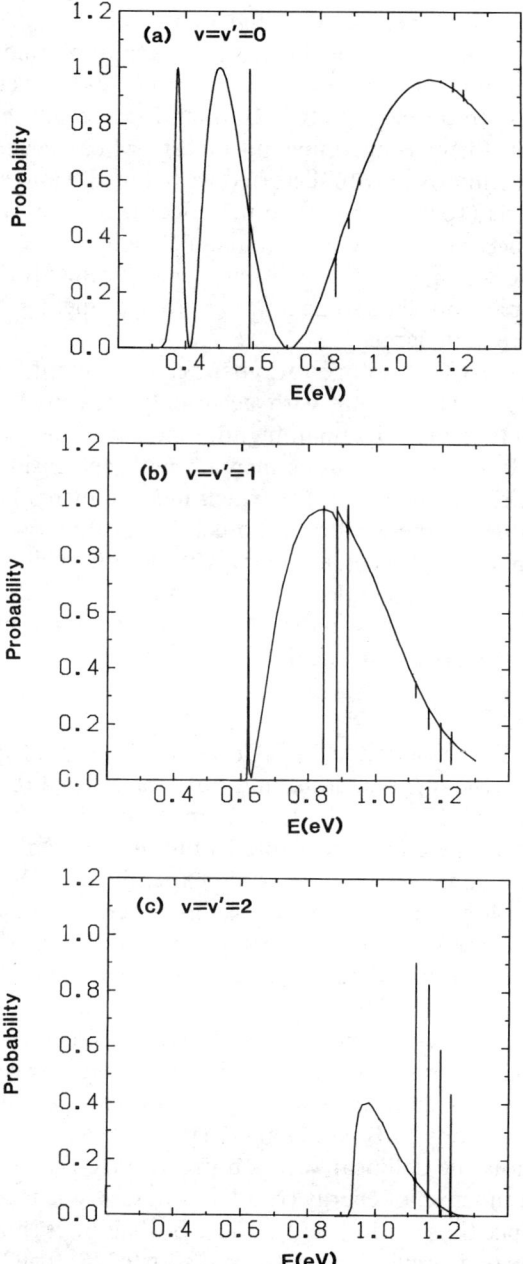

Figure 3 Reaction probability $P_{vv'}^R$ for (*a*) $v = v' = 0$, (*b*) $v = v' = 1$, and (*c*) $v = v' = 2$ versus E for collinear Cl+HCl, using potential surface of Bondi et al (173).

direct scattering), and sharp oscillations (spikes) at 0.588, 0.620, 0.845, 0.884, 0.916, 1.123, 1.162, 1.197, 1.226, and 1.232 eV. [The last of these spikes (1.232 eV) is not apparent in Figure 3 but does show up in other transition probabilities such as P_{33}^R.] The broad oscillations in Figure 3 are also seen in quasiclassical calculations, but the sharp oscillations are not. Bondi et al have shown that the sharp oscillations are due to resonances in which the Cl–H–Cl symmetric stretch mode v_3 is excited during reaction. The (asymmetric, symmetric) quantum numbers for these resonances are (0, 2), (1, 2), (0, 4), (1, 4), (2, 4), (0, 6), (1, 6), (2, 6), (3, 6), (4, 6), respectively. Notice that only positive even values of v_3 lead to resonances. In addition, the spacing between different symmetric stretch states is much larger than between different asymmetric stretch states, so the resonances come in three groups (at 0.60, 0.88, and 1.19 eV when averaged over v_1) corresponding to $v_3 = 2, 4$, and 6, respectively.

The peaks found by Metz et al occur at 0.45, 0.64, 0.86, and 1.13 eV on the same (total) energy scale as in Figure 3. The last three of these are in reasonable correspondence with where the $v_3 = 2, 4$, and 6 resonances are located. The first peak is apparently due to nonresonant photodetachment to the lowest VA potential (one that does not support a resonance), and it corresponds closely with where the threshold for reaction occurs in Figure 3. The excellent agreement between theory and experiment is probably somewhat fortuitous, since the threshold and resonance locations in three dimensions should be shifted by bending zero point energy (~ 0.06 eV) relative to one dimension. The potential surface used by Bondi et al is also known to be somewhat inaccurate (45), so perfect agreement is not expected. The experimental spectra do show sensible isotope effects when ClDCl$^-$ is used in place of ClHCL$^-$, so it seems likely that interpretation of the experiment as due to resonances is correct. Also, experiments on IHI$^-$, IHBr$^-$, BrHBr$^-$, and FHCl$^-$ show similar structure (178).

$H + CO$

RESONANCES   The H+CO collision system has received a great deal of attention in recent years, and from a variety of perspectives. Hot H atoms with 1–3 eV translational energy have been used to study vibration/rotation excitation (179–182), and a number of trajectory (182–184) and quantum scattering studies (185) of this have been done. Thermal addition and dissociation in the H+CO $\rightleftarrows$ HCO system has been studied both experimentally (157, 159, 186–190) and theoretically (190, 191) because of its importance in combustion. One important result of the theoretical analysis of addition/dissociation (191) is that tunneling must be excluded from an activated complex/RRKM treatment in order to define the reaction threshold properly. This somewhat disturbing conclusion suggests that refinements to the standard model are needed, and some work in this direction has been done (192). At very low temperatures there is interest in collisional energy transfer in the H+CO system because of its relevance in astrophysics (193–198). Finally, a recent electron photodetachment study of HCO^- (199) has produced detailed spectra of HCO in both bound and predissociative (i.e. resonant) states. This spectrum has been successfully interpreted theoretically through vibrational bound state calculations (200, 201) on a globally determined ab initio surface (184, 200).

Although the photodetachment spectrum provides the only direct measurement of resonances in H+CO to date, several of the other measurements probe them directly or indirectly. One possible direct measurement that has not been done would involve doing the hot atom experiments as a function of translational energy. Although there do not seem to be resonances associated with the HCO species in the 1–3 eV energy range (185), there are resonances associated with the metastable species HOC (185, 200), and by analogy with the $H+H_2$ resonance measurements (152) one might expect changes in the CO internal state distributions at energies where such resonances occur.

One experiment that is indirectly sensitive to the HCO resonances is thermal dissociation of HCO (190). Because there are only 15 bound states in nonrotating HCO (200), the density of states of HCO at its dissociation threshold is too small ($\sim 0.005/cm^{-1}$) for standard continuum theory implementations of RRKM theory to be correct. Wagner & Bowman (191) have developed a discretized theory that sums the contribution of each resonance to the dissociation rate. Results consistent with experiment have been obtained (190, 191).

CONCLUSION

Let us conclude by comparing the present status of quantum effects in chemical reactions with that 10 years ago and with what might be in the future. Ten years ago, tunneling was quantitatively understood only for $H+H_2$ (9), and the tunneling theories that were being used in the context of transition state theory were very inaccurate. However, the Marcus-Coltrin theory of tunneling had just been published (42) and accurate quantum dynamical studies of chemical reactions in three dimensions were starting to appear (60). Resonances had been seen in a number of theoretical calculations (68–71, 81), but the only experimental results were Lee's early data (83) on $F+H_2$.

Today we have a variety of high quality theoretical and experimental studies that show the importance of tunneling on both ground and excited state reaction rates and on isotope effects. This review has been confined to $H+H_2$, $O+H_2$, $Cl+HCl$, and $H+CO$, but other examples exist and have been reviewed (1, 2). Also, the importance of vibrational adiabaticity in determining reaction thresholds has been established, and a large number of accurate quantum methods in both full and reduced dimensionality exist for calculating quantitative tunneling factors. Resonances associated with the transition state region of reactive potential surfaces have been observed for several reactions just within the past two years and a large number of quantitative theoretical methods are now available for studying them accurately. At the same time, however, the reactions being studied are still among the simplest of chemical reactions. The sophisticated understanding that we have of these simple systems has only occasionally been used for more complex reactions (202, 203).

The future looks very promising for further experimental and theoretical discoveries in the area of quantum effects in chemical reactions. The types of experiments that led to the discovery of resonances in H_3 and ClHCl can be generalized to many other systems, and new classes of experiments such as those on precursor geometry limited reactions hold great promise for unlocking new features of quantum reaction dynamics. Advances in computer technology will provide continually improving theoretical capabilities for generating potential surfaces and doing quantum reaction dynamics calculations, and we may also see the emergence of new theoretical methods based on wavepackets (see, e.g. 204, 205) and path integrals (reviewed in 1, 206) that will play important roles in generating new discoveries.

ACKNOWLEDGMENTS

This research was supported by NSF Grant CHE-8715581.

Literature Cited

1. Schatz, G. C. 1987. *Chem. Rev.* 87: 81–89
2. Miller, W. H. 1987. *Chem. Rev.* 87: 19–28
3. Truhlar, D. G., Garrett, B. C. 1981. *Acc. Chem. Res.* 13: 440–48
4. Truhlar, D. G., Hase, W. L., Hynes, J. T. 1983. *J. Phys. Chem.* 87: 2664–82, 5523 (erratum)
5. Truhlar, D. G., Garrett, B. C. 1984. *Ann. Rev. Phys. Chem.* 35: 150–89
6. Truhlar, D. G., Isaacson, A. D., Garrett, B. C. 1985. In *Theory of Chemical Reaction Dynamics*, ed. M. Baer, 4: 65–137. Boca Raton: CRC
7. Kuppermann, A. 1981. In *Potential Energy Surfaces and Reaction Dynamics Calculations*, ed. D. G. Truhlar, pp. 375–420. New York: Plenum
8. Truhlar, D. G., ed. 1984. *Resonances in Electron-Molecule Scattering, Van der Waals Complexes and Reactive Chemical Dynamics Calculations, ACS Symp. Ser.* 263. Washington, DC: Am. Chem. Soc. 522 pp.
9. Truhlar, D. G., Wyatt, R. E. 1976. *Ann. Rev. Phys. Chem.* 27: 1–43
10. Schatz, G. C. 1986. In *Theory of Chemical Reaction Dynamics*, ed. D. C. Clary, pp. 1–26. Dordrecht: Reidel
11. Schatz, G. C. 1981. See Ref. 7, pp. 287–310
12. Connor, J. N. L. 1979. *Comput. Phys. Commun.* 17: 117–44

13. Bowman, J. M. 1985. *Adv. Chem. Phys.* 61: 115–67
14. Jellinek, J., Kouri, D. J. 1985. In *Theory of Chemical Reaction Dynamics*, ed. M. Baer, 1: 1. Boca Raton: CRC
15. Miller, W. H. 1975. *Adv. Chem. Phys.* 30: 77–136
16. Anderson, J. B. 1980. *Adv. Chem. Phys.* 41: 229–68
17. Shoemaker, C. L., Wyatt, R. E. 1981. *Adv. Quant. Chem.* 14: 169–240
18. Levine, R. D., Bernstein, R. B. 1987. *Molecular Reaction Dynamics and Chemical Reactivity*. New York: Oxford Univ. Press. 535 pp.
19. Kollman, P. 1987. *Ann. Rev. Phys. Chem.* 38: 303–16
20. Miller, W. H., Handy, N. C., Adams, J. E. 1980. *J. Chem. Phys.* 72: 99–112
21. Marcus, R. A., 1966. *J. Chem. Phys.* 45: 4493–4504
22. Garrett, B. C., Truhlar, D. G. 1979. *J. Phys. Chem.* 83: 1052–1112, 3058(E)
23. Garrett, B. C., Truhlar, D. G. 1980. *J. Phys. Chem.* 84: 805–12
24. Garrett, B. C., Truhlar, D. G., Grev, R. S., Magnuson, A. W. 1980. *J. Phys. Chem.* 84: 1730–52
25. Garrett, B. C., Truhlar, D. G. 1984. *J. Chem. Phys.* 81: 309–17
26. Carrington, T., Miller, W. H. 1984. *J. Chem. Phys.* 81: 3942–50
27. Carrington, T., Miller, W. H. 1986. *J. Chem. Phys.* 84: 4364–70
28. Schatz, G. C. 1983. *J. Chem. Phys.* 79: 5386–91
29. Abu-Salbi, N., Kouri, D. J., Baer, M., Pollak, E. 1985. *J. Chem. Phys.* 82: 4500–8
30. Garrett, B. C., Truhlar, D. G., Schatz, G. C. 1986. *J. Am. Chem. Soc.* 108: 2876–81
31. Shoemaker, C. L., Wyatt, R. E. 1982. *J. Chem. Phys.* 77: 4994–5008
32. Pollak, E., Wyatt, R. E. 1984. *Chem. Phys. Lett.* 110: 340–45
33. Garrett, B. C., Truhlar, D. G., Bowman, J. M., Wagner, A. F. 1986. *J. Phys. Chem.* 90: 4305–11
34a. Garrett, B. C., Truhlar, D. G. 1985. *J. Phys. Chem.* 89: 2204–8
34b. Connor, J. N. L., Jakubetz, W., Lagana, A. 1979. *J. Phys. Chem.* 83: 73–78
35. Karplus, M., Porter, R. N., Sharma, R. D. 1965. *J. Chem. Phys.* 43: 3259–87
36. Porter, R. N. 1974. *Ann. Rev. Phys. Chem.* 25: 317–55
37. Bunker, D. L. 1971. *Methods Comput. Phys.* 10: 287
38. Porter, R. N., Raff, L. M., Miller, W. H. 1975. *J. Chem. Phys.* 63: 2214–18
39. Schatz, G. C. 1983. In *Molecular Collision Dynamics*, ed. J. M. Bowman, *Top. Current Phys.* 25: 25–60. Heidelberg: Springer
40. Bowman, J. M., Kuppermann, A., Adams, J. T., Truhlar, D. G. 1973. *Chem. Phys. Lett.* 20: 229–32
41a. Weston, R. E. Jr., Schwarz, H. A. 1972. *Chemical Kinetics*. Englewood Cliffs, NJ: Prentice-Hall. 274 pp.
41b. Bell, R. P. 1980. *The Tunnel Effect in Chemistry*. London: Chapman & Hall
42. Marcus, R. A., Coltrin, M. E. 1977. *J. Chem. Phys.* 67: 2609–13
43. Miller, W. H. 1982. *J. Chem. Phys.* 76: 4904–8
44. Skodje, R. T., Truhlar, D. G., Garrett, B. C. 1981. *J. Phys. Chem.* 85: 3019–23
45. Garrett, B. C., Truhlar, D. G., Wagner, A. F., Dunning, T. H. 1983. *J. Chem. Phys.* 78: 4400–13
46. Babamov, V. K., Lopez, V., Marcus, R. A. 1983. *J. Chem. Phys.* 78: 5621–28
47. Garrett, B. C., Truhlar, D. G. 1983. *J. Chem. Phys.* 79: 4931–38
48a. Hiller, C., Manz, J., Miller, W. H., Römelt, J. 1983. *J. Chem. Phys.* 78: 3850–56
48b. Connor, J. N. L. 1976. *Chem. Soc. Rev.* 5: 125–48
49. Benson, S. W. 1976. *Thermochemical Kinetics*. New York: Wiley. 320 pp. 2nd ed.
50. Applebury, M. L., Peters, K. S., Rentzepis, P. M. 1978. *Biophys. J.* 23: 375–82
51. Ippen, E. P., Shank, C. V., Lewis, A., Markus, M. A. 1978. *Science* 200: 1279–81
52. Gochev, A. D. 1985. *Int. Agrophys.* 1: 121–33
53. Doba, T., Ingold, K. U., Siebrand, W., Wildman, T. A. 1984. *J. Phys. Chem.* 88: 3165–67
54. Gomer, R. 1973. *Surf. Sci.* 38: 373–93
55. Mitchell, N., LeRoy, D. J. 1973. *J. Chem. Phys.* 58: 3449–53
56. Katunin, A. Ya., Lukashevich, I. I., Orozmamatov, S. T., Sklyarevskii, V. V., Suraev, V. V., Filipov, V. V., Filipov, N. I., Shevtsov, V. A. 1982. *Sov. J. Low Temp. Phys.* 8: 240–42
57. Miyazaki, T., Lee, K.-P. 1986. *J. Phys. Chem.* 90: 400–7
58. Robie, D. C., Arepalli, S., Presser, N., Kitsopoulos, T., Gordon, R. J. 1987. *Chem. Phys. Lett.* 134: 579–82
59. Clary, D. C. 1985. *J. Chem. Phys.* 83: 1685–92
60. Schatz, G. C., Kuppermann, A. 1976. *J. Chem. Phys.* 65: 4668–92

61. Colton, M. C., Schatz, G. C. 1986. *Int. J. Chem. Kinet.* 18: 961–75
62. Schatz, G. C. 1985. *J. Chem. Phys.* 83: 5677–86
63. Schatz, G. C., Amaee, B., Connor, J. N. L. 1988. *J. Phys. Chem.* 92: 3190–95
64. Neumark, D. M., Wodtke, A. M., Robinson, G. N., Hayden, C. C., Lee, Y. T. 1985. *J. Chem. Phys.* 82: 3045–66
65. Neumark, D. M., Wodtke, A. M., Robinson, G. N., Hayden, C. C., Shobatake, K., Sparks, R. K., Schafer, T. P., Lee, Y. T. 1985. *J. Chem. Phys.* 82: 3067–77
66. Radhakrishnan, G., Buelow, S., Wittig, C. 1986. *J. Chem. Phys.* 84: 727–38
67. Hulbert, H. M., Hirschfelder, J. O. 1943. *J. Chem. Phys.* 11: 276–90
68. Truhlar, D. G., Kuppermann, A. 1972. *J. Chem. Phys.* 56: 2232–52
69. Wu, S.-F., Levine, R. D. 1971. *Mol. Phys.* 22: 881–97
70. Schatz, G. C., Kuppermann, A. 1973. *J. Chem. Phys.* 59: 964–65
71. Schatz, G. C., Bowman, J. M., Kuppermann, A. 1975. *J. Chem. Phys.* 63: 674–84
72. Bondi, D. K., Clary, D. C., Connor, J. N. L., Garrett, B. C., Truhlar, D. G. 1982. *J. Chem. Phys.* 76: 4986–95
73a. Walker, R. B., Stechel, E. B., Light, J. C. 1978. *J. Chem. Phys.* 69: 2922–23
73b. Bondi, D. K., Connor, J. N. L. 1985. *J. Chem. Phys.* 82: 4383–84
74. Garrett, B. C., Schwenke, D. W., Skodje, R. T., Thirumalai, D., Thompson, T. C., Truhlar, D. G. 1984. See Ref. 8, pp. 375–400
75. Kaye, J. A., Kuppermann, A. 1981. *Chem. Phys. Lett.* 77: 573–79
76. Römelt, J. 1986. See Ref. 10, pp. 77–104
77. Römelt, J. 1983. *Chem. Phys.* 79: 197–209
78. Aquilanti, V. 1986. See Ref. 10, pp. 383–413
79. Micha, D., Kuruoglu, Z. C. 1984. See Ref. 8, pp. 401–19
80. Schwenke, D. W., Truhlar, D. G. 1987. *J. Chem. Phys.* 87: 1095–1106
81. Schatz, G. C., Kuppermann, A. 1975. *Phys. Rev. Lett.* 35: 1266–69
82. Redmon, M. J., Wyatt, R. E. 1979. *Chem. Phys. Lett.* 63: 209–12
83. Sparks, R. K., Hayden, C. C., Shobatake, K., Neumark, D. M., Lee, Y. T. 1980. In *Horizons in Quantum Chemistry*, ed. K. Fukui, B. Pullman, pp. 91–105. Dordrecht: Reidel
84. McNutt, J. F., Wyatt, R. E., Redmon, M. J. 1984. *J. Chem. Phys.* 81: 1692–1703
85. McNutt, J. F., Wyatt, R. E., Redmon, M. J. 1984. *J. Chem. Phys.* 81: 1704–15
86. Zhang, J. Z. H., Abu Salbi, N., Baer, M., Kouri, D. J., Jellinek, J. 1984. See Ref. 8, pp. 457–77
87. Abu Salbi, N., Shoemaker, C. L., Kouri, D. J., Jellinek, J., Baer, M. 1984. *J. Chem. Phys.* 80: 3210–22
88. Shoemaker, C. L., Kouri, D. J., Jellinek, J., Baer, M. 1983. *Chem. Phys. Lett.* 94: 359–62
89. Hayes, E. F., Walker, R. B. 1984. See Ref. 8, pp. 493–513
90a. Muckerman, J. T. 1981. *Theor. Chem. Adv. Perpect.* A6: 1
90b. Jakubetz, W., Connor, J. N. L. 1977. *Faraday Discuss. Chem. Soc.* 62: 324–25
91. Truhlar, D. G., Garrett, B. C., Blais, N. C. 1984. *J. Chem. Phys.* 80: 232–40
92. Brown, F. B., Steckler, R., Schwenke, D. W., Truhlar, D. G., Garrett, B. C. 1985. *J. Chem. Phys.* 82: 188–201
93. Schwenke, D. W., Steckler, R., Brown, F. B., Truhlar, D. G. 1987. *J. Chem. Phys.* 86: 2443–44
94a. Bauschlicher, C. W., Taylor, P. R. 1987. *J. Chem. Phys.* 86: 858–61
94b. Zhang, J. Z. H., Miller, W. H. 1988. *J. Chem. Phys.* 88: 4549–50
94c. Takayanagi, T., Sato, S. 1988. *Chem. Phys. Lett.* In press
95. Gibson, L. L., Schatz, G. C., Ratner, M. A., Davis, M. J. 1987. *J. Chem. Phys.* 86: 3263–72
96. Brown, R. C., Wyatt, R. E. 1986. *Phys. Rev. Lett.* 57: 1–4
97. Brown, R. C., Wyatt, R. E. 1986. *J. Phys. Chem.* 90: 3590–99
98. Kuppermann, A., Schatz, G. C., Baer, M. 1976. *J. Chem. Phys.* 65: 4596–4623
99. Doll, J. D., George, T. F., Miller, W. H. 1973. *J. Chem. Phys.* 58: 1343–51
100. Miller, W. H. 1969. *J. Chem. Phys.* 50: 407–18
101. Saxon, R. P., Light, J. C. 1972. *J. Chem. Phys.* 56: 3885–95
102. Choi, B. H., Poe, R. T., Sun, J. C., Tang, K. T. 1983. *J. Chem. Phys.* 78: 5590–5605
103. Schultz, W. R., LeRoy, D. J. 1965. *J. Chem. Phys.* 42: 3869–73
104. Garrett, B. C., Truhlar, D. G. 1979. *Proc. Natl. Acad. Sci. USA* 76: 4755–59
105. Mead, C. A. 1980. *J. Chem. Phys.* 72: 3839–40
106. Mead, C. A., Truhlar, D. G. 1979. *J. Chem. Phys.* 70: 2284–96

107. Aharonov, Y., Bohm, D. 1959. *Phys. Rev.* 115: 485–91
108. Webster, F., Light, J. C. 1986. *J. Chem. Phys.* 85: 4744–45
109. Pack, R. T, Parker, G. A. 1987. *J. Chem. Phys.* 87: 3888–3921
110. Parker, G. A., Pack, R. T, Archer, B. J., Walker, R. B. 1987. *Chem. Phys. Lett.* 137: 564–68
111. Kuppermann, A., Hipes, P. G. 1986. *J. Chem. Phys.* 84: 5962–64
112. Schwenke, D. W., Haug, K., Truhlar, D. G., Sun, Y., Zhang, J. Z. H., Kouri, D. J. 1987. *J. Phys. Chem.* 91: 6080–82
113a. Miller, W. H., Jansen op de Haar, B. M. D. D. 1987. *J. Chem. Phys.* 86: 6213–20
113b. Zhang, J. Z. H., Miller, W. H. 1987. *Chem. Phys. Lett.* 140: 329–37
113c. Zhang, J. Z. H., Chu, S.-I., Miller, W. H. 1988. *J. Chem. Phys.* 88: 6233–39
114. Baer, M. 1987. *J. Phys. Chem.* 91: 5846–47
115. Linderberg, J. 1986. *Int. J. Quant. Chem. Symp.* 19: 467–76
116. Liu, B. 1984. *J. Chem. Phys.* 80: 581
117. Ceperley, D. M., Alder, B. J. 1984. *J. Chem. Phys.* 81: 5833–44
118. Barnett, R. N., Reynolds, P. J., Lester, W. A. 1985. *J. Chem. Phys.* 82: 2700–7
119. Blomberg, M. R. A., Liu, B. 1985. *J. Chem. Phys.* 82: 1050–51
120. Varandas, A. J. C., Brown, F. B., Mead, C. A., Truhlar, D. G., Blais, N. C. 1987. *J. Chem. Phys.* 86: 6258–69
121. Reid, I. D., Garner, D. M., Lee, L. Y., Senba, M., Arseneau, D. J., Fleming, D. G. 1987. *J. Chem. Phys.* 86: 5578–83
122. Garner, D. M., Fleming, D. G., Mikula, R. J. 1982. *Chem. Phys. Lett.* 121: 80–88
123. Schatz, G. C. 1985. *J. Chem. Phys.* 83: 3441–48
124. Blais, N. C., Truhlar, D. G., Garrett, B. C. 1983. *J. Chem. Phys.* 78: 2363–67
125. Iskovskikh, A. S., Katunin, A. Ya., Lukashevich, I. I., Sklyarevskii, V. V., Shevtsov, V. A. 1985. *JETP Lett.* 42: 30–34
126. Miyazaki, T., Lee, K.-P., Fueki, K., Takeuchi, A. 1984. *J. Phys. Chem.* 88: 4959–63
127. Lee, K.-P., Miyazaki, T., Fueki, K., Gotoh, K. 1987. *J. Phys. Chem.* 91: 180–82
128. Takayanagi, T., Nobuyuki, M., Nakamura, K., Okamoto, M., Sato, S., Schatz, G. C. 1987. *J. Chem. Phys.* 86: 6133–39
129. Schwenke, D. W., Truhlar, D. G. 1985. *J. Chem. Phys.* 83: 3454–61
130. Wolfrum, J. 1987. *Faraday Discuss. Chem. Soc.* 84: In press
131. Dreier, Th., Wolfrum, J. 1986. *Int. J. Chem. Kinet.* 18: 919–35
132. Mayne, H. R. 1979. *Chem. Phys. Lett.* 66: 487–92
133. Gordon, E. B., Ivanov, B. I., Perminov, A. P., Balalev, V. E., Ponomarev, A. V., Filatov, V. V. 1978. *Chem. Phys. Lett.* 58: 425–30
134. Osherov, V. I., Ushakov, V. G., Lomakin, L. A. 1978. *Chem. Phys. Lett.* 55: 513–14
135. Bowman, J. M., Lee, K.-T. 1979. *Chem. Phys. Lett.* 64: 291–94
136. Khare, V., Kouri, D. J., Jellinek, J., Baer, M. 1981. See Ref. 7, pp. 475–93
137. Bowman, J. M., Ju, G.-Z., Lee, K.-T. 1982. *J. Phys. Chem.* 86: 2232–39
138. Pollak, E. 1985. *J. Chem. Phys.* 82: 106–12
139. Schatz, G. C. 1983. *Chem. Phys. Lett.* 94: 183–87
140. Mayne, H. R., Toennies, J. P. 1981. *J. Chem. Phys.* 75: 1794–1803
141. Abu Salbi, N., Kouri, D. J., Shima, Y., Baer, M. 1985. *J. Chem. Phys.* 82: 2650–61
142. Sun, J. C., Choi, B. H., Poe, R. T., Tang, K. T. 1980. *Phys. Rev. Lett.* 44: 1211–14
143. Walker, R. B., Hayes, E. F. 1983. *J. Phys. Chem.* 87: 1255–63
144. Bowman, J. M., Lee, K.-T., Walker, R. B. 1983. *J. Chem. Phys.* 79: 3742–45
145. Garrett, B. C., Truhlar, D. G., Varandas, A. J. C., Blais, N. C. 1986. *Int. J. Chem. Kinet.* 18: 1065–77
146. Glass, G. P., Chaturvedi, B. K. 1982. *J. Chem. Phys.* 77: 3478–84
147. Rozenshtein, V. B., Gershenzon, Yu. M., Ivanov, A. V., Kucheryavii, S. I. 1984. *Chem. Phys. Lett.* 105: 423–26
148. Haug, K., Schwenke, D. W., Shima, Y., Truhlar, D. G., Zhang, J. Z. H., Kouri, D. J. 1986. *J. Phys. Chem.* 90: 6757–59
149. Götting, R., Herrero, V., Toennies, J. P., Vodegel, M. 1987. *Chem. Phys. Lett.* 137: 524–32
150. Götting, R., Toennies, J. P., Vodegel, M. 1986. *Int. J. Chem. Kinet.* 18: 949–60
151. Buntin, S. A., Giese, C. F., Gentry, W. R. 1987. *J. Chem. Phys.* 87: 1443–45
152. Nieh, J.-C., Valentini, J. J. 1988. *Phys. Rev. Lett.* 60: 519
153. Flynn, G. W., Weston, R. E. Jr. 1986. *Ann. Rev. Phys. Chem.* 37: 551–85
154. Colton, M. C., Schatz, G. C. 1986. *Chem. Phys. Lett.* 124: 256–59

155. Bowman, J. M. 1986. *Chem. Phys. Lett.* 124: 260–63
156. Hipes, P. G., Kuppermann, A. 1987. *Chem. Phys. Lett.* 133: 1–7
157. Baulch, D. L., Drysdale, D. D., Horne, D. G. 1973. In *Evaluated Kinetic Data for High Temperature Reactions*, Vol. 1. London: Butterworth. 433 pp.
158. Cohen, N., Westberg, K. R. 1983. *J. Phys. Chem. Ref. Data* 12: 531–90
159. Warnatz, J. 1984. In *Combustion Chemistry*, ed. W. C. Gardiner, Chap. 5. New York: Springer-Verlag
160. Marshall, P., Fontijn, A. 1987. *J. Chem. Phys.* 87: 6988–94
161. Presser, N., Gordon, R. J. 1985. *J. Chem. Phys.* 82: 1291–97
162. Walch, S. P. 1987. *J. Chem. Phys.* 86: 5670–75
163. Lee, K. T., Bowman, J. M., Wagner, A. F., Schatz, G. C. 1982. *J. Chem. Phys.* 76: 3563–82
164. Johnson, B. R., Winter, N. W. 1977. *J. Chem. Phys.* 66: 4116–20
165. Garrett, B. C., Truhlar, D. G. 1984. *J. Chem. Phys.* 81: 309–17
166. Garrett, B. C., Truhlar, D. G. 1986. *Int. J. Quant. Chem.* 29: 1463–82
167a. Haug, K., Schwenke, D. W., Truhlar, D. G., Zhang, Y., Zhang, J. Z. H., Kouri, D. J. 1987. *J. Chem. Phys.* 87: 1892–94
167b. Zhang, J. Z. H., Zhang, Y., Kouri, D. J., Garrett, B. C., Haug, K., Schwenke, D. W., Truhlar, D. G. 1987. *Faraday Discuss. Chem. Soc.* 84: In press
168. Bowman, J. M. 1987. *Chem. Phys. Lett.* 141: 545–47
169. Garrett, B. C., Truhlar, D. G., Bowman, J. M., Wagner, A. F., Robie, D., Arepalli, S., Presser, N. 1986. *J. Am. Chem. Soc.* 108: 3515–16
170. Bowman, J. M., Wagner, A. F. 1987. *J. Chem. Phys.* 86: 1967–75
171. Bowman, J. M., Wagner, A. F. 1987. *J. Chem. Phys.* 86: 1976–81
172. Furue, H., Pacey, P. D. 1987. *J. Phys. Chem.* 91: 4132–37
173. Bondi, D. K., Connor, J. N. L., Manz, J., Römelt, J. 1983. *Mol. Phys.* 50: 467–88
174. Schatz, G. C., Amaee, B., Connor, J. N. L. 1986. *Chem. Phys. Lett.* 132: 1–5
175. Schatz, G. C., Amaee, B., Connor, J. N. L. 1987. *Comp. Phys. Commun.* 47: 45–53
176. Amaee, B., Connor, J. N. L., Whitehead, J. C., Jakubetz, W., Schatz, G. C. 1987. *Faraday Discuss. Chem. Soc.* 84: 387–403
177. Metz, R. B., Kitsopoulos, T., Weaver, A., Neumark, D. M. 1988. *J. Chem. Phys.* 88: 1463–65
178. Kitsopoulos, T., Metz, R. B., Weaver, A., Neumark, D. M. 1988. In *Proc. 1987 Int. Laser Sci. Conf., Atlantic City, Oct. 1987. (Adv. Laser Sci.* Vol. 3). Am. Inst. Phys. In press
179. Wood, C. F., Flynn, G. W., Weston, G. W. 1982. *J. Chem. Phys.* 77: 4776–77
180. Wight, C. A., Leone, S. R. 1983. *J. Chem. Phys.* 78: 4875–86
181. Wight, C. A., Leone, S. R. 1983. *J. Chem. Phys.* 79: 4823–29
182. Chawla, G. K., McBane, G. C., Houston, P. L., Schatz, G. C. 1988. *J. Chem. Phys.* 88: 5481–88
183. Geiger, L. C., Schatz, G. C. 1984. *J. Phys. Chem.* 88: 214–21
184. Geiger, L. C., Schatz, G. C., Harding, L. B. 1985. *Chem. Phys. Lett.* 114: 520–25
185. Geiger, L. C., Schatz, G. C., Garrett, B. C. 1984. See Ref. 8, pp. 421–40
186. Hucknall, D. J. 1985. *Chemistry of Hydrocarbon Combustion*. New York: Chapman & Hall. 304 pp.
187. Hikida, T., Iyre, J. A., Dorfman, L. M. 1971. *J. Chem. Phys.* 54: 3422–28
188. Wang, H. Y., Eyre, J. A., Dorfman, L. M. 1973. *J. Chem. Phys.* 59: 5199–5200
189. Ahumada, J. J., Michael, J. V., Osborne, D. T. 1972. *J. Chem. Phys.* 57: 3736–45
190. Timonen, R. S., Ratajczak, E., Gutman, D., Wagner, A. F. 1987. *J. Phys. Chem.* 91: 5325–32
191. Wagner, A. F., Bowman, J. M. 1987. *J. Phys. Chem.* 91: 5314–24
192. Bowman, J. M. 1986. *J. Phys. Chem.* 90: 3492–95
193. Caracciolo, G., Ellis, T. H., Este, G. O., Ruffolo, A., Scoles, G., Valbura, V. 1979. *Astrophys. J.* 229: 451–54
194. Chu, S., Dalgarno, A. 1975. *Proc. R. Soc. London Ser. A* 342: 191–207
195. Green, S., Thaddeus, P. 1976. *Astrophys. J.* 205: 766–85
196. Romanowski, H., Lee, K.-T., Bowman, J. M., Harding, L. B. 1986. *J. Chem. Phys.* 84: 4888–93
197. Lee, K.-T., Bowman, J. M. 1987. *J. Chem. Phys.* 86: 216–25
198. Lee, K.-T., Bowman, J. M. 1986. *J. Chem. Phys.* 85: 6225–26
199. Murray, K. K., Miller, T. M., Leopold, D. G., Lineberger, W. C. 1986. *J. Chem. Phys.* 84: 2520–25
200. Bowman, J. M., Bittman, J. S., Harding, L. B. 1986. *J. Chem. Phys.* 85: 911–21
201. Christoffel, K. M., Bowman, J. M., Bittman, J. S. 1987. *Chem. Phys. Lett.* 133: 525–30
202. Kreevoy, M. M., Ostovic, D., Truhlar,

D. G., Garrett, B. C. 1986. *J. Phys. Chem.* 90: 3766–74
203. Kreevoy, M. M., Truhlar, D. G. 1986. In *Rates and Mechanisms of Reactions*, ed. C. F. Bernsaconi, Chap. 1. New York: Wiley. 4th ed.
204. Kouri, D. J., Mowrey, R. C. 1987. *J. Chem. Phys.* 86: 2087–94
205. Heather, R., Metiu, H. 1987. *J. Chem. Phys.* 86: 5009–17
206. Gubernatis, J. E., ed. 1986. *Proc. Conf. Frontiers of Quantum Monte Carlo, Los Alamos. J. Stat. Phys.* 43: 729–1238

THE NATURE OF SIMPLE PHOTODISSOCIATION REACTIONS IN LIQUIDS ON ULTRAFAST TIME SCALES[1]

A. L. Harris

AT&T Bell Laboratories, Murray Hill, New Jersey 07974

J. K. Brown and C. B. Harris

Department of Chemistry, University of California at Berkeley, and Materials and Chemical Sciences Division of Lawrence Berkeley Laboratory, Berkeley, California 94720

INTRODUCTION

The photodissociation and recombination of iodine in liquids has been studied for 50 years as a simple example of a bimolecular solution reaction. Franck & Rabinowitch first proposed in 1934 that solvents could trap the initial atom pair (the *geminate* pair) from a photodissociated molecule and force them to recombine in a process they termed the "cage effect" (1–3). Noyes and co-workers carried out quantitative studies in the 1950s of the geminate recombination yield as a function of the solvent viscosity and wavelength of the photodissociating light and estimated the nongeminate recombination rate (4–8). Extensive studies of atom recombination in the gas phase have been done (9), and the density dependence of geminate recombination yields and *non*geminate recombination rates have been investigated in fluids across the density range from gases to liquids (10–14).

In spite of this history, an accurate microscopic picture of geminate atom recombination did not emerge until very recently. Careful picosecond

[1] The US Government has the right to retain a nonexclusive, royalty-free license in and to any copyright covering this paper.

spectroscopy and several important theoretical reaction simulations have been required to elucidate the reaction steps and their rates. In this review, we describe the dramatic experimental progress of the past five years, and we relate the experimental studies to reaction models of geminate atom recombination and theoretical approaches to such reactions. One theme that emerges from the recent picosecond studies is the surprising complexity of iodine geminate atom recombination. A detailed picture is necessary to understand the actual dynamics of the recombination process. This picture takes account of the multiple electronic potential surfaces involved in any recombination process, and of the molecular energy relaxation that must accompany a highly energetic reaction.

The possible steps in a solution photolysis event are outlined in Figure 1 on a set of potential energy surfaces similar to those of iodine. Light absorption by the ground state molecule populates a dissociative or a predissociative electronic level, causing the atoms to separate on a repulsive potential. The separation may be delayed somewhat if a predissociative level is populated. The separating atoms encounter solvent molecules, and may be forced to recombine or may escape. We can say that recombination has occurred when a molecule is reformed and stabilized to some level, for instance $k_B T$, below the dissociation limit. Recombination may occur within a few collisions, or alternatively the atomic velocities could thermalize and the geminate recombination might occur on a much slower time scale, by relative atom diffusion.

Whether the recombination is collisional or diffusive, its rate is dependent upon the actual diatomic potential surfaces, which affect trapping rates and pathways. The geminately recombined molecules may form in

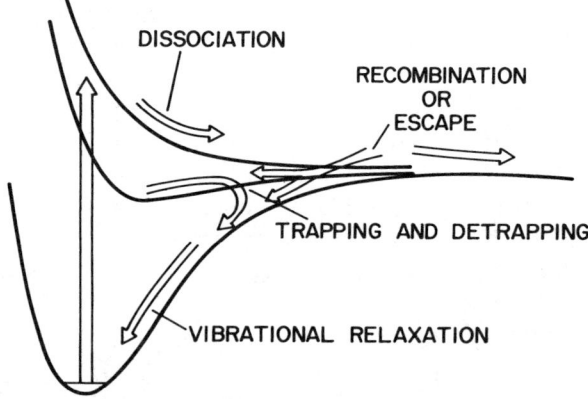

Figure 1 Processes of photodissociation and geminate recombination in solution. *Vertical arrow* designates initial light absorption.

an excited state if the excited state potential surface has an energetically stable potential well. These trapped excited state molecules must relax to the ground state by radiative or nonradiative transitions. Ground state molecules may thus be formed either immediately upon recombination or after surface hopping from excited states. In either case, they will be formed in highly vibrationally excited levels. The final step in the recombination process is therefore vibrational relaxation to an equilibrium distribution of vibrational levels of the ground electronic state.

The caging dynamics of polyatomic radical pairs in solution must also involve many of the reaction steps just discussed. The microscopic dynamics of polyatomic radical pair recombination, however, are not easily probed. Some detail on polyatomic radical caging is obtained through chemically induced dynamic nuclear polarization (CIDNP) spectra (15, 16), but detailed dynamics must be inferred from final results in the CIDNP spectrum. Iodine recombination dynamics, in contrast, can now be probed directly by picosecond spectroscopy, utilizing detailed knowledge of the iodine electronic states and the allowed optical transitions. The dynamics are also accessible to relatively simple theoretical simulations, and can serve as a simple model of more general caging processes.

Theories of solution reactions, many of which appertain to the studies discussed here, were recently reviewed by Hynes (17). The rapid development of ultrafast spectroscopy and its application to a number of fundamental questions in reaction dynamics has been described by Fleming (18). Both authors briefly reviewed the studies of geminate atom recombination dynamics that were available at that time. The present review updates the experimental picture of the geminate reaction process and focuses on comparisons of experimental and theoretical results for the various reaction steps.

We develop in the second section an accurate picture of iodine geminate recombination dynamics based upon the picosecond experiments of recent years. The third section compares the experimental recombination time scale to several theoretical descriptions of the reactive dynamics. The final section discusses vibrational energy transfer processes from the highly vibrationally excited molecule that is formed upon recombination and compares the experimental data with theories and simulations of vibrational relaxation in liquids.

THE EXPERIMENTAL PICTURE— CAGING DYNAMICS

Iodine Spectroscopy and Potential Surfaces

The key tool in understanding fast caging dynamics of the iodine molecule has been picosecond transient spectroscopy. The interpretation of transient spectral measurements depends upon detailed knowledge of the iodine molecule electronic potential energy surfaces that is available from gas phase studies. Figure 2 illustrates the most important potential surfaces for iodine photolysis and recombination. Seven of the ten Hund's case c states that correlate with ground state iodine atoms are known (19–25). Five of these are illustrated in Figure 2. The five states not shown are

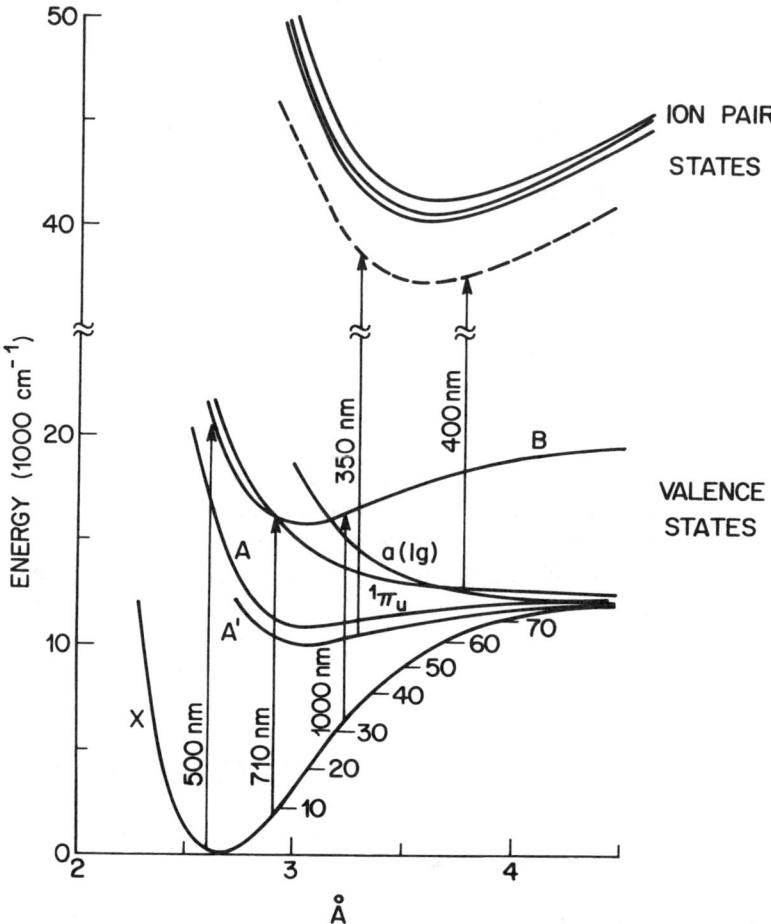

Figure 2 Relevant electronic potential energy surfaces of iodine. Five of the ten states correlating with ground state atoms are shown. *Solid lines* represent gas phase potentials. *Dashed line* shows approximate solvent shift of ion pair states. *Vertical arrows* designate optical transitions used to probe geminate recombination dynamics. *Numbers* and *tick marks* along the X-state curve label vibrational energy levels.

known to be, or presumed to be, primarily repulsive (20, 25). Of the molecular states that correlate with one excited state atom, only the B state is well studied (26), and is illustrated in Figure 2. In addition to these valence levels, a number of ion pair states of the iodine molecule are known at higher energies (27–29). The lowest energy group of ion pair states determine the transient ultraviolet absorption after photolysis in inert solvents, and are illustrated in Figure 2.

The geminate recombination dynamics have been studied in inert solvents, where the gas phase potentials are not strongly perturbed (30a–c), and where charge-transfer interactions (31, 32) of iodine molecules (33–35) and atoms (36) with the solvent are weak. A good assumption appears to be that the dynamics take place on potentials that are very similar to the gas phase potentials in Figure 2. The ion pair states, on the other hand, are affected even in relatively inert solvents (35). The approximate shift of the lowest set of ion pair states is illustrated in Figure 2 by the *dashed line*.

The equilibrium molecular iodine absorption spectrum peaks in the visible around 500 nm, and extends to 700 nm. Excitation within this absorption populates the A, $1u(^1\Pi_u)$ or B levels: wavelengths shorter than 600 nm populate primarily the B state, while wavelengths in the range 660–700 nm populate the A state above its dissociation limit (37, 38). Although the B state is bound, it is known to undergo collisional predissociation in the gas phase, probably by curve crossing to the $a(1_g)$ potential (25). A similar mechanism appears to lead to very efficient predissociation of the B state in liquids.

Following dissociation, the subsequent dynamics of molecules that recombine are followed by transient absorption spectra on optical transitions among the molecular valence levels or from the valence levels to molecular ion pair levels. Free atoms that escape are not observed directly, and recombine with other free atoms on a microsecond time scale. (The fraction that escape ranges from 30 to 90%, depending upon the solvent.) Most of the observed optical absorptions are known from the gas phase (23, 24, 37, 39), and are shown in Figure 2. Absorption from the X state is dominated by the optical transition to the B state (37, 38). The wavelength of this absorption is strongly dependent upon the X state vibrational level, as is indicated by the *vertical arrows* in Figure 2, which represent the classically allowed Franck-Condon transitions. This dependence on X state vibrational level is used to derive the X state vibrational dynamics. Ultraviolet absorption on transitions to ion pair states occurs from two regions; the stable A' state potential well, and regions of the X and A' states near their dissociation limits. The stable A' state is also probed by absorptions around 600–800 nm (40). The transition accounting for this red absorption is not illustrated in Figure 2, since the upper state has not been characterized in gas phase studies (40).

Picosecond Spectroscopy

Having drawn a picture of the relevant iodine molecular potential surfaces, we turn to the picosecond spectroscopic studies that have probed the nuclear motion on these surfaces following photodissociation. The first picosecond study of iodine geminate recombination dynamics used 532 nm light pulses to excite the iodine primarily into the B state (41). The recovery of the ground state absorption was also monitored at 532 nm. The experiment resolved an initial bleach of the molecular absorption in the first 20 ps, and partial recovery of the molecular absorption after 100–200 ps. The experiment was interpreted as showing that relatively fast predissociation of the B state (~ 20 ps) was followed by diffusion-controlled geminate recombination over a period of 100–200 ps.

This pioneering picosecond laser experiment provided the first glimpse of a process that had only been guessed at previously. The interpretation of the results, however, was biased by earlier notions of diffusive geminate dynamics. It failed to take into account molecular relaxation processes that might affect the ground state recovery time. Several subsequent picosecond experiments were interpreted in a similar fashion (42, 43).

An alternate explanation of the observed picosecond dynamics was suggested in 1982 by Nesbitt & Hynes (44) and Wilson and co-workers (48), who noted that the picosecond experimental wavelength of 532 nm probed only the absorption of lower vibrational levels of the X state (see Figure 2). Fast recombination, followed by slow X state vibrational relaxation, might account for the 100–200 ps recovery time at 532 nm. Molecular dynamics simulations of iodine recombination supported this idea, showing that geminate recombination was dominated by rapid collisional trapping rather than diffusion (45–48). Both Nesbitt & Hynes (44) and Wilson and co-workers (48) carried out simulations of iodine vibrational relaxation in the X state that demonstrated that a 100–200 ps relaxation time is reasonable. Wilson and co-workers also showed that as vibrational relaxation proceeded on the X state surface, the molecular absorption spectrum should shift from the near infrared region to the equilibrium peak near 500 nm (48), as is shown in Figure 2.

The recognition of this new aspect of the recombination dynamics prompted a new series of picosecond experiments designed to observe the transient spectrum of the recombining molecules. Wilson and co-workers observed transient red absorptions whose spectrum shifted to shorter wavelengths over a 50–200 ps period, as predicted by simulations of vibrationally relaxing iodine (48–51). In contrast, however, Kelley & Abul-

Haj observed a transient red absorption whose spectrum did not shift to shorter wavelengths with time (40, 52). Kelley & Abul-Haj also found that the transient absorption, as well as the recovery of the ground state bleach, had a recovery ranging from 100 ps to as long as 3 ns in simple solvents such as hydrocarbons or CCl_4. These recovery times were substantially longer than had been previously observed. Kelley & Abul-Haj assigned the red absorption to molecules trapped in the A or A' potential wells (see Figure 2), and concluded that the long time bleach recovery was determined primarily by the trapping time in the excited molecular states. Such trapping onto excited state potentials has been observed in rare gas matrices (53–57). The contrasting experimental results, however, left uncertain the relative roles of electronic trapping and vibrational relaxation in iodine geminate recombination.

Subsequent detailed picosecond studies by Harris and co-workers with improved time resolution, probing transient absorptions in the range 300 nm to 1000 nm, clarified the geminate recombination dynamics (58–60). Red absorptions that shifted from 1000 nm toward 500 nm over the first 40–200 ps in molecular solvents were assigned to X state vibrational relaxation. These absorptions were, however, superimposed on a fixed wavelength red absorption whose dynamics were assigned to the A/A' trapping proposed by Kelley & Abul-Haj (40). Ultraviolet absorptions from the A' level were also first observed. The recovery of the equilibrium molecular absorption near 500 nm was determined by both X state and A/A' state dynamics, and by the fraction of molecules that recombined onto each potential surface. Further picosecond studies of both I_2 and Br_2 recombination in molecular solvents have supported this assignment of slow vibrational and electronic relaxation (61, 62).

There has been disagreement on the time scale of the initial recombination event itself. Harris and co-workers concluded that geminate recombination occurred in less than 10–15 ps (59, 60). Kelley and co-workers argued that a substantial fraction of atoms might recombine on a slower, 50–100 ps timescale (62). Recent experiments by Harris and co-workers appear to resolve the issue, showing strong evidence for recombination within 2 ps after photodissociation (63).

Summary of Geminate Caging Dynamics

We summarize the present understanding of geminate recombination dynamics, with an emphasis on results for the most carefully studied solvent, CCl_4. Following short pulse excitation, the iodine molecule dissociates in a liquid within 2 ps, independent of the initial state populated (63). This result is based on strong transient ultraviolet absorptions, which are assigned to recombining molecules as they relax through the upper regions of the X and A' potential wells (see Figure 2). These absorptions are identical, within the experimental time resolution of 1–2 ps, after photoexcitation of either the A or B states in CCl_4 solvent. Since the A state is excited above its dissociation limit, it is expected to dissociate immediately (<1 ps). The B state predissociation time must therefore be less than 2 ps as well. Previous assignments of a 15 ps B state predissociation time (41, 60) were incorrect. Fast B state predissociation in the liquid is consistent with large collisional predissociation cross-sections measured in the gas phase for iodine (63–65).

As the iodine molecules dissociate, the solvent cage dynamics rapidly partition the atoms to recombine or escape. This conclusion is based upon transient absorption features assigned to recombining molecules in near-dissociation regions of the X or A' states (350–400 nm) and in intermediate vibrational levels of the X state (1000 nm), which rise and decay within 10–15 ps (see the experimental absorption curve at 1000 nm in CCl_4 solvent in Figure 3) (60, 63). Residual population in both of these regions must therefore be small after 15 ps, thus indicating that the reservoir of recombining atoms is depleted. Recombination into the A' state potential also occurs within 15 ps, based upon the rise times of the red and ultraviolet absorptions assigned to the bottom of the A' potential (60, 63). This is consistent with fast recombination, and provides an additional argument against slow diffusive recombination.

The detrapping rate from the A' potential is found to be strongly solvent dependent, varying from approximately 60 ps in alkane solvents to 2.7 ns in CCl_4 (62, 59, 60, 40). Trapping onto the A' potential appears to account for a large fraction (30–60%) of the recombined population. Dynamics of relaxation between the A and A' levels have not been directly observed.

The vibrational relaxation of molecules that recombine onto the X state potential is monitored through the B–X absorption spectrum, which shifts continuously from the near infrared to its equilibrium peak at 500 nm as the molecule vibrationally relaxes. The A' absorption in the red appears in the same wavelength region, but can be subtracted from the total absorption curve, leaving the X state component (60). The resulting X state absorption transients at wavelengths from 1000 nm to 500 nm in CCl_4 are shown in Figure 3. The absorption shift to shorter wavelength over a 150 ps period is apparent. Complete vibrational relaxation of iodine in other molecular solvents occurs in 50–150 ps (60). Somewhat faster relaxation rates were found for recombined bromine molecules in similar solvents (61). Iodine vibrational relaxation in xenon fluid has recently been estimated to take 5 ns (66), in contrast to early measurements (67). Experimental results for the vibrational distribution as a function of time in

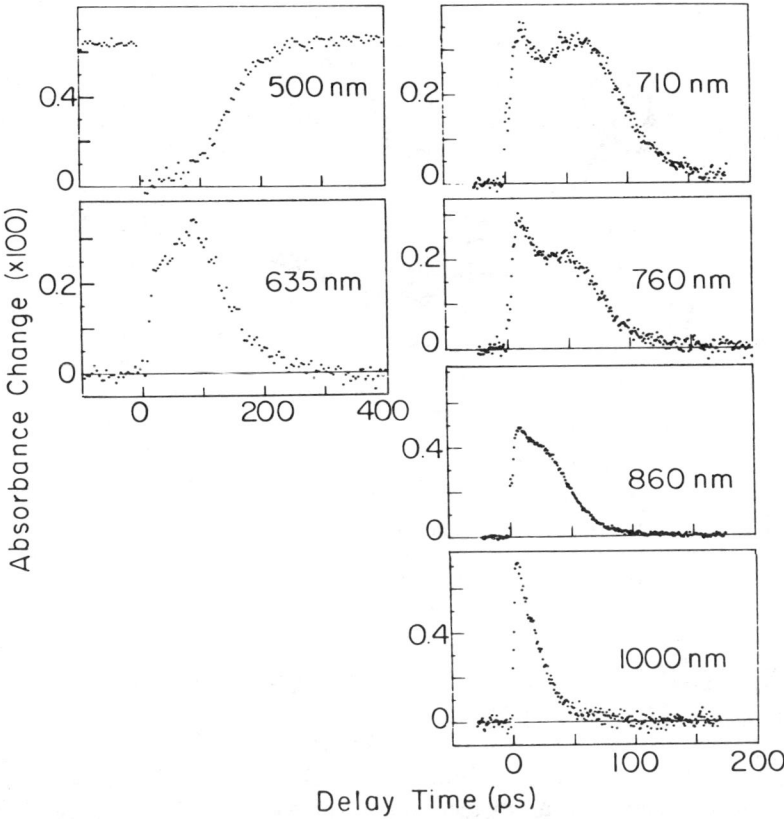

Figure 3 Transient absorptions in the red and near infrared assigned to vibrationally relaxing X state iodine after recombination in CCl_4 solvent. A' state red absorption has been subtracted from the spectra.

several solvents are discussed in connection with the theory of vibrational relaxation in a later section.

Figure 4 summarizes the geminate recombination processes and time scales in molecular solvents. Very fast dissociation or predissociation (1–2 ps) is followed by rapid recombination onto the X, A, or A' potential surfaces, or escape into the solvent, within about 15 ps. Subsequent vibrational relaxation on the X state potential takes 50–200 ps in molecular solvents. Detrapping of the A/A' potentials onto the X state potential is strongly solvent dependent, occurring on a 60 ps to several ns time scale. The various reaction steps now appear to be well established.

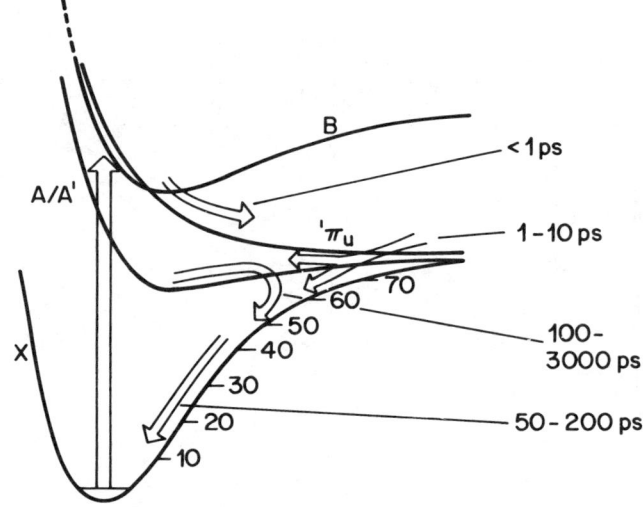

Figure 4 Summary of time scales for geminate recombination processes of iodine in molecular solvents (compare to Figure 1).

GEMINATE RECOMBINATION AND CURVE CROSSING DYNAMICS

Numerous reaction dynamics simulations have been applied to the caging dynamics of geminate atom recombination. Historically the cage effect was based on simple atom diffusion models, but they fail to account for important effects of the coupling of atomic and solvent motions. Approximate solute-solvent coupling and realistic atom-atom potential energy surfaces can be incorporated into Langevin simulations. More accurate assumptions about solvent-solute coupling can be incorporated into generalized Langevin simulations. Finally, molecular dynamics simulations of geminate recombination using reasonable iodine potential curves and iodine-solvent interactions present the most realistic view of recombination dynamics.

Diffusion

Diffusion models are simple to conceive and apply, and have been used frequently in discussions of caging dynamics. Treatments of polyatomic radical reactions frequently describe caging as a competition between "recombination" and "diffusion" (68). Simple calculations based upon Fick's Law diffusion were first applied to atom recombination by Noyes to estimate the geminate escape quantum yield as a function of solvent viscosity and wavelength of the photodissociating light (7, 8, 69–71). Early picosecond time-resolved measurements of the molecular signal recovery were also modeled by simple diffusion (41–43, 72, 73).

Diffusion models of recombination begin with a radially symmetric distribution of atoms representing the separation probability after photodissociation. The distribution evolves by Fick's Law diffusion to a central reactive sink, representing recombination. Appropriate boundary conditions at the surface of the sink give an analytic solution for the probability of recombination with time (73). Such models neglect details of potential surfaces and atom-atom forces. Since the calculated time scale and probability of geminate recombination are strongly dependent upon the assumed sink radius and initial distribution, these models cannot be predictive.

More serious problems arise because the diffusion models do not realistically model the early time dynamics of geminate recombination. Computer simulations (discussed in a later section), supported by the picosecond data, indicate that geminate recombination is complete within 5 ps and is dominated by fast collisional trapping rather than thermal diffusion (45–47). In addition, the diffusion models cannot treat the effects of multiple potential surfaces or excited state trapping. Finally, simple diffusion models do not correctly predict the viscosity dependence of recombination yields observed in careful pressure-dependent experiments (10–13). Such simple models thus offer little insight into the dynamics of solvent caging and recombination.

Langevin Dynamics

Langevin simulations characterize atomic motion on realistic potential energy surfaces. In such simulations, a single reaction trajectory is obtained by integrating the Langevin equation for motion along the reaction coordinate (74):

$$m\frac{d^2 r}{dt^2} = -m\gamma \frac{dr}{dt} + F(r) + R(t) \qquad 1.$$

where r is the position of the Brownian particle of mass m, $F(r)$ is the negative of the spatial derivative of the I–I potential, γ is the friction, and $R(t)$ represents solvent random forces. The random force and the friction are related by the fluctuation-dissipation theorem, which assures the ultimate equilibration of the system to the final temperature T:

$$\langle R(0)R(t) \rangle = 2k_B T \gamma \delta(t) \qquad 2.$$

where $\delta(t)$ is a delta function. To simulate a reaction event, the magnitude of the random force distribution is fixed by Eq. 2. An ensemble of trajectories, obtained by integrating Eq. 1, simulates the average behavior of the system.

Early Langevin simulations treated geminate recombination on a single potential surface. Hynes et al (75) found that at liquid viscosities, there were small differences between a Langevin simulation and the Smoluchowski equation (diffusion on the potential surface) in both time scale and quantum yield of escape. The results of either method were different from free diffusion to a reactive sink, and demonstrated that the I–I potential should be included in any realistic simulation. A comparison of the Fokker-Planck and BGK equations, different ways to incorporate the random forces in a simulation, also demonstrated similar results at viscosities where both methods approach the Smoluchowski behavior (76).

Multiple electronic potentials have been incorporated into Langevin simulations to account for the role of electronic curve crossing in recombination dynamics. The earliest such simulation of iodine recombination included a ground state potential and a repulsive potential curve to represent the state initially populated by photoexcitation (77). The crossing rate from the repulsive state to the ground state was varied arbitrarily. The reappearance of molecules in the lower vibrational levels of the ground state was controlled by the curve crossing rate, because vibrational relaxation on the ground state potential was fast, in contrast with experiment.

Other Langevin simulations have addressed electronic curve crossing dynamics in a less arbitrary fashion, based upon physical models of collision-induced curve crossing. Ali & Miller (78) compared the rate of recombination with and without multiple iodine electronic surfaces, using the Miller-George formulation of collision-induced Tully-Preston curve crossing (79, 80). Recombination was defined as relaxation to a vibrational level more than $k_B T$ below the dissociation limit. Langevin simulations of recombination showed that the multiple potential curves slowed recombination onto the ground state surface by a factor of 3–4.

Dawes & Sceats used a different surface-hopping model based upon collisional reorientation of electronic angular momentum (81) in a Langevin simulation of recombination onto X, A, and A' levels, and of detrapping from the A and A' levels (82). Trapping on the shallow well of the $1u(^1\Pi_u)$ state slowed recombination, which occurred on a time scale of 100 ps onto the A or A' potential. Relaxation from the A state to the A' or X states took about 1 ns, and detrapping from the A' level occurred over a 1–10 ns period. All of the rates are somewhat slower than experimental values. The viscosity dependence of the A' detrapping rate was estimated and compared to experimental results. The Langevin simulations qualitatively predicted the viscosity dependence of the A' lifetime in some solvents, but failed to predict the lack of viscosity dependence in hydrocarbon solvents. Dawes & Sceats proposed that the high frequency viscosity might control the A' dynamics, and that this viscosity might be very similar in a series of hydrocarbons. Small changes in the potential energy surfaces may also affect the relative rates in various solvents (40). At present, however, the source of the solvent dependence of the A' detrapping rate remains uncertain.

Langevin simulations incorporate realistic potentials and multiple electronic surfaces into reaction simulations. They overestimate the efficiency of coupling between atomic and solvent motions, however. Because the random force in Eq. 2 is delta-function correlated, the power spectrum of bath forces acting on the atomic motion contains all frequencies equally, overemphasizing the high frequency components. This affects relaxation of high frequency motions of the atoms particularly strongly (83), as in vibrational relaxation. Adelman, for instance, has shown that a Langevin simulation typically leads to vibrational relaxation on the iodine X state that is about two orders of magnitude too fast (84). General discussion of the formal limitations of the Langevin equation is found in (83, 85).

Generalized Langevin Dynamics

Memory effects can be included in reaction trajectory simulations to introduce temporal correlations into the bath motions. This damps the high frequency components of the bath forces acting on the reaction trajectory. This can be accomplished using a generalized Langevin equation (86, 87):

$$m \frac{d^2 r}{dt^2} + \int_0^t \gamma(t-\tau) \frac{dr(\tau)}{d\tau} d\tau = F(r) + R(t) \qquad 3.$$

where $\gamma(t)$ is a memory function that decays with the relaxation time of the bath and describes the dissipation of energy from the system. The random force and the memory function are related by the second fluctuation-dissipation theorem:

$$\langle R(0) R(t) \rangle = k_B T \gamma(t). \qquad 4.$$

To simulate the reactive motion, $R(t)$ and $\gamma(t)$ must be determined. These quantities are not readily accessible from experimental data, and must be guessed or calculated by other means. For certain forms of $\gamma(t)$, Eq. 3 can be written as a set of coupled linear differential equations that may be solved numerically. Examples of such techniques are Mori's continued fraction approximation (88–90) and the molecular time scale generalized Langevin equation (MTGLE) discussed by Adelman (91). The parameters in these expansions may be chosen based on experimental data or as "best" fits to molecular dynamics simulations of the system or pure solvent.

In the MTGLE approach, a set of sequentially coupled differential equations is numerically integrated to produce the solute trajectory. The equations represent a chain of coupled harmonic oscillators whose parameters are chosen to represent the solvent. The last oscillator is driven by delta-correlated random forces, but only the first is coupled to the solute coordinate. The chain generates solute-solvent coupling, but filters out unrealistic high frequency components of the random force (91).

Adelman has applied MTGLE simulations to several aspects of iodine recombination. The MTGLE simulation can treat vibrational relaxation, and such simulations and their limitations are discussed below. Adelman has shown that caging of the iodine atoms motion in simulations with CCl_4 as the solvent is very efficient in the MTGLE approach, but almost nonexistent in a Langevin simulation (92). The difference was related to the strong restoring forces that appear as a result of harmonic coupling between the first chain coordinate and the reaction coordinate. The same method applied to simulate recombination onto the ground state after photodissociation in liquid xenon (93) gave a very high escape yield, in agreement with recent experiments (66). These results, as well as applications of the MTGLE equations to vibrational relaxation, have been described in a recent article by Adelman (94). One weakness of such simulations is the need to fix the values of parameters that are not related to solvent properties in a simple way. The efficiency of the trapping in the CCl_4 simulation, for example, was strongly dependent upon an arbitrary "lag" time for response of the first shell (92).

Ali & Miller also applied MTGLE methods to recombination on multiple iodine electronic surfaces (95), by using the same surface-hopping model as in (78). They found that the slowing of recombination by the multiple potential surfaces was a factor of two in the generalized Langevin simulation. In addition, 10–20% of the recombination was onto the A or A' surfaces.

Molecular Dynamics Simulations

The primary result of computer simulations is that fast collisional caging dynamics can account for most of the geminate recombination yield in dense gas and liquid environments. Bunker & Jacobsen carried out one of the first molecular dynamics simulations of geminate recombination in liquids (45). Morse potential iodine interacted with 26 particles by Lennard-Jones 6–12 potentials. The mass and size of the particles were designed to simulate CCl_4. Summarizing the finding that most recombination events occurred in less than 1–2 ps, the authors stated, "Our most striking finding is surely the relative unimportance of truly diffusive effects, involving solvent intervention, under all conditions (45)." They concluded that once a solvent molecule intervenes between the two photodissociating atoms, the likelihood of recombination is considerably reduced. Similar conclusions were reached for larger model systems (101, 102), and fast recombination was also observed in simulations of dense gaseous krypton (46, 47) and nitrogen (96).

These results confirm the experimental conclusion that the partitioning between escaped and recombined atoms is very fast. The discrete nature of the solvent, and highly energetic, nonthermal collisions must dominate this process. The results emphasize that diffusive models of geminate recombination are not appropriate.

VIBRATIONAL RELAXATION

Fast geminate recombination creates a highly vibrationally excited ensemble of diatomic iodine molecules on the X state potential. In solvents where the A/A' state detrapping time does not interfere, such as CCl_4, the X state vibrational relaxation may be monitored over many vibrational levels. This presents a unique opportunity to examine vibrational level scaling of relaxation rates, and to study vibrational energy transfer between a low frequency anharmonic oscillator and its solvent bath. We begin with a discussion of the energy transfer mechanisms that are likely to dominate the vibrational dynamics, and of the role of molecular dynamics simulations in understanding these mechanisms. We then examine more generally two theoretical approaches to vibrational relaxation that have been applied to iodine vibrational relaxation: isolated binary collision models and generalized Langevin models. The discussion focuses on the problem of a low frequency anharmonic diatomic oscillator in a dense solvent medium. More general discussion of vibrational relaxation in liquids is found in recent reviews by Oxtoby (97, 98).

Vibrational Relaxation of X State Iodine

Calculations of vibrational relaxation on the iodine X state potential by Nesbitt & Hynes (44, 99, 100) raised the first questions about the relative importance of energy transfer from the iodine vibrational coordinate to either solvent translation (V–T) or vibrational (V–V) degrees of freedom. On the basis of binary collision trajectory calculations, they predicted that V–T energy transfer was efficient from upper vibrational levels of the X state, but inefficient at lower levels, where the vibrational spacing was larger and the amplitude of the vibrational motion was smaller. Complete vibrational relaxation was estimated to take about a nanosecond in xenon solvent, where only V–T energy transfer is possible. They proposed that near-resonant V–V energy transfer to a CCl_4 vibrational mode accounted for the faster (~ 150 ps) experimental relaxation time in that solvent.

The relative importance of V–V or V–T energy transfer is approximately determined by the power spectrum of solvent motions due to each degree of freedom. In terms of a simple perturbation picture of vibrational relaxation (97), the transition rate from one vibrational level to another is proportional to the magnitude of the power spectrum of perturbing forces at the transition frequency. Translational motion is predominantly a low frequency motion, so that energy transfer from a high frequency oscillator is often dominated by V–V energy transfer, when solvent vibrational degrees of freedom are available (97). But iodine is perhaps the lowest frequency oscillator for which good vibrational relaxation rates are available. Is it in the high frequency regime?

Experimental and molecular dynamics investigations have addressed this question. An experimental relaxation time of 5 ns has recently been estimated for complete relaxation of the iodine X state in xenon fluid ($\rho = 1.8$ g/cm^3) (66), more than one order of magnitude slower than in molecular solvents. V–V energy transfer in the molecular solvents might be invoked to account for this difference. The experimental relaxation rate in xenon was also within a factor of 3 or 4 of the V–T transfer rate predicted by Nesbitt & Hynes (44), a finding that perhaps supports the general accuracy of their conclusions.

On the other hand, a molecular dynamics simulation suggests that at least part of the difference between relaxation rates in CCl_4 and xenon may arise from differences in solvent density (101, 102). Simulations in a xenon solvent interacting via Lennard-Jones potentials at a Lennard-Jones reduced density ($\rho^* = 0.95$, 3.0 g/cm^3) comparable to CCl_4 showed X state vibrational relaxation was four times faster than in the same Lennard-Jones solvent held at the lower xenon experimental density ($\rho^* = 0.57$, 1.8 g/cm^3). A relatively small steepening of the interaction potentials in the simulation of CCl_4 relative to xenon could increase this by another factor of two, which would account for most of the rate difference between xenon and CCl_4 without invoking vibrational degrees of freedom.

Other experimental evidence in molecular solvents does not support the dominant role of near-resonant V–V energy transfer. Iodine relaxation in most of the molecular solvents examined (60) has been faster than in CCl_4, yet the vibrational modes of the other solvents are typically farther from resonance with the iodine oscillator than is CCl_4. In particular, the chlorinated methane solvents give iodine relaxation rates that are faster in the order $CH_2Cl_2 > CHCl_3 > CCl_4$ (60). The lowest frequency mode in these solvents is above the iodine oscillator frequency, and the gap is greater in the solvents with faster relaxation. An opposite trend is expected if near-resonant V–V energy transfer dominates the rate. V–T energy transfer, on the other hand, may be able to account for the trend of relaxation rates. Molecular dynamics simulations of iodine relaxation rates in Lennard-Jones solvents show that the relaxation rate is two times faster when the solvent is two times lighter with identical potential parameters (101, 102). This is consistent with the approximate 2:1 mass ratio and 1:2 relaxation rate ratio for CCl_4 solvent versus CH_2Cl_2 solvent.

When the X state vibrational dynamics are analyzed in more detail, V–T energy transfer can qualitatively account for the vibration level dependence of the relaxation rate, as well. The experimental vibrational energy as a function of time in CCl_4 solvent was approximately extracted from the transient X state absorptions, using calculated Franck-Condon factors (60). The time dependence of the vibrational populations is shown in Figure 5, *top*. The mean vibrational energy as a function of time is plotted on the lower part of Figure 5, calculated from this distribution. The trend of very fast relaxation at the top of the well, with slower relaxation near the bottom of the well, is typical in the solvents examined (60, 63). In terms of V–T energy transfer, the trend is due to smaller vibrational energy spacing and larger amplitude vibrational motion at the top of the well, which lead to more efficient V–T energy transfer there. The results from molecular dynamics simulations for a Lennard-Jones solvent at several densities are shown on the *upper* part of Figure 6 (101–103). They show behavior that is qualitatively similar to the experimental data.

To summarize, V–T energy transfer can be an important component of iodine vibrational relaxation in molecular solvents, although whether it is the dominant mechanism remains uncertain. Hindered rotational degrees of freedom may also help to account for the faster relaxation rates in molecular solvents. Further studies to clarify the roles of vibrational and rotational degrees of freedom in the vibrational relaxation of iodine are needed.

Theory of V–T Energy Transfer

We next consider more general aspects of the vibrational relaxation theories that have been applied to V–T energy transfer in the iodine system. Isolated binary collision trajectory calculations and generalized Langevin simulations have been particularly useful in studying iodine vibrational relaxation. These general models are evaluated, in part, by comparison with molecular dynamics simulations. The simulations themselves may or may not exactly reproduce experimental data, in part due to uncertainties in the appropriate interaction potentials, but they provide a reference system where the interaction potentials are exactly known. The iodine

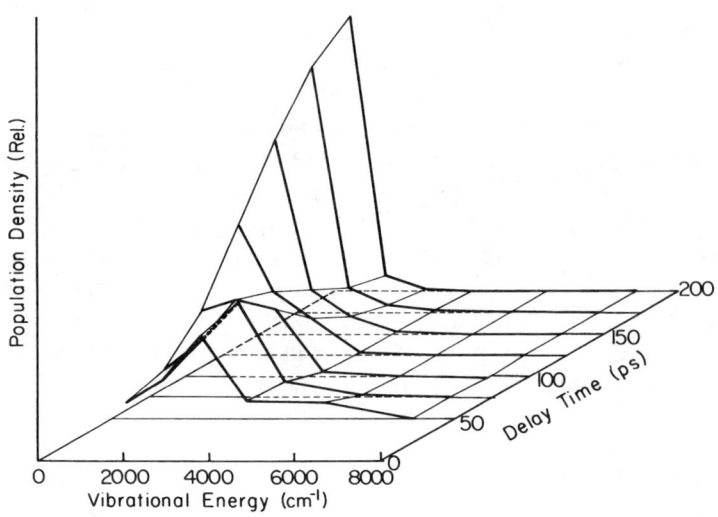

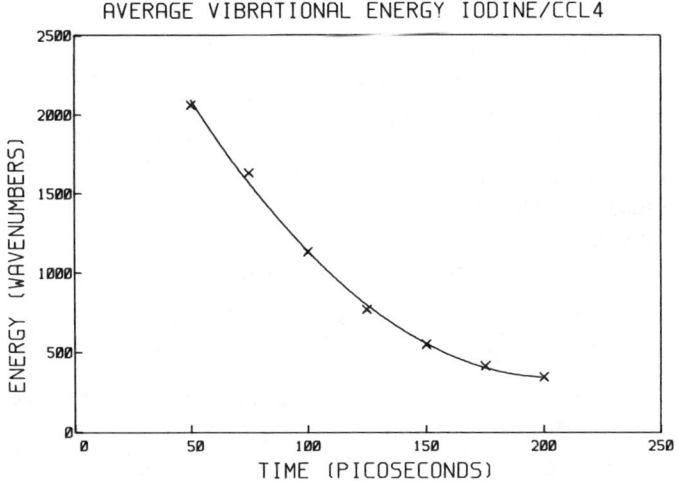

Figure 5 Top: Experimental vibrational energy distribution as a function of time for recombined X state iodine in CCl_4 solvent. *Bottom*: Mean vibrational energy as a function of time, from same data.

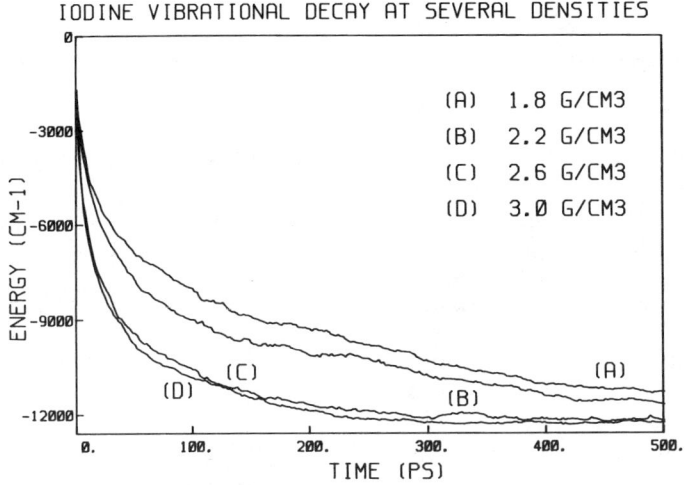

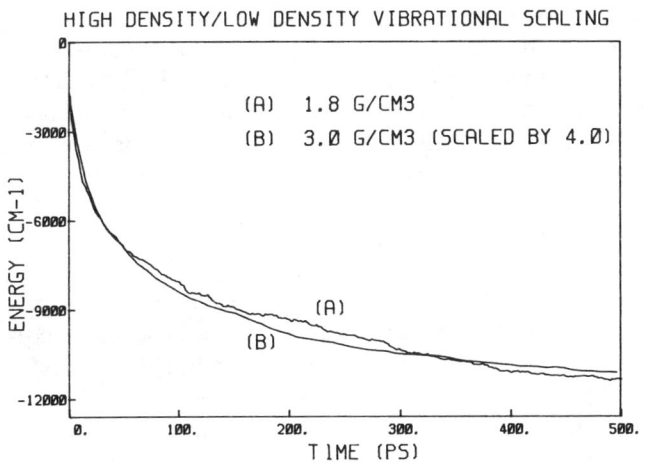

Figure 6 Top: Molecular dynamics simulation results for mean vibrational energy of X state iodine in xenon at four different densities; 1.8, 2.2, 2.6, 3.0 g/cm³. *Bottom*: Simulation data from lowest and highest density, scaled in time to illustrate similar behavior.

relaxation problem is amenable to the use of such classical mechanics simulations, because the iodine vibrational transition energies are less than the thermal energy at room temperature.

IBC MODELS OF VIBRATIONAL RELAXATION   The simplest theories of vibrational relaxation describe the relaxation process by a set of rate constants, k_{ij}, that determine the transition rate from vibrational state i to vibrational state j of the oscillator. One of the simplest models for the rate constants is based on the extension of gas phase isolated binary collision (IBC) models to densities characteristic of liquids (104). IBC models decompose the state-to-state transition rates into the product of two terms:

$$k_{ij} = \Gamma_{ij} * v_{ci} \qquad\qquad 5.$$

where v_{ci} is the rate of collisions by the solvent with the oscillator in state i, and Γ_{ij} is the probability for changing from state i to state j in a collision. The Γ_{ij} may be determined, for example, from experimental gas phase data, from trajectory calculations of a solvent molecule colliding with an excited oscillator (99, 100), or from a semiclassical model such as SSH theory (103, 105).

The determination of the collision frequencies is more complex since the tight liquid packing continuously keeps solvent molecules within the range of the oscillator-solvent interaction potential. There has been some success, however, in interpreting experimental data by using IBC models with a collision frequency determined from cell models (104) or from radial distribution functions (106, 107). Collision frequencies also may be calculated from molecular dynamics trajectories (108), although, except for hard sphere simulations, the results are sensitive to the chosen definition of a collision.

The validity of the IBC model for liquids has been much discussed (97). The model assumes that collisions are both binary and uncorrelated. It is possible and perhaps likely that two or more solvent molecules may "collide" with the oscillator simultaneously in a liquid, or that correlated collisions may occur, and the effects of these multiple collisions may not be additive (109). On the other hand, only the rare, very hard collisions of the solvent with the oscillator may be important in the vibrational relaxation process for high frequency oscillators, $\hbar\omega \gg k_B T$, thus decreasing collisional correlations (106, 110, 111). The low frequency iodine oscillator may be, in any case, much more susceptible than a high frequency oscillator to correlated motions of the solvent.

Recent molecular dynamics simulations have addressed the validity of IBC models for a low frequency oscillator such as iodine (101, 102, 112). The density dependence of the simulated iodine vibrational relaxation in liquid xenon is illustrated in Figure 6, *top* drawing. The functional forms of the vibrational decays calculated were approximately the same over a fairly broad density range, as shown by the scaled relaxation curves from the lowest and highest densities in Figure 6, *bottom*. This is consistent with very simple IBC models. However, the density scaling of the rate was stronger than would be predicted by many simple models of the liquid binary collision rate. Collision frequencies were also estimated directly from the crossing of solvent atoms through surfaces surrounding the iodine atoms in the simulation. The frequencies showed a strong dependence on the iodine vibrational amplitude, because the iodine molecule created collisions by its own vibrational motion. The IBC model thus appears to underestimate the density scaling of the relaxation rate, and to neglect important correlations in the oscillator-solvent motions (112).

Because of their simplicity IBC models can be useful for qualitative predictions. The IBC calculation of Nesbitt & Hynes, for instance, has been influential by elucidating the role of vibrational relaxation in geminate recombination dynamics (44). IBC models may, however, be poor choices for quantitatively modeling V–T energy transfer from a low frequency oscillator in a dense solvent, because of the limitations we have discussed.

GENERALIZED LANGEVIN MODELS: VIBRATIONAL RELAXATION   Generalized Langevin simulations can incorporate oscillator-solvent correlations that are missing in the IBC models (and in simple Langevin simulations, see the section on *Langevin Dynamics*). They account for the frequency response of the solvent through the form of the friction integral and the random force applied to the iodine.

One example of this approach is the MTGLE method described above. Adelman has shown that an MTGLE simulation of iodine relaxation from high vibrational levels leads to reasonable rates that are fast near the top of the potential well but that slow considerably as the vibrational frequency increases and the vibrational amplitude decreases further down the well (84). Adelman has extended to formal MTGLE method to include solvent vibrational modes, but has not made comparisons to experimental data (113).

Direct comparisons of the generalized Langevin Eq. 3 and molecular dynamics results for iodine vibrational relaxation in a simulated xenon fluid have also been done (103, 112). The memory function, $\gamma(t)$, in Eq. 3 was a simple single or double exponential function (103, 112) whose parameters were chosen to approximately reproduce the velocity autocorrelation function of xenon atoms in the simulation fluid. The generalized Langevin equation was thus made to fit as closely as possible the conditions of the simulated fluid. The equation gave relaxation times that were too fast when a single exponential fit was used for the memory function. This is due to an unphysical cusp in the memory function at $t = 0$ that leads to artificially large high frequency components in the random force, $R(t)$. This cusp was removed using a two exponential fit for the memory function, and the comparison to molecular dynamics simulations was more favorable. These results are summarized in Figure 7, which also demonstrates the failure of a simple Langevin simulation to model the relaxation rate. The generalized Langevin equation accounts, in this case, for many of the important solvent forces that determine vibrational relaxation. However, the results for vibrational relaxation are still noticeably faster than molecular dynamics, perhaps reflecting the necessity to better represent the iodine local environment. Moreover, it is not yet apparent how effectively these GLE models will be when extended to more complex solvents with low frequency internal vibrational modes.

SUMMARY

The microscopic processes of iodine geminate recombination are somewhat more complex than was supposed five years ago, but appear to be well understood. Very fast dissociation and predissociation is followed by recombination or escape within 5 ps. Subsequent slow vibrational relax-

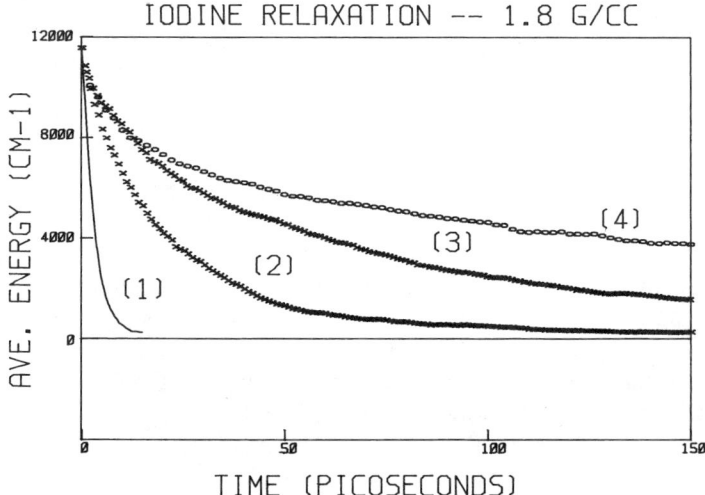

Figure 7 Molecular dynamics simulation results for mean vibrational energy of X state iodine in xenon at low density [4], compared to Langevin simulation [1]; Generalized Langevin simulation with single exponential memory function [2], and double exponential memory function [3].

ation on the X state potential occurs in 50–200 ps in molecular solvents, and in several nanoseconds in liquid xenon. Detrapping times from excited state potentials range from 50 ps to 3 ns in various solvents.

The elucidation of the recombination processes at a microscopic level in a simple system has presented opportunities to evaluate theories of reaction dynamics and energy transfer in liquids. Simple diffusion calculations do not explain the time scale of recombination. Molecular dynamics simulations provide a much better microscopic picture of the very rapid, nonthermal caging dynamics that lead to geminate recombination. Langevin and generalized Langevin simulations provide insight into the role of multiple electronic surfaces in recombination dynamics. Such simulations have not, however, accounted for the wide range of excited state trapping times observed in various solvents.

Several experiments and simulations indicate that V–T energy transfer may be able to account for the X state vibrational relaxation rate, although uncertainties remain. Isolated binary collision models of vibrational relaxation have been important in understanding the qualitative vibrational relaxation dynamics on the X state surface. Such models do not, however, account quantitatively for correlations in the solvent-molecule interactions that appear to be particularly important for a low frequency oscillator such as iodine. Generalized Langevin simulations allow a more accurate representation of the solvent environment, and can reproduce many features of vibrational relaxation rates of the low frequency iodine oscillator.

The coupling of detailed experimental results with reaction theories has elucidated the reaction steps in the classic iodine atom recombination reaction, and has probed reaction models. Improvements in ultrafast spectroscopy, particularly in the ability to probe molecular structure directly through transient infrared or Raman spectroscopy, should soon provide a similar level of detail about polyatomic reaction dynamics in solution, and should further challenge and refine current theoretical models of photodissociation reaction dynamics.

ACKNOWLEDGMENTS

The authors wish to acknowledge the many contributions made at the University of California at Berkeley by M. Berg, R. L. Hoff, M. P. Paige, D. J. Russell, and D. E. Smith. J.K.B. and C.B.H. acknowledge support by the National Science Foundation for this work. We also acknowledge the US Department of Energy, Office of Basic Energy Sciences, Chemical Sciences Division under Contract No. DE-AC03-76SF00098 for some of the specialized equipment used in some of the experiments reviewed in this article.

Literature Cited

1. Franck, J., Rabinowitch, E. 1934. *Trans. Faraday Soc.* 30: 120–31
2. Rabinowitch, E., Wood, W. C. 1936. *Trans. Faraday Soc.* 32: 547–56
3. Rabinowitch, E., Wood, W. C. 1936. *Trans. Faraday Soc.* 32: 1381–87
4. Zimmerman, J., Noyes, R. M. 1950. *J. Chem. Phys.* 18: 658–66
5. Marshall, R., Davidson, N. 1953. *J. Chem. Phys.* 21: 2086
6. Lampe, F. W., Noyes, R. M. 1954. *J. Am. Chem. Soc.* 76: 2140–44
7. Booth, D., Noyes, R. M. 1960. *J. Am. Chem. Soc.* 82: 1868–71
8. Meadows, L. F., Noyes, R. M. 1960. *J. Am. Chem. Soc.* 82: 1872–76
9. Troe, J. 1978. *Ann. Rev. Phys. Chem.* 29: 223–50
10. Hippler, H., Luther, K., Troe, J. 1973. *Ber. Bunsenges. Phys. Chem.* 77: 1104–14
11. Luther, K., Schroeder, J., Troe, J., Unterberg, U. 1980. *J. Phys. Chem.* 84: 3072–75
12. Otto, B., Schroeder, J., Troe, J. 1984. *J. Chem. Phys.* 81: 202–13
13. Hippler, H., Schubert, V., Troe, J. 1984. *J. Chem. Phys.* 81: 3931–41
14. Sceats, M. G., Dawes, J. M., Millar, D. P. 1985. *Chem. Phys. Lett.* 114: 63–70
15. Ward, H. R. 1973. In *Free Radicals*, ed. J. K. Koch, 1: 239–73. New York: Wiley. 713 pp.
16. Glarum, S. H. 1973. In *Chemically Induced Magnetic Polarization*, ed. A. R. Lepley, G. L. Closs, pp. 7–39. New York: Wiley. 416 pp.
17. Hynes, J. T. 1985. *Ann. Rev. Phys. Chem.* 36: 573–97
18. Fleming, G. R. 1986. *Chemical Applications of Ultrafast Spectroscopy*. New York: Oxford Univ. Press. 262 pp.
19. LeRoy, R. J. 1970. *J. Chem. Phys.* 52: 2683–89
20. Mulliken, R. S. 1971. *J. Chem. Phys.* 55: 288–309
21. Tellinghuisen, J. 1973. *J. Chem. Phys.* 58: 2821–34
22. Churassy, S., Martin, F., Bacis, R., Verges, J., Field, R. W. 1981. *J. Chem. Phys.* 75: 4863–68
23. Viswanathan, K. S., Sur, A., Tellinghuisen, J. 1981. *J. Mol. Spectrosc.* 86: 393–405
24. Tellinghuisen, J. 1982. *J. Mol. Spectrosc.* 94: 231–52
25. Tellinghuisen, J. 1985. *J. Chem. Phys.* 82: 4012–16
26. Barrow, R. F., Yee, K. K. 1973. *J. Chem. Soc. Faraday Trans. 2* 69: 684–700
27. Brand, J. C. D., Hoy, A. R., Kalkar, A. K., Yamashita, A. B. 1982. *J. Mol. Spectrosc.* 95: 350–58
28. Perrot, J. P., Broyer, M., Chevaleyre, J., Femelat, B. 1983. *J. Mol. Spectrosc.* 98: 161–67
29. Viswanathan, K. S., Tellinghuisen, J. 1983. *J. Mol. Spectrosc.* 101: 285–99
30a. Sension, R. J., Strauss, H. L. 1986. *J. Chem. Phys.* 85: 3791–3806
30b. Sension, R. J., Kobayashi, T., Strauss, H. L. 1987. *J. Chem. Phys.* 87: 6221–32
30c. Sension, R. J., Kobayashi, T., Strauss, H. L. 1987. *J. Chem. Phys.* 87: 6233–39
31. Mulliken, R. S., Person, W. B. 1969. *Molecular Complexes. A Lecture and Reprint Volume*. New York: Wiley. 498 pp.
32. Tamres, M., Strong, R. L. 1979. In *Molecular Association*, ed. R. Foster, 2: 331–456. New York: Academic. 497 pp.
33. Evans, D. F. 1955. *J. Chem. Phys.* 23: 1424–26
34. Orgel, L. E., Mulliken, R. S. 1957. *J. Am. Chem. Soc.* 79: 4839–46
35. Julien, L. M., Person, W. B. 1968. *J. Phys. Chem.* 72: 3059–61
36. Bonneau, R., Joussot-Dubien, J., Fournier de Violet, P. 1975. *J. Chem. Soc. Faraday Trans. 1* 71: 2148–55
37. Tellinghuisen, J. 1982. *J. Chem. Phys.* 76: 4736–44
38. Bergsma, J. P., Berens, P. H., Wilson, K. R., Fredkin, D. R., Heller, E. J. 1984. *J. Phys. Chem.* 88: 612–19
39. Tellinghuisen, J., Whyte, A. R., Phillips, L. F. 1984. *J. Phys. Chem.* 88: 6084–87
40. Kelley, D. F., Abul-Haj, N. A., Jang, D. J. 1984. *J. Chem. Phys.* 80: 4105–11
41. Chuang, T. J., Hoffman, G. W., Eisenthal, K. B. 1974. *Chem. Phys. Lett.* 25: 201–5
42. Langhoff, C. A., Moore, B., DeMeuse, M. 1982. *J. Am. Chem. Soc.* 104: 3576–79
43. Langhoff, C. A., Moore, B., DeMeuse, M. 1983. *J. Chem. Phys.* 78: 1191–99
44. Nesbitt, D. J., Hynes, J. T. 1982. *J. Chem. Phys.* 77: 2130–43
45. Bunker, D. L., Jacobsen, B. S. 1972. *J. Am. Chem. Soc.* 94: 1843–48
46. Murrell, J. N., Stace, A. J., Dammel, R. 1978. *J. Chem. Soc. Faraday Trans. 2* 74: 1532–39
47. Lipkus, A. H., Buff, F. P., Sceats, M. G. 1983. *J. Chem. Phys.* 79: 4830–38
48. Bado, P., Berens, P. H., Wilson, K. R.

1982. *Proc. Soc. Photo-Opt. Instrum. Eng.* 322: 230–36
49. Bado, P., Wilson, K. R. 1984. *J. Phys. Chem.* 88: 655–57
50. Bado, P., Dupuy, C. G., Bergsma, J. P., Wilson, K. R. 1984. In *Ultrafast Phenomena IV*, ed. D. H. Auston, K. B. Eisenthal, pp. 296–99. Berlin/Heidelberg/New York: Springer-Verlag. 509 pp.
51. Bado, P., Dupuy, C., Magde, D., Wilson, K. R., Malley, M. M. 1984. *J. Chem. Phys.* 80: 5531–38
52. Kelley, D. F., Abul-Haj, N. A. 1984. See Ref. 50, pp. 292–95
53. Beeken, P. B., Hanson, E. A., Flynn, G. W. 1983. *J. Chem. Phys.* 78: 5892–99
54. Mandich, M. L., Beeken, P. B., Flynn, G. W. 1982. *J. Chem. Phys.* 77: 702–13
55. Beeken, P. B., Mandich, M. L., Flynn, G. W. 1982. *J. Chem. Phys.* 76: 5995–6001
56. Bondybey, V. E., Bearder, S. S., Fletcher, C. 1976. *J. Chem. Phys.* 64: 5243–46
57. Bondybey, V. E., Fletcher, C. 1976. *J. Chem. Phys.* 64: 3615–20
58. Berg, M., Harris, A. L., Brown, J. K., Harris, C. B. 1984. See Ref. 50, pp. 300–3
59. Berg, M., Harris, A. L., Harris, C. B. 1985. *Phys. Rev. Lett.* 54: 951–54
60. Harris, A. L., Berg, M., Harris, C. B. 1986. *J. Chem. Phys.* 84: 788–806
61. Abul-Haj, N. A., Kelley, D. F. 1985. *Chem. Phys. Lett.* 119: 182–87
62. Abul-Haj, N. A., Kelley, D. F. 1986. *J. Chem. Phys.* 84: 1335–44
63. Smith, D. E., Harris, C. B. 1987. *J. Chem. Phys.* 87: 2709–15
64. Steinfeld, J. I. 1966. *J. Chem. Phys.* 44: 2740–49
65. Capelle, G. A., Broida, H. P. 1973. *J. Chem. Phys.* 58: 4212–22
66. Paige, M. E., Russell, D. J., Harris, C. B. 1986. *J. Chem. Phys.* 85: 3699–3700
67. Kelley, D. F., Rentzepis, P. M. 1982. *Chem. Phys. Lett.* 85: 85–90
68. Lowry, T. H., Richardson, K. S. 1976. *Mechanism and Theory in Organic Chemistry*, pp. 489–90. New York: Harper & Row. 748 pp.
69. Noyes, R. M. 1954. *J. Chem. Phys.* 22: 1349–59
70. Noyes, R. M. 1960. *Z. Electrochem.* 64: 153–56
71. Noyes, R. M. 1961. *Progr. React. Kinet.* 1: 129–60
72. Evans, G. T., Fixman, M. 1976. *J. Phys. Chem.* 80: 1544–48
73. Shin, K. J., Kapral, R. 1978. *J. Chem. Phys.* 69: 3685–96
74. Chandrasekhar, S. 1943. *Rev. Mod. Phys.* 15: 1–89
75. Hynes, J. T., Kapral, R., Torrie, G. M. 1980. *J. Chem. Phys.* 72: 177–88
76. Schell, M., Kapral, R. 1981. *Chem. Phys. Lett.* 81: 83–86
77. Martire, B., Gilbert, R. G. 1981. *Chem. Phys.* 56: 241–48
78. Ali, D. P., Miller, W. H. 1983. *J. Chem. Phys.* 78: 6640–45
79. Tully, J. C., Preston, R. K. 1971. *J. Chem. Phys.* 55: 562–72
80. Miller, W. H., George, T. F. 1972. *J. Chem. Phys.* 56: 5637–52
81. Sceats, M. G. 1985. *Chem. Phys.* 96: 299–313
82. Dawes, J. M., Sceats, M. G. 1985. *Chem. Phys.* 96: 315–26
83. Mazur, P., Oppenheim, I. 1970. *Physica* 50: 241–58
84. Brooks, C. L., Balk, M. W., Adelman, S. A. 1983. *J. Chem. Phys.* 79: 784–803
85. Deutch, J. M., Oppenheim, I. 1971. *J. Chem. Phys.* 54: 3547–54
86. Mori, H. 1965. *Progr. Theor. Phys.* 33: 423–55
87. Kubo, R. 1966. *Rep. Progr. Theor. Phys.* 29: 255–84
88. Mori, H. 1965. *Progr. Theor. Phys.* 34: 399–416
89. Vesely, F. J. 1984. *Mol. Phys.* 53: 505–24
90. Ciccotti, G., Ryckaert, J.-P. 1981. *J. Stat. Phys.* 26: 73–82
91. Adelman, S. A. 1979. *J. Chem. Phys.* 71: 4471–86
92. Balk, M. W., Brooks, C. L., Adelman, S. A. 1983. *J. Chem. Phys.* 79: 804–15
93. Brooks, C. L., Adelman, S. A. 1984. *J. Chem. Phys.* 80: 5598–5609
94. Adelman, S. A. 1985. *J. Phys. Chem.* 89: 2213–21
95. Ali, D. P., Miller, W. H. 1984. *Chem. Phys. Lett.* 105: 501–5
96. Stace, A. J. 1981. *J. Chem. Soc. Faraday Trans 2* 77: 2105–10
97. Oxtoby, D. W. 1981. *Adv. Chem. Phys.* 47: 487–519
98. Oxtoby, D. W. 1981. *Ann. Rev. Phys. Chem.* 32: 77–101
99. Nesbitt, D. J., Hynes, J. T. 1981. *Chem. Phys. Lett.* 82: 252–54
100. Nesbitt, D. J., Hynes, J. T. 1982. *J. Chem. Phys.* 6002–14
101. Brown, J. K. 1987. *Molecular dynamics simulations of simple liquid phase chemical reactions.* PhD thesis. Univ. Calif., Berkeley
102. Brown, J. K., Harris, C. B., Tully, J. C. 1988. *J. Chem. Phys.* 88: In press
103. Harris, C. B., Brown, J. K., Paige, M. E., Smith, D. E., Russell, D. J. 1986. In *Ultrafast Phenomena V*, ed. G.

R. Fleming, A. E. Siegman, pp. 326–29. Berlin/Heidelberg/New York: Springer-Verlag. 551 pp.
104. Madigosky, W. M., Litovitz, T. A. 1961. *J. Chem. Phys.* 34: 489–97
105. Schwartz, R. N., Slawsky, Z. I., Herzfeld, K. F. 1952. *J. Chem. Phys.* 20: 1591–99
106. Davis, P. K., Oppenheim, I. 1972. *J. Chem. Phys.* 57: 505–17
107. Delalande, C., Gale, G. M. 1977. *Chem. Phys. Lett.* 50: 339–43
108. Alder, B. J., Wainwright, T. E. 1960. *J. Chem. Phys.* 33: 1439–51
109. Zwanzig, R. 1961. *J. Chem. Phys.* 34: 1931–35
110. Herzfeld, K. F. 1962. *J. Chem. Phys.* 36: 3305–7
111. Zwanzig, R. 1962. *J. Chem. Phys.* 36: 2227
112. Brown, J. K., Russell, D. J., Smith, D. E., Harris, C. B. 1987. *Rev. Phys. Appl.* 22: 1787–92
113. Adelman, S. A., Balk, M. W. 1986. *J. Chem. Phys.* 84: 1752–61

KINETICS OF RADICAL REACTIONS IN THE ATMOSPHERIC OXIDATION OF CH_4[1]

A. R. Ravishankara

Aeronomy Laboratory, National Oceanic and Atmospheric Administration, Boulder, Colorado 80303

INTRODUCTION

Thermal gas phase reactions of free radicals play central roles in many practical systems. Historically, the growth in gas kinetics was closely linked to the understanding of hydrocarbon combustion. The same is true, to some extent, today. The recognition that the Earth's atmosphere is being affected by human activity and that a large fraction of the chemistry in the atmosphere is due to free radicals stimulated laboratory studies on atmospheric free radical reactions in the 1970s and 1980s. Free radical reactions in the oxidation of methane constitute one of the areas that has received a great deal of attention and, consequently, it is better understood than most. I have attempted to review the reactions taking place in the atmospheric oxidation of methane.

Methane is the simplest hydrocarbon and the most abundant hydrocarbon in the Earth's atmosphere. Its concentration is at least a thousand times greater than the next most abundant hydrocarbon. Atmospheric methane concentration has been increasing since the onset of the industrial era (1a,b) and the growth has accelerated in the last two decades (2a,b). The reasons for the increase are not clear. The CH_4 flux into the atmosphere may be increasing or its loss rate from the atmosphere may be decreasing. The oxidation of methane generates nearly 30% of the atmospheric CO

---

[1] The US Government has the right to retain a nonexclusive, royalty-free license in and to any copyright covering this paper.

(3). Methane is a greenhouse gas. It can augment global temperature changes due to increases in atmospheric CO_2 (4a,b). Methane oxidation produces O_3 and, directly or indirectly, many other oxidants such as OH. OH is particularly important because it initiates oxidation of reduced sulfur and nitrogen compounds as well as hydrocarbons. Methane transported into the stratosphere produces a large fraction of the water vapor present in this region. Water vapor is critical to the chemistry of the stratosphere. Because of these reasons, CH_4 oxidation has received a great deal of attention.

The extensive laboratory work carried out to understand CH_4 oxidation has also uncovered interesting information on free radical reactions. Previously accepted mechanisms of some of the reactions were shown to be incorrect. More importantly, it has shown us that we do not fully understand the magnitudes and the negative temperature dependences of some of the rate coefficients in terms of the current theories of reaction rates. Even the mechanisms of some of the reactions involving CH_3O, HCO, and CH_3O_2 are not understood. Contributions of more than one electronic state to the thermal rate coefficients, for example the spin orbit states of Cl and multiple paths in CH_3+O_2 reaction, are suggested but not understood. However, the rate coefficients and product yields data on reactions in the CH_4 oxidation scheme are known well enough for atmospheric purposes. Atmospheric chemists are likely to stop working on these reactions. Therefore, some of the questions raised by these studies may be ignored unless people interested in basic chemical kinetics continue this work. This review is partly aimed at providing a background for such studies.

There are a great many similarities between the oxidation of methane in the atmosphere and in combustion. Even though the differences and similarities between these two oxidation processes were not well known at the time, because of a lack of data at atmosphere temperatures, most of the reaction mechanisms and rate data for the CH_4 oxidation schemes developed in the early 1970s were taken from combustion studies. Extrapolation of some of the data obtained at the high temperatures of combustion to the low temperatures (190–300K) of the atmosphere was not very successful, as shown by the last 15 years of laboratory measurements. However, the basic scheme put together by Levy in the early 1970s (5, 6) based on such extrapolations is valid today and is shown in Figure 1.

In this review I discuss the reactions responsible for conversion of CH_4 to CO. As seen in Figure 1, the oxidation of CH_4 to CO goes through five major intermediates shown in highlighted boxes: CH_3, CH_3O_2, CH_3O, CH_2O, and HCO. I address the reactions that generate and remove these species. There are a large number of species in addition to the five intermediates that enter the picture. The important ones are OH, $O(^1D)$, Cl,

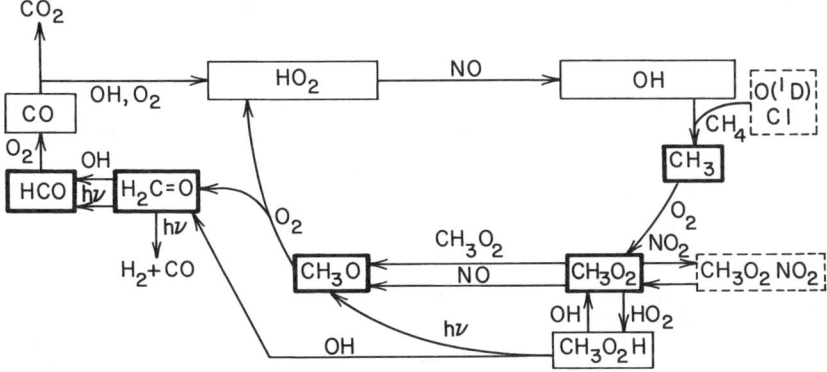

Figure 1 Schematic outline of the methane oxidation cycle. The *dashed boxes* and the *arrow* connecting CH_3OOH to CH_2O were not included in the original discussions by Levy (5, 6). The chemistry of the species in the *high-lighted boxes* are discussed in this article.

NO, NO_2, HO_2, and O_2. I do not discuss the chemistry of these species except as they relate to CH_4 oxidation. I do not discuss the photochemistry that is necessary for a thorough understanding of methane oxidation. In the cases of CH_2O and CH_3OOH where photochemical removal is important, I briefly summarize the present understanding of their photochemistry so as not to leave a gap in the description of the conversion scheme.

EXPERIMENTAL METHODS

I discuss kinetics data on specific reactions. Therefore, an understanding of the experimental methods used to obtain the rate coefficients is necessary to evaluate the data. I do not provide a description of the methods. This section is included to help interested readers find a few sources of descriptions on modern experimental methods.

The basic techniques most often used today to study elementary gas phase reactions of free radicals have been in existence for the last 15 to 20 years. These methods are the descendents of the classic flash photolysis-kinetic spectroscopy and the discharge flow techniques. Kaufman (7) in his last *Annual Reviews* article discussed these methods. Concise descriptions are given by Howard (8) on flow tube methods and Michael & Lee (9) on flash photolysis methods. Hancock (10) has given brief descriptions of most techniques currently used in thermal reaction rate constant measurements. Recent improvements have been in utilizing laser-based spectroscopic methods to detect very low concentrations of free radicals, modern optics to improve absorption methods, and lasers to selectively generate free radicals (11, 12). Laser-induced fluorescence (LIF) spectra of polyatomics have been discovered and successfully applied to kinetics and photochemical studies, e.g. CH_3O (13), NO_3 (14a,b), and CH_3S (15). Because of the short (10 to 20 ns) pulse duration of Q-switched and excimer lasers, the time scales accessible for kinetics studies have been reduced to nanoseconds in pulsed photolysis experiments. Using these lasers requires caution. When studying thermal reactions using quantum state specific detectors, care has to be taken to ensure that the internal degrees of freedom of the photolytically created reactant are thermalized prior to reaction, and excited products are thermalized prior to detection.

In conjunction with flow tubes, chemical ionization and photoionization mass spectrometry have been used to detect reactants and products. Such methods enable detection of some free radicals and stable molecules that were either undetectable or detected with poor sensitivity by other methods (12, 71, 73).

DETAILS OF CH_4 OXIDATION

Loss of CH_4

The major methane loss process in the atmosphere is its reaction with OH. In the stratosphere, its reactions with Cl and $O(^1D)$ also contribute. The photolysis of methane is not important below the mesosphere, where the methane concentration is very low. Therefore, CH_4 photolysis is not discussed here.

$$OH + CH_4 \rightarrow CH_3 + H_2O; \quad k_1 \qquad \qquad 1.$$

Reaction 1 initiates CH_4 oxidation in the Earth's troposphere and, to a large extent, in the stratosphere. It is also an extremely important reaction in flames and combustion. Greiner (16, 17), in his classic studies on hydrocarbon-OH reactions, was the first to measure k_1 under pseudo–first-order conditions in OH. Subsequently, Reaction 1 has been extensively studied between 240 and 1000K (16–30). Many shock tube and flame studies have also been carried out at higher temperatures relevant to combustion. A detailed compilation of data on k_1 is given by Baulch et al (31).

Figure 2 shows a plot of k_1 (on a logarithmic scale) vs $1/T$ for all the data obtained by monitoring the concentration of OH in an excess of CH_4. It does not show data obtained only at 298K. The average of all the 298K data yields $k_1 = 7.7 \times 10^{-15}$ cm^3 s^{-1}. A plot of $\ln(k_1)$ vs $1/T$ is not linear, i.e. it is non-Arrhenius. However, the data in the temperature range 240–400K can be fit to $k_1 = 2.3 \times 10^{-12} \exp(-1700/T)$ cm^3 s^{-1}. This expression, most commonly used in atmospheric calculations, is heavily

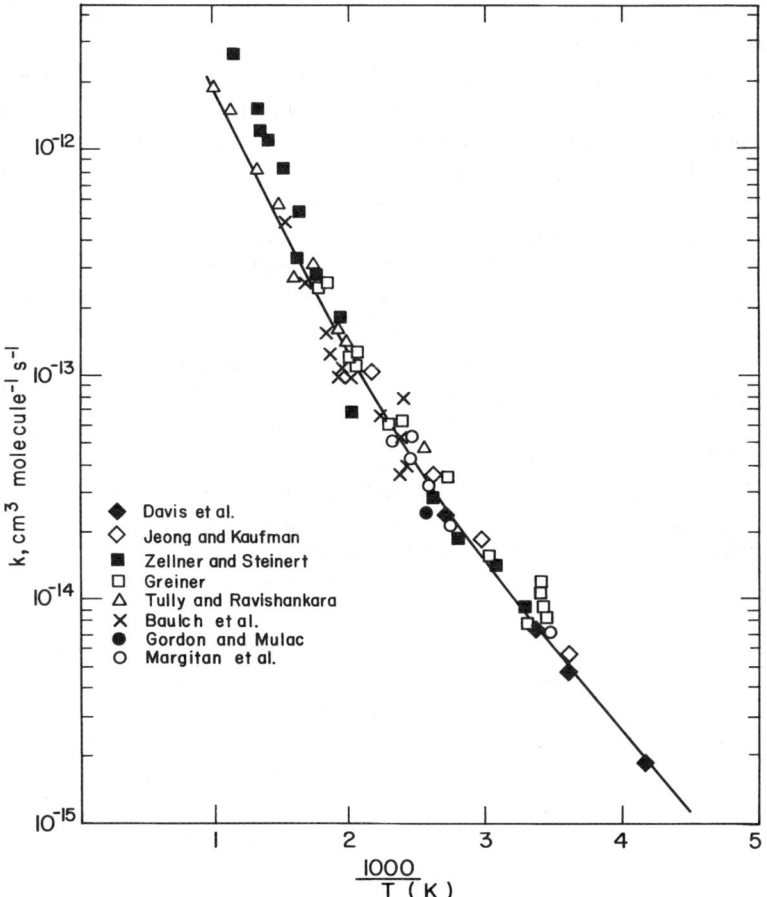

Figure 2 A plot of k_1 (on log scale) vs $1/T$. Only k_1 values measured as functions of temperature are included. The *line* is a three parameter fit given by Tully & Ravishankara (27).

weighted by three studies in which k_1 was measured at $T < 298$K (21, 26, 28). In only one study was k_1 measured down to 240K (26). Thus, there are only a few data points at temperatures of the upper troposphere and the lower stratosphere. At atmospheric temperatures k_1 is extremely small and requires very large concentrations of CH_4 to obtain measurable increases in OH loss rate due to Reaction 1 over those due to background losses. Reactive impurities in methane need to be removed so as not to influence the measured rate coefficients. Slow reactions such as $O(^3P) + H_2/D_2$ have been studied using pulsed photolysis experiments (32), and they show that such measurements are feasible on Reaction 1. A careful study of k_1 down to 180K would be very beneficial.

Efforts are currently underway to determine the sources of atmospheric methane. The ratio of $^{13}CH_4$ to $^{12}CH_4$ concentrations is different in different sources, i.e. natural gas, biogenic emission, etc (33). The atmospheric lifetime of each isotopic species of methane depends on its reactivity with OH. Therefore, to use the measured $^{13}C/^{12}C$ ratio in atmospheric CH_4 to estimate the contributions from different sources, one needs to know the ratio of rate coefficients for reactions of OH with $^{12}CH_4$ and $^{13}CH_4$ extremely accurately. This ratio has been measured by Rust & Stevens (34) and Davidson et al (35), yielding an average value of 1.010 ± 0.007. A more accurate value for this ratio would allow a better assessment of the source strengths. Similar CH_4 source assessments may be made by using deuterated methanes. However, kinetics data on OH + deuterated methanes at atmospheric temperatures are not available. The only rate coefficients data on OH + CH_3D, CH_2D_2, CHD_3, and CD_4 are the 416K values from Gordon & Mulac (23). Measurements of rate coefficients for OH + deuteromethanes would be beneficial not only to atmospheric chemistry but also to test calculations of the rate coefficients by the TST method or trajectory calculations on ab initio potential energy surfaces. A set of rate coefficient data on OH + H_2 and D_2 reactions obtained using one apparatus over a wide temperature range proved to be extremely useful in the past (36); once the barrier for the OH + H_2 reaction was fixed to match the measured rate constant, no other adjustable parameter remained in TST for calculations of the OH + D_2 reaction rate constant. Jeong & Kaufman (28) carried out similar temperature dependence measurements on reactions of a series of halogen substituted methanes with OH for comparison with TST calculations. For further discussions the reader is referred to Kaufman (37).

$$Cl + CH_4 \rightarrow HCl + CH_3; \quad k_2 \qquad 2.$$

In the stratosphere, the reaction of Cl with CH_4 is a minor path for CH_4 oxidation. Its main role is to provide a temporary sink for active chlorine by forming HCl. The rate coefficient for this reaction has been extensively studied using various direct and indirect methods (38–51) and the agreement at 298K among these studies is excellent, yielding an average value of k_2 (298K) = 1.0×10^{-13} cm^3 s^{-1}. However, at temperatures less than 230K the competitive chlorination experiments (41) and the low pressure flow tube measurements (45, 46) have consistently produced slightly lower values than the pulsed photolysis studies. Ravishankara & Wine (48) have hypothesized that the discrepancy is due to different rate coefficients for the reactions of $Cl(^2P_{1/2})$ and $Cl(^2P_{3/2})$ with CH_4 and populations of the spin orbit states that deviate from the thermal distribution in the flow tube and competitive experiments. This hypothesis needs verification. The reaction of $Cl(^2P_{3/2})$ with CH_4 is endothermic by ~ 1 kcal mole^{-1}, and it is intriguing that the activation energy for this process is ~ 2.5 kcal/mole, the same as the spin-orbit splitting in Cl. The Arrhenius plots obtained by most investigators at temperatures greater than 350K are curved upward. For atmospheric purposes all data at temperatures less than 350K are averaged to obtain $k_2 = 1.1 \times 10^{-11} \exp(-1400/T)$ cm^3 s^{-1}.

$$O(^1D) + CH_4 \rightarrow PRODUCTS; k_3 \qquad 3.$$

This reaction is of marginal importance as a methane loss process in the stratosphere and is unimportant in the troposphere. However, it is a source of OH in the stratosphere. At 298K, k_3 has been measured by monitoring the decay of $O(^1D)$ via its 630 nm emission (52) or 115.2 nm resonance absorption (53) and the growth of $O(^3P)$ via resonance absorption (54). The $O(^1D)$ quenching rate coefficients measured by using the 115.2 nm absorption are 1.7 times higher than those obtained by using other $O(^1D)$ detection methods for all reactions studied, except for that with O_3. This deviation has been attributed to the $O(^1D)$ concentration being linearly dependent on the measured absorbance, while the experimenters assumed a nonlinear relationship (174). After accounting for this artifact, the agreement among various studies is good. The average value of k_3 at 298K is 1.5×10^{-10} cm^3 s^{-1}. k_3 is independent of temperature (52), and hence the 298K value is valid at the lower atmospheric temperatures.

The interaction of $O(^1D)$ with CH_4 can produce $OH + CH_3$; (3a), CH_3OH; (3b), $CH_2O + H_2$; (3c), $CH_2OH + H$ or $CH_3O + H$; (3d), $O(^3P) + CH_4$; (3e), and $CH_2 + H_2O$; (3f). Lin & DeMore (56) have carried out end product analysis studies and suggest that the yield of OH is $\sim 90\%$ and that of CH_2O is $\sim 10\%$. Wine & Ravishankara (55) have measured the yield of $O(^3P)$ to be less than 4.3% by directly monitoring $O(^3P)$ production. In a molecular beam study, Casavecchia et al (57) found evidence for the occurrence of channel (3d) but no quantitative information on branching ratios was obtained. The mechanism for the reaction seems to be insertion of $O(^1D)$ into the C–H bond to generate very hot CH_3OH, which decomposes to various products, as suggested by the work of Cvetanovic and co-workers (58–61) and others (62–64). Rotational and vibrational distributions of nascent OH produced by Reaction 3 have been measured (65a,b). The rotational level populations in $v' = 0$ and 1 levels exhibit bimodal distributions with one component that is very hot. Vibrational levels up to $v' = 4$ are populated and they are not highly inverted. These results do not contradict the insertion mechanism. There is no evidence to suggest the formation of CH_2 in Reaction 3.

Loss of CH_3

$$CH_3 + O_2 \xrightarrow{M} CH_3O_2; k_4 \qquad 4.$$

The only important reaction of CH_3 in the Earth's atmosphere is that with O_2. Until the early 1970s Reaction 4 had been studied mainly via end product analysis (66, 67). CH_3 radicals were photolytically generated and the quantum yields for stable end products were measured in various concentrations of O_2 (68, 69) to obtain k_4. For good reviews on work carried out prior to 1975 the reader is referred to McMillan & Calvert (66) and Walker (67). From the end product analysis studies it was known that CH_3 added to O_2 to form CH_3O_2, and that at atmospheric pressures and temperatures Reaction 4 is in the fall off region between second and third order. There were conflicting data on the contribution of the bimolecular path, $CH_3 + O_2 \rightarrow CH_2O + OH$, at 298K to the overall reaction.

The UV absorption spectrum of CH_3 was discovered by Herzberg & Shoosmith (70) in 1956 but was not exploited for direct measurement of k_4 in flash photolysis systems until the early 1970s (74–80). Mass spectrometry (71–73) and IR absorption (81) were also used to detect CH_3 for measuring k_4. The results obtained by various investigators are not in very good agreement. There are three discrepancies in the 298K data: (a) the value of k_4^∞, (b) the value of k_4'' at a given pressure in the fall-off region,[2] and (c) the significance of the $CH_3 + O_2 \rightarrow CH_2O + OH$ reaction.

Pilling and co-workers (79, 80) have minimized the secondary reactions that plague the measurement of k_4. They pointed out that the measured values of k_4 would be in error due to secondary reactions such as $CH_3 + CH_3O_2 \rightarrow 2CH_3O$ and $CH_3 + CH_3 \xrightarrow{M} C_2H_6$ (79) unless the ratio of $[CH_3]_0$ to $[O_2]$ is kept very low. Pilling & Smith calculated $k_4^\infty = 1.05 \times 10^{-12}$ cm^3 s^{-1} at 298K, by extrapolating their subatmospheric pressure data using Troe's formalism (82, 83). Their k_4^∞ value is approximately one half that measured by Cobos et al (78) at 100 atmospheres of Ar and N_2, where the reaction is expected to be at the high pressure limit. Pilling & Smith (79) have hypothesized that more than one reaction channel leading to CH_3O_2 formation is open at higher pressures. The same suggestion was made earlier by Selzer & Bayes (73) to explain the fall-off

---

[2] For a third-order reaction, $A + B \xrightarrow{M} AB$, when the rate coefficient is measured under pseudo–first-order conditions in [A], $-d[A]/dt = k'[A]$ where $k' = k''[B]$. In the third-order regime, k'' varies linearly with $[M]$, the concentration of the third body, such that $k'' = k[M]$, where k is the third-order rate coefficient. In the fall-off regime where the reaction is between third- and second-orders, $-d[A]/dt \propto [M]^n$ where $0 < n < 1$. At high pressures when $n = 0$, $k'' = k^\infty$, the high pressure limiting rate constant.

behavior at pressures less than 5 torr. Better data is required to confirm the existence of multiple channels.

Low pressure measurements by Plumb et al (72) and Selzer & Bayes (73) carried out using flow tube reactors are in reasonable agreement and show that Reaction 4 is in the fall-off region even at pressures as low as 2 torr. These two groups agree with Baldwin & Golden (84) and Klais et al (85) that the direct channel leading to CH_2O+OH is negligible, since in these studies k_4'' vs pressure plots did extrapolate to zero at zero pressure after accounting for the fall-off at these low pressures. In the previous low pressure studies (71) it was incorrectly assumed that Reaction 4 was in the third-order regime and linear extrapolations were made in the k_4'' vs pressure plots. These extrapolations yielded positive intercepts that were interpreted as evidence for a bimolecular channel giving OH and CH_2O.

Keiffer et al (80) and Pratt & Wood (86) have studied Reaction 4 above 298K. Keiffer et al extrapolated the k_4'' vs pressure data using Troe's method to calculate k_4^∞. The k_4'' values at the higher temperatures are lower than those at 298K; therefore, the extrapolations are less certain. The k_4^∞ values increase with increasing temperature and suggest a small (~ 1 kcal/mole) barrier for the association. Pratt & Wood carried out their relative rate constant measurements at 2 to 10 torr between 230 and 568K and did not have sufficient data to calculate k_4^∞ by extrapolation.

Despite the noted discrepancies among the large number of studies, the fate of CH_3 radical in the Earth's atmosphere is very clear—it leads to CH_3O_2 within a fraction of a second. Therefore, it is unnecessary to recommend a rate constant, and no new measurements on this reaction are needed for atmospheric purposes.

Reactions of CH_3O_2

If the concentration of NO is more than ten times that of peroxy radicals (HO_2 being the most important), CH_3O_2 will react predominantly with NO. Otherwise, its reactions with peroxy species become competitive, and it is necessary to know the rate coefficients of all the CH_3O_2 loss processes accurately. The reaction of CH_3O_2 with O_3 is very slow (87) and so can be neglected.

$$CH_3O_2 + NO \rightarrow PRODUCTS; k_5 \qquad 5.$$

The early end product analyses provided valuable information on the mechanism and products of the reaction (88–93). The discovery, in the mid 1970s, of the UV absorption spectrum of CH_3O_2 (76, 94) enabled many direct measurements of k_5. In their flash photolysis apparatus, Anastasi et al (95) were unable to detect CH_3O_2 in the presence of NO 150 μs after the flash, and placed a lower limit of 1×10^{-12} cm^3 s^{-1} on k_5. Adachi & Basco

(96), Sander & Watson (97), and later Simonaitis & Heicklen (98) used the CH_3O_2 UV absorption to measure k_5 directly. The initial studies of Cox & Tyndall (99) provided a value of $\sim 6 \times 10^{-12}$ cm^3 s^{-1}; but as the authors acknowledged, their system was not in its optimal configuration for the measurement of such a fast reaction. Plumb et al (100, 101) have measured k_5 in a flow tube coupled to a mass spectrometer where CH_3O_2 was ionized by electron impact. Ravishankara et al (102) monitored the rise of product NO_2, while Zellner et al (103) have monitored both the decay of CH_3O_2 and the rise of CH_3O to obtain k_5. The values of k_5 obtained by all these investigators are shown in Table 1. It is clear that, except for the data of Adachi & Basco, k_5 values obtained using different experimental methods agree well.

The experiments of Adachi & Basco (96) yielded a low value of k_5 because CH_3ONO formed as a secondary product absorbed at 245 nm, the monitoring wavelength. The first study of Plumb et al (100) had problems with their source of CH_3, i.e. $O + C_2H_4 \rightarrow CH_3 + HCO$, which was converted into CH_3O_2. The $O + C_2H_4$ reaction is now known to be more complicated than previously believed, giving H atoms and vinoxy radicals in addition to CH_3 (104–108). Plumb et al obtained a value of 8.0×10^{-12} cm^3 s^{-1}. In a later investigation, Plumb et al (101) circumvented the source chemistry problems by using the $Cl + CH_4$ reaction to generate CH_3 radicals. The result of this study, $k_5 = (8.6 \pm 2.0) \times 10^{-12}$ cm^3 mol^{-1} s^{-1}, agrees with their previous value. The average value for k_5 (298K)

Table 1 Summary of k_5 measured at 298K by isolating the reaction

k_5, 10^{-12} cm^3 s^{-1}	Method[a]	Reference
>1.0	FP/KS	Anastasi et al (95)
3.0±0.2[b]	FP/KS	Adachi & Basco (96)
6.5±2.0	MMS	Cox & Tyndall (99)
8.0±2.0	DF/MS	Plumb et al (100)
7.1±1.4	FP/UVA	Sander & Watson (97)
8.1±1.6	LP/LIF(NO$_2$)	Ravishankara et al (102)
8.6±2.0	DF/MS	Plumb et al (101)
7.7±1.8	FP/UVA	Simonaitis & Heicklen (98)
7±2	LP/LIF-UVA[c]	Zellner et al (103)
7.8±1.2		Average of last six values

[a] FP: flash photolysis. KS: kinetic spectroscopy. MMS: molecular modulation spectroscopy. UVA: uv absorption. LP: laser photolysis. LIF: laser induced fluorescence. DF: discharge flow. MS: mass spectrometry.
[b] All errors are those quoted by the authors.
[c] LIF detection of CH_3O and UV absorption detection of CH_3O_2.

shown in Table 1 was obtained by excluding the results of the first three studies listed in the table.

Ravishankara et al (102) and Simonaitis & Heicklen (98) have measured k_5 in the temperature range 240–339K and 218–365K, respectively. At 218K, the lowest temperature of their study, Simonatitis & Heicklen observed a pressure dependence. Interference due to the $CH_3O_2 + NO_2$ reaction, which is faster at the lower temperatures, cannot be ignored in their system. Therefore, the recommended temperature dependence, $k_5 = 4.3 \times 10^{-12} \exp(+180/T)$ cm^3 s^{-1}, is derived from the data of Ravishankara et al and the higher temperature values of Simonaitis & Heicklen.

End product analysis studies by Heicklen's group (88) had suggested that CH_3ONO_2 and $CH_2O + HONO$ are the main products. However, later measurements by this group (90) and those by Pate et al (92) agree that $CH_3O + NO_2$ are produced. Direct measurements of the NO_2 yield by Ravishankara et al (102) and CH_3O yield by Zellner et al (103) have clearly shown that NO_2 and CH_3O are the major primary products at 298K and pressures less than 100 torr.

$$CH_3O_2 + NO_2 \rightarrow PRODUCTS; \ k_6 \qquad\qquad 6.$$

Addition of CH_3O_2 to NO_2 forms $CH_3O_2NO_2$, which in turn can thermally decompose back to reactants.[$\Delta H_r^0(298) = 22.4$ kcal mole^{-1}. Heats of formation are from CODATA evaluation (111). $\Delta H_f^0(298)$ for CH_3O_2 is from Slagle & Gutman (109) and Kachatryan et al (110)]. Therefore, at temperatures and pressures where $CH_3O_2NO_2$ decomposition is fast, Reaction 6 has little effect on the chemistry of the atmosphere. In the temperature (200–300K) and pressure (50–760 torr) ranges of the lower atmosphere, this reaction is in the fall-off region between second and third order. Depending on the lifetime of $CH_3O_2NO_2$, it could be transported to provide oxidants at locations other from where it is generated.

There have been four direct measurements of k_6. Cox & Tyndall (99) observed that k_6 was nearly independent of pressure at 298K, with values of 1.2×10^{-12} cm^3 s^{-1} at 50 torr (Ar + CH$_4$) and 1.6×10^{-12} cm^3 s^{-1} at 520 torr (N$_2$). This observation is in agreement with the results of Adachi & Basco (113), who obtained a value of 1.53×10^{-12} cm^3 s^{-1} independent of pressure. In contrast, Sander & Watson (97) and Ravishankara et al (114) observed the measured rate coefficients to vary with pressure between 50 and 700 torr N$_2$ at 298K. Their results over this pressure range are in excellent agreement. As pointed out by Sander & Watson, it is likely that thermal decomposition of $CH_3O_2NO_2$ and depletion of NO_2 were problems in the study by Cox & Tyndall. Adachi & Basco used very high concentrations of CH_3O_2 such that Reaction 9 was competing with Reaction 6. Also, they had to make corrections for the reaction of CH_3O_2 with NO produced by NO_2 photolysis. Ravishankara et al observed that as the temperature was lowered, at a given concentration of the bath gas, the rate coefficient increased. The shape of the observed fall-off curve [i.e. k_6'' vs (M)] was characteristic of an addition reaction. The rate coefficients for the thermal decomposition of $CH_3O_2NO_2$ investigated by Bahta et al (115) show a pressure dependence that is consistent with the pressure dependence for the forward reaction. These observations show that Reaction 6 is a pressure-dependent addition reaction in the temperature range 200–300K between 50 and 760 torr of N_2.

Cox & Tyndall (99) and Sander & Watson (97) have detected the $CH_3O_2NO_2$ product via its UV absorption. Based on thermochemistry Reaction 6 could produce CH_2O and $HONO_2$ and not CH_3O and NO_3. CH_2O and $HONO_2$ may be formed via a six-center reaction or by adduct formation followed by an H atom transfer and decomposition. Such reactions seem unlikely. Based on extrapolations of the measured k_6'' to zero pressure, the bimolecular channel, if taking place at all, contributes no more than a few percent to the overall CH_3O_2 loss due to Reaction 6 at atmospheric pressures and temperatures. Quantitative measurements of the yield of $CH_3O_2NO_2$ would be useful.

ABSORPTION CROSS SECTION OF CH_3O_2 All measurements of k_7 and k_9 have utilized CH_3O_2 detection by UV absorption. Reaction 7 cannot be easily studied under pseudo–first-order conditions in CH_3O_2. Because the self-reaction of HO_2 is rapid, one cannot maintain a constant excess concentration of HO_2. Reaction 9 is second-order in $[CH_3O_2]$ and hence one measures k_9/σ, which requires σ, the absorption cross section of CH_3O_2, to calculate k_9. Therefore, measurements of k_7 and k_9 require the absolute cross section of CH_3O_2. The wavelength-dependent absorption cross sections obtained by various groups (76, 77, 79, 116–120) are shown in Figure 3; unfortunately, the agreement is poor. Two causes for the discrepancies are errors in the CH_3O_2 concentration measurements and the presence of absorbing species other than CH_3O_2 in the cell. The CH_3O_2 concentration is calculated by using an actinometry experiment, and it assumes a thorough knowledge of the source reactions. In no study has CH_3O_2 been converted to a stable product of known UV absorption cross section to obtain its initial concentration. The lower part of Figure 3 shows the relative cross sections obtained by normalizing the value at the peak to 1. The peak wavelength was chosen for normalization because, in each measurement, this cross section should have the least error. This figure suggests interferences at wavelengths shorter than 250 nm. The agreement in the relative cross sections is better at longer wavelengths than at shorter wavelengths. Even though all of these studies utilized Reaction 4 to generate CH_3O_2,

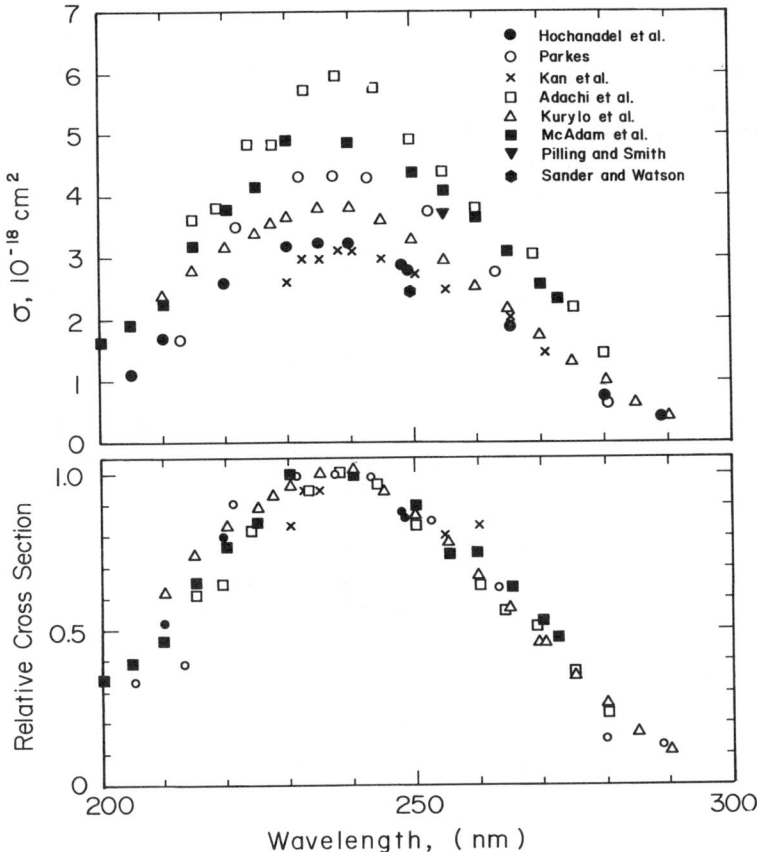

Figure 3 *Top:* The absorption cross sections of CH_3O_2 measured by various investigators in the wavelength range 200–300 nm. *Bottom:* The data from the top figure normalized to a value of 1 at 240 nm.

different interfering species may have been produced because of different reaction conditions.

To resolve this conflict, a CH_3O_2 source other than Reaction 4 should be used. Abstraction of an H atom from the oxygen in CH_3OOH could be such a reaction. Also, CH_3O_2 should be converted quantitatively to a stable molecule whose concentration is easy to measure accurately. The reaction of CH_3O_2 with NO followed by the conversion of CH_3O to CH_3ONO looks promising. NO_2 produced in the $CH_3O_2 + NO$ reaction can also be easily measured. The temperature dependence of the absorption cross section should also be measured. When such measurements are made the discrepancies are likely to be resolved.

$CH_3O_2 + HO_2 \rightarrow$ PRODUCTS; k_7 7.

Four direct studies have been carried out on Reaction 7 (120–124). In all cases the UV absorption was used to detect CH_3O_2 and HO_2 and the $CH_3 + O_2$ reaction was used to produce CH_3O_2. HO_2 was generated by either the $H + O_2$ or $CH_2OH + O_2$ reactions. The four studies disagree on the value of k_7. All the studies do agree that Reaction 7 is faster than the self-reactions of the reactants, i.e. Reaction 9 and $HO_2 + HO_2 \rightarrow H_2O_2 + O_2$.

In order to convert the observed temporal absorption profile at the monitoring wavelength into a rate coefficient, one should take into account the absorptions due to HO_2, CH_3O_2, CH_3OOH, H_2O_2, and perhaps others, e.g. CH_3OOCH_3, because all these species absorb in the 210–260 nm range. Since the absorption spectra of these species are different, the contributions to their measured absorption depends on wavelength. Hence, measurements at two or more wavelengths should help. In addition to the problem of multiple species absorption, Reaction 7 will be taking place in competition with the self-reactions of CH_3O_2 and HO_2. Therefore, the complete reaction sequence has to be modeled. The values obtained to date range from ~ 3 to 6×10^{-12} cm^3 s^{-1} with an average $k_7 = (4.5 \pm 3.0) \times 10^{-12}$ cm^3 s^{-1}. Cox & Tyndall (274–338K), and Dagaut et al (228–380K) (123) also measured the rate coefficient for this reaction as a function of temperature by assuming the cross sections of both CH_3O_2 and HO_2 to be temperature independent. They obtained E/R (E is the activation energy and R is the gas constant) values of -1300K and -720K, respectively. Dagaut et al quote an unpublished value of ≈ -1500K measured by B. Veyret.

To obtain more accurate values of k_7, Reaction 7 should be studied under isolated conditions. If 10^{11} cm^{-3} of HO_2 could be detected in a flash photolysis experiment, it would be feasible to measure k_7 under pseudo-first-order conditions in HO_2. To improve our understanding, it would be very useful to measure the end products accurately by using methods such as IR absorption. Different methods for generation and detection of CH_3O_2 and HO_2 may hold the key to understanding Reaction 7.

It appears that k_7 does not depend on pressure and water vapor concentration (122), unlike the $HO_2 + HO_2$ reaction; ordinarily this would be interpreted as an indication of H atom abstraction mechanism. However, the observed negative temperature dependence would support a long-range attraction and, perhaps, a complex formation route. Studies of the $CH_3O_2 + DO_2$ reaction and ^{18}O substitution for ^{16}O could shed more light on the mechanism. Direct accurate measurements of the product yields would be very helpful. In an end products analysis study, Moortgat et al (124) measured CH_3OOH produced in Reaction 7. The yield of CH_3OOH

was found to be lower than unity. This lower value could be due to problems in measuring CH_3OOH concentration or it could indicate the existence of a second reaction channel.

CH_3OOH REMOVAL The role CH_3OOH plays in the atmosphere depends on its lifetime, the products of its reactions with OH, and products of its photolysis. CH_3OOH is not as soluble in water as H_2O_2, but it still could be removed via scavenging during precipitation. If CH_3OOH is taken up by liquid drops, it may oxidize SO_2 and other species in a manner similar to H_2O_2. The different pathways by which CH_3OOH is removed from the atmosphere affects ozone production and odd hydrogen removal differently.

The absorption cross-sections of CH_3OOH have been measured in the laboratory (121, 125), and they are similar to those of H_2O_2 (111, 126). The rate of photolysis of CH_3OOH in the atmosphere can be calculated assuming a quantum yield of 1 for dissociation. The energetically feasible paths along with the threshold wavelengths are:

$$CH_3OOH \rightarrow CH_3O + OH, \quad \lambda \leq 650 \text{ nm}$$
$$\rightarrow CH_3O_2 + H, \quad \lambda \leq 329 \text{ nm}$$
$$\rightarrow CH_3OH + O, \quad \lambda \leq 680 \text{ nm}$$
$$\rightarrow CH_3O + H + O, \quad \lambda \leq 195 \text{ nm}$$
$$\rightarrow CH_3 + HO_2, \quad \lambda \leq 408 \text{ nm}.$$

Preliminary results from Dr. Vaghjiani's work in our laboratory indicate that OH is the major product of photolysis at 248 nm. Further work on the photochemical pathways is needed.

The reaction of CH_3OOH with OH has not been studied under isolated conditions. In a competitive study, Niki et al (127) have measured $k_8 \approx 1.0 \times 10^{-11}$ cm^3 s^{-1}, ten times faster than that estimated by analogy with the $OH + H_2O_2$ (126) and $CH_3OH + OH$ (128a–f) reactions. Niki et al also estimated that the branching ratios for the two possible paths,

$$CH_3OOH + OH \rightarrow CH_3O_2 + H_2O; k_{8a} \quad \quad \quad 8a.$$
$$\rightarrow CH_2OOH + H_2O; k_{8b} \quad \quad \quad 8b.$$
$$\downarrow$$
$$CH_2O + OH$$

are 60% and 40%, respectively. CH_2OOH formed in Reaction 8b is unstable and decomposes to OH and CH_2O (127). Therefore, when OH disappearance is monitored, as is customary, only the rate coefficient for Reaction 8a is measured. However, when ^{18}OH or OD are reacted with CH$_3$OOH, only ^{16}OH is regenerated, and the loss rates of ^{18}OH or OD yield the rate coefficients for the overall reaction. Results of such studies using isotopically labeled species by Dr. Vaghjiani in our laboratory indicate values that are approximately half those of Niki et al.

The reaction of CH$_3$OOH with O(^3P) is slow (129) whereas that with Cl is expected to be extremely fast (130). However, in the atmosphere, Reaction 8 is expected to dominate all other radical reactions. Also, the loss rate of CH$_3$OOH due to Reaction 8 is similar in magnitude to that via photolysis and scavenging by water drops. It is interesting to note that Reaction 8b leads directly to CH$_2$O; this step was not considered by Levy (5, 6).

$$CH_3O_2 + CH_3O_2 \rightarrow PRODUCTS; \ k_9 \qquad 9.$$

Under conditions where the peroxy radical concentrations are very high due to low concentrations of NO$_x$ or high concentrations of organics, Reaction 9 is important. Unlike the reaction between two HO$_2$ radicals, which is believed to form mostly H$_2$O$_2$ (131), the reaction between two CH$_3$O$_2$ radicals has many possible products.

$$CH_3O_2 + CH_3O_2 \rightarrow 2\,CH_3O + O_2;$$

$$\Delta H_r^0(298) = +3.0 \text{ kcal mole}^{-1} \qquad 9a.$$

$$\rightarrow CH_2O + CH_3OH + O_2;$$

$$\Delta H_r^0(298) = -79.4 \text{ kcal mole}^{-1} \qquad 9b.$$

$$\rightarrow CH_3OOCH_3 + O_2;$$

$$\Delta H_r^0(298) = -34.6 \text{ kcal mole}^{-1} \qquad 9c.$$

$$\rightarrow CH_3OOH + CH_2O_2;$$

$$\Delta H_r^0(298) = -14 \text{ kcal mole}^{-1} \qquad 9d.$$

[All the values of $\Delta H_r^0(298)$ were calculated by using $\Delta H_f^0(298)$ of 2.7 kcal mole^{-1} for CH$_3$O$_2$ (109, 110) and 4.2 kcal mole^{-1} for CH$_3$O (112). Other heats of formation were taken from the CODATA review (111).] Kan et al (132) and Niki et al (133) used FT-IR instruments to estimate the branching ratios k_{9a}/k_9, k_{9b}/k_9, and k_{9c}/k_9 to be 0.35, 0.55, and 0.1, respectively. Parkes (77) had earlier measured k_{9a}/k_9 to be $\sim$0.35. Neither Kan et al nor Niki et al found evidence for Reaction 9d, which is an exothermic H atom abstraction reaction. Nangia & Benson (134) suggested this pathway to rationalize the rapidity of Reaction 9 compared to self-reactions of other peroxy radicals such as t-butyl peroxy radicals. Reaction 9 may proceed via the formation of a CH_3O_2-O_2CH_3 (tetraoxide) complex that undergoes different rearrangements to produce different products. There is no evidence to suggest that the tetraoxide intermediate, if formed, can be stabilized.

The overall rate coefficient k_9, defined by the equation,

$$-\frac{d[CH_3O_2]}{dt} = 2k_9[CH_3O_2]^2$$

has been measured by many groups who detected CH_3O_2 by UV absorption in the 250 nm region (76, 77, 94, 95, 97, 116–118, 120, 121, 135, 136). The quantity actually measured in all investigations is k_9/σ. The agreement on k_9/σ among all measurements is reasonable even though they do not agree on the value of σ. For example, the five measurements (97, 117, 120, 121, 136) at 250 nm yield a mean value of $k_9/\sigma = (1.24 \pm 0.20) \times 10^5$ cm s^{-1}, where the error is 1σ. Converting a value of k_9/σ into k_9 using one value of σ requires us to know σ. As mentioned above in the section on CH_3O_2 absorption cross section, the agreement on σ is not good and one reason for the disagreement could be the inaccuracies in measuring the CH_3O_2 concentration. If we assume that this is the sole reason for the different values of σ, we can convert the k_9/σ to k_9 by using a value of σ that we believe is correct. In the absence of a preferred value for σ, $k_9 = (5 \pm 3) \times 10^{-13}$ cm^3 s^{-1} can be recommended.

The negative temperature dependence of k_9, $E/R \approx -200$K, suggests the formation of a complex during this reaction. It is likely that the product distribution changes with temperature. Further studies on product identification as a function of temperature would be useful in understanding the mechanism of Reaction 9.

The values of k_9 measured by all the investigators could be in error due to the reactions of CH_3O product and HO_2 with CH_3O_2. To correct k_9 for the CH_3O reaction requires careful consideration of all CH_3O losses in the studied system. If CH_3O reacts with Cl_2, as could be the case in many studies, rather than with O_2, the corrections would be different. Measurements of k_9, where CH_3O_2 is observed by a method other than UV absorption and is generated by a different method, would be helpful in obtaining an accurate value of k_9.

In the atmosphere Reaction 9 is less important than Reaction 7 for the loss of CH_3O_2, since HO_2 concentration is high when that of CH_3O_2 is high and k_9 is ten times smaller than k_7. However, Reaction 9 is a source of CH_3OH, which is not formed in any other reactions in the methane oxidation scheme.

CH_3O Loss

$CH_3O + O_2 \rightarrow$ PRODUCTS; k_{10} 10.

Reactions of methoxy radicals were studied for many years at temperatures relevant to combustion using indirect methods (66, 67, 137–139). From these early studies it was known that k_{10} is quite small at 298K and that it has a large activation energy. Even after CH_3O was directly detected via the laser magnetic resonance method (140), Reaction 10 was not studied using this technique. The discovery by Inoue et al (13, 141) that CH_3O fluoresces upon laser excitation at ~ 320 nm enabled direct studies on k_{10}. Sanders et al (142) attempted to measure this rate coefficient at 298K, but obtained only an upper limit of 1.7×10^{-15} cm^3 s^{-1}. The difficulty in their study was the efficient quenching of the fluorescence signal by O_2. Later, Gutman et al (143) were able to measure k_{10} between 413 and 628K since lower concentrations of O_2 were needed than at 298K. They obtained an Arrhenius expression, $k_{10} = 1.0 \times 10^{-13} \exp(-1310/T)$ cm^3 s^{-1}, which extrapolated to a room temperature value of 1.3×10^{-15} cm^3 s^{-1}.

Lorenz et al (144) and Wantuck et al (145) improved the LIF detection sensitivity sufficiently to measure k_{10} at 298K, yielding an average value of $k_{10}(298K) = 1.9 \times 10^{-15}$ cm^3 s^{-1}. Lorenz et al measured k_{10} over the temperature range 298–450K and obtained a linear Arrhenius plot. Wantuck et al studied this reaction between 298 and 973K and showed the Arrhenius plot of k_{10} to be curved over this range. The expression, $k_{10} = 1.5 \times 10^{-10} \exp(-6028/T) + 3.6 \times 10^{-14} \exp(-880/T)$ cm^3 s^{-1}, fits their data well. The data from the three studies where k_{10} was directly measured under pseudo–first-order conditions in (CH_3O) are shown in Figure 4. The data in the temperature range 298–500K obtained by Gutman et al (143), Lorenz et al (144), and Wantuck et al (145) are fit by the Arrhenius form, $k_{10} = 3.9 \times 10^{-14} \exp(-900/T)$ cm^3 s^{-1}, which is essentially the second half of Wantuck et al's expression. Extrapolation of this expression to atmospheric temperatures yields k_{10} that is uncertain by no more than a factor of three. In the atmosphere, only Reaction 10 is important for CH_3O loss.

Zellner (146) has shown that the yield of CH_2O in Reaction 10 is unity, within the uncertainty of his measurement. So, the products of this reaction, $CH_2O + HO_2$, are those expected for an H atom abstraction. (The other possible products are $HCO + H_2O_2$ and $HCOO + H_2O$.) The lower temperature Arrhenius A factor, 3.9×10^{-14} cm^3 s^{-1}, is too low for a simple H atom abstraction reaction. It is conceivable that the $CH_3O + O_2$ reaction actually proceeds through the formation of a $CH_3O \cdot O_2$ complex. The reaction between HO_2 and CH_2O, the products of Reaction 10, is known

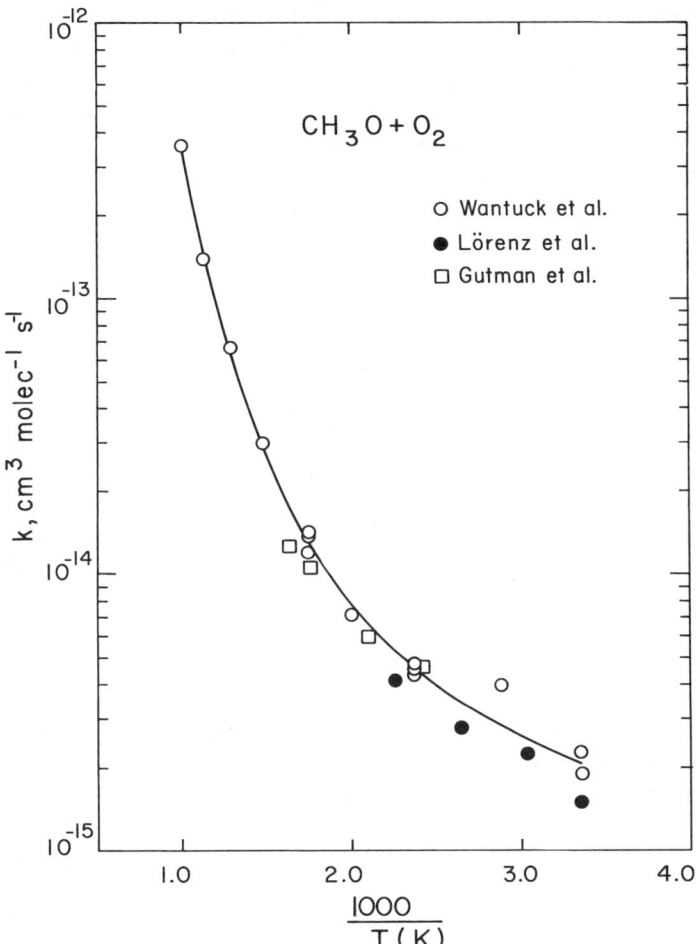

Figure 4 A plot of k_{10} (on log scale) vs $1/T$. Only k_1 values measured by direct detection of CH_3O are shown. The *line* is a fit to the data given by Wantuck et al (145).

to proceed through the formation of an adduct. Measurements of k_{10} at temperatures lower than 298K, determinations of the products of Reaction 10 as functions of pressure and temperature, and investigation of the $CD_3O + O_2$ reaction would help us understand this reaction better.

Comparison of Reaction 10 with the $CH_2OH + O_2$ reaction shows that the carbon centered radical reacts with O_2 much faster; $k(CH_2OH + O_2) = 9.6 \times 10^{-12}$ cm^3 s^{-1} at 298K (126). Therefore, Reaction 10 may proceed via isomerization of CH_3O to CH_2OH followed by reaction with O_2. Such a process is unlikely because the isomerization is expected to have a very large activation energy according to the electronic structure calculations (147). As discussed by Batt et al (148), CH_3O does not isomerize at low temperatures. In the atmosphere, isomerization of CH_3O, if it took place at all, would only speed up the loss of CH_3O. The isomerization and the thermal decomposition

$$(CH_3O \xrightarrow{O_2} CH_2O + H)$$

are two possible reasons for the observed non-Arrhenius behavior at temperatures greater than 500K.

Loss of CH_2O

In the atmosphere, CH_2O either reacts with OH or it is photolyzed. In the lower atmosphere, the OH reaction rate is of the same order of magnitude as the photolysis rate. Therefore, accurate photochemical parameters and kinetics data are required to understand the atmospheric chemistry of CH_2O.

CH₂O PHOTOLYSIS   The near UV absorption spectrum of CH_2O stretches from 260 to 370 nm and has discrete bands. The photolysis and photochemistry of CH_2O is extremely complex but has been thoroughly investigated. Excellent reviews by Moore & Weisshaar (149) and Calvert (150) cover the photophysics and photochemistry of CH_2O relevant to the atmosphere. Atmospheric photochemistry of CH_2O is controlled by three factors. First, there are two competing photodissociation channels. They are shown below along with the threshold wavelengths:

$CH_2O \xrightarrow{h\nu} H + HCO; \quad \lambda \leq 328$ nm at 298K   11a.

$\rightarrow H_2 + CO;$ at all λ.   11b.

The quantum yields for these two channels change with wavelength. Second, the quantum yield for Channel 11b decreases with increase in pressure at $\lambda > 330$ nm and is always less than 1 at $\lambda > 340$ nm. Third, the absorption cross-section itself is temperature dependent. To calculate the product yields in the atmosphere, the cross sections, which depend on temperature and wavelength, and quantum yields, which depend on pressure and wavelength, have to be convoluted with the wavelength-dependent solar flux. Therefore, quantum yields measured directly in air (151) are often used in atmospheric calculations. It is worth noting that

CH_2O photolysis is the main source of atmospheric H_2.

$$OH + CH_2O \rightarrow HCO + H_2O; \quad k_{12} \qquad \qquad 12.$$

Atkinson & Pitts (152) and Stief et al (153) measured k_{12} using pulsed photolytic production of OH that was detected by resonance fluorescence, over the temperature range 299 to 466K and 228 to 362K, respectively. They agree that k_{12} is independent of temperatures with a value of $\approx 1.0 \times 10^{-11}$ cm^3 s^{-1}. Niki and his co-workers measured k_{12} directly using a discharge flow mass spectrometer system (154) and indirectly, relative to the $OH + C_2H_4$ reaction, in a chamber (155, 156). Their results are in reasonable agreement with the above value. Smith (157) used a discharge flow apparatus to obtain $k_{12} \approx 6 \times 10^{-12}$ cm^3 s^{-1} at 298K and an E/R of ~ 700K. His 298K value is close to the value quoted above, but the temperature dependence disagrees with the direct measurements. Smith measured k_{12} relative to the rate coefficient for the $OH + OH$ reaction, which is not known accurately enough to be a reference reaction. Temps & Wagner (158), using a discharge flow tube equipped with a laser magnetic resonance (LMR) detector for OH, directly measured a value of $(8.1 \pm 1.7) \times 10^{-12}$ cm^3 s^{-1} at 298K, in agreement with the flash photolysis studies. A temperature independent value of $(1.0 \pm 0.2) \times 10^{-11}$ cm^3 s^{-1} is recommended.

Horowitz et al (159) have some evidence to suggest that $H + HCOOH$ could be the products of Reaction 12. Niki et al (156) showed that CH_2O is quantitatively converted to CO in air and also suggested that HCOOH seen by Horowitz et al is due to the formation of $HO_2 \cdot CH_2O$ rather than to Reaction 12. The upper limit for the addition channel leading to $CH_2O \cdot OH$, followed by production of HCOOH, was measured by Niki et al to be less than 2%. Temps & Wagner (158) directly measured the yield of HCO in Reaction 12 to be 1.00 ± 0.05 at 298K by detecting both OH and HCO. The observed temperature independence is similar to that observed for a few other $OH + $ aldehyde reactions (160), i.e. $E/R \approx 0$K or negative, and it suggests that complex formation may be taking place. The complex formation, if taking place at all, should still lead to HCO and H_2O.

Loss of HCO

$$HCO + O_2 \rightarrow PRODUCTS; \quad k_{13} \qquad \qquad 13.$$

HCO thermally decomposes to $H + CO$ in a time scale of seconds at 298K ($HCO \rightarrow H + CO$; $\Delta H_r^0(298) = 16.7$ kcal mole^{-1}) (11). However, in the atmosphere it reacts much faster with O_2. Various methods of HCO detection (161–168) have been used to measure k_{13} directly. The results at 298K

Table 2 Summary of rate coefficient data on $HCO+O_2$ reaction at 298K

k_{13}, 10^{-12} cm^3 mol^{-1} s^{-1}	Method[a]	Reference
5.7 ± 1.2[b]	DF/MS	Washida et al (161)
5.6 ± 0.9	FP/KS	Shibuya et al (162)
5.6 ± 0.6	FP/LRA	Veyret & Lesclaux (165)
4.65 ± 0.6	LP/LRA	Langford & Moore (167)
5.25 ± 1.0	DF/LMR	Temps & Wagner (168)
4.2 ± 0.7	FP/IDLA	Gill et al (166)
3.7 ± 0.8	FP/IDLA	Nadtochenko et al (164)
4.0 ± 0.8	LP/IDLA	Reilly et al (163)
5.4 ± 1.0	average of first five studies	

[a] DF/MS: discharge flow with mass spectrometry detection of HCO. FP/KS: flash photolysis with kinetic spectroscopy of HCO. LP/IDLA: laser photolysis with intracavity dye laser absorption. LP/LRA: laser photolysis with extracavity dye laser absorption. DF/LMR: discharge flow with laser magnetic resonance detection of HCO. FP/LRA: flash photolysis with extracavity dye laser absorption.
[b] All errors are those quoted by the authors.

are summarized in Table 2, and it shows that the intracavity dye laser absorption, IDLA (163, 164, 166), yields lower values of k_{13} than the other methods. In the IDLA method the measured absorption may not vary linearly with the concentration of the absorber. However, measurements made on the $HCO+NO$ reaction using the same intracavity absorption apparatus agree quite well with those made using other HCO detection methods. Therefore, it is unlikely that the detection method is the source of the problem. Veyret & Lesclaux (165) have suggested that in some of these experiments the O_2 concentration was not accurately known, since substantial amounts of O_2 were lost due to chain reactions. All the intracavity absorption studies were carried out by subjecting the reaction mixture to multiple flashes; therefore, the O_2 concentrations during the measurements of k_{13} were lower than that initially introduced into the reactors. In some studies, the HCO source chemistry was also not well defined (166). Therefore, only the first five measurements in Table 2 are averaged to give $k_{13} = 5.4 \times 10^{-12}$ cm^3 s^{-1} at 298K. Veyret & Lesclaux (165) measured k_{13} between 298 and 500K and found it to decrease slightly with increase in temperature. They recommend $k_{13} = 5.5 \times 10^{-11} T^{-0.4 \pm 0.3}$ cm^3 s^{-1}.

There are three energetically allowed channels:

$$HCO + O_2 \rightarrow HO_2 + CO \qquad \text{13a.}$$
$$\rightarrow OH + CO_2 \qquad \text{13b.}$$
$$\xrightarrow{M} HCO_3. \qquad \text{13c.}$$

In the end product measurements on Reaction 13 by Horner et al (169), Pearson (170), and Osif & Heicklen (171), the yield of CO_2 was less than 0.20, suggesting that Channel 13b is quite small. The same studies also indicate formation of HCO_3 to be significant. The recent study by Niki et al (156) on the $OH + {}^{13}CH_2O$ reaction showed that $HCO + O_2$ should lead to only CO since CH_2O was quantitatively converted to CO. Radford et al (172), in their attempts to identify the LMR spectra of HO_2, used the $HCO + O_2$ reaction to produce HO_2 but did not report its yield. Temps & Wagner (168) also used LMR detection of HCO and HO_2 to obtain an HO_2 yield of 1.01 ± 0.07 at 198K. They place an upper limit of 0.4% for Channel 13b.

Langford & Moore (167) found the rate coefficient for the $HCO + O_2$ reaction to be slightly smaller than that for the $DCO + O_2$ reaction and the rate coefficient for $HCO(010) + O_2$ to be smaller than that for $HCO(000) + O_2$. They suggested a complex formation route for Reaction 13. Electronic structure calculations by Winter et al (173) show that the product formed by HCO addition to O_2 is the ground state ($^2A''$) HCO_3, which has to convert to the $^2A'$ state to correlate with $OH + CO_2$ products. Reaction 13 is exothermic enough that the nascent HCO_3 ($^2A''$) can convert to HCO_3 ($^2A'$). The $^2A''$ state is ~ 9 kcal mole^{-1} more stable than the $^2A'$ state. Since OH and CO_2 are not formed, it must mean that the rate of formation of $HO_2 + CO$ is much faster than that for $OH + CO_2$ formation via the $^2A'$ state. The measurements of HO_2 yield by Temps & Wagner were carried out between 1 and 3 torr pressure. k_{13} is not pressure dependent. Therefore, it is unlikely that the HO_2 yield at higher pressures is less than unity.

CONCLUSIONS

The rate-determining step in CH_4 oxidation is the initial H atom abstraction by OH to generate CH_3. Methane has a chemical lifetime of eight years in the troposphere. Once CH_3 is formed, its conversion to CO takes place within a few hours. Species such as $CH_3O_2NO_2$ may live for a few days. In the process of converting CH_4 to CO, more HO_x radicals are produced than consumed. So, CH_4 oxidation acts as a free radical source.

Figure 1 shows our current understanding of the CH_4 oxidation cycle. It is nearly indistinguishable from the one Levy drew over 17 years ago.

The only exception is the arrow connecting CH_3O_2H to H_2CO! Was anything new discovered in the past 17 years regarding CH_4 oxidation? The answer is a resounding "yes." If we compare the reaction rate coefficients that Levy used with the critically evaluated tables of NASA/JPL (126) and the CODATA panel (111), we see that virtually every one of the rate constants has been revised. Some rate constants, such as k_{13}, have been changed by orders of magnitude. Now, we see many reactions with negative activation energies. The quality of the available data has increased tremendously. Figures 2 and 4 and Tables 1 and 2 were included to show some of the rate coefficients that have been measured by many people, using different methods to obtain values that are as free as possible from systematic errors.

The "sketchy" outline of 1971 has been now changed to a quantitative picture. During this 17 years, a great deal of information on basic free radical reactions has been obtained. Many unanswered questions regarding reaction mechanisms still remain.

ACKNOWLEDGMENT

I thank all my co-workers who studied with me many of the reactions discussed here. I am very grateful to C. J. Howard and G. S. Tyndall for innumerable discussions, suggestions, and critical readings of this manuscript and to T. P. Murrells and M. Trolier for thorough readings of the manuscript. I am indebted to Twyla Barrett for her patience in typing the many versions of this review.

Literature Cited

1a. Graedel, T. E., McRae, J. E. 1980. *Geophys. Res. Lett.* 7: 977
1b. Stauffer, B., Fischer, G., Neftel, A., Oeschger, H. 1985. *Science* 229: 1386
2a. Blake, D. R., Mayer, E. W., Tyler, S. C., Makide, Y., Montague, D. C., Rowland, F. S. 1982. *Geophys. Res. Lett.* 9: 477
2b. Rinsland, C. P., Levine, J. S., Miles, T. 1985. *Nature* 318: 245
3. Logan, J. A., Prather, M. J., Wofsy, S. C., McElroy, M. B. 1981. *J. Geophys. Res.* 86: 7210
4a. Ramanathan, V., Cicerone, R. J., Singh, H. B., Kiehl, J. T. 1985. *J. Geophys. Res.* 90: 5547
4b. Wang, W. Y., Molnar, G. 1985. *J. Geophys. Res.* 90: 12971
5. Levy, H. 1972. *Planet. Space Sci.* 20: 919
6. Levy, H. 1973. *Planet. Space Sci.* 21: 575
7. Kaufman, F. 1979. *Ann. Rev. Phys. Chem.* 30: 411
8. Howard, C. J. 1979. *J. Phys. Chem.* 83: 3
9. Michael, J. V., Lee, J. H. 1979. *J. Phys. Chem.* 83: 10
10. Hancock, G. 1987. In *Modern Gas Kinetics: Theory, Experiment, and Application*, ed. M. J. Pilling, I. W. M. Smith, pp. 137–62. London: Blackwell
11. Wahner, A., Ravishankara, A. R. 1987. *J. Geophys. Res.* 92: 2189
12. Timonen, R. S., Ratajczak, E., Gutman, D., Wagner, A. F. 1987. *J. Phys. Chem.* 91: 5325
13. Inoue, G., Akimoto, H., Okuda, M. 1979. *Chem. Phys. Lett.* 63: 213
14a. Ishiwata, T., Fujiwara, I., Naruge, Y.,

Obi, K., Tanaka, I. 1983. *J. Phys. Chem.* 87: 1349
14b. Nelson, H. H., Pasternack, L., McDonald, J. R. 1983. *J. Phys. Chem.* 87: 1286
15. Suzuki, M., Inoue, G., Akimoto, H. 1983. *J. Chem. Phys.* 81: 4505
16. Greiner, N. R. 1967. *J. Chem. Phys.* 46: 2795
17. Greiner, N. R. 1970. *J. Chem. Phys.* 53: 1070
18. Dixon-Lewis, G., Wilson, W. E., Westenberg, A. A. 1966. *J. Chem. Phys.* 44: 2877
19. Wilson, W. E., Westenberg, A. A. 1967. *11th Symp. (Int.) on Combust.*, p. 1143
20. Horne, D. G., Norrish, R. G. W. 1967. *Nature* 215: 1373
21. Margitan, J. J., Kaufman, F., Anderson, J. G. 1974. *Geophys. Res. Lett.* 1: 80
22. Overend, R., Paraskevopoulos, G., Cvetanovic, R. J. 1975. *Can. J. Chem.* 53: 3374
23. Gordon, S., Mulac, W. A. 1975. *Inst. J. Chem. Kinet. Symp.* 7(1): 289
24. Howard, C. J., Evenson, K. M. 1976. *J. Chem. Phys.* 64: 197
25. Zellner, R., Steinert, W. 1976. *Int. J. Chem. Kinet.* 8: 397
26. Davis, D. D., Fischer, S., Schiff, R. 1974. *J. Chem. Phys.* 61: 2213
27. Tully, F. P., Ravishankara, A. R. 1980. *J. Phys. Chem.* 84: 3126
28. Jeong, K. M., Kaufman, F. 1982. *J. Phys. Chem.* 86: 1808
29. Sworski, T., Hochanadel, C. J., Ogren, P. J. 1980. *J. Phys. Chem.* 84: 129
30. Husain, D., Plane, J. M. C., Slater, N. K. H. 1981. *J. Chem. Soc. Faraday Trans. 2* 77: 1949
31. Baulch, D. L., Bowers, M., Malcom, D. G., Tuckerman, R. T. 1986. *J. Phys. Chem. Ref. Data* 15: 465
32. Presser, N., Gordon, R. J. 1985. *J. Chem. Phys.* 82: 1291
33. Tyler, S. C. 1986. *J. Geophys. Res.* 91: 13232
34. Rust, F., Stevens, C. M. 1980. *Int. J. Chem. Kinet.* 12: 371
35. Davidson, J. A., Cantrell, C. A., Tyler, S. C., Shetter, R. E., Cicerone, R. J., Calvert, J. G. 1987. *J. Geophys. Res.* 92: 2195
36. Ravishankara, A. R., Nicovich, J. M., Thompson, R. L., Tully, F. P. 1981. *J. Phys. Chem.* 85: 2498
37. Kaufman, F. 1982. *Ber. Bunsenges. Phys. Chem.* 86: 362
38. Davis, D. D., Braun, W., Bass, A. M. 1970. *Int. J. Chem. Kinet.* 2: 101
39. Poulet, G., LeBras, G., Combourieu, J. 1974. *J. Chem. Phys.* 71: 101
40. Clyne, M. A. A., Walker, R. F. 1973. *J. Chem. Soc. Faraday Trans. 1* 69: 1547
41. Lin, C. L., Leu, M. T., DeMore, W. B. 1978. *J. Phys. Chem.* 82: 1772
42. Watson, R., Machado, G., Fischer, S., Davis, D. D. 1976. *J. Chem. Phys.* 65: 2126
43. Manning, R. G., Kurylo, M. J. 1977. *J. Phys. Chem.* 81: 291
44. Whytock, D. A., Lee, J. H., Michael, J. V., Payne, W. A., Stief, L. J. 1977. *J. Chem. Phys.* 66: 2690
45. Zahniser, M., Berquist, B. M., Kaufman, F. 1978. *Int. J. Chem. Kinet.* 10: 15
46. Keyser, L. F. 1978. *J. Chem. Phys.* 69: 214
47. Baghal-Vayjooee, M. H., Colussi, A. J., Benson, S. W. 1979. *Int. J. Chem. Kinet.* 11: 147
48. Ravishankara, A. R., Wine, P. H. 1980. *J. Chem. Phys.* 72: 25
49. Heneghan, S. P., Benson, S. W. 1983. *Int. J. Chem. Kinet.* 15: 1311
50. Michael, J. V., Lee, J. H. 1977. *Chem. Phys. Lett.* 51: 303
51. Dobis, O., Benson, S. W. 1987. *Int. J Chem. Kinet.* 19: 691
52. Davidson, J. A., Sadowski, C. M., Schiff, H. I., Streit, G. E., Howard, C. J., Jennings, D. A., Schmeltekopf, A. L. 1976. *J. Chem. Phys.* 64: 57; Davidson, J. A., Schiff, H. I., Streit, G. E., McAfee, J. R., Schmeltekopf, A. L., Howard, C. J. 1977. *J. Chem. Phys.* 67: 5021
53. Heidner, R. F., Husain, D. 1973. *Int. J. Chem. Kinet.* 5: 819
54. Amimoto, S. T., Force, A. P., Gulotty, R. G., Wiesenfeld, J. R. 1979. *J. Chem. Phys.* 71: 3640
55. Wine, P. H., Ravishankara, A. R. 1982. *Chem. Phys. Lett.* 69: 365
56. Lin, C. L., DeMore, W. B. 1973. *J. Phys. Chem.* 77: 863
57. Casavecchia, P., Buss, R. J., Sibener, S. J., Lee, Y. T. 1980. *J. Chem. Phys.* 73: 6351
58. Yamazaki, H., Cvetanovic, R. J. 1964. *J. Chem. Phys.* 41: 3703
59. Paraskevopoulos, G., Cvetanovic, R. J. 1969. *J. Chem. Phys.* 50: 590
60. Paraskevopoulos, G., Cvetanovic, R. J. 1970. *J. Chem. Phys.* 52: 5821
61. Michaud, P., Cvetanovic, R. J. 1972. *J. Phys. Chem.* 76: 1375
62. Jayanty, R., Simonaitis, R., Heicklen, J. 1976. *Int. J. Chem. Kinet.* 8: 107
63. DeMore, W. B., Raper, O. F. 1967. *J. Chem. Phys.* 46: 2500
64. Basco, N., Norrish, R. G. W. 1961. *Proc. R. Soc. Ser. A* 260: 293

65a. Luntz, A. C. 1980. *J. Chem. Phys.* 73: 1151
65b. Aker, P. M., O'Brien, J. J. A., Sloan, J. J. 1986. *J. Chem. Phys.* 84: 745
66. McMillan, G. R., Calvert, J. G. 1965. *Oxid. Combust. Rev.* 1: 83
67. Walker, R. W. 1975. In *Specialist Periodical Reports—Reaction Kinetics*, ed. P. G. Ashmore, 1: 161. London: The Chem. Soc.; 1977. In *Specialist Periodical Reports—Reaction Kinetics*, ed. P. G. Ashmore, R. J. Donovan, 2: 297. London: The Chemical Soc.
68. Laufer, A. H., Bass, A. M. 1975. *Int. J. Chem. Kinet.* 7: 639
69. Sokolova, N. A., Nikisha, L. V., Polyak, S. S., Nalbandyan, A. B. 1973. *Kinet. Catal.* 14: 721 (translated from Russian)
70. Herzberg, G., Shoosmith, J. 1956. *Can. J. Phys.* 34: 523
71. Washida, N., Bayes, K. D. 1976. *Int. J. Chem. Kinet.* 8: 777; Washida, N. 1980. *J. Chem. Phys.* 73: 1665
72. Plumb, I. C., Ryan, K. R. 1982. *Int. J. Chem. Kinet.* 14: 861
73. Selzer, E. A., Bayes, K. D. 1983. *J. Phys. Chem.* 87: 392
74. Basco, N., James, D. G. L., James, F. C. 1972. *Int. J. Chem. Kinet.* 4: 129
75. Van den Bergh, H. E., Callear, A. B. 1971. *J. Chem. Soc. Faraday Trans.* 67: 2017
76. Hochanadel, C. J., Ghormley, J. A., Boyle, J. W., Ogren, P. J. 1977. *J. Phys. Chem.* 81: 3
77. Parkes, D. A. 1977. *Int. J. Chem. Kinet.* 9: 451
78. Cobos, C. J., Hippler, H. H., Luther, K., Ravishankara, A. R., Troe, J. 1985. *J. Phys. Chem.* 89: 4334
79. Pilling, M. J., Smith, M. J. C. 1985. *J. Phys. Chem.* 89: 4713
80. Keiffer, M., Pilling, M. J., Smith, M. J. C. 1987. *J. Phys. Chem.* 91: 6028
81. Laguna, G. A., Baughcum, S. L. 1982. *Chem. Phys. Lett.* 88: 568
82. Troe, J. 1983. *Ber. Bunsenges. Phys. Chem.* 87: 161
83. Gilbert, R. G., Luther, K., Troe, J. 1983. *Ber. Bunsenges. Phys. Chem.* 87: 169
84. Baldwin, A. C., Golden, D. M. 1978. *Chem. Phys. Lett.* 55: 350
85. Klais, O., Anderson, P. C., Laufer, A. H., Kurylo, M. J. 1979. *Chem. Phys. Lett.* 66: 598
86. Pratt, G. L., Wood, S. W. 1984. *J. Chem. Soc. Faraday Trans. 1* 80: 3419
87. Simonaitis, R., Heicklen, J. 1975. *J. Phys. Chem.* 79: 298
88. Spicer, C. W., Villa, A., Wiebe, H. A., Heicklen, J. 1975. *J. Am. Chem. Soc.* 95: 13
89. Hanst, P. L., Calvert, J. G. 1959. *J. Phys. Chem.* 63: 2071
90. Simonaitis, R., Heicklen, J. 1974. *J. Phys. Chem.* 78: 2417
91. Cox, R. A., Derwent, R. G., Holt, P. M., Kerr, J. A. 1976. *J. Chem. Soc. Faraday Trans. 1* 72: 2444
92. Pate, C. T., Finlayson, B. J., Pitts, J. N. Jr. 1974. *J. Am. Chem. Soc.* 96: 6554
93. Simonaitis, R., Heicklen, J. 1979. *Chem. Phys. Lett.* 65: 361
94. Parkes, D. A., Paul, D. M., Quinn, C. P., Robson, R. C. 1973. *Chem. Phys. Lett.* 23: 425
95. Anastasi, C., Smith, I. W. M., Parkes, D. A. 1978. *J. Chem. Soc. Faraday Trans. 1* 74: 1093
96. Adachi, H., Basco, N. 1979. *Chem. Phys. Lett.* 63: 490
97. Sander, S. P., Watson, R. T. 1980. *J. Phys. Chem.* 84: 1664
98. Simonaitis, R., Heicklen, J. 1981. *J. Phys. Chem.* 85: 2946
99. Cox, R. A., Tyndall, G. S. 1979. *Chem. Phys. Lett.* 65: 357
100. Plumb, I. C., Ryan, K. R., Steven, J. R., Mulcahy, M. F. R. 1979. *Chem. Phys. Lett.* 63: 255
101. Plumb, I. C., Ryan, K. R., Steven, J. R., Mulcahy, M. F. R. 1981. *J. Phys. Chem.* 85: 3136
102. Ravishankara, A. R., Eisele, F. L., Kreutter, N. M., Wine, P. H. 1981. *J. Chem. Phys.* 74: 2267
103. Zellner, R., Fritz, B., Lorenz, K. 1986. *J. Atmos. Chem.* 4: 241
104. Hunziker, H. E., Kneppe, H., Wendt, H. R. 1981. *J. Photochem.* 17: 377
105. Buss, R. J., Baseman, R. J., He, G., Lee, Y. T. 1981. *J. Photochem.* 17: 389
106. Sridharan, U. C., Kaufman, F. 1983. *Chem. Phys. Lett.* 102: 45
107. Smalley, J. F., Nesbitt, F. L., Klemm, R. B. 1986. *J. Phys. Chem.* 90: 491
108. Endo, Y., Tsuchiya, S., Yamada, C., Hirota, E., Koda, S. 1986. *J. Chem. Phys.* 85: 4446
109. Slagle, I. R., Gutman, D. 1985. *J. Am. Chem. Soc.* 107: 5342
110. Khachatryan, L. A., Niazyan, O. M., Mantashyan, A. A., Vedeneev, V. I., Teitel'boim, M. A. 1982. *Int. J. Chem. Kinet.* 14: 1231
111. Baulch, D. L., Cox, R. A., Hampson, R. F., Kerr, J. A., Troe, J., Watson, R. T. 1984. *J. Phys. Chem. Ref. Data* 13: 1259
112. McMillen, D. F., Golden, D. M. 1982. *Ann. Rev. Phys. Chem.* 33: 493
113. Adachi, H., Basco, N. 1980. *Int. J. Chem. Kinet.* 12: 1

114. Ravishankara, A. R., Eisele, F. L., Wine, P. H. 1980. *J. Chem. Phys.* 73: 3743
115. Bahta, A., Simonaitis, R., Heicklen, J. 1982. *J. Phys. Chem.* 86: 1849
116. Kan, C. S., McQuigg, R. D., Whitbeck, M. R., Calvert, J. G. 1979. *Int. J. Chem. Kinet.* 11: 921
117. Adachi, H., Basco, N., James, D. G. L. 1980. *Int. J. Chem. Kinet.* 12: 949
118. Sander, S. P., Watson, R. T. 1981. *J. Phys. Chem.* 85: 2960
119. Kurylo, M. J., Wallington, T. J., Ouellette, P. A. 1987. *J. Photochem.* 39: 201
120. McAdam, K., Veyret, B., Lesclaux, R. 1987. *Chem. Phys. Lett.* 133: 39
121. Cox, R. A., Tyndall, G. S. 1980. *J. Chem. Soc. Faraday Trans. 2* 76: 153
122. Kurylo, M. J., Dagaut, P., Wallington, T. J., Neuman, D. M. 1987. *Chem. Phys. Lett.* 139: 513
123. Dagaut, P., Wallington, T. J., Kurylo, M. J. 1988. *J. Phys. Chem.* In press
124. Moortgat, G. K., Burrows, J. P., Schneider, W., Tyndall, G. S., Cox, R. A. 1986. *Proc. 4th Eur. Symp. on Physico-Chem. Behav. Atmos. Pollutants*, Stresa, Italy, pp. 271–81
125. Molina, M. J., Arguello, G. 1979. *Geophys. Res. Lett.* 6: 953
126. DeMore, W. B., Margitan, J. J., Molina, M. J., Watson, R. T., Golden, D. M., Hampson, R. F., Kurylo, M. J., Howard, C. J., Ravishankara, A. R. 1985. *Evaluation #7*, NASA Panel for Data Eval. Pasadena: JPL Publ. 85-37
127. Niki, H., Maker, P. D., Savage, C. M., Breitenbach, L. P. 1983. *J. Phys. Chem.* 87: 2190
128a. Campbell, I. M., McLaughlin, D. F., Handy, B. J. 1976. *Chem. Phys. Lett.* 38: 362
128b. Overend, R., Paraskevopoulos, G. 1978. *J. Phys. Chem.* 82: 1329
128c. Ravishankara, A. R., Davis, D. D. 1978. *J. Phys. Chem.* 82: 2852
128d. Hagele, J., Lorenz, K., Rhasa, D., Zellner, R. 1983. *Ber. Bunsenges. Phys. Chem.* 87: 1023
128e. Meier, U., Grotheer, H. H., Just, Th. 1984. *Chem. Phys. Lett.* 106: 97
128f. Greenhill, P. G., O'Grady, B. V. 1986. *Aus. J. Chem.* 39: 1775
129. Slemr, F., Warneck, P. 1977. *Int. J. Chem. Kinet.* 9: 267
130. Michael, J. V., Nava, D. F., Payne, W. A., Stief, L. J. 1981. *Chem. Phys. Lett.* 77: 110
131. Niki, H., Maker, P. D., Savage, C. M., Breitenbach, L. P. 1980. *Chem. Phys. Lett.* 73: 43
132. Kan, C. S., Calvert, J. G., Shaw, J. H. 1980. *J. Phys. Chem.* 84: 3411
133. Niki, H., Maker, P. D., Savage, C. M., Breitenbach, L. P. 1981. *J. Phys. Chem.* 85: 877
134. Nangia, P. S., Benson, S. W. 1980. *Int. J. Chem. Kinet.* 12: 43
135. Sanhueza, E., Simonaitis, R., Heicklen, J. 1979. *Int. J. Chem. Kinet.* 11: 907
136. Kurylo, M. J., Wallington, T. J. 1987. *Chem. Phys. Lett.* 138: 543
137. Barker, J. R., Benson, S. W., Golden, D. M. 1977. *Int. J. Chem. Kinet.* 9: 31
138. Batt, L., Robinson, G. N. 1979. *Int. J. Chem. Kinet.* 11: 1045
139. Cox, R. A., Derwent, R. G., Kearsey, S. V., Batt, L., Patrick, K. G. 1980. *J. Photochem.* 13: 149
140. Radford, H. E., Russell, D. K. 1977. *J. Chem. Phys.* 66: 222
141. Inoue, G., Akimoto, H., Okuda, M. 1980. *J. Chem. Phys.* 72: 1769
142. Sanders, N., Butler, J. E., Pasternack, L. R., McDonald, J. R. 1980. *Chem. Phys.* 48: 203
143. Gutman, D., Sanders, N., Butler, J. E. 1982. *J. Phys. Chem.* 86: 66
144. Lorenz, K., Rhasa, D., Zellner, R., Fritz, B. 1985. *Ber. Bunsenges. Phys. Chem.* 89: 341
145. Wantuck, P. J., Oldenborg, R. C., Baughcum, S. L., Winn, K. R. 1987. *J. Phys. Chem.* 91: 4653
146. Zellner, R. 1986. *9th Int. Symp. on Gas Kinet.*, Bordeaux, France, July 20–25
147. Saebo, S., Radom, L., Schaefer, H. F. III. 1983. *J. Chem. Phys.* 78: 845
148. Batt, L., Burrows, J. P., Robinson, G. N. 1981. *Chem. Phys. Lett.* 78: 467
149. Moore, C. B., Weisshaar, J. C. 1983. *Ann. Rev. Phys. Chem.* 34: 525
150. Calvert, J. G. 1980. *The Homogeneous Chemistry of Formaldehyde Generation and Destruction within the Atmosphere.* Washington DC: US Dept Transport., FAA HAP Program, Rep. No. FAA-EE-80-20
151. Moortgat, G. K., Seiler, W., Warneck, P. 1983. *J. Chem. Phys.* 78: 1185
152. Atkinson, R., Pitts, J. N. Jr. 1978. *J. Chem. Phys.* 68: 3581
153. Stief, L. J., Nava, D. F., Payne, W. A., Michael, J. V. 1980. *J. Chem. Phys.* 73: 2254
154. Morris, E. D. Jr., Niki, H. 1971. *J. Chem. Phys.* 55: 1991
155. Niki, H., Maker, P. D., Savage, C. M., Breitenbach, L. P. 1978. *J. Phys. Chem.* 82: 132
156. Niki, H., Maker, P. D., Savage, C. M., Breitenbach, L. P. 1984. *J. Phys. Chem.* 88: 1185
157. Smith, R. H. 1978. *Int. J. Chem. Kinet.* 10: 519

158. Temps, F., Wagner, H. Gg. 1984. *Ber. Bunsenges. Phys. Chem.* 88: 415
159. Horowitz, A., Su, F., Calvert, J. G. 1978. *Int. J. Chem. Kinet.* 10: 1099
160. Semmes, D. H., Ravishankara, A. R., Gump-Perkins, C. A., Wine, P. H. 1985. *Int. J. Chem. Kinet.* 17: 303
161. Washida, N., Martinez, R. I., Bayes, K. D. 1974. *Z. Naturforsch. Teil A* 29: 251
162. Shibuya, K., Ebata, T., Obi, K., Tanaka, I. 1977. *J. Phys. Chem.* 81: 2292
163. Reilly, J. P., Clark, J. H., Moore, C. B., Pimentel, G. C. 1978. *J. Chem. Phys.* 69: 4381
164. Nadtochenko, V. A., Sarkisov, O. M., Vedeneev, V. I. 1979. *Dokl. Akad. Nauk USSR* 244: 152
165. Veyret, B., Lesclaux, R. 1981. *J. Phys. Chem.* 85: 1918
166. Gill, R. J., Johnson, W. D., Atkinson, G. H. 1981. *Chem. Phys.* 58: 29
167. Langford, A. O., Moore, C. B. 1984. *J. Chem. Phys.* 80: 4220
168. Temps, F., Wagner, H. Gg. 1984. *Ber. Bunsenges. Phys. Chem.* 88: 410
169. Horner, E. C. A., Style, D. W. G., Summers, D. 1954. *Trans. Faraday Soc.* 50: 1201
170. Pearson, G. S. 1963. *J. Phys. Chem.* 67: 1686
171. Osif, T. L., Heicklen, J. 1976. *J. Phys. Chem.* 80: 1526
172. Radford, H. E., Evenson, K. M., Howard, C. J. 1974. *J. Chem. Phys.* 60: 3178
173. Winter, N. W., Goddard, W. A., Bender, C. F. 1975. *Chem. Phys. Lett.* 33: 25
174. Phillips, L. F. 1976. *Chem. Phys. Lett.* 37: 421

THE SEMICLASSICAL WAY TO MOLECULAR DYNAMICS AT SURFACES[1]

J. W. Gadzuk

Center for Chemical Physics, National Bureau of Standards, Gaithersburg, Maryland 20899

INTRODUCTION

The dynamics of atomic motion at solid surfaces provides not only a fertile area of intellectual inquiry in its own right, but also a formidable challenge to be understood if selective control of chemical events, based on molecular-level manipulations (1), is to be realized in surface-assisted chemistry (2). Various aspects of surface dynamics have been surveyed both in past (3a–g) and present (4) volumes of this series as well as in other review articles (5–11). These works should provide the reader with not only a cross-section of current wisdom and ignorance, but also access to the increasing literature.

Since the unifying thread linking the topics discussed here is the focus on molecular-level modeling of time-dependent phenomena in which a solid surface is one of the participants, it is reasonable to look first toward the more established gas phase studies of time-dependent quantum behavior in molecular systems for guidance (1, 12–19). Of unique importance to the present study is the semiclassical wavepacket methodology developed by Heller (17). The history is nicely summarized in the following direct quote from Sawada et al (20):

> In a series of papers Heller (17, 21–25a,b) developed a scheme for computing and interpreting time dependent quantum mechanical processes by representing the wave function as a superposition of Gaussian wave packets. Since each packet is characterized by several parameters (the position and the momentum of the packet's center, a complex

[1] The US Government has the right to retain a nonexclusive, royalty-free license in and to any copyright covering this paper.

width and complex phase) the calculation of the time evolution of the wave function is reduced to that of the time evolution of these parameters.

In his *applied work* Heller used a version of his method [which we call here the simplest Heller method (SHM)] which is based on two simplifying assumptions. 1. The first assumes that if we must represent the wave function by a sum of Gaussians, we can propagate each Gaussian independently. . . . 2. It is further assumed that throughout the process (i.e. collision or photon absorption) the width of each Gaussian is smaller than the length over which the potential changes. This allows . . . second order Taylor expansion of the potential around the instantaneous center of the Gaussian . . . the locally harmonic approximation.

SHM was used successfully by Heller to analyze a variety of time dependent processes such as atom-diatomic collision (21), photodissociation (23a–d), photoabsorption (23a–d), Raman scattering (24a–e), and atom diffraction by surfaces (25a,b). The method provides accurate results as well as a novel and beautiful interpretation of quantum dynamics in terms of a classical language. A common feature of these applications is that they all deal with the short time dynamics of localized quantum degrees of freedom; in a way their success reflects mostly Heller's skill in identifying important problems that fit the SHM validity conditions, rather than the generality of the method.

The ultimate purpose here is to present and analyze a number of different phenomena in surface spectroscopies, scattering, and dynamics that can be considered as surface analogues of the gas phase "applied work" of Heller. As a first step toward this goal, consider the following provocative question directed toward surface dynamics. What do

1. photoemission lineshapes and/or satellites,
2. stimulated desorption energy distributions,
3. vibrational overtone losses in resonance Electron Energy Loss Spectroscopy (EELS), and
4. internal state excitation in molecular beam scattering.

have in common?

Accepting the promise that relevant details of each of these processes will soon be put forth, a plausible answer is that all four phenomena involve following the evolution of the nuclear motion of an initially prepared, nonstationary state where "preparation" is considered to be a fast electronic transition. From the point of view of nuclear motion, this is equivalent to switching on a new potential energy surface (PES). In the energy domain, the subsequent nuclear motion or relaxation can be thought of as dephasing or loss of coherence in the initial superposition of eigenstates (26). Alternatively, in the time domain the relaxation can be viewed as the spread and displacement of the initially prepared wave packet. As is seen below, the methodology of Heller (17, 21–25a,b) is a most enlightening and practical means for treating this relaxation.

We hope to show that a new and worthwhile perspective on the physics and chemistry of surface processes such as **1–4** can be attained through inquiries focused on common aspects of dynamics. This approach fits in with Hamann's list of "Real Challenges for Surface Theory," which stresses the need to "transcend the case-by-case mentality" characterizing orthodox surface science and, further, to "generate some intellectual excitement," which seems to be often lacking (27). Indeed, this has previously been identified by Comsa & David, who note that, "For surface scientists the interest in the fulfillment of general laws seems to be nowadays rather limited" (28).

The present paper on the topic of semiclassical wave-packet modeling in surface dynamics is offered in the spirit of a response to these concerns. It demonstrates that a wide class of system-independent and seemingly unrelated surface phenomena in fact show an aesthetically pleasing unity when considered as problems in time-dependent quantum mechanics. In order to achieve this goal, the remainder of the paper is structured as follows. The next section, called "Methodology," presents the abstract models upon which the theoretical constructions are based. Sufficient mathematical and algebraic details are included here to make the paper self-contained. The applications section presents an analysis of common aspects of photoemission lineshapes from adsorbed atoms, stimulated desorption distributions, resonance EELS, and vibrational excitation in molecular scattering from surfaces, based on the unified wave packet picture. Final thoughts are then offered.

MODELING AND METHODOLOGY

Almost all tractable analytic theories involving the time development of an initially excited or prepared system, possibly with subsequent decay, have built into them the equivalent of the following reduction scheme.

1. Born-Oppenheimer separation of "fast" and "slow" degrees of freedom;
2. Franck-Condon or sudden approximation;
3. Average transition moment approximation in which the excitation matrix elements of the fast system are evaluated at some mean value for the coordinates of the "slow" systems.

With these simplifications, an electronic (fast system) transition appears as a change in the Hamiltonian or potential energy surface (PES) of the slow systems that allows for motion and redistribution of energy among the various "slow" degrees of freedom (e.g. translational, vibrational, rotational = T, V, R).

The theoretical method of choice here is semiclassical wave packet dynamics (17, 21–25a,b), chosen because it: (a) optimizes the use of the information content in classical trajectories; (b) requires knowledge of the potential energy surfaces only within (and close to) the regions that influence the classical dynamics; (c) implicitly carries out the averaging over initial conditions required in a strictly classical calculation; (d) automatically builds in interference effects between the various neighboring trajectories; (e) is computationally simple and efficient; and (f) is physically transparent. This technique, Gaussian wave packet propagation, is based on the premise that the classical trajectory steers the wave packet(s) that are "prepared" as well-defined nonstationary states during the initial excitation process. Although the method is most appropriate for motion over a smoothly varying PES in which there are no curve-crossings or bifurcations (1, 5c, 15, 29a,b), means do exist for treating such situations (20, 30a,b, 31).

The basic ideas are as follows. Upon excitation of a localized mode initially in its ground state $\equiv |\phi_0\rangle$ (say a Gaussian harmonic oscillator wavefunction) onto a new PES $\equiv V(\mathbf{R})$, the subsequent time-dependent wavefunction is required to retain the simple Gaussian form (or superposition of possibly displaced Gaussians), as illustrated in Figure 1a (21)

$$\phi(y;t) = \exp[i/\hbar\{\alpha_t^y(y-y_t)^2 + p_t^y(y-y_t) + \gamma_t^y\}] \qquad 1.$$

with $\phi(\mathbf{R};t) = \prod_{i=1}^{3} \phi(r_i;t)$. With the insertion of Eq. 1 into the time-dependent Schrödinger equation and retention of terms up to quadratic in deviations of the PES from y_t, the value at the instantaneous position of the wave packet center [i.e. $V(y;t) \approx V_0(y_t) + (y-y_t)V' + \frac{1}{2}(y-y_t)^2 V'']$ leads to four equations of motion for the wave packet dynamics (per degree of freedom)

$$\dot{y}_t = \partial H/\partial p_t^y$$

$$\dot{p}_t^y = -\partial H/\partial y_t$$

$$\dot{\alpha}_t^y = -(2/m)(\alpha_t^y)^2 - \frac{1}{2}\partial^2 V/\partial y^2|_{y=y_t}$$

$$\dot{\gamma}_t^y = i\hbar\alpha_t^y/m + p_t^y \dot{y}_t - E. \qquad \text{2a–d.}$$

In Eqs. 1 and 2, y_t and p_t are the time-evolving position and momentum of the wave packet center that follows the classical trajectory, α_t^y is the width, γ_t^y the phase and normalization along the trajectory, and E is the conserved energy that follows from the Hamiltonian $H(p_t, y_t) = p_t^2/2m + V(y_t) = E$. Equations 1 and 2 are the embodiment of the SHM. Although refinements upon the SHM exist (20, 32a–c), they are not required for present purposes.

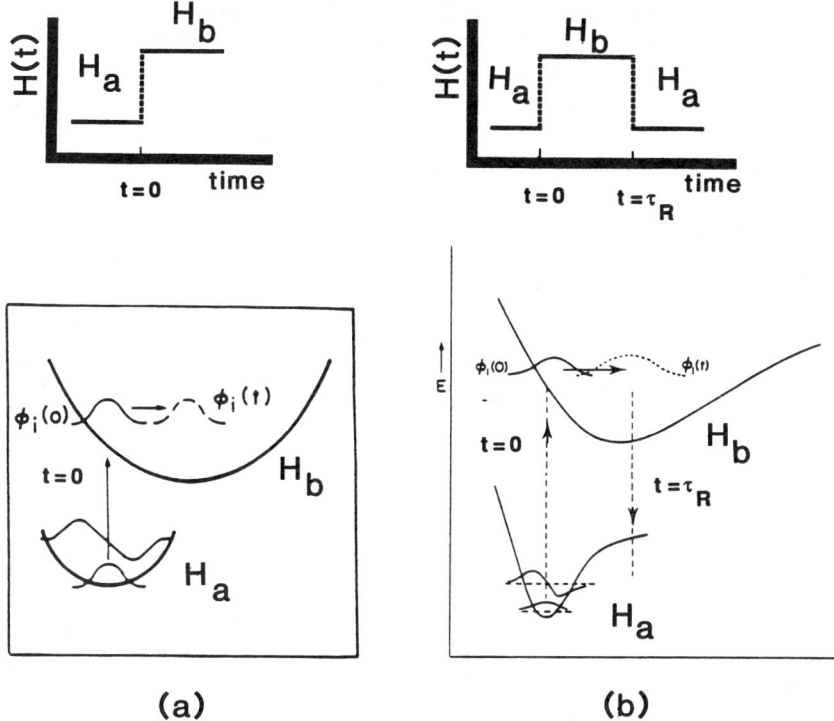

Figure 1 Time dependence of intramolecular Hamiltonian (*top*) and wave packet evolution (*bottom*) for (*a*) single electronic transition; (*b*) double transition (adapted from Ref. 24a).

Single Electronic Transitions

Wave packet propagation is ideally suited to deal with the type of excitation traditionally viewed as a Franck-Condon transition involving electronic excitations and hence instantaneous changes of the PES entering into Eq. 2. For an event characterized by an $a \to b$ electronic transition, it is straightforward to demonstrate that the slow system (e.g. vibrational) excitation spectrum

$$I(\omega) = \sum_n \delta(\omega - \omega_n^b) |\langle n; b|0; a\rangle|^2$$

where $\hbar\omega_n^b$ is the nth vibrational eigenvalue on PES b and $\langle n; b|0; a\rangle$ the overlap integral of the ground state wavefunction on V_a with the vibrational state n on V_b, is equivalent to

$$I(\omega) = \frac{1}{2\pi} \int_{-\infty}^{\infty} dt e^{i\omega t} \langle \phi_0 | \phi(t) \rangle. \qquad 3.$$

In Eq. 3, $|\phi_0\rangle \equiv |0;a\rangle$ and $|\phi(t)\rangle \equiv e^{-iH_b t}|\phi_0\rangle$, which says that the vibrational spectrum is the Fourier transform of the overlap between the initial wavepacket defined by H_a and the wavepacket that time evolves according to H_b on the final state PES. Within the "simplest" Heller picture, this $|\phi(t)\rangle$ is given by Eq. 1 and the Gaussian parameters by solution of the system of Eq. 2.

The connections between the temporal structure of $\langle\phi_0|\phi(t)\rangle$ and the vibrational spectrum are nicely displayed in Figure 2, taken from Heller (33). Shown in Figure 2a are prototypical realizations of not only V_b but also the initial wave packet drawn as the full circular structure labeled 1, the classical trajectory followed on V_b, and "snapshots" at various times after initial excitation. The time-dependent overlap is displayed below the PES. In this illustration, there are three distinctly different time scales, which show up in three different ways in the vibrational spectrum. First is T_a, which is a measure of the initial departure rate, determined by the slope of V_b along the segment 1–2 of the trajectory. Next is T_b, the time required for the resurrection of the overlap (in simple terms, the "vibrational period") along the segment 1–4. Last in this picture is T_c, which measures the "decay rate" of the "vibrational amplitude." In addition, if a reflecting wall on V_b turned the trajectory around somewhere to the right of the drawn portion in Figure 2a, then a recurrence pattern could be possible that would introduce yet other time scales. A schematic Fourier transform of this overlap evolution is shown in Figure 2b. The width of the envelope is set by $1/T_a$ (i.e. steep potentials yield broad vibrational bands), the "discrete" vibrational intervals by $1/T_b$, the "resurrection" frequency, and the "discrete" width by $1/T_c$, the amplitude decay rate. Actual numerical implementation of this procedure is much less demanding than calculating Franck-Condon factors, which require knowledge of the vibrational (or scattering) wave functions over the entire PES.

Multiple Electronic Transitions

Another important class of vibrational excitation that is usefully addressed within the wave packet framework is resonance and Raman excitation in which the excitation process involves temporary propagation over some intermediate state PES as depicted in Figure 1b. Since the amplitude for exciting the nth vibrational state on V_a is $A_n = \langle n;a|\phi(t)\rangle$, it can be demonstrated that the vibrational spectrum is given by

$$I_{on}(\omega) = |\mu|^2 \left| \int_0^\infty dt\, e^{i\omega t} \langle n;a|\phi(t)\rangle \right|^2, \qquad 4.$$

which depends upon the overlap of the vibrational state in question with

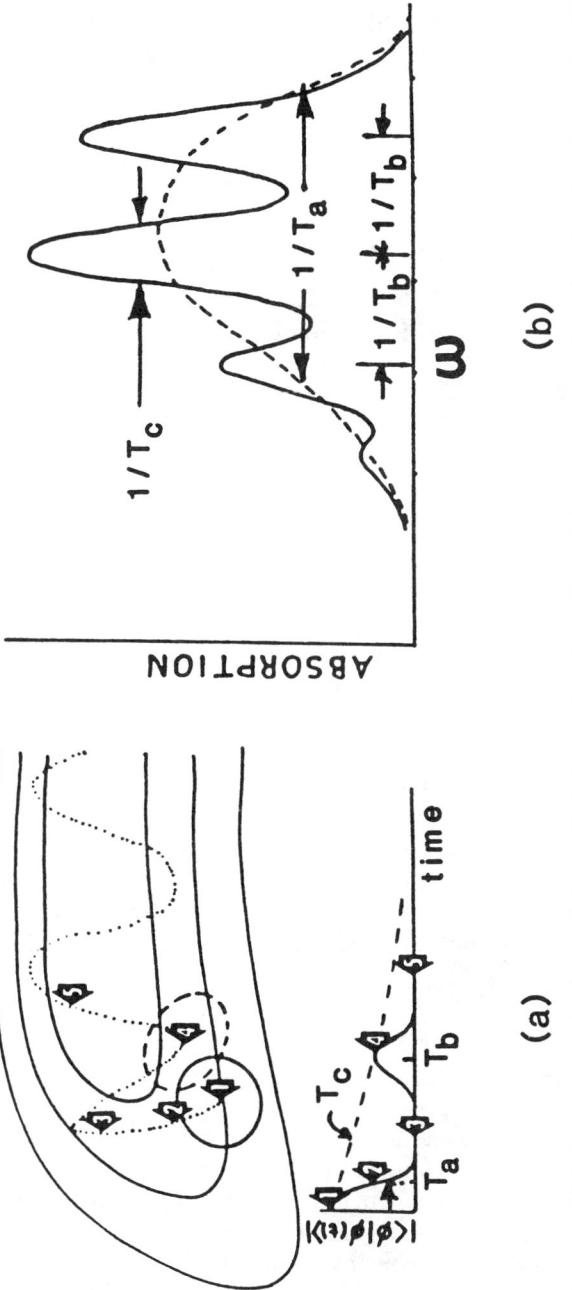

Figure 2 (a) Path of wave packet on excited state PES. *Points* along the paths are labeled and shown also in the time dependent overlap. The packet at times "T_1" and "T_4" are outlined. (b) Equivalent absorption spectrum. (Ref. 33.)

the time-dependent wave packet (24a–e). The "proportionality constant" $|\mu|^2$ includes matrix element effects for the electronic transitions.

A particularly useful adaptation of Eq. 4 is the case shown in Figure 1b in which a transition from $a \to b$ occurs at $t = 0$, followed by the return $b \to a$ at $t = \tau_R$. Since

$$|\phi(t > \tau_R)\rangle = \exp(-iH_a(t-\tau_R))\exp(-iH_b\tau_R)|0;a\rangle,$$

the probability for excitation of the nth state on V_a is

$$P_n = \lim_{t \to \infty} |\langle n;a|\phi(t)\rangle|^2$$

and thus the excited state distribution is

$$I(\omega) \sim \sum_n \delta(\omega - \omega_n^a)|\langle n;a|\phi(\tau_R)\rangle|^2 \qquad 5.$$

where $|\phi(\tau_R)\rangle = \exp(-iH_b\tau_R)|0;a\rangle$, the excited wave packet at τ_R. This is a Franck-Condon factor between the *moving* wave packet and the nth vibrational state.

Short Time Dynamics

An attractive aspect of the wavepacket picture is that one needs only as much information to calculate a dynamic observable as demanded by the resolution of the experiments. Thus, knowledge of the total time evolution of the prepared wave packet is only necessary in Eqs. 3, 4, or 5, when an infinitely precise spectrum is required. In fact, the short time limit (STL) is often sufficient, particularly for setting spectral envelope structure as in Figure 2b.

It is instructive and useful to consider the algebraic details of the STL in the frozen Gaussian approximation (34), for processes such as those shown in Figures 1 and 2. Equations 2a and b, the classical equations of motion for the wave packet center, are

$$\dot{y}_t = p_t^y/m \quad \text{and} \quad \dot{p}_t^y = -\partial V(y;t)/\partial y|_{y=y_t},$$

which in the STL integrate to

$$y_t = y_0 + 0(t^2) \quad \text{and} \quad p_t^y = -V'|_{y=y_0}t + 0(t^2).$$

Note that the initial phase space displacement is in momentum and not position. In the frozen Gaussian limit, the width in Eq. 1 is replaced by $\alpha_t = i\hbar\alpha_0$ and $\alpha_0 = (2\bar{u})^{-2}$ where $\bar{u}$ is determined by the confining potential in the initial state, as is discussed below. Finally, in the STL, $\gamma_t = -\gamma_0 t + 0(t^2)$ where γ_0 is a constant. Thus the STL propagating Gaussian from Eq. 1 is

$$|\phi_0(t)\rangle \simeq \exp\left[-\alpha_0(y-y_0)^2 - \frac{i}{\hbar}V't(y-y_0) - \frac{i}{\hbar}\gamma_0 t\right]$$

and the resulting overlap in Eq. 3 integrates to

$$\langle\phi_0|\phi_0(t)\rangle \sim \exp\left(-i\omega't - \frac{V'^2 t^2}{8\alpha_0 \hbar^2}\right) \qquad 6.$$

with $\omega' \equiv \gamma_0/\hbar$. Insertion of Eq. 6 into Eq. 3 yields the Gaussian spectrum

$$I(\omega) \sim \exp\left(-\frac{1}{2}\left(\frac{\hbar(\omega-\omega')}{\Delta}\right)^2\right) \qquad 7a,b.$$

with

$$\Delta \equiv \bar{u}\, dV/dy|_{y=y_0}.$$

Thus excitation of an initial state of width $\bar{u}$ onto a steep PES (V' large) produces a spectrum whose intrinsic width scales as the product of $\bar{u} \times V'$. The initial wave packet width $2\bar{u}$ is derived from a thermal average of y over harmonic oscillator states as

$$\bar{u}^2 = \langle y^2 \rangle_T$$

which for $T=0$ is just the ground state expectation value $\bar{u}^2 = (\hbar/2m\omega) \equiv u_0^2$. It is an easy exercise in Boson algebra to demonstrate that for finite temperature,

$$\bar{u}(T)^2 = u_0^2(1+2n)$$

where the Bose-Einstein factor is

$$n = (e^{\hbar\omega/kT} - 1)^{-1}.$$

Consequently,

$$\bar{u}(T) = (\hbar/2m\omega)^{1/2}[\coth(\hbar\omega/2kT)]^{1/2}, \qquad 8.$$

a result that will be useful below.

Operationally, as can be seen from this simple example, the algebra of Gaussian wave packet dynamics is quite neat, since one generates many Gaussian integrals, convolutions, and Fourier transforms, all of which usually produce more Gaussians.

Harmonic Oscillators

The driven harmonic oscillator (HO) is at the heart of almost any *analytic* model of a time-dependent process in (surface) chemical physics (35a–c).

Not only discrete, localized, intramolecular-vibrational modes but also the delocalized continua of elementary excitations (phonons, pairs, plasmons, etc) of a (metallic) solid are usually considered within the harmonic limit. A classical HO, subjected to a time-dependent force $\equiv -\lambda(t)$ [i.e. $V_{int}(y,t) = y\lambda(t)$] will gain an amount of energy

$$\Delta\varepsilon_{classical} = \frac{1}{2}k\frac{|\lambda(\omega)|^2}{(m\omega)^2} = \frac{|\lambda(\omega)|^2}{2m}.$$

Here $\lambda(\omega)$ is the ωth Fourier component of $\lambda(t)$. The correspondence principle provides an elegant and exact connection between the energy gain of the forced classical harmonic oscillator and the vibrational excitation probability distribution of the equivalent quantum mechanical harmonic oscillator subjected to the same forcing function. In terms of the parameter

$$\beta = \Delta\varepsilon_{class}/\hbar\omega = |\lambda(\omega)|^2/2m\hbar\omega, \qquad 9.$$

it has been demonstrated many times that the probability for a $0 \to n$ HO transition is

$$P_{0 \to n} = e^{-\beta}(\beta^n/n!), \qquad 10.$$

a Poisson distribution (36a,b). Furthermore, when $\beta \gg 1$, the Poisson distribution becomes a Gaussian whose width $\Delta = \beta^{1/2}\hbar\omega$ is identical with Eq. 7b, the STL expression. The dynamics of a specific process enter into Eq. 10 solely through Eq. 9, which expresses the functional dependence of β on the Fourier transform of $\lambda(t)$.

It is instructive to consider the motion of a classical point particle, initially at rest, throughout the time sequence shown in Figure 1b. The classical Hamiltonians for the two oscillators are

$$H_a = (p^2/2m) + \frac{1}{2}ky^2 \qquad 11.$$

and

$$H_b = (p^2/2m) + \frac{1}{2}\tilde{k}(y-y_0)^2$$

where $k \neq \tilde{k}$, in general. For $0 \leq t \leq \tau_R$, the harmonic motion over V_b is conveniently given in a displaced coordinate system $y' = y - y_0$ with $dy'/dt = dy/dt$, as $y'(t) = y_0\cos(\tilde{\omega}t + \pi)$ and from $\ddot{y}' = \partial H_b/\partial p$, $p(t) = -m\tilde{\omega}y_0\sin(\tilde{\omega}t + \pi)$. At $t = \tau_R$, the particle is returned to V_a at the position

$$y(\tau_R) = y'(\tau_R) + y_0 = y_0(1 - \cos\tilde{\omega}\tau_R) \qquad 12.$$

with momentum

$$p(\tau_R) = m\tilde{\omega} y_0 \sin \tilde{\omega}\tau_R. \qquad 13.$$

The energy delivered to the A oscillator as a result of this sequence is

$$\Delta\varepsilon_{\text{class}} = H_a(t \geq \tau_R) - H_a(t \leq 0),$$

which with Eqs. 11–13 can be reduced to

$$\Delta\varepsilon_{\text{class}} = ky_0^2\left(1 - \cos\tilde{\omega}\tau_R - \left(\frac{k-\tilde{k}}{2k}\right)\sin^2\tilde{\omega}\tau_R\right). \qquad 14.$$

Making the identification $\beta_0 \equiv ky_0^2/2\hbar\omega = (\bar{u}V'/\hbar\omega)^2$, the usual Poisson parameter associated with a permanent switch of HO potentials (Figure 1a), a dynamic parameter follows from Eqs. 9 and 14 as:

$$\beta(\tau_R; k, \tilde{k}) = 2\beta_0\left(1 - \cos\tilde{\omega}\tau_R - \left(\frac{k-\tilde{k}}{2k}\right)\sin^2\tilde{\omega}\tau_R\right). \qquad 15.$$

Further, if and only if $k = \tilde{k}$ then from Eq. 15

$$\beta(\tau_R) = 2\beta_0(1 - \cos\omega\tau_R) \qquad 16.$$

and the exact final excited state distribution is the Poisson distribution given by Eqs. 10 and 16. If τ_R is such that $\omega\tau_R$ is an odd (even) integral-multiple of π, maximum (minimum) oscillator excitation occurs where from Eqs. 9 and 16, $\Delta\varepsilon^{\max} = 4\beta_0\hbar\omega$. Thus a commensurability between the oscillator frequency and the intermediate state lifetime is a crucial factor.

For the more general case in which $k \neq \tilde{k}$, the final state distribution is not expected to be precisely Poisson since a necessary condition is that the form of the time-dependent change in the Hamiltonian is $H'(t) = y\lambda(t)$. Equation 11 can be cast into this form only when $k = \tilde{k}$. Otherwise, Eq. 11 would imply $H'(t) = (y + (1 - \tilde{k}/k)y^2/2y_0)\lambda(t)$. The quadratic term gives rise to a distinctly quantum mechanical effect when wave packet rather than point particle motion is considered; namely, wave packet spreading or contraction. It is true that once a Gaussian, always a Gaussian when propagating on a quadratic potential. However, only in the case for which the curvature (frequency) of the displaced harmonic potential is identical with the confining potential determining the initial width of the Gaussian wave packet does the width remain constant (minimum uncertainty states) (21, 22a–c, 37a–c).

APPLICATIONS

Having invested a substantial effort in understanding the fundamentals of a unified theoretical apparatus for time-dependent quantum dynamics, we

Adsorbate Photoemission Lineshapes

Some of the most important and detailed knowledge of the electronic structure and chemical bonding at both "clean" and adsorbate covered metallic surfaces has come from Ultraviolet and X-ray Photoelectron Spectroscopy (UPS and XPS) studies (38–40). From a comparison between the photoelectron spectrum (possibly angle-resolved, photo-ejected electron energy distribution) of the isolated gas phase atom or molecule and the spectrum of the same object adsorbed on a surface or condensed into a solid, one can learn about the nature of the interaction of the atom or molecule with its condensed phase environment. The observed differences show up as both initial state chemical and final state relaxation shifts (40) in electron binding energies or "ionization potentials," as extra-atomic shakeup satellites (38, 41), and as lineshapes due to both electron-hole pair (42a,b, 43) and phonon (43, 44) excitation. Since most of these effects are due to some sort of background electron response to the creation of a quasilocalized positive hole potential in the photoionization process, they would not be expected to exhibit a significant temperature dependence. Conversely, if a significant temperature dependence is observed, then nuclear motion and displacement are almost certainly involved and in these cases, wave packet modeling should be useful for analysis.

As an example, consider the UPS ($h\nu = 21.2$ eV, He I light) studies of Xe physisorbed upon Cu[110] shown in Figure 3 (44). As expected, the spectrum (*insert*) shows a spin-orbit split doublet from the closed Xe $5p$ shell with a binding energy roughly 2 eV less than in atomic Xe due to the extra-atomic screening or relaxation shift. The linewidth [full width at half maximum (FWHM)] of the Xe $5p$ levels were measured as a function of temperature (150 meV energy resolution) and the pronounced temperature dependence is also shown in Figure 3.

The origins of this phonon-induced photoemission line broadening are displayed in Figure 4. The lowest-lying potential-energy curve describing the neutral Xe-surface interaction is characterized by some equilibrium position z_0 and curvature $d^2V_a/dz^2|_{z=z_0} = M\omega^2$. Photoionization creates a quasilocalized positive hole on the Xe that induces a negative screening (image) charge within the substrate (45, 46). The interaction between the hole and its image provides an additional, attractive term augmenting $V_a(z)$, the neutral curve, which results in the Xe$^+$-metal interaction shown as the *upper curve*, whose equilibrium distance from the surface is $z^+ < z_0$ and whose curvature is such that ω^+ is not necessarily equal to ω.

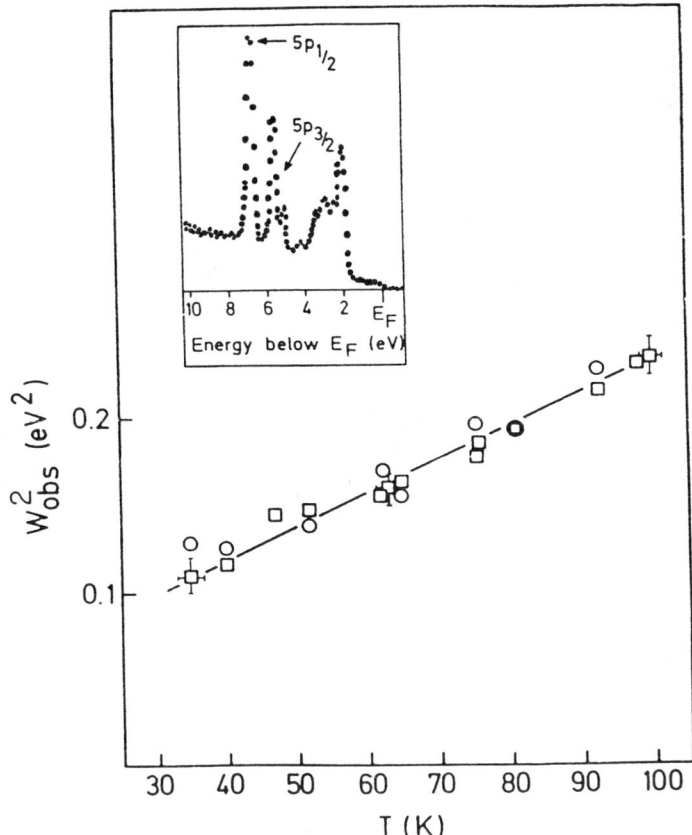

Figure 3 Square of the linewidth W_{obs} vs layer temperature T of the Xe $5p_{1/2}$ (*squares*) and $5p_{3/2}$ (*circles*) photoemission peaks for a Xe layer adsorbed on Cu[110]; a sample spectrum, recorded in normal emission, is shown in the *inset*. (Ref. 44.)

As a reasonable first approximation,

$$V_+(z) \simeq V_a(z) - V_{\text{image}}(z) + V_i,$$

with $V_a(z)$ the atom+metal curve, V_i the gas phase ionization potential, and $V_{\text{image}}(z) = e^2/4(z-d_0)$ where d_0 is the effective location of the image plane (47). Since $dV_a/dz|_{z=z_0} = 0$, the relevant slope of the Xe$^+$+metal curve is

$$dV_+(z)/dz|_{z=z_0} = e^2/4(z_0-d_0)^2. \qquad 17.$$

If the variation with final state electron energy of the Xe dipole matrix element is weak over an energy scale of order the observed linewidths,

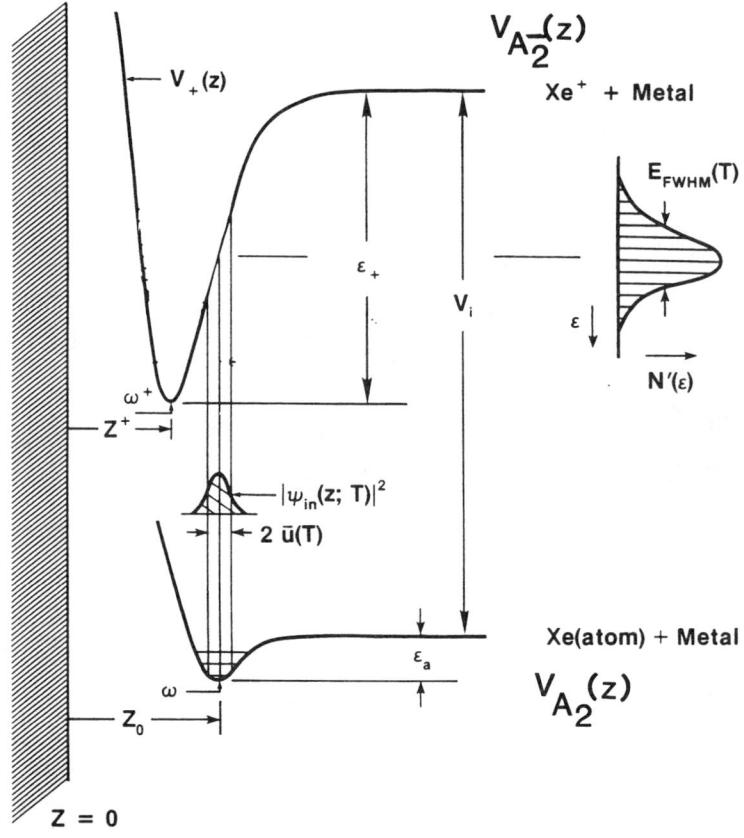

Figure 4 Potential-energy curves for atomic and ionic Xe interaction with a surface. $\varepsilon_a - \hbar\omega/2$ and $\varepsilon_+ - \hbar\omega^+/2$ are atom and adsorbed-ion desorption energies. The structure labeled $|\psi_{in}(z;T)|^2$ represents the initial-state charge distribution. (Ref. 44.)

then the photoemission lineshape is given by Eqs. 7 and 8 where ω_0 is the frequency of the Xe-metal bond. Combining Eqs. 8 and 17, the (FWHM) linewidth, $W = 2.35\Delta$, is given by

$$W(T) = \frac{2.35e^2}{4(z_0-d_0)} \left(\frac{(\hbar/2M\omega_0)^{1/2}}{(z_0-d_0)} \right) \left(\coth\left(\frac{\hbar\omega_0}{2kT}\right) \right)^{1/2}$$

$$= \frac{2.35e^2}{4(z_0-d_0)} \left(\frac{\bar{u}(T)}{z_0-d_0} \right), \qquad 18.$$

where $\bar{u}(T)$ is the vibrational amplitude of the initial state of the physisorbed Xe atom. For Xe on $r_s = 2$ jellium (48), $z_0 \simeq 2.5$ Å, $d_0 \simeq 0.85$ Å, and thus $e^2/4(z_0-d_0) \simeq 2$ eV, which is consistent with other experimental results (44). For the temperature range covered in Figure 3, with $\hbar\omega_0 = 2.35$ meV, $\coth(\hbar\omega_0/2kT) \simeq 2kT/\hbar\omega_0$, and thus

$$W(T)^2 = aT$$

where $a = 0.00194$ eV$^2/K$ was determined from Eq. 18. The observed linewidth is given by

$$W_{obs}(T) = [W_{other}^2 + W(T)^2]^{1/2},$$

where W_{other} includes temperature-independent broadening mechanisms such as experimental resolution and hole "lifetime" effects. Both the theoretical and experimental data are shown in Figure 3. The linear theoretical curve implies that $W_{other} \simeq 0.2$ eV, which estimates the sum of instrumental resolution and (surprisingly long) hole-lifetime effects.

This particular study suggests how one could use the temperature dependence of a photoemission lineshape to determine parameters characterizing the absorbate-surface bond such as ω_0 and z_0. By having the wave packet methodology at hand, lineshape analysis could be carried out immediately.

Stimulated Desorption Distributions

Based on the success of wave packet modeling in providing intuitively appealing accounts of gas phase photodissociation dynamics due to electronic excited states (23a–d, 49–51), a similar level of success might reasonably be expected with regards to the closely related phenomena of (electron or photon) stimulated desorption (52–54). Perhaps this expectation has inspired what seems to be more recent and varied theoretical studies in the area of stimulated desorption than in any other single area of surface dynamics (52a,b, 55–63a,b). From the point of view of simplified one-dimensional nuclear dynamics, the situation in stimulated desorption is modeled in terms of "desorbate"-surface potential energy curves, as shown in Figure 5. The initial state of the particle that will be desorbed is thermally distributed among the bound states of the atom/molecule-surface PES, as in the case of physisorbed Xe, which was shown in Figure 4. According to both the long-standing Menzel-Gomer-Redhead (MGR) model (64, 65) as well as the more recent Knotek-Feibelman (KF) model (66a,b), desorption is initiated by first creating an electronically excited state of the surface complex, either with an electron or photon beam. Usually MGR is meant to imply something like a bonding-to-antibonding transition among valence electrons. In contrast, KF is initiated by core-hole creation, then two-hole production via Auger decay, and finally, Coulomb repulsion. In either case, the excited electronic state provides a new PES for subsequent nuclear motion. Depending upon specifics of the excited state, this

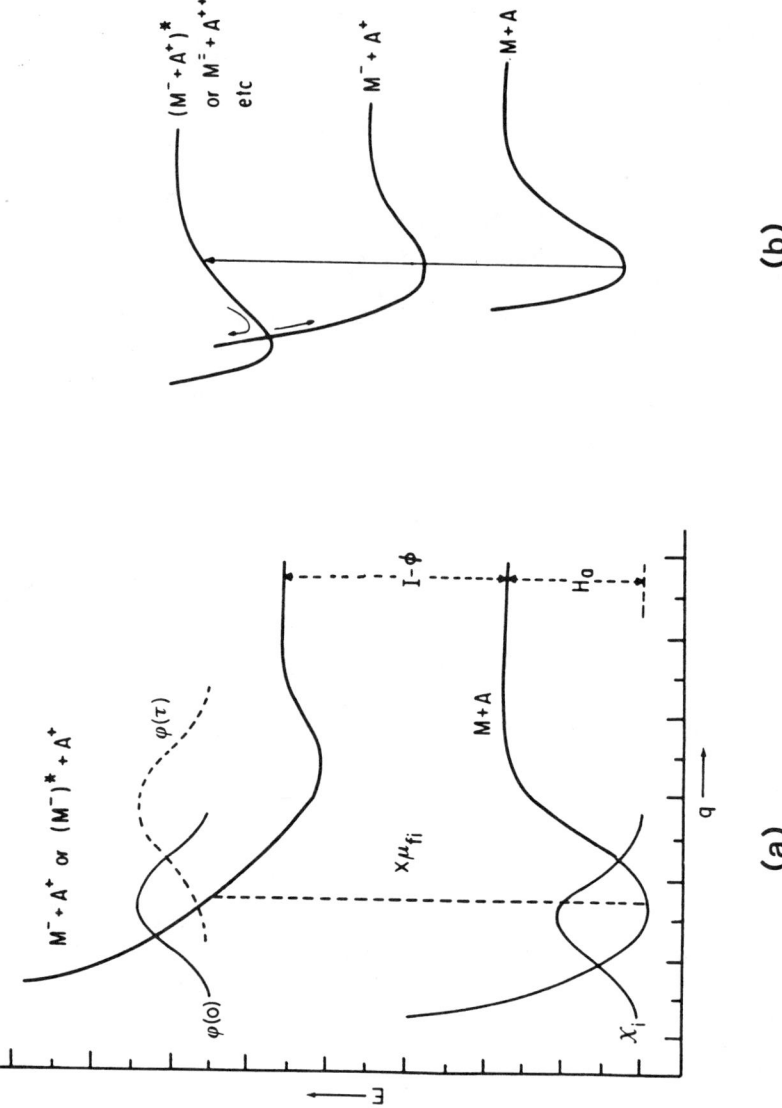

Figure 5 Potential curves illustrating various mechanisms for stimulated desorption. (*a*) Direct excitation onto a repulsive desorptive curve such as MGR or KF. (*b*) Excitation onto an intermediate bound ion curve prior to desorption, as suggested by Antoniewicz (56).

PES may be repulsive, as in Figure 5a, in which case desorption can occur if the excited state is sufficiently long-lived to survive surface quenching. Although arguments go on within the stimulated desorption community as to whether desorption occurs via a MGR or KF mechanism, from the point of view of nuclear dynamics the most relevant question pertains to the nature of the final state PES, whether repulsive or not, and not on the particular electronic properties responsible for the repulsion. Finally, we note that although most stimulated desorption experiments have observed ionic desorption due to ease of detection (52–54), recent studies have also measured quantities such as yields, kinetic energy distributions, and internal quantum state distributions of desorbed neutral molecules (63a, 67, 68a,b). This seems well advised, since the yield of neutrals greatly exceeds that of charged particles.

As an example of the usefulness of the wave packet dynamics here, consider the results of a study on the system O on W[110] (69). Among other things, the desorbed O^+ energy distribution and the temperature dependence of the fractional change of the width was measured, as shown in Figure 6. These results are easily understood, since the energy distribution is given by Eq. 3, which in the STL is the Gaussian of Eq. 7. Further, with $x \equiv \hbar\omega_0/2kT$, Eqs. 7b and 8 give the width

$$\Delta(T) = u_0 V'|_{z=z_0} \coth^{1/2} x$$

$$\simeq u_0 V'(1 + \exp(-2x)).$$

Thus, the semi-log slope of $[\Delta(T) - \Delta(300)]/\Delta(300)$ vs $1/T$, plotted in Figure 6b, yields the value of $\hbar\omega_0 = 0.158$ eV, for the vibrational energy of the O–W bond. Thermal broadening of a similar origin has also been observed in the angular distributions of desorbing ions (54).

Another interesting stimulated desorption scenario, shown in Figure 5b, was suggested by Antoniewicz (56). He realized that if the initially excited state were ionic, then, due to image potential screening (45), the forces determining initial wave packet propagation could be directed toward the surface, similar to the Xe photoemission example shown in Figure 4. Closer to the surface, additional electronic transitions and thus curve crossings could be experienced onto repulsive states, in analogy with predissociative processes in gas phase.

Stimulated desorption of this sort, with curve crossings, is particularly interesting when the internal state distributions of desorbed molecules are considered. As an illustrative example, let τ_R be the time interval between initial placement on the upper curve and the transition onto the final repulsive curve. Further suppose that the molecule is a homo-nuclear diatomic and the sequence of electronic transitions corresponds to

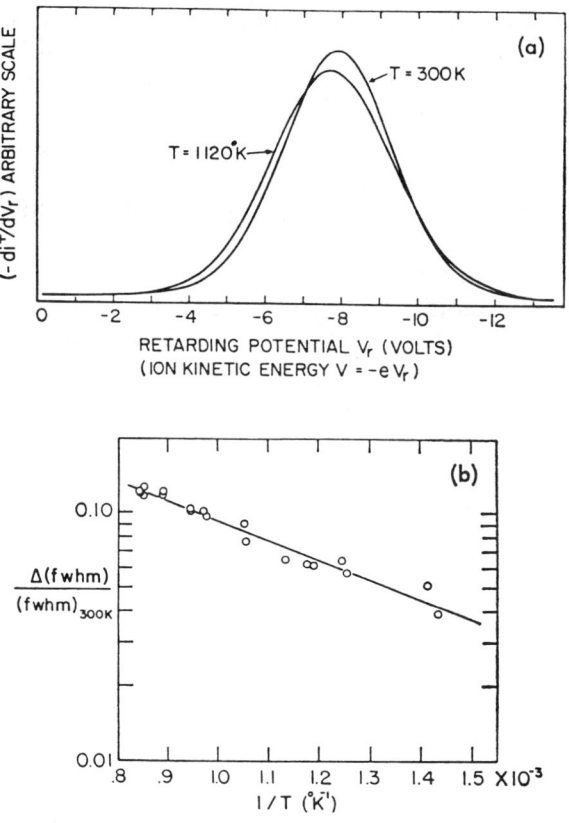

Figure 6 (*a*) Temperature dependent stimulated desorption ion energy distributions for O on W[110]; (*b*) Boltzman plot of fractional change in broadening of ion energy distribution. (Ref. 69.)

$$A_2(t = 0) \to A_2^* \to A_2(t = \tau_R).$$

Under these circumstances, the desorbed molecule should emerge vibrationally hot with a Poisson internal state distribution given by Eqs. 10 and 16. This possibly inverted distribution has absolutely nothing to do with any thermal equilibrium distributions characterized by the surface temperature. Similar effects have been anticipated (70a,b) and observed (68a,b) in hot rotational distributions due to hindered-to-free rotor transitions in stimulated desorption (63a,b). At present, conceptualizing and modeling along these lines is being pursued most seriously and effectively by Stechel and associates (59, 63a).

Resonance EELS

A number of recent electron-energy-loss spectroscopy (EELS) studies of molecules adsorbed on surfaces have shown greatly enhanced overtone excitations of intramolecular modes of the adsorbate, due to the formation of temporary negative-ion resonances (3e, 71–73). With regards to simple diatomic molecules, the equivalent gas-phase phenomenon has long been understood in terms of the so-called "boomerang model" (74–76). In the widely considered case of N_2, incident electrons with ~ 2 eV kinetic energy are trapped within a molecular-ion shape resonance characterized by an equilibrium internuclear separation greater than that of the neutral molecule. Throughout the negative-ion lifetime τ_R, the nuclei separate toward the new equilibrium point as in Figure 1b. Once the electron exits the shape resonance, the nuclei are extended relative to the neutral equilibrium separation, and the molecule is left in a vibrationally excited state. Models for such resonance EELS processes have been developed in which certain surface modifications to the resonance excitation of intramolecular modes can be taken into account (9, 77–79).

An illuminating example of overtone excitation due to the formation of temporary negative molecular ions is the EELS study of N_2 physisorbed and condensed on Ag surfaces by Demuth, Schmeisser & Avouris (3e, 71). The N_2 loss spectrum for a primary electron energy = 1.5 eV is shown in Figure 7a, where substantial overtone excitation of the 0.29 eV, $0 \to 1$ fundamental is to be noted. The gas phase $0 \to 1$ cross-section, as a function of primary energy, is shown in Figure 7b. It is argued that the loss spectrum shown in Figure 7a is due to the formation of a temporary negative ion, as in the gas phase, but with the threshold moved lower in energy due to image potential reductions in the electronic bound state energies. This is supported by the observed cross-sections for both the physisorbed and condensed N_2, also shown in Figure 7b. The physisorbed cross-section appears as a smoothed gas phase version shifted down in primary energy ~ 1.5 eV, which would be a reasonable image shift. In the case of condensed multilayers that are not in intimate contact with the Ag, the resulting cross-sections are just those of the gas phase, appropriately broadened.

The observed intensity distribution versus overtone excitation number shown in Figure 7c for both gas phase and physisorbed N_2 display a notable attenuation for physisorbed relative to free N_2. It was suggested that the presence of the surface could decrease the negative ion resonance lifetime, and in some way this could be responsible for the reduced overtone excitation of the physisorbed N_2 compared to that of the free molecule (3e, 71).

Resonance vibrational excitation in EELS is a very nice realization of

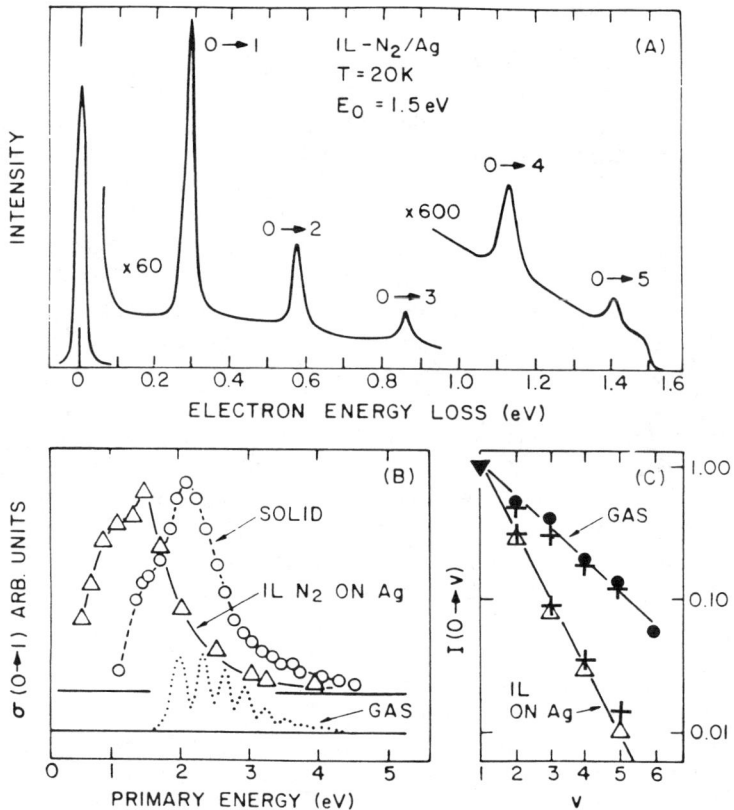

Figure 7 (a) The vibrational spectrum (fundamental and several overtones) of N_2 on a silver surface at 20K obtained via resonance electron scattering. (b) Resonance profiles for the fundamental vibrational excitation of N_2 in the gaseous, solid, and adsorbed (on Ag) phases. (c) Normalized intensity of the vibrational overtones of a monolayer of N_2 on Ag ($\triangle$) compared to the corresponding intensities for gaseous N_2 ($\bigcirc$), and theoretical values ($+$) from Eqs. 10 and 16. (Ref. 3e.)

the double switching process shown in Figure 1b in which the excited state distribution is given by Eq. 5. Further, if the intramolecular potential curves of N_2 and N_2^- were displaced HOs with identical frequency, then the excitation probabilities would be Poisson distributed according to Eqs. 10 and 16. Such distributions have been calculated by using parameters appropriate to the free molecule obtained from theoretical potential energy curves (80), specifically $\hbar\omega = 0.29$ eV, $\beta_0 \simeq 3$, and $\hbar/\tau_R \simeq 0.5$ eV, and the results are shown as the $+$ signs in Figure 7c. The physisorbed distribution was obtained by reducing the negative ion resonance lifetime to a value

~80% of the free space value. Certainly the frozen wave packet propagation picture is able to account for the observed resonance loss intensities in a plausible way.

For the sake of completeness, two possible improvements should be noted. First, since $\hbar\omega_{N_2^-} \simeq 0.26$ eV, Eq. 15 rather than 16 might be used. However, as mentioned in the HO discussion, when $\omega \neq \tilde{\omega}$, then the excited-state distribution is not necessarily Poisson. Under these conditions, the vibrational state distribution can be obtained either from explicit calculation with Eqs. 1, 2, and 4, or from a more traditional Franck-Condon picture in which

$$P_{0 \to n}(\tau_R) = |\langle n | e^{-iH_{N_2^-}\tau_R} | 0 \rangle|^2$$

$$= \left| \sum_{\tilde{m}} \langle n | \tilde{m} \rangle e^{-i\hbar\tilde{\omega}\tilde{m}\tau_R} \langle \tilde{m} | 0 \rangle \right|^2 \quad 19.$$

where $|0\rangle$ and $|n\rangle$ are the ground and nth excited state vibrational wave functions for N_2, and $\{|\tilde{m}\rangle\}$ is the manifold of vibrational wave functions of the intermediate, displaced oscillator state (77). The objects $\langle n|\tilde{m}\rangle$ and $\langle \tilde{m}|0\rangle$ are overlap integrals. The coherent sum over paths connecting $|0\rangle$ with $|n\rangle$ in Eq. 19 leads to interference effects depending upon the intermediate state lifetime τ_R relative to $\tilde{\omega}$, i.e. the commensurability effect already noted. The exact determination of the vibrational state distribution, Eq. 19 is facilitated by using HO recursion relations such as those due to Manneback (81). Head-to-head comparison between the approximate Poisson distributions and those obtained from Eq. 19 have been made, and, at least for cases in which $0.85 \lesssim \tilde{\omega}/\omega \lesssim 1.15$, the differences are not significant for present purposes (82). Second, if the negative ion decays exponentially rather than according to the "top-hat" structure shown in Figure 1b, then $P_{0 \to n}(\tau_R)$ should be averaged over all times, weighted by the existence probability

$$\langle P_{0 \to n}(\tau_R) \rangle = \int_0^\infty \frac{dt}{\tau_R} e^{-t/\tau_R} P_{0 \to n}(t).$$

This was the procedure followed elsewhere (77), where a somewhat different value for the ratio $\tau_R^{gas}/\tau_R^{phys}$ was required to fit the data of Figure 7b.

Yet another potentially observable feature in an EELS spectrum that should accompany temporary negative ion formation is based on the necessary existence of a usually lower frequency vibrational mode associated with the adsorption bond between the molecule and the substrate (e.g. $\hbar\omega \sim 50$ meV for adsorbed CO, N_2, etc) (79). To illustrate the point, reconsider the potential-energy curves shown in Figure 4, in which the energy of the molecule-surface bond, as a function of separation, is depicted. The neutral molecule curve labeled $V_{A_2}(z)$ is characterized by a relatively shallow well with a minimum at $z = z_0$. In analogy with ionic states formed in photoemission, when the incident electron becomes trapped in the shape resonance, an additional image-potential type of attraction is turned on. This attraction results in the augmented molecular-ion-surface curve $V_{A_2^-}(z) \simeq V_{A_2}(z) - e^2/4(z-d_0) + V_0$, where V_0 is a constant whose value is irrelevant for the present argument. During the life of a negative ion, the molecule's center of mass is displaced on $V_{A_2^-}$, and is ultimately returned to V_{A_2} at a position $z(\tau_R) \neq z_0$. Hence (low-frequency) surface-molecule modes should show an enhancement in excitation probability as the primary electron energy is tuned through a negative-ion shape resonance associated with the high-frequency intramolecular stretch mode.

A quantitative measure of this effect follows from the no-loss probability obtained from Eqs. 10 and 16 as

$$P_0(\tau_R) = \exp[-2\beta_0(1 - \cos \omega_0 \tau_R)], \qquad 20.$$

which in the typical $\omega_0 \tau_R < 1$ limit reduces to

$$P_0(\tau_R) \simeq \exp(-\beta_0 \omega_0^2 \tau_R^2).$$

From the STL of Eq. 20, it is apparent that large-scale depletion of the no-loss line requires a resonance lifetime long enough for the molecule to displace, which it does on a time scale set by ω_0^{-1}. The magnitude of the effect is characterized by the quantity

$$Q \equiv \beta_0 \omega_0^2 \tau_R^2 = \frac{(V'_{A_2^-} \tau_R)^2}{2\hbar M \omega_0}, \qquad 21.$$

obtained from Eqs. 7b, 8, and 17. Typically, $e^2/4(z_0 - d_0) \simeq 2$ eV, so $z_0 - d_0 \approx 1.8$ Å, and $V'_{A_2^-} \simeq 1.1$ eV/Å. The resonance lifetime corresponds to a width in the 0.1–1.0 eV range (10^{-15} sec $\lesssim \tau_R \lesssim 10^{-14}$ sec). Aside from H_2, $M \approx 28 \, M_{\text{proton}}$ sets a lower limit on the molecular mass. For given τ_R, the only remaining "free variable" is the molecule-surface-vibrational frequency ω_0. The softer this mode, the smaller is P_0, thus enhancing the loss features. For instance, using the parameters just mentioned and $\hbar\omega_0 \simeq 0.05$ eV, Eqs. 20 and 21 give $P_0(\tau_R = 10^{-15}$ sec$) = 0.9982$ and $P_0(\tau_R = 10^{-14}$ sec$) \simeq 0.83$. Since unitarity requires that $1 = P_0 + \Sigma_{n>0} P_n$, it is reasonable to take $P_{n=1} \simeq 1 - P_0$. Although the resulting excitation probability per resonance event is not large, looking for loss features whose intensities are only a small fraction of the elastic peak is standard procedure in EELS. The intriguing aspect of this particular loss mechanism is that it can be tuned into by varying the primary electron energy through a negative-molecular-ion resonance associated with an entirely different bond (the intramolecular A_2 bond), and, as such, its intensity variation with energy could provide additional information concerning the dynamics of both shape resonances in molecules adsorbed on metal surfaces, and also molecular motion over the potential-energy curve associated with the adsorption bond.

Molecule Surface Collisions

State-to-state molecular beam scattering is emerging as a major experiment of choice for elucidation of the microscopic mechanisms of relevance in molecular chemical dynamics at surfaces (3–8). Among the fundamental issues being addressed experimentally are the role of (a) energy redistribution between the translational and rotational (83, 84) or vibrational (85) (T, R, V) degrees of freedom of the participating molecules; (b) nonadiabatic electronic transitions in atomic (86) and molecular (87) beams; and (c) dissipative effects due to the continuum of both electron-hole pair (88) and phonon (89a,b) excitations of the solid. As a result of these studies, it is becoming apparent that charge transfer or harpooning (5b, 15, 87) frequently occurs during a molecule-surface collision. Furthermore, significant internal vibrational excitation is expected in such collisions (90a–h). This is particularly important since placement of energy within the vibrational mode is a necessary first step in dissociative sticking and other surface reactions (91a–c, 92).

The essential features of the charge-transfer-induced vibrational excitation mechanism are illustrated in Figure 8. Suppose that the molecule far removed from the surface is prepared in its electronic and vibrational ground state and then is allowed to interact with the surface, subject to the constraint that it remain in the prepared state. The total energy of the coupled system, as a function of the position of the molecular center-of-mass, might be given as the repulsive curve labeled A_2 in Figure 8a. Now imagine that the same molecule is converted into a negative molecular ion by extracting an electron from the substrate (at a cost of energy = ϕ, the work function) and attaching it onto the molecule (with energy gain = A, the electron affinity) and then is allowed to interact with the substrate, again subject to the frozen-internal-state constraint. A possible potential curve, labeled A_2^- in Figure 8a, shows a strongly attractive potential well due to the image potential between the molecular ion and the surface. If $V_{A_2}(z)$ is reasonably constant prior to the short-range repulsion, then the curve crossing point is given by

$$\phi - A = e^2/4R_c, \qquad 22.$$

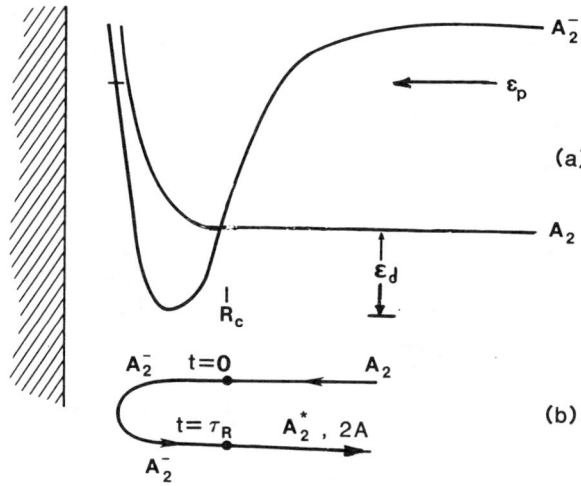

Figure 8 (a) Diabatic potential-energy curves as a function of z, the normal distance from the surface of an incident A_2 molecule with center-of-mass kinetic energy ε_p. At the curve crossing ($z = R_c$), passage to the A_2^- negative molecular ion curve is possible. (b) Time line showing the state of the molecule at various times along its scattering trajectory.

in analogy with gas phase harpooning reactions (15). Although far from the surface, A_2^- corresponds to an electronically excited state of the coupled system; when $z \leq R_c$, it becomes the ground state. At the curve crossing where $V_{A_2}(z) = V_{A_2^-}(z)$, passage from one diabatic state to another can occur at no cost in energy. Thus a possible scattering scenario is one in which an incident neutral molecule with center-of-mass translational energy ε_p propagates on $V_{A_2}(z)$ up to $z = R_c$ (time $t = 0$) where an electron from the surface tunnels into the molecule with some probability $\equiv P_{LZ}$, essentially a surface Landau-Zener probability (5b, 15, 29a,b, 30a,b). The resulting molecular ion then translates on $V_{A_2^-}(z)$ to the classical turning point, reflects, and passes through $z = R_c$ (at $t = \tau_R$) proceeding left to right, after which the electron can be returned to the surface. The hairpin time line, Figure 8b, illustrates the fact that a negative-molecular ion existed for a time duration τ_R.

The intramolecular dynamics is that of Figure 1b. Electron capture by the molecule has the effect of altering the intramolecular potential from a neutral to a negative-molecular ion curve. If the donation is into an antibonding orbital, then the equilibrium atomic spacing increases and, as a result, the atoms separate. Subsequently, when the electron departs, the intramolecular dynamics occurs over the neutral potentials. Since the atoms are now displaced relative to the neutral equilibrium configuration, the molecule ends up vibrationally excited, exactly as in the resonance electron scattering discussed in the previous section. The significant factor determining the ultimate degree of $T \to V$ energy redistribution is, from Eqs. 9 and 15 or 16, the negative ion lifetime (with respect to $\tilde{\omega}$, the intramolecular frequency). This is just the time duration for translation on $V_{A_2^-}(z)$ in Figure 8a, which is expressed as

$$\tau_R = \int_{z_{in}}^{z_{out}} \frac{dz}{v_z(z)}, \qquad 23.$$

where the integral is performed along the trajectory, and, as drawn in Figure 8a, $z_{in} = z_{out} = R_c$ (90d,e). The velocity as a function of z is obtained from

$$\varepsilon_{tot} = \frac{1}{2} M v_z^2(z) + V_{A_2^-}(z), \qquad 24.$$

which requires a functional expression for $V_{A_2^-}(z)$. From Eqs. 23 and 24, the molecular ion lifetime is thus

$$\tau_R(\varepsilon_p) = (M/2\varepsilon_{tot})^{1/2} \int_{z_{in}}^{z_{out}} [1 - V_{A_2^-}(z)/\varepsilon_{tot}]^{-1/2} \, dz, \qquad 25.$$

and it is here that the kinetic energy dependence enters the formulation.

An important and exciting aspect of molecular collisions with surfaces involving intermediate charge-transfer states is that τ_R, the intermediate state lifetime, can in principle be controlled. The obvious method simply depends upon changing the kinetic energy of the incident beam. A more dramatic way, with a broad dynamic range, involves changing the work function of the surface, say by cesiation (93a,b). From Eq. 22, it is apparent that the location of the curve crossing, $R_c = e^2/4(\phi - A)$, can easily be shifted by varying ϕ. From Eq. 25, this is equivalent to setting τ_R. Thus a new degree of control over any process that is a function of τ_R, such as vibrational excitation (90a–h), dissociation (90e, 94), or selectivity in branching ratios of final states (92, 95), has been introduced.

Two aspects of the theoretical vibrational distributions are illustrated in Figure 9. The probability that the molecule will remain unexcited throughout the double curve crossing, the so-called no-loss probability given by Eq. 20, is shown in Figure 9a as a function of $\omega \tau_R$ taking β_0 parametrically as labeled. [Note: $\beta_0 \simeq 1-3$ and $\hbar \omega \sim 0.2$ eV for C, N, O diatomics (Poisson required), whereas $\beta_0 \simeq 50$ and $\hbar \omega \sim 0.02$ eV for halogens (Gaussian adequate).] Since P_0 is periodic, if $\omega \tau_R$ is an odd (even) integral multiple of π, maximum (minimum) oscillator excitation occurs. Thus commensurability between the oscillator frequency and the

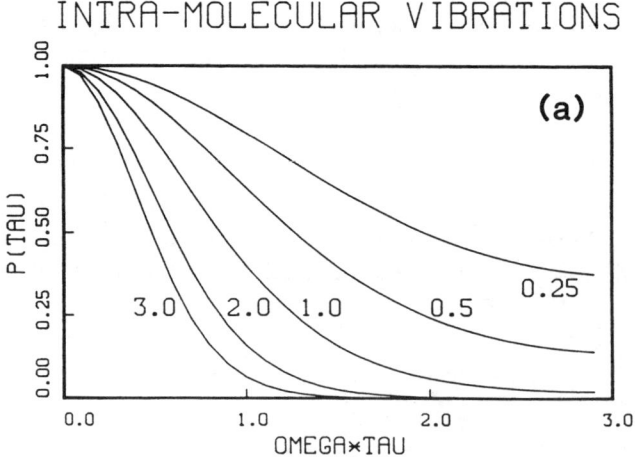

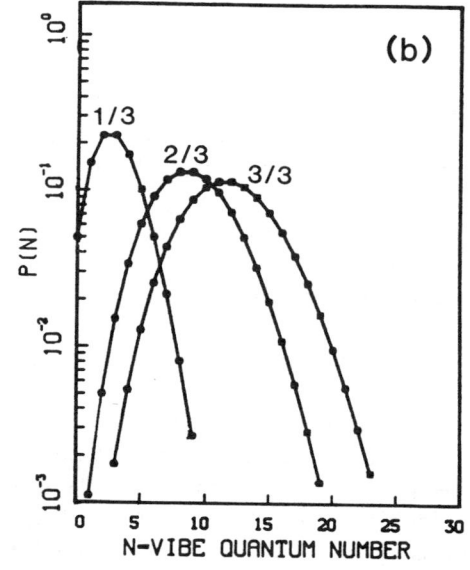

Figure 9 (*a*) No loss intensity vs $\omega\tau_R$, from Eq. 20, with β_0 varied parametrically. Due to periodicity, $P_0(\tau_R)$ is symmetric about $\omega\tau_R = n\pi$ with $n =$ integer. (*b*) Vibrational excitation probability as a function of vibrational quantum number with $\omega\tau_R/\pi = 1/3, 2/3, 3/3$.

time delay between curve crossings is a crucial factor in determining the degree of excitation of the scattered molecule.

Vibrational state distributions obtained from Eqs. 10 and 16 with $\beta_0 = 3$, as appropriate for O_2, and $\omega\tau_R = \pi/3, 2\pi/3$, and $3\pi/3$ are shown in Figure 9b. The mean vibrational energy carried by the scattered molecules is $\Delta\varepsilon(\tau_R) = \beta(\tau_R)\hbar\omega$, which, as is apparent from Figure 9b, can in principle provide a controllable source of molecules with extreme vibrational population inversion. Such a controlled inversion in scattered polyatomic molecules could provide a feasible alternative (92) to a two-photon pump-dump proposal for controlled selectivity in reactivity (95).

Finally, we consider the current experimental situation with regard to the scattering scenario put forth in this section. First, greatly enhanced non-thermal vibrational excitation in beams of NO scattered from Ag[111] (85) and in recombinatively desorbed H_2 from Cu (96) has been observed. Second, charge transfer/harpooning in $O_2^+ \to O_2, O_2^-$, O scattering from Ag[111] has been documented (87). Last, O_2 is known to nondissociatively physisorb as O_2^- on several metal surfaces (97). Thus every component in the proposed model has been observed, and it only seems like a matter of time before all three parts will combine in a single experiment.

FINAL THOUGHTS

A major purpose of this survey on applications of semiclassical wave packet techniques in surface dynamics has been to demonstrate that a wide class of seemingly unrelated phenomena occurring in chemical physics research involving surfaces, can in fact fit under a single "intellectual umbrella" when considered as problems in time-dependent quantum mechanics. The position was then taken that for present purposes, the wave packet techniques developed by Heller and associates (17, 21–25a,b) can provide a unified and insightful single framework for modeling the surface phenomena considered here. The essential features of what Metiu has called the "simplest Heller method" have been presented. Technical aspects related to Gaussian algebra, forced harmonic oscillator mechanics, and the SHM in the short time limit have been presented in enough detail for this article to be self-contained, once analysis of the specific surface phenomena was undertaken. Those processes considered here, namely photoemission, stimulated desorption, resonance electron energy loss spectroscopy, and charge-transfer-induced-vibrational-excitation in molecule/surface collisions all involve fast electronic transitions and thus "preparation" (and perhaps decay) of nonstationary states. The subsequent atomic motion of the prepared state, described so clearly within the SHM, gives rise to experimental observables characterizing both the potential energy surface over which the wave packets propagate and the inherent time-dependences of the excited electronic states. By using only the theory presented here in the Methodology section, each of the four examples has been modeled in ways in which nontrivial information on atom or molecule interactions with surfaces was able to be extracted immediately from data, once one invested the effort needed to "transcend the case-by-case mentality" of surface science (27) and to develop an "interest in the fulfillment of general laws" (28). I hope that this philosophical approach to surface dynamics will prove to be as stimulating and informative to others as it has been to me.

Acknowledgments

It is a pleasure to thank John Harris, Eric Heller, Aart Kleyn, Alan Luntz, Ted Madey, Horia Metiu, and Ellen Stechel for useful discussions over the years, on topics related to this work. Special thanks go to Steve Holloway, who has been a lively collaborator on a good deal of the research discussed here.

Literature Cited

1. Levine, R. D., Bernstein, R. B. 1987. *Molecular Reaction Dynamics and Chemical Reactivity.* New York/Oxford: Oxford Univ. Press
2. Goodman, D. W. 1986. *Ann. Rev. Phys. Chem.* 37: 425
3a. Saltsburg, H. 1973. *Ann. Rev. Phys. Chem.* 24: 493
3b. Ceyer, S. T., Somorjai, G. A. 1977. *Ann. Rev. Phys. Chem.* 28: 477
3c. Tully, J. C. 1980. *Ann. Rev. Phys. Chem.* 31: 319
3d. Cardillo, M. J. 1981. *Ann. Rev. Phys. Chem.* 32: 331
3e. Avouris, P., Demuth, J. 1984. *Ann. Rev. Phys. Chem.* 35: 49
3f. Madey, T. E., Ramaker, D. E., Stockbauer, R. 1984. *Ann. Rev. Phys. Chem.* 35: 215
3g. Lin, M. C., Ertl, G. 1986. *Ann. Rev. Phys. Chem.* 37: 587
4. Ceyer, S. T. 1988. *Ann. Rev. Phys. Chem.* 39: 479–510
5a. Kasemo, B., Lundqvist, B. I. 1984. *Comm. Atom. Mol. Phys.* 14: 229
5b. Gadzuk, J. W. 1985. *Comm. Atom. Mol. Phys.* 16: 219
5c. Drolshagen, G. 1985. *Comm. Atom. Mol. Phys.* 17: 47
6. Pullman, B., Jortner, J., Nitzan, A., Gerber, R. B., eds. 1984. *17th Jerusalem Symp. Devoted to Dynamics on Surfaces.* Dordrecht: Riedel
7. Barker, J. A., Auerbach, D. J. 1985. *Surf. Sci. Rep.* 4: 1
8. Yoshimori, A., Tsukada, M., eds. 1985. *Dynamical Processes and Ordering on Solid Surfaces.* Berlin: Springer-Verlag
9. Gadzuk, J. W. 1986. *J. Elec. Spectrosc.* 38: 233
10. Luntz, A. C. 1987. *Phys. Scripta* 35: 193
11. Gerber, R. B. 1987. *Chem. Rev.* 87: 29
12a. Miller, W. H. 1974. *Adv. Chem. Phys.* 25: 69
12b. Miller, W. H. 1975. *Adv. Chem. Phys.* 30: 77
13. Miller, W. H., ed. 1976. *Dynamics of Molecular Collisions.* New York: Plenum
14. Bernstein, R. B., ed. 1979. *Atom-Molecule Collision Theory.* New York: Plenum
15. Truhlar, D. G., ed. 1981. *Potential Energy Surfaces and Dynamics Calculations.* New York: Plenum
16. Noid, D. W., Koszykowski, M. L., Marcus, R. A. 1981. *Ann. Rev. Phys. Chem.* 32: 267
17. Heller, E. J. 1981. *Acc. Chem. Res.* 14: 368
18. Kleyn, A. W., Los, J., Gislason, E. A. 1982. *Phys. Rep.* 90: 1

19. Bang, J., de Boer, J., eds. 1985. *Semiclassical Descriptions of Atomic and Nuclear Collisions*. Amsterdam: Elsevier
20. Sawada, S., Heather, R., Jackson, B., Metiu, H. 1985. *J. Chem. Phys.* 83: 3009
21. Heller, E. J. 1975. *J. Chem. Phys.* 62: 1544
22a. Heller, E. J. 1976. *J. Chem. Phys.* 65: 1289
22b. Heller, E. J. 1976. *J. Chem. Phys.* 65: 4979
22c. Heller, E. J. 1977. *J. Chem. Phys.* 67: 3339
23a. Heller, E. J. 1978. *J. Chem. Phys.* 68: 2066
23b. Heller, E. J. 1978. *J. Chem. Phys.* 68: 3891
23c. Heller, E. J. 1980. *J. Chem. Phys.* 72: 1337
23d. Heller, E. J., Stechel, E. B., Davis, M. J. 1980. *J. Chem. Phys.* 73: 4720
24a. Lee, S.-Y., Heller, E. J. 1979. *J. Chem. Phys.* 71: 4777
24b. Tannor, D. J., Heller, E. J. 1982. *J. Chem. Phys.* 77: 202
24c. Myers, A. B., Mathies, R. A., Tannor, D. J., Heller, E. J. 1982. *J. Chem. Phys.* 77: 3857
24d. Heller, E. J., Sundberg, R. L., Tannor, D. 1982. *J. Phys. Chem.* 86: 1822
24e. Sundberg, R. L., Heller, E. J. 1982. *Chem. Phys. Lett.* 93: 586
25a. Drolshagen, G., Heller, E. J. 1983. *J. Chem. Phys.* 79: 2072
25b. Drolshagen, G., Heller, E. J. 1984. *Surf. Sci.* 139: 260
26. Lefebvre, R., Mukamel, S., eds. 1987. *Stochasticity and Intramolecular Redistribution of Energy*. Dordrecht: Reidel
27. Hamann, D. 1988. *Proc. Solvay Conf. on Surface Sci.* Berlin: Springer-Verlag. In press
28. Comsa, G., David, R. 1982. *Surf. Sci.* 117: 77
29a. Tully, J. C., Preston, R. K. 1971. *J. Chem. Phys.* 55: 562
29b. Nikitin, E. E., Zulicke, L. 1978. *Selected Topics of the Theory of Chemical Elementary Processes*. Berlin: Springer-Verlag
30a. Sawada, S.-I., Nitzan, A., Metiu, H. 1985. *J. Chem. Phys.* 32: 851
30b. Sawada, S.-I., Metiu, H. 1986. *J. Chem. Phys.* 84: 6293
31. Coalson, R. D. 1987. *J. Chem. Phys.* 86: 6823
32a. Coalson, R. D., Karplus, M. 1982. *Chem. Phys. Lett.* 90: 301
32b. Tal-Ezer, H., Kosloff, R. 1984. *J. Chem. Phys.* 81: 3967
32c. Heather, R., Metiu, H. 1987. *J. Chem. Phys.* 86: 5009
33. Heller, E. J. 1981. See Ref. 15, p. 103
34. Heller, E. J. 1981. *J. Chem. Phys.* 75: 2923
35a. Lucas, A. A., Šunjić, M. 1972. *Surf. Sci.* 32: 439
35b. Leung, K. M., Schön, G., Rudolph, P., Metiu, H. 1984. *J. Chem. Phys.* 81: 3307
35c. Gadzuk, J. W., Holloway, S. 1988. *Prog. Surf. Sci.* In press
36a. Rapp, D. 1971. *Quantum Mechanics*. New York: Holt, Rhinehart & Winston
36b. Gentry, W. R. 1979. See Ref. 14, p. 391
37a. Nieto, M. M., Gutschick, V. P. 1981. *Phys. Rev. D* 23: 922
37b. Nieto, M. M., Simmons, L. M., Gutschick, V. P. 1981. *Phys. Rev. D* 23: 927
37c. Stechel, E. B., Schwartz, R. N. 1981. *Chem. Phys. Lett.* 83: 350
38. Feuerbacher, B., Fitton, B., Willis, R. F., eds. 1978. *Photoemission and the Electronic Properties of Surfaces*. Chichester/New York: Wiley
39. Plummer, E. W., Eberhardt, W. 1982. *Adv. Chem. Phys.* 49: 533
40. Egelhoff, W. F. Jr. 1987. *Surf. Sci. Rep.* 6: 253
41. Gadzuk, J. W. 1987. *Phys. Scripta* 35: 171
42a. Doniach, S., Šunjić, M. 1970. *J. Phys. C* 3: 285
42b. Gadzuk, J. W., Šunjić, M. 1975. *Phys. Rev. B* 12: 524
43. Citrin, P., Wertheim, G. K., Baer, Y. 1977. *Phys. Rev. B* 16: 4256
44. Gadzuk, J. W., Holloway, S., Mariani, C., Horn, K. 1982. *Phys. Rev. Lett.* 48: 1288
45. Gadzuk, J. W. 1976. *Phys. Rev. B* 14: 2267
46. Lang, N. D., Williams, A. R. 1977. *Phys. Rev. B* 16: 2408
47. Lang, N. D., Kohn, W. 1973. *Phys. Rev. B* 7: 3541
48. Lang, N. D. 1981. *Phys. Rev. Lett.* 46: 842
49. Shapiro, M., Bersohn, R. 1982. *Ann. Rev. Phys. Chem.* 33: 409
50. Imre, D., Kinsey, J. L., Sinha, A., Krenos, J. 1984. *J. Phys. Chem.* 88: 3956
51. Brumer, P., Shapiro, M. 1985. *Adv. Chem. Phys.* 60: 371
52a. Tolk, N. H., Traum, M. M., Tully, J. C., Madey, T. E., eds. 1983. *Desorption Induced by Electronic Transitions DIET I*. Berlin: Springer-Verlag
52b. Brenig, W., Menzel, D., eds. 1985. *Desorption Induced by Electronic Transitions DIET II*. Berlin: Springer-Verlag
53. Knotek, M. L. 1984. *Rep. Prog. Phys.* 47: 1499
54. Madey, T. E. 1986. *Science* 234: 316
55a. Brenig, W. 1976. *Z. Phys. B* 23: 361
55b. Brenig, W. 1982. *J. Phys. Soc. Jpn.* 51: 1914

55c. Brenig, W. 1983. See Ref. 52a, p. 90
56. Antoniewicz, P. R. 1980. *Phys. Rev. B* 21: 3811
57. Freed, K. F. 1982. *Surf. Sci.* 122: 317
58. Ueba, H. 1983. *Phys. Rev. B* 27: 7389
59. Stechel, E. B. 1985. See Ref. 52b, p. 32
60. Cini, M. 1985. *Phys. Rev. B* 32: 1945
61. Gortel, Z. W., Kreuzer, H. J., Feulner, P., Menzel, D. 1987. *Phys. Rev. B* 35: 8951
62. Clinton, W. L., Pal, S., Jutila, R. E. 1987. *Phys. Rev. B* 36: 4123
63a. Burns, A. R., Stechel, E. B., Jennison, D. R. 1987. *Phys. Rev. Lett.* 58: 250
63b. Avouris, Ph., Kawai, R., Lang, N. D., Newns, D. M. 1987. *Phys. Rev. Lett.* 59: 2215
64. Menzel, D., Gomer, R. 1964. *J. Chem. Phys.* 41: 3311
65. Redhead, P. E. 1964. *Can. J. Phys.* 42: 886
66a. Feibelman, P. J., Knotek, M. L. 1978. *Phys. Rev. B* 18: 6531
66b. Feibelman, P. J. 1983. See Ref. 52a, p. 61
67. Feulner, P., Menzel, D., Kreuzer, H. J., Gortel, Z. W. 1984. *Phys. Rev. Lett.* 53: 671
68a. Burns, A. R. 1985. *Phys. Rev. Lett.* 55: 525
68b. Burns, A. R. 1986. *J. Vac. Sci. Tech. A* 4: 1499
69. Madey, T. E., Yates, J. T. 1969. *J. Chem. Phys.* 51: 1264
70a. Gadzuk, J. W., Landman, U., Kuster, E. J., Cleveland, C. L., Barnett, R. N. 1982. *Phys. Rev. Lett.* 49: 426
70b. Landman, U., Kleiman, G. G., Cleveland, C. L., Kuster, E., Barnett, R. N., Gadzuk, J. W. 1984. *Phys. Rev. B* 29: 4313
71. Demuth, J. E., Schmeisser, D., Avouris, Ph. 1981. *Phys. Rev. Lett.* 47: 1166
72. Sanche, L., Michaud, M. 1983. *Phys. Rev. B* 27: 3856
73. Kesmodel, L. L. 1984. *Phys. Rev. Lett.* 53: 1001
74a. Herzenberg, A. 1968. *J. Phys. B* 1: 548
74b. Dube, L., Herzenberg, A. 1975. *Phys. Rev. A* 11: 1314
75. McCurdy, C. W., Turner, J. L. 1983. *J. Chem. Phys.* 78: 6773
76. Domcke, W., Berman, M., Estrado, H., Mündel, C., Cederbaum, L. S. 1984. *J. Phys. Chem.* 88: 4862
77. Gadzuk, J. W. 1983. *J. Chem. Phys.* 79: 3982
78. Gerber, A., Herzenberg, A. 1985. *Phys. Rev. B* 31: 6219
79. Gadzuk, J. W. 1985. *Phys. Rev. B* 31: 6789
80. Krauss, M., Mies, F. H. 1970. *Phys. Rev. A* 1: 1592
81. Manneback, C. 1951. *Physica* 17: 1001
82. Waldenstrom, S., Razi Nagvi, K. 1982. *Chem. Phys. Lett.* 85: 81
83. Kleyn, A. W., Luntz, A. C., Auerbach, D. 1981. *Phys. Rev. Lett.* 47: 1169
84. Andersson, S., Wilzen, L., Harris, J. 1985. *Phys. Rev. Lett.* 55: 2591
85. Rettner, C. T., Kimman, J., Fabre, F., Auerbach, D. J., Morawitz, H. 1987. *Surf. Sci.* 192: 107
86. Newns, D. M., Makoshi, K., Brako, R., van Wunnik, J. N. M. 1984. *Phys. Scripta* T6: 5
87. Haochang, P., Horn, T. C. M., Kleyn, A. W. 1986. *Phys. Rev. Lett.* 57: 3035
88. Lim, C., Tully, J. C., Amirav, A., Trevor, P., Cardillo, M. J. 1987. *J. Chem. Phys.* 87: 1808
89a. Toennies, J. P. 1987. *J. Vac. Sci. Tech. A* 5: 440
89b. Persson, M., Harris, J. 1987. *Surf. Sci.* 187: 67
90a. Gadzuk, J. W. 1983. *J. Chem. Phys.* 79: 6341
90b. Gadzuk, J. W., Nørskov, J. K. 1984. *J. Chem. Phys.* 81: 2828
90c. Holloway, S., Gadzuk, J. W. 1985. *J. Chem. Phys.* 82: 5203
90d. Gadzuk, J. W., Holloway, S. 1985. *Phys. Scripta* 32: 413
90e. Gadzuk, J. W., Holloway, S. 1986. *J. Chem. Phys.* 84: 3502
90f. Newns, D. M. 1986. *Surf. Sci.* 171: 600
90g. Gadzuk, J. W. 1987. *Surf. Sci.* 184: 483
90h. Gadzuk, J. W. 1987. *J. Chem. Phys.* 86: 5196
91a. Gadzuk, J. W., Holloway, S. 1985. *Chem. Phys. Lett.* 114: 314
91b. Harris, J., Andersson, S. 1985. *Phys. Rev. Lett.* 55: 1583
91c. Chiang, C.-M., Jackson, B. 1987. *Chem. Phys.* 87: 5497
92. Gadzuk, J. W. 1987. *Chem. Phys. Lett.* 136: 402
93a. van Wunnik, J. N. M., Los, J. 1983. *Phys. Scripta* T6: 27
93b. Yu, M. L., Lang, N. D. 1983. *Phys. Rev. Lett.* 50: 127
94. Kolodney, E., Amirav, A., Elber, R., Gerber, R. B. 1984. *Chem. Phys. Lett.* 111: 366
95. Tannor, D. J., Kosloff, R., Rice, S. A. 1986. *J. Chem. Phys.* 85: 5805
96. Kubiak, G. D., Sitz, G. O., Zare, R. N. 1985. *J. Chem. Phys.* 83: 2538
97. Backx, C., de Groot, C. P. M., Biloen, P. 1981. *Surf. Sci.* 104: 300

CHAIN MOLECULES AT HIGH DENSITIES AT INTERFACES

Ken A. Dill, J. Naghizadeh, and J. A. Marqusee

Departments of Pharmaceutical Chemistry and Pharmacy, University of California, San Francisco, California 94143

INTRODUCTION

Properties of polymeric systems often depend on the interactions of chain molecues with surfaces or interfaces. Chains may be constrained by walls, they may adsorb to surfaces or interfaces, or they may be "anchored" to surfaces; that is, they may be attached through one or a small number of chain segments that have a strong affinity for the interface. One example of an anchored phase is that of the stationary phases used in reversed-phase liquid chromatography, wherein short alkyl chains, generally of lengths ranging from 1 to 18 carbons, have one end covalently attached to a silica surface. Another example is that of the interphase region in semicrystalline polymers; in this case, the chains emanate from the surface of a crystallite, a crystalline region of parallel chains. They emerge into a region that becomes increasingly disordered with distance from the crystallite, the chains utimately adopting random bulk organization. A third example of anchored chains is that of aggregates of amphiphilic chain molecules. For example, detergents, fatty acids, and lipids are comprised of a polar headgroup attached to a longer hydrophobic chain tail. They have affinity for hydrophobic/hydrophilic interfaces; some amphiphiles localize at air/water interfaces, with their polar heads in the water and their hydrocarbon tails in the air. In these systems, interfacial forces are sufficiently strong that surface fluctuations are typically less than 1–2 times the chain cross section (5–10 Å); hence these may often be regarded as sharp interfaces to which the chains are effectively anchored. Amphiphilic aggregates include Langmuir-Blodgett and surfactant monolayer films; bilayers and biomembranes and their curved counterparts, vesicles and liposomes; and micelles and microemulsions of various geometries.

The complete subject of interfacial chain molecules is of greater breadth than could be covered in this brief review. It divides naturally, however, into subfields, depending on the relative concentrations of polymer and solvent at the interface, and depending on whether the interface is sharp, as for chains anchored to solid surfaces, or diffuse, as for chains at some liquid/liquid interfaces. Sharp interfaces are defined here as those with spatial variations smaller than 1–2 chain widths (cross section), the shortest possible correlation distance among molecules. We consider here only chains at sharp interfaces at high surface densities, where solvent within the chain environment is completely absent or is in low concentration. In parallel with the recent rapid development of experimental methods in this area, much theoretical work has emerged during the past decade to predict the physical properties of these high-density interfacial chain systems, which have been referred to as "interphases" (1–3); see Figure 1. Our

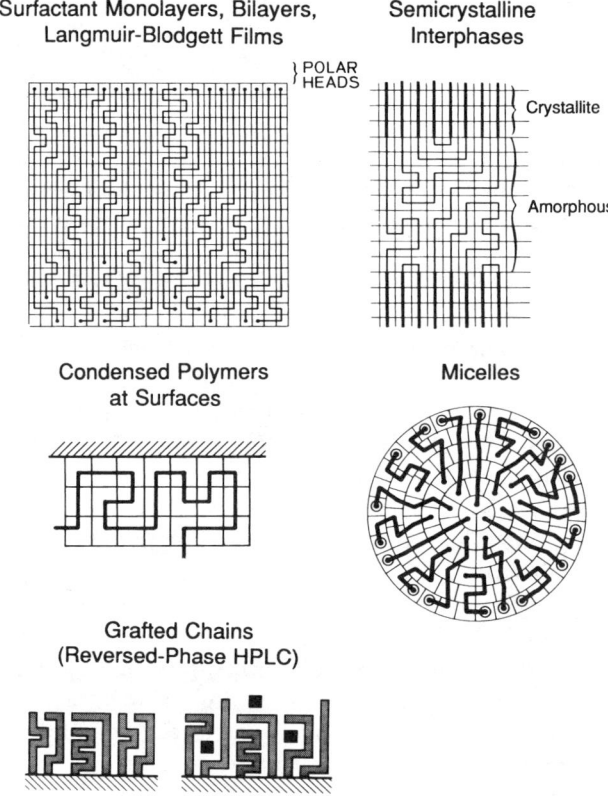

Figure 1 Chain molecule interphases.

purpose here is to review the theories and the principles of interfacial constraint on polymer conformations in these systems, and to consider experiments and computer simulations insofar as they bear on these issues.

As brief introduction, we consider the broader context of theory for chains at interfaces. The properties of chain molecules depend on their accessible configurations. Except for crystalline materials, the configurations of chains are statistical, due to the low energy barriers to bond rotation, and thus due to the large number of degrees of freedom. Hence statistical mechanical methods are required to account for their properties. Classical theories of long chains at sharp interfaces at low surface densities have treated the chains as freely-jointed lattice walks (4, 5), and have taken solvent interactions into account (6, 7) through use of Bragg-Williams/Flory-Huggins χ parameters (8). Whereas those methods are appropriate for chains in theta solvents (i.e. random flights), they do not suitably account for perturbations of the long-ranged intrachain interactions in other solvents. For that purpose, the self-consistent field method of Edwards (9) has been applied to the prediction of the lowest order corrections for long-ranged intrachain excluded-volume interactions among molecules in good solvents near interfaces (10, 11), and scaling law theories have been applied for chains in good or poor solvents (12–16). These theories aim to address properties of polymer adsorption and colloid stabilization, for example, in the dilute and semidilute regimes in non-theta solvents. We do not further consider these problems here; a broad insightful review has recently been given by deGennes (17).

In contrast, at high polymer densities, intrachain interactions become ideal because the long-ranged interactions are screened by neighboring interpenetrating chains (18). For diffuse interfaces, such as those in block copolymers, a principal contribution to the conformational and thermodynamic properties is due to the fluctuations in the longitudinal displacement of each chain relative to the mean interfacial position. Helfand & Tagami (19) modeled this situation through a self-consistent field diffusion equation approach. Since that treatment, many similar functional integral theories have been applied to interfacial structure and microphase separation equilibria (20–26). These problems of diffuse interfaces are also not further considered here.

Our focus is on polymers at sharp concentrated interfaces. Following a lattice treatment by Roe (27, 28), Helfand developed an elegant theory for long chains at various interfaces, diffuse or sharp (29–31), in which he introduced a symmetry constraint missing in the Roe formulation. Weber & Helfand applied the theory to bulk polymers at interfaces (31). A related problem is that of chain conformations in the semicrystalline interphase. The simplest early models of the interphase are as random flights from a fixed wall (32), but the density constraint is important and significantly perturbs the conformations; that constraint has recently been taken into account (33–37a) through methods described below.

For chains at high densities at sharp interfaces, it is useful to distinguish between "long" chain systems (ends are rare, and properties of segments are independent of their position along the chain) and "finite" chain systems (chain segments, including ends, can have properties dependent upon their position in the chain). With great power and generality, properties due to short-ranged interactions along finite chains have been treated by the elegant matrix method of Kramers & Wannier for the one-dimensional Ising model (38). This approach was applied by Zimm & Bragg to the helix-coil transition (39), by Flory to the general problem of bond-configurational properties, the so-called "rotational isomeric state" theory (40), and by Rubin and Di Marzio to conformations of chains at low densities constrained by impenetrable interfaces (41–43).

In this spirit, Dill & Flory (DF) (2, 3) developed theory for finite chains at high densities; subsequent related treatments are those of Ben-Shaul, Szliefer & Gelbart (44–50) and Naghizadeh & Dill (51). Independently, Scheutjens & Fleer (52–54) and Levine et al (55) adapted the method of Rubin & DiMarzio to problems of polymer adsorption over the full range of chain densities. However, the neglect in the latter treatments of the intrachain long-ranged interactions may be unwarranted (17) except at the highest densities, as noted above. Theodorou has recently extended these approaches to copolymers (56–59). Our aim is to present here the simplest unified description of chains of all lengths at high densities at sharp interfaces; the discussion below most nearly resembles the treatments of Helfand (29, 30) and Marqusee & Dill (37) for long chains, and Naghizadeh & Dill (51) for finite chains.

LONG CHAINS CONDENSED AT INTERFACES

Theory

Relative to chains in the bulk, the configurational freedom of interfacial polymers is limited if the interface is impenetrable. Hence the configurational freedom will increase with distance from the interface until it ultimately reaches its bulk value. Therefore, interfacial systems are characterized by gradients of chain organization. In general, the bond orientations will depend on distance from the interface. This dependence can be represented by considering the chains to be configured on a lattice of layers, numbered $l = 1, 2, 3, \ldots, L$, from the interface. If the interface is planar, then layers are parallel to it; if the interface is curved, then the layers are concentric with it. The number of lattice sites per layer is

$$N_l = \begin{cases} N_0 \text{ for planar interfaces} \\ \dfrac{4\pi}{3}(3(r_0-l+1)(r_0-l)+1) \text{ for chains directed radially} \\ \quad \text{inward from spherical interfaces, of radius } r_0 \\ \pi h(2(r_0-l)+1) \text{ for chains directed radially inward from} \\ \quad \text{cylindrical interfaces, of radius } r_0, \text{ and length } h \gg r_0. \end{cases} \qquad 1.$$

The reduced surface density of the N chains is $\sigma = N/N_1$.

From one layer l to the next $l+1$, the bond configurations may differ, but they are assumed to be uniform within a single layer. The configurations of the chains can be characterized by the number of bonds in each layer with each possible relative orientation. Let $v_{l,c}^+$ represent the number of segments per chain of "forward" bonds (directed away from the interface, from layer l to $l+1$) in configuration c, and $v_{l,c}^0$ and $v_{l,c}^-$ represent likewise the number of "lateral" bonds (parallel to the interface, from one site to another within layer l), and "reverse" bonds (back toward the interface, from layer l to $l-1$), respectively. The properties do not depend on which of the $(z-2)$ directions a lateral bond takes, where z is the coordination number of the lattice. Hence a chain configuration need only be characterized by the number of steps of each type taken by the chain, $c = \{v_{1,c}^+, v_{1,c}^0, v_{1,c}^-, v_{2,c}^+, v_{2,c}^0, v_{2,c}^-, \ldots, v_{L,c}^+, v_{L,c}^0, v_{L,c}^-\}$. The number of conformations g_c, represented by c is given by

$$g_c = \prod_{l=1}^{L} (z-2)^{v_{l,c}^0}. \qquad 2.$$

If the number of chains in conformation c is N_c, then the total number of chains, N, is

$$N = \sum_c N_c. \qquad 3.$$

The multiplicity, W, of chain configurations is given by

$$W = \prod_c g_c^{N_c} \left[\frac{N!}{\prod_c N_c!} \right]. \qquad 4.$$

In terms of the probability of configuration c,

$$P_c = \frac{N_c}{N} \qquad 5.$$

the normalized partition function is

$$Z = \sum_c P_c. \qquad 6.$$

The use of Stirling's approximation and the Boltzmann-Planck equation lead to the entropy of the system, $S = k \ln W$,

$$\frac{S}{Nk} = -\sum_c P_c \ln \left(\frac{P_c}{g_c}\right) \qquad 7.$$

where k is Boltzmann's constant. The average number of steps of each type per chain in layer l is

$$\langle v_l^+ \rangle = \sum_c v_{l,c}^+ P_c$$

$$\langle v_l^0 \rangle = \sum_c v_{l,c}^0 P_c$$

$$\langle v_l^- \rangle = \sum_c v_{l,c}^- P_c. \qquad 8.$$

These average bond conformational properties can be calculated if the equilibrium distribution, P_c, is known. The equilibrium distribution is that for which the entropy S is a maximum subject to the imposed constraints. For the problem at hand of chains of infinite length, the system is subject to two constraints: 1. packing, and 2. symmetry.

1. For a system of pure chains, the volume in each layer must be fully filled by chain segments, which are: (a) forward steps from layer $l-1$, (b) lateral steps within layer l, or (c) reverse steps from layer $l+1$,

$$N(\langle v_{l-1}^+ \rangle + \langle v_l^0 \rangle + \langle v_{l+1}^- \rangle) = N_l. \qquad 9.$$

This constraint on the mean bond occupancies is of the same nature as that of the canonical ensemble wherein the total energy (mean energy per particle) is held constant.

2. For long chains, it is impossible to distinguish segment k from $k+1$; hence the chain cannot have directionality (29, 30, 51). Thus there must be symmetry of forward and reverse bonds across each interface between layers l and $l+1$,

$$\langle v_l^+ \rangle = \langle v_{l+1}^- \rangle. \qquad 10.$$

Combined with Eqs. 8, these two sets of constraints can be expressed as:

$$\sum_c \sum_{l=1}^{L} (v_{l-1,c}^+ + v_{l,c}^0 + v_{l+1,c}^-) \, dP_c = 0 \qquad 11.$$

$$\sum_c \sum_{l=1}^{L} (v_{l,c}^+ - v_{l+1,c}^-) \, dP_c = 0. \qquad 12.$$

Subject to packing and symmetry constraints (Eqs. 11 and 12), maximization of the entropy, Eq. 7, using the standard method of Lagrange multipliers, leads to

$$dS = \sum_c \left\{ \ln\left(\frac{P_c}{g_c}\right) + 1 + \sum_{l=1}^{L} [\lambda_l(v^+_{l-1,c} + v^0_{l,c} + v^-_{l+1,c}) + \xi_l(v^+_{l,c} - v^-_{l+1,c})] \right\} dP_c = 0 \qquad 13.$$

where λ_l and ξ_l are undetermined multipliers. According to the Lagrange method, each variation is assumed to be independent, hence Eq. 13 is solved by

$$P_c = (g_c/e) \prod_{l=1}^{L} p_l^{v^+_{l,c}} q_l^{v^0_{l,c}} u_l^{v^-_{l,c}} \qquad 14.$$

where the quantities p_l, q_l, u_l are given by

$$p_l = \exp[-\lambda_{l+1} - \xi_l]$$
$$q_l = \exp[-\lambda_l]$$
$$u_l = \exp[-\lambda_{l-1} + \xi_{l-1}]. \qquad 15.$$

Ratios of these quantities lead to the set of L equations,

$$\frac{p_l}{q_l} = \frac{q_{l+1}}{u_{l+1}}. \qquad 16.$$

It is clear from the form of Eq. 14 that p_l, q_l, and u_l represent the *a priori* statistical weights for forward, lateral, and reverse steps, respectively (51). Hence the averges over chain configurations can alternatively be expressed as averages over the segment configurations,

$$\langle v^+_l \rangle = \frac{\partial \ln Z}{\partial \ln p_l} = \sum_k v_l(k) p_l(k)$$

$$\langle v^0_l \rangle = \frac{\partial \ln Z}{\partial \ln q_l} = \sum_k v_l(k) q_l(k)$$

$$\langle v^-_l \rangle = \frac{\partial \ln Z}{\partial \ln u_l} = \sum_k v_l(k) u_l(k) \qquad 17.$$

where $v_l(k)$ is the number of kth segments per chain in layer l. For long chains, p_l, q_l, and u_l are independent of k. Inasmuch as $v_l(k)$ is also independent of k for long chains, and independent of l for interphases without curvature (51), then the use of Eq. 17 with Eqs. 9 and 10, can be expressed more succinctly as

$$p_l = u_{l+1} \qquad 18.$$

$$p_{l-1} + (z-2)q_l + u_{l+1} = 1. \qquad 19.$$

Substitution of the L Eqs. 18 and L Eqs. 19 into Eqs. 16 lead to the set of L coupled equations,

$$(z-2)^2 u_l^2 = (1 - u_l - u_{l+1})(1 - u_l - u_{l-1}) \qquad l = 1, 2, 3, \ldots, L \qquad 20.$$

which are solved simultaneously, subject to the relevant boundary conditions. For the semicrystalline interphase, the boundary condition is that all N chains cross from layer 0 to 1; for bulk polymers bounded by surfaces, the boundary condition is that no chains cross from layer 0 to layer 1.

The Semicrystalline Interphase

This theory predicts that immediately adjacent to the interface there is a high degree of "order," i.e. alignment along the axis defined by the crystalline stems (i.e. normal to the interface) (37). The order rapidly diminishes to the bulk, fully disordered value, within about 1–2 chain widths, i.e. 5–10 Å from the interface, the interface thickness increasing only slightly with chain stiffness. This predicted interphase thickness is approximately the same as that determined by Raman (60, 61) and NMR spectroscopy (62, 63).

The conformations of chains in the semicrystalline interphase have been the subject of much controversy (32, 64–68). The "flux" of chains, i.e. the number of chains per unit area passing through a plane normal to the chain axes, is the inverse of the area per chain. Flory first showed that the requirement for chain randomness in the amorphous phase is that the flux must have only half the value in the amorphous bulk phase as in the crystallite (1). Hence, in the absence of ends, chains must fold back and return to the crystallite in order to dissipate this flux and to progress from a region of spatial order to disorder. This condition, however, makes no prescription for the specific molecular conformations of the foldbacks. Two opposing views have arisen.

1. Chain foldbacks may be due predominantly to "adjacent re-entry," whereby each chain re-enters the crystallite at a site immediately adjacent to its exit site. In a lattice model, this would be represented as a hairpin loop (which has the specific configuration of a forward, then lateral, then reverse bond).
2. Alternatively, chain foldbacks may be due predominantly to "random

re-entry," whereby chains do not re-enter often in neighboring sites (68).

Several recent calculations along the lines described above, and others (33, 35–37a), are in general agreement on the following qualitative predictions of re-entrant chains. First, a decrease in chain flux cannot commence without at least a small number of hairpin loops in the first layer. Second, for chains that are flexible (i.e. 0 bending energy), constrained by volume-filling requirements, the incidence of adjacent foldbacks is approximately 73% (35, 37, 54). In contrast, for chains that are flexible but unconstrained by density requirements, which obey random flight statistics, strict adjacencies constitute about 17% of the total (32). Third, when realistic models for the interphase chains are considered, wherein account is taken of: (*a*) volume-filling constraints, (*b*) chain-bending energies between bond pairs, and (*c*) excess hairpin energies, between bond triplets (33–35, 37a), then the incidence of strictly adjacent foldbacks is a relatively small fraction of chain re-entrants. For example, using reasonable choices for bending and hairpin energies in a Monte Carlo simulation, Mansfield predicts that the incidence of strict adjacent re-entrants ranges from 9.3 to 31.9%. Figure 1*a* shows the fraction of adjacent re-entrants as a function of bending energy when the hairpin energy is assumed

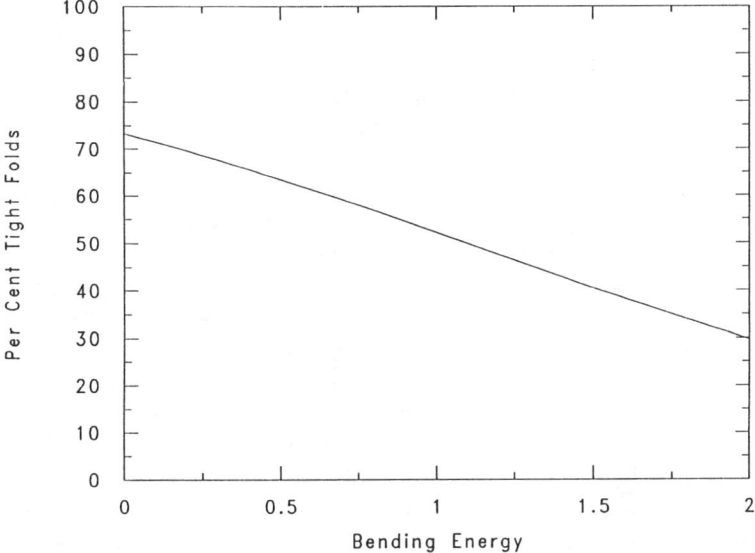

Figure 1a Percentage adjacent re-entrant chains in semicrystalline polymers vs bending energy (with hairpin loop energy = 0), from Marqusee (37a).

to equal zero, according to the calculations of Marqusee (37a). Fourth, inasmuch as the high density in the interphase is the principal factor favoring shorter loops, then factors that diminish the number of chains in the interphase lead to reduced numbers of adjacent re-entrants. For example, tilted chains at the crystal interface (tilting appears in the theory as reduced flux from layer 0 to 1) lead to fewer adjacent re-entrants (34). Fifth, there is a broad distribution of loop sizes, with smaller ones substantially favored. For example, Leermakers et al (36) predict for their stiff chains that the *most probable* loop size is approximately second-nearest-neighbor, but because the distribution is so broad, the *average* loop length is about 6–8 lattice segments, or about 25 methylene units. The results of Leermakers et al may err somewhat due to use of a model of bulk chains terminated by the planar interfaces, which therefore contains too high a concentration of chain ends in the interphase. Nevertheless, in sum, the present conclusions from theory are that: (*a*) the fraction of adjacent re-entrants decreases strongly with increasing bending and hairpin energies, and is relatively small for realistic energies, and (*b*) there is a broad distribution of loop sizes, with shorter ones being favored by the constraint of high density.

Certain effects of the interfacial constraints are propagated over long distances away from the interface. For example, simple random-flight theory predicts that the number of chain loops should equal the number of ties for large separation between crystallites (32). "Loops" exit and re-enter the same crystallite; "ties" exit and enter neighboring crystallites. In contrast, the interphase theory shows that even for amorphous regions 40 layers in thickness, the ratio of loops to ties (per layer) is approximately 0.81. The average lengths of loops and ties are greater in the interphase than predicted by the random-flight model (37). Neutron scattering measurements of the number of ties (69) may permit tests of these predictions. Using an elegant model, Lacher (70–72) has shown that significant numbers of loops from opposing crystallites should be interlinked. Both links and ties contribute to the physical connection of crystallites, and undoubtedly to the mechanical strength of the material. Even though there are fewer loops in the interphase than predicted by the random-flight model, Lacher's model predicts more links in the interphase (72). It is particularly interesting that his model predicts that the ratio of links to ties increases with separation of the crystallites.

Vonk (73) has argued that lattice theories for the semicrystalline interphase are flawed inasmuch as the "flux" normal to the interface is 1/3 from one lattice layer to the next in the bulk, but must be 1/2 according to the Flory argument. The flaw in this view is that it does not average properly over the discreteness of lattice layer planes; the average flux, when correctly summed over planes between layers and planes slicing through mid-layers, equals 1/2. Hence, in this regard at least, the lattice models introduce no artifact.

FINITE CHAINS CONDENSED AT INTERFACES

Theory

In contrast to infinitely long chains considered above, finite chains are subject to a length constraint,

$$\sum_{l=1}^{L} v_{l,c}^{+} + v_{l,c}^{0} + v_{l,c}^{-} = n \qquad 21.$$

where n is the number of chain bonds. A convenient way to enumerate all the configurations that satisfy this constraint (2, 3) is to use the matrix method originated by Kramers & Wannier for the one-dimensional Ising model (38), and first applied to isolated chains at interfaces by Rubin & DiMarzio (43). The distribution of segment k throughout all layers can be represented by a vector,

$$\mathbf{v}(k) = [v_1(k), v_2(k), v_3(k), \ldots, v_L(k)]. \qquad 22.$$

For chains that are constrained only by the impenetrable interface and are not subject to packing constraints due to neighboring chains, the distribution of segment k can be written in terms of the distribution of the preceding segment $k-1$,

$$\mathbf{v}(k) = \mathbf{v}(k-1)\mathbf{G} \qquad 23.$$

where

$$\mathbf{G} = \begin{bmatrix} z-2 & 1 & 0 & 0 & 0 & 0 & \\ 1 & z-2 & 1 & 0 & 0 & 0 & \\ 0 & 1 & z-2 & 1 & 0 & 0 & \\ 0 & 0 & 1 & z-2 & 1 & 0 & \cdots \\ & & \cdots & & 0 & 1 & z-2 \end{bmatrix} \qquad 24.$$

is a matrix that generates the configurations of each succeeding chain segment from its predecessor. Iterative application of Eq. 23 leads to

$$\mathbf{v}(k) = \mathbf{v}(1)\mathbf{G}^{k-1} \qquad 25.$$

where $\mathbf{v}(1) = [v_1(1), v_2(1), \ldots, v_L(1)]$ accounts for the intrinsic distribution of segment 1 throughout the layers. If chains all have their first segments anchored to the interface, then $\mathbf{v}(1) = \mathbf{A} = [1\,0\,0\,0\cdots 0]$. The single-chain partition function, Z_1, is the sum over all such configurations for chains of length n,

$$Z_1 = v(n)\mathbf{B} = v(1)\mathbf{G}^{n-1}\mathbf{B} \qquad 26.$$

where $\mathbf{B} = \text{col}\,[1\,1\,1\ldots 1]$. While this method enumerates all the configurations for chains of length n that do not penetrate the interface, it does so only for chains that are free of constraint from neighboring molecules.

For chains that are also subject to constraint by neighboring molecules, the generating matrix will be modified by introduction of the *a priori* conditional weights, p_l, q_l, and u_l as defined in Eq. 14:

$$\mathbf{G} = \begin{bmatrix} (z-2)q_1 & p_1 & 0 & 0 & 0 & 0 \\ u_2 & (z-2)q_2 & p_2 & 0 & 0 & 0 \\ 0 & u_3 & (z-2)q_2 & p_3 & 0 & 0 \\ 0 & 0 & u_4 & (z-2)q_4 & p_4 & 0 \\ & & & \ldots & 0 & u_L(z-2)q_L \end{bmatrix}$$
$$27.$$

(see Figure 2).

Condensed phases of chains at interfaces are subject to the packing constraint, Eq. 9. Finite chains are not, however, subject to the symmetry constraint, Eq. 10, because for finite chains, segments k and $k+1$ are readily distinguishable; chains have directionality. For example, more bonds may be directed (from segments k to $k+1$) forward away from the interface than toward it. Hence for finite chains, the equilibrium configurations of the system are predicted by the state of maximum entropy subject only to the packing constraint (i.e. Eq. 13 with $\xi_l = 0$) with the result,

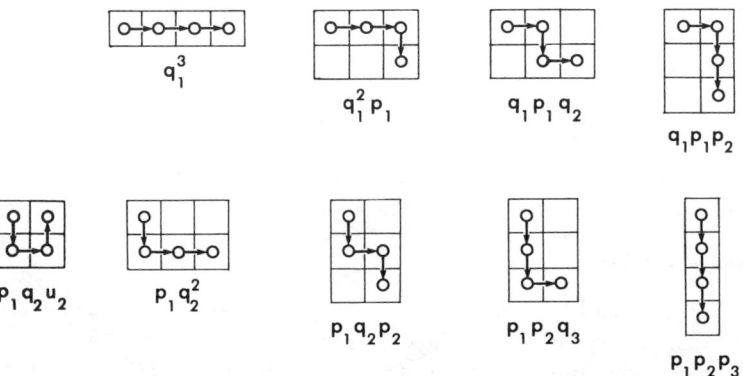

Figure 2 A priori statistical weights for the configurations of a four-segment chain.

$$p_{l-1} = q_l = u_{l+1}.$$
28.

These relationships are substituted into **G** to give

$$\mathbf{G} = \begin{bmatrix} (z-2)q_1 & q_2 & 0 & 0 & 0 & 0 \\ q_1 & (z-2)q_2 & q_3 & 0 & 0 & 0 \\ 0 & q_2 & (z-2)q_3 & q_4 & 0 & 0 \\ \vdots & & \cdots & & 0 & q_{L-1} & (z-2)q_L \end{bmatrix}.$$
29.

This quantity **G** is then substituted into the expressions,

$$\langle v_l^+ \rangle = \frac{\partial \ln Z_1}{\partial \ln p_l} = \frac{p_l}{Z_1} [v(1) \{ \mathbf{G}'_{p_l} \mathbf{G}^{n-2} + \mathbf{G}\mathbf{G}'_{p_l} \mathbf{G}^{n-3}$$
$$+ \cdots + \mathbf{G}^{n-2} \mathbf{G}'_{p_l} \} \mathbf{B}]$$

$$\langle v_l^0 \rangle = \frac{\partial \ln Z_1}{\partial \ln q_l} = \frac{q_l}{Z_1} [v(1) \{ \mathbf{G}'_{q_l} \mathbf{G}^{n-2} + \mathbf{G}\mathbf{G}'_{q_l} \mathbf{G}^{n-3}$$
$$+ \cdots + \mathbf{G}^{n-2} \mathbf{G}'_{q_l} \} \mathbf{B}]$$

$$\langle v_l^- \rangle = \frac{\partial \ln Z_1}{\partial \ln u_l} = \frac{u_l}{Z_1} [v(1) \{ \mathbf{G}'_{u_l} \mathbf{G}^{n-2} + \mathbf{G}\mathbf{G}'_{u_l} \mathbf{G}^{n-3}$$
$$+ \cdots + \mathbf{G}^{n-2} \mathbf{G}'_{u_l} \} \mathbf{B}],$$
30.

which arise from the relevant derivatives of the matrix Eq. 26 using Eq. 27 (51), and $\mathbf{G}'_{p_l} = \partial \mathbf{G}/\partial p_l$ for example. Finally, substitution of these equations into the packing constraint, Eq. 9, leads to a set of L coupled equations in L unknowns, dependent on interphase geometry through N_l. Initial estimates are made for the q_l values in **G**, then these equations can be solved in matrix form by standard iterative methods for nonlinear equations in L variables. Various moments of this probability distribution function can then be readily computed for comparison with experiments.

Although the theory described above follows Naghizadeh & Dill (51) (ND), it summarizes many treatments (2, 3, 29–32, 37, 37a, 44–49, 51–53, 56–59), all of which derive from the same physical premises, and which differ only in minor stylistic or technical details. The few principal differences are described briefly here. In the method of Scheutjens & Fleer (52, 53) (SF), the principal constraint is a condition on the chain connectivity. Although it can be shown (51) to be identical to the packing constraint, Eq. 9, our preference for the present approach follows from what we feel is a greater transparency of the physical premise. Applications of the SF theory were initially for polymers in solution, hence they have also incorporated solvent/polymer contact interactions from the outset through Flory-Huggins χ parameters, and in some cases have considered the generalization to chains with longitudinal fluctuations (96). Dill & Flory (2, 3) (DF) originally neglected chain steps in the reverse direction, which are relatively rare except in the lowest-density interphases. They used an additional symmetry condition, rather than an entropy maximization procedure, to establish closure of the equations. The symmetry condition is identical to the entropy maximization result under most circumstances, but it introduces some error where chain ends are concentrated in highly curved interphases (51), for spherical micelles, for example. Where figures are shown below of comparisons of experiments with the DF model, it implies that those results have been tested with the approach above (51), and that they do not differ significantly from the DF predictions. The only significant difference is in the bond order parameter in spherical micelles, for which DF predicts an increase toward the chain ends, whereas the treatment above predicts decreased bond order along the chain (*vide infra*). The treatment of Theodorou follows those of Scheutjens & Fleer and of Dill & Flory, but it is also generalized for copolymers of any sequence (56–59). Theodorou has suggested that the SF and DF models differ in their emphasis on "sites" or on "bonds" (site-pairs), respectively, but this difference is only apparent; the statistical weights can be shown to have the same meaning (51). The treatment of Ben-Shaul et al (44–49) does not apply a systematic method for counting chain conformations, introduced above through use of the **G** matrix. Conformations are counted by hand in their procedure, which restricts its use to chains no longer than about 4–5 segments. They have established, however, the important result that theories of this type are not dependent on use of a regular lattice. Szleifer et al (48) have considered fluctuations of the interface, but more empirically than the preferable treatment of Leermakers et al (96). Gruen (74–76) has developed a numerical simulation method for generating the ensemble of conformations and for iteratively changing the statistical weights until the packing constraint is satisfied. This method permits relatively realistic representation of the chain conformations and includes separate statistical weights for the hydrophobic interaction at the surface. However, in our opinion, the far greater computational complexity of this method is not justified by any real improvement in predictive capability.

The predictions of theory have been tested by experiments on a variety of systems, as described below.

Chain Configurations in Bilayer Membranes

The interfacial nature of bilayer membranes was first observed in the electron spin resonance experiments of Hubbell & McConnell (77), who demonstrated a gradient of chain disorder. This type of experiment is now widely performed using ^{2}H-NMR (78). The NMR experiment measures the frequency splitting of the quadrupolar interaction due to the electric field of a C–D bond; if that bond is at the jth position along the chain, then the experiment determines directly the bond order parameter $S_j = 1/2(3\langle\cos^2\theta_j\rangle - 1)$, where θ is the angle between the local chain direction and the interfacial normal. Different positions along the chains are deuterated in separate experiments, and the orientational order parameter is determined at each position. Figure 3 shows a comparison of the theoretical prediction of the disorder gradient, with experiments on three different systems: a soap bilayer, with a single chain emanating from each head group; a phospholipid bilayer, with a pair of chains emanating from each head group; and a biomembrane from *Acholeplasma laidlawii*, compared with the theory of Dill & Flory (2). Figure 4 shows predictions of theories of Gruen (74) and Ben-Shaul et al (48), compared with molecular dynamics simulations of van der Ploeg & Berendsen (79) and experiments on a bilayer of mixed surfactants (80). It is clear that the predictions from these models show that the chains have the highest orientational order (are most solid-like) near the head groups, and the lowest (most liquid-like) near the terminal methyl groups of the tails.

Whereas these NMR experiments monitor the bond orientational disorder, the positional disorder may also be measured by small-angle neutron

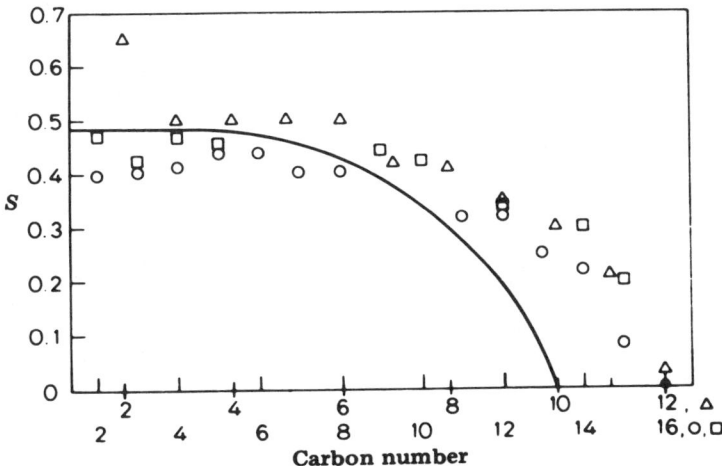

Figure 3 Calculated disorder gradient for planar monolayers of $n = 4$, $\sigma = 0.63$ from DF (2) compared with experiments on bilayers of potassium laurate ($\triangle$ upper abscissa), dipalmitoylphosphatidylcholine ($\square$ lower abscissa), and *A. laidlawii* ($\bigcirc$ lower abscissa).

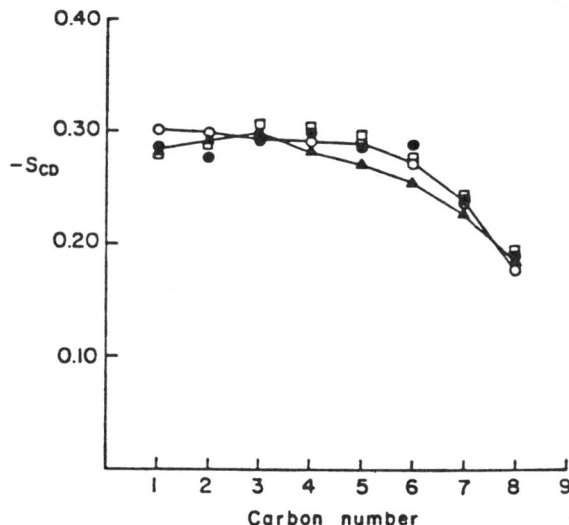

Figure 4 Disorder gradient in bilayers: comparison of theories of Ben-Shaul et al (46) (○) and Gruen (75) (□) with molecular dynamics simulations of van der Ploeg & Berendsen (▲) (79) and experiments of Niederberger & Seelig on mixed bilayers (●) (80).

scattering on deuterated systems. By contrasting the deuterated and protonated species, the distribution of distances of a few selected chain segments along the bilayer normal relative to the position of the head group plane have been determined (82). Figure 5 shows that the predictions of theory (81) are in good agreement with these experiments.

Chain Configurations in Micelles

Among the simplest realistic models of the chain organization in micelles is that their head groups are fixed at an interface of simple geometry, such as a sphere or cylinder. In that approximation, fluctuations of the interfacial geometry are neglected, and only the internal conformational degrees of freedom of the chains are taken into account.

The disorder gradient in planar short-chain interphases, such as bilayers and monolayers, arises principally from the distribution of the terminal methyl groups of the chains. Curvature gives rise to an additional contribution to the chain flux, for example in micelles. As the chains tend toward the center of a curved interphase, the area diminishes through which the chains pass. Hence the area per chain decreases radially inward; correspondingly, the flux increases. The balance between the effects due to the chain ends, and to the interphase geometry, determines the chain

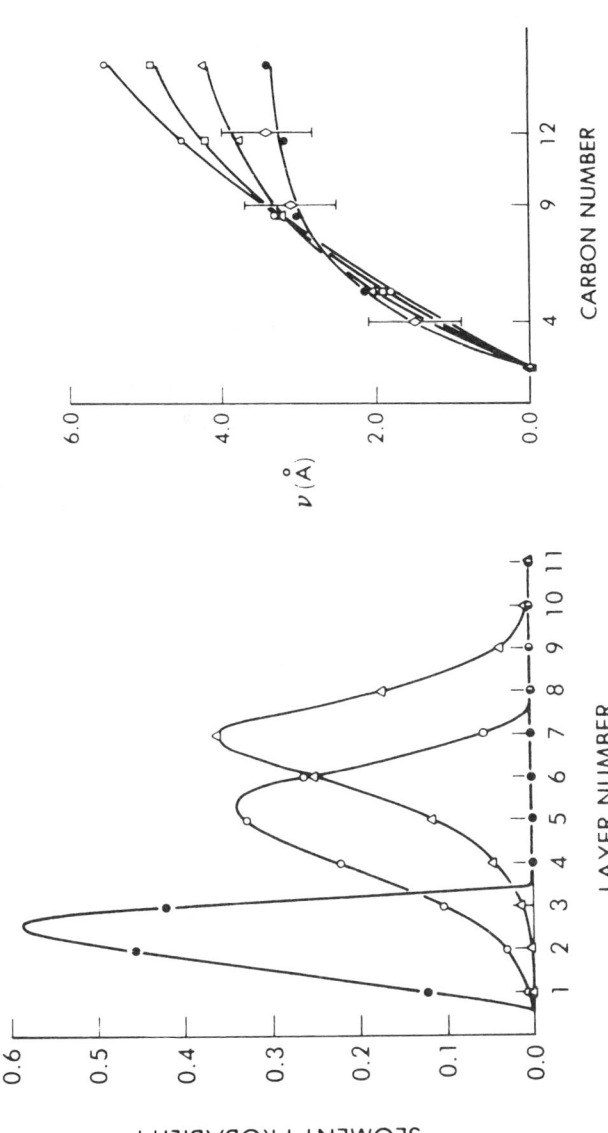

Figure 5 Left: predicted distribution of chain segments along an axis normal to the head-group plane of the bilayer, from Cantor & Dill (81). *Right*: Widths of those distributions vs segment position along the chain, compared with neutron scattering experiments of Zaccai et al (82). Calculated results are for $n = 4$, $\sigma = 0.64$, for different values of ω, the statistical weight for a bent relative to an unbent bond: (●)$\omega = 1$, (△)$\omega = 0.5$, (□)$\omega = 0.3$, (○)$\omega = 0.2$.

organization. Figure 6 shows that a disorder gradient, similar to that in planar bilayers, is observed in both spherical and cylindrical micelles. However the magnitude of the chain anisotropy is smaller in these systems of higher curvature. Figure 7 is reproduced from the work of Woods, Haile & O'Connell (83, 84), who have performed molecular dynamics simulations on small spherical micelles, using a parabolic restraining potential for the head groups along the radius. They show comparisons with the DF theory for the chain end distribution (*top*, Figure 7) and the layer order parameters (*bottom*, Figure 7). The layer order parameter refers to the order at fixed spatial positions relative to the micellar center, in contrast to the bond order parameter, shown in Figure 6, which is a function of bond position along the chain.

Paramagnetic ions in micellar solutions can be used to shift the NMR relaxation times of the various methylene groups along the micellar chains, for the purpose of measuring the distance of each group from the interface (85, 86). What is determined is the configurational average of the inverse sixth power of the distance from the paramagnetic ion at the surface to the position of the given methylene group. Figure 8 shows a comparison of the data of Cabane (85) with (*a*) the interphase theory, (*b*) a "radial chain" model, in which the chains are all-*trans* and directed radially inward like the spokes of a wheel, and (*c*) an "oil-droplet" model, in which the chains are maximally random, and not necessarily connected to head groups (86, 87). These latter are the two extremes of possible molecular ordering, for comparison. The interphase theory is in good agreement with the experiments.

^{19}F-NMR chemical shifts can be used to determine the nature of the solvent environment in perfluoro-octanoate micelles (88, 89). These chemical shifts can be compared to those of reference systems in which chain segments are in pure water or in pure perfluorinated hydrocarbon. The fluorinated methylene groups along the chain are observed to be located in different solvent environments; see Figure 9. Groups near the head are in an environment similar to that of ethanol, as is expected for groups at the interface between water and the micellar hydrocarbon core. Groups near the methyl termini are in a more hydrocarbon-like environment. However, for the short octanoate chains, the environment of the chain ends, including the terminal methyls, is not that of the pure fluorocarbon; there is a small degree of water-like character experienced by all the tail segments. This and related experiments have been interpreted by some authors to imply that the micellar hydrocarbon core is porous and is substantially penetrated by water (see references in 90).

A better explanation for the solvation of hydrocarbon chain segments, however, is provided by the interphase theory, shown also in Figure 9

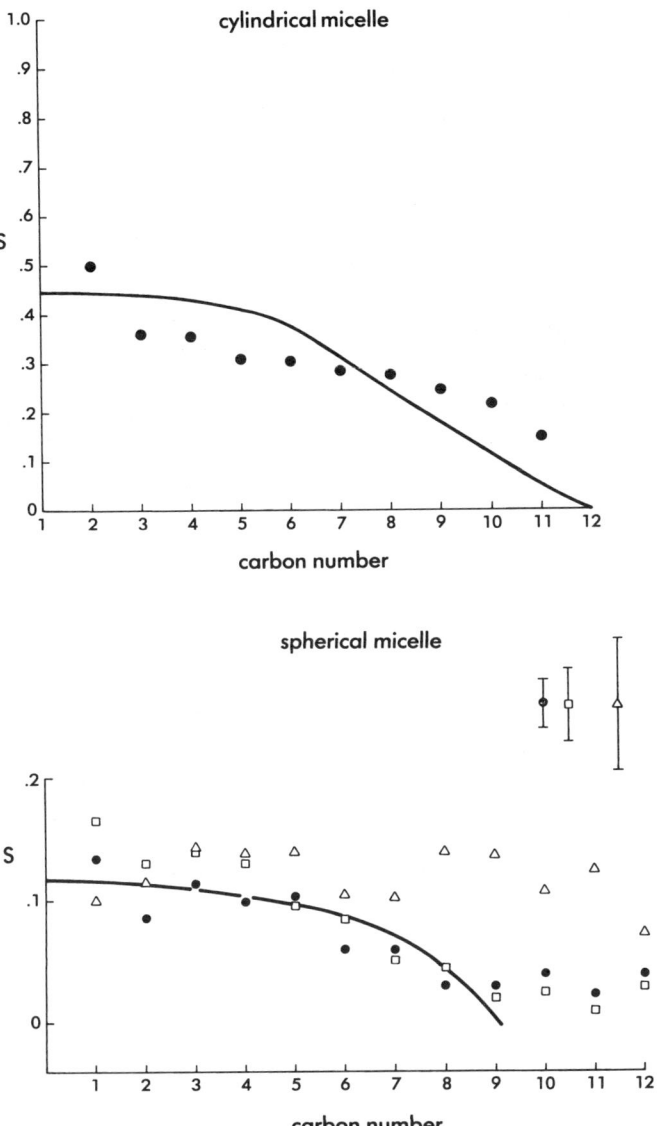

Figure 6 Disorder gradient for cylindrical (*top*) and spherical micelles (*bottom*). Bond order parameter predictions of Naghizadeh & Dill (51) are compared with data of Mely & Charvolin (122) for cylinders; and compared with molecular dynamics simulations (□, ●) (83, 84) and experimental data on sodium dodecyl sulfate spherical micelles (△) (123).

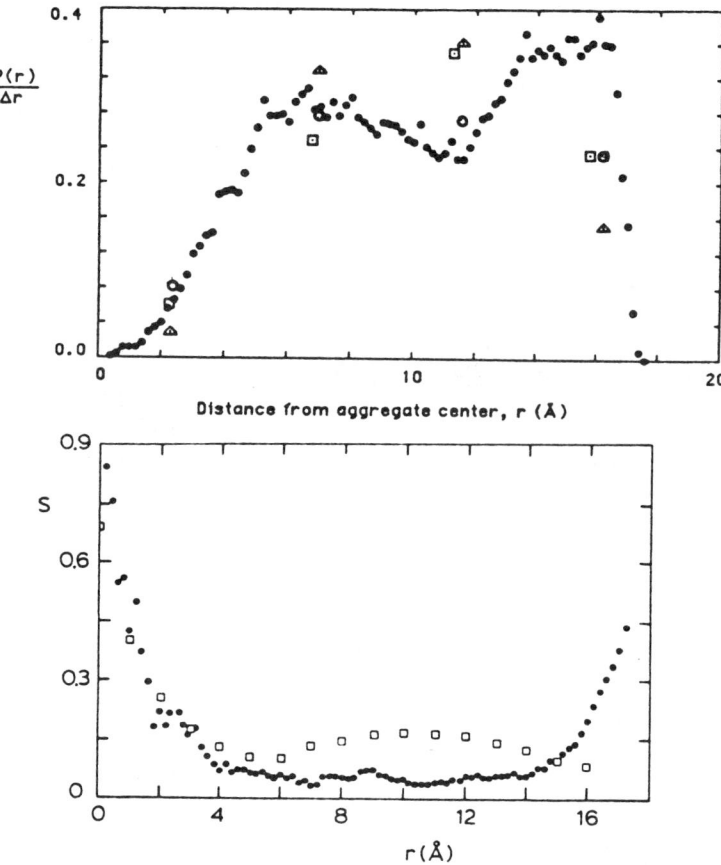

Figure 7 Top: Probability of chain end incidence in a spherical shell of radius r from the center of the micelle. *Bottom*: Layer order parameters for spherical micelle. Note the order increases toward the micellar center. From Woods, Haile & O'Connell (84), comparisons of their molecular dynamics simulations (●) with the theory of Dill & Flory (□). Reproduced with permission.

(89, 90). In that calculation it is assumed that no water penetrates the hydrocarbon core, a process that would be expensive in free energy. It is assumed that the only hydrocarbon/water contact occurs at the micellar interface. The chain conformational statistics are used to calculate the probability that any chain segment, k, is located on the surface. For short chains, the theory predicts that all chain segments have some incidence at the surface, due to the conformational freedom to configure relatively randomly. This probability of surface incidence decreases along the chain.

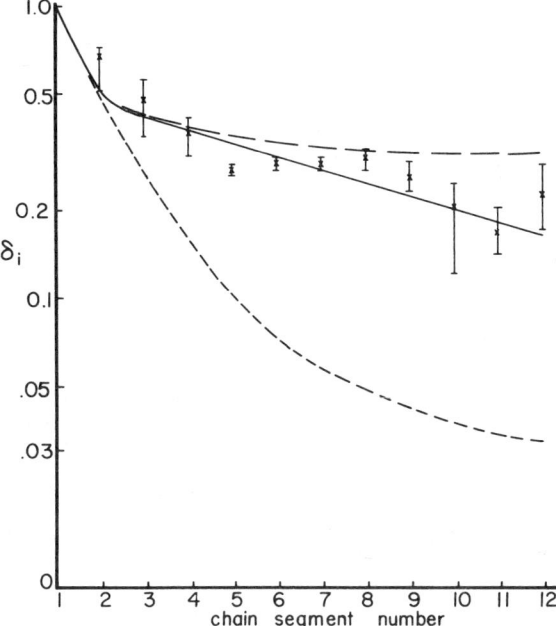

Figure 8 ^{13}C-NMR experiment of Cabane (85) on radial alignment of the chains in spherical micelles, compared with radial chain model (----), oil droplet model (---), DF theory (——) (87).

Hence the theory predicts that the substantial contact of water with hydrocarbon segments is not principally due to water penetrating the core, but to the disordered configurations of the core chains causing the various chain segments to contact the interface. For surfactant chains longer than about 12 carbons, however, the incidence of chain ends at the micelle/water interface is predicted to be relatively small. If, in addition, account is taken of the different sizes and chemical nature of methyls and methylene groups, then even somewhat longer chains will be expected to have a measurable incidence of chain ends at the interface, since methyl groups will be somewhat surface-active in the medium of methylene groups (76, 91).

Small-angle neutron scattering (SANS) experiments have been performed on small, nearly spherical, lithium dodecylsulfate micelles of chains that have been deuterated at the methyl ends (92). The radial distribution function of the chain ends has been determined from those experiments (90, 92); its Fourier transform, the scattering amplitude, is shown in Figure 10. These small-angle experiments are sensitive to the coarse features of the chain distribution. Although the interphase theory is in good agreement

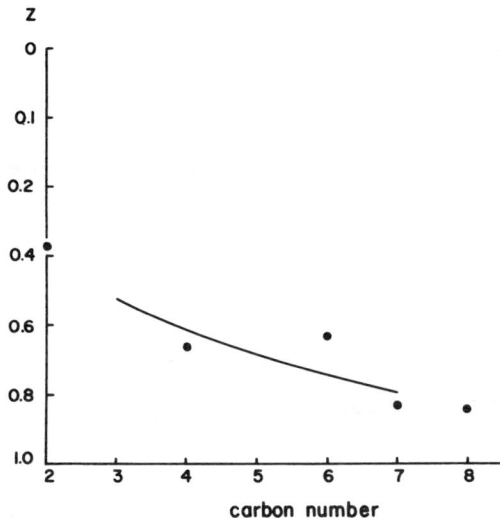

Figure 9 ^{19}F-NMR experiments of Muller & Simsohn (88) on the environment of the chain segments in perfluoro octanoate micelles. $z = 0$ implies a water-like environment, $z = 1$ implies that the chain segment is fully surrounded by other chain segments. Compared with the DF theory of the probability that each segment, k, resides at the surface (89). All segments of these short chains have some incidence at the micellar surface.

with the data, it is clear from Figure 10 that the data rule out the the radial chain and oil droplet models.

These SANS experiments are limited, however, and can resolve only relatively large spatial features. Higher resolution experiments have recently been performed by Cabane et al (93) on sodium dodecylsulfate micelles to intermediate scattering angles, approximately 0.6 Å^{-1}. In some experiments, they have specifically deuterated certain chain segments, including the termini. They conclude that micellar cores are "dry" (the water concentration inside is less than one molecule per surfactant molecule) but that the micelles undergo significant fluctuations in aggregation number and shape.

From the scattering function, they have determined the distribution of pairwise distances between CD_3 groups. For comparison, they have calculated the corresponding distributions from various theoretical models. Their comparison of experiments with the DF theory is shown in Figure 10a. The discrepancy between the theory and experiments is not improved by any other model in which the interface is assumed to have fixed geometry. Hence, this suggests that any model that assumes fixed geometry will err at this level of resolution for the distribution of the ends.

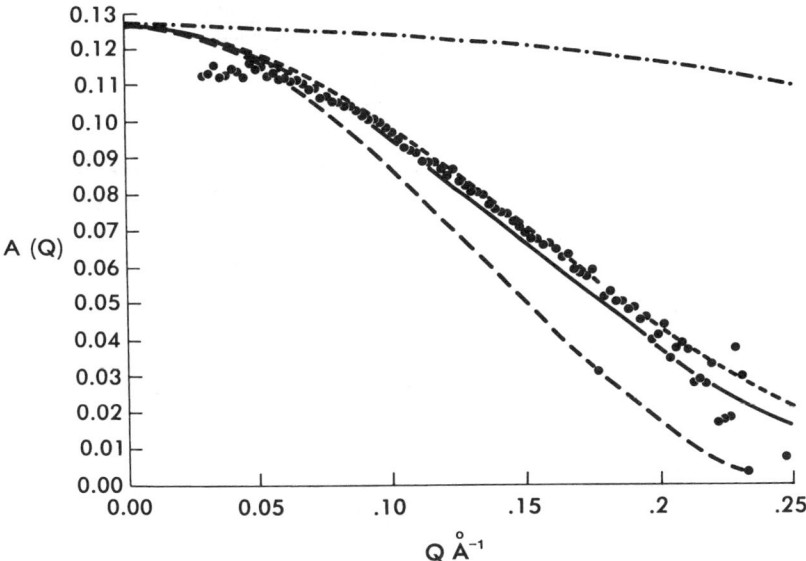

Figure 10 SANS scattering amplitude vs wavevector, Q. Experiments of Bendedouch & Chen (92) compared with radial chain model (-·-·-), oil droplet model (---), theory of DF for 0 chain stiffness (——), and Cantor & Dill for $\omega = 0.5$ (----) (81, 90).

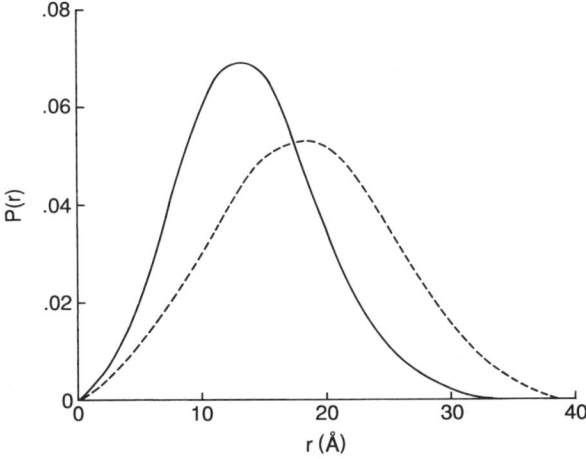

Figure 10a Probability distribution function for the pairwise distances between chain end CD_3 groups: from intermediate-angle neutron scattering experiments (----), compared to the DF theory (——), from Cabane et al (93).

There are several simulations that attempt a realistic accounting for interfacial fluctuations (94–97), and in that regard they do not assume the simplified interfacial structure of the models described above. Predictions from these models are not unequivocal, since there are difficulties with choice of the dielectric constant in the molecular dynamics calculations, and there are some unresolved issues with the definitions and choices of the adjustable parameters in the lattice models in that group. Nevertheless, they suggest that interfacial fluctuations can be as large as 5–10 Å, about 1–2 chain widths. Szleifer et al (48) have pointed out, however, that since the interfacial fluctuations derive from relatively short-ranged interactions, their magnitude is likely to be relatively unaffected by micellar size; hence the surface fluctuations should have less overall importance for larger micelles of more common experimental interest. Nevertheless, combined with the high resolution neutron scattering experiments, this suggests that the assumption of the fixed interface may be the weakest remaining approximation in all the interphase models. The differences among the interphase models are otherwise relatively minor.

Alkane Crystals

Crystals of short alkane molecules have significant numbers of chain ends at the crystal surface. Through infrared measurements of rocking vibrations at 622 cm^{-1} and 658 cm^{-1}, Maroncelli, Strauss & Snyder have shown that rotator phases of C21 and C29 alkanes have a gradient of increasing numbers of *gauche* bonds toward the chain ends (see Figures 11 and 12) (98). They have successfully modeled this disorder gradient through a modification of the theory above. The surface densities in the crystal are near maximal ($\sigma \geq 0.95$), much higher than in the bilayers discussed above, for example, hence they have argued that it is the lack of longitudinal register of the chains that is a principal contributor to the disorder gradient in these systems. This premise has been confirmed by

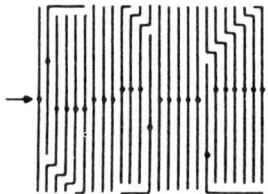
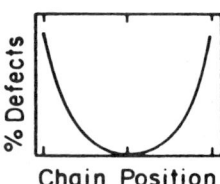

Figure 11 Alkane Crystal model of Maroncelli et al (98). Midpoints of the chains can fluctuate longitudinally. Reproduced with permission.

molecular dynamics simulations of the rotator phase of C23 alkanes by Ryckaert et al (99). Thus in the model of Maroncelli et al, the midpoints of the chains, which they identify with the interfacial plane of an interphase, are permitted to fluctuate longitudinally parallel to the chain axis. In their generalization, the probability that a chain midpoint is displaced longitudinally by Δl bonds relative to the midplane is

$$d(\Delta l) = \begin{cases} d(0) & \text{if } \Delta l = 0 \\ ce^{-\varepsilon/\Delta l} & \text{if } \Delta l \neq 0 \end{cases} \qquad 31.$$

where $d(0)$ and ε are parameters chosen in comparison with the data, and c is a normalization constant. For $T = 39°C$, for example, $\sigma = 0.9985$,

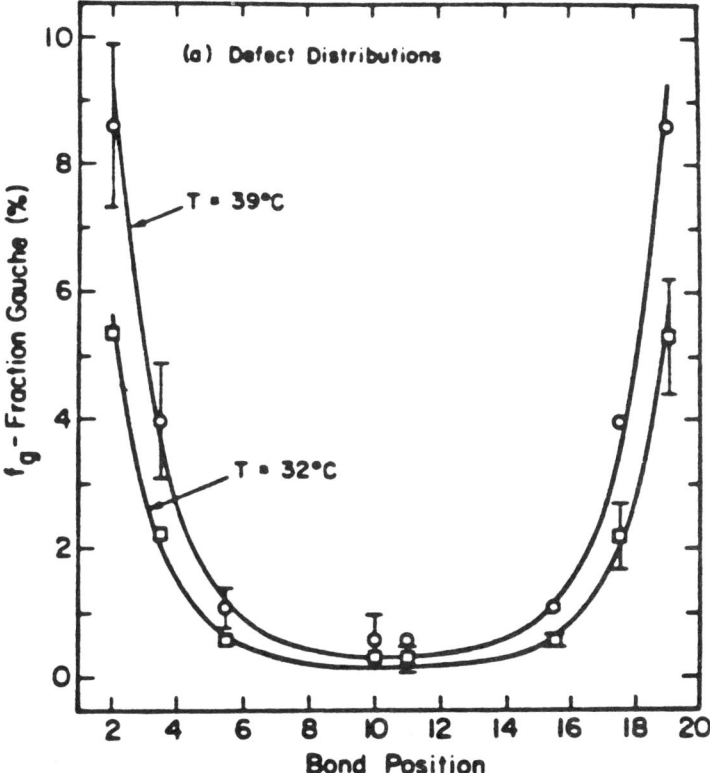

Figure 12 Theory of Maroncelli et al (98) compared with infrared vibrational experiments. Reproduced with permission.

$d(0) = 0.87$, and $\varepsilon = 0.67$: 87% of the chains are centered at the midplane, and for C21 only 1–2% of the centers are displaced by four or more bonds from the midplane. They conclude that in these alkane crystals, the chain end distribution is much sharper, and there are fewer lateral steps, than in bilayers.

Polymers Adjacent to Walls

Condensed chains at surfaces have been treated using the long-chain interphase theory (31, 67) and the finite-chain model (53, 54, 57, 58). Also, deGennes (14) has treated the case of pure chains at a surface where some are anchored and some are mobile. The appropriate boundary condition is that no chains pass through the interface, from layer 0 to layer 1. The theories predict an enhanced probability of lateral steps at the interface. For homopolymers, the order is predicted to be $S = -0.226$ in the first layer adjacent to the surface, depending slightly on chain length, and diminishing by one order of magnitude per each of the next four layers (58, 59). (If all bonds were perfectly aligned within the layer plane, then $S = -0.5$.) Moreover, all the segment statistical weights reach within about 2% of their bulk values in the second layer. Hence the presence of the surface leads to short-ranged order, of about one molecular diameter from the interface, 4–5 Å. Scheutjens & Fleer have shown that adsorbed chains tend to have the central 1/3 of the molecule in contact with the surface, with the remainder of the chain in two equal-length tails (53, 54). The midsection is comprised of short trains and longer loops.

This treatment has also been applied to the interfacial tension of pure polymeric melts. The entropic contribution to the interfacial free energy per chain segment of pure homopolymer at an interface is predicted to be small, 0.184 kT, for infinitely long chains, decreasing for shorter chains. Theodorou (58) notes that the interfacial free energy is predicted to depend on n^{-1} whereas experiments indicate the dependence should be $n^{-2/3}$ (100, 101). Measured interfacial free energies, however, will also have contributions due to the cohesive energy, and may depend significantly on the density, which is known to increase with chain length (102).

Theodorou has developed a generalization of the matrix **G** in Eq. 27, which permits consideration of copolymer molecules at interfaces (58, 59). The sequence of monomeric units is taken into account through a statistical weight factor in each element of the matrix corresponding to the appropriate Flory χ parameter (8) for segment/segment interactions. In this manner, any specific sequences of monomers can be readily described. Predictions of this model of Theodorou for triblock and random copolymers are in good agreement with experiments of surface tension as a function of copolymer composition; see Figure 13.

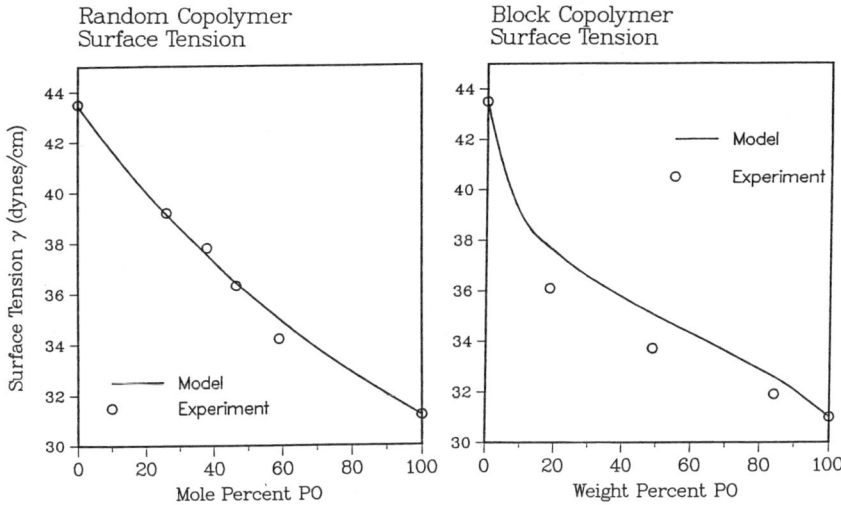

Figure 13 Surface tensions of copolymers, as a function of the ethylene oxide, propylene oxide monomer composition. Experiments of Rastogi & St. Pierre (124), compared with the theory of Theodorou (59), reproduced with permission.

SOLUTE PARTITIONING INTO INTERPHASES

Theory

We now consider mixing properties, principally of small molecules within interphases, for example the partitioning of solutes at infinite dilution from an external medium into the interphase. We limit the discussion to solute molecules of a size commensurate with that of a single chain segment, and an interphase that remains at fixed surface density throughout the partitioning process; see Figure 14.

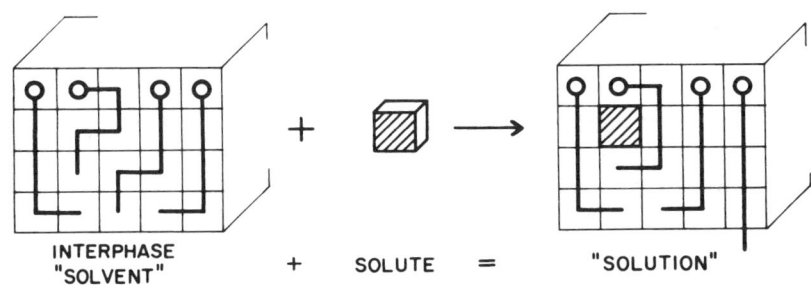

Figure 14 Partitioning of solutes into interphases, reproduced from (108).

For the unmixed interphases of pure homopolymer chains described above, we have intentionally omitted a constant factor irrelevant for the conformational properties. We now consider this factor, however, because it is relevant to the mixing process. In the pure system, the partition function can be constructed as a product of "insertion factors" (8, 51, 53, 108).

$$W_{\text{insertion}} = \prod_{l=1}^{L} \prod_{j=0}^{N_l-1} (1-j/N_l) = \prod_{l=1}^{L} N_l^{-N_l} N_l!. \qquad 32.$$

This constant factor has no bearing on conformations of the pure interphase.

Now consider the mixing within the interphase of S_l solute molecules, each of size equal to that of a single lattice site and having no internal degrees of freedom. Through use of the insertion process, the number of distinguishable configurations of the mixed system is

$$W_{\text{insertion}} = \prod_{l=1}^{L} \frac{1}{S_l!} \prod_{j=0}^{N_l-1} (1-j/N_l)$$

$$= \prod_{l=1}^{L} \frac{1}{S_l!} N_l^{-N_l} N_l!. \qquad 33.$$

Hence the total entropy of the system is a sum of a mixing contribution, determined by use of the Boltzmann equation and Stirling's approximation,

$$\frac{\Delta S_{mix}}{k} = -\sum_{l=1}^{L} S_l \ln\left(\frac{S_l}{N_l}\right) \qquad 34.$$

and a configurational entropy term, calculated as in Eq. 7, but now with the packing constraint, Eq. 9, replaced by

$$N(\langle v_{l-1}^+ \rangle + \langle v_l^0 \rangle + \langle v_{l+1}^- \rangle) = N_l - S_l \qquad 35.$$

(108). In this fashion, the solute chemical potential is calculated for each layer. The solute distribution within the interphase can be determined by setting these chemical potentials equal. Furthermore, contact interactions among solute, chains, and external medium are readily taken into account in the Bragg-Williams approximation using Flory χ parameters to predict the partition coefficient, the equilibrium constant for partitioning into the interphase from an external medium (108).

Partitioning into Bilayer Membranes

The theory predicts that there are three principal respects in which solutes partition differently into interphases than into bulk phase polymers (108). First, because of their interfacial nature, in which properties vary with distance from the interface, it is expected that solute at equilibrium will not be uniformly distributed throughout the interphase; see Figure 15. Instead, for solutes of the same chemical character as the chains, the solute is predicted to be most concentrated in the region where chain disorder is greatest, i.e. in the most "liquid-like" region. Hence the theory predicts that solute should be more concentrated in the midplane of bilayer membranes than near the head groups; this prediction is confirmed by the small-angle neutron experiments of White et al with deuterated hexane in dioleoyl lecithin bilayers (103) (see Figure 16).

Second, the partial chain ordering in the interphase should disfavor solute partitioning due to the chain conformational entropy, relative to partitioning into bulk amorphous polymeric phases (108; J. Naghizadeh,

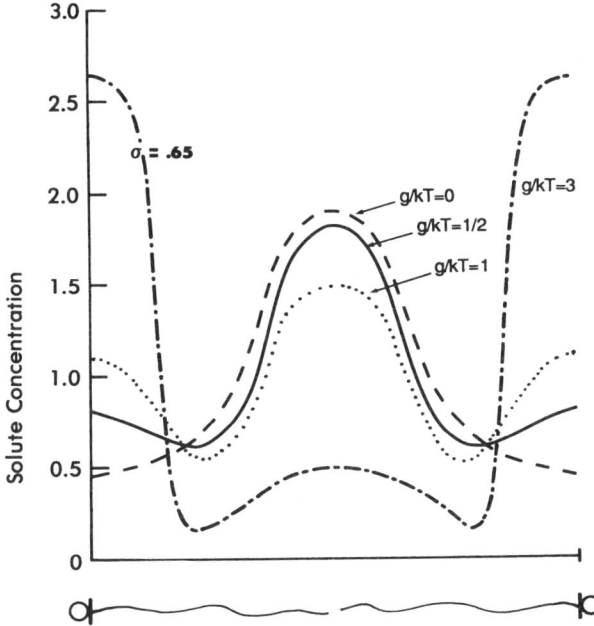

Figure 15 Equilibrium solute distribution within planar bilayers, predictions of Marqusee & Dill (108), for various values of g/kT, where g is the free energy cost of moving a solute molecule from the surface to a deeper layer. Solutes identical to chain segments ($g = 0$) partition to mid-bilayer, whereas more surface-active solutes distribute between mid-bilayer and the surface.

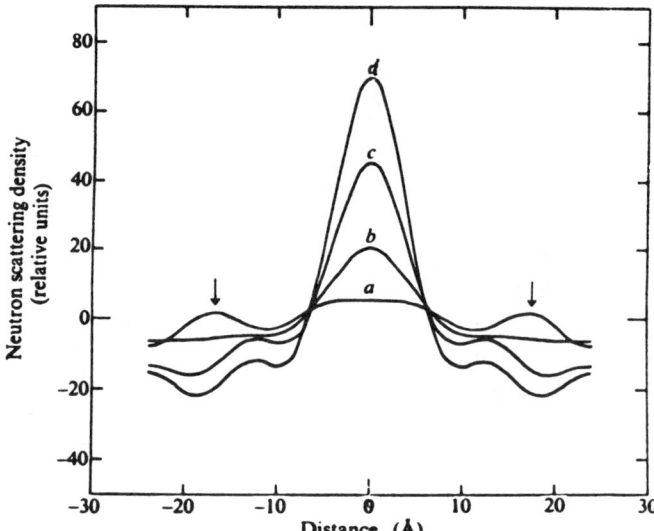

Figure 16 SANS experiments on the distribution of deuterated hexane in dioleoyllecithin bilayers, reproduced from White et al (103) with permission. Increasing hexane concentrations from *a* to *d*. *Arrows* indicate the positions of the headgroups. Hexane localizes at mid-bilayer.

K. A. Dill, in preparation). There is some suggestive, but indirect, evidence to support this prediction. The partition coefficients of anesthetics into biomembranes is about five-fold less than into olive oil (104). Entropies of transfer of short-chain hydrocarbons (105, 106) and noble gases (107) into bilayers are more negative than transfer into amorphous hydrocarbon.

Third, the solute uptake is predicted to decrease significantly as the surface density of the interphase chains increases (108) (see Figure 17), provided the surface density is sufficiently high ($\sigma \gtrsim \frac{1}{3}$). Indirect evidence has been available that supports this conclusion. Factors that cause increased surface density of the bilayer, such as decreasing the temperature through the main phase transition of the phospholipid, or incorporation of cholesterol above that temperature (see references in 109), lead to lower partition coefficients. Recently, more direct gas-phase partitioning experiments of benzene and hexane have been performed on phosphatidylcholine bilayers of chain lengths C12, C14, and C16 (109; L. De Young, K. A. Dill, in preparation). Surface density was varied by changing temperature or cholesterol incorporation, and measured by ^{2}H-NMR. The results show that partitioning decreases by nearly an order of magnitude as the surface density increases from 50 to 90% of its maximum value (see Figure 18).

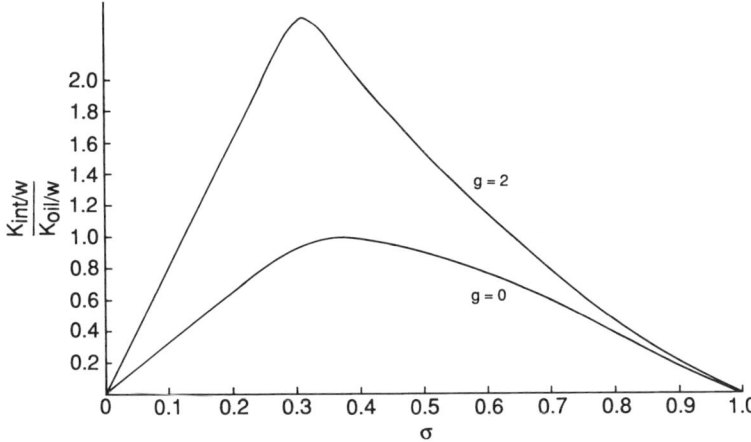

Figure 17 Predictions of Marqusee & Dill theory (108, 126) for partitioning of solutes vs surface density of the interphase chains. At high densities, chain ordering causes entropic expulsion of the solutes; at low densities, partitioning is linearly proportional to surface coverage by the chains.

Partitioning in Reversed-Phase Liquid Chromatography

An important class of interphases involves chains covalently grafted onto solid surfaces. One example is that of the silica beads used as stationary phases in reversed-phase liquid chromatography, onto which short alkyl chains are typically attached (110). Limited information is currently available on the conformations of these grafted chains. One problem is that surface densities are not commonly measured, and may vary widely, hence conformational properties may also vary. FTIR experiments that monitor IR absorption bands characteristic of *gauche* bonds show that there is much conformational disorder (111). NMR measurements of cross-polarization (112) and T1 relaxation rates and line-widths (113–115) show that the ends of these chains are in more rapid motion than segments near the points of covalent attachment to the silica surfaces. ^{2}H–NMR experiments are consistent with the view that in some stationary phases the chains are largely independent of each other: at temperatures above about 140K, the spectrum shows only an isotropic peak (116). Substantial quadrupolar splitting occurs only at temperatures below 140K, presumably where most of the chains are in their nearly all-*trans* conformations, interacting little with each other and aligned normal to the surface.

The partitioning of solutes into these grafted-chain stationary phases is predicted to depend on the surface density of the chains (117); see Figure

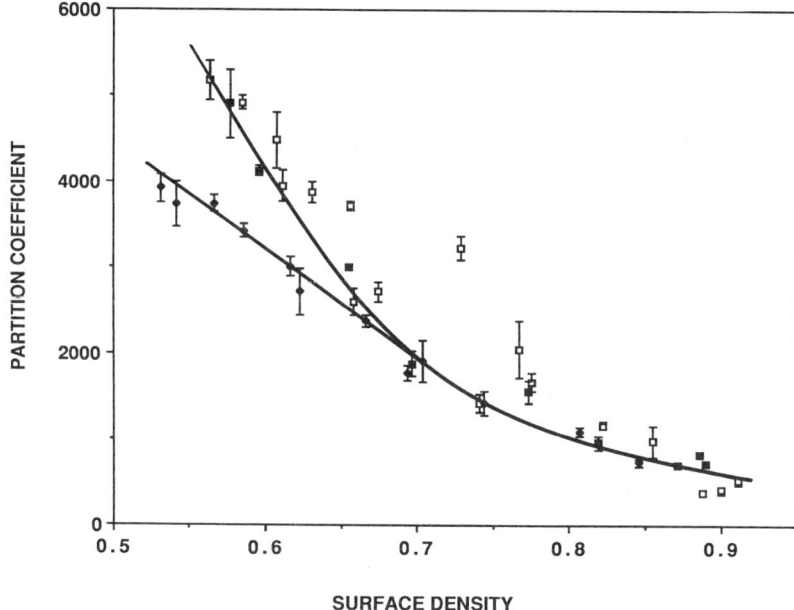

Figure 18 Experiments of De Young & Dill (109) on partitioning of benzene into bilayers of dilaurel, dimyristoyl, and dipalmitoyl lecithin, vs surface density, σ, of the bilayer chains, at high surface densities. Surface density is varied by changing the temperature and/or the concentrations of incorporated cholesterol, and is measured using ^{2}H-NMR.

17. As with the partitioning into bilayer membranes, solute concentration in the interphase should decrease with increasing surface density, provided the surface density is sufficiently high that chain conformations are constrained by neighbor interactions ($\sigma \gtrsim \frac{1}{3}$). Below this concentration, the partitioning of nonpolar solutes should depend simply on the total surface coverage by the chains, which increases linearly with surface density. These predictions, shown in Figure 17, are confirmed by the partitioning experiments of Sentell & Dorsey (118) on naphthalene into a series of C18 stationary phases of different bonding densities; see Figure 19.

Partitioning into Micelles

The theory predicts that partitioning should depend on the curvature of interphases, such as that of spherical and cylindrical micelles. Whereas the theory predicts that solute concentration per unit volume should be greatest at the micellar centers, where the chain disorder is the greatest, nevertheless the volume per radial layer increases radially outward. These two contributions to chain flux compete (toward the center, chain ends

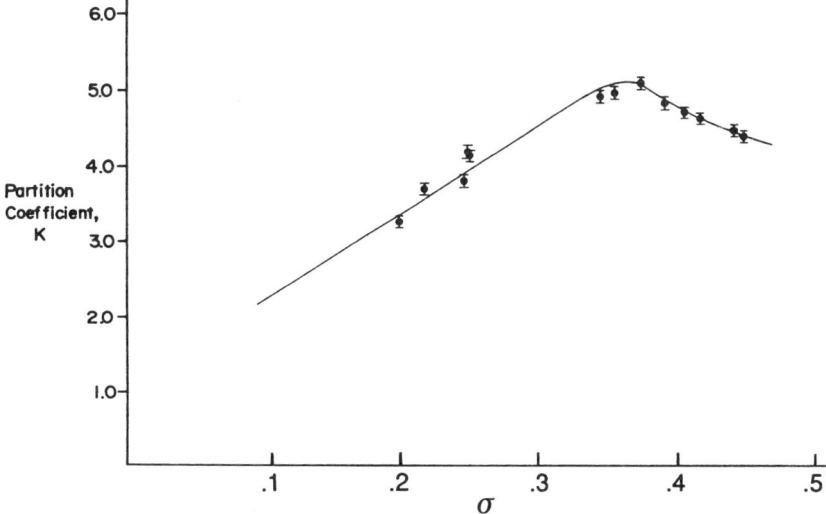

Figure 19 Experiments of Sentell & Dorsey (118) on partitioning of naphthalene into C18 reversed-phase liquid chromatography stationary phases vs surface density, σ, at low surface densities. Surface coverage is varied by changing the grafting density.

increase but the volume decreases), with the result that the largest number of molecules of small hydrocarbon solutes should be at the centers of cylindrical micelles but at the outer interface of the hydrocarbon cores of spherical micelles (108); see Figure 20. This latter prediction suggests that even very hydrophobic probes and solutes in solutions of spherical micelles should be in a relatively water-like environment, because most of them should reside at the micellar surface, due to the large volume available there. Molecules that have some surface activity should be even more concentrated at the core/water interface. Mukerjee & Cardinal (120) have observed spectroscopically that benzene, for example, at low concentrations localizes at the core/water interface. The situation may differ at high concentrations. For example, Rehfeld has observed that benzene at high concentrations is predominantly in the micellar center (121); these structures may be significantly perturbed, perhaps being essentially oil droplets surrounded by a surfactant monolayer.

SUMMARY

We have reviewed the general principles of interfacial constraint on highly concentrated polymers near sharp interfaces. First, chains are constrained by their inability to penetrate the boundary. Second, at high concentration,

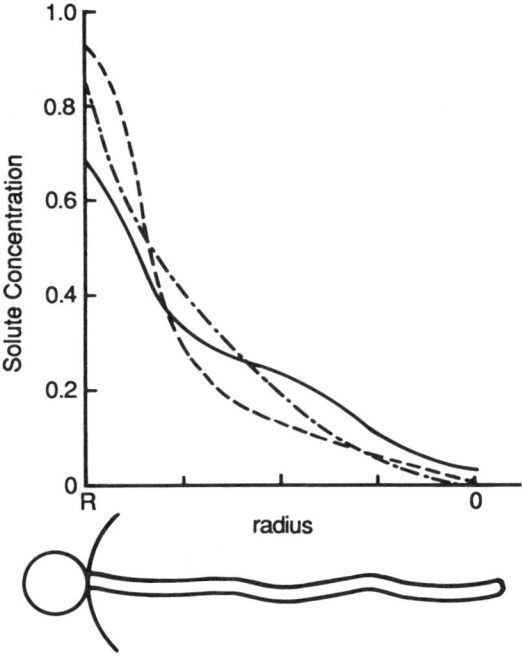

Figure 20 Theory of Marqusee & Dill (108) for the solute distribution within spherical micelles. Numbers of lattice sites per radial layer (-··-), number of solute molecules per radial layer for solutes with no surface activity ($g = 0$) (———) and for solutes with surface activity ($g = 0.75$ kT) (---).

polymers are also constrained by interactions with neighboring chains. Third, one additional constraint depends on the chain length: (*a*) for long chains, a symmetry condition arises from the indistinguishability of segments k and $k+1$, whereas (*b*) for shorter molecules, wherein the segments are distinguishable, the length of the chains is fixed. Subject to these restraints, chains at equilibrium will be configured to maximize their entropy, and hence their configurational disorder. The physical properties of chains at interfaces are often quite different from those of bulk polymers. In most such systems, the conformational ordering is dissipated within only 5–10 Å from the interface, but some physical properties depend on effects that are propagated over much longer distances. The currently available theory is found to be in quite good general agreement with a large number of conformational and mixing properties of polymers at interfaces, in semicrystalline polymers, in alkane crystals, in stationary phases used in reversed-phase liquid chromatography, and in amphiphilic aggregates including bilayer membranes and micelles.

Acknowledgments

We thank the National Institutes of Health and the University Research Initiative program of the Defense Advanced Research Projects Agency for support of this work, and the PEW Scholars Foundation for support to K.A.D.

Literature Cited

1. Flory, P. J. 1962. *J. Am. Chem. Soc.* 84: 2857–67
2. Dill, K. A., Flory, P. J. 1980. *Proc. Natl. Acad. Sci. USA* 77: 3115–19
3. Dill, K. A., Flory, P. J. 1981. *Proc. Natl. Acad. Sci. USA* 78: 676–80
4. Silberberg, A. 1962. *J. Phys. Chem.* 66: 1872
5. DiMarzio, E. A., Peyser, P. 1965. *J. Chem. Phys.* 42: 2558
6. Silberberg, A. 1968. *J. Chem. Phys.* 48: 2853
7. Hoeve, C. A. J. 1966. *J. Chem. Phys.* 44: 1505
8. Flory, P. J. 1953. *Principles of Polymer Chemistry.* Ithaca: Cornell Univ.
9. Edwards, S. F. 1965. *Proc. Phys. Soc.* 85: 613
10. Dolan, A. K., Edwards, S. F. 1975. *Proc. R. Soc. London Ser. A* 393: 427
11. Jones, I. S., Richmond, P. 1977. *J. Chem. Soc. Faraday Trans. 2* 73: 1062–70
12. deGennes, P.-G. 1976. *J. Phys. Paris* 37: 1445
13. deGennes, P.-G. 1977. *J. Phys. Paris* 38: 426
14. deGennes, P.-G. 1980. *Macromolecules* 13: 1069–75
15. deGennes, P.-G. 1981. *Macromolecules* 14: 1637–44
16. Klein, J., Pincus, P. 1982. *Macromolecules* 15: 1129–35
17. deGennes, P.-G. 1987. *Annal. Chim.* 77: 389–410
18. deGennes, P.-G. 1979. *Scaling Concepts in Polymer Physics.* Ithaca: Cornell Univ. Press
19. Helfand, E., Tagami, Y. 1972. *J. Chem. Phys.* 56: 3592–3601
20. Helfand, E. 1975. *Acct. Chem. Res.* 8: 295
21. Helfand, E., Wasserman, Z. R. 1982. In *Developments in Block Copolymers*, ed. I. Goodman, p. 99. New York: Applied Science
22. Fredrickson, G. H., Helfand, E. 1987. *J. Chem. Phys.* 87: 697–705
23. Leibler, L. 1980. *Macromolecules* 13: 1602–17
24. Hong, K. M., Noolandi, J. 1981. *Macromolecules* 14: 727–36
25. Noolandi, J., Hong, K. M. 1982. *Macromolecules* 15: 482–92
26. Whitmore, M. D., Noolandi, J. 1985. *Macromolecules* 18: 2486–97
27. Roe, R. J. 1974. *J. Chem. Phys.* 60: 4192
28. Roe, R. J. 1975. *J. Chem. Phys.* 62: 490–99
29. Helfand, E. 1975. *J. Chem. Phys.* 63: 2192–98
30. Helfand, E. 1976. *Macromolecules* 9: 307–10
31. Weber, T. A., Helfand, E. 1976. *Macromolecules* 9: 311–16
32. Guttman, C.-M., DiMarzio, E. A., Hoffman, J. D. 1981. *Polymer* 22: 1466–79
33. Flory, P. J., Yoon, D. Y., Dill, K. A. 1984. *Macromolecules* 17: 862–68
34. Flory, P. J., Yoon, D. Y. 1984. *Macromolecules* 17: 869–71
35. Mansfield, M. L. 1983. *Macromolecules* 16: 914–20
36. Leermakers, F. A. M., Scheutjens, J. M. H. M., Gaylord, R. 1984. *Polymer* 25: 1577–88
37. Marqusee, J., Dill, K. A. 1986. *Macromolecules* 19: 2420–26
37a. Marqusee, J. 1988. *Macromolecules.* Submitted
38. Kramers, H. A., Wannier, G. H. 1941. *Phys. Rev.* 60: 252–62, 263–76
39. Zimm, B. H., Bragg, J. K. 1959. *J. Chem. Phys.* 31: 526–35
40. Flory, P. J. 1969. *Statistical Mechanics of Chain Molecules.* New York: Wiley
41. Rubin, R. J. 1965. *J. Chem. Phys.* 43: 2392–2407.
42. DiMarzio, E. A. 1965. *J. Chem. Phys.* 42: 2101
43. DiMarzio, E. A., Rubin, R. J. 1971. *J. Chem. Phys.* 55: 4318–36
44. Ben-Shaul, A., Szleifer, I., Gelbart, W. M. 1984. *Proc. Natl. Acad. Sci. USA* 81: 4601–5
45. Ben-Shaul, A., Szleifer, I., Gelbart, W. M. 1985. *J. Chem. Phys.* 83: 3597–3611

46. Ben-Shaul, A., Szleifer, I., Gelbart, W. M. 1985. *J. Chem. Phys.* 83: 3612–20
47. Ben-Shaul, A., Gelbart, W. M. 1985. *Ann. Rev. Phys. Chem.* 36: 179–211
48. Szleifer, I., Ben-Shaul, A., Gelbart, W. M. 1986. *J. Chem. Phys.* 85: 5345–58
49. Szleifer, I., Ben-Shaul, A., Gelbart, W. M. 1987. *J. Chem. Phys.* 86: 7094–7109
50. Viovy, J. L., Gelbart, W. M., Ben-Shaul, A. 1987. *J. Chem. Phys.* 87: 4114–25
51. Naghizadeh, J., Dill, K. A. 1988. *J. Chem. Phys.* Submitted for publication
52. Scheutjens, J. M. H. M., Fleer, G. J. 1979. *J. Phys. Chem.* 83: 1619–35
53. Scheutjens, J. M. H. M., Fleer, G. J. 1980. *J. Phys. Chem.* 84: 178–90
54. Scheutjens, J. M. H. M. 1985. Dissertation. Agricultural University, Wageningen, The Netherlands
55. Levine, S., Thomlinson, M. M., Robinson, K. 1978. *Discuss. Faraday Soc.* 65: 202
56. Theodorou, D. N. 1988. *Macromolecules.* 21: 1391–1400
57. Theodorou, D. N. 1988. *Macromolecules.* 21: 1400–10
58. Theodorou, D. N. 1988. *Macromolecules.* 21: 1411–21
59. Theodorou, D. N. 1988. *Macromolecules.* 21: 1422–36
60. Strobl, G. R., Hagedorn, W. J. 1978. *J. Polym. Sci. Polym. Phys. Ed.* 16: 1181
61. Glotin, M., Mandelkern, L. 1982. *Colloid Polym. Sci.* 260: 182–92
62. Kitamura, R., Horii, F. 1978. *Adv. Polym. Sci.* 26: 137
63. Bergmann, K. 1978. *J. Polym. Sci. Polym. Phys. Ed.* 16: 1611
64. 1979. *Faraday Discuss. Chem. Soc.* Vol. 68
65. DiMarzio, E. A., Guttman, C. M. 1980. *Polymer* 21: 733–44
66. DiMarzio, E. A., Guttman, C. M., Hoffman, J. D. 1980. *Polymer* 21: 1374–84
67. Yoon, D. Y., Flory, P. J. 1977. *Polymer* 18: 509–13
68. Flory, P. J., Yoon, D. Y. 1978. *Nature* 272: 226
69. Fischer, E. W., Hahn, K., Kugler, S., Struth, U., Born, R. 1984. *J. Polymer Sci. Polym. Phys. Ed.* 22: 1491–1513
70. Lacher, R. C., Bryant, J. L., Howard, L. N., Summers, D. W. 1986. *Macromolecules* 19: 2639–43
71. Lacher, R. C., Bryant, J. L., Howard, L. N. 1986. *J. Chem. Phys.* 85: 6147–52
72. Lacher, R. C. 1987. *Macromolecules* 20: 3054
73. Vonk, C. G. 1986. *J. Polym. Sci. C. Polym. Lett.* 24: 305–9
74. Gruen, D. W. R. 1981. *J. Colloid Interface Sci.* 84: 281–83
75. Gruen, D. W. R. 1985. *J. Phys. Chem.* 89: 146–53
76. Gruen, D. W. R. 1985. *J. Phys. Chem.* 89: 153–63
77. Hubbell, W. L., McConnell, H. M. 1971. *J. Am. Chem. Soc.* 93: 314–26
78. Seelig, J. 1977. *Q. Rev. Biophys.* 10: 353–418
79. van der Ploeg, P., Berendsen, H. J. C. 1983. *Mol. Phys.* 49: 233–48
80. Seelig, J., Niederburger, W. 1974. *Biochemistry* 13: 1585
81. Cantor, R. S., Dill, K. A. 1984. *Macromolecules* 17: 384–88
82. Zaccai, G., Buldt, G., Seelig, A., Seelig, J. 1979. *J. Mol. Biol.* 134: 693
83. Haile, J. M., O'Connell, J. P. 1984. *J. Phys. Chem.* 88: 6363–66
84. Woods, M. C., Haile, J. M., O'Connell, J. P. 1986. *J. Phys. Chem.* 90: 1875–85
85. Cabane, B. 1981. *J. Phys. Paris* 42: 847
86. Zemb, T., Chachaty, C. 1982. *Chem. Phys. Lett.* 88: 68–73
87. Dill, K. A. 1982. *J. Phys. Chem.* 86: 1498–1500
88. Muller, N., Simsohn, H. 1971. *J. Phys. Chem.* 75: 942
89. Dill, K. A. 1983. *Surfactants in Solution,* ed. K. L. Mittal, B. Lindman, 1: 307–19. New York/London: Plenum
90. Dill, K. A., Koppel, D. E., Cantor, R. S., Dill, J. A., Bendedouch, D., Chen, S.-H. 1984. *Nature* 309: 42–45
91. Czarniecki, M. F., Breslow, R. 1979. *J. Am. Chem. Soc.* 101: 3675–76
92. Bendedouch, D., Chen, S.-H., Koehler, W. C. 1983. *J. Phys. Chem.* 87: 153–59, 2621–28
93. Cabane, B., Duplessix, R., Zemb, T. 1985. *J. Phys. Paris* 46: 2161–78
94. Owensen, B., Pratt, L. R. 1984. *J. Phys. Chem.* 88: 2905–15
95. Pratt, L. R., Owensen, B., Sun, Z. 1986. *Adv. Colloid Int. Sci.* 26: 69–97
96. Leermakers, F. A. M., Scheutjens, J. M. H. M., Lyklema, J. 1983. *Biophys. Chem.* 18: 353–60
97. Jonsson, B., Edholm, O., Teleman, O. 1986. *J. Chem. Phys.* 85: 2259–71
98. Maroncelli, M., Strauss, H. L., Snyder, R. G. 1985. *J. Chem. Phys.* 82: 2811–24
99. Ryckaert, J.-P., Klein, M. L., McDonald, I. 1987. *Phys. Rev. Lett.* 58: 698
100. Legrand, D. G., Gaines, G. L. 1969. *J. Coll. Int. Sci.* 31: 162
101. Legrand, D. G., Gaines, G. L. 1975. *J. Coll. Int. Sci.* 50: 272–79
102. Orwoll, R. A., Flory, P. J. 1967. *J. Am. Chem. Soc.* 89: 6814

103. White, S. H., King, G. I., Cain, J. E. 1981. *Nature* 290: 161–63
104. Seeman, P. 1972. *Pharmacol. Rev.* 24: 583–655
105. Miller, K. W., Hammond, L., Porter, E.-G. 1977. *Chem. Phys. Lipids* 20: 229–41
106. Simon, S. A., Stone, W. L., Busto-Latorre, P. 1977. *Biochim. Biophys. Acta* 645: 327–38
107. Katz, Y., Diamond, J. M. 1974. *J. Memb. Biol.* 17: 101–20
108. Marqusee, J. A., Dill, K. A. 1986. *J. Chem. Phys.* 85: 434–44
109. De Young, L., Dill, K. A. 1988. *Biochemistry*. In press
110. Melander, W. R., Hovarth, C. S. 1980. In *High Performance Liquid Chromatography*, ed. C. S. Hovarth, 2: 113. New York: Academic
111. Sander, L. C., Callis, J. B., Field, L. R. 1983. *Anal. Chem.* 55: 1068–75
112. Sindorf, D. W., Maciel, G. E. 1983. *J. Am. Chem. Soc.* 105: 1848
113. Gilpin, R. K., Gangoda, M. E. 1984. *Anal. Chem.* 56: 1470
114. Gilpin, R. K. 1984. *J. Chromatogr. Sci.* 22: 371
115. Gilpin, R. K., Gangoda, M. E. 1983. *J. Chromatogr. Sci.* 21: ?52
116. Kelusky, E. C., Fyfe, C. A. 1986. *J. Am. Chem. Soc.* 108: 1746–49
117. Dill, K. A. 1987. *J. Phys. Chem.* 91: 1980–88
118. Sentell, K. B., Dorsey, J. G. 1988. *Anal. Chem.* Submitted
119. Sentell, K. B. 1988. PhD thesis. Univ. Florida, Gainsville
120. Mukerjee, P., Cardinal, J. R. 1978. *J. Phys. Chem.* 82: 1620–27
121. Rehfeld, S. J. 1971. *J. Phys. Chem.* 75: 3905
122. Mely, B., Charvolin, J., Keller, P. 1975. *Chem. Phys. Lipids* 15: 161–73
123. Ellena, J. F., Dominey, R. N., Cafiso, D. S. 1987. *J. Phys. Chem.* 91: 131–37
124. Rastogi, A. K., St. Pierre, L. E. 1969. *J. Coll. Int. Sci.* 31: 168–75

THEORY OF PURE DEPHASING IN CRYSTALS

J. L. Skinner

Department of Chemistry, Columbia University, New York, New York 10027

INTRODUCTION

We have come a long way in the last 25 years from the notion that the width of an absorption line is determined solely by population lifetimes. In particular, for a transition between the ground state and some excited state of a system, it is well known that if the excited state lifetime in seconds is T_1, then the absorption linewidth (FWHM) in Hz, Δv, is given by $\Delta v = 1/2\pi T_1$. Indeed, such a relation can be viewed as a direct manifestation of the time-energy Heisenberg uncertainty principle. However, it has become clear in recent years that the process of pure dephasing makes an additional, and often dominant, contribution to the absorption linewidth.

To understand the origin of pure dephasing, it is convenient to divide the Hilbert space of the universe into two subspaces—one for the "system" and one for the "bath." One can then decompose the Hamiltonian for the universe into three terms—a system Hamiltonian, a bath Hamiltonian, and a system-bath coupling. The actual specification of the system subspace is chosen such that the spectroscopy of interest measures transitions between eigenstates of the system Hamiltonian. Coupling terms that are off-diagonal in the system states will lead to population relaxation (e.g. coupling to the electromagnetic field produces radiative relaxation), whereas those terms that are diagonal in the system states lead only to pure dephasing (they do not produce population relaxation). In the interaction picture, these diagonal coupling terms can be viewed as providing time-dependent perturbations to the system eigenenergies. Thus it is not surprising that these additional "uncertainties" in the system energies lead to line broadening over and above the Heisenberg uncertainty lifetime broadening. The origin of the term "dephasing" comes from the time-domain viewpoint: if a linear superposition state of the system evolves in time, these fluctuating energies cause the phase relation between amplitudes in different states to become uncorrelated, or "dephased."

For the most part, this review concerns the pure dephasing of vibrational or electronic transitions of dilute molecular or ionic impurities in crystals. In all cases, it is the collective vibrations of the crystal lattice, or phonons, that serve as the bath, and are responsible for dephasing. Due to the impurity-phonon coupling, the introduction of impurities into crystals can provide a useful probe of both the host crystal phonons and the coupling itself. Inasmuch as it is precisely this interaction that is invoked in theories of intermolecular exciton or vibrational energy transfer, it is very important to determine the nature and strength of this coupling. Furthermore, the study of these interactions in relatively simple crystalline hosts will surely be useful (and indeed, already has been useful) in trying to understand spectroscopy and transport in complex systems such as glasses and proteins.

The usual theoretical approach to the problem of pure dephasing involves second-order perturbation theory in the system-bath coupling. This approach was pioneered by McCumber & Sturge more than 20 years ago (1), and their results have been rederived many times (2–6). While their weak coupling result has been tremendously useful for analyzing experiment, it should be recognized that this approach makes an uncontrolled approximation by truncating the perturbation series at second order, without knowing whether the remaining terms are really "small" or not. Recently Osad'ko (7–9), and Hsu and myself (10–16), have discovered that the standard model in the field can be solved exactly, and our general results are in agreement with each other. Two other workers have also pursued an exact solution, but their results differ from ours (17, 18). Our own approach is in fact perturbative in nature, but we sum the perturbation series to all orders. Thus, in this review I discuss our (nonperturbative) results for the pure dephasing problem. A number of excellent reviews of the more traditional theoretical approaches, and of some relevant experiments, have appeared within the last decade or so (19–24). In addition, we have recently reviewed our own work (25, 26). Therefore, the present article, while attempting to be self-contained, focuses on our new results of the last two years. These include an extension of the usual model of one collective bath coordinate in the system-bath interaction, to the most general case of arbitrary (quadratic in the bath) system-bath interaction (15); a unified microscopic model for the creation of pseudo-local phonons and the concomitant dephasing they produce (14); and a description of how our theory, originally derived for the case of optical dephasing, can be applied equally well to the problem of vibrational dephasing (16). I also describe the application of our results to several recent experiments, without, however, attempting to provide a comprehensive survey of the experimental literature.

PURE DEPHASING OF A TWO-LEVEL SYSTEM

Before I discuss the physical problems of interest, it is convenient to quote some exact results for the general problem of dephasing of a two-level system. In subsequent sections I describe how the problems of optical and vibrational dephasing relate to our general result, and assign physical meaning to the various terms below.

We begin by considering the problem of the interaction of a two-level system (TLS) with a collection of harmonic oscillators (the bath). Since we are only interested here in pure dephasing, we need only consider diagonal terms in the TLS-bath interaction, and so we can write the most general Hamiltonian as:

$$H = E_0|0\rangle\langle 0| + E_1|1\rangle\langle 1| + \Delta_0|0\rangle\langle 0| + \Delta_1|1\rangle\langle 1| + H_P, \qquad 1.$$

$$H_P = \sum_\alpha \hbar\omega_\alpha^P \left(a_\alpha^\dagger a_\alpha + \frac{1}{2}\right). \qquad 2.$$

In the above, $|0\rangle$ and $|1\rangle$ are the ground and excited state kets of a TLS with energies E_0 and E_1, H_P is the Hamiltonian for the N harmonic oscillators, and $a_\alpha^\dagger$, a_α, and ω_α^P are the creation and annihilation operators and frequency for the αth mode. In principle, Δ_0 and Δ_1 are general operators in the Hilbert space of H_P. In what follows, however, we assume that they depend only on the coordinates, $q_\alpha \equiv a_\alpha^\dagger + a_\alpha$. Each of these operators can then be expanded in a Taylor series, which to second order we write as:

$$\Delta_0 = \Delta E_0 + \sum_\alpha g_{0\alpha} q_\alpha + \frac{1}{2}\sum_{\alpha\beta} g_{0\alpha\beta} q_\alpha q_\beta, \qquad 3.$$

$$\Delta_1 = \Delta E_1 + \sum_\alpha g_{1\alpha} q_\alpha + \frac{1}{2}\sum_{\alpha\beta} g_{1\alpha\beta} q_\alpha q_\beta. \qquad 4.$$

For such a TLS coupled to a bath, the pure dephasing rate, $1/T_2'$, and the frequency shift, $\Delta\omega$, of the TLS transition, are defined by the Bloch equation for the off-diagonal reduced density matrix element, $\sigma_{10}(t)$ (26):

$$\dot\sigma_{10}(t) = -\left(i(\Omega+\Delta\omega) + \frac{1}{T_2'}\right)\sigma_{10}(t), \qquad 5.$$

where $\Omega = (E_1 + \Delta E_1 - E_0 - \Delta E_0)/\hbar$. For the above Hamiltonian we have found exact analytic (up to quadrature) expressions for $1/T_2'$ and $\Delta\omega$ (15). While this is a substantial theoretical achievement, it is, at this time, not particularly useful for understanding experiments, because of the very large number of parameters involved. We have found that, in fact, a very simple special case of the above is adequate for explaining most of the experimental data that have been examined to date. First of all, except in the limit of an Ohmic bath (27), which is not realistic for molecular systems interacting with phonons, the linear terms in Eqs. 3 and 4 lead only to a temperature-independent contribution to $\Delta\omega$ and can therefore be neglected or transformed away (26). Second, the quadratic terms are simplified by making the assumptions that $g_{0\alpha\beta} = U h_\alpha^P h_\beta^P$, and $g_{1\alpha\beta} = (U+W) h_\alpha^P h_\beta^P$. With these assumptions, we write

$$\Delta_0 = \Delta E_0 + \frac{U}{2}\phi^2, \qquad 6.$$

$$\Delta_1 = \Delta E_1 + \frac{(U+W)}{2}\phi^2, \qquad 7.$$

$$\phi = \sum_\alpha h_\alpha^P q_\alpha. \qquad 8.$$

Thus the only TLS-bath coupling involves a single collective bath coordinate, which is a linear combination of the bath modes.

With the above simplified model we have found that $1/T_2'$ and $\Delta\omega$ are given (exactly) by (10–16)

$$\frac{1}{T_2'} = \frac{1}{4\pi} \int_0^\infty d\omega \ln\{1 + 4n(\omega)[n(\omega)+1] W^2 \Gamma_0(\omega) \Gamma_1(\omega)\}, \qquad 9.$$

$$\Delta\omega = \frac{1}{2\pi} \int_0^\infty d\omega \arctan\left\{\frac{2n(\omega) W \Gamma_0(\omega)[1 - W\Omega_0(\omega)]}{[1 - W\Omega_0(\omega)]^2 + [2n(\omega)+1]^2 W^2 \Gamma_0(\omega)^2}\right\}, \qquad 10.$$

$$n(\omega) = \frac{1}{\exp(\hbar\omega/kT) - 1}, \qquad 11.$$

$$\Gamma_0(\omega) = \frac{\Gamma_P(\omega)}{[1 - U\Omega_P(\omega)]^2 + U^2 \Gamma_P(\omega)^2}, \qquad 12.$$

$$\Gamma_1(\omega) = \frac{\Gamma_P(\omega)}{[1 - (U+W)\Omega_P(\omega)]^2 + (U+W)^2 \Gamma_P(\omega)^2}, \qquad 13.$$

$$\Gamma_P(\omega) = \frac{\pi}{\hbar} \sum_\alpha h_\alpha^{P2} \delta(\omega - \omega_\alpha^P), \qquad 14.$$

$$\Omega_P(\omega) = \frac{2}{\pi} \int_0^\infty dx\, \Gamma_P(x)\, P\!\left(\frac{x}{\omega^2 - x^2}\right). \qquad 15.$$

(In Eq. 15 P denotes the principal value.) Both $1/T_2'$ and $\Delta\omega$ are strongly temperature dependent and vanish as $T \to 0$. Thus all temperature-independent contributions to the frequency shift have been included in Ω.

OPTICAL DEPHASING OF IMPURITIES

Let me now discuss how the problem of the optical dephasing of substitutional impurities in crystals can be cast into the formalism of the previous section. We begin by considering a perfect crystal without impurities, whose Hamiltonian describes acoustic or optical phonons, and an isolated impurity molecule (or ion or atom) in the gas phase, with ground and excited state electronic energies E_0 and E_1. Now imagine replacing a host molecule (ion, atom) by an impurity in its ground electronic state. The potential energy of the crystal coordinates will be perturbed, and this perturbation can be expanded in a Taylor series in the perfect crystal coordinates. Within the harmonic approximation, one keeps terms up to quadratic order. The constant term in the expansion corresponds to an energy shift for the substitution without allowing the crystal coordinates to change. The term linear in phonon coordinates results from the fact that, in general, the lattice will relax around the impurity, leading to a lower energy equilibrium state. The quadratic term describes a change in the force constant matrix of the lattice coordinates. The linear term can be eliminated by transforming to a new set of coordinates with shifted equilibrium positions. The energy shift that occurs with this transformation, along with the constant term in the expansion described above, is the *solvent shift* for the impurity in the ground state. We also note that the form of the perfect crystal Hamiltonian is unchanged by this transformation. The most general quadratic perturbation of the phonons is unnecessarily complicated for our purposes. A simpler model assumes that this quadratic term can be expressed in terms of a single *collective* coordinate that is a linear combination of the perfect crystal normal modes. Examples of such a linear combination are the strain field of acoustic or optical phonons, or a particular nearest neighbor displacement. Thus if H_P is taken to be the perfect crystal phonon Hamiltonian, but for the shifted coordinates, and ΔE_0 is the solvent shift, then Eqs. 1, 2, 6, and 8 are consistent with the above description.

Now imagine repeating the argument, but replacing a host molecule with the impurity in its excited electronic state. Again, there will be constant, linear, and quadratic terms in the expansion of the perturbation potential in the perfect crystal coordinates, which, in general, will be different from the terms in the ground state. One can again eliminate the linear term, which leads to an excited state solvent shift, ΔE_1, and a phonon Hamiltonian and quadratic perturbation in terms of the shifted coordinates. Of course the coordinate shifts necessary to eliminate this linear term will not be the same shifts that eliminate the linear term when the impurity is in its ground state. This difference is in fact responsible for the production of phonon sidebands in the absorption spectrum. However, for the zero-phonon line (ZPL) we have shown that except for the (in this case) unphysical Ohmic bath (27), this difference in the coordinate shifts is irrelevant (26). Therefore, in what follows we neglect this difference, writing the quadratic Hamiltonians for the ground and excited states in terms of the same phonon coordinates. If we invoke again a collective coordinate, and further make the not completely unreasonable assumption that the *same* collective coordinate perturbs the ground and excited state potentials, then we recover Eqs. 1, 2, and 6–8. In summary then, H_P is the acoustic or optical phonon Hamiltonian for the perfect crystal without impurities, $\Delta E_1 - \Delta E_0$ is the total solvent shift, $(U/2)\phi^2$ is the perturbation in the ground state, and $(W/2)\phi^2$ is the *additional* perturbation when the impurity is excited. Different models for the phonons, the collective coordinate, and the coupling constants lead to different dephasing "mechanisms." Several examples are discussed below.

Acoustic Phonons

In the simplest model of acoustic phonons the density of states is given by a Debye spectrum. However, in the dephasing theory, the central quantity is not the density of states itself, but the weighted density of states, or spectral density, given by $\Gamma_P(\omega)$ in Eq. 14. If one chooses the collective coordinate to be the strain field, then within the deformation potential approximation we have (11, 28)

$$\Gamma_P(\omega) = \frac{3\pi}{2}\left(\frac{\omega}{\omega_D}\right)^3 \theta(\omega)\theta(\omega_D - \omega), \qquad 16.$$

$$\Omega_P(\omega) = -\left\{1 + 3\left(\frac{\omega}{\omega_D}\right)^2 + \frac{3}{2}\left(\frac{\omega}{\omega_D}\right)^3 \ln\left(\frac{1-\omega/\omega_D}{1+\omega/\omega_D}\right)\right\}, \qquad 17.$$

where ω_D is the Debye cutoff. Implicit in the assumption of dephasing by acoustic phonons is the stipulation that $U = 0$ in Eqs. 6, 12, and 13. In other words, one assumes that while the *introduction* of the ground state impurity into the crystal does not significantly perturb the phonons, the electronic *excitation* does significantly perturb the phonons, so that $U = 0$

and $W \neq 0$. Cases where $U \neq 0$ are discussed below. From Eqs. 9–13, 16, and 17 one can then calculate the temperature-dependent linewidth and lineshift.

Optical Phonons

In a similar vein, to describe dephasing by optical phonons, we set $U = 0$, and take a simple model for the optical phonon spectral density (11):

$$\Gamma_P(\omega) = \frac{3\pi\omega_{op}}{8\bar{\omega}^3}[\bar{\omega}^2 - (\omega - \omega_{op})^2]\theta(\omega - \omega_{op} + \bar{\omega})\theta(\omega_{op} + \bar{\omega} - \omega), \qquad 18.$$

$$\Omega_P(\omega) = -\frac{3\omega_{op}}{8\bar{\omega}^3}\left\{[\bar{\omega}^2 - (\omega + \omega_{op})^2]\ln\left[\frac{\bar{\omega} + \omega_{op} + \omega}{\omega_{op} + \omega - \bar{\omega}}\right]\right.$$

$$\left. + [\bar{\omega}^2 - (\omega - \omega_{op})^2]\ln\left(\frac{\bar{\omega} + \omega_{op} - \omega}{\bar{\omega} - \omega_{op} + \omega}\right) + 4\bar{\omega}\omega_{op}\right\}, \qquad 19.$$

where ω_{op} is the center frequency of the phonon dispersion, and $2\bar{\omega}$ is the width.

Pseudolocal Phonons

Often when an impurity is introduced substitutionally into a crystal, the phonon perturbation is substantial, and pseudolocal modes are created, typically with a frequency within the acoustic phonon spectrum of the host (29, 30). One can describe the creation of these modes theoretically by considering a quadratic perturbation of Debye acoustic phonons by the strain field (14). Thus the spectral density for the ground state impurity is given by $\Gamma_0(\omega)$ in Eq. 12, with Eqs. 16 and 17. We have found that for $-1 < U \leq -0.8$, $\Gamma_0(\omega)$ is sharply peaked at a frequency $\omega_0 \ll \omega_D$, which is given approximately by (14)

$$\omega_0 = \omega_D\sqrt{\frac{1+U}{-3U}}, \qquad 20.$$

and has a full-width at half maximum, $1/\tau_0$, given by

$$\frac{1}{\tau_0} = \frac{\pi\omega_0^2}{2\omega_D}, \qquad 21.$$

where τ_0 is the local model lifetime. When the impurity molecule is excited, if $-1 < (W+U) \leq -0.8$, then the excited state spectral density, $\Gamma_1(\omega)$ (from Eq. 13), will also be sharply peaked, with

$$\omega_1 = \omega_D \sqrt{\frac{1+U+W}{-3(U+W)}}, \qquad 22.$$

$$\frac{1}{\tau_1} = \frac{\pi \omega_1^2}{2\omega_D}, \qquad 23.$$

where ω_1 and τ_1 are the frequency and lifetime of the local mode when the impurity is in its excited state. Thus, if both the ground and excited state spectral densities are sharply peaked, then one can legitimately speak of dephasing by pseudolocal modes. The linewidth and shift of the impurity ZPL can be calculated from Eqs. 9 and 10. We have also shown that at low temperatures, the dephasing rate reduces to a simple and interesting bi-Arrhenius analytic form (26):

$$\frac{1}{T_2'} = \frac{(\delta\omega\tau^*)^2}{1+(\delta\omega\tau^*)^2}\left[\frac{1}{2\tau_0}\exp(-\hbar\omega_0/kT) + \frac{1}{2\tau_1}\exp(-\hbar\omega_1/kT)\right], \qquad 24.$$

$$\delta\omega = \omega_1 - \omega_0, \qquad 25.$$

$$\frac{1}{\tau^*} = \frac{1}{2}\left(\frac{1}{\tau_0} + \frac{1}{\tau_1}\right). \qquad 26.$$

We have also examined the dephasing by pseudolocal modes with a more phenomenological model (12), simply choosing $\Gamma_0(\omega)$ to be a particular sharply peaked function, rather than deriving its form from a microscopic Hamiltonian. Again, however, we have derived Eq. 24 in the low temperature limit.

The subject of dephasing by low-frequency modes has also been addressed from a still more phenomenological viewpoint, considering a four-level system consisting of only two vibrational levels of the low-frequency mode for each electronic level (31–39). Our formula, Eq. 24, in fact reduces to the "exchange" theory result of Harris et al (31a,b) in the limits of $\tau_0 = \tau_1$ and $\hbar\delta\omega/kT \ll 1$, and reduces to the "uncorrelated phonon scattering" theory of deBree & Wiersma (39) in the limit $\delta\omega\tau^* \gg 1$. Thus our result generalizes these two previous theories, and provides an expression valid for the complete range of parameters (in the low temperature limit). Our microscopic model for the creation of pseudolocal modes also furnishes relations among the parameters (see Eqs. 20–23).

Weak Coupling Limit

The traditional theory of dephasing in crystals treats the electron-phonon coupling $(W/2)\phi^2$ as a perturbation (1–6). Thus these results can be recovered from our formulae by expanding to second order in W. For the dephasing rate and the lineshift one finds from Eqs. 9 and 10:

$$\frac{1}{T'_2} = \frac{W^2}{\pi} \int_0^\infty d\omega\, n(\omega)[n(\omega)+1]\Gamma_0(\omega)^2, \qquad 27.$$

$$\Delta\omega = \frac{W}{\pi} \int_0^\infty d\omega\, n(\omega)\Gamma_0(\omega). \qquad 28.$$

These expressions have been used to describe, for example, dephasing by acoustic phonons, by setting $U = 0$ and thus taking $\Gamma_0(\omega) = \Gamma_P(\omega)$, where the latter is given by Eq. 16. This leads to the famous T^7 low temperature dependence, first quantified by McCumber & Sturge (1). They have also been used to describe dephasing by optical phonons, again by setting $U = 0$ and taking $\Gamma_0(\omega) = \Gamma_P(\omega)$, where $\Gamma_P(\omega)$ can be modeled (for example) by Eq. 18 (3, 12, 40). Finally, as an approach to dephasing from pseudolocal modes, one can describe $\Gamma_0(\omega)$ phenomenologically by a Lorentzian (2, 3, 12, 40). This leads (12) to results that are completely equivalent to the fast exchange limit of Harris' theory (31a,b).

VIBRATIONAL DEPHASING

We first consider a perfect crystal with acoustic and optical phonons, and an impurity molecule whose ground vibrational state, $|0\rangle$, has energy E_0, and whose first vibrational excited state (for a particular coordinate q), $|1\rangle$, has energy E_1 (16). The molecular vibrational potential for the impurity may be anharmonic. Next we imagine substituting the *rigid* (with $q = 0$) impurity into the crystal. As discussed above, the potential energy of the phonons will be perturbed, and we can expand this perturbation in the old phonon coordinates, keeping terms up to quadratic order. As before, one then eliminates the linear term by adding a shift to the phonon coordinates, which leaves the form of the phonon Hamiltonian unchanged. We will also make the collective coordinate assumption, so that the crystal vibration Hamiltonian for the (rigid) impurity–crystal system is $\Delta E_0 + H_P + (U_1/2)\phi^2$, where ΔE_0 is the solvent shift, and U_1 is the quadratic coupling constant. Now one imagines turning on the coupling between the coordinate q and the phonons. If we assume that the same collective coordinate couples to q, we can write this interaction to second order in the phonons as $A(q)\phi + (B(q)/2)\phi^2$. We should emphasize that, in general, $A(q)$ and $B(q)$ are nonlinear functions of q, so that this coupling is anharmonic. To discuss dephasing, we need the diagonal matrix elements of this interaction for the two vibrational states. As in the case of optical dephasing, the terms linear in ϕ lead only to a temperature-independent frequency shift and the appearance of phonon sidebands but do not contribute to dephasing, so they can be neglected. Therefore, defining $U \equiv U_1 + \langle 0|B(q)|0\rangle$, the total Hamiltonian for the ground vibrational state is $E_0 + \Delta E_0 + H_P + (U/2)\phi^2$, and defining $W \equiv \langle 1|B(q)|1\rangle - \langle 0|B(q)|0\rangle$, the Hamiltonian for the excited vibrational state is $E_1 + \Delta E_1 + H_P + ((U+W)/2)\phi^2$, in agreement with Eqs. 1, 2, 6, and 7.

The above discussion shows that the vibrational dephasing problem can be discussed quite naturally within the same formalism as the optical dephasing problem. Thus one can consider vibrational dephasing by acoustic, optical, or pseudolocal phonons as described above. The only real difference between the two cases is that in the optical dephasing problem, U and W are independent, while in vibrational dephasing they are not. That is, in the optical problem, U describes the distortion caused by the introduction of the impurity, while W describes the distortion upon excitation. However, in the vibrational problem, both U and W involve the vibration–phonon coupling function $B(q)$. If, for example, one chooses the molecular vibration to be harmonic, and takes $B(q)$ to be quadratic in q, then one finds (16) that

$$U = U_1 + W/2. \qquad 29.$$

Vibrational Dephasing in a Pure Crystal

Molecular vibrations in a pure molecular crystal are in general coupled, and form vibrational exciton bands. If, however, this coupling is sufficiently weak (compared to static or dynamic site energy fluctuations), then one can treat the molecular vibrations independently, and can understand the vibrational spectroscopy by considering the motion of only a single molecule. This case differs from that of an impurity vibration in a crystal, in that here $U = 0$. That is, when all the molecules are in the ground state, the lattice Hamiltonian is exactly H_P, which describes the acoustic or optical phonons (16).

OPTICAL DEPHASING EXPERIMENTS

In this section I would like to illustrate how the theory described above has been applied to experiments on optical transitions of impurities in crystals. What follows, however, is not intended to be a comprehensive or critical discussion of the experimental literature—simply a summary of the efforts we and others have made to analyze experiment with our theory.

The theory that we have developed is for the pure dephasing contribution to the homogeneous linewidth, $1/T'_2$. As discussed in the introduction, there is also a lifetime contribution. The total dephasing rate, $1/T_2$, is given by $1/T_2 = 1/T_2' + 1/2T_1$, which is related to the homogeneous linewidth in cm^{-1}, $\Delta\tilde{v}$, by $\Delta\tilde{v} = 1/c\pi T_2$. In general, the lifetime is nearly temperature independent at low temperatures, while the pure dephasing rate is a very strong function of temperature, and vanishes in the limit $T \to 0$. Thus T_1 can be determined either independently or from the low temperature value of the linewidth, in either case allowing one to extract the pure dephasing contribution. Then there is the matter of inhomogeneous broadening. This refers to a distribution of transition energies, resulting from a distribution of microscopic environments for the impurity, which can be considered static on the experimental timescale (T_2). At 1 or 2K the observed absorption lineshape is usually inhomogeneous, crossing over to a Lorentzian homogeneous lineshape at higher temperatures, as T_2 decreases. Therefore, at these higher temperatures one can obtain T_2 directly from the absorption spectrum, at intermediate temperatures one can obtain T_2 by deconvoluting the observed partially inhomogeneously broadened spectrum, while at the lowest temperatures one must use a line-narrowing technique such as photon echoes or hole burning. In fact, the data that we discuss in this section were all obtained by absorption spectroscopy.

In Table 1 are listed the different optical experiments that have been analyzed with our theoretical results for dephasing by either acoustic or pseudolocal phonons with Eqs. 9, 12, 13, 16, and 17. Thus there are three theoretical parameters: the Debye temperature, $T_D \equiv \hbar\omega_D/k$, U, and W. T_D can usually be obtained from heat capacity or phonon density of states measurements or calculations on the pure host crystal. As discussed above,

Table 1 Parameters for thermal zero-phonon line broadening of optical impurity spectra in crystals. The last two columns indicate the references for the experiment and for the theoretical analysis, respectively

Impurity	Host	Transition	T_D (K)	W	U	Exp. Ref.	Th. Ref.
Cr^{3+}	Al$_2$O$_3$	R$_1$	935	$-0.31, 3.01$	0	1	13
		R$_2$	935	$-0.28, 1.39$	0	1	13
Cr^{3+}	BeAl$_2$O$_4$	R$_{1i}$	500	$-0.2, 0.79$	0	41	41
1,3-d	naph.	(0, 0)	113	-0.785	0	42	13
		539 cm^{-1}	113	-0.489	0	42	13
		904 cm^{-1}	113	-0.639	0	42	13
		1070 cm^{-1}	113	-0.50	0	42	13
		1158 cm^{-1}	113	-0.62	0	42	13
Cu$^+$	NaF	$^1A_{1g} - {}^1T_{2g}$	492	-0.154	-0.819	43	14
		$^1A_{1g} - {}^1E_g$	492	-0.174	-0.819	43	14
3,4,6,7-d	n-octane	3962 Å	84	0.0455	-0.869	44	14

U is a measure of the distortion of the potential energy function when the impurity is substitutionally introduced into the crystal, while W is a measure of the additional distortion when the impurity is electronically excited. Information about U and W can be obtained from the phonon sideband.

If there are no pseudolocal mode peaks in the sidebands, then one can assume that U is small, and pseudolocal modes are not created when the impurity is introduced into the crystal. In this case one analyzes the experiments by setting $U = 0$ and treating W as an adjustable parameter. This was the approach taken for the first three systems in Table 1 (in the third system 1,3-d stands for 1,3-diazaazulene, naph. for naphthalene, and the five vibronic transitions are all for $S_0 \rightarrow S_1$). In all cases a satisfactory one-parameter fit to the experimental data could be obtained (T_D was determined independently), and in all cases, with the possible exception of Cr^{3+} in $BeAl_2O_4$, the value of W obtained shows that our nonperturbative result is significantly different from the weak coupling result of Eq. 27.

The last two systems in Table 1 (3,4,6,7-d is 3,4,6,7-dibenzopyrene) exhibit pronounced pseudolocal mode peaks in the absorption and emission phonon sidebands. Thus for example for Cu^+ in NaF, since T_D can be obtained from heat capacity measurements, the parameters U and W can then be determined from Eqs. 20 and 22. This is a particularly nice system since one can make a *no-free-parameter* comparison with experiment. For both electronic transitions the agreement is reasonably good, although there are discrepancies on the order of a factor of three between the measured and predicted (from Eqs. 21 and 23) local mode lifetimes (14). In the last example in Table 1, the Debye temperature is not known, so instead we used the measured local model frequencies and the lifetime in the ground state to determine the three parameters U, W, and T_D through Eqs. 20–22. This again provides a no-free-parameter test of the theory, and again the results are quite satisfactory. Both of these last two systems have also been analyzed with our phenomenological model (12) with similar results (13, 43). Another approach to the analysis of experimental systems with pseudolocal modes is to use our low temperature result of Eq. 24, or one of its limits (31a,b, 39). While this works satisfactorily for the above two systems at the lowest temperatures (13, 43), above 10 or 15K it underestimates the amount of line broadening. This is simply a result of using the expression outside of its (low temperature) range of validity.

Attempts to understand two other experiments on free-base porphin in n-decane (45) and pentacene in benzoic acid (46) with our model have been somewhat less successful (13). Perhaps this is due to the presence of anharmonicity, which is not considered in our theory.

VIBRATIONAL DEPHASING EXPERIMENTS

In Table 2 are shown the two vibrational systems that we have attempted to analyze (16) with our theoretical results. For the first system (47), the v_3 transition of ReO_4^- in KI, we have no evidence for the creation of pseudolocal modes, so following the discussion above, we set $U_1 = 0$. That is, we assume that the perturbation from the introduction of the rigid impurity into the crystal is negligible. Next we must assume a specific form for the vibration-phonon coupling, in order to relate U to W. As discussed above, if this (anharmonic) interaction is taken to be biquadratic, and the vibration itself is harmonic, then from Eq. 29 we have that $U = W/2$. With this identification, the dephasing rate is calculated from Eqs. 9, 12, and 13. Since at the highest temperatures of the experiment both the acoustic and optical phonons are populated, we assumed a model of independent dephasing from these two modes, with the models of Eqs. 16–19. The parameters for the acoustic phonons (T_D) and the optical phonons ($\tilde{v}_{op} = \omega_{op}/2\pi c$ and $\tilde{v} = \bar{\omega}/2\pi c$) were estimated from the phonon density of states, and the coupling parameters for acoustic and optical phonons respectively, W_a and W_o, were treated as adjustable parameters. These five parameters are shown in Table 2. Although the fit to the experimental data was not too bad, there were systematic deviations at high temperatures, the origin of which are still unclear (16).

This system was originally analyzed by Moerner et al (47) using the model of weak coupling to acoustic phonons, Eq. 27, with $U = 0$ and $\Gamma_P(\omega)$ given by Eq. 16. (If $U_1 = 0$, and if W is very small, then for any reasonable model of vibration-phonon coupling U will be very small, and so the above procedure is consistent with the weak coupling assumption.) They analyzed the data with two adjustable parameters, T_D and W, and in fact found better agreement than our two-parameter (W_a and W_o) fit. However, this analysis is also somewhat unsatisfactory because their value of T_D does not correspond to either the heat capacity or the cutoff in the phonon density of states, and further, their value of W is not consistent with the weak coupling limit.

The second system shown in Table 2 allowed a much more satisfactory theoretical analysis. Since the experiments (48, 49a,b) are on pure naph-

Table 2 Parameters for the analysis of vibrational dephasing experiments

Impurity	Host	Transition	T_D (K)	W_a	$\tilde{v}_{op}$ (cm^{-1})	$\tilde{v}$ (cm^{-1})	W_o
ReO_4^-	KI	923 cm^{-1}	87	−0.048	109	13	0.011
naph.	naph.	766 cm^{-1}	72	0.011	123	17.5	0.026

thalene, following the discussion above, we set $U = 0$. Here again we assumed dephasing by both acoustic and optical phonons, and therefore calculated the dephasing rate from Eqs. 9, 12, 13, and 16–19. The parameters for the acoustic and optical phonon models were estimated from the phonon density of states, and the acoustic and optical coupling constants were treated as adjustable parameters (Table 2). The theoretical fit to the experimental data (48, 49a,b) was excellent (16). In this case we had an additional no-parameter test of the theory, since the thermal lineshift had also been measured as a function of temperature (48) and had been corrected for thermal expansion. Thus the calculation of $\Delta\omega$ from Eq. 10 with the same parameters agreed quite well with the experimental data (16). Although this does not prove that the model is correct, and indeed other explanations (49b, 50) have been advanced (16), it does provide a consistent theoretical interpretation.

A third experimental system, the v_3 transition of SF_6 in xenon, has also been analyzed with our theory (51). It was found that the dephasing above and below 60K was dominated by acoustic and pseudolocal phonons, respectively. Although these conclusions are probably correct, because of the large number of adjustable parameters (four), we do not consider this analysis to be a definitive test of the theory (16).

CONCLUSIONS

The theory that I have outlined above involves a relatively simple Hamiltonian for modeling electronic or vibrational transitions of molecules or ions in crystals, and it provides exact results for the pure dephasing contribution to linewidths and lineshifts. Thus this work extends previous perturbative and phenomenological results for the same problems. Our approach also has the flexibility to consider different phonon mechanisms for dephasing within the same theoretical framework. I have also discussed how the study of phonon sidebands can play an important role in determining dephasing mechanisms. In that regard, I have described a microscopic model for pseudolocal phonons that provides a direct and useful connection between dephasing and sideband spectroscopy.

It is ironic that despite the recent technological advances in short-pulse and narrow-band lasers and the concomitant interest in photon echo and hole-burning techniques, all of the optical dephasing data that we have analyzed were obtained from conventional absorption spectroscopy [although some of the vibrational dephasing data were obtained by hole-burning (47) or psec CARS (49a,b)]. We chose these particular experimental systems for analysis simply because they are able to provide the most unambiguous tests of theory, since auxiliary information can be obtained cleanly from the phonon sidebands. On the other hand, these absorption experiments cannot probe adequately low-temperature dynamics because of inhomogeneous broadening. In fact, there have been some ground-breaking nonlinear experiments performed in the last several years to investigate the low-temperature dephasing of optical transitions [see especially the reviews by Hesselink & Wiersma (19) and Fayer (21)]. Thus it would be particularly useful to have low-temperature photon echo data on systems that have simple sideband structure. This, coupled to high-temperature broadening studies in absorption spectroscopy, will provide us with a more stringent test of theory, and hence will enable us to learn a great deal more about dephasing in crystals.

What have we learned from the studies that have been performed so far? We have learned that, in general, electronic transitions are strongly coupled to phonons in the sense that second-order perturbation theory in the electron-phonon coupling is inadequate to describe dephasing, and also that simple harmonic phonon models are usually adequate for understanding experiments. We have also seen that, often, pseudolocal phonons play a major role in dephasing at low temperatures. This then supports the conjecture that for impurities in glasses, pseudolocal modes may also be the dominant dephasing mechanism (52, 53). Finally, we have seen that for vibrational dephasing, the vibration-phonon coupling is, in general, weaker than the electron-phonon coupling for optical transitions (compare the W values in Tables 1 and 2). This is, of course, quite reasonable, since one would expect that the perturbation to the phonons upon effecting a 1000 cm^{-1} vibrational transition would be substantially smaller than the perturbation for a 20,000 cm^{-1} electronic transition.

ACKNOWLEDGMENTS

The author thanks Dr. David Hsu for his invaluable contributions to the theory of dephasing over the last several years. The author also acknowledges support from the National Science Foundation (Grant No. DMR-86-03394 and the Presidential Young Investigator program), the Camille and Henry Dreyfus Foundation, and the Alfred P. Sloan Foundation.

Literature Cited

1. McCumber, D. E., Sturge, M. D. 1963. *J. Appl. Phys.* 34: 1682
2. Krivoglaz, M. A. 1964. *Fiz. Tverd. Tela* 6: 1707 (*Sov. Phys. Solid State* 6: 1340); 1965. *Zh. Eksp. Teor. Fiz.* 48: 310 (*Sov. Phys. JETP* 21: 204)
3. Jones, K. E., Zewail, A. H. 1978. In *Advances in Laser Chemistry*, ed. A. H. Zewail. New York: Springer
4. Halperin, B. 1985. *Chem. Phys.* 93: 39
5a. Diestler, D. J. 1976. *Chem. Phys. Lett.* 39: 39
5b. Diestler, D. J., Zewail, A: H. 1979. *J. Chem. Phys.* 71: 3103; 3113
6. Mukamel, S. 1980. *J. Chem. Phys.* 73: 5322
7. Osad'ko, I. S. 1979. *Usp. Fiz. Nauk* 128: 31 [*Sov. Phys. Usp.* 22: 311]

8. Osad'ko, I. S. 1983. In *Spectroscopy and Excitation Dynamics of Condensed Molecular Systems*, ed. V. M. Agranovich, R. M. Hochstrasser, p. 437. Amsterdam: North-Holland
9. Osad'ko, I. S., Zaitsev, N. N. 1985. *Chem. Phys. Lett.* 121: 209
10. Hsu, D., Skinner, J. L. 1984. *J. Chem. Phys.* 81: 1604
11. Hsu, D., Skinner, J. L. 1984. *J. Chem. Phys.* 81: 5471
12. Hsu, D., Skinner, J. L. 1985. *J. Chem. Phys.* 83: 2097
13. Hsu, D., Skinner, J. L. 1985. *J. Chem. Phys.* 83: 2107
14. Hsu, D., Skinner, J. L. 1987. *J. Chem. Phys.* 87: 54
15. Hsu, D., Skinner, J. L. 1987. *J. Lumin.* 37: 331
16. Skinner, J. L., Hsu, D. 1988. *Chem. Phys.* In press
17. Abram, I. I. 1977. *Chem. Phys.* 25: 87
18. Hizhnyakov, V. 1987. *J. Phys. C* 20: 6073
19. Hesselink, W. H., Wiersma, D. A. 1983. See Ref. 8, p. 249
20. Burns, M. J., Liu, W. K., Zewail, A. H. 1983. See Ref. 8, p. 301
21. Fayer, M. D. 1983. See Ref. 8, p. 185
22. Wiersma, D. A. 1981. *Adv. Chem. Phys.* 47.2: 421
23. Sapozhnikov, M. N. 1973. *Phys. Status Solidi B* 56: 391; 1976. *Phys. Status Solidi B* 75: 11
24. Dubost, H. 1984. In *Inert Gasses: Potentials, Dynamics and Energy Transfer in Doped Crystals*, ed. M. L. Klein. Berlin: Springer-Verlag
25. Skinner, J. L., Hsu, D. 1986. *Adv. Chem. Phys.* 65: 1
26. Skinner, J. L., Hsu, D. 1986. *J. Phys. Chem.* 90: 4931
27. Leggett, A. J., Chakravarty, S., Dorsey, A. T., Fisher, M. P. A., Garg, A., Zwerger, W. 1987. *Rev. Mod. Phys.* 59: 1
28. DiBartolo, B. 1968. *Optical Interactions in Solids*. New York: Wiley
29. Barker, A. S. Jr., Sievers, A. J. 1975. *Rev. Mod. Phys.* 47(Suppl. 2): S1
30. Böttger, H. 1983. *Principles of the Theory of Lattice Dynamics*. Weinheim: Physik
31a. Harris, C. B., Shelby, R. M., Cornelius, P. A. 1977. *Phys. Rev. Lett.* 38: 1415; 1978. *Chem. Phys. Lett.* 57: 8; Harris, C. B. 1977. *J. Chem. Phys.* 67: 5607; Shelby, R. M., Harris, C. B., Cornelius, P. A. 1979. *J. Chem. Phys.* 70: 34
31b. Marks, S., Cornelius, P. A., Harris, C. B. 1980. *J. Chem. Phys.* 73: 3069
32. Abbott, R. J., Oxtoby, D. W. 1979. *J. Chem. Phys.* 70: 4703
33. Wertheimer, R. K. 1980. *Chem. Phys.* 45: 415
34. Hiroike, E. 1981. *Chem. Phys. Lett.* 78: 323; 1982. *J. Phys. Soc. Jpn.* 51: 958; 1983. *Chem. Phys. Lett.* 103: 49
35. Arimitsu, T., Shibata, F. 1982. *J. Phys. Soc. Jpn.* 51: 1070
36. Gadzuk, J. W., Luntz, A. C. 1984. *Surf. Sci.* 144: 429
37a. Persson, B. N. J., Ryberg, R. 1985. *Phys. Rev. Lett.* 54: 2119; *Phys. Rev. B* 32: 3586
37b. Nitzan, A., Persson, B. N. J. 1985. *J. Chem. Phys.* 83: 5610
38. Kosloff, R., Rice, S. 1980. *J. Chem. Phys.* 72: 4591
39. deBree, P., Wiersma, D. A. 1979. *J. Chem. Phys.* 70: 790
40. Small, G. J. 1978. *Chem. Phys. Lett.* 57: 501
41. Powell, R. C., Xi, L., Gang, X., Quarles, G. J., Walling, J. C. 1985. *Phys. Rev. B* 32: 2788
42. Burke, F. P., Small, G. J. 1974. *J. Chem. Phys.* 61: 4588
43. Pack, D. W., McClure, D. S. 1987. *J. Chem. Phys.* 87: 5161
44. Korotaev, O. N., Kaliteevskii, M. Yu. 1980. *Zh. Eksp. Teor. Fiz.* 79: 439 [*Sov. Phys. JETP* 52: 220]
45. Dicker, A. I. M., Dobkowski, J., Völker, S. 1981. *Chem. Phys. Lett.* 84: 415
46. Molenkamp, L. W., Wiersma, D. A. 1984. *J. Chem. Phys.* 80: 3054
47. Moerner, W. E., Chraplyvy, A. R., Sievers, A. J. 1984. *Phys. Rev. B* 29: 6694
48. Hess, L. A., Prasad, P. N. 1980. *J. Chem. Phys.* 72: 573
49a. Dlott, D. D., Schosser, C. L., Chronister, E. L. 1982. *Chem. Phys. Lett.* 90: 386
49b. Schosser, C. L., Dlott, D. D. 1984. *J. Chem. Phys.* 80: 1394
50. Righini, R. 1984. *Chem. Phys.* 84: 97
51. Swanson, B. I., Jones, L. H., Ekberg, S. A., Fry, H. A. 1986. *Chem. Phys. Lett.* 126: 455
52. Jackson, B., Silbey, R. 1983. *Chem. Phys.* 99: 331
53. Walsh, C. A., Berg, M., Narasimhan, L. R., Fayer, M. D. 1987. *J. Chem. Phys.* 86: 77

DISSOCIATIVE CHEMISORPTION: DYNAMICS AND MECHANISMS

Sylvia T. Ceyer

Department of Chemistry, Massachusetts Institute of Technology, Cambridge, Massachusetts 02139

INTRODUCTION

The last 20 years have witnessed a blossoming of knowledge about the structure and stability of molecules and reaction intermediates adsorbed on single crystal surfaces. From Auger electron spectroscopy and low energy electron diffraction to high resolution electron energy loss spectroscopy and scanning tunneling microscopy, the development and refinement of new techniques has facilitated the acquisition of an immense data base for the chemical identity of the adsorbate, its binding site, structure, and reactivity. With this storehouse of information as a solid foundation, the time is now ripe for asking questions not only about the structure and stability of adsorbates but also about their dynamics and mechanisms.

For example, consider the collision of a gas phase molecule with a surface and its subsequent adsorption on this surface. Five years ago, there was little appreciation for the dynamics of this initial collision and the mechanism by which the incident molecule finds a chemisorption site. In particular, there was scant understanding of the dominant features of the molecule-surface potential energy hypersurface and how they dictate the dynamics and mechanism of both dissociative and molecular chemisorption. This state of affairs obtained because our knowledge of dynamics and mechanism was derived from measurements of adsorption kinetics (1–3) such as the behavior of the adsorption probability as a function of surface temperature and coverage. Although these types of studies are an essential first step, the information derived from them is averaged over a limited area of the potential energy surface and is insufficient for developing a unique picture of dynamics and mechanism. Lacking was a sensitive and specific probe of the major features of the potential surface. One such probe is the effect of a molecule's incident energy on adsorption. Such experiments were not executed because it was commonly believed, despite some early and elegant experiments to the contrary (4, 5), that the surface was the all-important source of energy in a molecule-surface interaction. The majority of surface studies had been undertaken after adsorption of gas molecules from the ambient background where molecules strike the surface from all directions and with thermally distributed energies. It is correct that the surface is the sole energy source when the molecule is bound on the surface. However, in the adsorption process, especially those adsorptions that involve dissociation, the interaction time during which the molecule collides with the surface is usually too short for substantial amounts of energy to be absorbed by the molecule and transferred into the modes that lead to dissociation. Therefore, the energy of the incoming adsorbate can be the decisive factor in the mechanism and dynamics for chemisorption. It follows that, in general, all the initial conditions of the incident adsorbate such as its vibrational phase and orientation relative to the atomic structure of the surface also play a major role in the interaction. Given the sensitivity of the reaction probability to the energy and orientation of two molecules colliding in the gas phase, these assertions come as no surprise.

Another notion prevalent in earlier surface studies is that the mechanism and dynamics of surface reactions involved only molecules adsorbed on the surface. Support for this view stems from the *lack* of clear evidence for the direct participation of gas phase molecules in surface reactions. For example, the Eley-Rideal mechanism (6, 7) whereby a gas phase molecule is incident on and reacts directly with an adsorbed molecule has been discussed for over 50 years. Despite the longevity of this proposed mechanism, there is no unambiguous evidence for it. Yet, it is a very reasonable mechanism. The translational and internal energy of an incident gas phase molecule is another source of energy to activate the reaction with the adsorbate or to activate the adsorbate's dissociation or desorption. As with the collision of an incoming adsorbate with the bare surface, the collision energy can also be the decisive factor in the mechanism and dynamics of interaction with an adsorbed species and can also be exploited as a powerful probe of the adsorption dynamics.

The purpose of this discussion is to illustrate how experiments carried out in the last five years have succeeded in establishing the collision energy in its rightful and major role in molecule-surface interactions, especially reactive ones (8, 9). Many of these studies have confirmed hypotheses of the presence of certain dominant features of the potential hypersurface that dictate the chemisorption dynamics. The results of other studies suggest that the fine details of the shape of the interaction potential are also important in chemisorption. The translational and internal energy and the angle of incidence of the incoming molecules are employed as probes of these features. Molecular beam techniques are predominantly used to vary these parameters while measuring the probability for chemisorption of the incident molecule or of an adsorbed species hit by the incident molecule. Molecular beam techniques are usually coupled with electron spectroscopies to monitor the result of the adsorption event.

With one exception, this discussion is limited to molecule-surface interactions that involve dissociative chemisorption. The importance of this new knowledge about the microscopic details of the interaction for understanding macroscopic phenomena in heterogeneous catalysis is emphasized. The unexpected bonus from these studies is the recognition of a synthetic tool for production of novel adsorbates. These results are also described. Another focus of this discussion is the recent investigations using molecular beam techniques to identify unambiguously an adsorbate's dissociation and desorption induced by the collision with a gas phase species. These observations represent the discovery of new kinds of mechanisms for dissociative chemisorption and desorption called *collision-induced dissociation and desorption*. A comparison of the dynamics of collision-induced dissociation to that of established mechanisms for dissociation is presented in detail.

EXPERIMENTAL TECHNIQUE

The collision energy of a molecule with a surface is an effective probe of the dominant features of the interaction potential that dictates the dissociative chemisorption dynamics. Molecular beam techniques allow the translational energy and the direction of the incoming adsorbate to be varied over a wide range and the vibrational energy over a limited range. The detector of the result of the chemisorption event must be capable of chemically identifying the adsorbate and quantitatively measuring its coverage with high sensitivity. These requirements are met in an experimental arrangement that couples molecular beam techniques with electron spectroscopies (9). Electron spectroscopies, especially those with vibrational resolution, provide the most definitive and sensitive detection of the result of the chemisorption event.

The molecular beam-ultrahigh vacuum apparatus shown in Figure 1 is specifically designed for measurements of the adsorption probability as a function of the energy of the incident molecules (10, 11). The stainless steel ultrahigh vacuum chamber houses a quadrupole mass spectrometer, a single pass cylindrical mirror electrostatic energy analyzer (CMA), a low

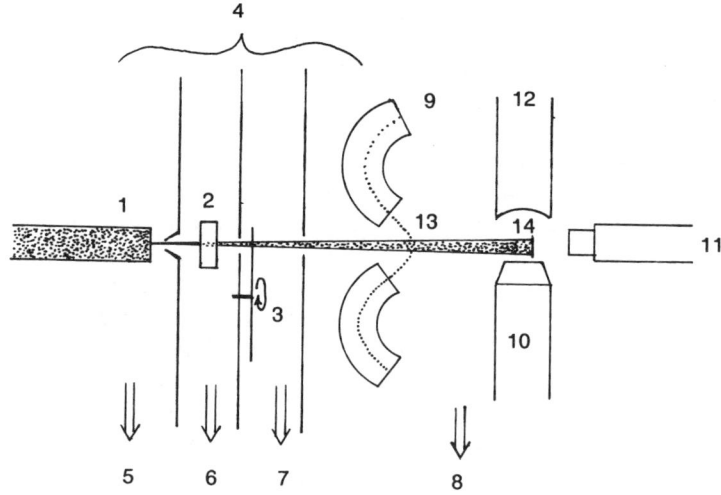

Figure 1 Schematic drawing of a molecular beam-UHV apparatus: 1. nozzle; 2. electronic shutter; 3. chopper; 4. differential pumping stages; 5. to 10" diffusion pump; 6. to 6" diffusion pump; 7. to 4" diffusion pump; 8. to liquid nitrogen trapped 10" diffusion pump; 9. HREEL spectrometer; 10. single pass CMA Auger spectrometer; 11. quadrupole mass spectrometer; 12. LEED apparatus; 13, 14. possible crystal positions.

energy electron diffraction apparatus, a 127° cylindrical deflector high resolution electron energy loss spectrometer (HREELS), a crystal manipulator, an ion sputter gun, and an Alpert-Bayard type nude ion gauge. The main chamber is precisely coupled to the beam source chamber through two differential pumping chambers so that the molecular beam is aligned exactly along the axis of the main chamber.

A supersonic molecular beam source, mounted in the source chamber and triply differentially pumped, is used as the monoenergetic source of incident molecules. The beam passes through a skimmer and through a series of circular apertures mounted in the walls of the differential stages that collimate the beam to about 10^{-4} sr. It then continues directly through the center of the high resolution electron energy loss spectrometer, in front of the cylindrical mirror analyzer Auger spectrometer and LEED apparatus and directly into the line-of-sight of the quadrupole mass spectrometer. The translational temperature and the collision energy of the beam are measured by a time-of-flight technique that utilizes a slotted disk chopper in the second differential stage. An electronic shutter mounted in the first differential chamber controls the exposure time of the beam on the surface.

The surface is mounted on a manipulator that allows the sample to be cooled to 8K, heated, rotated 360° around an axis parallel to its surface, and translated in three mutually perpendicular directions. The entire manipulator is mounted eccentrically on a 35.6 cm diameter differentially pumped, Teflon® sealed, rotary lid. Rotation of this lid by 180° with a chain and sprocket arrangement translates the crystal horizontally a distance of 19 cm from the center of the EEL spectrometer to a position in front of the CMA Auger spectrometer. A 0.5 mm thick sheet of Conetic alloy, which shields the electron optics from the earth's magnetic field, lines the inside of the UHV chamber. The resultant field at the position of the EEL spectrometer is ~30 mG. The main chamber is pumped by a liquid nitrogen trapped 10-inch diffusion pump containing polyphenyl ether oil. The operating base pressure is 4×10^{-11} torr.

In these experiments, the key measurement is the probability of dissociative chemisorption. This probability is simply the ratio of the number of molecules dissociatively chemisorbed on the surface to the total number of molecules incident on the surface. The latter is evaluated from the absolute beam flux, which is equal to the leak rate of the beam into the main chamber divided by the cross-sectional area of the beam on the crystal. The leak rate of the beam is determined by the total pressure rise (as measured by an ionization gauge) in the main chamber with the beam on times the experimentally determined pumping speed in the main chamber.

The number of dissociatively chemisorbed molecules is determined from the absolute coverage measured by an electron spectroscopy. Auger electron spectroscopy is used as quantitative detector of the result of an adsorption event when elemental analysis is sufficient to distinguish between adsorbates. The Auger signal is easily calibrated for absolute coverage by noting its intensity at a coverage that results in a LEED pattern. However, elemental analysis afforded by Auger measurements is often insufficient to identify chemically an adsorbed species. Therefore, high resolution electron energy loss spectroscopy (HREELS), a vibrational spectroscopy, is employed as a detector. This spectroscopy allows the species produced from dissociative chemisorption to be differentiated from molecularly chemisorbed species. Novel species produced as a result of extremes in incident energy can be identified. Thermal desorption spectroscopy, which employs a quadrupole mass spectrometer as a detector, may also be used as a quantitative probe of the adsorbate coverage. The use of thermal desorption as a probe of the chemical identity of the adsorbate is predicated on the lack of dissociation of the adsorbate during the temperature ramp and the absence of desorption of two distinct chemical species with the same mass.

The lowest dissociation probability detectable is determined by the length of exposure of the beam on the crystal. For a beam with an effective pressure of 10^{-5} torr, this dissociation probability is 10^{-6}. The accuracy of the absolute values of the dissociative chemisorption probabilities is limited to a factor of 2–3 due to inaccuracies in beam flux and coverage calibration procedures.

MECHANISMS FOR DISSOCIATIVE CHEMISORPTION: DIRECT VS PRECURSOR

Two kinds of mechanisms, the direct and precursor state models, are used to describe the dissociation of a molecule on a surface. In direct dissociative chemisorption, the incident molecule dissociates into adsorbed fragments immediately upon collision with the surface. In dissociative chemisorption through a precursor state, the molecule is adsorbed intact for as few as a single or very many vibrational periods before dissociating. Trapping of the molecule in a molecular chemisorption or physisorption state that serves as a precursor state occurs as a result of the loss of the molecule's incident kinetic energy to the surface upon its collision. The molecule remains in this state until enough energy is supplied from the solid to reorient it into an energetically favorable configuration for dissociation and/or for it to diffuse onto an energetically favorable site for dissociation whereupon it dissociates. Alternatively, the molecule remains in this state until enough energy is supplied from the solid for it to desorb without dissociating. Which mechanism for dissociation the molecule follows is dictated by the potential energy hypersurface of the interaction. The goal of recent molecular beam experiments in which the collision energy and incident angle are varied is to determine the dominant features of the interaction potential that distinguish these mechanisms.

Direct Dissociative Chemisorption

Molecule-surface interactions in which only direct dissociative chemisorption has been observed to be operative can all be classified as activated dissociative chemisorption processes. This term refers to the presence of a potential barrier above the zero of energy and along the dissociative reaction coordinate. The zero of energy corresponds to the energy of the system when the molecule is infinitely far from the surface. When the translational energy of the incident molecule is greater than the barrier height, the molecule dissociates and its fragments adsorb directly upon impact with the surface. At incident translational energies below the barrier energy, the molecule simply reflects from the surface. The temperature of the surface has no effect on the probability of dissociative chemisorption.

The presence of a potential barrier to dissociative chemisorption was first discussed by Lennard-Jones almost 60 years ago (12). One general origin (13, 14) of the barrier is the following: as a molecule approaches a surface, its long-range attractive van der Waals interaction with the surface gives way to a short-range repulsive interaction. Unlike the case of the interaction with an inert gas atom, the repulsion does not continue to grow at shorter distances. Instead, because of filling of the antibonding orbitals of the approaching molecule, the interaction first becomes less repulsive and then turns strongly attractive. This inner, deeply attractive well accounts for the dissociative state. The potential maximum between the outer and inner well forms the barrier between the molecular adsorption state and the dissociative chemisorption state. Another view considers the barrier to arise from the repulsion of two-orbital, four electron interactions (15, 16). As a molecule approaches the surface, the repulsion continues to grow as the filled antibonding component is pushed up in energy. However, as the antibonding component passes through the Fermi level, the electrons are dumped into the empty metal bonding levels. The interaction, which suddenly turns attractive, results in a maximum or a barrier along the potential energy surface.

This discussion and the majority of previous discussions have considered the barrier to be one-dimensional (17) where the potential energy depends only on the distance between the molecule and the macroscopic surface. Such a model predicts the probability for dissociative chemisorption to depend solely on the molecule's incident energy in the normal direction. This type of energy scaling has become known as a "normal energy scaling." The energy in the normal direction is given by $E_n = E_i \cos^2 \theta_i$, where E_i is the incident total energy and θ_i is the angle of incidence measured from the surface normal. Although this one-dimensional picture provides a convenient framework in which to discuss these processes, it is a gross over-simplification of the actual potential energy surface, even in cases where normal energy scaling is observed. The true potential energy surface is multidimensional and depends on the point of impact of the molecule on the surface lattice, the orientation of the molecule, the coordinates of the atoms forming and breaking bonds, and the coordinates of the spectator atoms. The multidimensional nature therefore gives rise to the distribution of barrier energies that is observed experimentally.

ACTIVATED DISSOCIATIVE CHEMISORPTION OF CH₄ ON Ni[111]   The dynamics of the activated dissociative chemisorption of CH_4 are the most studied and best understood of all direct dissociative chemisorption systems. Methane dissociative chemisorption on Ni[111] (8, 18–20), Ni[100] (21), W[110] (22, 23) and Ir[110] (24) has been investigated recently using molecular beam techniques. Earlier work includes the measurements of the dissociation of methane on polycrystalline tungsten (25) and rhodium (26) and Rh[111] (27). An investigation into the effect of translational and vibrational energy on the dissociative chemisorption of CH_4 on Ni[111] is discussed in the next section to illustrate the kinds of detailed dynamical and mechanistic information that are attainable, the implications of this information for the pressure gap in catalytic reactivity, and the development of a synthetic tool.

Mechanism and dynamics of the dissociation The probability for dissociative chemisorption of CH_4 on Ni[111] has been measured as a function of the translational energy of the incident CH_4 molecule (18-20) and is shown in Figure 2. Below 12 kcal mol^{-1}, there is no dissociative chemisorption above the sensitivity limit for carbon detection, but as the translational energy is increased to 17 kcal mol^{-1}, the dissociation probability increases exponentially by two orders of magnitude. The dissociation probabilities shown in Figure 2 were measured with the molecular beam incident at the normal angle to the crystal. Dissociation probabilities measured at other angles of incidence and plotted versus the translational energy in the normal direction fall on the straight line in Figure 2. Therefore, the dissociation probability correlates not with the total energy of the incident methane molecule but with the kinetic energy only in the normal direction. The temperature of the surface has no effect on the dissociation probability. These results show that there is a barrier to the direct dissociative chemisorption of CH_4 which only the translational energy in the normal direction is effective in overcoming.

The effectiveness of vibrational energy in surmounting the barrier was also investigated. The vibrational energy is varied over a limited range by using molecular beam techniques. Because significant vibrational relaxation does not occur in the beam expansion, the vibrational energy distribution of the expanded CH_4 is characterized by the temperature of the nozzle and is varied by changing the nozzle temperature. The translational energy is maintained at a constant value by adjusting the ratio of CH_4 to the carrier gas to compensate for the change in nozzle temperature. Measurements of the dissociation probability as a function of nozzle temperature for a fixed translational energy indicate that vibrational energy, most of it concentrated in v_4 (umbrella mode) and v_2 (bending mode), is at least as effective as translational energy if not slightly more effective.

These observations are assimilated into a coherent model for the physical origin of the barrier to dissociation and how translational and vibrational excitation allows the molecule to overcome it. The telling clue is the

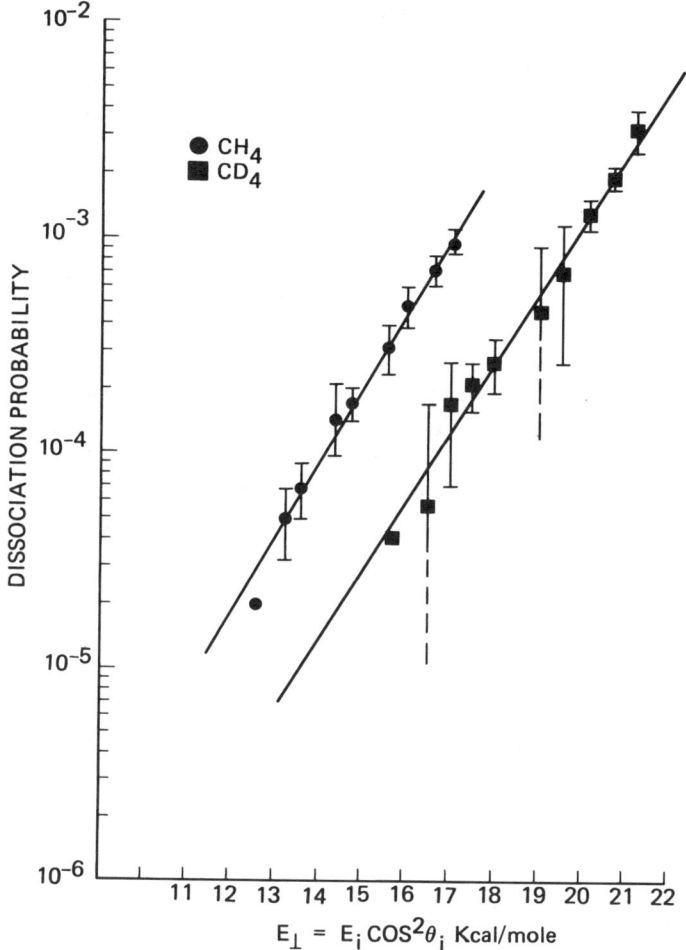

Figure 2 The absolute dissociation probability of CH_4 and CD_4 as a function of the normal component of translational energy. The lines are linear least squares fit to the data and the error bars are 95% confidence limits of a series of 6–8 measurements for CH_4 and of a series of three measurements for CD_4.

equivalent effectiveness of the normal component of the translational energy and of the vibrational energy, which suggests that translational and vibrational excitation lead to the same motion of the nuclei passing over the barrier along the reaction coordinate. This is possible if the role of translational energy is the distortion or deformation of the CH_4 molecule upon impact with the repulsive wall of the surface. The deformation that occurs upon impact is essentially a conversion of translational energy into vibrational motion associated with the v_2 and v_4 vibrational modes in CH_4. These are also the modes excited to the largest degree in the vibrational excitation measurements. Therefore, since both kinds of excitation lead to the identical motion of the nuclei in close proximity to the surface, the effectiveness of translational and vibrational energy are equivalent. However, only the normal component of the translational energy promotes dissociative chemisorption, because for a spherical molecule like CH_4 incident on a flat surface, deformation can arise only from the motion perpendicular to the surface. The parallel velocity component of the molecule's motion encounters no repulsive interaction from which conversion of translational to vibrational energy can take place.

The fact that the methane molecule must be distorted from its tetrahedral configuration in order for dissociation to occur is reasonable upon consideration of the energetic requirements for C–H bond cleavage. The C–H bond energy in methane is about 100 kcal mol^{-1}. In order for there to be sufficient energy release to break the C–H bond, a Ni–H bond, which is worth about 63 kcal mol^{-1}, and a Ni–C bond, worth approximately 40 kcal mol^{-1}, must be formed. To a flat Ni surface with no protrusions, the methane molecule appears spherical; that is, the interaction between the flat surface and the methane molecule is isotropic (28). The hydrogen atoms effectively bury the carbon atom, preventing a strong attractive interaction between the Ni atoms of the surface and the carbon atom. A slowly moving methane molecule incident on the surface interacts primarily through the shielding hydrogen atoms. No C–H bond cleavage occurs because the carbon atom cannot move close enough to the Ni surface to interact strongly. However, as the translational energy of the incident molecule is increased, the methane molecule begins to suffer substantial deformation upon collision with the repulsive wall of the surface due to its increasing impact. This deformation serves to push the hydrogen atoms out from between the surface and the carbon atom in the same way as vibrational excitation of the v_4 and v_2 modes, thereby exposing the carbon atom to the Ni surface. In this configuration, both a Ni–C bond and a Ni–H bond can be formed, breaking a C–H bond.

This picture for the effect of translational and vibrational energy on the probability of dissociative chemisorption depicts the barrier as the energy required to deform the molecule into the proper configuration for the transition state. If the deformation energy is the sole contribution to the barrier, then, in principle, the height of the barrier for each incident configuration of the molecule can be calculated from the known force field of the methane molecule (29, 30). This calculation requires that some assumptions be made about the unknown geometry or geometries of the transition state. A pyramidal configuration for the transition state is possible for a methane molecule incident on the surface with three of its hydrogen atoms pointing toward the surface. When this molecule hits the surface, the three hydrogen atoms move into one plane with the carbon atom, thereby clearly exposing the carbon atom to the surface. The energy required to distort the tetrahedral methane molecule into this pyramidal configuration is 16.3 kcal/mole calculated from the harmonic force field. The absolute value of this calculated deformation energy is significant because it lies in the range of the translational energies over which probabilities for dissociative chemisorption are large. The exponential dependence of the dissociation probability on energy can also be obtained if the dissociation probability is considered to be proportional to the degree of deformation of the v_4 mode. These observations along with recent molecular orbital calculations (31) lend support to the idea that translational energy leads to deformation of the molecule and that the pyramidal configuration of the molecule is the geometry of the transition state with the lowest energy that leads to dissociation. Therefore, the energy required to distort the molecule is a major part of the barrier to dissociative chemisorption.

However, deformation of the molecule upon impact is only part of the mechanism for dissociation. The remainder of the picture comes into view upon comparison of the dissociation probabilities of CH_4 to CD_4, also shown in Figure 2. The same exponential dependence of the dissociation probability on incident energy is observed, but the dissociation probabilities for CD_4 are consistently about an order of magnitude below those of CH_4. This is a large difference in the dissociation probabilities compared to the classical kinetic isotope effect of 4.8 expected at a nozzle temperature of 640K, the lowest temperature used in the expansion. This value is calculated based on the assumption that the transition state is essentially an adsorbed CH_3 species and that its vibrational frequencies are essentially those determined in this study for the adsorbed CH_3 product of the dissociative chemisorption event. Therefore, the magnitude of this isotope effect cannot be explained solely by the difference in zero point energies. Rather, quantum mechanical tunneling of the light hydrogen atom is indicated. This is a reasonable mechanism because in order for the C–H bond to break, there must be motion of the nuclei along the C–H coordinate once the methane molecule is sufficiently deformed. The light hydrogen atom tunnels through the remaining barrier along the C–H coordinate before the molecule is deformed completely. The dissociation probability then is directly proportional to the tunneling probability. The difference in the tunneling probability between the light H atom and the heavy D atom can easily explain the order of magnitude difference in the dissociation probabilities. In fact, the dissociative chemisorption of methane on tungsten has previously been discussed and the exponential dependence explained solely in terms of a tunneling mechanism (22, 25). However, a tunneling mechanism coupled with deformation of the molecule upon impact provides a more consistent explanation for the observation that translational and vibrational excitation induce the identical motion of the nuclei over the barrier along the reaction coordinate.

The process by which the dissociative chemisorption of CH_4 occurs as indicated by these results is readily summarized. As the methane molecule approaches the surface more closely than the equilibrium position for molecularly adsorbed methane, it experiences a repulsive interaction that can be overcome by increasing the normal component of the translational energy or the vibrational energy. This increased translational or vibrational energy results in deformation of the molecule, thereby moving the hydrogen atoms out of the way of the Ni–C attractive interaction. However, the molecule does not need to be deformed completely because at some distance the attractive part of the potential, through interaction with the antibonding orbitals, tempers the barrier so that the barrier becomes sufficiently narrow for the light hydrogen atom to tunnel through to the product regime. This picture is supported by very similar effects observed in a study of the dissociative chemisorption of CH_4 on W[110] (22, 23). Vibrational energy was also observed to be at least as effective as translational energy. This fact, plus the similar range of translational energies over which significant dissociation of CH_4 occurs on W[110] and the similar effectiveness of translational energy, suggests that this deformation model is quite general. This is expected if the main contribution to the barrier is the energy required to distort the molecule, as suggested by this model. The chemical nature of the metal or, more specifically, the strength of the attractive part of the potential may play a secondary role in the energy requirements for dissociation. If the attractive part of the potential is relegated to a secondary role, then it is also clear why previous attempts to promote dissociative chemisorption of CH_4 on Rh by direct laser excitation of the vibrational modes were not successful (27). Two quanta of v_4 excitation (7.5 kcal mol^{-1}) do not provide enough energy to deform the molecule to the point where dissociation is detectable. This model for the dynamics of the dissociative chemisorption process that couples deformation with tunneling provides the most consistent explanation for the origin of the approximate equal effectiveness of translational and vibrational energy toward dissociation, the exponential dependence of the dissociation probability on energy, and the large kinetic isotope effect.

Barriers: an origin of the pressure gap in heterogeneous catalytic reactivity The above study provides not only a detailed understanding of the interaction between a methane molecule and a Ni surface but also a means to understand a macroscopic phenomenon known as the *pressure gap* in catalysis (32, 33). The pressure gap refers to the absence of reactivity at the lower adsorbate pressures readily attainable in ultrahigh vacuum systems. In short, the catalytic activity observed under high pressure conditions common in industrial heterogeneous processes is often absent in the low pressure, laboratory surface science experiments.

This phenomenon is manifested in the dissociative chemisorption of methane. Methane is observed not to adsorb dissociatively when the CH_4 is introduced above the Ni surface as a room temperature gas at a pressure below 10^{-4} torr. The reason for this lack of dissociative adsorption is the very low dissociative adsorption probability of CH_4 on Ni. Several studies (34–36) have reported a dissociative adsorption probability for CH_4 on a Ni catalyst or Ni single crystal of less than 10^{-9}. This low probability for dissociative chemisorption makes it very difficult to observe the dissociatively chemisorbed products.

However, as the pressure of methane above the surface is increased, the flux of CH_4 molecules incident on the surface is increased. The higher fluxes can overcome the low probability of a dissociative chemisorption event. Therefore, dissociative chemisorption appears to occur only at higher adsorbate pressures. Based on the results of the above experiment, the explanation for the low probability for dissociative chemisorption, and hence for the effect of pressure on the observation of dissociative chemisorption, is the presence of a barrier along the dissociative reaction coordinate. A potential barrier above the zero of energy can be overcome by those molecules incident on the surface with energies greater than the barrier height. If the barrier is sufficiently high such that a very small fraction of molecules incident with thermally distributed energies has the requisite energy, then the rate of dissociative chemisorption will be very slow. An increase in the pressure simply increases the absolute number of molecules with energies sufficient to overcome the barrier, thereby increasing the rate and allowing the products of dissociative chemisorption to be observed.

This conclusion is corroborated in the following way. Thermal rate constants were calculated from the convolution of the low pressure dissociation probabilities as a function of energy (18–20) with a Maxwell-Boltzmann distribution of energies. These rate constants were then compared to those measured for methane decomposition on a Ni[111] crystal under high pressure conditions as a function of the temperature of the system (37). The temperature of the system refers to the equivalence of the gas and surface temperatures. Excellent agreement is found to within a factor of two to three between the rate constants measured at high pressure

(37) and those calculated from the low pressure dissociation probability measurements (18–20). This result clearly establishes the barrier to the dissociative chemisorption of CH_4 as the origin of the pressure gap in the reactivity. The presence of this barrier along the dissociative reaction coordinate establishes a link between experiments carried out in ultrahigh vacuum environments where the adsorbate pressures are $<10^{-5}$ torr and heterogeneous catalysis where the reactant pressures may be as high as hundreds of atmospheres. This implies that for catalytic reactions that have a reasonable rate only at high pressures because the dissociative chemisorption of one or both of the reactants is activated, the high pressure requirement can now be bypassed by increasing the energy of the incident molecule. Establishment of this link means that high pressure reactions can now be carried out at low pressure, where the entire arsenal of surface science techniques is available to study these practically important catalytic reactions.

Synthesis of adsorbed methyl radicals A complete picture of the dissociative chemisorption of CH_4 on Ni[111] requires that the product of the dissociative chemisorption event be spectroscopically identified. This is accomplished by measuring the high resolution electron energy loss spectrum, shown in Figure 3, of methane after deposition on the surface at 140K with a translational energy of 17 kcal mol^{-1}. The crystal temperature is maintained at a low value in order to trap the nascent product of the dissociative chemisorption event rather than a species produced by thermal decomposition of the nascent product. From the absolute exposure and dissociation probability for CH_4, the carbon coverage is calculated to be 0.04 monolayer. By analogy to the vibrational frequencies of metal alkyls (38) and to the intensities of the loss features for NH_3 adsorbed on Ni[111] (39), the products of the dissociation are identified as an adsorbed methyl radical and adsorbed H atom (18–20). The feature in the loss spectrum observed at 370 cm^{-1} is assigned to the Ni–CH_3 stretch, the feature at 1220 cm^{-1} to the –CH_3 symmetric deformation mode, and the feature at 2660 cm^{-1} to a "soft" C–H stretching mode. The –CH_3 rocking and –CH_3 degenerate deformation modes are nondipole allowed transitions observed in spectra measured away from the specular angle.

Adsorbed methyl radicals have long been invoked as reaction intermediates in a wide variety of hydrocarbon-surface reactions carried out both in an ultrahigh vacuum environment and under high pressure industrial conditions. Despite their importance as proposed reaction intermediates, this result represents the first time that adsorbed methyl radicals have been produced cleanly on a single crystal metal surface and unam-

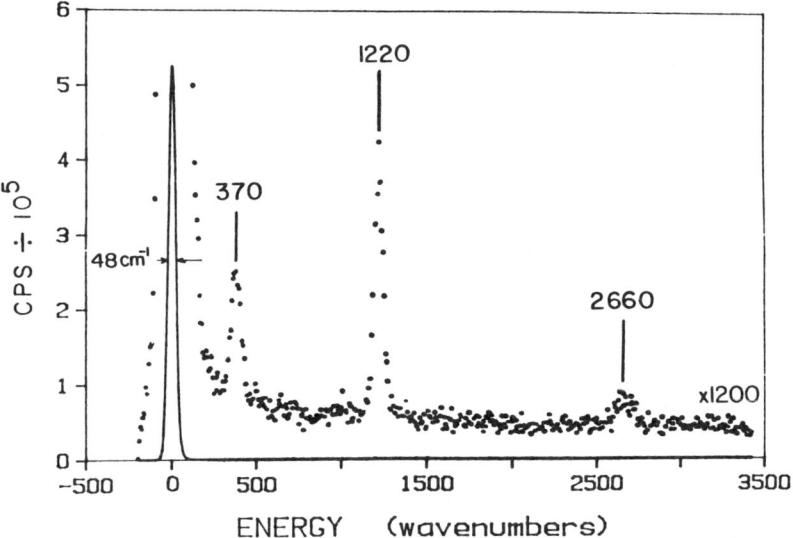

Figure 3 HREEL spectrum of CH_3 adsorbed on Ni[111] measured at the specular angle. The methyl radical is synthesized by deposition of CH_4 at a normal incidence with an energy of 17 kcal/mole on a Ni[111] surface at 140K. The loss feature at 370 cm^{-1} is assigned to a Ni–CH_3 stretch, 1220 cm^{-1} to a –CH_3 symmetric deformation mode, and 2660 cm^{-1} to a convolution of the symmetric and asymmetric C–H stretching modes.

biguously identified by vibrational spectroscopy. The reason for the difficulty in methyl radical production is that there is no simple way to produce them. The commonly studied adsorbates, C_2H_2 and C_2H_4, which dissociate readily under UHV conditions, have not been observed to produce methyl radicals on metallic surfaces. Methane and ethane, natural candidates for the clean production of methyl radicals, are completely unreactive with most metal surfaces under low pressure conditions of the adsorbing gas. However, with the work described above, it is now clear both why methane is unreactive at low pressures and how to activate it. The bonus derived from the experiments probing the dynamics of methane decomposition is that the products of the activated dissociative chemisorption event are an adsorbed CH_3 radical and an adsorbed H atom. The methyl radical is a nonequilibrium species that cannot be produced thermally by adsorption of a parent species at high surface temperatures because some process, usually desorption, competes more effectively. For example, desorption of molecular methane from Ni[111] occurs at much lower surface temperatures than those necessary to cleave the C–H bond. This difficulty can be overcome with molecular beam techniques that provide the necessary energy to produce adsorbed methyl radicals. The variability of the collision energy afforded by molecular beams makes them a tool with which novel adsorbates can be synthesized.

With adsorbed CH_3 species conveniently and cleanly synthesized, their stability and reactivity has been probed by monitoring the vibrational spectrum as a function of surface temperature (19, 40). The methyl radicals adsorbed on Ni[111] are stable below surface temperatures of 150K. Above this temperature, the methyl radicals dissociate into CH_2 species that recombine and dissociate at 230K into C_2H_2, a rehybridized form of acetylene. At temperatures above 450K, the C_2H_2 species dissociates into a CH species. The CH species are stable up to 700K, at which point they dissociate and dissolve into the bulk nickel.

OTHER SYSTEMS The direct mechanism for dissociative chemisorption is followed in several other molecule-surface interactions. These interactions include the dissociation of H_2 on Ni[100] (41, 42) and on Ni[111] (42–45), the dissociation of ethane, propane, and butane on Ni[100] (21), and ethane on Ir[110] (46). A few specific examples are discussed in detail in the following sections.

Normal energy scaling—CO_2 on Ni[100] Normal kinetic energy scaling of the probability for dissociative chemisorption has been established for the dissociation of CO_2 into CO and O adsorbed on Ni[100] (47). The probability for dissociation increases by three orders of magnitude as the translational energy in the normal direction is increased from 2 to 25 kcal mol^{-1}. One or more quanta of vibrational excitation in the bending mode is also observed to result in a significantly higher dissociation probability. The authors conclude that the configuration of the transition state for dissociation is bent in a similar fashion to the bent transition state known for the recombinative desorption of CO and O on Pt[111] (48–51). As in the case of the activation of CH_4 on Ni[111] (20), it is likely that the role of translational and vibrational energy is to deform CO_2 molecules into the proper bent configuration for the transition state that leads to dissociative chemisorption.

Total energy scaling—N_2 on W[110] The dynamics of the dissociative chemisorption of N_2 on W[110] are unique (52–55). The probability for dissociation scales with the total translational energy of the incident N_2 molecule. That is, the dissociation probability is independent of the incident angle. Total energy scaling means that the energy in the parallel direction is equally efficacious in overcoming the potential barrier to chemisorption as the energy in the normal direction. Two explanations have been offered for this observation. If the N_2 molecule collides with the surface in a region of a highly corrugated interaction potential, then its energy or momentum in the parallel and perpendicular directions readily mix upon the initial collision, leading to the equivalent efficacy in promoting dissociation. Such a strong corrugation is suggested by a "chemical" interaction of N_2 with W[110] where the binding energies vary greatly across the unit cell, yielding a highly corrugated potential.

Alternatively, a N_2^- negative ion state has been suggested (56, 56a) as an intermediate state to dissociation. In this case, the affinity level of the N_2 molecule falls below the Fermi level at some critical distance. From this distance inward, the electron can tunnel onto the incoming molecule to form a negative ion without requiring any activation. The negative ion state is coupled to the dissociative reaction coordinate but a potential barrier lies along the pathway to dissociation. It is this barrier that is overcome by translational activation. However, the negative ion state allows the energy in the parallel direction to be mixed with normal kinetic energy via vibrational excitation of the N_2. Therefore, it is total translational energy of the N_2 that is effective in overcoming the barrier. This mechanism requires that the coupling between the components and the subsequent dissociation occur before the translational energy is dissipated to phonons. This in turn requires that the potential energy surface for the ion-surface interaction be highly corrugated.

Translational energy also affects the saturation coverage of atomic nitrogen. Nitrogen atom saturation coverage increases from 0.25 monolayer at 2 kcal mol^{-1} to 0.5 monolayer at 23 kcal mol^{-1}. New features are observed in the thermal desorption spectra as the result of the higher coverage attainable with higher translational energies. These findings are consistent with direct dissociative chemisorption where the higher potential barrier for dissociation at high coverage can be overcome with higher translational energies. This behavior indicates that the effective number of sites for chemisorption is energy dependent.

Lack of translational activation of CO on Ni[111] and Ni[100] These systems are included as examples of dissociations that are not activated by translational energy. In the case of CO on Ni[111] (57), translational and vibrational energies as high as 45 and 18 kcal mol^{-1}, respectively, do not result in detectable dissociation. The minimum detectable dissociation probabilities are 2×10^{-6} and 9×10^{-4} at the maximum translational and vibrational excitation, respectively. No increase in the probability for CO dissociation on Ni[100] (58) was apparent in measurements carried out with translational energies up to 20 kcal mol^{-1}. Like the dissociative chemisorption of CH_4 on Ni[111], CO dissociation on Ni[111] is not observed at pressures attainable in a UHV system, but dissociation is observed at high pressure. It is concluded that translational energies greater than 45 kcal mole^{-1} do not contribute to the CO dissociation rate at high pressures. Rather, the highly anisotropic potential energy surface for the CO–Ni[111] interaction likely mandates vibrational excitation greater than the amount that could be achieved in this experiment for activation of the C≡O bond.

Centrifugal mechanism—I_2 on MgO   Measurements and analysis of the dissociation probability of I_2 on a MgO surface as a function of collision energy up to 8 eV indicate that the dissociation takes place solely through the impulsiveness of the collision (59–61). The surface is chemically inert to I_2 and atomic iodine and serves only as a repulsive wall that converts part of the translational energy of the incident I_2 molecule into internal energy. If sufficient energy is transferred to the internal degrees of freedom along the reaction coordinate such that the binding energy is exceeded, the molecule dissociates. In the case of I_2, it is shown that dissociation to two gas phase iodine atoms takes place by a centrifugal mechanism as a result of the high rotational excitation attained upon impact. The dissociation of Ar clusters on metal surfaces (62) and the isomerization of cyclopropane on mica (63) may also take place through transfer of translational energy to vibrational energy. A similar energy transfer mechanism is believed to be operative in overcoming the barrier to the dissociation of CH_4 on Ni[111] where the surface first serves as a repulsive wall to convert translational energy into vibrational energy necessary to deform the molecule (20). This mechanism represents the most elementary and fundamental process for dissociation of molecules at surfaces.

Precursor Mechanism

Consider the dependence on collision energy of the probability for dissociative chemisorption via a precursor state mechanism. Because the dissociation step is virtually decoupled from the adsorption of the incident molecule into the precursor state, the dependence of the dissociation probability on energy is actually a measure of the energy dependence of adsorption into the molecular precursor state. Scattering experiments in which both angular and velocity distributions are measured have shown that the probability for adsorption into a molecular chemisorption or physisorption state is a strong function of the ratio of the translational energy in the normal direction to the depth of the attractive well (64). The larger this ratio, the lower the probability for adsorption. Therefore, the precursor state model predicts the dissociative chemisorption probability to decrease as the incident translational energy in the normal direction is increased. The dissociative chemisorption probability increases as the incident angle is increased because glancing incident trajectories contain less energy in the normal direction.

Unlike dissociation through a direct mechanism, the temperature of the surface is a critical parameter in a precursor state mechanism. This is because the surface temperature controls the population of molecules bound in the precursor state by regulating both the rate of desorption of the precursor molecules and their rate of dissociation. The temperature dependence of the dissociative chemisorption probability can be complex because, whereas higher temperatures enhance the dissociation rate, they decrease the supply of precursor molecules through desorption. The important point here is that any effect of surface temperature on the magnitude of the dissociation probability is usually viewed as evidence for dissociation via a precursor state.

IDENTIFICATION OF AND REASONS BEHIND A PRECURSOR MECHANISM With these expected trends as the basis for identification, dissociative chemisorption via a precursor state has recently been identified in several systems: the dissociative chemisorption of N_2 on W[100] (65), O_2 on W[110] (66, 67), H_2 on Ni[997] (42, 68), propane and butane on Ir[110]-(1 × 2) (24, 69), and O_2 on Pt[111] (70). For a fixed angle of incidence, the probabilities of dissociative chemisorption are observed to decrease by as much as a factor of four as the translational energy is increased from 1 to about 10 kcal mol^{-1}. Information is not yet available on the relative importance of the translational energies in the normal and parallel directions. The dissociative chemisorption probabilities are observed to decrease rapidly with increasing surface temperature.

In each of these systems, the first step of the precursor mechanism is the initial collision of the molecule with the surface. It is a revealing endeavor to consider the reason for the failure of the molecule to dissociate directly upon its initial collision. Presumably, either the orientation of the incident molecule or the surface site on which it is incident is energetically unfavorable for direct dissociation. The operation of a precursor mechanism on a large scale implies that there are very few orientations of the incident molecule or few surface sites that are energetically favorable for direct dissociation. Therefore, the molecule requires time to find the site or achieve the orientation that offers a low energy pathway for dissociation. The time is provided by its adsorption into a precursor state, and the low energy pathway is one where the barrier along the potential surface to the dissociative state is below the zero of energy. Although the probability is low, had the molecule impacted the surface at this low energy site with an energetically favorable orientation, the direct dissociative chemisorption would have occurred.

The dissociation probability exhibits intriguing behavior as the translational energy is increased further: it reaches a minimum and then increases (65–70). This reversal is interpreted as a change from a precursor mechanism to direct dissociative chemisorption. The interpretation is supported by the insensitivity of the dissociation probability at high translational energies to surface temperature. This is an important observation. It shows that, at low energies, a dominant feature of the potential energy surface prevents the molecule from dissociating directly and instead forces it into the precursor state. However, this feature can be overcome at higher energies. These trends are most consistent with a barrier high above the zero of energy as that dominant feature. Molecules incident with low energies cannot overcome the barrier but, as the energy is increased, the barrier is surmountable and direct chemisorption becomes the dominant mechanism. Therefore, the barrier prevents direct dissociative chemisorption and forces the population of the precursor state.

There are now collectively enough observations to provide grist for speculation as to the major distinctions between the potential energy surfaces that support direct dissociation and those that support dissociation via a precursor state. The simple difference that emerges is that the precursor state surface has holes or low energy pathways to the dissociative state. That is, the two surfaces are identical in that each has a distribution of barriers rising high above the zero of energy that correspond to individual positions on the surface lattice and to different orientations of the molecule and that each potential surface has a distribution of wells corresponding to the molecular precursor states. But on a surface supporting a precursor mechanism, some of the barriers lie below the zero of energy, creating low energy pathways or holes to the dissociative state. A molecularly adsorbed species eventually finds these low energy pathways by reorienting itself or by diffusing onto a different site and then dissociates. The dissociative chemisorption of CH_4 on Ni[111] is an example of a system that has no low energy pathway and therefore only direct dissociation occurs.

This picture for the distinctions between the potential energy surfaces is naive in the sense that the dominant features, the barriers, are considered to be the sole distinguishing elements. In reality, the fine details of the shape of the potential surface along the entire reaction coordinate also determine the mechanistic fate of the incident molecule. This is because the shape of the potential surface dictates the dynamical constraints to dissociation. It determines the volume of phase space accessible to the molecule. It is possible that direct chemisorption fails in some cases not because there are potential barriers above the zero of energy but because dynamical constraints, such as the incorrect vibrational phase of the molecule, prevent it. In this case, the molecule then proceeds to collide with the repulsive wall of the potential energy surface, where it loses energy and becomes trapped in the precursor state. The collision with the repulsive wall of the potential surface is now responsible for forcing the molecule into the precursor state. At higher translational energies, these dynamical constraints may be lifted to allow direct chemisorption. Likewise, there may be cases where a low energy pathway to dissociation exists, but the trapped precursor molecule never finds it because the precursor molecule cannot access that region of phase space due to dynamical constraints.

However, even with this caveat concerning the shape of the potential energy surface clearly in mind (71), the dramatic increase in the dissociation probability with collision energy, as observed in many experiments, is most consistent with the surmounting of barriers. Barriers and potential wells are indeed extrema along the potential energy surface of the interaction. Their shapes and the shape of the entire potential are secondary features. Therefore, until appropriate spectroscopic techniques are developed to probe the secondary details of the interaction, these simple models for the types of potential energy surfaces constitute the essence of the distinctions between them and provide a foundation from which to launch further investigations.

BARRIERS TO A MOLECULAR PRECURSOR STATE   Recently, evidence has been presented for a distribution of potential barriers rising above the zero of energy to molecular adsorption of N_2 into the α-N_2 state on Fe[111] (72). The barriers manifest themselves as an increase in the probability of adsorption into the α-N_2 state from 5×10^{-2} to 1.2×10^{-1} as the translational energy of N_2 is increased from 23 to 90 kcal/mole. Vibrational excitation is shown to be half as effective as translational energy (73, 74). This represents the first observation of a potential barrier to molecular adsorption in any interaction between a molecule and a surface and should provide sufficient reason to relinquish the traditional notion that barriers to molecular chemisorption do not exist.

The important point here is that the α-N_2 molecule has long been known to be a precursor molecule to dissociatively adsorbed nitrogen on Fe[111] (75). Therefore, even though nitrogen dissociates via a precursor mechanism, the probability for dissociative chemisorption increases with increasing translational energy because higher energies provide access to the precursor state. However, as expected for a precursor mechanism, surface temperature is an important variable in the extent of dissociation. The dissociative chemisorption probabilities at all translational energies decrease at higher temperatures because desorption of the precursor molecule competes effectively with its dissociation. Further evidence for a precursor mechanism is found in the independence of the nitrogen atom saturation coverage on incident energy. Because the nitrogen molecule is adsorbed in the precursor state, it loses all memory of its initial conditions prior to its dissociation into two adsorbed nitrogen atoms. This contrasts with the direct dissociative chemisorption of N_2 on W[110] for which the atomic saturation coverage is dependent on the energy of the incident N_2 molecule (52–54).

The results of this molecular beam study are consistent with and provide explanations for many puzzling phenomena observed in previous studies (75) in which the adsorption of N_2 on Fe[111] was carried out from a room temperature ambient gas. The consistently low probability for dissociative chemisorption ($\sim 10^{-6}$) can now be understood as arising from the barrier to the precursor state. The observed decreasing dissociation probability with increasing temperature is consistent with the precursor mechanism verified by the molecular beam experiments. It is clear how macroscopic adsorption properties can be understood from a knowledge of the dominant features of even a very complex interaction potential.

ROTATIONAL RESONANCES AND PRECURSOR STATES It is well known that a diatomic molecule can suffer a sufficiently inelastic collision with the surface such that all of its kinetic energy in the normal direction is transferred to rotation (76). This loss of kinetic energy results in trapping of the molecule near the surface in a selective adsorption resonance. The resonances are observed as discrete losses in the scattered intensity from diffraction channels. The resonance condition for the transition to the bound level can be effected by selecting the appropriate incident angle for a given kinetic energy, thereby trapping the resonant molecules temporarily in vibrational states of a laterally averaged, hindered rotor-surface interaction potential. Such a trapped molecule, tumbling across the surface, fulfills the general idea of a precursor molecule to dissociative chemisorption.

An experiment was recently carried out to explore whether these rotationally mediated selective adsorption resonances provide a pathway for dissociative chemisorption via a precursor state. The probability for dissociative chemisorption of HD on W[110] was monitored as a function of the incident angle (77, 78). This procedure allows several resonances with varying rotational energies to be accessed. No correlation between the dissociative chemisorption probability and the population of the selective adsorption resonances was found. The probability of desorption of the trapped HD molecule apparently greatly exceeds the probability of dissociation. The low probability for dissociation is believed to be due not to a substantial barrier between the molecular and dissociative states but rather to a dynamical bottleneck or a restriction of phase space arising from stringent steric requirements for dissociation relative to desorption.

In the following section, evidence for a similar dynamical bottleneck between a precursor state and a molecular chemisorption state is discussed.

PRECURSOR MOLECULES TO MOLECULAR CHEMISORPTION

Because of the importance of the precursor mechanism to dissociative chemisorption, the following section discusses recent findings of a similar mechanism to molecular chemisorption.

Molecularly chemisorbed molecules must also attain a minimum energy orientation over some minimum energy site. It is certainly not possible for every molecule to be incident exactly at this site with the proper orientation, yet the probability for molecular chemisorption is often one! The explanation for this dilemma has long been attributed to a weakly bound precursor molecule (1, 2), whose high mobility and unhindered rotation allow it to find the binding site and chemisorb there with the energetically favorable orientation.

Evidence has recently been presented for a long-lived ($\tau > 1$ μs at $T_S = 300$K) CO precursor molecule to molecularly chemisorbed CO on Ni[111] (11, 79). The CO molecule is trapped with unit probability as a precursor molecule when its incident energy is less than 4 kcal mol^{-1}. At slightly higher collision energies, the probability for molecular chemisorption drops to 0.5 and remains constant at 0.5 up to 30 kcal mol^{-1}. Only direct molecular chemisorption occurs at these higher translational energies. This behavior is reminiscent of that observed for dissociative chemisorption via a precursor mechanism (65–70).

Stabilization and spectroscopic identification of the CO precursor molecule at a surface temperature of 8K was not successful (80). This result enables maximum values for the apparent activation energy barrier along the coordinate connecting the CO precursor state to the molecular chemisorption state and preexponential factor for the unimolecular conversion of the CO precursor molecule to the chemisorbed molecule to be determined. These values are 0.3 kcal mol^{-1} and 10^6 sec^{-1}. The very low value of the preexponential factor for a unimolecular process indicates the presence of a dynamical bottleneck for conversion of the precursor molecule to the chemisorbed molecule.

A possible simple explanation (11) for the physical origin of the precursor molecule and the dynamical bottleneck lies in the very anisotropic interaction of CO with Ni. The interaction is strongly attractive (~ 30 kcal mol^{-1}) with the carbon end oriented toward the Ni but weakly attractive ($\lesssim 4$ kcal mol^{-1}) toward the oxygen end. When CO is incident with the carbon end toward the surface and with high kinetic energies ($E_i > 4$ kcal mol^{-1}) it will directly chemisorb as long as enough energy can be transferred to surface phonons upon its collision. The amount of energy transferred has been shown to be a constant fraction of the sum of the well depth and the incident translational energy (64). The value of the constant fraction is determined by the kinematics of the collision and is mainly a ratio of the masses of the CO and the effective mass of the Ni surface atom. For CO on Ni[111], this fraction is at least 0.5. Therefore, even if the collision energy is as high as 30 kcal mol^{-1}, sufficient energy will be transferred to trap CO in the chemisorption state. However, for CO incident at 7 kcal mol^{-1} and with the O end toward the surface, the acceleration of the incident molecule due to the attractive part of the potential (~ 4 kcal mol^{-1}) is so small compared to its incident translational energy that a loss of a half of this total energy (11 kcal mol^{-1}) is not sufficient to trap CO with the oxygen end toward the surface. Therefore, CO molecules incident with this approximate configuration, which comprise half of the incident molecules, do not trap and the chemisorption probability drops to 0.5 as the energy is increased much above 4 kcal mol^{-1}.

When CO is incident with the O end toward the surface and with an energy below 4 kcal mol^{-1}, sufficient energy can be transferred to trap CO with the oxygen end down. Once trapped, the anisotropy of the interaction produces a large torque on the molecule. The molecule begins to tumble end over end, exchanging its energy in the parallel direction for rotational energy. This tumbling molecule is the precursor molecule, which moves fast ($\sim 10^4$ cm/sec) and lives a long time before it dissipates its energy completely in the proper orientation over the correct site for chemisorption. The dissipation of the energy is the bottleneck to chemisorption. In a subsequent investigation, a similar "hot" precursor molecule has been suggested for CO adsorption on Ni[100] (81). A different kind of mechanism has recently been proposed for a hot H_2 precursor molecule (82). Verification of these proposals for the nature of the precursor molecule awaits development of the technological capability for its spectroscopic detection on at least a microsecond time scale.

COLLISION OF A GAS PHASE ATOM WITH AN ADSORBATE: NEW MECHANISMS FOR DISSOCIATIVE CHEMISORPTION AND DESORPTION

Consider a surface at low temperature on which a molecule is adsorbed. As the temperature is raised, the adsorbed molecule may dissociate into adsorbed fragments or desorb intact. The energy required to overcome the barrier to decomposition or desorption is acquired from the surface. Molecular beam techniques have recently unveiled a different kind of energy source available to adsorbates for dissociation or desorption. Gas phase atoms incident on an adsorbate covered surface have been shown to provide sufficient energy to either dissociate or desorb the adsorbate (28, 83, 84). This process represents a new kind of mechanism for dissociative chemisorption and desorption called *collision-induced dissociation and desorption*.

Collision-Induced Dissociation of CH_4 on $Ni[111]$

The barrier to the dissociative chemisorption of CH_4 on $Ni[111]$ was identified in a previous section as arising largely from the energy required to deform methane. If this is correct, it should be possible to supply this energy to a physisorbed methane molecule by collision with a gas phase inert gas atom. The impact of the inert gas atom pounds the molecularly adsorbed methane into the distorted shape of the transition state that leads to dissociation.

The experiment is performed by measuring the dissociation rate as an argon atom beam impinges on a layer of methane physisorbed on a $Ni[111]$ surface at 46K. The dissociation rate is measured as a function of the energy of the incident Ar atoms. Since the dissociation rate is equal to the product of the collision-induced dissociation cross-section times the argon atom flux times the methane coverage, the cross-section is easily calculable. Figure 4 shows a plot of the cross-section for dissociation of the physisorbed methane versus the normal component of the kinetic energy of the incident Ar atoms for several total kinetic energies of Ar. It is clear from this plot that the strict adherence to normal kinetic energy scaling observed in the case of translational activation of methane (18–20) and shown in Figure 2 is not observed in collision-induced activation.

To understand this complex energy dependence, it is necessary to divide the mechanism for the collision-induced dissociation process into two steps. In the first step, some fraction of the kinetic energy of the incident Ar atom is transferred by collision to the physisorbed methane. The effectiveness of the collision in transferring energy depends on the type of collision or impact parameter. That is, the Ar atom can collide with the adsorbed methane molecule with an impact parameter as large as the hard sphere collision diameter or as small as zero in a head-on collision. The amount of energy transferred as a function of the impact parameter and incident angle is easily calculable using a hard sphere collision model. However, the energy crucial to the dissociation of methane is the energy in the normal direction, E_n. Therefore E_n is given by

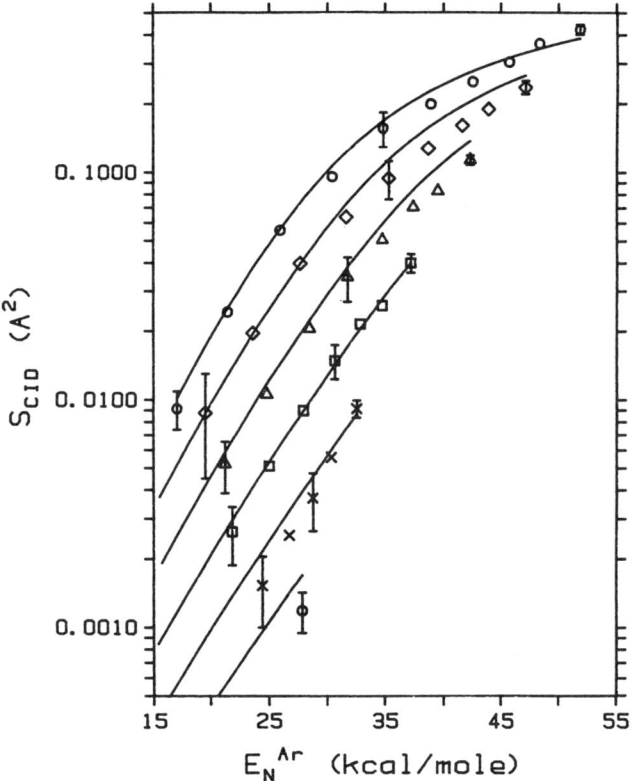

Figure 4 Cross-section for dissociation of CH_4 physisorbed on Ni[111] at 47K induced by the impact of Ar atoms vs the kinetic energy of the Ar atom in the normal direction. Methane coverage is 0.3 monolayers. Each point at the same total energy represents a different incident angle from 0° to 55°. (*Top circle*) 51.8 kcal mol^{-1} total energy, (*diamond*) 47.1, (*triangle*) 42.3, (*square*) 37.2, (*X*) 32.5, (*bottom circle*) 27.8. The *solid lines* are the result of the model calculation discussed in the text.

$$E_n(E_i, \theta_i, b, \phi) = \frac{4m_A m_C}{(m_A + m_C)^2} E_i \left[\left(1 - \frac{b^2}{d^2}\right) \cos\theta_i - \frac{b}{d}\left(1 - \frac{b^2}{d^2}\right)^{1/2} \right.$$

$$\left. \times \sin\theta_i \cos\phi \right]^2 \quad \text{for} \quad b \leq d$$

$$= 0 \quad \text{for} \quad b > d$$

where E_i is the total energy and θ_i is the incident angle of the Ar atom, b is the impact parameter and ϕ is the azimuthal angle of the collision, m_A

and m_C are the masses of Ar and CH_4, and d is the hard sphere collision diameter.

Once the methane has acquired the energy E_n, the second step in the mechanism is dissociation. The probability that the methane molecule dissociates is given as a function of E_n by the previous measurements of the dissociation probability, P, versus the normal component of the kinetic energy of the incident methane molecule. These data, which are shown in Figure 2, are described by the functional form

$$P(E_n) = \frac{A}{1 - e^{-\alpha(E_n - V)}}$$

where α, V, and A are treated as fitting parameters. This expression is combined with that derived from the hard sphere collision model for the energy transferred to CH_4, to yield the probability for collision-induced dissociative chemisorption, P_{CID}, as a function of the incident energy and angle of the Ar atom and the impact parameter and azimuthal angle of the collision

$$P_{CID}(E_i, \theta_i, b, \phi) = P[E_n(E_i, \theta_i, b, \phi)].$$

Although E_i and θ_i are fixed in the experimental measurements, the impact geometry is not. Therefore, it is necessary to integrate P_{CID} over all values of b and ϕ to yield a calculated value for the collision-induced cross section for dissociation, σ

$$\sigma(E_i, \theta_i) = \frac{1}{\cos \theta_i} \int_0^{2\pi} d\phi \int_0^\infty b \, db \, P_{CID}(E_i, \theta_i, b, \phi).$$

The results of this calculation are shown as the *solid lines* in Figure 4. The agreement between the model and data is excellent. The model closely reproduces the complicated dependence of the cross-section on the normal kinetic energy of the incident Ar atoms. The excellent agreement strongly suggests that the breakdown in normal kinetic energy scaling in the Ar energy is the result of the range of impact parameters that contribute to the cross-section and that a hard sphere collision model accurately describes the energy transfer.

The physical picture that emerges is straightforward. The Ar atom collides with the methane molecule, transfers some fraction of its energy, depending on the kinematics of the collision, and then reflects from the surface. The subsequent methane-surface collision is the same as in translational activation. The methane molecule, with its newly acquired energy, is accelerated into the surface, deforms and dissociates. As in the case of translational activation, the products of collision-induced dissociation are an adsorbed methyl radical and an adsorbed hydrogen atom (28). Therefore, dissociation of methane after translational activation and after collision-induced activation are completely consistent with each other. They are simply different methods for providing the energy to deform the molecule, but the mechanism for dissociation of methane is the same.

Collision-induced dissociation is not limited to weakly bound species. Methyl radicals adsorbed on Ni[111] dissociate to a mixture of CH_2 and CH under impact of a Xe atom beam (83, 84). The collision-induced dissociation of CO to C and O atoms adsorbed on Ni[111] has also been achieved (Q. Y. Yang, A. D. Johnson, S. T. Ceyer, unpublished).

Collision-Induced Desorption of CH_4 from Ni[111]

In competition with dissociation induced by the impact of the argon atom, collision-induced desorption also occurs (28, 83, 84). That is, once the argon atom transfers energy to the methane, the methane molecule collides with the surface and can rebound into the gas phase if the site on which it is physisorbed or if its orientation is not energetically favorable for dissociation. Desorption induced by collisions is roughly one to two orders of magnitude more probable than collision-induced dissociation.

The dynamics of collision-induced desorption are studied by measuring the desorption cross-section as a function of the energy and incident angle of the argon atoms. A plot of the desorption cross-section versus incident angle for several total kinetic energies is shown in Figure 5. The absolute magnitudes of the cross-sections are about an order of magnitude smaller than hard sphere, gas kinetic cross sections. The desorption cross section is observed to approximately double as the incident angle is increased from 0° to 70°. However, the magnitude of the increase in the desorption rate is dependent on and is largest for high total energies of the incident argon atom. This is the result of two competing effects. As the incident angle is increased, the normal component of the kinetic energy, which is the component responsible for desorption, decreases. Therefore, the desorption cross-section should decrease at glancing incidence because there is less energy in the normal direction to push the molecule away from the surface. However, as the incident angle increases the collision cross-section of an incoming Ar atom with the adsorbed molecule increases. The complicated dependences of the desorption cross-section on energy and incident angle are the result of these competing factors. Results of classical trajectory calculations treating CH_4 and Ar as hard spheres on a rigid lattice, reproduce well these trends and show that only direct collisions of Ar with CH_4 contribute to desorption. Surface-mediated processes, such as the production of hot spots, are not important mechanisms for collision-induced desorption.

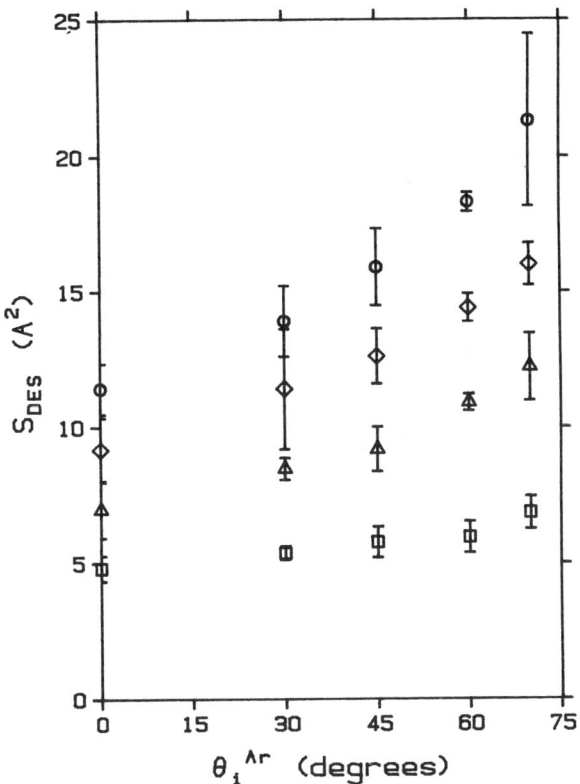

Figure 5 Cross-section for desorption of CH_4 physisorbed on Ni[111] at 40K induced by the impact of Ar atoms vs the incident angle of the Ar atom for several Ar total kinetic energies: (*circle*) 51.8 kcal mol^{-1}, (*diamond*) 42.3, (*triangle*) 32.5, (*square*) 27.8.

Importance of Collision-Induced Chemistry and Desorption for Catalysis

Perhaps more important than the physics behind these processes is the fact that they have been observed and what the knowledge of their existence means for understanding the complex environment of high pressure heterogeneous catalysis. Collision-induced chemistry and desorption likely play significant roles in a high pressure catalytic reaction, since, under these conditions, the catalyst is covered with adsorbate and is continually bombarded by gas phase molecules. With the observation of these processes, no mechanism for a high presure reaction can now be considered complete without an assessment of the role of collision-induced chemistry and desorption as potential major steps. These observations are now cause for reexamination of the mechanisms of catalytic reactions in which inert gas effects on reaction rates have been noted (85). Collision-induced chemistry and desorption are additional contributors to the pressure gap in the reactivity of heterogeneous catalysis. They are additional reasons why surface chemistry at high pressures is often very different from surface chemistry at low pressures.

CONCLUDING REMARKS

Many examples discussed here illustrate how complex a potential energy surface can be for even the most elementary surface chemical reaction: the dissociative chemisorption of a diatomic molecule. Despite the complexity of the multidimensional interaction potential, studies carried out in the last few years of the collision energy dependence of the dissociation probability have successfully singled out the major features of the potential that are responsible for the dynamics of and mechanism for dissociation. They have provided a structural framework on which future refinements will be confidently constructed. From the dissociation of CH_4 on Ni and W, CO_2 on Ni, and I_2 on MgO, the importance of the surface as a repulsive wall that serves to convert translational to internal energy is an emerging common thread. Direct and precursor mechanisms for dissociative chemisorption are becoming distinguishable by their potential energy surfaces. The newly discovered collision-induced dissociation and desorption mechanisms potentially offer unprecedented probes of the molecular motions leading to dissociation and desorption. But even more important, the type of understanding of microscopic interactions gained from these studies allows macroscopic phenomena such as the pressure gap in heterogeneous catalytic reactivity and inert gas effects on catalytic reaction rates to be rationalized and provides the opportunity for them to be advantageously modified.

ACKNOWLEDGMENTS

Michelle T. Schulberg is gratefully acknowledged for her critical reading of this manuscript.

This work is supported in part by the National Science Foundation (CHE-8508734), the Energy Laboratory at the Massachusetts Institute of Technology, and the Joint Services Electronics Program (DAAL-03-86-K-0002). S. T. Ceyer thanks the Dreyfus Foundation for a Teacher-Scholar Award and the A. P. Sloan Foundation for a Fellowship.

Literature Cited

1. Taylor, J. B., Langmuir, I. 1933. *Phys. Rev.* 44: 423
2. Kisliuk, P. J. 1957. *J. Phys. Chem. Solids* 3: 95–101; 1958. *J. Phys. Chem. Solids* 5: 78–84
3. King, D. A. 1978. *CRC Crit. Rev. Solid State Mat. Sci.* 7: 167–208
4. King, D. A., Wells, M. G. 1974. *Proc. R. Soc. London Ser. A* 339: 245–69
5. Balooch, M., Cardillo, M. J., Miller, D. R., Stickney, R. E. 1974. *Surf. Sci.* 46: 358–92
6. Rideal, E. K. 1938. *Proc. Cambridge Philos. Soc.* 35: 130
7. Eley, D. D., Norton, P. R. 1966. *Discuss. Faraday Soc.* 41: 135
8. Ceyer, S. T., Beckerle, J. D., Lee, M. B., Tang, S. L., Yang, Q. Y., Hines, M. A. 1987. *J. Vac. Sci. Technol. A* 5: 501–7
9. Ceyer, S. T., Gladstone, D. J., McGonigal, M., Schulberg, M. T. 1988. In *Physical Methods of Chemistry*, ed. J. F. Hamilton, R. C. Baetzold. New York: Wiley. In press
10. Tang, S. L., Lee, M. B., Yang, Q. Y., Beckerle, J. D., Ceyer, S. T. 1986. *J. Chem. Phys.* 84: 1876–83
11. Tang, S. L., Beckerle, J. D., Lee, M. B., Ceyer, S. T. 1986. *J. Chem. Phys.* 84: 6488–6506
12. Lennard-Jones, J. E. 1932. *Trans. Faraday Soc.* 28: 333–60
13. Lundqvist, B. I., Gunnarsson, O., Hjelmberg, H., Norskov, J. K. 1979. *Surf. Sci.* 89: 196
14. Norskov, J. K., Holloway, S., Lang, N. D. 1984. *Surf. Sci.* 137: 65
15. Hoffman, R. 1988. *Rev. Mod. Phys.* In press
16. Garfunkel, E. L., Feng, X. 1986. *Surf. Sci.* 176: 445
17. Van Willigen, W. 1968. *Phys. Lett. A* 28: 80
18. Lee, M. B., Yang, Q. Y., Tang, S. L., Ceyer, S. T. 1986. *J. Chem. Phys.* 85: 1693-94
19. Ceyer, S. T., Lee, M. B., Yang, Q. Y., Beckerle, J. D., Johnson, A. D. 1988. In *Methane Conversion*, ed. D. Bibby, C. Chang, R. Howe, S. Yurchak, 36: 51–66. Amsterdam: Elsevier Science. 741 pp.
20. Lee, M. B., Yang, Q. Y., Ceyer, S. T. 1987. *J. Chem. Phys.* 87: 2724–41
21. Hamza, A. V., Madix, R. J. 1987. *Surf. Sci.* 179: 25–46
22. Rettner, C. T., Pfnür, H. E., Auerbach, D. J. 1985. *Phys. Rev. Lett.* 54: 2716–19
23. Rettner, C. T., Pfnür, H. E., Auerbach, D. J. 1986. *J. Chem. Phys.* 84: 4163–67
24. Hamza, A. V., Steinrück, H.-P., Madix, R. J. 1987. *J. Chem. Phys.* 86: 6506–14
25. Winters, H. F. 1975. *J. Chem. Phys.* 62: 2454–62; 1976. *J. Chem. Phys.* 64: 3495–3500
26. Brass, S. G., Reed, D. A., Ehrlich, G. 1979. *J. Chem. Phys.* 70: 5244–50
27. Yates, J. T. Jr., Zinck, J. J., Sheard, S., Weinberg, W. H. 1979. *J. Chem. Phys.* 70: 2266–72
28. Beckerle, J. D., Yang, Q. Y., Johnson, A. D., Ceyer, S. T. 1987. *J. Chem. Phys.* 86: 7236–37
29. Herzberg, G. 1945. *Molecular Spectra and Molecular Structure II. Infrared and Raman Spectra of Polyatomic Molecules.* New York: van Nostrand Reinhold. 632 pp.
30. Gray, D. L., Robiette, A. G. 1979. *Mol. Phys.* 37: 1901–20
31. Anderson, A. B., Maloney, J. J. 1988. *J. Phys. Chem.* 92: 809–12
32. Somorjai, G. A. 1979. *Surf. Sci.* 89: 496; 1981. In *Chemistry in Two Dimensions*. Ithaca: Cornell Univ. Press. 576 pp.
33. Stoltze, P., Norskov, J. K. 1985. *Phys. Rev. Lett.* 55: 2502–5
34. Gaidai, N. A., Babernich, L., Guczi, L. 1974. *Kinet. Catal.* 15: 868
35. Frennet, A., Lienard, G. 1974. *Catal. Rev. Sci. Eng.* 10: 37
36. Schouten, F. C., Gijzeman, O. L. J., Bootsma, G. A. 1979. *Bull. Soc. Chim. Belg.* 88: 541–47; 1979. *Surf. Sci.* 87: 1
37. Beebe, T. P. Jr., Goodman, D. W., Kay, B. D., Yates, J. T. Jr. 1987. *J. Chem. Phys.* 87: 2305–15
38. Nakamoto, K. 1978. *Infrared and Raman Spectra of Inorganic and Coordination Compounds.* New York: Wiley
39. Fisher, G. B., Mitchell, G.E. 1983. *J. Electron Spectros. Rel. Phenom.* 29: 253
40. Yang, Q. Y., Ceyer, S. T. 1988. *J. Vac. Sci. Technol. A* 6: 851–52
41. Hamza, A. V., Madix, R. J. 1985. *J. Chem. Phys.* 89: 5381–86
42. Rendulic, K. D., Winkler, A., Karner, H. 1987. *J. Vac. Sci. Technol. A* 5: 488–91
43. Steinrück, H.-P., Rendulic, K. D., Winkler, A. 1985. *Surf. Sci.* 154: 99–108
44. Robota, H. J., Vielhaber, W., Lin, M. C., Segner, J., Ertl, G. 1985. *Surf. Sci.* 155: 101–20
45. Steinrück, H.-P., Luger, M., Winkler, A., Rendulic, K. D. 1985. *Phys. Rev. B* 32: 5032–37
46. Steinrück, H.-P., Hamza, A. V., Madix, R. J. 1986. *Surf. Sci.* 173: L571–75
47. D'Evelyn, M. P., Hamza, A. V., Gdow-

ski, G. E., Madix, R. J. 1986. *Surf. Sci.* 167: 451–73
48. Becker, C. A., Cowin, J. P., Wharton, L., Auerbach, D. J. 1977. *J. Chem. Phys.* 67: 3394–95
49. Campbell, C. T., Ertl, G., Kuipers, H., Segner, J. 1980. *J. Chem. Phys.* 73: 5862–73
50. Segner, J., Campbell, C. T., Doyen, G., Ertl, G. 1984. *Surf. Sci.* 138: 505–23
51. Mantell, D. A., Kunimori, K., Ryali, S. B., Haller, G. L., Fenn, J. B. 1986. *Surf. Sci.* 172: 281–302
52. Lee, J., Madix, R. J., Schlaegel, J. E., Auerbach, D. J. 1984. *Surf. Sci.* 143: 626–38
53. Auerbach, D. J., Pfnür, H. E., Rettner, C. T., Schlaegel, J. E., Lee, J., Madix, R. J. 1984. *J. Chem. Phys.* 81: 2515–16
54. Pfnür, H. E., Rettner, C. T., Lee, J., Madix, R. J., Auerbach, D. J. 1986. *J. Chem. Phys.* 85: 7452–66
55. Kara, A., DePristo, A. E. 1988. *Surf. Sci.* 193: 437–54
56. Holloway, S., Gadzuk, J. W. 1986. *J. Chem. Phys.* 85: 7452–66
56a. Gadzuk, J. W. 1988. *Ann. Rev. Phys. Chem.* 39: 395–424
57. Lee, M. B., Beckerle, J. D., Tang, S. L., Ceyer, S. T. 1987. *J. Chem. Phys.* 87: 723
58. Steinrück, H.-P., D'Evelyn, M. P., Madix, R. J. 1986. *Surf. Sci.* 172: L561–67
59. Kolodney, E., Amirav, A., Elber, R., Gerber, R. B. 1984. *Chem. Phys. Lett.* 111: 366–71
60. Kolodney, E., Amirav, A. 1984. In *Dynamics on Surfaces*, ed. B. Pullman et al, pp. 231–42. Amsterdam: Reidel
61. Gerber, R. B., Amirav, A. 1986. *J. Phys. Chem.* 90: 4483–91
62. Xu, G. Q., Bernasek, S. L., Tully, J. C. 1988. *J. Chem. Phys.* 88: 3376–84
63. Tsou, L., Haller, G. L., Fenn, J. B. 1987. *J. Phys. Chem.* 91: 2654–58
64. Barker, J. A., Auerbach, D. J. 1984. *Surf. Sci. Rep.* 4: 1–99
65. Auerbach, D. J., Rettner, C. T. 1987. In *Kinetics of Interface Reactions*, ed. M. Grunze, H. J. Kreuzer, p. 125. Berlin: Springer-Verlag
66. Rettner, C. T., DeLouise, L. A., Auerbach, D. J. 1986. *J. Vac. Sci. Technol. A* 4: 1491–92
67. Rettner, C. T., DeLouise, L. A., Auerbach, D. J. 1986. *J. Chem. Phys.* 85: 1131–49
68. Karner, H., Luger, M., Steinrück, H.-P., Winkler, A., Rendulic, K. D. 1985. *Surf. Sci.* 163: L641–44
69. Hamza, A. V., Steinrück, H.-P., Madix, R. J. 1986. *J. Chem. Phys.* 85: 7494–95
70. Williams, M. D., Bethune, D. S., Luntz, A. C. 1988. *J. Chem. Phys.* 88: 2843–45
71. Doren, D. J., Tully, J. C. 1988. *Langmuir* 4: 256–68
72. Rettner, C. T., Stein, H. 1987. *Phys. Rev. Lett.* 59: 2768
73. Rettner, C. T., Stein, H. 1987. *J. Chem. Phys.* 87: 770–71
74. Holloway, S., Hodgson, A., Halstead, D. 1988. *Chem. Phys. Lett.* 147: 425–29
75. Ertl, G., Lee, S. B., Weiss, S. 1982. *Surf. Sci.* 114: 515
76. Cowin, J. P., Yu, C. F., Sibener, S. J., Hurst, J. E. 1981. *J. Chem. Phys.* 75: 1033–34
77. Rettner, C. T., DeLouise, L. A., Cowin, J. P., Auerbach, D. J. 1985. *Faraday Discuss. Chem. Soc.* 80: 127–36
78. Rettner, C. T., DeLouise, L. A., Cowin, J. P., Auerbach, D. J. 1985. *Chem. Phys. Lett.* 118: 355–58
79. Tang, S. L., Lee, M. B., Beckerle, J. D., Hines, M. A., Ceyer, S. T. 1985. *J. Chem. Phys.* 82: 2826–27; 1985. *J. Vac. Sci. Technol. A* 3: 1665
80. Beckerle, J. D., Yang, Q. Y., Johnson, A. D., Ceyer, S. T. 1988. *Surf. Sci.* 195: 77–93
81. D'Evelyn, M. P., Steinrück, H.-P., Madix, R. J. 1987. *Surf. Sci.* 180: 47–76
82. Muller, J. E. 1987. *Phys. Rev. Lett.* 59: 2943–46
83. Beckerle, J. D., Johnson, A. D., Yang, Q. Y., Ceyer, S. T. 1988. *J. Vac. Sci. Technol. A* 6: 903–4
84. Beckerle, J. D., Johnson, A. D., Yang, Q. Y., Ceyer, S. T. 1988. In *Proc. Solvay Conf. on Surface Sci.*, ed. G. Ertl. Berlin: Springer-Verlag. In press
85. Hudgins, R. R., Silveston, P. L. 1975. *Catal. Rev. Sci. Eng.* 11: 167

HIGH-RESOLUTION SOLID-STATE NMR OF PROTEINS

Steven O. Smith and Robert G. Griffin

Francis Bitter National Magnet Laboratory, Massachusetts Institute of Technology, Cambridge, Massachusetts 02139

INTRODUCTION

In the past decade, nuclear magnetic resonance (NMR) methods have become well established for determining the structure and dynamics of biological molecules. In particular, the introduction and development of *two-dimensional* (2D) NMR methods has provided a versatile approach for resolving and assigning resonances in complex solution NMR spectra, such as those exhibited by proteins. These methods and their applications have been described in detail in several reviews and monographs (1–5). 2D NMR techniques have been used extensively for studying small proteins (MW < 15,000) in solution, where secondary and in some cases tertiary structures can now be determined solely on the basis of NMR data (6). However, for larger proteins, protein complexes, and membrane proteins, problems involving spectral line broadening are encountered, owing to reduced tumbling rates and correspondingly longer rotational correlation times.

The development of NMR methods for determining the structure of large biological molecules has taken two approaches. The first involves some novel extensions of high-resolution solution NMR techniques. For example, specific ^{13}C or ^{15}N isotopic labeling, in combination with indirect detection via protons, has taken advantage of the large ^1H-heteronuclear *J*-couplings and heteronuclear chemical shift dispersions to edit resonances and increase the spectral resolution in 2D spectra (7). In addition, deuteration has long been utilized to simplify protein spectra. Random deuteration has been employed to narrow proton resonances by reducing proton-

proton dipolar interactions (8a,b), while selective deuteration has been coupled with double difference spectroscopy to yield ^{1}H spectra of selected sites in antibody-hapten complexes (9).

The second approach for obtaining structural information on large biomolecules and membrane proteins is the use of solid-state NMR. In these systems, molecular motion is restricted and the NMR resonances broaden due to *anisotropy* in the dipolar, quadrupolar, and/or chemical shift interactions. In solution, rapid molecular motion serves to average these spin interactions, and the NMR spectrum exhibits only isotropic chemical shifts and scalar spin-spin couplings. For noncrystalline proteins (protein aggregates, frozen solutions, lyophilized powders) and polycrystalline proteins, solid-state NMR methods such as magic angle spinning (MAS) and multiple pulse techniques have been developed for averaging anisotropic interactions (10, 11). These methods restore high-resolution to the NMR spectrum and permit detailed investigations of the structural and dynamical properties of proteins, protein-complexes, and crystals. One significant benefit of MAS is that information on the spatial anisotropy of the dipolar couplings and chemical shift can still be extracted from the high-resolution spectrum, and related to molecular structure.

The focus of this review is high-resolution, solid-state NMR methods, particularly magic angle spinning, and recent applications to proteins. We provide an overview of the types of protein systems that are especially suited for solid-state NMR investigations, and describe the methods that have been developed for utilizing the anisotropy of the chemical shift and dipolar interactions for structural studies. Solid-state NMR spectroscopy of *static* protein samples represents a related area that for the most part is not covered. Several reviews of solid-state NMR of biological systems that include static NMR studies have recently appeared (12–17).

SOLID-STATE NMR METHODS

The use of solid-state NMR to investigate molecular structure has been correlated with the development of methods to manipulate various terms in the nuclear spin Hamiltonian. In this section we discuss the dominant spin interactions in solid-state NMR spectroscopy and draw comparisons with solution NMR. For high-resolution structural studies of proteins with solid-state methods, the "magnetically dilute" spin 1/2 nuclei such as ^{13}C, ^{15}N, ^{31}P, and ^{113}Cd have thus far found the widest application. Consequently, much of our discussion focuses on the solid-state NMR methods used to study these spin systems. In this regard, the most widely applicable technique is magic angle spinning for resolving isotropic and anisotropic chemical shifts. Finally, methods are discussed for measuring dipolar couplings and chemical shift anisotropies, and for relating these interactions to molecular structure.

Spin Hamiltonians

The nuclear spin Hamiltonian describes the interactions that determine the NMR spectrum, and has the general form (11),

$$\mathcal{H} = \mathcal{H}_Z + \mathcal{H}_{RF} + \mathcal{H}_{CS} + \mathcal{H}_D^{II} + \mathcal{H}_D^{IS} + \mathcal{H}_J + \mathcal{H}_Q \qquad 1.$$

where

$$\mathcal{H}_Z = -\gamma \mathbf{H}_0 \cdot \mathbf{I}$$

$$\mathcal{H}_{RF} = -2\gamma H_1 (\cos \omega t) I_x$$

$$\mathcal{H}_{CS} = -\gamma \mathbf{H}_0 \cdot \hat{\sigma} \cdot \mathbf{I}$$

$$\mathcal{H}_D^{II} = \sum_{i \neq j} (\gamma_i \gamma_j \hbar) I_i \cdot \mathcal{D} \cdot I_j$$

$$\mathcal{H}_D^{IS} = \sum_i (\gamma_I \gamma_S \hbar) I_i \cdot \mathcal{D} \cdot \mathbf{S}$$

$$\mathcal{H}_J = \sum_{i \neq j} I_i \cdot \mathcal{J} \cdot I_j$$

$$\mathcal{H}_Q = \frac{(eQ)}{2I(2I-1)\hbar} \mathbf{I} \cdot \mathcal{V} \cdot \mathbf{I}. \qquad 2.$$

The Zeeman Hamiltonian ($\mathcal{H}_Z$) and radiofrequency Hamiltonian ($\mathcal{H}_{RF}$) describe the interaction of the nuclear spins with the external magnetic field ($\mathbf{H}_0$) and with the rf field [$H_1 (\cos \omega t)$]. Both the size of the Zeeman Hamiltonian, determined by $\mathbf{H}_0$, and the rf irradiation scheme are under the control of the experimenter. The remaining terms in Eq. 1 are due to *interactions* that exist at the molecular level, which are in general anisotropic and can be described by second rank tensors ($\mathcal{D}, \hat{\sigma}, \mathcal{J}, \mathcal{V}$). The "effective size" of these interactions can change dramatically in going from solution to the solid-state (Table 1) due to motional averaging. Specifically,

Table 1 Comparison of the size of various spin interactions in liquids and solids, and the methods used for solid-state NMR studies

Interaction	Liquids (Hz)	Solids (Hz)	Techniques
σ	$10–10^4$	$10–10^4$	Decoupling, MAS
J_{IS}	10^2	10^2	Decoupling
D_{II}	0	10^4	Multiple-pulse
D_{IS}	0	10^4	Decoupling, MAS
V	0	$10^5–10^6$	Quad echo, MAS

rapid isotropic motion in solution reduces each interaction to its isotropic value, which is one third the trace of the interaction tensor. The trace of the dipolar ($\mathscr{D}$) and quadrupolar ($\mathscr{V}$) interactions is zero, and these terms contribute to the solution NMR spectrum only through relaxation. In contrast, the chemical shift ($\hat{\sigma}$) and J-coupling ($\mathscr{J}$) tensors have non-vanishing traces and thus display an isotropic component in solution NMR spectra.

The chemical shift Hamiltonian ($\mathscr{H}_{CS}$) accounts for the "shielding" of the nuclear spins from the external magnetic field due to an opposing magnetic field induced by the motion of surrounding electrons. In a principal axis system (PAS) where $\hat{\sigma}$ is diagonal, the tensor is completely described by its three principal values ($\sigma_{11}, \sigma_{22}, \sigma_{33}$) and three vectors that orient the tensor within a molecular reference frame. Figure 1 illustrates

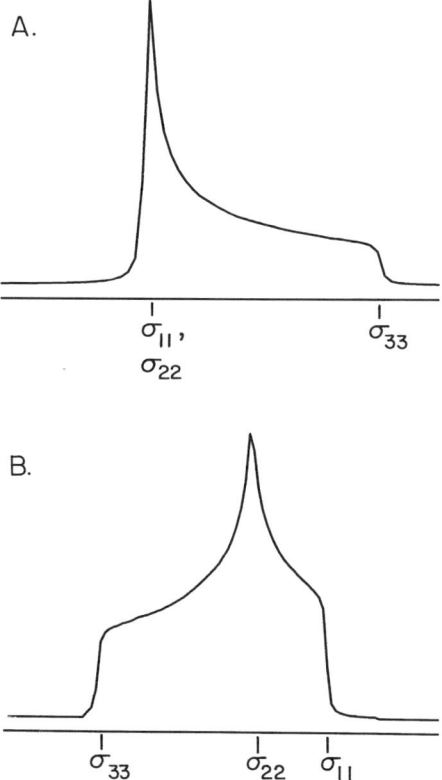

Figure 1 Powder lineshapes due to the chemical shift anisotropy observed in solid-state NMR spectra of static samples. (*A*) Axially symmetric lineshape. (*B*) Axially asymmetric lineshape. The principal values ($\sigma_{11}, \sigma_{22}, \sigma_{33}$) of the chemical shift tensor are indicated.

characteristic "powder lineshapes" that are observed in static samples and result from anisotropy in the chemical shift when the shift tensor is axially symmetric ($\sigma_{33} \neq \sigma_{22} = \sigma_{11}$) and axially asymmetric ($\sigma_{33} \neq \sigma_{22} \neq \sigma_{11}$). Lineshapes such as these are observed for dilute spin-1/2 nuclei when heteronuclear interactions are attenuated by proton decoupling. The chemical shift anisotropy is an *inhomogeneous* interaction (18), and as such can be averaged by MAS even when $\omega_r \ll |\mathcal{H}_{CS}|$, where ω_r is the spinning speed in a MAS experiment and $|\mathcal{H}_{CS}|$ is the size of the chemical shift anisotropy. Thus, powder lineshapes such as those illustrated in Figure 1 break up into a centerband at the isotropic shift, flanked by rotational sidebands spaced at $\omega_r/2\pi$. One important advantage of solid-state NMR is that the chemical shift anisotropy can in principle be directly related to molecular structure and environment once the axis system of the shift tensor is related to a molecular-fixed system. In solution, only the isotropic component of $\hat{\sigma}$ is retained and the *directional* information encoded in the shift anisotropy is lost.

The methods used for restoring high-resolution in solid-state NMR spectra depend on the spin species being observed. The magnetically dilute spin-1/2 nuclei, such as ^{13}C, ^{15}N, and ^{31}P, are the most amenable to high-resolution studies, since chemical shift anisotropy is the dominant spin interaction that is expressed in the NMR spectrum. Spectra of these nuclei are not complicated by homonuclear dipolar and quadrupolar effects (see below). Also, the heteronuclear interactions between dilute spin-1/2 nuclei and protons can be attenuated with proton decoupling. An additional advantage of dilute spin-1/2 nuclei that have a very low natural abundance, such as ^{13}C and ^{15}N, is that selective isotopic enrichment can provide site specific probes in biological macromolecules. The standard approach to obtain high-resolution spectra of dilute spin-1/2 nuclei is to use MAS in combination with cross-polarization (CP) and proton decoupling. The pulse sequence commonly used for obtaining solid-state NMR spectra is illustrated in Figure 2. CP involves a transfer of magnetization from the

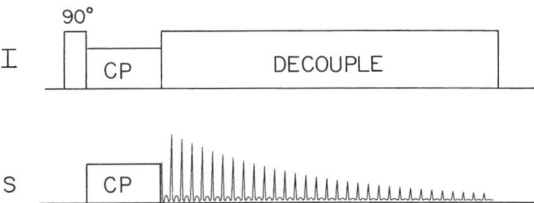

Figure 2 Standard CP-MAS pulse sequence for obtaining high-resolution NMR spectra of dilute spins (S) in the solid-state. The 90° pulse together with the cross-polarization (CP) pulses are employed to transfer magnetization to the S spins. Proton decoupling is applied during acquisition of the free induction decay and results in a high-resolution spectrum.

¹H spins to the dilute spin being observed and results in an *increase in sensitivity* (19). Also, the recycle delay can generally be *decreased* in this pulse sequence, since the T_1 of the protons in the sample is generally much shorter than the T_1 of the dilute spin. MAS methods are discussed extensively in the next section.

The homonuclear ($\mathcal{H}_D^{II}$) and heteronuclear ($\mathcal{H}_D^{IS}$) dipolar Hamiltonians describe the direct dipole-dipole coupling between like and unlike spins, respectively. The strength of these couplings (D) depends on the internuclear distance (r_{ij}), and the angle (θ_{ij}) that the internuclear vector makes with respect to the external field

$$D_{ij} = (1/r_{ij}^3)(3\cos^2\theta_{ij} - 1). \qquad 3.$$

The homonuclear dipolar Hamiltonian is an example of a *homogeneous* interaction, and is not averaged by MAS unless the condition $\omega_r \gg |\mathcal{H}_D^{II}|$ is satisfied. The protons in a protein (and in most molecular solids) are homogeneously broadened spin systems, since they are strongly coupled owing to their high (isotopic) abundance and large gyromagnetic ratio. This strong coupling results in broad lines (25–50 kHz) that overwhelm the smaller (~5 kHz) but more interesting chemical shifts. Multiple pulse methods have been developed to attenuate the homonuclear dipolar interaction (20–22), and in combination with MAS (23, 24) can yield proton linewidths on the order of 1–2 ppm. Although this represents a tremendous improvement in ¹H spectral resolution, it only begins to approach the resolution attained in solution. As a result, ¹H solid-state NMR spectroscopy is not as successful as high-resolution solution NMR for structural studies. The heteronuclear dipolar term ($\mathcal{H}_D^{IS}$) is inhomogeneous and therefore behaves in the same manner as the chemical shift. A high-resolution "dipolar spectrum" can be analyzed to determine heteronuclear bond distances and the mutual orientation of shift and dipolar tensors. The techniques for obtaining dipolar spectra and determining the heteronuclear dipolar couplings are discussed in more detail below.

The electron mediated or J-coupling Hamiltonian ($\mathcal{H}_J$) has a nonzero trace and consequently isotropic J-couplings are observed in *solution* NMR spectra. In 2D spectroscopy, the J-couplings provide a mechanism for establishing connectivities between atoms in a molecule. Both $^{13}C-^1H$ and $^{13}C-^{13}C$ J-couplings have been observed in solid-state NMR spectra (25–28). However, in general the J-couplings are not resolved, since their size is often less than the solid-state NMR linewidth.

The quadrupolar Hamiltonian ($\mathcal{H}_Q$) is nonzero for nuclei with spin $I \geq 1$. The spin 1 nuclei most often encountered in biological NMR studies are ^2H and ^{14}N and both of these spin species exhibit quadrupolar couplings that can be much larger than either the chemical shift of dipolar interactions. The first-order quadrupole interaction ($\mathcal{H}_Q^{(1)}$) is inhomogeneous, and in some cases MAS has been used to narrow quadrupole powder patterns. However, sample spinning has not been generally applicable to these spin species, since molecular motion often leads to very short ^{2}H and ^{14}N T_2s, and in many cases ^{14}N quadrupole couplings are large enough (~ 3–5 MHz) that setting the magic angle accurately is difficult. In contrast, *static* ^{2}H-NMR has been widely exploited for investigating dynamics in proteins (15, 16, 29), membrane phospholipids (12, 30), and nucleic acids (31, 32). These studies have taken advantage of the sensitivity of the deuterium lineshape to molecular motion by using quadrupolar echo spectroscopy (33). Recently, ^{14}N overtone spectroscopy has been introduced as an approach for obtaining ^{14}N spectra of oriented and polycrystalline samples (34).

There are many spin $I > 1$ nuclei that are potentially very interesting for studies of proteins, including ^{17}O ($I = 5/2$), ^{23}Na ($I = 3/2$), ^{39}Mg ($I = 5/2$), ^{35}Cl ($I = 3/2$), ^{43}Ca ($I = 7/2$), and ^{67}Zn ($I = 5/2$). In most of these cases, the higher order transitions (e.g. $-3/2 \to -1/2$) are out of the frequency range of the rf excitation, and, in addition, they are generally broadened by lattice imperfections and dynamic effects. As a result, the central $1/2 \to -1/2$ transition is selectively excited and observed. This transition is free from the first-order quadrupolar interaction, but is still broadened by second-order quadrupole effects, with the broadening being proportional to the square of the quadrupole coupling and inversely proportional to the field. In contrast to $\mathcal{H}_Q^{(1)}$, the second-order quadrupole term ($\mathcal{H}_Q^{(2)}$) behaves homogeneously, and its size is such that high spinning speeds are necessary to achieve significant narrowing in MAS experiments (35–37). Furthermore, the "magic angles" for $\mathcal{H}_Q^{(2)}$ differ from those for the chemical shift, and to date schemes for narrowing both interactions have not been implemented. However, because of its potentially wide applicability, this remains an area of intense interest.

Magic Angle Spinning

The dominant broadening mechanisms in solid-state spectra of dilute spin-1/2 nuclei are chemical shift anisotropy and heteronuclear dipolar interactions. In solution, these terms are averaged by rapid isotropic motions, whereas in the solid-state inhomogeneous interactions such as these can be averaged by *mechanical* spinning of the sample at an angle of $\theta_m = 54.7°$ (the magic angle) relative to the external magnetic field (18, 38–41). The resulting spectra exhibit centerbands at the isotropic resonance frequency, flanked by sidebands spaced at the spinning frequency. As is seen below, these sideband intensities can be analyzed to determine the shift anisotropies and heteronuclear dipolar couplings.

CHEMICAL SHIFT ANISOTROPY   MAS imparts a *periodic* time dependence to the orientation of each "crystallite" in a polycrystalline or powdered sample. Sample rotation causes the chemical shift of each crystallite to sweep through a range of frequencies each rotor period; the average frequency experienced over a rotor cycle corresponds to the isotropic chemical shift. The effect of sample rotation depends on the relationship between the spinning speed (ω_r) and the anisotropy ($\delta = \sigma_{33} - (1/3)\text{Tr}(\hat{\sigma})$) of the interaction being averaged. In the regime where $\omega_r \gg \delta$, the broad chemical shift powder spectrum collapses into a narrow line at the isotropic chemical shift. For $\omega_r \ll \delta$, the spectrum contains additional lines (rotational sidebands) spaced at the spinning speed that result from refocusing of the magnetization each rotor period. This refocusing generates "rotational echoes" in the time domain NMR signal (the free induction decay), which, when Fourier-transformed, produce the observed sidebands in the frequency domain. Figure 3 illustrates the effect of MAS on spectra of zinc acetate. The methyl and carboxyl groups in zinc acetate both exhibit nearly axially symmetric static spectra, but with very different anisotropies. Figures 3A–C show how the number and intensity of the rotational sidebands change as the spinning speed is increased from 0 to 4.2 kHz.

The relative intensities of the rotational sidebands can be analyzed to obtain the principal values of the chemical shift tensor (18, 42). As mentioned briefly above, both the isotropic and anisotropic components of the chemical shift provide information on molecular structure. A change in the isotropic chemical shift in response to a change in molecular structure or environment, such as hydrogen-bonding, may be *localized* in one or more of the three principal tensor elements. Such a localized shift provides a means for establishing the *origin* of a change in the isotropic shift. The first evidence of localized shifts in the individual tensor elements was reported by Waugh and co-workers (19) in a series of ^{13}C spectra of carbonyl compounds.

When only the isotropic chemical shift is of interest, it is advantageous to suppress the rotational sidebands in MAS spectra, particularly in spectra that contain several isotropic resonances. The simplest method for achieving this is to increase ω_r so that the condition $\omega_r \gg \delta$ is satisfied. However, in many situations this is impractical and, consequently, two approaches have been developed for controlling sideband intensities with rf pulse trains. The first method, *chemical shift scaling*, reduces the effective size of δ by application of a train of paired pulses having opposite phases (43, 44). A reduction in δ transfers spectral intensity to the centerband and attenuates the sidebands. An interesting but potentially annoying feature of chemical shift scaling methods, or multiple pulse MAS experiments in general, is the appearance of spurious "rotor lines" that result when a

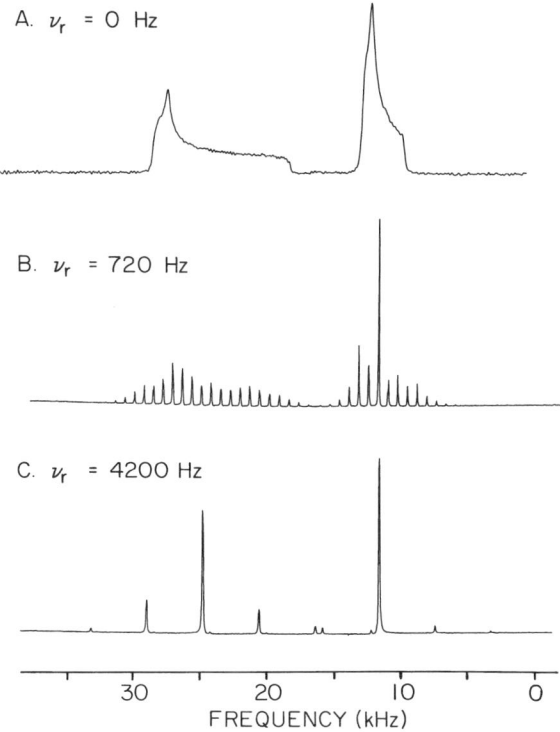

Figure 3 Static (*A*) and MAS (*B* and *C*) ^{13}C spectra of zinc acetate showing the effect of MAS on chemical shift anisotropy powder patterns. The spectra were obtained with the pulse sequence shown in Figure 2. The rotational sidebands observed in (*B*) and (*C*) are spaced at the spinning frequency. The breadths of the powder patterns are 10.2 kHz for the carboxyl and 3.2 kHz for the methyl resonances. Note that the narrower methyl pattern is almost completely averaged at $v_r = 4.2$ kHz.

genuine resonance is located near the carrier frequency or multiples of the spinning frequency. The second method, *total suppression of sidebands, or TOSS*, modulates the phase of the sidebands from individual crystallites through a four (45) or six π pulse sequence (46), but keeps the phase of the centerband constant. The net effect of a TOSS pulse sequence is to cancel the sideband intensities while retaining the centerband. A hybrid experiment involving both scaling and TOSS has been described that has the advantage of attenuating sidebands while transferring intensity to the centerband (47).

HETERONUCLEAR DIPOLAR EFFECTS IN CHEMICAL SHIFT SPECTRA MAS is also effective in averaging heteronuclear dipolar interactions between dilute spin-1/2 nuclei such as ^{13}C and ^{15}N. In these systems, the lines are narrowed, since $\mathcal{H}_D^{IS}$ is an inhomogeneous interaction. However, the sideband intensities are no longer a function solely of the chemical shift anisotropy, but also contain contributions from the heteronuclear dipolar coupling. Coupling between a dilute spin-1/2 species and a dilute quadrupolar spin ($I \geq 1$) also influences the sidebands, but in a slightly different fashion. For these spin systems, the quadrupole coupling is often not small compared to the Zeeman Hamiltonian for the quadrupolar spin, and consequently the NMR energy levels are an admixture of the Zeeman and quadrupolar terms. In this situation, the MAS centerbands and sidebands appears as miniature powder patterns rather than sharp Lorentzian resonances. These effects were first noticed in ^{13}C spectra of amino acids and peptides at low magnetic fields where ^{13}C–^{14}N dipolar coupling produced substantial broadenings (48). Dramatic examples of these effects are observed when the quadrupole couplings are larger as in ^{31}P–^{63}Cu systems (49). Finally, for coupling between a dilute spin and strongly dipole-coupled ^{1}H spins, MAS is ineffective, since the strongly coupled ^{1}H spins introduce a *homogeneous* component to the heteronuclear dipolar interaction. In this case, it is necessary to use homonuclear proton decoupling to narrow the NMR resonances completely.

The strong coupling that exists between dilute nuclei and directly bonded protons has been exploited by Opella & Frey (50) as a means for distinguishing protonated from nonprotonated nuclei. They inserted a delay without proton decoupling between the cross-polarization pulse and data acquisition in the standard CP-MAS pulse sequence. The delay broadens the resonances of the protonated nuclei, and, if sufficiently long (normally 40–50 μsec), produces a spectrum of only the nonprotonated resonances. When sidebands are present, it is necessary to introduce a refocusing pulse with the sequence. These methods can be combined with TOSS sequences to yield only the centerband resonances of nonprotonated nuclei (46). A similar method uses a delay without decoupling as a "T_1 filter" to remove resonances with short T_1s (51, 52). Such methods are an aid to spectral simplification and assignment.

We mentioned above that large proton linewidths in MAS spectra greatly limit the utility of ^{1}H NMR in the solid-state. One approach for regaining resolution that may be of benefit in some circumstances is through heteronuclear correlation spectroscopy. 2D heteronuclear chemical shift correlation techniques with MAS have been applied to ^{13}C–^{1}H systems (53). The superior dispersion and resolution of the carbon chemical shifts effectively separate overlapping proton resonances in the second dimension. In this way, the overall proton resolution is increased over the 1D spectra. These experiments may aid in the assignments of proton and carbon chemical shifts, and allow one to determine the *connectivities* between spins. These methods can also be extended to $^{13}\text{C}-^{15}\text{N}$ heteronuclear correlation.

CHEMICAL EXCHANGE   A third source of line broadening in spectra of spin $I = 1/2$ nuclei arises from chemical exchange or molecular motion. In MAS spectra, the linewidths are narrow in the fast and slow exchange regimes, but broaden when the spins are in intermediate exchange. Intermediate exchange broadening occurs when the timescale of the motion (τ_c) is comparable to the rotational period of the sample (ω_r), the shift anisotropy (δ), or the difference in chemical shift between exchanging sites. Exchange broadening can also occur if the motion is on the timescale of the decoupling field (54). The dynamics of the exchange process are customarily studied by recording the lineshapes as a function of temperature and by simulating these with appropriate models (18, 55, 56). In biological systems, the thermal stability of the samples often precludes narrowing resonances by increases of temperature. Thus, the approach to removing exchange broadening effects from the spectra has typically involved low-temperature spectroscopy. MAS spectra can routinely be obtained down to ~150K by using cooled N_2 as the spinning gas, down to 77K if the sample is in contact with liquid nitrogen (57), and as low as 5K by using cooled He gas (58).

Methods for Measuring Dipolar Interactions

Direct dipole-dipole couplings are through-space interactions between nuclear spins. In many circumstances these couplings can be extracted from the solid-state NMR spectrum in the form of sideband intensities or direct splittings in the spectra. These data then provide information on bond orientations and internuclear distances.

DIPOLAR-CHEMICAL SHIFT SPECTROSCOPY   The methods for obtaining high-resolution solid-state NMR spectra generally involve the suppression of dipolar interactions in favor of the underlying chemical shifts and J-couplings. In some instances dipolar splittings can be directly observed (59), although more often the chemical shift and dipolar interactions are of the same magnitude and are not resolved. The separation of chemical shift and heteronuclear dipolar interactions is possible by 2D methods, similar in principle to the methods used in solution NMR (60–63). One pulse sequence used to record dipolar/chemical shift spectra from spin pairs, such as $^{13}\text{C}-^{1}\text{H}$ and $^{15}\text{N}-^{1}\text{H}$, is shown in Figure 4. In this experiment, a variable delay, t_1, is inserted between cross-polarization and acquisition of the spectrum, and during the delay, proton decoupling is replaced by a multiple pulse sequence (MREV-8). The MREV-8 cycle attenuates *homonuclear* dipolar coupling and allows the spins to evolve under a scaled

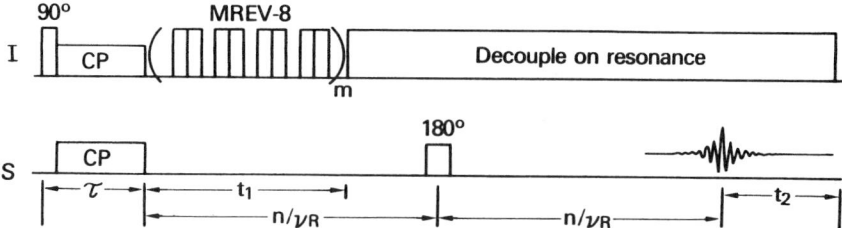

Figure 4 General 2D dipolar-chemical shift pulse sequence used to obtain heteronuclear bond distances. The timing of the 180° pulse is synchronized with the rotor period as indicated.

heteronuclear dipolar Hamiltonian. Fourier transformation with respect to both the t_1 and t_2 (acquisition period) time variables produces the two-dimensional spectrum, consisting of sidebands produced by the chemical shift anisotropy and the heteronuclear dipolar coupling. The sideband intensities in the 2D spectrum provide bond lengths for C–H and N–H bonds that are accurate under favorable conditions to within 0.005 Å (62), and also information on the relative orientation of the chemical shift and dipolar tensors. The relative orientation of tensorial interactions can often aid in the characterization of molecular structure (64). The pulse sequence shown in Figure 4 uses a refocusing π pulse to separate chemical shift and dipolar interactions completely in the two frequency dimensions. An alternate version of this sequence that omits the refocusing pulse has been suggested. This approach provides better signal-to-noise ratios at the sacrifice of incomplete separation of the interactions (65).

The dipolar-chemical shift methods work well for strong dipolar couplings as in N–H and C–H bonds. However, small heteronuclear couplings between dilute heteronuclear spin 1/2 pairs are averaged by MAS. Recently, two methods have been proposed for measuring weak heteronuclear couplings in rotating solids (66). In the first method, the dipolar coupling is "reintroduced" into the MAS spectrum by matching rf irradiation with the spinning frequency. The result is a high-resolution MAS spectrum where the centerbands and rotational sidebands exhibit splittings due to the heteronuclear coupling. The second method utilizes the dephasing properties of π pulse trains and difference methods to obtain spectra of only the spins dipole coupled to other dilute spin $\frac{1}{2}$ nuclei.

MAGNETIZATION TRANSFER   Magnetization transfer measurements (often referred to as spin exchange or spin diffusion) are a possible solid-state approach for obtaining distance information between spins not directly bonded, similar to the nuclear Overhauser effect (NOE) measurements in solution NMR. There are both 1D and 2D experiments for measuring magnetization transfer in MAS experiments. Magnetization transfer in the 1D experiments is monitored by changes in peak intensities following inversion of a selected resonance (67, 68), while magnetization transfer in the 2D experiment is exhibited by cross-peaks between dipolar coupled spins (69–71). These methods have been used to observe dipolar coupling in amino acids and cyclic peptides (72). The magnetization transfer measurements require that the two exchanging nuclei have different chemical shifts, and in many instances the difference in chemical shift is larger than the dipolar coupling. In this case, the transfer of magnetization is not energy conserving, and consequently the rate of transfer is slow. Under these conditions it is difficult to measure magnetization transfer rates accurately over long distances. Recently, a method for selectively enhancing magnetization transfer has been developed that relies upon matching the rotational frequency of the sample in a MAS experiment with the difference in chemical shift between the two sites (28, 73). When "rotational resonance" is achieved, the "flip-flop" term in the dipolar Hamiltonian is effectively reintroduced and the transfer rate is enhanced. For ^{13}C and ^{15}N, these methods work best by selectively labeling the two sites of interest. This approach has thus far been tested experimentally on di^{13}C-labeled zinc acetate, and computer simulations suggest that ^{13}C...^{13}C distances can be measured between nuclei up to ~4–6 Å apart.

PROTEINS

Model Systems

The first stage in structural studies of proteins often involves characterizing the chemical shift anisotropy and dipolar interactions in amino acids, oligopeptides or prosthetic groups. The principal values of the shift tensor can readily be extracted from the discontinuities in the static spectrum (Figure 1) or from the intensities of the rotational sidebands in a MAS spectrum. In fact, the shift tensors for a wide range of model compounds incorporating important functional groups have been tabulated (11). These studies yield the principal values of the chemical shift tensor, but not its orientation relative to a molecular axis system. Determining both the principal values of the shift tensor and their orientation *relative to a molecular-fixed axis system* is necessary to characterize the chemical shift anisotropy of a specific site completely (74, 75). The traditional approach to this problem involves single crystal studies of the molecule of interest (76, 77). An indirect approach that can yield information on relative tensor orientation involves the dipolar-chemical shift methods outlined above.

Single crystal tensor studies have been reported for several amino acids (78–80) and the dipeptide glycylglycine (81). The single crystal analysis of the ^{15}N chemical shift and ^{15}N–^{13}C dipolar tensor for the peptide bond in glycylglycine is particularly relevant, since the unique shift tensor element does not lie along the N–H bond vector as might be assumed, but is offset by 21°. These measurements are in agreement with the 2D dipolar-chemical shift methods (82). A more extensive analysis of the magnitude and relative orientations of chemical shift tensors in dipeptides has exploited the approach of ^{13}C–^{15}N labeling the peptide bond (83–86). In this case, measurements can be made on polycrystalline samples, since the unique axis of the dipolar tensor is known to lie along the ^{13}C–^{15}N bond. These studies have shown that the lattice environment has a strong influence on the chemical shift tensors of peptide carbonyl carbons.

Studies on amino acids and polypeptides (87, 88) have also been undertaken to characterize the chemical shift dependence on pH and secondary structure. These studies have recently been reviewed (89). Extensive studies of polypeptides labeled with ^{13}C and ^{15}N have shown that the peptide resonances are sensitive to local conformation and are not strongly affected by the amino acid sequence (90–93). High-resolution spectra have also been obtained of *cyclic* peptides as model systems to study local conformations, hydrogen-bonding between peptide amino acids, and the dynamics of sidechain functional groups (89, 94). Figure 5 presents ^{13}C-NMR spectra of the cyclic peptide cyclo(gly-pro-gly)$_2$ in the solid-state (polycrystalline) with MAS (A) and in solution (B). An interesting aspect of these spectra is the splitting of the C^β resonance of the proline residues in the solid-state. This is attributed to asymmetry in hydrogen bonding and local conformation of the proline residues in which one of the prolines forms an intramolecular hydrogen bond and has a ψ angle of 126°, while the other proline does not form such a bond and has a ψ angle of 36° (89).

Model compound studies are also an important element in solid-state NMR studies of co-enzymes and metal ions bound to proteins. Ellis and co-workers have characterized the shift tensors in Cd complexes, particularly those containing oxygen ligands, that serve as model systems for Cd-binding proteins (95–97). These workers have developed tensor element–structure correlations that describe general symmetry features of oxygen ligands. Three such relationships are (97):

1. The least shielded tensor element is aligned perpendicular to the plane containing water oxygens.
2. If the coordination sphere is devoid of water oxygens, then the deshielded element is oriented to maximize the short-bond oxygen shielding contribution.
3. The most shielded tensor element is directed perpendicular to the longest cadmium-oxygen bond.

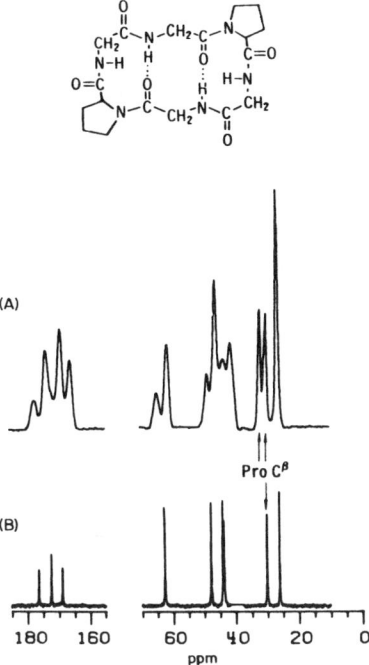

Figure 5 ^{13}C CP-MAS (*A*) and solution (*B*) NMR spectra of the cyclic peptide cyclo (gly-pro-gly)$_2$ (89).

These relationships give an idea of the type of information that can be extracted from shift tensor studies on single crystals.

Model studies have also been undertaken on quinones (98), porphyrins (99, 100) and chlorophyll a (101), components of photosynthetic systems, and on retinal Schiff bases, the chromophore in visual pigments and bacteriorhodopsin (102). A recent example of the use of chemical shift anisotropy in the structural analysis of a membrane protein is illustrated in Figure 6. The structure in Figure 6 is a portion of the photoreactive retinal chromophore in the membrane protein bacteriorhodopsin (bR) (52, 102). The isotropic ^{13}C chemical shift of the C-5 position of the chromophore in bR is 16 ppm downfield of the 6-s-*cis* retinal model compounds. Comparison of the individual shift tensor elements of 6-s-*trans* and 6-s-*cis* model compounds shows that isomerization about the C_6–C_7 single bond results in a localized shift in the σ_{33} shift tensor element from 217 and 237 ppm. The observation of the σ_{33} element at 237 ppm *in the protein* is the signature of a 6-s-*trans* bond. Further comparison of the

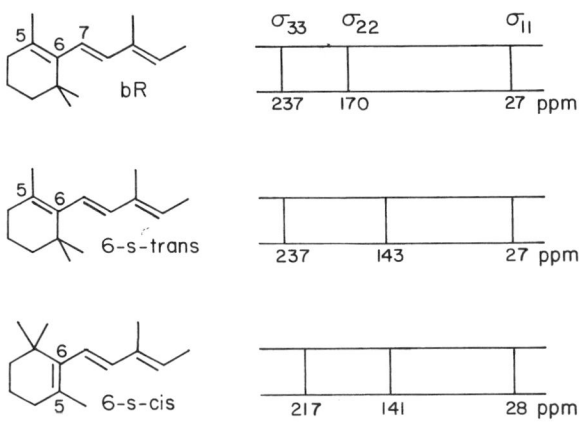

Figure 6 Chemical shift tensor elements of the 5-^{13}C resonance of the retinal chromophore in bacteriorhodopsin (bR), compared with those of 6-s-*trans* and 6-s-*cis* retinoic acid (52). Isomerization of the C_6–C_7 single bond in the model compounds leads to a 20 ppm shift in the σ_{33} tensor element. The shift in σ_{22} from ~142 ppm in the model compounds to 170 ppm in bR is attributed to a negative *protein* charge near the 5-position in bR.

protein with the 6-s-*trans* model compound shows a localized shift in the σ_{22} element that is attributed to a negative protein charge near the C-5 position of the chromophore. These results suggest that the large isotropic shift observed in the protein originates from two distinct sources, chromophore isomerization and a nearby protein charge, that have localized effects on the shift tensor elements.

Membrane Proteins

Membrane proteins are well suited for structural studies by solid-state NMR. Biological membranes greatly restrict the motion of embedded proteins and consequently are difficult to study by solution NMR methods. Also, only a few membrane proteins have been crystallized and studied by X-ray diffraction methods. High-resolution MAS studies have been undertaken on bacteriorhodopsin and the visual pigment rhodopsin by using specific isotopic labels to enhance the sensitivity of selected protein sites. Several natural abundance ^{13}C studies of membrane proteins have also been reported, including one that deals with bacterial photosynthetic reaction centers (103).

Bacteriorhodopsin and rhodopsin are both integral membrane proteins that contain the vitamin A aldehyde retinal as a photoreactive chromophore. High-resolution MAS studies have focused predominantly on the structure and environment of the chromophore. In these studies the retinal chromophore is removed and the apoprotein is regenerated with specifically ^{13}C-labeled retinal. In bR, the retinal functions as a light-driven proton-ion pump. Absorption of light initiates a photochemical reaction, and the pigment passes through a series of intermediates. Solid-state spectra of the dark-adapted pigment have yielded results on the structure of the retinal and specific protein charges in the retinal binding site (52, 104–106). These studies indicate that the two components of dark-adapted bR, bR_{568} and bR_{548}, contain all-*trans* and 13-*cis* retinals, respectively, in agreement with other experiments. In addition, the spectra show that the C=N bond in bR_{568} is *anti* and in bR_{548} is *syn*, and that in both components the C_6–C_7 bond is in the s-*trans* conformation. Recently, spectra have been obtained of one of the photointermediates of bR by blocking the thermal decay of the intermediate at 250K (106). Figure 7 shows MAS spectra of bR and the photointermediate M_{412} regenerated with 13-^{13}C retinal. The two components of dark-adapted bR are observed as two sharp resonances at 165.7 and 169.7 ppm in Figure 7*A*, while the M412 resonance is observed at 146.7 ppm in Figure 7*B*. A difference spectrum between 13–^{13}C M_{412} and *unlabeled* bR in Figure 7*C* highlights the ^{13}C label. The 13–^{13}C M_{412} chemical shift indicates that the retinal chromophore has an unprotonated retinal-lysine Schiff base linkage. These studies illustrate the use of low-temperature solid-state NMR methods for obtaining spectra of reaction intermediates, and difference spectroscopy (107, 108) for removing natural abundance backgrounds. Analogous experiments may prove to be useful in trapping enzymatic intermediates.

Solid-state NMR spectra have also been obtained of rhodopsin containing ^{13}C-labeled retinal chromophores (108–110). In the case of rhodopsin, rotational diffusion of the protein in the membrane significantly broadens the protein resonances. Two approaches for restricting protein motion have been used in solid-state NMR work, low-temperature and lyophilization in diphytanoylglycerophosphocholine. These studies have revealed the structure of the C_6–C_7 and C=N bonds of the chromophore. Furthermore, comparison of the chemical shifts of the chain positions with retinal model compounds have provided support for a negative protein charge near the C–12...C–13 region of the chromophore. A negative protein charge has previously been implicated in the mechanism for red-shifting the visible absorption bond of rhodopsin.

An interesting observation in the rhodopsin studies is that diffusion of the protein within the membrane results in loss of the protein NMR signal above $-10°$C. Motion on the timescale of the diffusive process ($\tau_c = 20$ μsec) might interfere with ^{1}H–^{13}C decoupling during data acquisition and

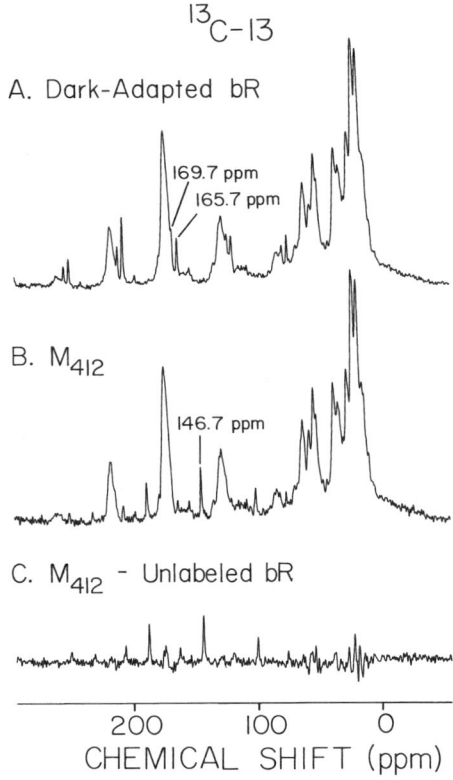

Figure 7 ^{13}C CP-MAS spectra of bacteriorhodopsin containing a retinal chromophore ^{13}C-labeled at position 13. The spectrum in (*A*) is of the dark-adapted pigment and exhibits two sharp resonances (and their associated sidebands) corresponding to two distinct forms of the pigment. The spectrum in (*B*) is of the M_{412} photoreaction intermediate that has been trapped at 250K. A difference spectrum is shown in (*C*) between M_{412} and a natural abundance spectrum of the protein.

may also reduce cross-polarization efficiency by reducing the proton $T_{1\rho}$ of the protein (24, 46). Whether this is a general problem (and low temperature methods are a "standard" solution) in MAS studies on membrane proteins remains to be seen.

Enzymes and Soluble Proteins

Solid-state NMR provides a unique method for structural studies of enzymes and soluble proteins that may complement and extend structural methods such as X-ray diffraction and solution NMR. High-resolution MAS experiments have been reported on metalloproteins (111–113), heme-proteins (114), α-lytic protease (115, 116), alanine racemase (117), and ribonuclease A (118).

Solid-state NMR studies of metal coordination sites in proteins have taken advantage of the sensitivity of the chemical shift tensor to metal-ligand coordination. Several proteins, including parvalbumin and concanavilin A, that coordinate calcium and zinc have been studied by using MAS methods. The approach in these systems is to substitute calcium or zinc with divalent cadmium cations. Cadmium is a spin 1/2 nucleus and exhibits a large chemical shift dispersion in complexes with oxygen, nitrogen, and sulfur. Unfortunately, in many of these systems the ^{113}Cd lines are tremendously broadened. Ellis, Bryant and co-workers (111) have shown that there is considerable disorder at the metal binding site in lyophilized metalloprotein samples and that hydration can significantly narrow broad ^{113}Cd resonances. Solution NMR studies have demonstrated facile metal-ligand exchange in several metalloproteins, such as carboxypeptidase A, where a ^{113}Cd resonance can only be observed in the presence of enzyme inhibitor (119). Low-temperature MAS methods offer a potentially attractive approach for halting the exchange process and for studying the metal coordination site in such cases.

The heme-containing proteins myoglobin and hemoglobin have been studied by both MAS and static NMR methods. ^{13}C- and ^{2}H-labeling of sperm whale myoglobin (114) was used to probe the dynamics of various amino acids. Met-55 and Met-131 were shown to exhibit different motions based on cross-polarization transfer rates and proton $T_{1\rho}$ measurements. The ^{13}C chemical shifts for these positions were similar in solution and crystalline solid-state spectra, suggesting that the average solution and crystal conformations were the same.

Alanine racemases are a group of pyridoxal-5-phosphate containing enzymes that catalyze the racemization of L- and D-alanine, the latter being an essential component of bacterial cell walls. Biochemical studies have shown that alanine racemase can be inactivated by the alanine analogue, 1-(aminoethyl)phosphonic acid (Ala-P). The protein is a dimer (MW = 78,000) in the molecular weight range suitable for MAS studies. Low-temperature ^{15}N-MAS NMR spectra of ^{15}N-Ala-P complexed with alanine racemase demonstrated that inactivation was due to transaldimation of the native enzyme, forming an Ala-P-pyroxidal phosphate Schiff base linkage. These studies resolved a conflict in the interpretation of biochemical studies that suggested that the Ala-P enzyme complex did not form an imine.

Solid-state NMR spectra of *protein crystals* of α-lytic protease have been obtained in order to resolve conflicting results from X-ray diffraction and solution NMR studies. α-Lytic protease is one of the serine proteases whose catalytic mechanism involves the sidechain groups of serine, histidine, and aspartate, known as the "catalytic triad." X-ray diffraction studies have concluded that a hydrogen bond does not exist between the serine and histidine groups in the resting enzyme. However, ^{15}N chemical shifts of the catalytic histidine residue in high-resolution solution spectra of α-lytic protease argue for a Ser-His hydrogen bond (120). The possibility that the solution and crystal structures of the protein are different was addressed by solid-state NMR studies of protein crystals containing ^{15}N-labeled histidine (Figure 8). These studies supported the solution NMR

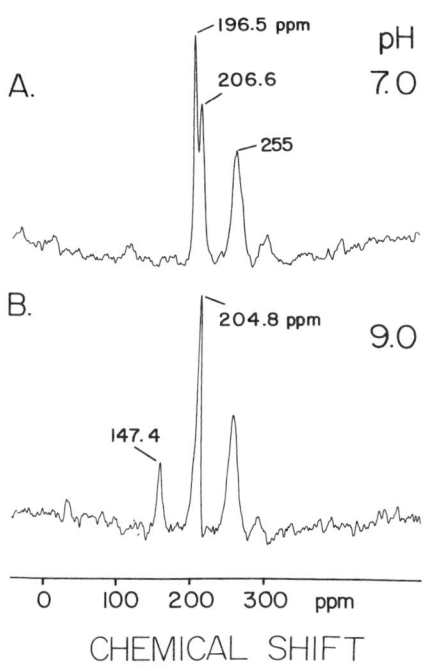

Figure 8 ^{15}N CP-MAS spectra of α-lytic protease containing a single histidine ^{15}N-labeled at the two nitrogens on the imidazole ring. At the high pH (B), where the imidazole ring is neutral and catalytically active, the ^{15}N chemical shifts of the two nitrogens are 204.8 and 147.4 ppm, respectively. The observation of a 147.4 ppm chemical shift provides evidence for a strongly hydrogen-bonded nitrogen. The chemical shift of this nitrogen hydrogen bonded to H$_2$O in model compounds is at ∼128 ppm (120).

results by revealing a strong Ser-His hydrogen bond in the catalytic active site.

Ribonuclease A is another enzyme that has been well-characterized by X-ray analysis and solution NMR. These studies use enzyme inhibitors, such as 2′- and 3′-cytidine monophosphate (CMP), as models of enzyme-substrate complexes, and have generated models of the nucleotide-RNase complex in which the phosphate group interacts with lysine 41 and histidine 119. Solid-state ^{31}P-NMR studies have been undertaken to compare the solution and crystal structures of the complex directly, and to examine the dynamics of the substrate bound to the protein. Both the 2′-CMP (118) and the 3′-CMP complexes (T.-H. Huang, W. Gilbert, G. A. Petsko, R. G. Griffin, unpublished results) have been studied and the results argue that the phosphate is in the dianionic state. In addition, studies on the dynamics of the 3′ CMP-RNase complex show that the motion of the phosphate can be modeled by a two-site hopping mechanism.

Structural Proteins

The structural components of cells are generally large macromolecules or polymers. Both MAS and stationary methods have been used to investigate structural proteins, such as peptidoglycan (121, 122), viral coat proteins (15, 16, 123), collagen (124, 125), elastin (126), and insect cuticle (127).

Peptidoglycan is formed from alternating N-acylated residues of glucosamine and muramic acid, and short peptide chains cross-linked together. Schaefer and co-workers (121) were able to label alanine and lysine groups in the glycoprotein with ^{13}C and ^{15}N and to use double cross-polarization (cross-polarization from ^{1}H to ^{15}N to ^{13}C) experiments to measure cross-linking in intact cell walls. They found that there are no terminal D-Ala or D-Ala-D-Ala units in uncross-linked chains. In a related study, the extent of cross-linking between the chitin and protein components in insect cuticle was investigated by these workers with CP-MAS methods (127). In this case, they found that covalent bonds form between the ring nitrogens of histidine residues and the ring carbons of catecholamines.

The outer protein casing of the filamentous bacteriophage (fd) is composed of many copies of a 50 amino acid coat protein. In a comprehensive study of the fd coat protein, Opella and co-workers have used solid-state NMR methods to determine both the structure and dynamics of the protein in the virus particle and in a membrane-bound form. A major structural difference in the protein between these two states involves the dynamics of the C-terminal residues, which are rigid when the protein forms the outer protein casing of the virus but are highly mobile when the protein is free in lipid bilayers. An important aspect of these studies is the development of solid-state NMR approaches for investigating the tertiary structure of a protein (16).

Future Developments

High-resolution solid-state NMR methods have become increasingly important in studying a range of biological systems. This trend should continue as biochemists and biophysicists become more familiar with the type of problems best suited for solid-state NMR studies. Although this review has been limited to proteins, the methods discussed are equally applicable to carbohydrates, nucleic acids, lipids, and other biomolecules. At present, solid-state NMR works well for systems that can be isotopically labeled at specific sites. Single sites enriched with ^{13}C or ^{15}N can be observed in proteins as large as 150 kD, although the quantity of protein necessary (50–150 mg) is generally much greater than for other spectroscopic techniques. Several developments should further boost the power and versatility of these methods. Dynamic nuclear polarization (DNP) can lead to dramatic increases in sensitivity by transferring polarization from electron to nuclear spin systems. To date, enhancements of ~ 25 have been achieved in favorable cases at room temperature (J. Schaefer, E. O. Stejskal, personal communication), and larger polarizations can be expected in low-temperature experiments. Efforts are underway to move DNP experiments from the current 60 MHz to 210 MHz ^{1}H-NMR fields. High-speed spinning systems (>15 kHz) are being developed that will greatly facilitate the averaging of spin interactions by MAS. Finally, advances in genetic engineering should increase the availability of proteins in large quantities and aid in the incorporation of specific isotopic labels.

ACKNOWLEDGMENTS

We acknowledge the support of the National Institutes of Health (GM-23403, GM-23289, GM-25505, and RR-00995). The authors wish to thank Dan Raleigh, Terry Oas, Andrew Kolbert, Valérie Copié, Huub DeGroot, and Ann McDermott for critical comments, and Jim Roberts, Julia Speyer, and Dan Raleigh for several of the figures. S.O.S. was the recipient of an NIH Postdoctoral Fellowship (GM-10502-02).

Literature Cited

1. Wemmer, D. E., Reid, B. R. 1985. *Ann. Rev. Phys. Chem.* 36: 105–37
2. Ernst, R. R., Bodenhausen, G., Wokaun, A. 1987. *Principles of Nuclear Magnetic Resonance in One and Two Dimensions.* Oxford: Clarendon Press. 610 pp.
3. Bax, A. 1984. *Two-Dimensional Nuclear Magnetic Resonance in Liquids.* Dordrecht: Delft Univ. Press. 200 pp.
4. Turner, D. L. 1985. *Progr. NMR Spectrosc.* 17: 281–358
5. Morris, G. A. 1986. *Magnet. Reson. Chem.* 24: 371–403
6. Wuthrich, K. 1986. *NMR of Proteins and Nucleic Acids.* New York: Wiley. 165 pp.
7. Griffey, R. H., Redfield, A. G. 1987. *Q. Rev. Biophys.* 19: 51–82
8a. LeMaster, D. M., Richards, F. M. 1988. *Biochemistry* 27: 142–50

8b. Torchia, D. A., Sparks, S. W., Bax, A. 1988. *J. Am. Chem. Soc.* 110: 2320–21
9. Frey, T., Anglister, J., McConnell, H. M. 1984. *Biochemistry* 23: 6470–73
10. Haberlen, U. 1976. *High-Resolution NMR in Solids: Selective Averaging.* New York: Academic. 190 pp.
11. Mehring, M. 1983. *High-Resolution NMR in Solids.* New York: Springer-Verlag. 342 pp.
12. Griffin, R. G. 1981. *Methods Enzymol.* 72: 108–74
13. Opella, S. J. 1982. *Ann. Rev. Phys. Chem.* 33: 533–62
14. Torchia, D. A. 1984. *Ann. Rev. Biophys. Bioeng.* 13: 125–44
15. Opella, S. J. 1986. *Methods Enzymol.* 131: 327–61
16. Opella, S. J., Stewart, P. L., Valentine, K. G. 1987. *Q. Rev. Biophys.* 19: 1–49
17. Ganesh, K. N. 1984. *Appl. Spectrosc. Rev.* 20: 107–57
18. Maricq, M. M., Waugh, J. S. 1979. *J. Chem. Phys.* 70: 3300–16
19. Pines, A., Gibby, M. G., Waugh, J. S. 1973. *J. Chem. Phys.* 59: 569–90
20. Waugh, J. S., Huber, L. M., Haeberlen, U. 1968. *Phys. Rev. Lett.* 20: 180–82
21. Rhim, W. K., Elleman, D. D., Vaughan, R. W. 1973. *J. Chem. Phys.* 58: 1772–73
22. Burum, D. P., Rhim, W. K. 1979. *J. Chem. Phys.* 71: 944–56
23. Gerstein, B. C. 1981. *Philos. Trans. R. Soc. London A* 299: 521–46
24. Rosenberger, H., Schnabel, B. 1978. *Wiss. Z. Friedrich-Schiller Univ. Jena Math. Naturwiss. Reihe* 27: 257
25. Miura, H., Terao, T., Saika, A. 1986. *J. Magnet. Reson.* 68: 593–96
26. Zilm, K. W., Grant, D. M. 1982. *J. Magnet. Reson.* 48: 524
27. Meier, B. H., Earl, W. L. 1987. *J. Am. Chem. Soc.* 109: 7937–42
28. Raleigh, D. P., Harbison, G. S., Neiss, T. G., Roberts, J. E., Griffin, R. G. 1987. *Chem. Phys. Lett.* 138: 285–90
29. Smith, R. L., Oldfield, E. 1984. *Science* 225: 280–88
30. Seelig, J., MacDonald, P. M. 1987. *Acc. Chem. Res.* 20: 221–28
31. Brandes, R., Vold, R. R., Vold, R. L., Kearns, D. R. 1986. *Biochemistry* 25: 7744–51
32. Vold, R. R., Brandes, R., Tsang, P., Kearns, D. R., Vold, R. L. 1986. *J. Am. Chem. Soc.* 108: 302–3
33. Davis, J. H., Jeffery, K. P., Bloom, M., Valic, M. F., Higgs, T. P. 1976. *Chem. Phys. Lett.* 42: 390–94
34. Tycko, R., Opella, S. J. 1986. *J. Am. Chem. Soc.* 108: 3531–32
35. Kundla, E., Samoson, A., Lippmaa, E. 1981. *Chem. Phys. Lett.* 83: 229–32
36. Samoson, A., Kundla, E., Lippmaa, E. 1982. *J. Magnet. Reson.* 49: 350–57
37. Ganapathy, S., Schramm, S., Oldfield, E. 1982. *J. Chem. Phys.* 77: 4360–65
38. Lowe, I. J. 1959. *Phys. Rev. Lett.* 2: 285–87
39. Andrews, E. R., Bradbury, D., Eades, R. G. 1958. *Nature* 182: 1659–61
40. Schaefer, J., Stejskal, E. O. 1976. *J. Am. Chem. Soc.* 98: 1031–32
41. Lippmaa, E., Alla, M., Tuherm, T. 1976. *Proc. Colloq. Ampere, 19th, Heidelberg*, pp. 113–18
42. Herzfeld, J., Berger, A. 1980. *J. Chem. Phys.* 73: 6021–30
43. Ellet, J. D. Jr., Waugh, J. S. 1969. *J. Chem. Phys.* 51: 2851–58
44. Aue, W. P., Ruben, D. J., Griffin, R. G. 1984. *J. Chem. Phys.* 80: 1729–38
45. Dixon, W. T., Schaefer, J., Sefcik, M. D., Stejskal, E. O., McKay, R. A. 1982. *J. Magnet. Reson.* 49: 341–45
46. Raleigh, D. P., Olejniczak, E. T., Vega, S., Griffin, R. G. 1987. *J. Magnet. Reson.* 72: 238–50
47. Raleigh, D. P., Olejniczak, E. T., Vega, S., Griffin, R. G. 1984. *J. Am. Chem. Soc.* 106: 8302–3
48. Hexem, J. G., Frey, M. H., Opella, S. J. 1982. *J. Chem. Phys.* 77: 3847–56
49. Menger, E. M., Veeman, W. S. 1982. *J. Magnet. Reson.* 46: 257–68
50. Opella, S. J., Frey, M. H. 1979. *J. Am. Chem. Soc.* 101: 5854–56
51. Torchia, D. A. 1978. *J. Magnet. Reson.* 30: 613–16
52. Harbison, G. S., Smith, S. O., Pardoen, J. A., Courtin, J. M. L., Lugtenburg, J., Herzfeld, J., Mathies, R. A., Griffin, R. G. 1985. *Biochemistry* 24: 6955–62
53. Roberts, J. E., Vega, S., Griffin, R. G. 1984. *J. Am. Chem. Soc.* 106: 2506–12
54. Rothwell, W. P., Waugh, J. S. 1981. *J. Chem. Phys.* 74: 2721–32
55. Schmidt, A., Smith, S. O., Raleigh, D. P., Roberts, J. E., Griffin, R. G., Vega, S. 1986. *J. Chem. Phys.* 85: 4248–53
56. Schmidt, A., Vega, S. 1987. *J. Chem. Phys.* 87: 6895–6907
57. Zilm, K. W., Merrill, R. A., Greenburg, M. M., Berson, J. A. 1987. *J. Am. Chem. Soc.* 109: 1567–69
58. Macho, V., Kendrick, N., Yannoni, C. S. 1983. *J. Magnet. Reson.* 52: 450–56
59. Pake, G. E. 1948. *J. Chem. Phys.* 16: 327–36
60. Munowitz, M. G., Griffin, R. G., Bodenhausen, G., Huang, T. H. 1981. *J. Am. Chem. Soc.* 103: 2529–33
61. Waugh, J. S. 1976. *Proc. Natl. Acad. Sci. USA* 73: 1394

62. Roberts, J. E., Harbison, G. S., Munowitz, M. G., Herzfeld, J., Griffin, R. G. 1987. *J. Am. Chem. Soc.* 109: 4163–69
63. DiVerdi, J. A., Opella, S. J. 1982. *J. Am. Chem. Soc.* 104: 1761–62
64. Linder, M., Hoehener, A., Ernst, R. R. 1980. *J. Chem. Phys.* 73: 4959–70
65. Zilm, K. W., Webb, G. G. 1986. *Exp. NMR Conf. Abstr.* A81
66a. Oas, T. G., Griffin, R. G., Levitt, M. H. 1988. *J. Chem. Phys.* 88: In press
66b. Gullion, T., Schaefer, J. 1988. *J. Magnet. Reson.* 77: In press
67. Caravatti, P., Bodenhausen, G., Ernst, R. R. 1983. *J. Magnet. Reson.* 55: 88–103
68. Linder, M., Henrichs, P. M., Hewitt, J. M., Massa, D. J. 1985. *J. Chem. Phys.* 82: 1585–98
69. Szeverenyi, N. M., Sullivan, M. J., Maciel, G. E. 1982. *J. Magnet. Reson.* 47: 462–75
70. Harbison, G. S., Raleigh, D. P., Herzfeld, J., Griffin, R. G. 1985. *J. Magnet. Reson.* 64: 284–95
71. DeJong, A. F., Kentgens, A. P. M., Veeman, W. S. 1984. *Chem. Phys. Lett.* 109: 337–42
72. Frey, M. H., Opella, S. J. 1984. *J. Am. Chem. Soc.* 106: 4942–45
73. Raleigh, D. P., Levitt, M. L., Griffin, R. G. 1988. *Chem. Phys. Lett.* 146: 71–76
74. Veeman, W. S. 1984. *Prog. Nucl. Magnet. Reson. Spectrosc.* 16: 193–235
75. Facelli, J. C., Grant, D. M., Michl, J. 1987. *Acc. Chem. Res.* 20: 152–58
76. Lauterbur, P. 1958. *Phys. Rev. Lett.* 1: 343–44
77. Pausak, S., Pines, A., Waugh, J. S. 1973. *J. Chem. Phys.* 59: 591–95
78. Haberkorn, R. A., Stark, R. E., van Willigen, H., Griffin, R. G. 1981. *J. Am. Chem. Soc.* 103: 2534–39
79. Naito, A., Ganapathy, S., Akasaka, K., McDowell, C. A. 1981. *J. Chem. Phys.* 74: 3190–97
80. James, N., Ganapathy, S., Oldfield, E. 1983. *J. Magnet. Reson.* 54: 111–21
81. Harbison, G. S., Jelinski, L. W., Stark, R. E., Torchia, D. A., Herzfeld, J., Griffin, R. G. 1984. *J. Magnet. Reson.* 60: 79–82
82. Munowitz, M., Aue, W. P., Griffin, R. G. 1982. *J. Chem. Phys.* 77: 1686–89
83. Oas, T. G., Hartzell, C. J., McMahon, T. J., Drobny, G. P., Dahlquist, F. W. 1987. *J. Am. Chem. Soc.* 109: 5956–62
84. Oas, T. G., Hartzell, C. J., Dahlquist, F. W., Drobny, G. P. 1987. *J. Am. Chem. Soc.* 109: 5962–66
85. Valentine, K. G., Rockwell, A. L., Gierasch, L. M., Opella, S. J. 1987. *J. Magnet. Reson.* 73: 519–23
86. Hartzell, C. J., Whitfield, M., Oas, T. G., Drobny, G. P. 1987. *J. Am. Chem. Soc.* 109: 5966–69
87. Munowitz, M., Bachovchin, W. W., Herzfeld, J., Dobson, C. M., Griffin, R. G. 1982. *J. Am. Chem. Soc.* 104: 1192–96
88. Frey, M. H., Opella, S. J. 1986. *J. Magnet. Reson.* 66: 144–47
89. Opella, S. J., Gierasch, L. M. 1985. *Peptides: Anal., Synth. Biol.* 7: 405–36
90. Ando, S., Yamanobe, T., Ando, I., Shoji, A., Ozaki, T., Tabeta, R., Saito, H. 1985. *J. Am. Chem. Soc.* 107: 7648–52
91. Shoji, A., Ozaki, T., Fujito, T., Deguchi, K., Ando, I. 1987. *Macromolecules* 18: 2135–40
92. Kricheldorf, H. R., Müller, D. 1984. *Coll. Polym. Sci.* 262: 856–61
93. Kricheldorf, H. R., Hull, W. E., Müller, D. 1985. *Macromolecules* 18: 2135–40
94. Frey, M. H., Opella, S. J., Rockwell, A. L., Gierasch, L. M. 1985. *J. Am. Chem. Soc.* 107: 1946–51
95. Marchetti, P. S., Honkonen, R. S., Ellis, P. D. 1987. *J. Magnet. Reson.* 71: 294–302
96. Honkonen, R. S., Marchetti, P. S., Ellis, P. D. 1986. *J. Am. Chem. Soc.* 108: 912–15
97. Honkonen, R. S., Ellis, P. D. 1984. *J. Am. Chem. Soc.* 106: 5488–97
98. Scheffer, J. R., Wong, Y.-F., Patil, A. O., Curtin, D. Y., Paul, I. C. 1985. *J. Am. Chem. Soc.* 107: 4898–4904
99. Okazaki, M., McDowell, C. A. 1984. *J. Am. Chem. Soc.* 106: 3185–90
100. Frydman, L., Olivieri, A. C., Diaz, L. E., Frydman, B., Morin, F. G., Mayne, C. L., Grant, D. M., Alder, A. D. 1988. *J. Am. Chem. Soc.* 110: 336–42
101. Brown, C. E., Spencer, R. B., Burger, V. T., Katz, J. J. 1984. *Proc. Natl. Acad. Sci. USA* 81: 641–44
102. Harbison, G. S., Mulder, P. P. J., Pardoen, J. A., Lugtenburg, J. A., Herzfeld, J., Griffin, R. G. 1985. *J. Am. Chem. Soc.* 107: 4809–16
103. Nozawa, T., Nishimura, M., Hatano, M., Hayashi, H., Shimada, K. 1985. *Biochemistry* 24: 1890–95
104. Harbison, G. S., Herzfeld, J., Griffin, R. G. 1983. *Biochemistry* 22: 1–5
105. Harbison, G. S., Smith, S. O., Pardoen, J. A., Winkel, C., Lugtenburg, J., Herzfeld, J., Mathies, R. A., Griffin, R. G. 1984. *Proc. Natl. Acad. Sci. USA* 81: 1706–9

106. Smith, S. O., Courtin, J., van den Berg, E., Winkel, C., Lugtenburg, J., Herzfeld, J., Griffin, R. G. 1988. *Biochemistry* 27: In press
107. De Groot, H. J. M., Copié, V., Smith, S. O., Allen, P. J., Winkel, C., Lugtenburg, J., Herzfeld, J., Griffin, R. G. 1988. *J. Magnet. Reson.* 77: 251–57
108. Mollevanger, L. C. P. J., Kentgens, A. P. M., Pardoen, J. A., Courtin, J. M. L., Veeman, W. S., Lugtenburg, J., de Grip, W. J. 1987. *Eur. J. Biochem.* 163: 9–14
109. Smith, S. O., Palings, I., Copié, V., Raleigh, D. P., Courtin, J., Pardoen, J. A., Lugtenburg, J., Mathies, R. A., Griffin, R. G. 1987. *Biochemistry* 26: 1606–11
110. Smith, S. O., Friedlander, J., Palings, I., Courtin, J., Pardoen, J. A., De Groot, H., Lugtenburg, J., Mathies, R. A., Griffin, B. G. 1988. *Biochemistry* 27: In press
111. Marchetti, P. S., Ellis, P. D., Bryant, R. G. 1985. *J. Am. Chem. Soc.* 107: 8191–96
112. Ellis, P. D. 1983. *The Multinuclear Approach to NMR Spectroscopy*, ed. J. B. Lambert, F. G. Riddell, pp. 457–523. New York: Reidel
113. Armitage, I. M., Boulanger, Y. 1983. *NMR of Newly Accessible Nuclei*, Vol. 2, pp. 337–65. New York: Academic
114. Keniry, M. A., Rothgeb, T. M., Smith, R. L., Gutowsky, H. S., Oldfield, E. 1983. *Biochemistry* 22: 1917–26
115. Huang, T.-H., Bachovchin, W. W., Griffin, R. G., Dobson, C. M. 1984. *Biochemistry* 23: 5933–37
116. Smith, S. O., Farr-Jones, S., Griffin, R. G., Bachovchin, W. W. 1988. *Science.* In press
117. Copié, V., Faraci, W. S., Walsh, C., Griffin, R. G. 1988. *Biochemistry.* In press
118. Dobson, C. M., Lian, L.-Y. 1987. *FEBS Lett.* 225: 183–87
119. Gettins, P. 1986. *J. Phys. Chem.* 261: 15513–18
120. Bachovchin, W. W. 1986. *Biochemistry* 25: 7751–59
121. Schaefer, J., Garbow, J. R., Jacob, G. S., Forrest, T. M., Wilson, G. E. Jr. 1986. *Biochem. Biophys. Res. Commun.* 137: 736–41
122. Jacob, G. S., Schaefer, J., Wilson, G. E. Jr. 1983. *J. Biol. Chem.* 258: 10824–26
123. Colnago, L. A., Valentine, K. G., Opella, S. J. 1987. *Biochemistry* 26: 847–54
124. Batchelder, L. S., Sullivan, C. E., Jelinski, L. W., Torchia, D. A. 1982. *Proc. Natl. Acad. Sci. USA* 79: 386–89
125. Torchia, D. A., Hiyama, Y., Sarkar, S. K., Sullivan, C. E., Young, P. E. 1985. *Biopolymers* 24: 65–75
126. Torchia, D. A., VanderHart, D. L. 1979. *Topics in Carbon-13 NMR Spectroscopy*, ed. G. C. Levy, 3: 325–60. New York: Wiley
127. Schaefer, J., Kramer, K. J., Garbow, J. R., Jacob, G. S., Stejskal, E. O., Hopkins, T. L., Speirs, R. D. 1987. *Science* 235: 1200–4

UV RESONANCE RAMAN STUDIES OF MOLECULAR STRUCTURE AND DYNAMICS: Applications in Physical and Biophysical Chemistry

Sanford A. Asher

Department of Chemistry, University of Pittsburgh, Pittsburgh, Pennsylvania 15260

INTRODUCTION

A fundamental understanding of molecular structure and dynamics requires a close coupling between theory and experiment. One ultimate objective is to utilize the understanding of molecular structure and dynamics to predict both molecular properties such as chemical reactivity and bulk properties such as phase behavior and material strengths. Often the major driving force for progress in science is the development of new experimental techniques capable of incisive glimpses into simple phenomena in complex systems. This is especially important for studies of large polyatomic molecules, and is crucial for the macromolecules studied in biology and biochemistry that control life processes.

This review outlines the recent progress in the development of UV resonance Raman spectroscopy (UVRR) as a new technique for the study of molecular structure and dynamics for both small molecules and larger molecules such as polypeptides and proteins. This recent work follows the pioneering UV Raman nucleic acid studies in 1975, which used frequency doubled Ar laser excitation at 257 nm (1–4), and Ziegler & Albrecht's pioneering nitrogen-dye laser preresonance Raman benzene derivative studies, which set the stage for true resonance Raman investigations of benzene derivatives and other aromatics (5–8). This work was followed by

Ziegler & Hudson's UV resonance Raman benzene derivative studies excited at 212.8 nm using a quintupled Yag laser (9, 10). Intense activity began in 1983 as tunable Nd-Yag based excitation sources became available (*vide infra*).

Raman spectroscopy monitors the coupling between electromagnetic radiation and the component of the molecular polarizability that is modulated by molecular vibrations (11). Thus, the Raman effect derives directly from dynamical coupling between electronic and vibrational motion. Excitation in resonance within a particular electronic transition results in Raman scattering that is dominated by the dynamical vibrational-electronic coupling present within the resonant rovibronic excited state. UV excitation (below 300 nm) occurs in the spectral region that is dominated by the electronic transitions of frontier orbitals of both small molecules and of small chromophoric segments of large macromolecular systems. Thus, UV Raman scattering can monitor the ro-vibronic-electronic dynamics of the excited states of small molecules such as ammonia, methyl iodide, and benzene. These small molecules are the modern candidates for theoretical electronic studies that attempt to calculate excited state geometries, dynamics, and potential surfaces. These theoretical studies have over the last 30 years remained at the frontier of physical chemistry as the sophistication of the calculations have increased and as the size of the molecules examined have increased. UV resonance Raman spectroscopy offers a major new experimental probe of the dynamics within the excited states of interest.

The selective examination of isolated segments of macromolecules is essential for progress in biochemistry and polymer science. UV resonance Raman spectroscopy can be used to excite particular chromophoric segments of proteins (12, 13). To the extent that the protein structure, conformation, and enzymatic reactivity are direct consequences of specific intermolecular interactions between amino acid residues, it may be possible to excite selectively those specific amino acid residues directly involved in the reaction coordinate in order to obtain information to help elucidate the enzymatic mechanism. As discussed in this review, rapid progress is being made in biophysical applications of UVRR. Many of these studies probe aromatic amino acids in proteins and take advantage of the numerous previous studies of simpler aromatic molecules. In addition, other recent UVRR biophysical studies take advantage of UV photochemistry to examine transient photochemically generated excited states. These studies probe the dynamics of return of these excited state species to the ground state.

In this review I examine UVRR instrumentation, the basic resonance Raman scattering phenomenology, and discuss studies that probe dynamics in the excited states of small molecules such as ammonia, ethylene, benzene and N-methyl acetamide. I also review some recent biophysical applications, and discuss two new emerging techniques, Saturation Raman Spectroscopy and Hyper-Raman Spectroscopy, which will have significant impact in the future.

INSTRUMENTATION

Excitation Source

The recent progress in UV Raman spectroscopy has been made possible by the development of UV laser excitation sources. The ideal excitation source for most UV Raman measurements would be a CW laser source that is conveniently tunable over the spectral region between 150 nm and 800 nm. In contrast, the conventional excitation source until recently was a high power, low duty cycle Nd-Yag based laser system. Figure 1, which shows a block diagram of our Raman spectrometer (14), indicates that we utilize a 20 Hz, 6 nsec Yag laser, which delivers ca 250 mJ per pulse at 1064 μm. By using efficient harmonic generation crystals such as KDP, the Yag fundamental is frequency doubled or tripled to 532 or 355 nm in order to excite a dye laser. The dye laser output is frequency doubled down to ca 260 nm. Light between 217 and 260 nm is obtained by mixing the 1064 nm Yag fundamental with the doubled dye output. Excitation below 217 nm is obtained by anti-Stokes Raman shifting the quadrupled Yag excitation of 266 nm. In this way, light between 190 to 800 nm is generated. The pulse energy (average power) will vary from a minimum of less than 0.05 mJ (1 mW average power) below 217 nm to 1.0 mJ (20 mW) between 217–260 nm and to 10 mJ (200 mW) above 260 nm. In all cases the nonlinear optical efficiencies are high due to the short temporal pulse widths and the fact that high efficiency nonlinear crystals are used. Unfortunately the short pulse widths result in extraordinarily high peak power or energy fluxes if the excitation source is tightly focused in the sample in order to optimize collection efficiency of the scattered light. For example, 6 nsec, 1 mJ pulses focused to a 0.1 mm diameter in the sample have peak powers and peak energy fluxes of ca 2 GigaWatts cm^{-2} and 10 J cm^{-2}. This commercially available laser system is easily and continuously tunable over the 217–800 nm spectral region. The major inconvenience derives from the necessity for changing dyes if excitation is to span a UV spectral region larger than ca 10–20 nm.

Other laboratories have used a single high power Yag laser to generate numerous discreet UV excitation lines between 320 and 184 nm (15, 16). These lines are generated by anti-Stokes Raman shifting the doubled, tripled, or quadrupled Yag beam in hydrogen gas. The high peak power

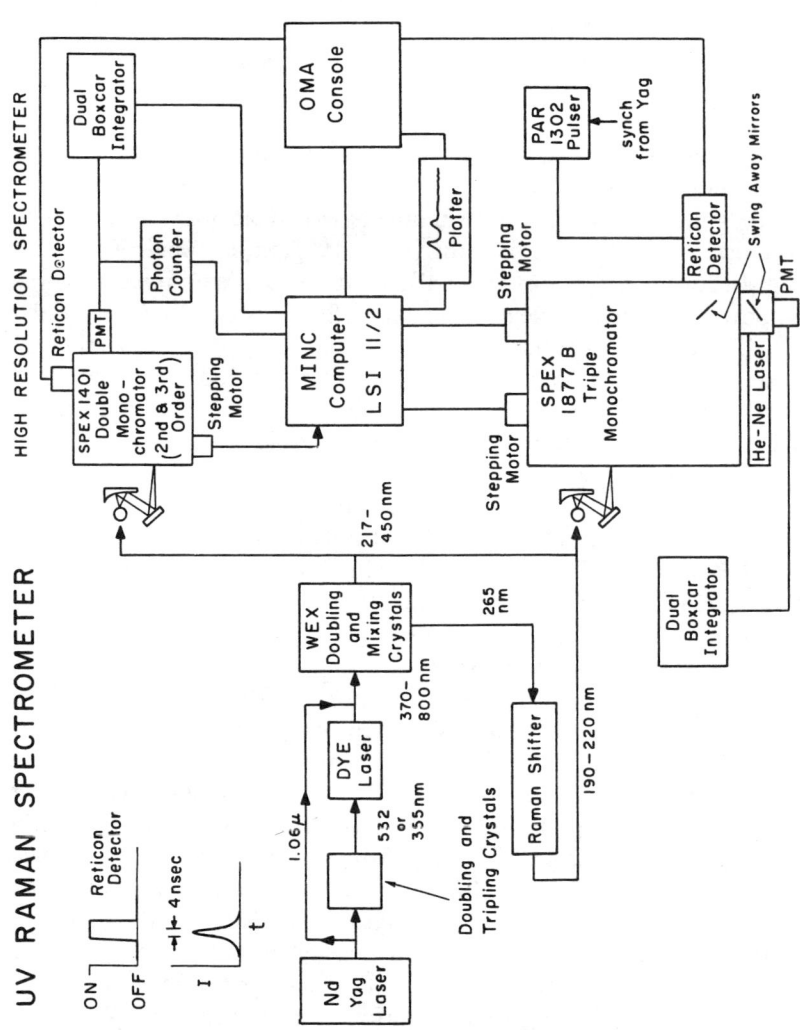

Figure 1 Block diagram of a tunable UV resonance Raman spectrometer. After Ref. (14).

Yag beam, when focused into a tube of hydrogen gas pressurized between two and five atmospheres, will stimulate Raman scattering to generate a number of anti-Stokes wavelengths that are shifted by harmonics of 4155 cm^{-1}, the hydrogen-stretching frequency. The efficiency of Raman shifting is very high for the high Yag peak powers and is often greater than 10% for the first anti-Stokes line. Ziegler (17) has developed a continuously tunable UV excitation source by anti-Stokes Raman shifting a Yag-pumped dye laser into the deep UV spectral region close to 200 nm.

The high pulse energies and the peak powers of the low duty cycle Yag excitation sources often induce optical phenomena that can interfere with the Raman spectral measurements (18, 19). Saturation of the Raman intensities can occur due to depletion of the ground state species during the laser excitation pulse. This saturation makes measurements of Raman cross-sections and relative intensities difficult and presents insidious potential artifacts for protein studies. For example, different protein aromatic amino acids show different ground state recovery rates and, thus, show different amounts of saturation; the protein aromatic amino acid relative intensities become power dependent. Interferences can also occur as a result of Raman scattering from high concentrations of photochemical intermediates. In addition, other photochemical processes can occur, such as two-photon ionization processes, to cause sample decomposition. Stimulated Raman scattering can also complicate the Raman studies.

Recently we demonstrated the utility of a high repetition rate excimer laser for decreasing the pulse energies to permit excitation at higher average powers (19). This lower pulse energy excitation helps to avoid saturation and other photochemical artifacts. The recent discovery that β-barium borate crystals can be used to frequency-double light efficiently into the 200–300 nm spectral region should make high repetition rate (greater than 500 Hz) excimer-pumped dye lasers the standard UV excitation source (20, 21); 500 Hz, 16 nsec pulse duration excimer laser sources are commercially available. The high efficiency (greater than 25% at 240 nm) of these β-barium borate frequency doubling crystals permits dramatic reductions in the required dye laser output, with concomitant reductions in the required excimer pump power. The delayed acceptance of excimer based UV Raman excitation sources derived mainly from the extraordinarily high projected costs for dye replacement (22). The poor doubling efficiencies below 260 nm of the previously available frequency doubling crystals necessitated extraordinarily high excimer pump powers to generate sufficient dye laser output to achieve modest levels (ca 20 mW average) of UV power below 260 nm. The development of the highly efficient β-barium borate crystals should dramatically decrease the dye replacement costs.

Other groups have attempted to use an injection-locked excimer laser as an excitation source. The narrow excimer excitation was to be used both directly and Raman shifted to obtain other wavelengths. Reliable and stable injection locking that prevents excitation wavelength hopping over the excimer gain curve has not as yet become routine and represents the main impediment for this laser excitation source.

Sample Handling and Collection Optics

Standard optics are used to focus the laser excitation into the sample and to collect and focus the scattered light into the spectrometer. We utilized a MgF_2 overcoated aluminized ellipsoidal mirror as the achromatic collection optic for our spectrometer to avoid the chromatic aberations that lead to artifacts in intensity and excitation profile measurements. However MgF_2 or quartz lenses are more convenient and may be more efficient if scattering occurs from nonpoint-like illuminated sample volumes; the ellipsoidal mirror is efficient for light collection from point sources.

The sample handling accessories are simple for gases, and typically involve flowing the gas through a nozzle. The gas stream is intersected by the laser excitation beam prior to the evacuation to a fume hood. For combustible mixtures, an annulus of flowing nitrogen surrounds the sample gas flow. Simple quartz cells filled with static sample gases have also been used when photochemistry did not interfere with the spectral measurements.

Liquid sample handling systems are more complex and are designed to replenish the sample volume between laser pulses. Sample solutions are circulated through thin quartz capillaries by using a peristaltic pump. It is important to avoid interference from plasticizers used in the rubber tubing, which can leach into the sample solutions. Other liquid sample handling systems attempt to avoid the nonlinear optical phenomena that often occur at surfaces to degrade the sample; opaque material deposits on the capillary walls to prevent transmission of the excitation beam into the sample. A thin (200 μm) free lamellar jet of liquid can be prepared by recirculating sample solutions through a dye laser jet nozzle by using a high pressure pump (14). An exterior cylindrical quartz tube can be used to isolate the liquid stream in order to establish vapor equilibrium to avoid solvent evaporation. The jet stream is rectangular and is ca 0.2 μm wide and 3 mm high. The sides of the stream are optically flat and parallel. Excitation can occur either at an angle to the flat surface, or 90° scattering can be accomplished by focusing into the narrow edge. The rapid flow and the unconstrained walls make this sampling system useful for high repetition laser excitation. Further, the thinness of the jet prevents significant self-absorption.

Another approach utilizes gentle flow through a capillary that has a hole drilled in the wall (23). For gentle lamellar flow with minimal back pressure, surface tension constrains the liquid, and a free surface can be excited by backscattering from the hole in the side. A free flowing liquid surface has also been prepared by flowing sample solutions down a pair of separated wires (24), or down a teflon tube that was cut and shaped such that the surface tension of the liquid formed a flat surface (25).

Solid samples have not generally been studied in the UV. The potential difficulties for studying them with pulsed laser excitation are obvious; the illuminated solid volume element must be replaced between laser pulses and the sample must have sufficient thermal conductivity (via linear or nonlinear processes) to remove the energy deposited during each pulse without significant thermal decomposition during the laser pulse.

Spectrometer and Detection System

The spectrometer and detection systems typically used fall into two categories: Either a complex spectrometer is used that contains a predispersing system to reject the Rayleigh scattering, followed by a spectrograph and multichannel detector, or a single high dispersion, long pathlength monochromator is used combined with a solar blind photomultiplier and boxcar detection. The single monochromator/solar blind photomultiplier detection system has a high dynamic range and high throughput efficiency (as high as 60–70%).

The photomultiplier detector is shot-noise limited. The use of a high dispersion grating and adjustable slits permits high resolution gas phase studies (to the bandwidth of the laser excitation, typically 1 cm^{-1}). The instrumentation is relatively inexpensive and rugged. Unfortunately, stray light can interfere in low frequency Raman measurements due to the inability of the single grating to reject all of the Rayleigh scattered light. Another disadvantage of the scanning systems is that only a small fraction of the spectrum is examined at any time; the signal-to-noise limitations are not generally shot-noise limited. Most of the noise in pulsed laser measurements derive from pulse-to-pulse fluctuations in Raman scattering due to fluctuations in energy, and spacial and temporal beam shapes. A portion of these fluctuations can be normalized out by using dual channel boxcar detection, where the detected Raman intensity for each pulse is normalized to the incident energy.

The multichannel Raman spectrometer shown in Figure 1 utilies a Spex Triplemate spectrometer and an intensified Reticon Array (EG&G OMA II Model 1420). The predispersing stage uses a subtractive double spectrometer to remove the Rayleigh scattering and transfers the Raman scattered light to a 0.64 m spectrograph, which disperses and images the light onto the detector. Although the transfer efficiency of this spectrograph has been measured to be low (ca 3%), the multichannel advantage more than compensates for this throughput loss. For the Reticon detector, a large fraction of the Raman spectrum is simultaneously accumulated (ca 1500 cm^{-1}), and this automatically normalizes out pulse-to-pulse fluctuations. Unfortunately, the ultimate low signal limit for the intensified Reticon detector derives from background electronic noise. This means that signal strengths below the detector noise levels cannot be observed; increasing the integration time proportionately increases both the signal and the noise. New multichannel detectors, such as charge coupled devices (CCDs), are emerging that have high sensitivities and lower noise levels (26, 27). Multichannel spectrometers such as shown in Figure 1 show high sensitivity and reasonably low stray light above 400 cm^{-1}, but have relatively modest resolution in the UV; the typically useful resolution is ca 5 cm^{-1} at 240 nm, which is determined by the spectrograph grating and the effective Reticon element pixel width. However, this resolution is adequate for most condensed phase studies. The multichannel spectrograph is ideal for monitoring small frequency shifts, since no mechanical changes occur between separate scans. Obviously, the ideal spectrometer would be a single high dispersion spectrometer designed to minimize stray light equipped with a multichannel detector.

RAMAN PHENOMENOLOGY

The Raman effect derives from the inelastic scattering of electromagnetic radiation by matter. For the purposes of this review the energy exchange is mainly limited to vibrational energy quanta. Figure 2 shows an energy level diagram for production of Stokes Raman scattering, where a molecule in its vibronic ground state is excited by light of frequency v_0. Raman shifted light is Stokes scattered at frequencies $v_0 - v_a$ and $v_0 - v_b$, where v_a and v_b are vibrational mode frequencies. Obviously the molecule is pro-

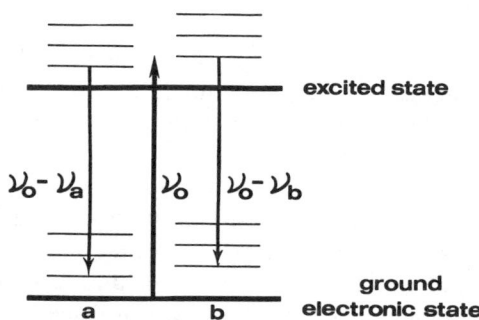

Figure 2   Illustration of Raman vibrational transitions.

moted to higher vibrational quantum levels of the ground electronic state during this process. Molecular spectroscopic information is conveyed by the vibrational frequency shift, by the efficiency (intensity) of the Raman scattering process, and by the polarization of the Raman scattered light relative to the incident scattering.

The vibrational frequency shifts derive from the nuclear vibrational dynamics that occur within the ground state of the molecule. Every molecule (nonlinear) has 3N-6 normal modes of vibration, each of which has a characteristic frequency of oscillation that depends upon interatomic force constants, atomic masses, and the couplings with the motion of both adjacent and distant atoms. The ability to calculate these normal mode frequencies and motions represents one of the great successes of physical chemistry. The Raman frequencies directly report upon the ground state molecular structure and intermolecular interactions, such as the presence of hydrogen bonding, for example. The Raman bandwidths in the condensed phase depend upon dynamical processes that control lifetimes of the ground state vibronic levels (homogeneous linewidth) and/or the distribution of molecular environments (inhomogeneous linewidth) (28, 29a,b).

The Raman band intensities at a particular excitation frequency depend upon the degree to which a particular vibration modulates the molecular polarizability. For excitation at frequency v_0, the Raman intensity, I_{mn}, observed over a unit solid angle for a Raman transition between vibronic levels m and n in the electronic ground state manifold is (30–34)

$$I_{nm} = \sigma_{nm}^R N I_0 W(\Omega) \qquad 1.$$

where σ_{nm}^R (cm$^2 \cdot$ mol$^{-1} \cdot$ str^{-1}) is the total differential Raman cross-section for a single gas phase molecule integrated over the Raman peak bandwidth. I_0 is the incident excitation intensity (photons $\cdot$ cm^{-2} sec^{-1}) into a particular sample volume element of area dA and length dl, N is the number of molecules within that volume element, and $W(\Omega)$ is a parameter that details the optical geometry and includes factors such as the collected solid angle. For a single orientationally averaged molecule in vacuum, the differential Raman scattering cross-section may be written

$$\sigma_{nm}^R = 4\pi^2 \left(\frac{e^2}{\hbar c}\right)^2 v_0(v_0 - v_{nm})^3 g f(T) \left| \sum_{\rho\sigma} \overline{\mathbf{e}_{s\rho}^* \alpha_{\rho\sigma} \mathbf{e}_{\sigma 0}} \right|^2 \qquad 2.$$

where $e^2/\hbar c$ is the fine structure constant, and v_0 and v_{nm} are the frequencies (cm^{-1}) of the excitation source and the Raman active vibration, respectively. g is a factor specifying the degree of degeneracy of the vibration, while $f(T)$ is the Boltzmann weighting factor specifying the thermal occupancy of the initial state, m. $\alpha_{\rho\sigma}$ is the $\rho\sigma$th element of the polarizability tensor where ρ and σ are molecular fixed coordinates, s and 0 are lab fixed coordinates, and $\mathbf{e}_{\sigma 0}$ and $\mathbf{e}_{sp}$ represent unit vectors that point in the polarization directions of the excitation and scattered light, respectively. The expression

$$\left| \sum_{\rho\sigma} \overline{\mathbf{e}_{s\rho}^{*} \alpha_{\rho\sigma} \mathbf{e}_{\sigma 0}} \right|^{2}$$

indicates that the value of $\alpha_{\rho\sigma}$ is averaged over all possible molecular orientations, given particular values of $\mathbf{e}_{\sigma 0}$ and $\mathbf{e}_{sp}$.

Typically, the differential Raman cross-section, σ_{nm}^{R}, is measured in units of Barns·str^{-1} (1 Barn = 10^{-24} cm^{2}·mol^{-1}), and Raman cross-sections of species such as hexane, benzene, and acetone have cross-sections of ca 1.0×10^{-6} Barns str^{-1} with visible wavelength excitation (31, 32). The strongest *resonance* Raman cross-sections observed are greater than 60 Barns str^{-1}.

The total differential Raman cross-section includes light polarized both parallel and perpendicular to the incident polarization (Figure 3). For the typical 90° scattering geometry where the molecule occurs randomly oriented in the gas, solution, or solid phase, and where we collect both parallel and perpendicular scattering components, it is convenient to write the Raman differential cross-section in terms of the Raman polarizability tensor invariants $\Sigma^{0}, \Sigma^{1}, \Sigma^{2}$ (34).

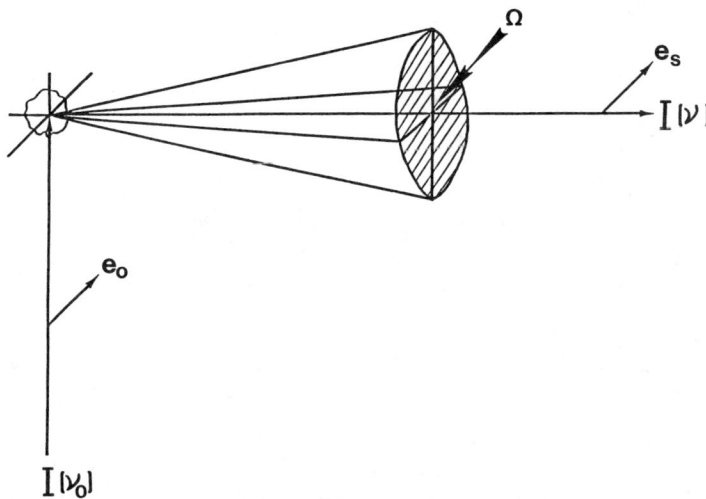

Figure 3 Raman 90° scattering geometry.

$$\sigma_{nm} = \sigma_{nm}^R(90°, \| + \bot) = 4\pi^2 \left(\frac{e^2}{\hbar c}\right)^2 v_0(v_0 - v_{nm})^3 g f(T) L(v_0)$$

$$\times (10\Sigma^0 + 5\Sigma^1 + 7\Sigma^2) \qquad 3.$$

where $L(v_0)$ details the increased local electromagnetic field strength within the sample medium over that in vacuum. The local field correction is suggested to be modeled accurately as (31, 32, 35)

$$L(v_0) = (n_s/n_0)(n_s^2+2)^2(n_0^2+2)^2/81 \qquad 4.$$

where n_s and n_0 are the refractive indices for light at the incident and Raman scattered frequency, respectively.

The Raman tensor invariants are those linear combinations of Raman tensor elements which turn out to be independent of the relative orientation between the molecular and laboratory frames.

$$\Sigma^0 = \frac{1}{3}|\alpha_{xx} + \alpha_{yy} + \alpha_{zz}|^2$$

$$\Sigma^1 = \frac{1}{2}\{|\alpha_{xy} - \alpha_{yx}|^2 + |\alpha_{xz} - \alpha_{zx}|^2 + |\alpha_{yz} - \alpha_{zy}|^2\}$$

$$\Sigma^2 = \frac{1}{2}\{|\alpha_{xy} + \alpha_{yx}|^2 + |\alpha_{xz} + \alpha_{zx}|^2 + |\alpha_{yz} + \alpha_{zy}|^2\}$$

$$+ \frac{1}{3}\{|\alpha_{xx} - \alpha_{yy}|^2 + |\alpha_{xx} - \alpha_{zz}|^2 + |\alpha_{yy} - \alpha_{zz}|^2\} \qquad 5.$$

Σ^0, Σ^1, and Σ^2 are known as the isotropic (trace), the antisymmetric anisotropic, and the symmetric anisotropic Raman invariants, respectively.

For nonresonance Raman scattering, Σ^1, the antisymmetric tensor component is zero, since the tensor is symmetric. Although for resonance Raman scattering Σ^1 may become large for particular vibrational symmetries in particular point groups, it vanishes for the examples discussed here.

The Raman depolarization ratio is the relative intensity ratio of Raman scattering observed perpendicular, $I_\bot$, and parallel, $I_\|$, to the incident polarization. The depolarization ratio can be calculated directly from the relative values of the Raman invariants

$$\rho = \frac{(I)_\bot}{(I)_\|} = \frac{5\Sigma^1 + 3\Sigma^2}{10\Sigma^0 + 4\Sigma^2}. \qquad 6.$$

If Σ^1 is zero, ρ will vary between a value of 0.00 and 0.75, since the invariants are positive definite quantities. If only one tensor element, say α_{xx}, is greater than zero, $\rho = 0.33$. For $\alpha_{xx} = \alpha_{yy}$ and $\alpha_{zz} = \alpha_{xy} = \alpha_{xz} = \alpha_{yz} = 0$, $\rho = 0.125$. For $\alpha_{xx} = \alpha_{yy} = \alpha_{zz}$, with all of the off-diagonal elements equal to zero, $\rho = 0.00$. Finally if $\alpha_{xx} = \sigma_{yy} = \alpha_{zz} = 0$ and off-diagonal elements are different from zero, then $\rho = 0.75$. Obviously, the depolarization ratio is very sensitive to the relative values of the Raman polarizability tensor elements and can be used to examine both electronic and vibrational dynamics for a Raman scattering process where the scattering time is essentially instantaneous relative to molecular rotation periods. If the scattering time becomes long compared to the rotational period, ρ will approach a value of 1.0 since the molecule rotates prior to emission and the polarization becomes isotropic.

We must specify $\alpha_{\rho\sigma}$ in terms of the molecular electronic and vibrational states. For this review I mainly rely on the familiar sum over states given by the Kramer Heisenberg-Dirac (KHD) formalism. It should be noted, however, that significant advances have recently been made in applying the propagating wave packet formalism of Lee & Heller (36, 37) to relate observed Raman cross sections to excited state potential surfaces. In fact, the wave packet formalism derives directly from the KHD expression through an inverse Laplace transform. The time dependent formalism, however, has been extraordinarily useful in describing dissociative excited state surfaces for molecules like CH_3I where the observed Raman cross-sections for the molecular vibrational normal modes can be related to the dynamical dissociation coordinates (38–40). However, most of the recent UV Raman studies have utilized the KHD approach for physical insight.

The KHD perturbation theory description of the Raman polarizability tensor elements is given by (30, 41)

$$\alpha_{\rho\sigma} = \sum_{ev} \frac{\langle n|\mathbf{r}_\rho|ev\rangle \langle ev|\mathbf{r}_\sigma|m\rangle}{v_{ev}-v_0-i\Gamma_{ev}} + \frac{\langle n|\mathbf{r}_\sigma|ev\rangle \langle ev|\mathbf{r}_\rho|m\rangle}{v_{ev}+v_0-i\Gamma_{ev}} \qquad 7.$$

where $\mathbf{r}_\rho$ is an internal molecular coordinate specifying the position of each electron, where $\mathbf{r}_\rho = \Sigma_i \mathbf{r}_{i\rho}$ and the summation occurs over all of the electrons. $\alpha_{\rho\sigma}$ has a contribution from all of the rovibronic molecular excited states as indicated by the summation over the excited states $|ev\rangle$ in Eq. 7. The weighting of an individual $|ev\rangle$ excited state contribution is determined by the values of the transition moment matrix elements in the numerator, and by the values of energy denominators. The energy denominators contain information on Γ_{ev}, the homogeneous linewidth for the transition between the ground state m and the excited state $|ev\rangle$, and the detuning of excitation from resonance, $(v_{ev}-v_0)$.

For normal Raman scattering, where excitation is far from resonance, the denominator only weakly depends upon excitation frequency, and all molecular rovibronic transitions contribute essentially in proportion to their transition moments. As resonance with an electronic transition is approached the first term of Eq. 7 begins to dominate, and the resonant excited state dominates the sum over states of the Raman polarizability expression. If we display the dependence of the transition moment matrix elements upon vibrational motion we can show that Eq. 7 will display three distinct scattering mechanisms (9, 42):

$$\alpha_{\rho\sigma} = A + B + C:$$

$$A = \sum_{ev} \frac{\langle g|\mathbf{r}_\rho|e\rangle \langle e|\mathbf{r}_\sigma|g\rangle}{v_{ev} - v_0 - i\Gamma_{ev}} (f|v)(v|i)$$

$$B = \sum_{ev} \left\{ \left[\frac{\partial \langle g|\mathbf{r}_\rho|e\rangle}{\partial Q_a}\right]_0 \cdot \frac{\langle e|\mathbf{r}_\sigma|g\rangle}{v_{ev} - v_0 - i\Gamma_{ev}} \right\} (f|Q_a|v)(v|i)$$

$$+ \left(\frac{\langle g|\mathbf{r}_\rho|e\rangle}{v_{ev} - v_0 - i\Gamma_{ev}} \cdot \left[\frac{\partial \langle e|\mathbf{r}_\sigma|g\rangle}{\partial Q_a}\right]_0\right) (f|v)(v|Q_a|i)$$

$$C = \sum_{ev}\sum_a\sum_b \left(\left[\frac{\partial \langle g|\mathbf{r}_\rho|e\rangle}{\partial Q_a}\right]_0 \left[\frac{\partial \langle e|\mathbf{r}_\sigma|g\rangle}{\partial Q_b}\right]_0\right) (f|Q_a|v)(v|Q_b|i). \qquad 8.$$

The transition moment matrix elements have been factored into separate electronic and vibrational integrals. The transition moment integrals between the ground and excited electronic states are denoted by angular brackets. Curved brackets denote the integrals between vibrational levels, where $|i)$ labels the initial vibrational level of the normal mode a in the ground electronic state, and $|v)$ labels the vibrational level of mode a in the excited electronic state. $|f)$ labels the final vibrational level of mode a in the ground electronic state ($f = i+1$ for Stokes Raman scattering from mode a).

A pictoral representation for these enhancement mechanisms is shown in Figure 4. Since the Raman cross-section is proportional to the modulus squared of the Raman polarizability, the A-term will dominate enhancement in strongly allowed transitions. The Raman cross-sections are porportional to the fourth power of the electronic transition moment to the resonant excited state. Thus, enhancement by a single electronic transition scales with the square of the molar absorptivity. The Franck-Condon overlap factors $(f|v)(v|i)$ will differ from zero only if the excited state equilibrium geometry is displaced along a symmetric normal mode coordinate relative to the ground state. This assumes identical *ortho*normal vibrational wavefunctions in the ground and excited states. A-term, Condon enhancement may also occur due to strong Franck-Condon overlaps

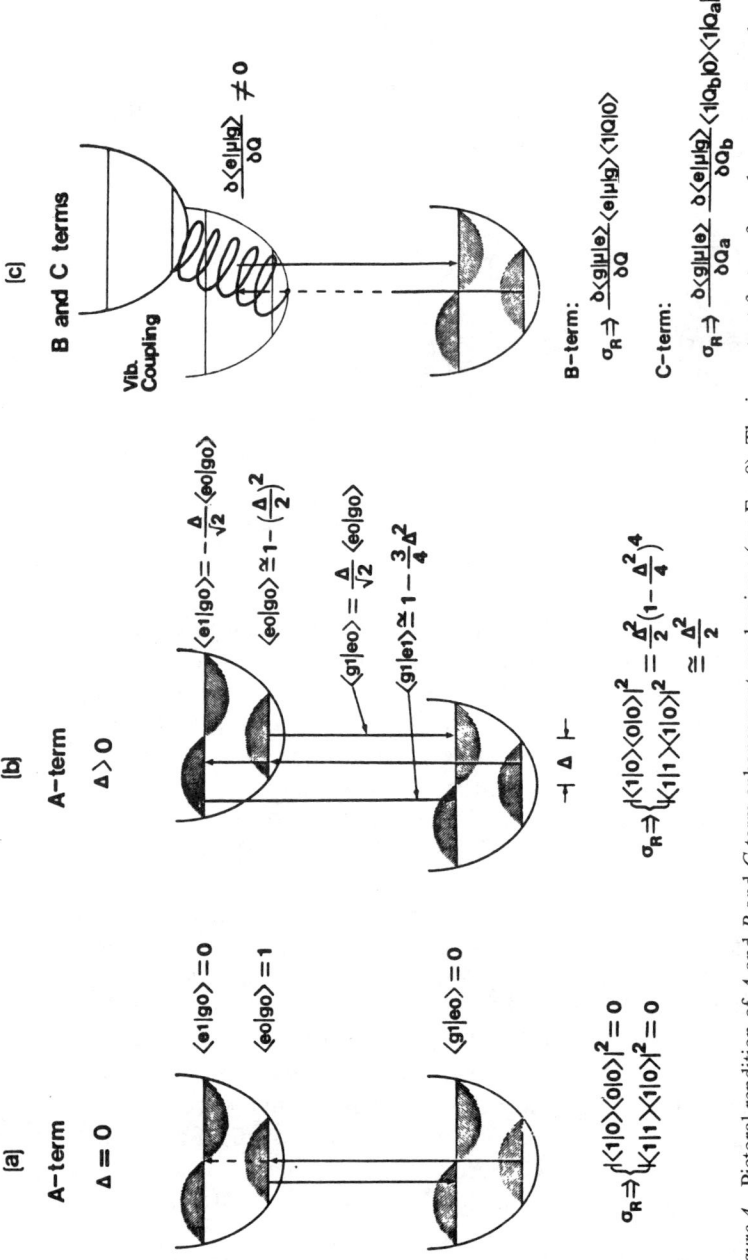

Figure 4 Pictoral rendition of *A* and *B* and *C*-term enhancement mechanisms (see Eq. 8). The important factors for enhancement are shown at *bottom* of figure where only a small excited state displacement is assumed. *A*-term enhancement derives from large Franck-Condon overlaps in the upward and downward transitions, and dominates for symmetric vibrations in strongly allowed transitions. As indicated in panels (*a*) and (*b*), *A*-term enhancement requires a displacement, Δ, for the excited state potential function relative to the ground state. If $\Delta = 0$, the Raman cross-section $\sigma_R = 0$. Otherwise $\sigma_R \approx \Delta^2/2$ for small excited state displacements. Non-Condon *B*-term enhancement (*c*) derives from the dependence of the transition moments upon vibrational motion. Both symmetric and nonsymmetric vibrations may show *B*-term enhancement. Panel (*c*) displays the Herzberg-Teller vibronic coupling mechanism where the enhanced vibration couples in an adjacent strongly allowed transition. *C*-term enhancement occurs for forbidden electronic transitions that show no oscillator strength at the equilibrium internuclear geometry. Non-Condon coupling must occur for both the upward and downward electronic transition moment matrix elements.

when the excited state vibrational states are solutions to a different Hamiltonian than that of the ground state. For the more common case of similar vibrational mode compositions in the ground and excited states, where the excited state geometry either expands or contracts relative to the ground state, the enhancement of totally symmetric vibrations scales roughly as $\Delta^2/2$, where Δ is the magnitude of displacement of the excited state potential surface along the Raman active normal coordinate. For small displacements only the fundamental shows significant enhancement. Larger displacements result in lengthy Franck-Condon progressions. Thus, excited state geometry changes can be deduced from the pattern of internal coordinates active in the enhanced vibrations (43). Enhancement of symmetric vibrations by the A-term can also derive from vibrational force constant changes in the excited state, as well as by alterations in the composition of the excited state normal mode coordinates (Duschinsky effect). Although the A-term Condon mechanism dominates enhancement within strongly allowed transitions, Stallard et al (44) have demonstrated the importance of non-Condon contributions to enhancement of symmetric vibrations within the strong Soret band of ferro cytochrome c. Jones & Asher (45) also recently observed ca 15% non-Condon contributions to the enhancement of symmetric vibrations within the strongly-allowed S_4 electronic transition of pyrene.

Enhancement via the B-term derives from the non-Condon dependence of the electronic transition moment upon the vibrational coordinate. One example of a non-Condon enhancement mechanism is Herzberg-Teller vibronic coupling of different electronic transitions. Both symmetric and nonsymmetric fundamentals can be enhanced by a B-term mechanism. For a strongly allowed transition, the magnitude of B-term enhancement of symmetric vibrations is significantly below that for A-term enhancement. B-term enhancement will dominate only for the nonsymmetric vibrations. In contrast, B-term enhancement can dominate in transitions where significant oscillator strength derives from Herzberg-Teller coupling to adjacent strongly allowed transitions (Figure 4).

If the transition is rigorously forbidden at the equilibrium geometry, enhancement of fundamentals cannot occur via either the A or B-terms; however, C-term enhancement of overtones and combinations can occur, and will involve two quanta of vibrational mode a, or the combination of one quantum of mode a and one quantum of mode b. The C-term only dominates enhancement for resonance excitation within forbidden electronic transitions.

The energy denominators of the A, B, and C terms cause individual electronic transitions to dominate the Raman cross-sections upon resonance excitation, provided that the spacing between adjacent transitions is greater than the homogeneous linewidth, Γ_{ev}. Measurements of the dependence of the Raman cross-sections upon the excitation frequency, the Raman excitation profile (RREP), are particularly informative in studies of the dynamics of excited states.

A major simplification in the polarizability tensor expressions can occur in the preresonance region, where excitation occurs close to a dominating transition, but where the offset from resonance is large compared to Γ_{ev}. If one strong transition dominates, A-term enhancement will dominate and the preresonance excitation profile of the Raman cross-sections for totally symmetric vibrations can be written (41, 46):

$$\sigma_{nm}^{pre} = 4\Pi^2 \left(\frac{e^2}{\hbar c}\right)^2 v_0(v_0 - v_{nm})^3 g f(T) L(v_0)(10\Sigma^0 + 7\Sigma^2) \qquad 9.$$

where Σ^0 and Σ^2 are defined by Eq. 5, but

$$\alpha_{\rho\rho} = \frac{2}{\hbar c}|\langle g|r_\rho|e\rangle|^2 \langle i|Q_a|f\rangle k_e \cdot \Delta \left(\frac{v_e^2 + v_0^2}{(v_e^2 - v_0^2)^2}\right)$$

$$I_{nm} = K v_0(v_0 - v_{nm})^3 \left(\frac{v_e^2 + v_0^2}{(v_e^2 - v_0^2)^2}\right)^2 \qquad 10.$$

where k_e is the vibrational force constant, Δ is the displacement of the potential surface along the vibrational mode coordinate, and K is a collection of frequency-independent factors.

Equation 10 indicates that a measurement of the dispersion of the preresonance enhancement can be used to locate the transition frequency v_e. The magnitude of preresonance enhancement of a vibration is indicative of the magnitude of the displacement Δ of the excited state potential surface along the vibrational mode coordinate.

Occasionally, one excited state does not completely dominate enhancement for excitation both far from resonance and well within the preresonance regime. This occurs because of the limited discrimination of the resonance energy denominators. Thus, the Albrecht A-term expression can be phenomenologically modified by adding a term to account for the contribution of states further from resonance (47, 48):

$$\alpha_{\rho\rho} = \frac{2}{\hbar c}|\langle g|r_\rho|e\rangle|^2 \langle i|Q|f\rangle k_e \cdot \Delta \left\{\frac{v_e^2 + v_0^2}{(v_e^2 - v_0^2)^2}\right\} + K_2. \qquad 11.$$

Although this additional term (K_2), alone, gives a frequency-independent contribution to $\alpha_{\rho\rho}$, it results in an additional frequency dependence in

$|\alpha_{\rho\rho}|^2$ due to the interference term with the first frequency-dependent term of Eq. 11 (48).

The expressions above are used to model the measured Raman cross-sections. The RREP data, the depolarization ratios, and the normal mode compositions of the enhanced vibrations can be combined to determine excited state geometry changes, and to monitor dynamical coupling among electronic, vibrational, and rotational motions.

APPLICATIONS

The applications discussed are chosen from studies that incisively utilize the unique capabilities of UVRR spectroscopy. I attempt to review the current activities in UV Raman spectroscopy and conclude this review by discussing two new techniques that are likely to make major contributions to the understanding of molecular structure and dynamics in the near future. These techniques are Saturation Raman Spectroscopy and Resonance Hyper-Raman Spectroscopy.

As discussed in detail toward the end of the review, Saturation Raman Spectroscopy (49a,b) examines the rate of recovery of molecules back to the ground state after photon absorption transfers these molecules into an excited state during a resonance Raman measurement. The rate of ground state recovery depends upon molecular structure and environment.

Resonance Hyper-Raman Spectroscopy (50–52) is a second-order process where excitation with light of frequency v_0 results in Raman scattering at $2v_0 - v_v$, where v_v is a vibrational mode frequency. A two-photon resonance occurs with electronic transitions centered at $2v_0$, to enhance the cross-sections of the Hyper-Raman bands. Although visible wavelength excitation is used, the technique probes far UV electronic transitions.

The intense activity in UV Raman studies of molecular excited states and dynamics can be categorized into three different types of studies. (*a*) studies that utilize Raman intensities and bandwidths to resolve excited state dynamics, especially within photodissociating electronic transitions; (*b*) studies that probe excited state structure by relating the Raman cross-sections to changes in excited state geometry; (*c*) studies that utilize the selectivity of UV Raman enhancement to examine molecular structure in complex systems.

Dynamical Studies

The recent study of CH_3I photodissociation by Imre et al (38–40) examined Raman scattering excited in resonance within a dissociating electronic transition. By using Heller et al's (36, 37) time-dependent formalism, the relative Raman intensities of the fundamentals and overtones were used to gain quantitative information upon molecular excited state structure, the nature of the electronic potential surfaces, and the dynamics of the photodissociation process. This time-dependent formalism derives from the inverse LaPlace Transform of the KHD expressions and is simply a restatement of second order perturbation theory in the time domain. This formalism, which is not reviewed here, has the advantage that it is cast in the time domain, and unique physical insight can be obtained on probabilistic short-time dynamics of a photodissociative process.

The CH_3I studies of Imre and the Kinsey group (38–40), and the NO_2 studies of Rohlfing & Valentini (53), clearly illustrate the dynamical information available from the relative Raman intensities. The relative CH_3I fundamental and overtone Raman intensities were used to determine the magnitudes of the slopes (the forces) of the potential surface along the stretching and bending molecular coordinates. The Raman intensities report upon the shape of the excited state potential surfaces and the probability density surface for photodissociation. Indeed, because of the dependence of the Franck-Condon overlap factors on the relative shifts between potential surfaces, the intensities of the higher overtones are determined selectively by the dissociating edge of the excited state potential surface. In contrast, the intensities of the fundamentals are dominated by the potential surface located around the ground state equilibrium nuclear geometry. As a result, the fundamentals and overtone intensities report on short and longer time dynamics, respectively (i.e. approximately less than and greater than one ground state vibrational period). These types of studies are uniquely informative of the likely trajectories of photodissociating molecules. It is likely that these methods will be applied to study other photodissociating systems.

Ziegler et al (54–58) adopted a different approach to examine dynamics of photodissociation in gaseous and condensed phase solutions of NH_3. They carefully examined the resonance Raman spectra and excitation profiles of the 160–220 nm $A \leftarrow X$ NH_3 transition (Figure 5). This Rydberg transition, which is associated with promotion of a nitrogen lone pair electron to a $3s$ Rydberg orbital, shows a long progression in the gas phase in the v_2 umbrella bending coordinate. The gas phase absorption bands are broad due to lifetime broadening (greater than 100 cm^{-1} linewidths), and no rotational fine structure is observed. In the condensed phase these bands severely weaken and broaden. Excitation into this state in the gas phase results in photodissociation to NH_2 and H with a quantum yield of unity; the homogeneous linewidth or dephasing rate, $(1/\Gamma_{ev})$, for this transition is dominated by the photodissociation rate. Resonance Raman excitation into the vibronic subbands show progressions in the v_2 out-of-plane umbrella bending coordinate, since the pyramidal ground state

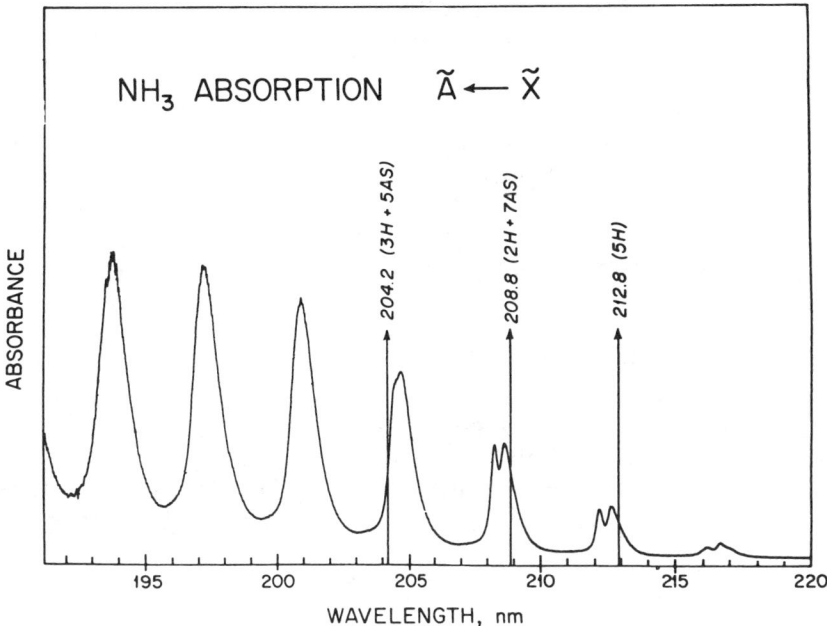

Figure 5 Absorption spectrum of NH_3 gas at room temperature. *Arrows* indicate excitation wavelengths. From Ref. (56).

becomes planar in the photodissociating excited state. In addition strong enhancement is observed for v_1, the symmetric stretching coordinate.

Careful examination of the rotational structure in the UVRR spectra clearly reveals that the lifetime of the dissociating excited state is a function of both the vibronic and rotational energy levels. As indicated in Eq. 8, the Raman cross-section depends upon the homogeneous linewidth parameter, Γ_{ev}, which is related to the reciprocal of the excited state lifetime. These expressions were written for vibrational Raman scattering, and can be readily extended to rovibronic scattering (where the rotational states are separable) by including the rotational states and by recognizing the rotational state dependence of the energy denominators. Ziegler has calculated the required expressions that can be used to model the relative rovibronic Raman intensities (56). Since NH_3 can be modeled well as a symmetric top, the only unknown parameter within the rigid rotor approximation is the homogeneous linewidth, which, as described above, is related to the dissociation rate. Qualitatively, the shorter the lifetime the more rotational transitions will be resonantly enhanced and observed, since their offset from resonance will become comparable to or less than Γ_{ev}, the bandwidth. In contrast, longer lifetimes will result in a smaller Γ_{ev}

and a sharper resonance, which will give rise to resonance Raman spectra with fewer calculated rotational transitions. Indeed, Figure 6 illustrates the dependence of the rotational structure upon excited state lifetime for 208.8 nm excitation of NH_3, and Figure 7 shows the agreement possible between the observed and calculated spectra. Obviously, the spectra are extraordinarily sensitive to lifetime, which permits Ziegler to conclude that the ammonia excited state lifetime decreases from 140 fsec in the $v_2 = 1$ vibronic level to 56 fsec in the $v_2 = 3$ vibronic level.

Ziegler also demonstrated (57) a rotational state dependence for photodissociation by examining the UV rotational excitation profiles through the $v'_2 = 2$ vibronic absorption band. The rotational state dependence of the dissociation rate is obtained directly from the excitation profile, since the only unknown parameter is the excited state lifetime. Figure 8 shows the R and S rotational transitions accompanying the $v_1 + v_2$ combination band as a function of excitation within the $v'_2 = 2$ vibronic subband; Figure 9 shows the measured and calculated excitation profiles. As indicated in Table 1 the rovibronic lifetimes and rates strongly depend upon the

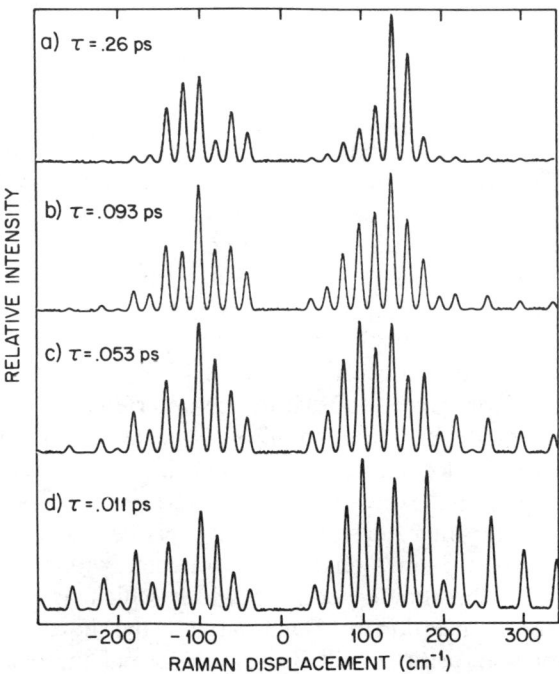

Figure 6 Dependence of the calculated NH_3 resonance Raman rotational structure upon excited state lifetime for 208.8 nm excitation where $\tau = (2\pi c\Gamma_{ev})^{-1}$. From Ref. (56).

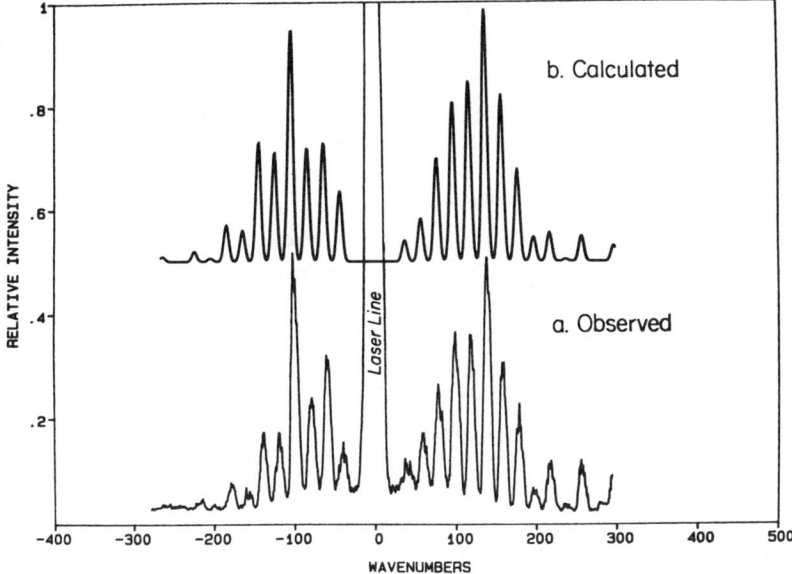

Figure 7 Comparison of observed and calculated resonance rotational Raman spectrum of NH_3 with 208.8 nm excitation. The calculated spectrum derives from the use of an excited state lifetime of 93 fsec ($\Gamma = 55$ cm^{-1}). From Ref. (56).

Table 1 Rovibronic lifetimes and photodissociation rates determined by S and O RREPs for the $v'_2 = 2$ absorption band of the $X \to A$ transition of NH_3

Raman transition	Initial J	Resonant J'	Γ (fsec)[a]	Rate $\times\ 10^{-12}\ s^{-1}$
S	1	2	141	7.1
	2	3	100	10.0
	3	4	91	11.0
	4	5	78	12.7
	5	6	75	13.3
	6	7	70	14.3
	7	8	68	14.7
O	3	2	139	7.2
	4	3	109	9.2
	5	4	111	9.0
	6	5	102	9.8
	7	6	91	11.0
	8	7	82	12.2

[a] Absolute lifetime error is estimated to be $\pm 10\%$. Relative lifetime error is $\pm 5\%$. From Ref. (57).

rotational levels involved. The lifetimes decrease from 151 fsec for $S1$ ($J = 1$, $J' = 2$) to 68 fsec for $S7$ ($J = 7$, $J' = 8$). Similar lifetime decreases are observed for the O, P, and R Raman transitions. It is interesting that the photodissociation rates appear to depend upon how the upper state is populated; the rates differ between S and O transitions that terminate in identical J' levels.

Thus, the intramolecular dissociation rates are strongly dependent upon the angular momentum of the populated states. Ziegler suggests that this rotational state dependence derives mainly from centrifugal effects on the quasibound A state surface and that higher resolution RREP studies will be used in the future to explore the relative contribution of Coriolis coupling to increasing the dissociation rate. These studies monitor the dynamics of photodissociation and detail the role of rotations in determining the dissociation rate constants.

The dissociation dynamics present in the gas phase are expected to be greatly modified in the condensed phase. Indeed, Rydberg transitions become weaker and more diffuse when a molecule such as NH_3 is placed in solution. Recently Ziegler & Roebber examined NH_3 dissolved in hexane and observed that the v_2 stretching Raman cross-section was decreased relative to that in the gas phase (58). This decrease was suggested to derive from an increased dephasing rate of the Rydberg transition due to coupling with low frequency solvent bath modes. These types of studies are crucial to establish connections between the vibronic transitions of gas phase molecules and those in solution. Careful establishment of these connections is especially important for higher energy transitions where host matrix transitions exist at energies similar to those of the solute.

We recently observed similar homogeneous linewidth differences upon excitation into the B_{2u} vibronic bands of gas and solution phase benzene (59–61). In solution, little enhancement derives from the B_{2u} vibronic components because of their large homogeneous linewidths. In the gas phase these transitions give strong resonance enhancement due to their narrow 3 cm^{-1} linewidths (60). Indeed, the resonance Raman cross-section maxima should show a Γ^{-2} dependence upon homogeneous linewidth.

By examining the details of the excitation profile of the v_1 band near one of these narrow vibronic resonances, we were able to probe the excited state dephasing dynamics (60). Figure 10 shows that benzene Raman spectra excited at 43,170, and 43,280 cm^{-1} are dominated by the symmetric v_1 ring and v_2 C–H stretching vibrations. The Raman cross-sections of benzene were measured relative to the known cross-sections of the internal standard methane (62); the C–H methane stretching band occurs at 2920

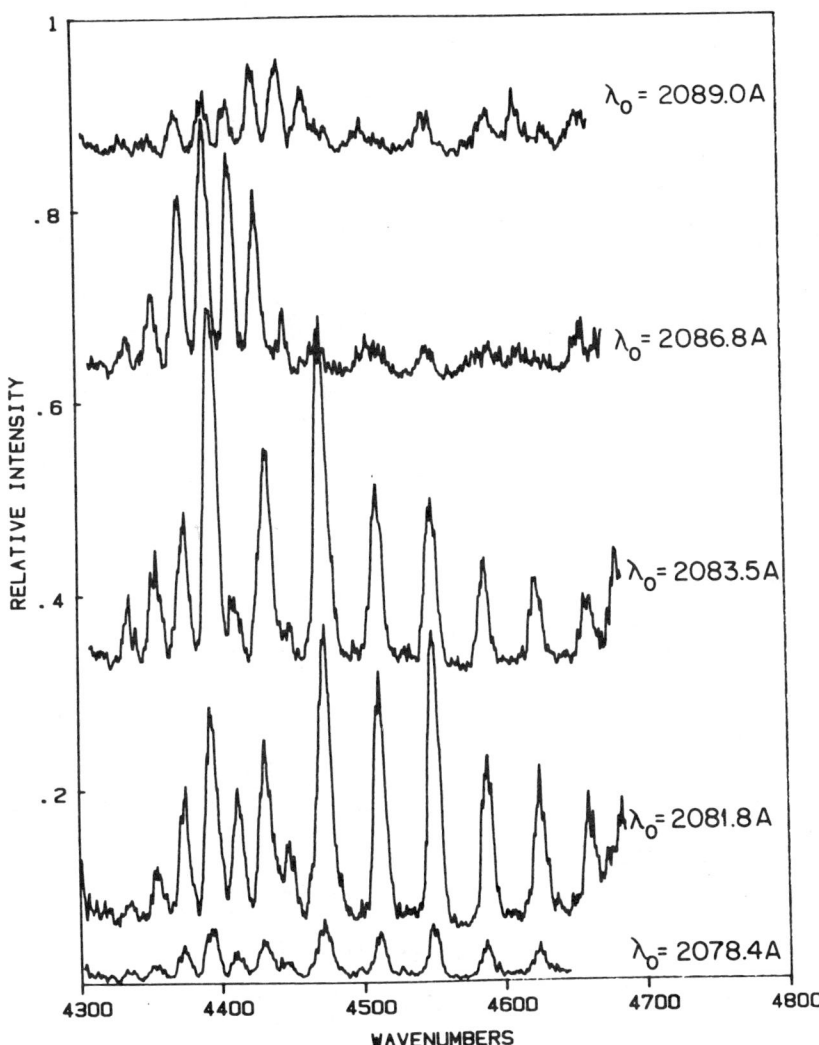

Figure 8 Resonance enhanced R and S transitions accompanying the $v_1 + v_2$ combination band of NH_3. The displayed intensities are normalized to the internal standard v_1 band of methane. From Ref. (57).

cm^{-1}. The Raman excitation profiles (Figure 11) measured between 19,000 and 45,000 cm^{-1}, excited away from resonance with B_{2u} vibronic features, monotonically increase with increasing excitation frequencies (Figure 11). The use of the modified A-term expression (Eq. 11) indicates that the

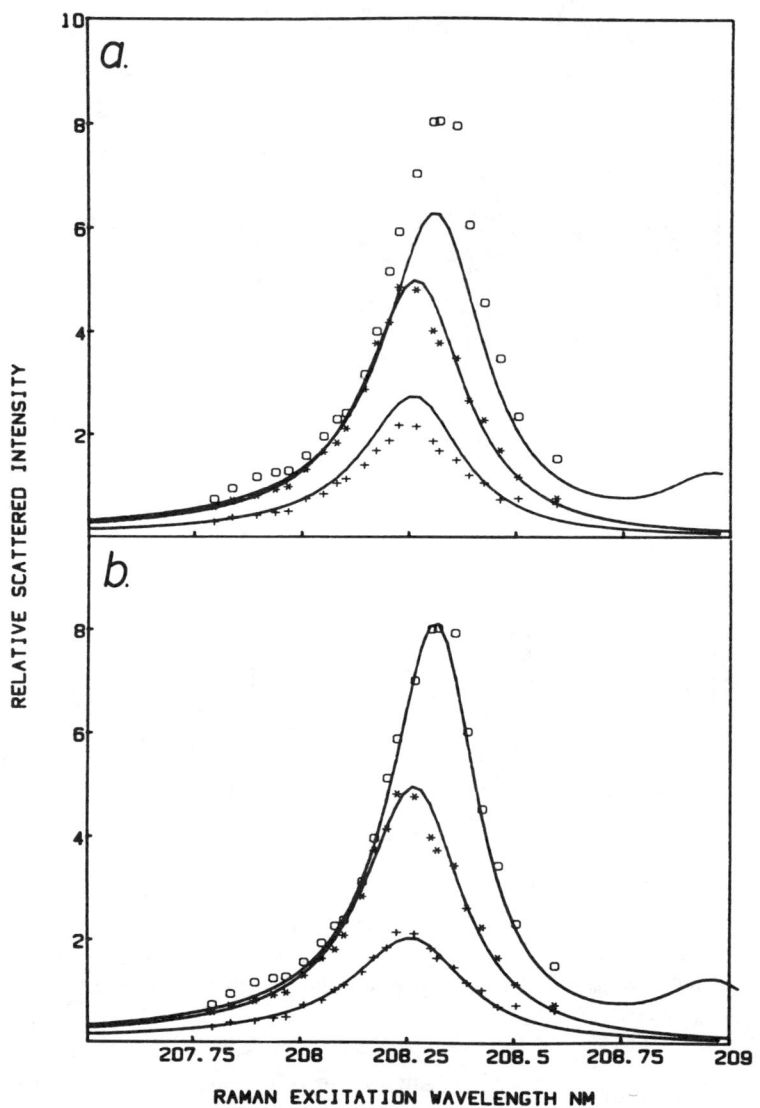

Figure 9 (a) RREP for the S3(○), S5(*), and S7(+) transitions of NH_3. *Solid curves* calculated with $\Gamma = 68$ cm^{-1}. (b) Best fit to the S3, S5, and S7 RREP give $\Gamma = 56$, 68, and 75 cm^{-1}, respectively. From Ref. (57).

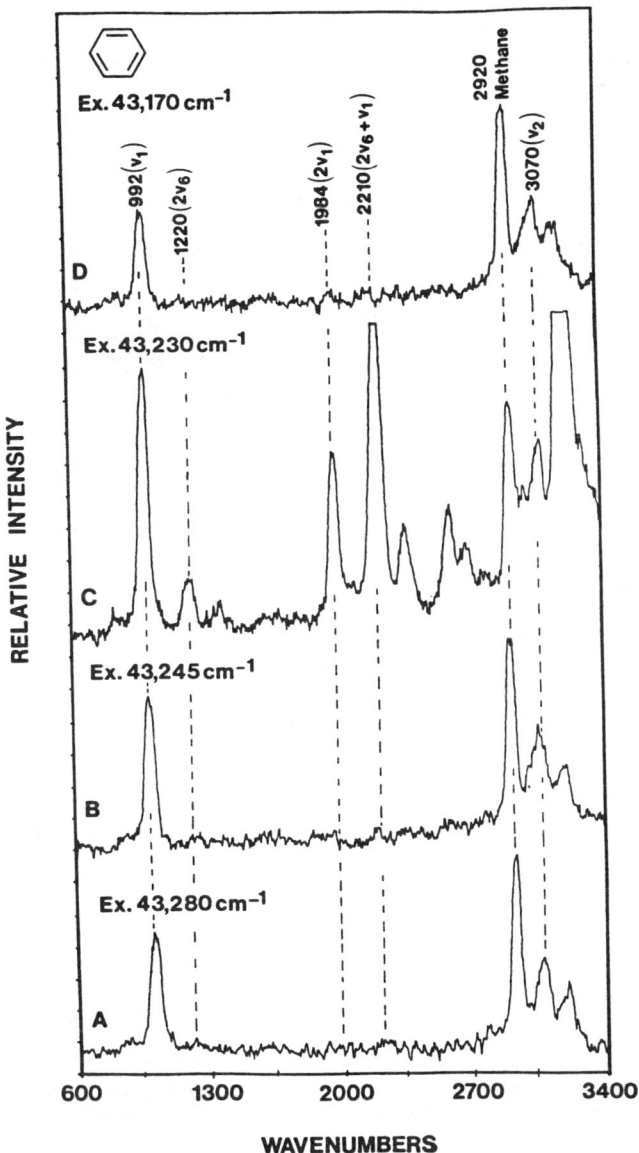

Figure 10 Raman spectra of gas phase benzene obtained with excitation in the vicinity of excitation of the B_{2u} $6_0^1 1_0^5$ transition. The benzene bands are labeled and the intensities are determined by reference to the methane internal standard band at 2920 cm^{-1}. The spectra were obtained at the excitation frequencies shown in the inset of Figure 11. From Ref. (60).

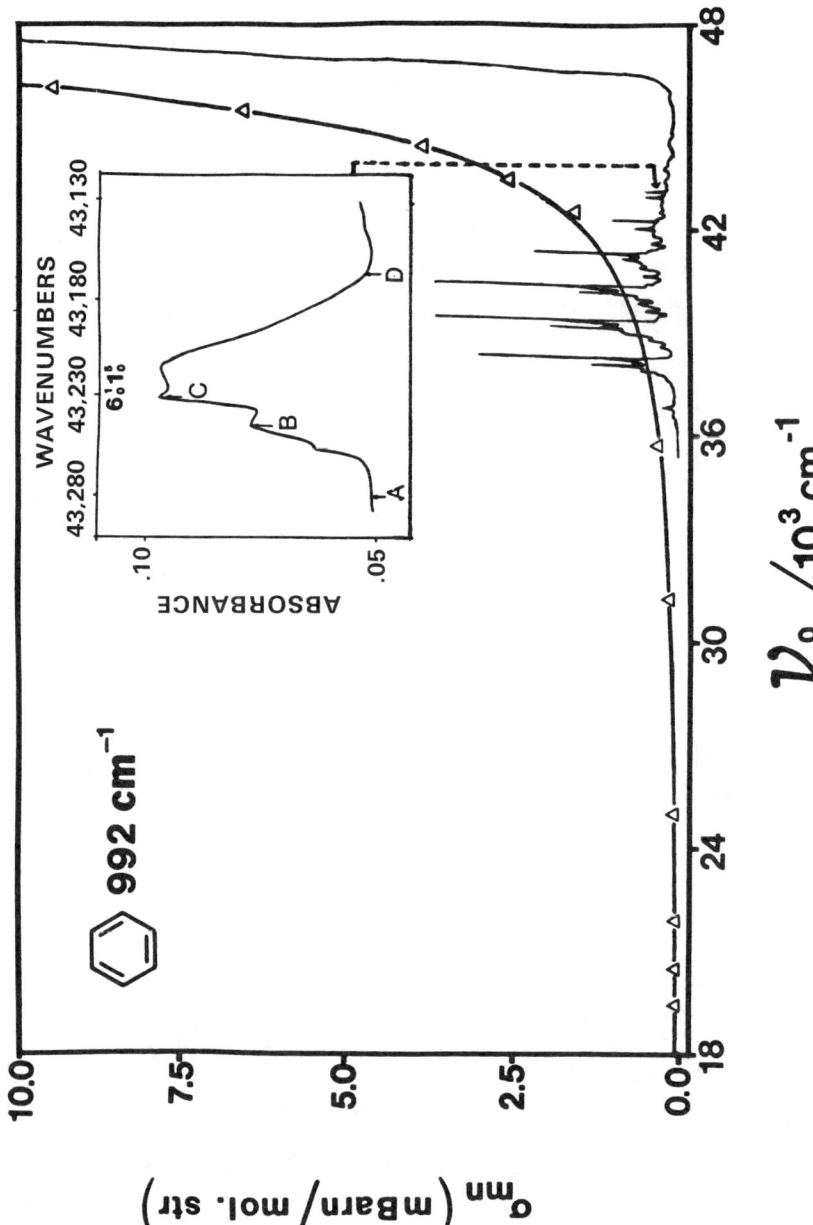

Figure 11 Gas phase benzene absorption spectrum and RREP of the 992 cm^{-1} symmetric ring breathing vibration. *Inset* shows details of the 43,200 cm^{-1} absorption region. From Ref. (61).

enhancement of the v_1 band derives almost entirely from the E_{1u} and B_{1u} transitions at higher energy. Excitation studies by Gerrity et al (63) directly into the E_{1u} transition show strong v_1 enhancement by the E_{1u} transition, but destructive interference with excitation between the B_{1u} and E_{1u} transitions (200 nm).

Figure 10, which shows the 43,230 cm^{-1} excited resonance Raman spectrum within the $6_0^1 1_0^5 B_{2u}$ vibronic transition, shows enhancement not only of v_1 but also of $2v_1$, and additional harmonics. In addition, a $2v_6$ band is observed as well as its combinations with v_1. The progressions in v_6 clearly derive from the C-term scattering mechanism, where a one-quantum change occurs for both the upward and downward Raman transitions; the strictly forbidden B_{2u} transition obtains oscillator strength mainly through Herzberg-Teller coupling by the v_6 vibration.

The excitation profile of the v_1 vibration can be used to examine dephasing, since the magnitude of interference between the preresonance Raman and resonance Raman scattering will decrease as the dephasing time shortens. In the absence of pure dephasing all of the emission is Raman and interference is a maximum. Fast pure dephasing results in resonance emission that derives mainly from hot luminescence (HL) (otherwise known as single vibrational level fluorescence, SVLF) that does not add at the amplitude level (polarizability level) to the preresonance Raman intensities. Prior to pure dephasing the resonance Raman excitation profile in the B_{2u} absorption feature will show constructive and destructive interference with the preresonance contribution as excitation passes through resonance; the energy denominator of Eq. 8 passes through zero to alter the sign of resonance Raman polarizability. In contrast, if emission occurs subsequent to dephasing and occurs from a molecular eigenstate, the SVLF or HL will contribute independently of the preresonance Raman intensity (see Figure 12).

Figure 13, which shows the RREP through the $6_0^1 1_0^5$ benzene vibronic absorption band, shows the presence of constructive and destructive interference in the RREP, and indicates that some resonance Raman emission occurs prior to pure dephasing. Quantitative modeling reveals that about 50% of the v_1 *resonance* emission intensity derives from SVLF or HL. We determined a ca 2 psec pure dephasing time. By examining the pressure dependence of the Raman intensities we were able to identify this ca 2 psec dephasing process as intramolecular vibrational redistribution (IVR) and were able to estimate the number of levels coupled by the IVR process. The presence of SVLF was also signaled by the observed depolarization ratios, which increased to $\rho = 1$ as the emission lifetime became larger than the rotational period.

As experimental resources increase for UV Raman measurements it is

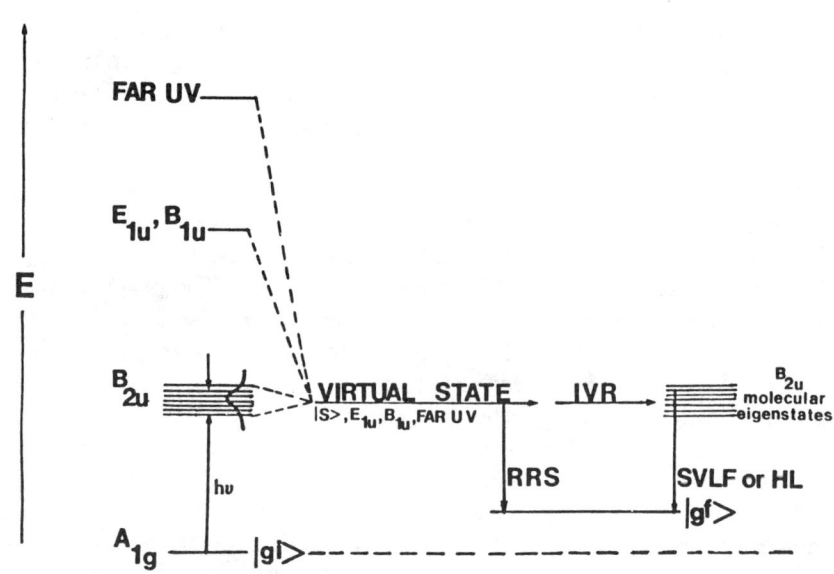

Figure 12 Schematic diagram of RRS, HL, and SVLF. Closely spaced *horizontal lines* represent B_{2u} molecular eigenstates. The superposition or virtual state, which gives rise to RRS and the interference effects, contains contributions from resonant and preresonant states. Intramolecular vibrational redistribution (IVR) dephases the superposition state and populates the B_{2u} molecular eigenstates. Emission from the eigenstates is HL (or SVLF). From Ref. (60).

certain that additional studies will examine dynamics of photodissociation and photochemistry and the dephasing of the coherent states prepared by the radiation field.

Studies of Electronic Excited States

The study of molecular excited states by Raman spectroscopy involves the investigation of the relative enhancement and the depolarization ratios of different vibrational modes. The depolarization ratios identify which invariants of the polarizability contribute. In resonance the depolarization ratio for symmetric vibrations will uniquely determine whether an electronic transition is polarized along a single direction, whether differently polarized electronic transitions overlap, or whether degeneracy is present. The relative enhancements of vibrations is directly diagnostic of geometry or force constant changes in the excited state. Since the magnitude of the Franck-Condon factors for the fundamental of a symmetric vibration is directly related to the displacement of the excited state potential surface along that vibrational coordinate, the magnitudes of enhancement can be

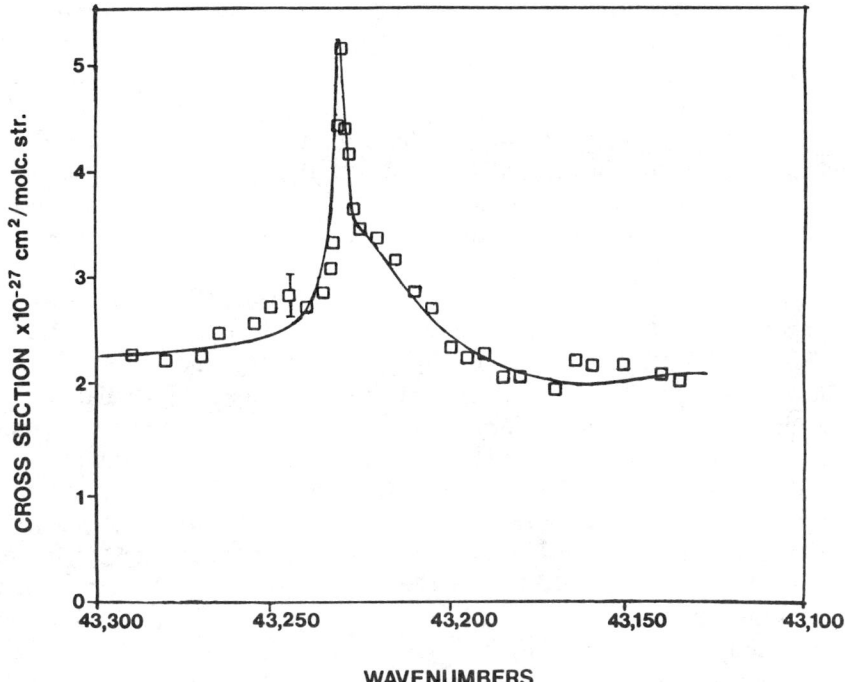

Figure 13 Raman excitation profile of the 992 cm^{-1} benzene ring vibration within the B_{2u} $6_0^1 1_0^5$ vibronic transition. The *solid curve* is the calculated excitation profile assuming almost equal contributions of resonance Raman and SVLF emission. This leads to a ca 2 psec dephasing time. From Ref. (60).

used to determine quantitatively the geometry change within the basis set of the symmetric normal mode coordinates (Figure 4).

RREP can also be used to identify underlying vibronic band structure or overlapping transitions. Obviously, dispersion of the depolarization ratios conveys additional information. Preresonance excitation profiles can be used to identify the location of higher energy transitions, and the relative preresonance enhancements of different vibrations can be used to probe qualitatively the geometry changes in preresonant excited state.

Preresonance Raman studies are especially useful for the identification of far UV electronic transitions. As clearly pointed out in the insightful volumes of Robin (64), the vacuum UV region is relatively unexplored, with few transitions clearly identified or characterized even for the simplest gas phase molecules. The information on vacuum UV transitions available for molecules in the condensed phase is truly "underwhelming."

Fitting the RREP data using the A-term expression (Eqs. 10 and 11) will extract the transition frequency of the preresonant transition. The maximum preresonance enhancement occurs for symmetric vibrations with major contributions from those internal vibrational coordinates that project along the coordinates of the excited state geometry change. Thus, excitation even out of resonance allows information to be obtained on transitions further in the UV. Often, direct absorption measurements of these transitions are impossible due to interference from overlapping solvent transitions. All solvents show absorption interferences below ca 180 nm.

Preresonance Raman studies have recently appeared for species such as acetone (48), CH_3CN (31), acetamide and N-methylacetamide (48), ClO_4^- (31), SO_4^{2-} (31), l-cystine (65), imidazole (66), and water (66). We developed a new technique to determine the dispersion of Raman cross-sections in the UV and examined species that were useful as internal intensity standards for future UV measurements (31). The absolute scaling of the cross-sections was obtained by referencing our relative cross-sections to the 514.5 nm benzene absolute differential Raman cross-sections of Abe et al (35). Measurements of gas phase Raman cross-sections are more straightforward, and Black & Bischel (62) have reported cross-section measurements for CH_4, D_2, H_2, N_2, and O_2 from the visible spectral region into the UV (below 200 nm)

Figure 14 shows the dispersion of the cross-sections of the C–C stretch (918 cm^{-1}) and the C–N stretch (2249 cm^{-1}) of CH_3CN, and the symmetric stretches of ClO_4^-, SO_4^{2-}, and NO_3^- ions in aqueous solution. The *solid lines* show A-term fits to the preresonance cross-section data. The NO_3^- preresonant cross-sections are accurately modeled by the simple A-term expression, which indicates that the enhancement derives from a transition at 191 nm, which is close to the well known 200 nm $\pi \to \pi^*$ transition. Thus, for NO_3^- preresonance enhancement derives solely from a UV state that undergoes an expansion along the N–O bonds.

In contrast, the C–C stretch of CH_3CN shows almost no preresonance enhancement, as evident from the extrapolation of its preresonant transition to 25 nm. Indeed, extraordinarily high energy preresonance transitions are derived from A-term fits for CH_3CN, sulfate, ClO_4^-, acetone, H_2O: 86 nm for the C≡N stretch of CH_3CN, 73 nm for the symmetric stretch of sulfate ion, 78 nm for the symmetric stretch of ClO_4, 80 nm for the carbonyl stretch of acetone, and 68 nm for the 1645 cm^{-1} bending vibration of water.

These extrapolated transition frequencies are surprisingly high relative to naive expectations, since most workers assume that strong acetonitrile and acetone C≡N and C=O $\pi \to \pi^*$ transitions occur in the 150–180 nm spectral region (31, 64). These transitions should result in significant C–N

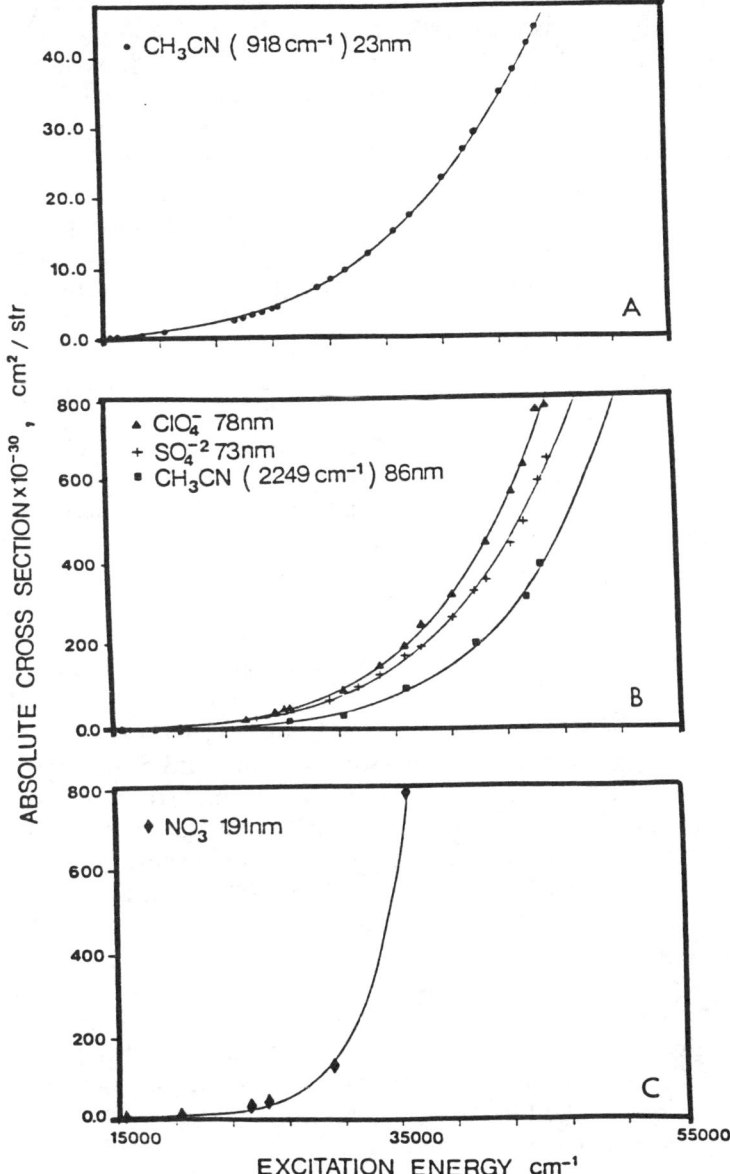

Figure 14 Total differential Raman cross-sections of (*A*) the 918 cm^{-1} C–C stretch of acetonitrile (*B*) The 2249 cm^{-1} C≡N stretch of acetonitrile, the 932 cm^{-1} symmetric stretch of ClO$_4^-$, and the 981 cm^{-1} symmetric stretch of SO$_4^{2-}$. (*C*) The 1045 cm^{-1} symmetric stretch of NO$_3^-$. The *solid lines* show *A*-term expression best fit to Eq. 10. From Ref. (31).

and C–O bond elongations, which would lead to strong enhancements. Similar expectations exist for SO_4^{2-} and ClO_4^-, which display absorption edges at ca 180 nm, and broad intense features are observed below 160 nm. Our perusal of the vacuum UV literature had lead us to the realization that no clear assignment exists for the acetone $\pi \rightarrow \pi^*$ transition, nor is there a clear understanding of the absorption spectra of SO_4^{2-} and ClO_4^- ions in water. Charge transfer transitions between the ions and water have been suggested to be important for SO_4^{2-} and ClO_4^- absorption bands. Indeed, the relevant question is whether molecular transitions are even likely to exist in the vacuum UV spectral region, where the density of states is so high and where intermolecular transitions (charge transfer) are likely to dominate the oscillator strengths. These UV Raman measurements are expected to serve as a sensitive probe for the existence and importance of vacuum UV electronic transitions.

In general, preresonance RREP of species with well-resolved, clearly assigned UV transitions show satisfactory A-term fits giving expected transition frequencies. Examples include imidazole (66), acetamide, and N-methylacetamide (48). Figure 15 shows the visible and UV Raman spectra of N-methylacetamide, while Figure 16 shows the preresonance RREP for these bands. Prominent bands observed include the Amide I, II, and III bands, which are dominated (see Figure 15) by carbonyl stretching (Amide I), and C–N stretching and C–H bending (Amide II and III). The 1162, 881 and 628 cm^{-1} bands each contain major contributions of either C–C stretching or N–CH$_3$ rocking in addition to NCO bending. As evident in the simple one-state A-term fits shown on the left side of Figure 16, an excellent modeling occurs for the Amide II and III bands which are dominated by C–N stretching. The preresonant state is extrapolated to occur at 185 nm, the location of the well-known amide $\pi \rightarrow \pi^*$ transition. In contrast, less acceptable fits occur for bands with appreciable C–C stretching and NCO bending. The inability to fit simultaneously the visible and UV Raman RREP for these bands indicates that more than one transition is important to their preresonance enhancement. The 185 nm $\pi \rightarrow \pi^*$ transition does not completely determine the enhancement of bands with appreciable C–C stretching and NCO bending. The RREP of the Amide I band, which is dominated by C–O stretching and shows little enhancement by the 185 nm transition, is not plotted because it overlaps the Amide II band. We conclude that the $\pi \rightarrow \pi^*$ transitions of amides involves large C–N bond elongation. This result is important because it argues against theoretical calculations that predict a larger C–O than C–N bond order change for the $\pi \rightarrow \pi^*$ transition (67, 68). Further, we have recently tentatively assigned the 1496 cm^{-1} N-methylacetamide band to the overtone of the Raman-forbidden Amide V torsional vibrational (68a).

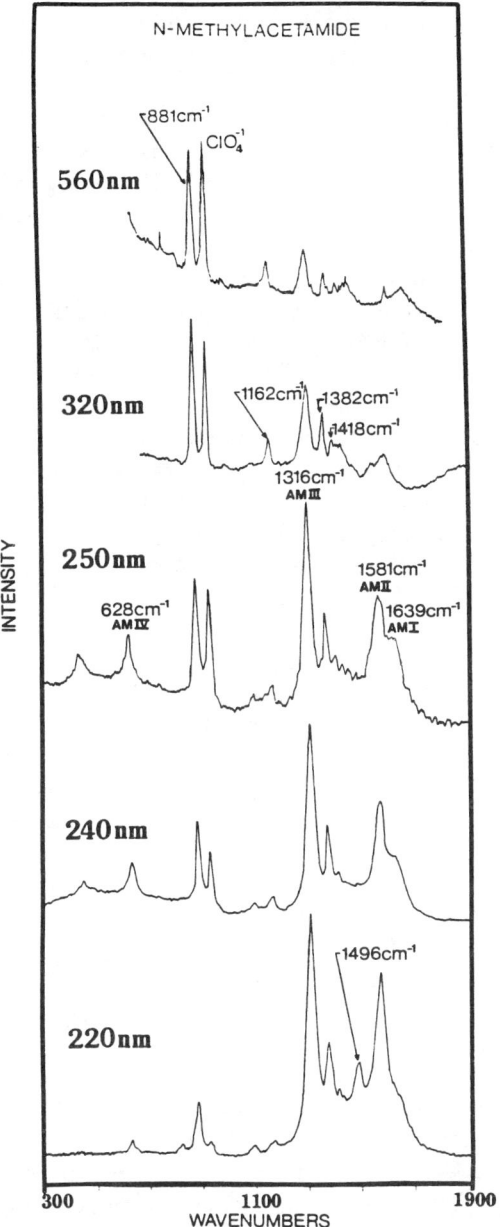

Figure 15 Raman spectra of aqueous solutions of *N*-methylacetamide. The internal standard perchlorate (0.2 M) contributes the 932 cm^{-1} band. *N*-methylacetamide concentrations were 1.0 M for 560 nm and 320 nm excitation, and 0.4 M for the other excitation wavelengths. From Ref. (48).

The anomalous enhancement of this band by the Amide $\pi \to \pi^*$ transition suggests that the excited state is twisted in a manner similar to that of ethylene (*vide infra*).

The enhancement pattern for *N*-methylacetamide is drastically changed by deuteration, as shown recently by Mayne et al (69). The increased deuterium mass results in decoupling of N–H bending from C–N stretching for the Amide II and III modes, and an Amide II' band replaces the Amide II and III bands. This Amide II' vibration is almost a pure C–N stretch, and shows a cross-section almost the sum of that of the Amide II and III bands. Thus, the Raman cross-sections quantitatively report upon the magnitude of the projection of the normal coordinate upon the excited state geometry changes. A similar decoupling of C–H bending and C–N stretching occurs for proline derivatives, which leads to dramatic enhancement of a similar Amide II' band for these molecules (70–72).

The excited states of numerous small gas phase molecules such as ethylene (76a,b), and derivatives such as dichloro and dimethylethylene (42), butadiene (77), CS_2(78), benzene (9, 10, 47, 60, 61, 63), and styrene (79), have been probed with UV Raman excitation. Studies of ethylene in the $N \to V$ transition (ca 190 nm) show enhancement of even overtones of the a_u torsional mode, which indicates that the excited state is twisted with respect to the planar ground state (76a,b). Large enhancements are also observed for C–C stretching and CH_2 scissoring motions, which indicate an increased C–C bond length and decreased CH_2 bond angles in the excited state. The increased C=C bond length presumably derives directly from a bond order decrease. The decreased C–H bond angle probably derives from the increased carbon sp^3-like hybridization in the excited state.

Studies of styrene with excitation in its S_2, S_3, and S_4 $\pi \to \pi^*$ transitions between 190 and 260 nm examined the relative enhancement of the ring vibrations compared to the vinyl group stretching, angle bending, and torsional motions (79). In contrast to ethylene, no vinyl group torsional modes were enhanced, thus indicating a much larger barrier to twisting in the excited state of styrene compared to ethylene. The enhancements observed for the ring modes and vinyl stretching were similar for excitation in the S_2 and S_3 transitions, a result that indicates the delocalized nature of the ground and/or the excited S_2 and S_3 states. The S_1 transition is known to be localized on the ring from absorption and fluorescence measurements. Excitation into S_4, however, results mainly in enhancement of bands that are predominantly ring vibrations; the nature of this excited state appears very reminiscent of the benzene E_{1u} transition.

We have recently concluded a study (80) of the electronic transitions of bis(imidazole) Fe(III) protoporphyrin IX-dimethylester, a model for the

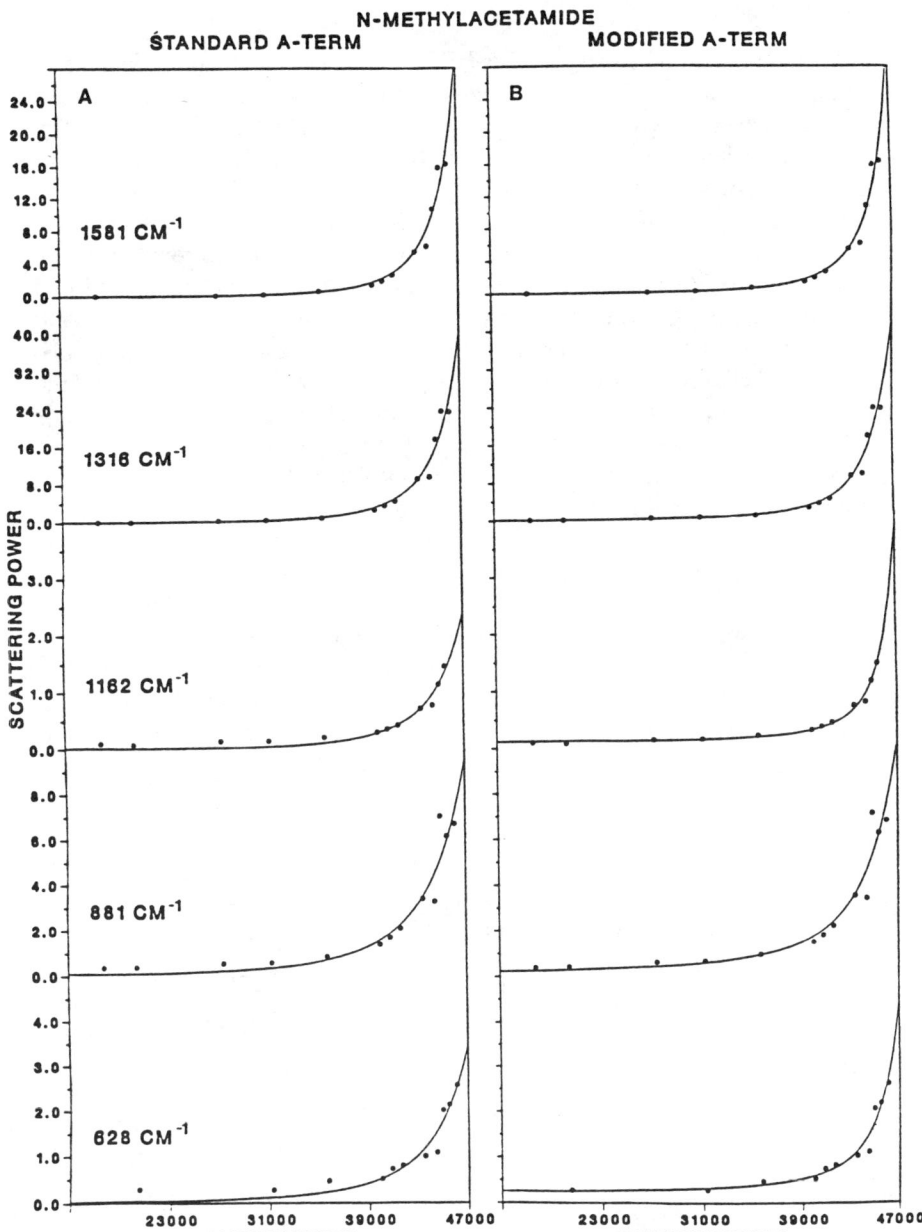

Figure 16 Preresonance RREP of *N*-methylacetamide. The *solid curve* in the left-hand panel corresponds to a best fit with the *A*-term expression given by Eq. 10, while the right-hand panel shows the best fit to the modified *A*-term expression given by Eq. 11. From Ref. (48).

prosthetic group of heme proteins (Figure 17). The electronic structure of hemes has been carefully investigated with a host of experimental techniques as well as by using numerous theoretical calculations (81). The visible wavelength transitions (Figure 18) are well understood and result from in-plane $\pi \rightarrow \pi^*$ transitions from the highest two occupied porphyrin orbitals of a_{1u} and a_{2u} symmetry to an unoccupied orbital of e_{2g} symmetry. Due to configuration interactions the transition dipoles add for the intense Soret transition centered at ca 400 nm and subtract for the Q band 0–0 and 0–1 transitions known as the α and β bands at ca 590 and 550 nm, respectively. Depending upon the central metal and its axial ligands, additional charge transfer transitions can occur to result in a rich electronic spectral region between 330 nm and 800 nm. The region below 330 nm is relatively unknown because of the paucity of experimental studies, which may be related to the fact that this region is rather featureless. Additional porphyrin $\pi \rightarrow \pi^*$ transitions are likely to occur in this region, as should the $\pi \rightarrow \pi^*$ transitions of the vinyl groups. A fundamental question is whether the vinyl groups are so conjugated with the porphyrin network that the vinyl group excited states dissolve in the sea of heme electron density changes, or whether they continue a separate existence. This question can be readily resolved by RREP measurements. Excitation within the Soret band strongly enhances both heme vibrations as well as the C=C vinyl stretching mode. Although the vinyl group bands appear in a con-

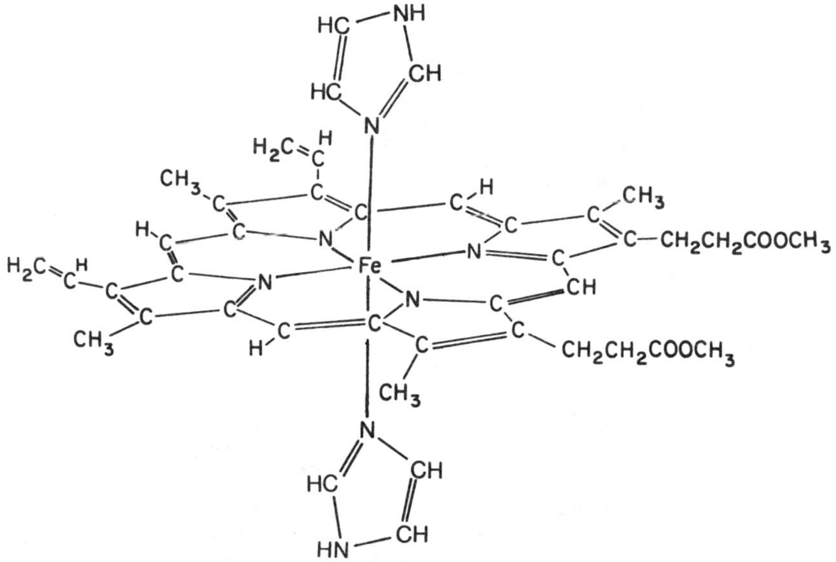

Figure 17 Structure of bis(imidazole) Fe (III) protoporphyrin IX dimethyl ester.

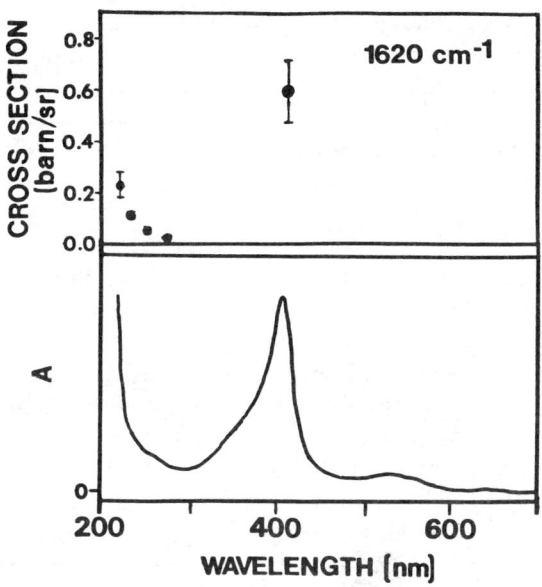

Figure 18 Absorption spectrum and RREP of the vinyl C=C stretching vibration of bis(imidazole) Fe (III) protoporphyrin-IX-dimethyl ester dissolved in CH_3CN.

gested region that shows numerous heme vibrations, the vinyl group band can be clearly assigned from the isotopic frequency shifts that occur upon deuterium substitution of the vinyl group. The observed Soret band maximum cross-sections can be estimated from the literature to be ca 0.6 Barns/str for vinyl stretching compared to ca 4.0 Barns/str for heme vibrations (82, 83). Thus, strong conjugation between the vinyl groups and the porphyrin orbitals is evident in the Soret transition. Excitation further in the UV results in cross-section decreases for all of the vibrational bands, both of the porphyin and the vinyl substituent. Indeed, only weak enhancement derives from the broad heme electronic transitions between 250 and 360 nm. However, excitation below 250 nm shows selective enhancement of the 1620 cm^{-1} vinyl stretching mode. The excitation profile clearly indicates the dominance of vinyl stretching, with a maximum in the RREP occurring somewhere below 220 (Figure 18). This selective enhancement of vinyl stretching proves the existence of an ethylenic $\pi \to \pi^*$ transition relatively uncoupled from the porphyrin ring. The stronger Soret band enhancement probably signals transfer of electronic density between the porphyrin and the vinyl group upon Soret excitation, or a significant contribution of porphyrin nuclear motion to this vibration. Because vinyl group electron density changes result in vinyl C=C bond length changes, vinyl stretching becomes enhanced via an A-term mechanism. Thus, these UV Raman measurements can determine the extent of delocalization of orbitals associated with substituents attached to aromatic ring systems and can be used to probe delocalization of the resonant excited state. Further, the extent of vinyl orbital delocalization should be sensitive to the vinyl group orientation with respect to the heme ring.

Biophysical and Analytical Applications

The Amide band frequencies and intensities are important markers of protein and peptide backbone folding and conformation (73, 74). Because of exciton interactions, the amide $\pi \rightarrow \pi^*$ transition dipoles can add constructively or destructively. For example, the α-helix conformations show a ca 40% hypochromic oscillator strength decrease compared to random coil or β-sheet. This oscillator strength decrease is accompanied by a greater than 60% decrease in the Amide II and III band cross-sections. In contrast, the Amide I band shows only a modest alteration in Raman cross-section. Spiro et al (71, 74, 75a,b) have begun to utilize this conformational dependence for the UV Raman intensities for protein structural studies. They recently demonstrated linear correlations between amide mode cross-sections and the protein peptide conformation (75a,b).

Numerous other studies have recently focused upon the protein aromatic amino acids (12, 13, 18, 19, 49a,b, 74, 84–89) in order to clarify the enhancement phenomena and to determine the selectivity available for enhancement of these residues in proteins. Important mechanistic information may derive from RR measurements that selectively excite those aromatic amino acid residues important in the protein enzymatic reaction coordinate. The vibrational mode frequencies and cross-sections will report on the aromatic amino acid residue environment. For example, the frequencies will depend upon hydrogen bonding. An environmental cross-section dependence occurs because the absorption band oscillator strength and its homogeneous and inhomogeneous linewidths depend upon environment. In addition, the Raman cross-sections depend upon the local field, which varies as the effective refractive index of the environment surrounding a residue changes. Careful studies of the environmental dependence of the aromatic amino acid band frequencies and cross-sections are just now underway (25, 88, 89).

The major impediment to these studies has been the facile formation of transient aromatic amino acid radicals (18) and the Raman saturation phenomena (19, 49a,b). The saturation phenomena derive from depopulation of the ground state and result in a power dependence of the Raman intensities. Because this power dependence strongly depends upon environment, it has been difficult to measure reliably protein aromatic amino acid cross-sections. The recent development of lower pulse energy sources promises to surmount the present obstacles (19).

Other studies have recently examined nucleic acid bases with the objective of utilizing this information to develop techniques for monitoring DNA conformation and nucleic acid-adduct interactions (23, 90-100). Extensive RREP have been reported for adenine, thymine, uracil, cytosine, and guanine, and attempts have been made to utilize the vibrational mode intensities to determine the excited state geometry changes. These studies have been aided by the high quality normal coordinate calculations and the electronic excited state calculations that exist for these molecules (93).

Recently studies examined complexation of the antitumor agent *cis*-dichloroammine platinum (II) with GMP (95, 98) and observed specific interactions that were associated with hydrogen bonding to the carbonyl group. These preliminary results illustrate that important intermolecular interactions can be monitored. It is clear that excitation wavelengths can be chosen to enhance specifically the Raman bands of particular classes of nucleic acid bases. The DNA nucleic acid Raman cross-sections depend upon DNA coiling and base stacking due to the exciton interactions between base pairs that cause hypochromism; the Raman cross-sections often will decrease with the square of the relative absorption decrease. However, hypochromism in the resonant band does not affect all bands identically, since some bands may derive their enhancement from adjacent electronic transitions that are not hypochromic (92).

The existing aromic amino acid and nucleic acid studies are just now being applied to examine protein and DNA conformation in in vivo systems (101-103). Nelson et al have recently used the relative intensities of the visible wavelength excited carotene bands (102) and the UV excited aromatic amino acid and nucleic acid bands (103) to speciate between different algae and bacteria. Obviously, these studies represent the most aggressive applications of UV resonance Raman spectroscopy and are at the forefront of bioanalytical chemistry.

UVRR studies have been reported for more extensively conjugated molecules such as naphthalene, triphenylene, and pyrene (19, 45, 104-115). These polycyclic aromatic hydrocarbons (PAH) represent a class of molecules of significant analytical and bioanalytical interest. Other members of this class are highly carcinogenic, and all are ubiquitous in the environment. In addition, these molecules are important for electronic structure theory because they represent test cases for new theoretical calculational techniques. These large planar aromatics have strong, allowed transitions that result in huge Raman enhancements for their symmetric in-plane ring modes. For example, pyrene shows a total differential Raman cross-section of 60 Barns/str for the 592 cm^{-1} pyrene ring mode with excitation at the maximum of its S_4 transition (240 nm) (19, 45). Recent studies of the S_1 (111–114) and S_4 (115) transitions of azulene and the S_4 transition of pyrene (45) have attempted to calculate the RREP by using Raman transform theory. The transform formalism utilizes the Franck-Condon information available in the absorption spectrum to predict the RREP of symmetric vibrations. Excellent agreement was found for the S_1 state of azulene within the framework of the transform formalism that assumes the Condon approximation. The transform theory, however, failed to model accurately the RREP of the S_4 excited state of azulene, even when the theory was extended to include excited state vibrational frequency shifts or non-Condon sources of scattering (115).

In contrast, we find excellent modeling of the RREP of the S_4 transition of pyrene by using the transform formalism, which includes non-Condon sources of scattering (45). Figure 19 shows the UV Raman spectra of pyrene excited in the S_4 transition, while Figure 20 shows the measured absorption spectrum and RREP for the most enhanced bands. The *solid line* is the best fit of the transform calculation to the experimental RREP within the Condon approximation, while the *dashed curve* derives from a calculation that includes a small linear non-Condon contribution. Obviously, acceptable modeling of the excitation profile requires inclusion of non-Condon activity. The extent of non-Condon contributions appears to increase with the vibrational frequency from a minimum of 3% for the 592 cm^{-1} band to a maximum of 13% for the 1632 cm^{-1} band. This systematic increase of the non-Condon contribution as the vibrational mode frequency increases can be diagnostic for nonadiabatic coupling between the S_4 and S_2 pyrene transitions.

These RREP studies examine the fundamental interactions between vibrational and electronic motion. Since much of our understanding of electronic spectroscopy derives from the numerous applications of molecular orbital calculations to rationalize the spectropic properties of PAHs, these studies continue to refine our understanding of excited state molecular structure and dynamics. Obviously, different transitions in PAHs exhibit different dynamical phenomena. It is striking that the RREP of the S_4 transitions of pyrene and azulene can be so different, since they both appear to involve isolated, fully allowed electronic transitions that show clear Franck-Condon vibrational substructure. Future studies are expected to clarify the major differences and the importance of non-Condon processes to electronic transitions.

Other, more applied studies have utilized the UVRR selectivity and sensitivity to empirically examine PAHs such as naphthalene, triphenylene, and pyrene in samples such as coal-derived liquids, coal-liquid distillates, and rat liver microsomes (104–108). These studies illustrate the analytical

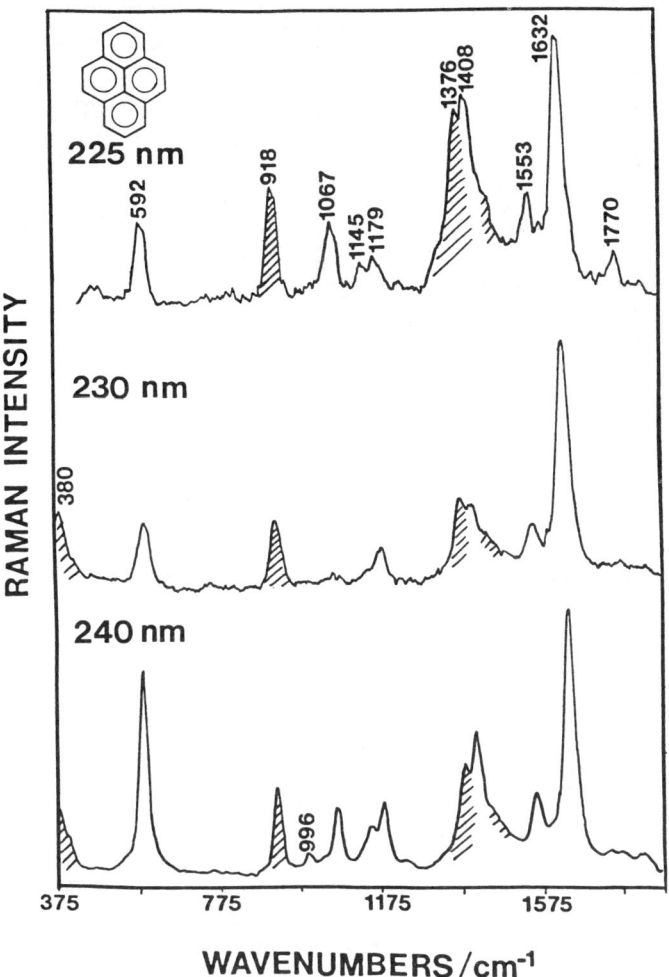

Figure 19 UV resonance Raman spectra of pyrene excited in the S_4 electronic transition. See S_4 absorption spectrum in Figure 20. The sample consists of a 5×10^{-4} M pyrene solution in acetonitrile. Note that the pyrene bands in the spectra show some Raman saturation. The cross-hatched bands are from the solvent. From Ref. (45).

and bioanalytical potential of UV Raman spectral measurements. In addition, these studies prove that fluorescence is not an impediment for Raman spectral measurements below 250 nm; species with their S_1 emitted state below 250 nm have vanishingly small fluorescent quantum yields in the condensed phase (105). Because most molecules emit only from S_1, any fluorescence occurs at much longer wavelengths. Analytical applications of

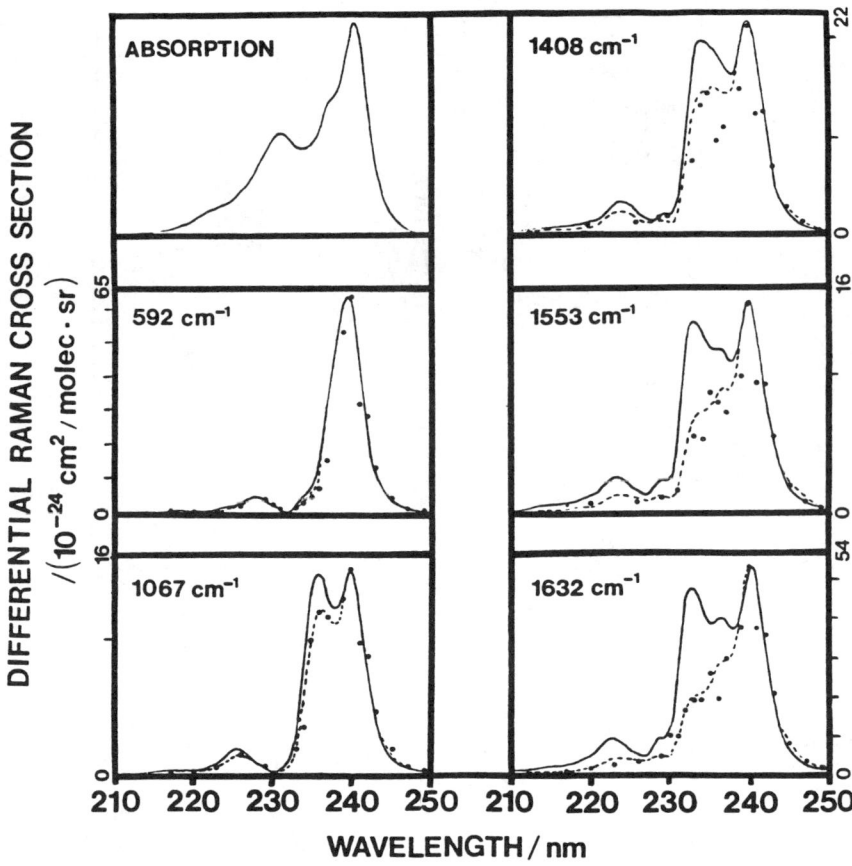

Figure 20 Absorption spectrum and RREP of symmetric vibrations of pyrene enhanced in the S_4 electronic transition. The *solid lines* show the transform best fit if only Condon sources of scattering are assumed. The *dashed lines* show a transform calculation that includes non-Condon sources of scattering.

UVRR spectroscopy are now just beginning and are expected to have impact in the fuel sciences, environmental studies, and bioanalytical studies of PAH metabolism.

SATURATION RAMAN SPECTROSCOPY

All of the studies described above monitor Raman bands deriving from vibrational transitions in the ground state manifold. The net Raman intensity is proportional to the incident intensity, the number of molecules in the sample volume, and the Raman cross-section for the band under study.

The high peak energy fluxes of the pulsed UV excitation source can induce photophysical processes that deplete the ground state population, leading to Raman intensities that do not increase linearly with incident photon flux (18, 19, 45, 49). In addition, either permanent photoproducts or transient intermediates can form, which may also be resonant with the incident excitation frequency (18, 19, 45).

For example, Figure 21 shows the UV Raman spectra of tyrosinate excited at 488 and 245 nm. Spectra obtained at 245 nm (B) excited at low pulse energy fluxes show bands that are also evident in the preresonance 488 nm spectrum (A). In contrast, the 245 nm spectrum (C) excited with higher pulse energies (corresponding to ca 100 photons $\cdot$ mol^{-1} $\cdot$ pulse^{-1}) shows new bands that are more evident in the difference spectrum (D) and are easily assigned to tyrosyl radical, which transiently forms during excitation due to monophotonic ionization of tyrosine. The power dependence of the intensity of the tyrosyl radical bands indicates the quantum yield for radical formation. Further, the power dependence of the tyrosinate ground state Raman bands is quantitatively related to the ground state recovery rate. High incident energy fluxes result in an apparent saturation of the ground state Raman band intensities due to ground state depopulation.

We recently described the kinetics associated with these processes and included terms related to the quantum yield of radical formation and kinetic parameters that are used to model ground state recovery rates (18, 19, 25). In the case that negligible recovery occurs during the excitation pulse, the expressions are particularly simple.

Raman saturation can also derive from depopulation of the ground state due to population of long-lived excited states. Indeed this is the mechanism of Raman saturation in pyrene (19, 45) and tryptophan (84) associated with excitation into their upper singlet states. Efficient fast internal conversion results in population of the long-lived S_1 state, which bottlenecks relaxation back to the ground state. Obviously the degree of saturation will depend upon intermolecular dynamical quenching processes that determine the lifetime of S_1.

The environmental dependence of the ground state recovery rate is the basis of a new spectroscopy we call Saturation Raman Spectroscopy that monitors relaxation back to the ground state. If relaxation is fast compared to the laser pulse length, a steady state can be built up during the excitation pulse. The steady state depopulation is directly related to the incident energy flux, the absorption cross-section (which measures the probability of transfer into the excited state), and the rate of relaxation back to the ground state. The spectroscopy monitors the dependence of the Raman intensities of ground state analyte bands upon the incident pulse energy flux.

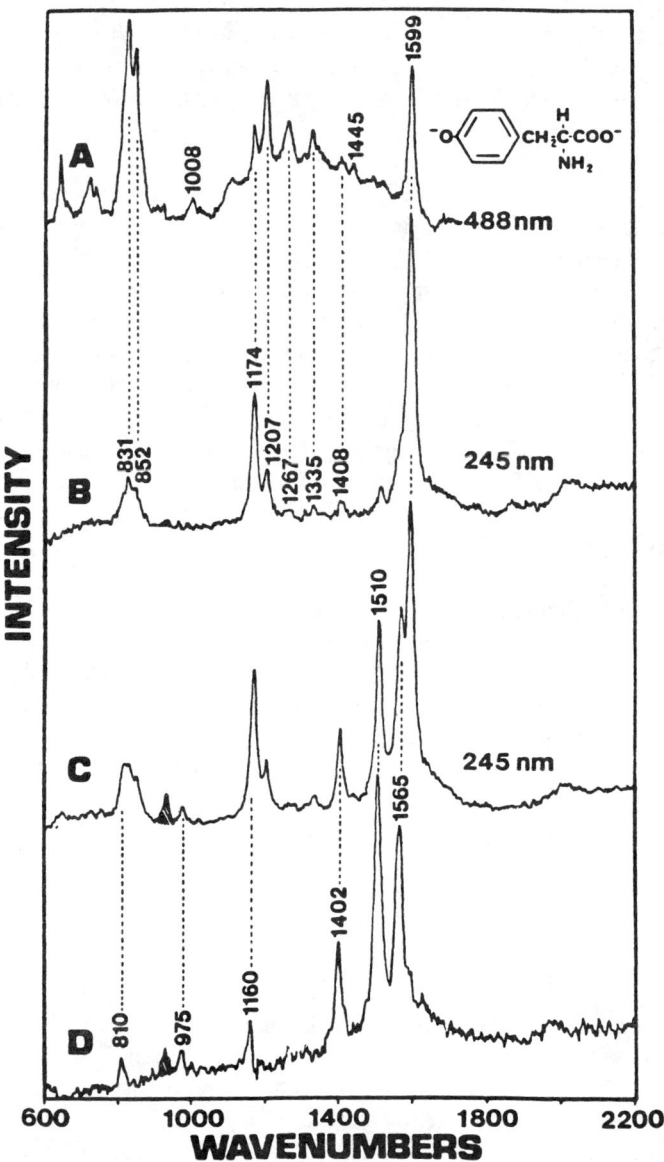

Figure 21 Raman spectra of tyrosinate in water containing 0.2 M ClO_4^- as an internal intensity standard (the shaded 932 cm^{-1} band). (*A*) Normal Raman spectrum excited at 488 nm with a CW laser. (*B*) UV Raman spectrum (245 nm) excited at low pulse energy. (*C*) Same as *B* but at higher pulse energy. (*D*) Difference spectrum: *C–B*. From Ref. (18).

Figures 22 and 23 illustrate one application of this spectroscopy to the study of the electronic structure of a dimer of tyrosine and tryptophan. Figure 22 shows Raman spectra of a mixture of monomeric tyrosine and tryptophan (*A* and *B*) and the Raman spectra of trp-tyr dimer (*C* and *D*) excited at 225 nm at high (*A* and *C*) and low (*B* and *C*) pulse energies. Figure 23 shows the intensity dependence of the tyrosine and tryptophan bands as a function of the incident pulse energy flux. The decreased relative tryptophan intensities (the 760 cm^{-1}, 1010 cm^{-1}, 1355 cm^{-1}, and 1550 cm^{-1} bands) in the dimer results from a hypochromic shift of the tryptophan absorption band due to the exciplex interactions. The degree of saturation differs between the tyr and trp residues in the monomers because of the different absorbances and relaxation rates of these residues.

The hypochromic absorption decreases in the dimer are clearly evident

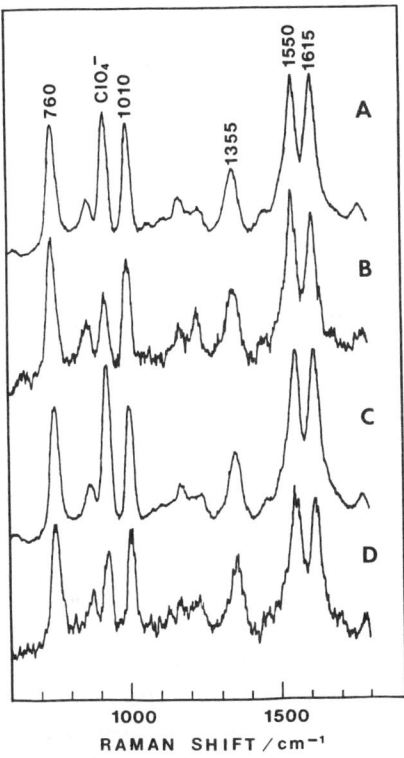

Figure 22 The pulse energy flux intensity dependence of the UV resonance Raman spectra of a stoichiometric mixture of tyrosine and tryptophan monomers, and that of the Trp-Tyr dimer excited at 225 nm. Note the 932 cm^{-1} ClO_4^- internal intensity standard band. From Ref. (25).

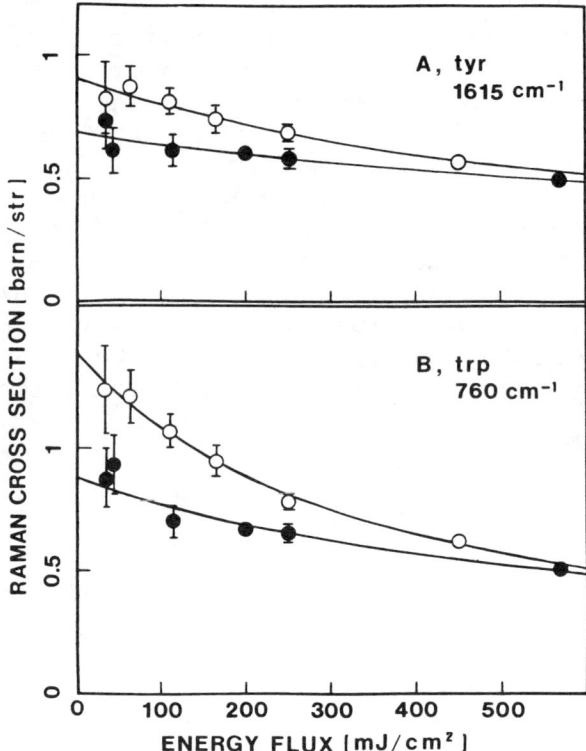

Figure 23 Dependence of Raman intensities of tyr and trp residues upon excitation pulse energy. Monomer saturation (○), dimer saturation (●). From Ref. (25).

by comparing the dimer and monomer absorption spectra. The dimer electronic transition occurs to an excited state that is a linear combination of the excited states of the tryptophan and tyrosine groups. This is evident from the saturation plot, which shows similar saturation behaviors for the tyrosine (1615 cm^{-1}) and the tryptophan bands in the dimer. This is also evident in the Raman spectra, since the tyr and trp relative intensities become independent of excitation power flux and the magnitude of saturation. The relaxation rate depends upon the intermolecular electronic interactions between these aromatic residues.

We expect more subtle interactions also to affect relaxation rates. For example, disulfide linkages in proteins are known to quench the fluorescence of adjacent tryptophans by increasing the rate of internal conversion. This will increase the relaxation rate back to the ground state and decrease the magnitude of Raman saturation. Indeed, we have already observed that the extent of aromatic amino saturation depends upon the protein studied. From these preliminary studies we are hopeful that Raman Saturation Spectroscopy will prove useful in studying the environments of aromatic molecules in complex environments. Indeed, it may be possible to simplify the spectra of proteins, for example, by using pump beams to saturate particular resonances in order to study others selectively.

HYPER-RAMAN SPECTROSCOPY

Recent studies of Ziegler et al (50–52) have demonstrated the potential utility of Resonance Hyper-Raman Spectroscopy to probe far UV excited states of molecules. Figure 24 shows that excitation at v_0 is two-photon resonant with a far UV transition and that a photon is inelastically scattered at frequency $2v_0-v_v$. For centrosymmetric molecules, resonance Raman and hyper-Raman vibrational transitions are mutually exclusive. In contrast, non-centrosymmetric molecules are expected to show hyper-Raman enhancement for the same symmetric modes as those enhanced in normal resonance Raman spectra excited at twice the frequency.

Hyper-Raman scattering appears at the $\chi^{(5)}$ level of the electrical susceptibility. Chung & Ziegler (52) have recently reviewed the theory for spontaneous hyper-Raman scattering and have expanded it in terms of the vibronic formalism typically used in discussions of normal resonance Raman scattering. They find similar enhancement mechanisms for hyper-Raman as for normal Raman scattering. The somewhat more complex expressions can be subdivided into A, B, and C-term-like contributions analogous to those displayed in Eq. 8 but that uniquely display the essence of the three-photon interaction. The A-term enhances symmetric vibrations in electronic transitions that are both one and two-photon allowed. The B-term involves a vibronic transition moment and can enhance both symmetric and nonsymmetric vibrations within electronic transitions that are either/both one- or two-photon allowed. The C-term

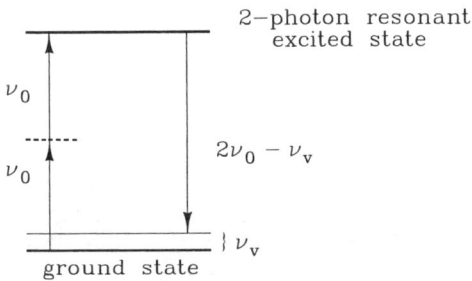

Figure 24 Energy level diagram showing hyper-Raman transition.

involves two vibronic transition moments and can be involved in Franck-Condon scattering of symmetric vibrations and overtones and combinations of nonsymmetric modes. The C-term is expected to be a weak source of resonance hyper-Raman enhancement.

Figure 25 shows the resonance hyper-Raman spectrum of ammonia excited at 425.6 nm and the calculated spectrum used to determine the excited state lifetime. Obviously the signal-to-noise ratios are comparable to those obtained by direct resonance Raman measurements of ammonia in the UV (see Figures 7 and 8), in spite of the dramatically decreased hyper-Raman cross-sections. This occurs because self-absorption of the incident excitation beam does not occur for the visible wavelength excitation; higher sample concentrations can be used and the beam can sample larger volume elements. A major advantage for hyper-Raman studies of far UV transitions is that the hyper-Raman excitation utilizes tunable visible wavelength excitation sources that have high power fluxes. In addition, little competing absorption and photochemistry occurs with visible excitation, in contrast to far UV excitation. Although not proven, it is possible that less stray-light spectral interferences may occur for hyper-Raman, since the cross-sections for the hyper-Rayleigh elastic scattering from particulates and refractive index inhomogeneities may be smaller relative to the vibrational hyper-Raman cross-sections than for normal Raman scattering.

It is likely that hyper-resonance Raman spectroscopy will prove to be an important new technique for studying far UV-excited states. Studies of ammonia, CH_3I, and CS_2 illustrate the A- and B-term like resonance hyper-Raman enhancement mechanisms (52). Future studies are likely to prove the utility of this technique for studying condensed phase systems as well.

CONCLUSIONS

This review was written in the winter of 1987, a period of active progress in fundamental and applied UV Raman studies. It is already clear that new information is being obtained that both confirms and leads the theoretical understanding of electronic structure and dynamics. The future directions of research will probe electronic-vibrational rotational coupling in small molecules as well as use this spectroscopy in applications in physical chemistry, physical organic, biophysical chemistry, analytical chemistry, and bioanalytical chemistry. At the moment the experimental limitations derive mainly from the inconvenience of the present high pulse energy flux UV excitations sources. These limitations are rapidly disappearing as new technology develops.

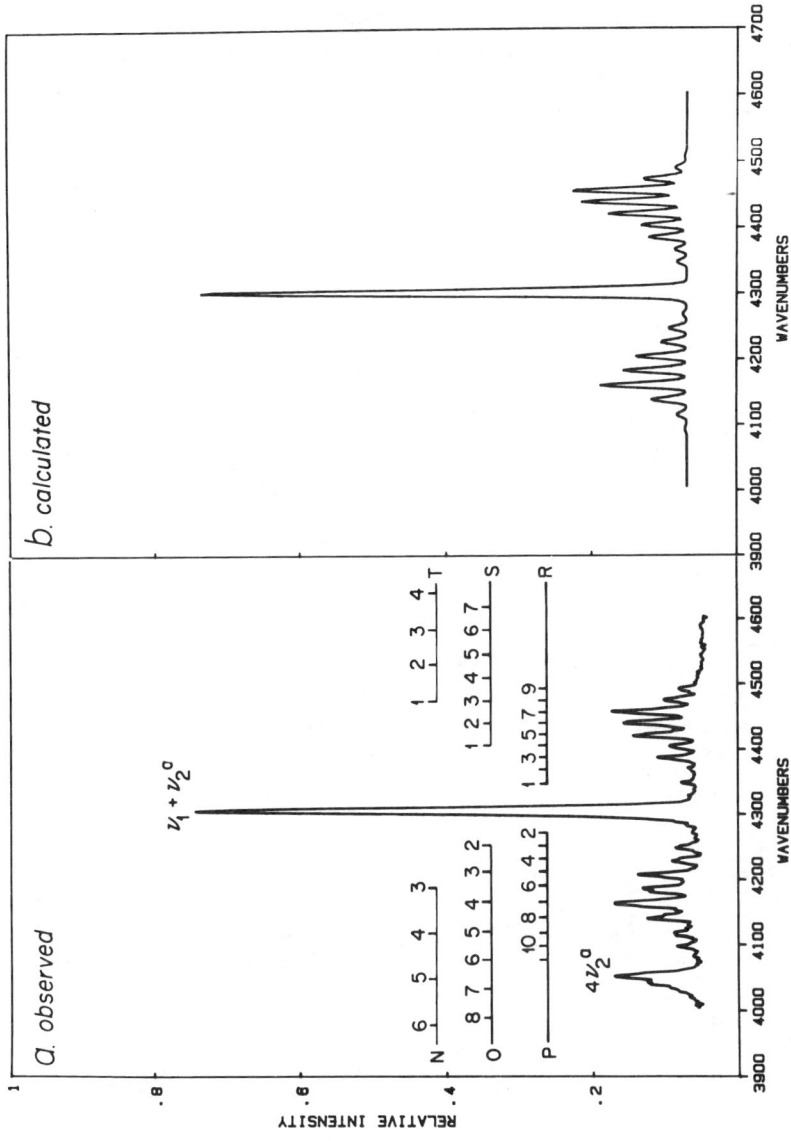

Figure 25 (a) Resonance Hyper-Raman rotationally resolved spectrum of NH_3 (ca 2 atm pressure) of the $\nu_1 + \nu_2$ band with 425.6 nm excitation. (b) Calculated spectrum, which gives an excited state lifetime of 130 fsec. From Ref. (50).

Acknowledgments

I gratefully acknowledge the work of the present and past members of my research group who made the progress reviewed here. Special thanks go to Dr. Colleen Jones, Val DeVito, Paul Harmon, and Dr. Junji Teraoka for their results that were filched prior to publication and for their helpful comments. I also gratefully acknowledge preprints and helpful conversations with Dr. Larry Ziegler, Dr. Bruce Hudson, and Dr. Tom Spiro. I also gratefully acknowledge support for this work from National Institute of Health Grant 1R01 GM30741-07. Sanford A. Asher is an Established Investigator of the American Heart Association; this work was done during the tenure of an Established Investigatorship of the American Heart Association, Pennsylvania affiliate.

Literature Cited

1. Hirakawa, A. Y., Tsuboi, M. 1975. *Science* 188: 359
2. Nishimura, Y., Hirakawa, A. Y., Tsuboi, M. 1978. *Adv. IR Raman Spectrosc.* 5: 217
3. Pezolet, M., Yu, T. J., Peticolas, W. L. 1975. *J. Raman Spectrosc.* 3: 55
4. Chinsky, L., Turpin, P. Y. 1982. *Biopolymers* 21: 277
5. Ziegler, L. D., Albrecht, A. C. 1979. *J. Chem. Phys.* 70: 2634, 2644
6. Ziegler, L. D., Albrecht, A. C. 1977. *J. Chem. Phys.* 67: 2753
7. Korenowski, G. M., Ziegler, L. D., Albrecht, A. C. 1978. *J. Chem. Phys.* 68: 1248
8. Ziegler, L. D., Albrecht, A. C. 1979. *J. Raman Spectrosc.* 8: 73
9. Ziegler, L. D., Hudson, B. S. 1981. *J. Chem. Phys.* 74: 982
10. Ziegler, L. D., Hudson, B. S. 1983. *J. Chem. Phys.* 79: 1134
11. Long, D. H. 1977. *Raman Spectroscopy.* New York: McGraw-Hill
12. Johnson, C. R., Ludwig, M., O'Donnell, S., Asher, S. A. 1984. *J. Am. Chem. Soc.* 106: 5008
13. Copeland, R. A., Spiro, T. G. 1985. *Biochemistry* 24: 4960
14. Asher, S. A., Johnson, C. R., Murtaugh, J. 1983. *Rev. Sci. Instr.* 54: 1657
15. Hudson, B. S. 1986. *Spectroscopy* 1: 22
16. Fodor, S. P. A., Rava, R. P., Spiro, T. G. 1986. *J. Raman Spectrosc.* 17: 471
17. Ziegler, L. D. 1987. *J. Chem. Phys.* 86: 1703
18. Johnson, C. R., Ludwig, M., Asher, S. A. 1986. *J. Am. Chem. Soc.* 108: 905
19. Jones, C. M., Devito, V. L., Harmon, P. A., Asher, S. A. 1987. *Appl. Spectrosc.* 41: 1268
20. Kato, K. 1986. *IEEE J. Quantum Electron.* QE-22: 1013
21. Hudson, B. S. 1987. *Spectroscopy* 2: 33
22. Asher, S. A. 1984. *Appl. Spectrosc.* 38: 276
23. Bajdor, K., Nishimura, Y., Peticolas, W. L. 1987. *J. Am. Chem. Soc.* 109: 3514
24. Caswell, D. S., Spiro, T. G. 1987. *J. Am. Chem. Soc.* 109: 2796
25. Teraoka, J., Harmon, P. A., Asher, S. A. 1988. *Appl. Spectrosc.* Submitted
26. Bilhorn, R. B., Sweedler, J. V., Epperson, P. M., Denton, B. M. 1987. *Appl. Spectrosc.* 41: 1114
27. Bilhorn, R. B., Epperson, P. M., Sweedler, J. V., Denton, M. B. 1987. *Appl. Spectrosc.* 41: 1125
28. Rousseau, D. L., Williams, P. F. 1976. *J. Chem. Phys.* 64: 3519
29a. Lee, D., Albrecht, A. C. 1985. *Adv. IR Raman Spectrosc.* 12: 179
29b. Mukamel, S. 1985. *J. Chem. Phys.* 82: 5398
30. Placzek, G. 1931. In *Handb. Radiol.* 6: 205 (Available in Engl. trans., UCRL-Trans-526. US Atomic Energy Comm., Div. Techn. Inform., 1962)
31. Dudik, J. M., Johnson, C. R., Asher, S. A. 1985. *J. Chem. Phys.* 82: 1732
32. Schrotter, H. W., Klockner, H. W. 1979. In *Topics in Current Physics,* ed. A. Weber, 11: 123. Berlin: Springer-Verlag
33. Eckhardt, G., Wagner, W. G. 1966. *J. Mol. Spectrosc.* 19: 407
34. Mortenson, O. S., Hassing, S. 1979. *Adv. IR Raman Spectrosc.* 6: 1

35. Abe, N., Wakayama, M., Ito, M. 1977. *J. Raman Spectrosc.* 6: 38
36. Lee, S.-Y., Heller, E. J. 1979. *J. Chem. Phys.* 71: 4777
37. Heller, E. J., Sundberg, R. L., Tannor, D. 1982. *J. Phys. Chem.* 86: 1822
38. Imre, D., Kinsey, J. L., Sinha, A., Krenos, J. 1984. *J. Phys. Chem.* 88: 3956
39. Sundberg, R. L., Imre, D., Hale, M. O., Kinsey, J. L., Coalson, R. D. 1986. *J. Phys. Chem.* 90: 5001
40. Hale, M. O., Galica, G. E., Glogover, S. G., Kinsey, J. L. 1986. *J. Phys. Chem.* 90: 4997
41. Tang, J., Albrecht, A. C. 1970. In *Raman Spectroscopy*, ed. H. Szymanski, 2: 33. New York: Plenum
42. Hudson, B. S., Kelly, P. B., Ziegler, L. D., Desiderio, R. A., Hess, W., Bates, R. 1986. *Advances in Laser Spectroscopy*, ed. B. A. Garetz, J. R. Lombardi, 3: 1. New York: Wiley
43. Myers, A. B., Mathies, R. A. 1987. In *Biological Applications of Raman Spectroscopy*, ed. T. B. Spiro. New York: Wiley
44. Stallard, B. R., Callis, P. R., Champion, P. M., Albrecht, A. C. 1984. *J. Chem. Phys.* 80: 70
45. Jones, C. M., Asher, S. A. 1988. *J. Chem. Phys.* In press
46. Albrecht, A. C., Hutley, M. C. 1971. *J. Chem. Phys.* 55: 4438
47. Asher, S. A., Johnson, C. R. 1985. *J. Phys. Chem.* 90: 1375
48. Dudik, J. M., Johnson, C. R., Asher, S. A. 1985. *J. Phys. Chem.* 89: 3805
49a. Ludwig, M., Asher, S. A. 1988. *J. Am. Chem. Soc.* 110: 1005
49b. Teraoka, J., Harmon, P., Asher, S. A. 1988. *Science.* In preparation
50. Ziegler, L. D., Chung, Y. C., Zhang, Y. P. 1987. *J. Chem. Phys.* 87: 4498
51. Ziegler, L. D., Roebber, J. L. 1987. *Chem. Phys. Lett.* 136: 377
52. Chung, Y. C., Ziegler, L. D. 1988. *J. Chem. Phys.* In press
53. Rohlfing, E. A., Valentini, J. J. 1985. *J. Chem. Phys.* 83: 521
54. Ziegler, L. D., Hudson, B. 1984. *J. Phys. Chem.* 88: 1110
55. Ziegler, L. D., Kelly, P. B., Hudson, B. 1984. *J. Chem. Phys.* 81: 6399
56. Ziegler, L. D. 1986. *J. Chem. Phys.* 84: 6013
57. Ziegler, L. D. 1987. *J. Chem. Phys.* 86: 1703
58. Ziegler, L. D., Roebber, J. L. 1988. *Chem. Phys. Lett.* In press
59. Asher, S. A., Johnson, C. R. 1985. *J. Phys. Chem.* 89: 1375
60. Harmon, P. A., Asher, S. A. 1988. *J. Chem. Phys.* 88: 2925
61. Harmon, P. A., Asher, S. A. 1988. *J. Chem. Phys.* Submitted
62. Black, G., Bischel, W. K. 1983. In *Excimer Lasers*, ed. C. K. Rhodes, H. Egger, H. Pummer. New York: Am. Inst. Physics
63. Gerrity, D. P., Ziegler, L. D., Kelly, P. B., Desiderio, R. A., Hudson, B. 1985. *J. Chem. Phys.* 83: 3209
64. Robin, M. B. 1985. In *Higher Excited States of Polyatomic Molecules*, Vols. 1–3. New York: Academic
65. Johnson, C. R., Asher, S. A. 1987. *J. Raman Spectrosc.* 18: 345
66. Asher, S. A., Murtaugh, J. L. 1988. *Appl. Spectrosc.* 42: 83
67. Robb, M. A., Csizmadia, I. G. 1968. *Theor. Chim. Acta* 10: 269
68. Del Bene, J., Jaffe, H. H., Ellis, R. L., Kuehnlenz, G. 1974. *Quantum Chem. Progr. Ex.* 10: 174
68a. Song, S., Asher, S. A., Krimm, S. 1988. *J. Am. Chem. Soc.* Submitted
69. Mayne, L. C., Ziegler, L. D., Hudson, B. 1985. *J. Phys. Chem.* 89: 3395
70. Mayne, L., Hudson, B. 1987. *J. Phys. Chem.* 91: 4438
71. Caswell, D. S., Spiro, T. G. 1987. *J. Am. Chem. Soc.* 109: 2796
72. Carey, P. R. 1982. *Biochemical Applications of Raman and Resonance Raman Spectroscopies*. New York: Academic
73. Tu, A. T. 1982. *Raman Spectroscopy in Biology*. New York: Wiley
74. Rava, R. P., Spiro, T. G. 1985. *Biochemistry* 24: 1861
75a. Copeland, R. A., Spiro, T. G. 1987. *Biochemistry* 26: 2134
75b. Copeland, R. A., Spiro, T. G. 1986. *J. Am. Chem. Soc.* 108: 1281
76a. Ziegler, L. D., Hudson, B. S. 1983. *J. Chem. Phys.* 79: 1197
76b. Sension, R. J., Mayne, L., Hudson, B. 1988. *J. Am. Chem. Soc.* Submitted
77. Chadwick, R. R., Gerrity, D. P., Hudson, B. S. 1985. *Chem. Phys. Lett.* 115: 24
78. Desiderio, R. A., Gerrity, D. P., Hudson, B. S. 1988. *Chem. Phys. Lett.* 115: 29
79. Ziegler, L. D., Varotsis, C. 1986. *Chem. Phys. Lett.* 123: 175
80. DeVito, V. L., Asher, S. A. 1988. *J. Am. Chem. Soc.* In preparation
81. Gouterman, M. 1978. In *The Porphyrins*, ed. D. Dolphin, 3: 1. New York: Academic
82. Choi, S., Spiro, T. G., Langry, K. C., Smith, K. M., Budd, D. L., La Mar, G. N. 1982. *J. Am. Chem. Soc.* 104: 4345
83. Bangcharoenpaurpong, O., Scho-

macker, K. T., Champion, P. M. 1984. *J. Am. Chem. Soc.* 106: 5688
84. Asher, S. A., Ludwig, M., Johnson, C. R. 1986. *J. Am. Chem. Soc.* 108: 3186
85. Rava, R. P., Spiro, T. G. 1984. *J. Am. Chem. Soc.* 106: 4062
86. Rava, R. P., Spiro, T. G. 1985. *J. Phys. Chem.* 89: 1856
87. Caswell, D. S., Spiro, T. G. 1986. *J. Am. Chem. Soc.* 108: 6470
88. Caswell, D. S., Spiro, T. G. 1986. *Biochim. Biophys. Acta* 873: 73
89. Hildebrand, P. G., Copeland, R. A., Spiro, T. G., Otlewski, J., Laskowski, M., Prendergast, F. G. 1988. *Biochemistry*. Submitted
90. Fodor, S. P. A., Rava, R., Hays, T. R., Spiro, T. G. 1985. *J. Am. Chem. Soc.* 107: 1520
91. Copeland, R. A., Spiro, T. G. 1986. *J. Phys. Chem.* 90: 6648
92. Fodor, S. P. A., Spiro, T. G. 1986. *J. Am. Chem. Soc.* 108: 3198
93. Tsuboi, M., Nishimura, Y., Hirakawa, A. Y., Peticolas, W. L. 1987. In *Biological Applications of Raman Spectroscopy*, Vol. 2, ed. T. G. Spiro. New York: Wiley
94. Wang, Y., Peticolas, W. L. 1987. *J. Phys. Chem.* 91: 3122
95. Ziegler, L. D., Hudson, B., Strommen, D. P., Peticolas, W. L. 1984. *Biopolymers* 23: 2067
96. Kubasek, W. L., Hudson, B., Peticolas, W. L. 1985. *Proc. Natl. Acad. Sci. USA* 82: 2369
97. Blazej, D. C., Peticolas, W. L. 1980. *J. Chem. Phys.* 72: 3134
98. Perno, J. R., Cwikel, D., Spiro, T. G. 1987. *Inorg. Chem.* 26: 400
99. Chinsky, L., Turpin, P. Y., Duquesne, M., Brahms, J. 1977. *Biochem. Biophys. Res. Commun.* 75: 766
100. Chinsky, L., Turpin, P. Y., Duquesne, M., Brahms, J. 1978. *Biopolymers* 17: 1347
101. Nocentini, S., Chinsky, L. 1983. *J. Raman Spectrosc.* 14: 9
102. Dalterio, R. A., Baek, M., Nelson, W. H., Britt, D., Sperry, J. F., Purcell, F. J. 1987. *Appl. Spectrosc.* 41: 241
103. Dalterio, R. A., Nelson, W. H., Britt, D., Sperry, J. F. 1987. *Appl. Spectrosc.* 41: 417
104. Asher, S. A. 1984. *Anal. Chem.* 56: 720
105. Asher, S. A., Johnson, C. R. 1984. *Science* 225: 311
106. Johnson, C. R., Asher, S. A. 1984. *Anal. Chem.* 56: 2258
107a. Jones, C. M., Naim, T. A., Ludwig, M., Murtaugh, J., Flaugh, P. L., Dudik, J. M., Johnson, C. R., Asher, S. A. 1985. *Trac-Trends Anal. Chem.* 4: 75
107b. Asher, S. A., Jones, C. M. 1986. In *New Applications of Analytical Techniques to Fossil Fuels*, ed. M. Perry, H. Retcofsky. ACS Div. Fuel. Chem. Preprints 31: 170
108. Rumelfanger, R., Asher, S. A., Perry, M. B. 1988. *Appl. Spectrosc.* 42: 267
109. Peticolas, W. L., Chinsky, L., Turpin, P. Y., Laigle, A. 1983. *J. Chem. Phys.* 78: 656
110. Koshihara, S., Kobayashi, T. 1986. *J. Chem. Phys.* 85: 1211
111. Liang, R., Schnepp, O., Warshel, A. 1976. *Chem. Phys. Lett.* 44: 394
112. Liang, R., Schnepp, O., Warshel, A. 1978. *Chem. Phys.* 34: 17
113. Brafman, O., Chan, C. K., Khodadoost, B., Page, J. B., Walker, C. T. 1984. *J. Chem. Phys.* 80: 5406
114. Chan, C. K., Page, J. B., Tonks, D. L., Brafman, O., Khodadoost, B., Walker, C. T. 1985. *J. Chem. Phys.* 82: 4813
115. Cable, J. R., Albrecht, A. C. 1986. *J. Chem. Phys.* 84: 1969

VIBRATIONAL SPECTROSCOPIC STUDIES OF THE STRUCTURE OF SPECIES DERIVED FROM THE CHEMISORPTION OF HYDROCARBONS ON METAL SINGLE-CRYSTAL SURFACES

Norman Sheppard

School of Chemical Sciences, University of East Anglia, Norwich, NR4 7TJ, England

INTRODUCTION

This review is primarily concerned with the identification of the structures of chemisorbed complexes formed between hydrocarbons and metal surfaces through the use of various forms of vibrational spectroscopy (1). The identification of the chemisorbed species and the study of their reactivities by spectroscopic methods provides an essential basis for the understanding of many classes of metal-catalyzed reactions of hydrocarbons such as hydrogenation/dehydrogenation, hydrogenolysis, isomerization, and metathesis.

The particular emphasis of this article is on the study of species formed on metal single crystals that are cut so that specific surfaces of known atomic arrangements, e.g. the [111], [100], and [110] surfaces of face-centered cubic (fcc) metals, are exposed to the hydrocarbon adsorbate. Such experiments have become feasible during the past 15 years or so. The principal vibrational spectroscopic techniques used for this purpose are high resolution electron energy loss spectroscopy (EELS) (2) and reflection/absorption infrared spectroscopy (RAIRS) (3).

In recent years it has been shown that results obtained with these simplified single-crystal systems are of great value in interpreting vibrational spectra obtained from hydrocarbons adsorbed on finely divided metal particles. The latter are usually studied in the form of oxide-supported metal catalysts. Their spectra are more complex than those from single-crystals because metal particles can exhibit a much wider range of adsorption sites, leading in turn to the coexistence of a multiplicity of chemisorbed species. On the other hand, the relatively high specific area of these finely-divided metals, *ca.* 1 to 20 m^2 g^{-1}, does lead to high spectroscopic sensitivity for adsorbed species in the accessible regions. For this reason transmission infrared spectroscopy has been applied to study such systems for more than 30 years, following the pioneering work of R. P. Eischens and his collaborators of the Texaco Research Laboratories in the USA (4, 5). It had originally been intended that this type of work would also be reviewed in this article. However, the extent of the literature on metal single-crystal work is such that this has not proved to be possible within the space allotted. Also it is now clear that a reinterpretation of many of the spectra obtained on finely divided metals has been necessitated by the collected single-crystal results.

Nevertheless I conclude this review with a "worked example" of the relationship between the two areas of research by illustrating a successful interpretation of the complex spectra from ethylene on finely divided Pt/SiO$_2$, which has been strongly aided by the results of single-crystal studies by EELS and RAIRS.

Other vibrational spectroscopic techniques that have been applied to finely divided or polycrystalline metal samples include Raman spectroscopy, inelastic neutron scattering, and inelastic electron tunneling spectroscopy. For silver, and to a lesser extent for copper and gold, much greater Raman sensitivity can be obtained through the mechanism—still incompletely explained—of surface-enhanced Raman spectroscopy (SERS) (6). This is applicable to the rough surfaces of cold-deposited metal films formed by evaporation. The inelastic neutron scattering method can also be applied to porous metals as such, which would be opaque to the other methods.

One reason for the comparatively great success of the vibrational spectroscopic methods in the structural diagnosis of chemisorbed species is the very extensive literature on vibrational spectra to which reference can be made (7). For the present purpose the latter has been very usefully extended during the past decade to include infrared, and sometimes Raman, spectra from specific hydrocarbon ligands on metal-cluster compounds whose structures have been determined unambiguously by x-ray crystallography (8). Not only do such spectra provide a "pattern recognition" method for the identification of likely adsorbed species, but also the spectra obtained from the unknown adsorbed species themselves exhibit absorption bands characteristic of particular hydrocarbon groupings. The latter provide a very useful initial guide to which general types of adsorbed species are likely to be worth considering. Luckily, with oxide-supported metal catalysts it is usually the high-frequency ends of the spectra of adsorbed species that are accessible, and this is particularly rich in group-frequency information (9).

A considerable number of other physical methods have also contributed to the identification of chemisorbed hydrocarbon species (10). These include low-energy electron diffraction (LEED), ultraviolet photoelectron spectroscopy (UPS), including the use of angle-resolved techniques (ARUPS), near-edge x-ray absorption fine structure (NEXAFS), nuclear magnetic resonance spectroscopy (NMR), and programmed thermal desorption (TPD), which is alternatively, less happily, named thermal desorption spectroscopy (TDS). Results from the use of these techniques are mentioned when they are pertinent to the interpretation of the vibrational spectra. However, I cannot claim a capability for critically evaluating these results to the same degree as for those from vibrational spectroscopy. If this leads to unduly heavy weight being given to the latter results I apologize in advance.

LEED techniques can often lead fairly straightforwardly to information about the relative pattern and spacings of adsorbed molecules when these form regular arrays on a metal surface (11). However, often the same pattern for the adsorbates can be superimposed on different positions with respect to the metal surface, implying alternative adsorption sites and often alternative adsorbed species. More detailed intensity/voltage measurements of the diffraction features have in some cases led to very valuable information about the specific adsorption sites and the orientation of the adsorbed species with respect to the surface atoms. A particularly valuable case in point was the evidence originally provided by this method that the C–C bond in, what is now known to be, the ethylidyne CH_3–C chemisorbed species was perpendicular to the metal surface (12). The diffraction and vibrational spectroscopic methods are complementary. The diffraction methods selectively give information about regular arrays of adsorbed species, and the diffraction patterns reflect the average situation over considerable areas of adsorbed species. On the other hand, the spectroscopic method provides information about local structure and is less sensitive to, and even does not require, the presence of regular adsorption arrays.

The NEXAFS technique, based on the carbon x-ray absorption-edge, provides information about the orientation of CC orbitals with respect of the metal surface and also about the degree of filling of π^* antibonding orbitals through back-donation from the metal (13). UPS also provides information about the filled orbitals of the hydrocarbon adsorbate and, by incorporating angle-resolved methods, about the symmetries of adsorbed species (14). For chemisorbed hydrocarbons the NMR technique to date has principally provided adsorbate C–C distances from sophisticated ^{13}C studies, which can be correlated with the degree of hybridization at the carbon atoms (15).

Temperature-programmed desorption provides information about hydrocarbon or hydrogen desorption as a function of temperature (16). Where, as is not infrequently the case, hydrocarbon adsorption occurs only at low temperatures from multilayer adsorption so that carbon from monolayer species is not desorbed, the amount of total hydrogen desorption at a particular temperature can be a very useful measure of the overall CH_x composition of the remaining adsorbed species. For example, that only 25% of hydrogen had been desorbed by room temperature from ethylene on Pt[111] implied an overall structure of $CH_{1.5}$ or C_2H_3 for the adsorbed species (17). This species was shown by LEED (12), EELS (18), and by comparison with the vibrational spectrum of a model cluster-compound (8), to correspond to the ethylidyne structure, CH_3–C.

This review deals in turn with the adsorbed species from chemisorbed alkenes, alkynes, aromatic hydrocarbons, and alkanes, including cyclic alkanes, usually studied on a wide variety of metals and, for a given metal, on several crystal faces. It is found, not unexpectedly, that the majority of studies have been concerned with the type-molecules ethylene (ethene), acetylene (ethyne), and benzene.

INTENSITIES AND SELECTION RULES IN THE VARIOUS SURFACE VIBRATIONAL SPECTROSCOPIES

These topics have been the subject of recent review articles (19, 20) and only the salient points are repeated here.

Infrared Spectroscopy

Metals are excellent electrical conductors in the infrared region. Under those conditions the generation of an oscillating vibrational dipole moment by absorption of an infrared photon leads to virtual images of the opposite sign in a flat metal surface. The practical result of this is that vibrational dipoles parallel to the surface are in effect cancelled out by corresponding motions of the image, and those perpendicular to the surface are enhanced by a factor of two (21). Hence only vibrations that give rise to oscillating dipoles or dipole-components perpendicular to the surface (these are the completely symmetrical modes of vibration of the complex) can be observed by absorption from an infrared beam reflected from the metal surface. This effect is termed the *metal surface selection rule*. Maximum absorption from an infrared beam polarized perpendicular to the surface can occur when the beam is reflected at near-grazing incidence (22).

Although the above consideration applies to a flat metal surface, in fact it can be shown to apply usefully to metal particles (21) down to about 2 nm diameter (23), which is still much larger than the adsorbate monolayer thickness of about 0.1 nm. The successful understanding of the spectra of hydrocarbon species on finely divided metal catalysts, which rarely have particles less than 2 nm in diameter, has depended a great deal on the realization of the effectiveness of this selection rule for such samples.

Electron Energy Loss Spectroscopy

This method utilizes a beam of monoenergetic electrons that is reflected from the metal surface. In the process, individual electrons lose discrete amounts of energy resulting from absorption of a vibrational quantum by the adsorbed monolayer. The reflected electrons are energy-analyzed so as to determine the numbers of electrons that have lost quanta of different magnitudes and can therefore be associated with the excitation of individual modes of molecular vibration (2).

Two mechanisms have been established for the absorption of energy from a monoenergetic electron beam reflected off a metal surface. The first one, termed *the dipolar mechanism* (24), depends on the absorption of energy by an interaction of the electric fields associated with the approaching or receding electron with a vibrational dipole of the adsorbed molecule. This is the same mechanism as applies to infrared absorption, and so the same metal surface selection rule applies once again, i.e. only vibrations with dipoles perpendicular to the surface will be detected by this means. Electrons interacting with the adsorbed layer by this mechanism are reflected close to the specular direction, i.e. where the angle of incidence equals the angle of reflection.

The fact that the dipolar mechanism in electron energy loss spectroscopy is the same as in infrared spectroscopy means that there ought to be a good correlation between the pattern of relative intensities of losses observed in EEL spectra in the specular direction and those observed by infrared absorption. As we shall see below, this has provided a most valuable method for correlating the spectra of the adsorbed species with the infrared bands from completely symmetrical modes of hydrocarbon ligands on metal clusters. Actually a slight deviation of the dipolar-excited electron-beam from the specular direction increases with the magnitude of the energy loss. Hence EEL spectrometers operating at the specular angle with narrow collecting angles for the electrons do systematically attenuate the intensities of particularly the higher energy electron losses (higher vibration frequencies).

A second electron-loss mechanism can give rise to electron reflections in many directions and is termed *the impact mechanism* (25). As its name implies, it arises from an intimate interaction of the electron with the adsorbed molecule. Because such electrons can be reflected over a whole hemisphere, only a small proportion will be detected close to the specular direction. The latter is therefore dominated by the electrons that interact by the dipolar mechanism.

Impact-excited vibrations are best evaluated by off-specular measurements, and it can be shown that these can also include vibrations associated with vibrational atomic motions parallel to the metal surface (2). Hence the off-specular measurements lead to additional vibrational information. It can also be shown that there are additional selection rules associated with electron losses caused by impact-excited modes. No such electrons should be observable in the specular direction from a vibration that is antisymmetric with respect to a two-fold axis perpendicular to the surface, or with respect to reflection planes perpendicular to the metal surface and either perpendicular or parallel to the plane of incidence (26). In the latter case, impact-scattered electrons can only be observed away from the plane of incidence, i.e. no off-specular electrons at all should be observed within the plane of incidence. These selection rules can also be used to determine the symmetry properties of the relevant vibrational modes, and can therefore provide indirect information about the overall symmetry of the adsorbed species. Because large energy losses excited by the dipolar mechanism are attenuated, it is usually found that the losses in the high frequency vCH bond-stretching region are mainly impact in character.

Inelastic Neutron Scattering

The intensities of the vibrational features in these spectra depend on the neutron scattering cross-sections of the nuclei and on the atomic amplitudes during the vibration mode in question. As the former is known and the latter can be approximately calculated, intensity considerations are useful in the interpretation of INS spectra (27). Because hydrogen ^{1}H nuclei have a much higher neutron scattering cross-section than other nuclei, vibrations involving hydrogenic motions caused either by CH bond-stretching or angle-bending within a group, or by the hydrogen atoms being "carried" during motions of carbon atoms, give strong features in the spectra of hydrocarbon groups. Further information is provided by the replacement of ^{1}H by ^{2}D, as the latter has a much-reduced scattering cross-section.

Raman Spectroscopy

Intensities within Raman spectra depend on induced dipole moments brought about the electric field associated with the monochromatic incident beam from a laser. The induced dipole moments in turn depend upon the electrical *polarizability* changes associated with the vibrations. Hence relative patterns of vibrational intensities are different between Raman spectra and infrared spectra. However, considerations similar to those described above in relation to infrared spectra again lead to the conclusion that the completely symmetrical modes will give the strongest features in Raman spectra from molecules adsorbed on a metal surface. However, because metals are not as good conductors at high visible frequencies as in the infrared, the above "selection rule" is less rigorous than in the infrared region (28).

SOME GENERAL CONSIDERATIONS

The literature in this field is now very voluminous and so this review has necessarily to be rather selective. For a given adsorbate, in addition to the obvious experimental variables of the metal and the crystal face, others are (*a*) the temperature of adsorption, (*b*) the temperature range, if any, covered with the adsorbed species before the spectral measurement, (*c*) the degree of exposure to the gas (to which is related the fractional coverage of the metal surface), (*d*) whether the spectral measurements are made on-specular or off-specular, and (*e*) if the latter, the angle of deviation from the specular direction.

With regard to (*a*) and (*b*) above it can be expected that, above the very low temperature at which physical adsorption is eliminated, the lowest-temperature form of chemisorption is most likely to be the one to occur without the breaking of C–C or CH bonds of the original adsorbate. In other words, the molecular formula for the adsorption complex is likely to be the same as that of the adsorbate itself, giving rise to what is termed nondissociative adsorption. Even so, we shall see that there is often more than one possibility for the mode of bonding of the carbon atoms to the metal surface.

It is found that with metal/hydrocarbon systems more and more hydrogen is driven off as the temperature is raised, but that the carbon from the original monolayer often remains on the surface, finally in the form of a graphitic layer. Immediately prior to this final stage, many hydrocarbons appear to give closely similar spectra, consisting principally of absorptions at between 750 and 840 and near 3030 cm^{-1}. These frequencies led to the original suggestions that this residual surface species is CH with the two absorptions representing the νCH and δCH modes, respectively. However, if, as seems likely for example on a [111] fcc face, the CH bond is perpendicular to the metal surface, the δCH mode should be forbidden by the metal surface selection rule. More recent analytical measurements have suggested that the spectrum described above occurs at an overall composition of C_2H (29, 30), although the surface complex may possibly be polymerized on the way to forming continuous patches of carbon overlayers at higher temperatures.

In between the low and high temperature extremes are to be found a variety of intermediate decomposition products, of which the best known is the ethylidyne CH_3C species derived from ethylene. Some of these have been identified and some not, the latter partly because of complexities from mixtures of species.

With regard to (c) above, the degree of exposure to the gas, it is sometimes found that a spectrum obtained at the lowest exposure is different from, or contains extra bands in comparison with, ones taken at higher coverage. Reasons for these differences may be that the very first dose of adsorption occurs at surface defects or steps and may be stronger and more dissociative in form than for adsorption on flatter terraces (31). In such a case the higher-coverage spectra will better reflect the species present on the principal crystal plane. Another possibility is that, even on areas of a well-formed crystal plane, one type of species may change into another with increasing coverage. In clear-cut cases of these types I discuss the structural implications of both the low-coverage and high-coverage spectra.

Under (d) above, because of the very useful correspondence, in intensities as well as in frequencies, between the infrared bands from completely symmetrical vibrational modes from hydrocarbon ligands on metal clusters, and the related features in EEL spectra from the same species on metal surfaces excited by the dipolar mechanism, I pay more attention in this review to on-specular than to off-specular EEL spectra. However, dipolar losses are rapidly reduced in intensity on going "off-specular" whereas those excited by the impact mechanism are not, and therefore a comparison of on-specular and off-specular spectra can be of great use in identifying the dipolar-active, completely symmetrical, vibrational modes in the on-specular spectrum. As expected, this method confirms that the majority of strong features in on-specular spectra are from dipolar-active modes. Observation of an absorption in a reflection/absorption infrared spectrum is itself evidence for the completely symmetrical nature of the vibration, because of the strict operation of the metal surface selection rule in the infrared region.

The resolution of infrared spectra is high (~ 1 cm^{-1}) so that normally all spectral features that are strong enough to be observed can be separately resolved, one from another. However, in EELS the achievable resolution is much lower (20 to 50 cm^{-1}) so that band overlap can occur and weak losses may not be separately distinguishable from neighboring strong ones.

Whereas it is not a prime aim of this article to give detailed vibrational assignments to the observed spectra, nevertheless some of these are mentioned when the features in question are well established and the assignment is convincing. The off-specular spectra frequently show additional frequencies, and such data are very valuable when a more complete assignment of vibrational modes is to be attempted.

In assembling spectral diagrams the metals in Groups VIII and IB that have been investigated are listed in their order in the Periodic Table. It should be noted that Fe is atypical in occurring in the body-centered cubic (bcc) rather than in the more normal face-centered cubic (fcc) form. Ruthenium also occurs in the hexagonal close-packed form (hcp), but the [001] face that has so far invariably been chosen is a close-packed one and very similar to [111] faces for fcc metals. For a given metal, spectra from different crystal faces will be given in the order [111], [100], and [110]. The arrangements of atoms on faces are illustrated in Figure 1 for fcc metals.

As the main emphasis in this article is on the identification of the chemical nature of an adsorbed species from the vibrational spectra, the most convenient way of summarizing and comparing the extensive experimental data seems to be by visual means. In the spectral diagrams that follow, on-specular EEL spectral data are represented schematically by "stick" diagrams, are labeled in terms of the metal surface on which adsorption occurs, and the temperatures where relevant and the appropriate numerical literature references are given. Where similar spectra have

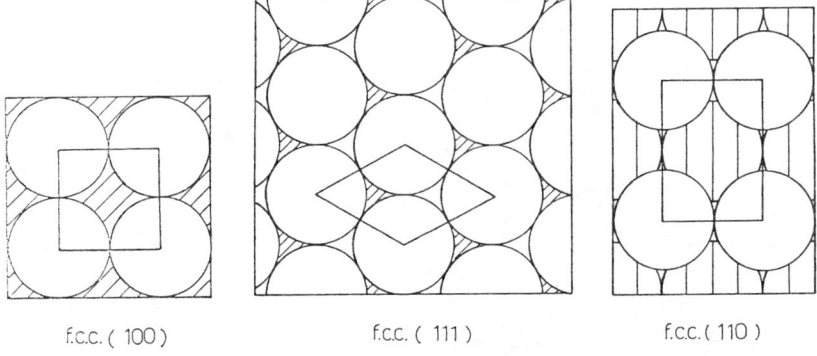

Figure 1 The arrangement of metal atoms in the [111], [100], and [110] faces of a face-centered-cubic metal.

been reported by more than one set of authors, the diagram represents the average data. Significantly dissimilar spectra are illustrated separately.

A band represented in outline only is considered to arise from the presence of a second minority species, the nature of which is indicated by a symbol. Spectral bands that have been shown to be substantially dipolar in origin by comparison with off-specular data are indicated by the symbol ●; these are "completely symmetrical" modes of vibration, i.e. those which are symmetrical with respect to all the symmetry elements associated with the surface complex. Some corresponding features in different spectra from a set, mostly completely symmetrical modes discussed in the text, are indicated by *dashed connecting lines*. Where spectra have been obtained at different coverages, those illustrated here are judged to be best related to monolayer coverage.

Relatively few RAIRS spectra are yet available. These are discussed individually in the text.

THE ALKENES

Ethylene (Ethene)

The adsorption of this hydrocarbon has now been studied on many different metal surfaces. The pioneering EELS study of ethylene on Pt[111] was carried out by Ibach & Lehwald in 1978 (32). The first RAIRS studies, also on Pt[111] were made very recently by Chesters & McCash (33) and by Trenary and colleagues (34), both in 1987.

THE LOW TEMPERATURE SPECTRA—GENERAL COMMENTS For most metals, two general types of low-temperature spectra have been reliably identified. These have been assigned to a di-σ bonded species (I) (32) and a π-bonded species (II) (35), respectively, both of a nondissociative nature, retaining

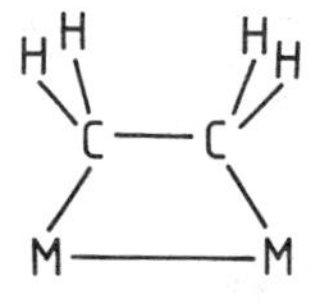

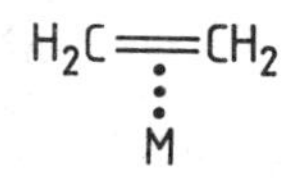

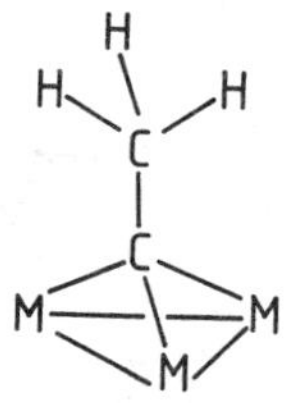

the hydrocarbon formula C_2H_4. The same numbering has previously been given to these two types of spectra (36–38) and is used once again in this article.

In a few cases the low temperature adsorption of hydrocarbons has been studied on metals such as tungsten at the other end of the transition metal period to the more widely studied Group VIII metals. Relatively poorly defined hydrocarbon spectra were obtained on W[100] and W[110] (39–41). It is clear that on this clean metal surface there is initial decomposition to dissociated species, and even to hydrogen and carbon, at very low temperatures.

TYPE I SPECTRA—THE DI-σ ADSORBED SPECIES   For clean metal surfaces the up-dated collection of Type I spectra taken from the literature until mid-1987 includes those on the crystal faces Ni[111] (42, 43); Ni[100] (44, 45a,b; H. Ibach, personal communication); Ni[110] (46a,b, 47); Ru[001] (48); Pt[111] (18, 32); Pt[100] (49); and Fe[110] (50). The spectra are summarized graphically in Figure 2A for adsorbed C_2H_4, and in Figure 2B for adsorbed C_2D_4. In general there is a striking family resemblance within sets of spectra, particularly down to 1000 cm^{-1}. The main exceptions are the spectrum on Fe[110], where additional absorptions of some prominence occur at 1250 cm^{-1} with C_2H_4 and at 700 cm^{-1} with C_2D_4 (50). However, unlike most of the other metals, iron has a body-centered cubic lattice, with the [110] face nearest to being close-packed. Also, as is described below, for a few spectra previously described as of Type I', there are some additional bands near 900 cm^{-1} for C_2H_4 and 680 cm^{-1} for C_2D_4.

At the structural level the assignment of a di-σ structure (I) to these species, with close to carbon sp^3 hybridization, is consistent with the views of most of the original authors. In individual cases, however, more detailed considerations have been discussed in the original papers concerning the precise locations or shape of this general type of adsorbed species, e.g. in the context of whether the C_2M_2 skeleton is coplanar. Nonplanarity is, for example, strongly indicated for Ni[111] (42, 43) and Ni[100] (44, 45; H. Ibach, personal communication), where "soft" νCH modes absorb between 2700 and 2850 cm^{-1} and indicate that some of the CH bonds are interacting more strongly with the metal surface than others.

Some general support for the di-σ formulation has also come from comparison with the infrared spectrum of such a ligand in the metal cluster compound $(C_2H_4)Os_2(CO)_8$, which is known to have this type of structure. Detailed considerations of the assignments of the spectra of the C_2H_4 and C_2D_4 versions of this complex are in hand (37, 51).

The principal exception to a di-σ compatible assignment from the original authors is the study on Ni[100] by Zaera & Hall (45a,b) where it was concluded that the spectrum is that of a π-complex. One reason for this different assignment was the observation of weak bands near 1575 cm^{-1} for both C_2H_4 and C_2D_4 adsorption, which were assigned the νC=C bond-stretching mode. However, in gas-phase C_2H_4 and C_2D_4 the νC=C fre-

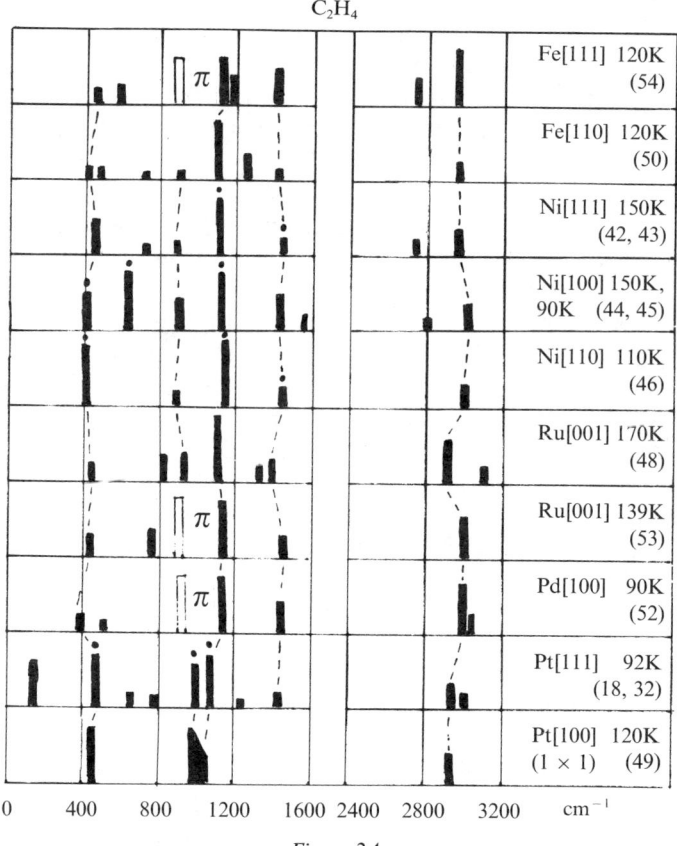

Figure 2A.

Figure 2 A schematic representation of on-specular EEL spectra of Types I and I' that have been assigned to di-σ adsorbed species and obtained by low-temperature nondissociative adsorption of (A) C_2H_4 and (B) C_2D_4. The Type I' spectra have additional bands, drawn in outline, which are considered to represent the presence of a proportion of π-bonded species (see Figure 3). For information on the symbols and conventions used, see the end of the section, Some General Considerations.

quencies have the very different values of 1623 and 1515 cm^{-1}. We conclude that the weak *ca.* 1575 cm^{-1} bands are experimental artifacts (possibly, see below, arising from a small fraction of vinylidene species) and that ethylene also has the di-σ structure on Ni[100]. The spectra on Ni[100] in both Figures 2A and 2B (44, 45a,b; H. Ibach, personal communication) also

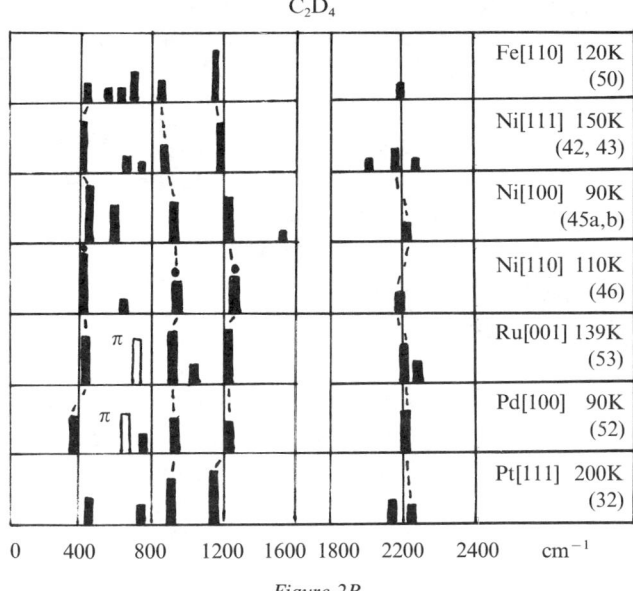

Figure 2B.

differ from the others in the set because of the presence of strong bands near 600 cm^{-1}. The other spectra in Figures 2A and 2B do show losses in this region but with much lower intensities. The small frequency shift between C_2H_4 and C_2D_4 shows that these bands arise from a nonhydrogenic mode, most probably νCM from another, possibly dissociated, species.

If an isolated di-σ complex had a planar C_2M_2 skeleton the overall symmetry would be C_{2v}. In the spectral diagrams the completely symmetrical vibrations identified as dipolar modes by off-specular data (with the exception of ν_1, νCH, which is mainly impact-excited for reasons given above), have been labeled ν_1 to ν_5. The frequency ranges derived from the collected data, together with an approximate description of these modes in group vibration terms, are as follows. Where the description of a normal mode involves contributions from more than one group vibration, the one that is considered to be more important is given first. Data or designations for C_2D_4 complexes are given in parentheses after those for C_2H_4 complexes.

ν_1, $\nu CH_2 s(\nu CD_2 s)$ 3000–2910 (2200–2150) cm^{-1}; ν_2, $\delta CH_2(\delta CD_2/\nu CC)$ 1470–1400 (1200–1100); ν_3, CH_2wag/νCC ($\nu CC/CD_2$ wag) 1170–1060 (950–850); ν_4, $\nu CC/CH_2$ wag (CD_2 wag) 920–830 (740–640); ν_5, νCM (νCM) 480–400 (450–400) cm^{-1}.

For nonplanar C_2M_2 skeletons, e.g. of symmetry C_s, C_2, or C_1, additional fully symmetrical modes would be dipolar allowed in the specular direction. However, those already allowed under C_{2v} symmetry are likely to remain the strongest.

TYPE I' SPECTRA I have included in Figures 2A and 2B a few spectra on the crystal faces Pd[100] (52); Ru[001] (53); Fe[111] (54) that have frequencies similar to those of Type I, but a notably different distribution of relative intensities of the bands. In an earlier survey they were designated Type I' spectra (36), although in this article I have already classified spectra on Ni[100] (44, 45a,b; H. Ibach, personal communication) as of Type I. It now seems very probable that the strong extra bands, near 900 cm^{-1} for the C_2H_4 spectra and near 680 cm^{-1} for the single C_2D_4 case, indicated in Figures 2A and 2B in unfilled outline only, are due to the strongest bands from the presence of a proportion of π-complexes (see below) coexistent with the di-σ species. On Pd[100] the bands in question fall at 920 and 660 cm^{-1}, respectively (52), while the π-complex alone (see below) has absorptions at 895 and 680 cm^{-1} (55). Also in a separate study on Ru[001] by Menzel et al cited above (48), the "extra" bands are much weaker so as to give a rather well-defined Type I spectrum (Figure 2A). Both spectra are illustrated in this case. We therefore now consider Type I' spectra to indicate the coexistence of di-σ and π species. In connection with acetylene spectra, to be discussed below, we shall find other examples of the presence of more than one type of adsorbed hydrocarbon species on a given crystal face. Mixtures of di-σ and π species from C_2H_4 probably also occur at low temperature on the stepped surface Ni 5[111] × [1$\bar{1}$0] where the spectra vary as a function of coverage (56).

TYPE II SPECTRA—THE π-ADSORBED SPECIES Figure 3 shows the Type II C_2H_4 and C_2D_4 spectra, respectively, on the clean metal surfaces Cu[100] (35); Pd[111] (57, 58); Pd[100] (55); Pd[110] (59). These can be assigned with confidence to the presence of π-adsorbed species as shown by the pattern of analogous, fully symmetrical, infrared absorptions for the authentic π-complexed ethylene ligand in Zeise's salt, $K^+[(C_2H_4)PtCl_3]^-$, H_2O (37, 60). The frequency regions characteristic of the completely symmetrical modes v_1 to v_4 for the π-complex are as follows:

v_1, $v CH_2 s(v CD_2 s)$ 3075–2990 (2250–2230); v_2, $vCC/\delta CH_2(vCC)$ 1560–1500 (1420–1350); v_3, $\delta CH_2/vCC(\delta CD_2)$ 1290–1225 (960–930); v_4, $\gamma CH_2 s/(\gamma CD_2 s)$ out-of-plane wag 915–900 (680–660) cm^{-1}.

For the π-complexes the frequencies of vCMs appear to be a sensitive function of the strength of bonding to the surface, but for the Pd complexes (55, 57–59) the values fall consistently close to 265 cm^{-1}.

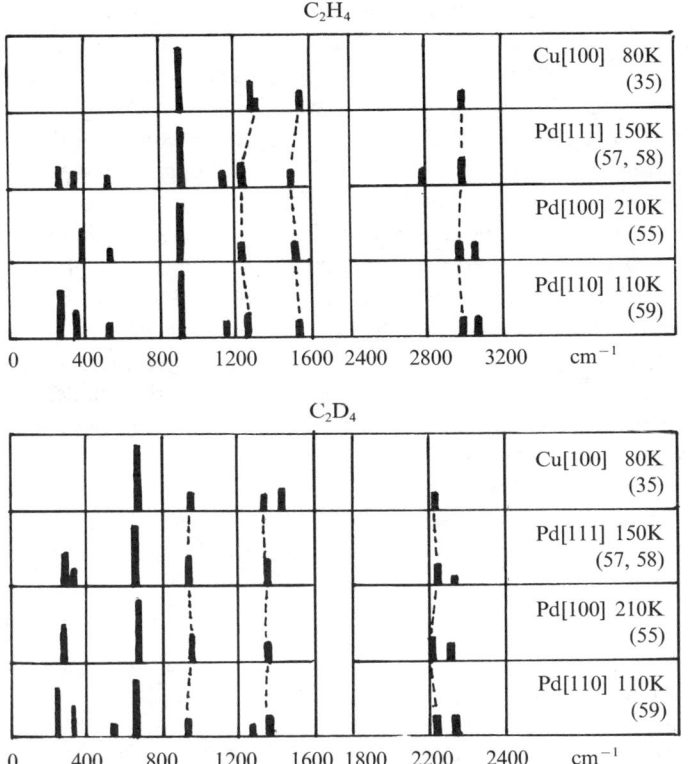

Figure 3 A schematic representation of on-specular EEL spectra of Type II that have been assigned to π-complex formation and obtained by low-temperature nondissociative adsorption of C_2H_4 and C_2D_4.

Perhaps the best measure of the vibration frequency of the CC bond in these π-complexes is the mean value of *ca.* 1385 cm^{-1} for the C_2D_4 complexes. In the C_2H_4 case there is well known to be a close coupling between νCC and δCH$_2$ modes, but in the C_2D_4 case the δCD$_2$ frequency is well removed so that νCC is a good group frequency (61). The corresponding pure νC=C double bond value for C_2D_4 gas is 1515 cm^{-1} (62) and, in conjunction with the νC–C single-bond value of 993 cm^{-1} for ethane (62), we can conclude that the bond order in the π complexes is about 1.7. Stuve & Madix (52) suggest an alternative way of deriving what they term a πσ parameter and reach a similar conclusion. We suspect that, because of coupling between CH$_2$ wagging and νCC modes in the spectra of the di-σ species, the πσ parameter is likely to be less reliable for assessing the CC bond order for the latter species than for the π-complexes.

The most remarkable finding among the collected data is that, unlike Ni and Pt, which readily form di-σ surface species, Pd has a strong preference for π-complex formation. That Cu does so (35) is less surprising as it is expected to adsorb ethylene more weakly.

Analogous to the case with adsorbed CO complexes (63), the di-σ and π complexes may be considered to differ in the degree of back-bonding from filled d-orbitals in the metal into the vacant π^* antibonding orbital of the parent ethylene molecule. For the di-σ species the back-donation must be substantial, but for the π-complex it is considerably weaker as reflected in both the νCC and νCH vibration frequencies. However, as shown by the Type I' spectrum for Pd[100] (52), it is at least possible for palladium to form a proportion of di-σ bonded surface species. Palladium atoms, unlike those of Ni and Pt, have filled d^{10} outer-shells. Although there is hence clearly no lack of d electrons, possibly the high stability of the completed d shell prevents back-donation.

Whereas relatively few clean metals seem to prefer π- to di-σ complexes, it seems to be true that all metals so far investigated revert to π-bonding to ethylene when appreciable amounts of either carbon or oxygen are preadsorbed on the surface. The numerous examples in the literature are listed below, the appropriate references being given in parantheses.

Ni[110]C (37); hcp Ru[001]O (49); Pd[100]O (52, 64); Ag[110]O (65, 66); Ir[111]O (67); Pt[111]O (18) and bcc Fe[111]O (54).

It is interesting to speculate whether in real catalytic systems, where it is difficult to exclude impurities, it might not be the π-bonded species that is the more active one of the nondissociatively adsorbed species.

A reason for the preference for π-bonding to metal surfaces with coadsorbed O or C might be that the latter withdraw electrons from the metal and hence reduce d-electron availability for back-donation to the adsorbed ethylene. It is alternatively possible that the coadsorbed atoms act in a steric manner by reducing the number of adjacent 2-atom metal sites that are needed for di-σ adsorption (69). The strongest bands from the ν_4, γCH$_2$ out-of-plane wagging modes of the π-complexes on surfaces with coadsorbed oxygen occur near 1000 (C$_2$H$_4$) and 720 cm^{-1} (C$_2$D$_4$) compared with the corresponding values of ca. 910 and 670 cm^{-1} for the clean metals.

ROOM TEMPERATURE SPECTRA: THE ETHYLIDYNE SPECIES (III) The general evidence from the spectra of adsorbed ethylene is that the low-temperature nondissociatively adsorbed species, discussed above, have been replaced at room temperature by one or another type of dissociatively adsorbed form.

One of the early successes of vibrational spectroscopy in this field, in conjunction with LEED, was to show than on Pt[111] at room temperature an ethylidyne, CH_3C species, is adsorbed in the center of a triangle of Pt atoms. The principal steps in the discovery were that Ibach & Lehwald originally suggested the presence of a methyl-containing species, probably CH_3CH, at room temperature from their EELS spectra (32). Somorjai and colleagues showed by LEED that the CC bond of this species was perpendicular to the surface (12); Demuth showed by TPD that the surface composition was C_2H_3 (17); and finally Skinner et al, by comparison of the EEL spectra with the infrared spectrum of the model compound $(CH_3C)Co_3(CO)_9$ (taking into account the simplifying effect of the metal surface selection rule), reinterpreted very successfully the spectra and off-specular features in the original EEL spectra (8). To my knowledge the presence of such a species had not even been suggested in the earlier literature prior to its discovery by the above means. This is probably because it was considered unlikely that on dissociation a hydrogen atom would migrate from one end to the other of the CC bond. Ethylidyne has, however, since proved to be a very stable species which has so far been found to occur over an appropriate temperature range on all the close-packed fcc [111] planes studied, with the exception of Ni[111], and on the similar close-packed hcp [001] plane of Ru. Spectra of this type occur on the following faces; Rh[111] (71, 72); Ru[001] (49, 53); Pd[111] (57, 58); Ir[111] (31); Pt[111] (18, 70); and Pt[100] 5 × 20 {reconstructed to be [111]-like} (48, 73). The spectra themselves are summarized in Figure 4, from which the characteristic frequency ranges of the completely symmetrical modes v_1 to v_4, based on C_{3v} symmetry for the complex, can be derived as follows:

v_1, $vCH_3s(vCD_3s)$ 2950–2880 (ca 2080); v_2, $\delta CH_3s(vC-C/\delta CD_3s)$ 1400–1330 (1160–1145); v_3, $vC-C(vC-C/\delta CD_3s)$ 1165–1080 (1050–950); v_4, $vCM_3s(vCM_3s)$ 460–410 (430–400) cm^{-1}.

It seems that the generation of the ethylidyne species from di-σ or π-bonded ethylene is likely to occur by the following reaction sequence suggested by Gates & Kesmodel (58):

surface $\underset{(\text{di-}\sigma \text{ or } \pi)}{CH_2=CH_2} \rightarrow \underset{(\text{di-}\sigma/\pi)}{C=CH_2} + H(ads) \rightarrow CCH_3$.

Although, as discussed in the next section, the $C=CH_2$ species has been identified in some circumstances, this is not yet so as an intermediate between adsorbed ethylene and ethylidyne. It may well be that the cogenerated surface H atoms lead to the immediate conversion of ethylidyne to CCH_3.

It is worth noting that, although ethylidyne species have also been found

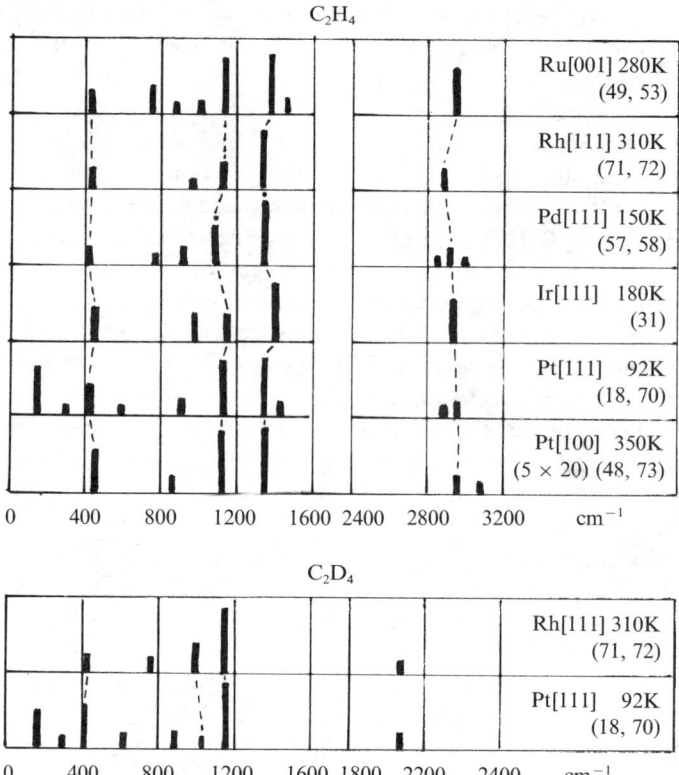

Figure 4 A schematic representation of on-specular EEL spectra that have been assigned on ethylidyne surface species and obtained by dissociative adsorption of C_2D_4 or C_2D_4 at intermediate temperatures.

bonded to a square of metal atoms in a metal cluster compound (74), as yet no such species has been observed on fcc [100] faces.

Finally, recent excellent reflection/absorption infrared spectra (RAIRS) between 4000 and 800 cm^{-1} on Pt[111] by Chesters & McCash (33) and by Trenary and his colleagues (34) have provided full confirmation for the assignment of the fully symmetrical modes of vibration of ethylidyne (the metal surface selection rule is very strict for the dipolar mechanism) and have provided extremely precise values for their frequencies, namely, $\nu CH_3 s$, 2884; $2 \times \delta CH_3 as$, 2795; $\delta CH_3 s$, 1339; νCC, 1118 cm^{-1}.

These pioneering RAIRS studies hold out much general hope of improved experimental data for hydrocarbons adsorbed on metal surfaces.

OTHER SPECIES DERIVED FROM ETHYLENE AT ROOM TEMPERATURE OR HIGHER: SURFACE ACETYLENE, VINYLIDENE, AND POSSIBLY VINYL It is well-established that by room temperature ethylene on Ni[111] has been converted to a Type A spectrum of adsorbed acetylene (42, 75) (see the discussion below of acetylene spectra). Also it seems most likely that on Pt[100] (1 × 1), i.e. unreconstructed Pt[100] (49), another acetylenic species with strong νCC absorptions close to 1130 cm^{-1} from both C_2H_4 and C_2D_4 is formed near 300K and persists even up to 650K. Additional medium to strong absorptions at 1589 (1573) cm^{-1} and 1008 (709) cm^{-1} near 300K, derived from adsorbed C_2H_4 (C_2D_4) on Pt[100] (1 × 1) suggest the additional presence of a vinylidene group, C=CH$_2$, di-σ bonded perpendicular to the surface, without π bonding, as discussed below.

Vinylidene species seem to be likely intermediates between adsorbed nondissociatively adsorbed ethylene at low temperatures and ethylidyne at room temperature or as dehydrogenation products of ethylidyne. Two very similar spectra from acetylene adsorbed on Pd[111] at 250K (58); and from ethylene adsorbed on Ru[001] with preadsorbed 0 at 350K (76), have been convincingly assigned to a di-σ bonded vinylidene group which is also π-bonded to a third metal atom. Figure 5 supports these structural assignments by comparison with the fully symmetrical modes in the infra-

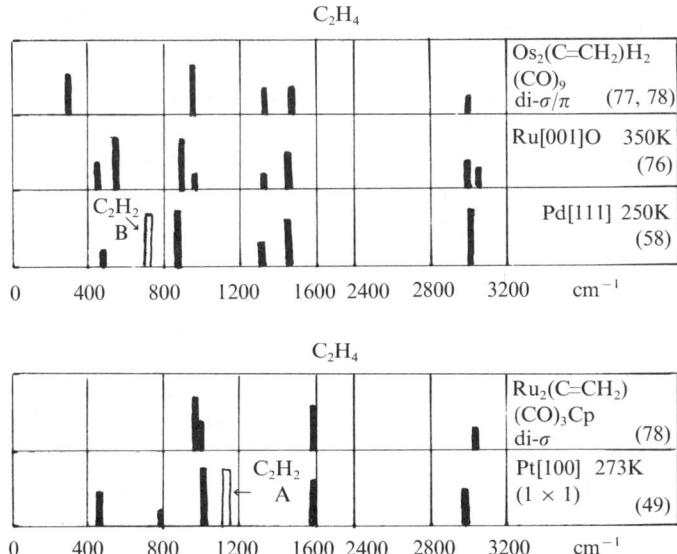

Figure 5 A comparison of infrared absorptions from fully symmetrical modes of vibrations of vinylidene ligands on metal clusters with analogous on-specular EEL spectra on metal surfaces. The *upper spectra* are from di-σ/π vinylidene species; the *lower* one from vinylidene species that are only di-σ bonded to metal atoms.

red spectra of the model compound $H_2Os_2(C=CH_2)(CO)_9$, which has the above structural features (77, 78). Figure 5 shows an equally convincing comparison between the infrared spectrum of $Ru_2(C_5H_5)_2(C=CH_2)(CO)_3$ (79), where the vinylidene group is only σ-bonded to the two Ru atoms, and the EEL spectrum from ethylene on unreconstructed Pt[100] (1 × 1) discussed above (49).

Finally with the help of spectra from deuterium-substituted isomers, and taking into account kinetic isotope effects, Zaera & Hall (45a,b) have suggested that a vinyl spectrum results from ethylene on Ni[100] heated to 175K. The spectra involved are coverage-dependent and the higher coverage version whose frequencies are not listed, rather than a lower coverage one, would appear to give better intensity/frequency agreement with the published spectrum of $\pi\sigma$ bonded vinyl in an Os_3 cluster compound (77, 78).

FURTHER THERMAL EVOLUTION OF THE SPECTRA   In several cases, heating ethylidyne surface species leads to a gradual conversion to C_nH or C_2H, which gives a simple spectrum consisting of a strong low frequency band between about 750 and 840 cm^{-1} (out-of-plane γCH deformation), often a weak broad band in the 1200–1300 cm^{-1} region (νCC), and a weak to medium νCH absorption between about 3050 and 2950 cm^{-1}. In the examples given below I list the individual νCH and γCH frequencies of the two strong bands.

On Ru[001] an ethylidyne species at 280K is gradually converted to C_nH at 360K (2960 and 750 cm^{-1}), and C_2H 500K (3010 and 810) (53). On Ir[111] (31) ethylidyne is formed at the much lower temperature of 180K and is gradually and finally completely converted to C_2H by 500K (3018 and 838 cm^{-1}). On Ni[110] (45a,b) there is a gradual conversion from the di-σ species at 80K to C_2H at 300K (2990 and 890 cm^{-1}). It is possible that the variable frequencies of these bands depend not only on the metal substrate but also on different degrees of polymerization and dehydrogenation of surface C_nH species as a final graphitic carbon layer is approached (29, 80).

Ethylene decomposition on Pt[100] (1 × 1) (49) is exceptional in that what may be a Type A acetylene spectrum is retained up to 650K. Also the thermal evolution of spectra on Ni[100] (45a,b) and Pd[100] (52) clearly take forms different from those originating from ethylidyne species on [111] faces.

STRUCTURAL EVIDENCE ON HYDROCARBON SURFACE COMPLEXES FROM OTHER SINGLE-CRYSTAL TECHNIQUES   Hammer et al (43) have carried out careful LEED studies on the di-σ species on Ni[111] at low temperatures, revealing several different diffraction patterns at different coverages but very similar

EEL spectra. They conclude that both ethylene and acetylene adsorb on rather similar sites and express a preference from the LEED data for bonding to adjacent two-fold bridged sites. This seems more reasonable for acetylene than for ethylene. On valency grounds two "on-top" sites would seem to be more likely for ethylene.

Koestner, Van Hove & Somorjai have summarised LEED work on the ethylidyne species on Rh[111] and Pt[111] (81); the CC bond lengths are given as 0.145 and 0.150 nm, respectively, values in line with the CC bond orders deduced approximately from vibrational data. They have also compared the reactivities of these two surfaces with respect to ethylene. Salmeron & Somorjai (16) have given a comprehensive discussion of TPD results for a range of unsaturated hydrocarbons on Pt[111]. Albert et al (82) and Lloyd & Netzer (83) have used angle-resolved UPS to confirm that in ethylidyne the C–C bond is perpendicular to the metal surface. Kruger & Benndorf (84) have used LEED, UPS, and TPD to confirm that ethylene is π-bonded on Ag[100]. Stöhr and colleagues (13, 85) have carried out NEXAFS studies that show that on Cu[100] at 60K the CC bond is parallel to the surface and only slightly elongated relative to the gas phase, as expected for the π-bonded species deduced from EELS. On Pt[111] the NEXAFS technique (86, 87) has shown that at low temperature the C–C bond of chemisorbed ethylene is parallel to the surface with a bond order of about 1.2 but that it is upright at 300K with a somewhat shorter bond, again in agreement with EELS/RAIRS results.

Although there are discrepancies, it can be concluded that in general the other physical methods provide rather good support to the pattern of conclusions derived from the more comprehensive series of EELS/RAIRS studies of ethylene chemisorption.

Higher Alkenes

Relatively little work has yet been done on the vibrational spectroscopy of alkyl-substituted ethylenes on metal surfaces. For noncyclic alkenes Avery & Sheppard (88–90) have studied propene, the four isomeric butenes, pent-1-ene, and buta-1:4-diene on Pt[111] by EELS and TPD. Somorjai et al (81, 91) have studied propene on Rh[111] and Pt[111] by LEED, EELS, and TPD.

On Pt[111] Avery & Sheppard (89) conclude that at low temperature (170K) propene adsorbs as the di-σ species (IV) and is converted to the propylidyne species (V) by 300K. Both products, and the reaction temperature, are analogous to those resulting from ethylene chemisorption. Somorjai et al (91) report a different type of low temperature spectrum from propene on Rh[111], involving less lowering of the C–C bond order compared with Pt[111], but have not specified its nature. On Rh also

(IV) (V) (VI)

conversion to propylidyne occurs on warming, but at 200K compared to 270K for Pt. Propylidyne on Rh[111] is in turn clearly shown by EELS to decompose to ethylidyne and probably methylidyne (CH) or C_2H at 270K (91), by breaking the C–C bond attached to the metal surface; the ethylidyne is further decomposed to C_2H (3025, 825 cm^{-1}) by 500K. On Pt[111] propylidyne appears to decompose directly, i.e. without intermediate conversion to ethylidyne, to give the same final products above 400K (88). However, there are substantial temperature gaps between the reported experiments. Very recently, E. M. McCash, P. Gardner, and M. A. Chesters (unpublished) have obtained a very good high resolution RAIRS spectrum of propylidyne on Pt[111] at room temperature that strongly confirms the structure of the adsorbed species.

In agreement with predictions by Salmeron & Somorjai (16) based on TPD(TDS) measurements, Avery & Sheppard also conclude that but-1-ene and pent-1-ene form the appropriate alkylidyne on warming from low-temperature di-σ adsorbed species on Pt[111]. They showed that *cis*- and *trans*-but-2-ene gave distinguishable di-σ spectra at low temperatures (88, 89). However, at room temperature they gave rise to the same species. On TPD evidence this is shown to be of molecular formula C_4H_6, which contrasts with the formula C_4H_7 for butylidyne. Analysis of the spectra, and comparison with the infrared spectrum of the model compound $(CH_3CCCH_3)Os_3(CO)_{10}$ (T. W. Matheson, personal communication) showed that the structure is of type (VI). This can be considered to be derived from coordinated dimethyl acetylene (but-2-yne) by the formation of σ-bonds with two Pt atoms and a π-bond with a third. Above 400K all the linear butenes gave similar spectra in two stages on Pt[111]. The spectrum at the first stage (>370K) was attributed tentatively to a C–CH=CH–C surface species (strong CH band at 720 cm^{-1}) and the second at >500K to the usual C_2H species (3070, 830).

Isobutene (2-methyl propene) on Pt[111] was also shown to give a di-σ low-temperature species and an isopropylidyne species at room temperature (89). Another rather simple spectrum was observed at 420K, prior to decomposition to smaller hydrocarbon fragments at higher temperatures. This was suggested to be a symmetrical $CH(CHM_2)_3$ species where the six surface metal atoms, M, are in the form of a hexagonal ring on the [111] Pt surface.

The proposed decomposition pathways for the linear butenes are shown schematically in Figure 6 (88, 89). Avery (94) has also made an interesting study of cyclopentadiene on Pt[111] and showed that at 300K it is converted to a symmetrical C_5H_5 cyclopentadienyl species.

THE ALKYNES

Acetylene (Ethyne)

THE LOW TEMPERATURE SPECTRA Once again two principal types of spectra have been identified. We shall adopt a previous designation of Type A and Type B (95, 36). By comparison with the infrared spectrum of an acetylene (CHCH) ligand in the cluster compound $(CHCH)Os_3(CO)_{10}$, it is con-

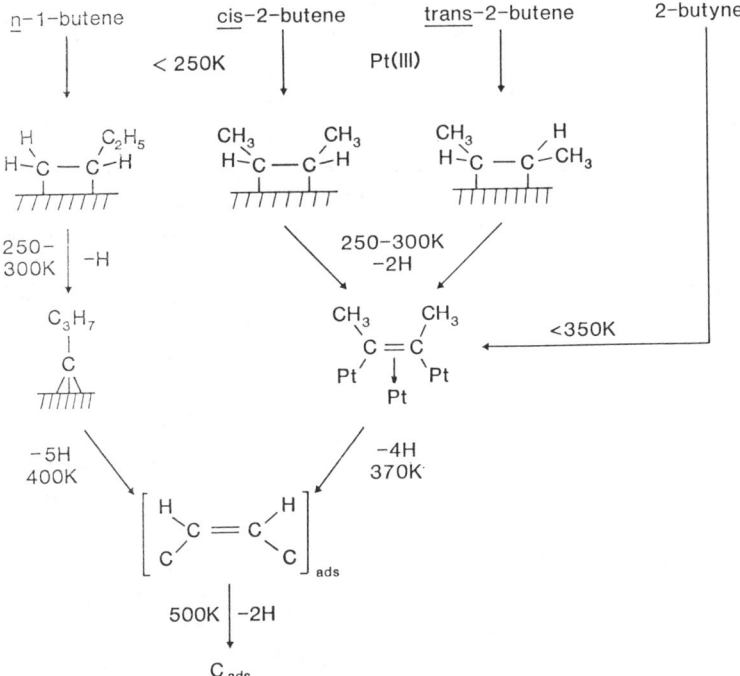

Figure 6 A schematic representation of the different species adsorbed at various temperatures on Pt[111] from but-1-ene and *cis* and *trans* but-2-enes as deduced from EEL spectra (89).

sidered that the Type B spectrum probably originates from a structure in which the acetylene ligand is σ-bonded to two metal atoms of a triangle

(VII)   (VIII)

and π-bonded to the third, as in (VII) above (96). It has been tentatively suggested that the species responsible for Type A spectra may involve bonding to four metal atoms, perhaps as in (VIII) for close-packed faces (36, 95, 106).

Until recently either a Type A or a Type B spectrum was found to occur on any particular crystal plane. However, Jakob, Cassuto & Menzel (97) have now observed a Type A spectrum (their designation was Type I) at low coverage on Ru[001] changing over to a Type B (Type II) spectrum at higher coverage. This type of situation had not previously been clearly found for hydrocarbon ligands, but is not uncommon among the many studies that have been made of CO adsorption. An example is provided by CO adsorption on Pd[111], which starts at low coverage with three-fold bridged sites being occupied when then convert to two-fold bridged sites at high coverage (98). A similar situation could be occurring with acetylene on Ru[001], with the first species occupying a four-atom metal site and subsequent ones three-atom sites. The conversion from one species to another merely by increase of coverage suggests, as in the CO cases, that there is only a relatively small stability difference between the two types of surface complex. It is probable that the increased bonding to the metal surface is accompanied by compensatingly weaker CC bonding within the hydrocarbon group, as has been suggested for ethylene (37).

In a study of acetylene adsorption on a stepped Ni[$5 \times (111)/(1\bar{1}0)$] surface, dissociative adsorption occurred even at the low temperature of 150K to give a strong band at 2220 cm^{-1} which seems best attributed to a hydrogen-free C$\equiv$C species.

THE TYPE A AND TYPE A′ SPECTRA   In a previous compilation (36) a distinction was made between Type A spectra, where the νCC stretching mode dominated the spectra, and Type A′ spectra, where the νCC mode still gave a strong band, but in addition lower frequency bands between 1000 and 900 cm^{-1} and 800 and 700 cm^{-1} for C$_2$H$_2$ and C$_2$D$_2$, respectively, had acquired comparable intensities. The up-dated sets of Type A spectrum on crystal faces Ni[111] (56, 99); Cu[111] (95); Ni[100] (101, 102); Ru[001] (97);

Fe[110] (50) and Fe[111] (54); and of Type A′ spectra on Ni[111] (43, 100); Cu[100] (103); Cu[110] (104); Pd[100] (55, 105), and Pt[100] (5 × 20) (49) are illustrated diagramatically in Figures 7 and 8, respectively.

There is now, however, an interesting cross-connection between the A and A′ spectra. A spectrum from adsorption on Ni[111] is present in each set, the one measured originally by Ibach et al (99) and the other more

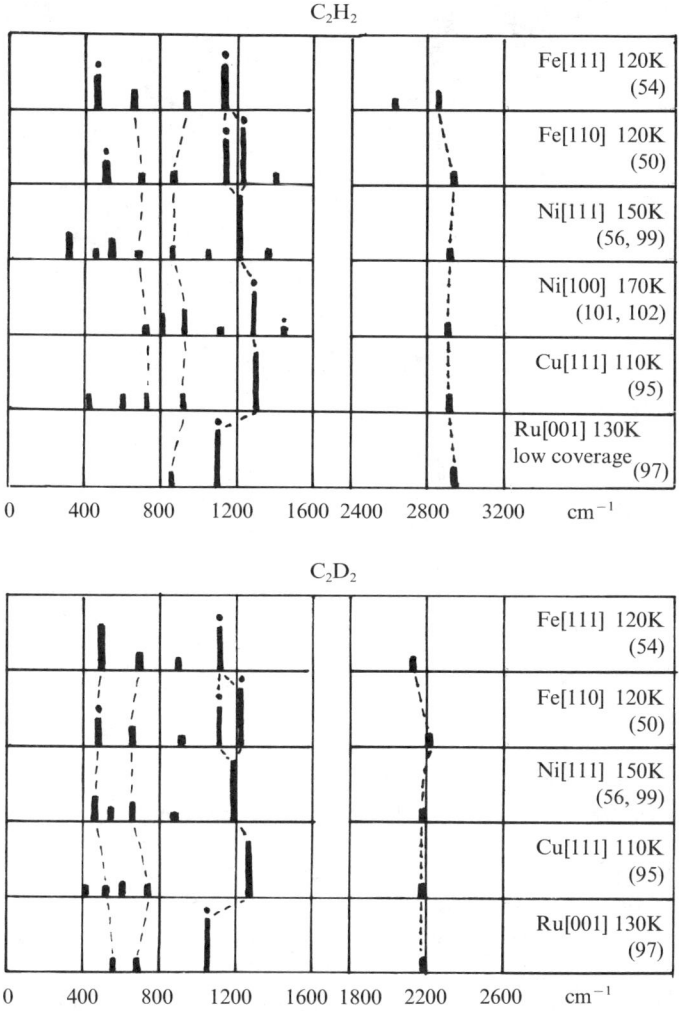

Figure 7 Type A on-specular EEL spectra from nondissociative adsorption of C_2H_2 and C_2D_2 on metal surfaces at low temperatures. For information on the symbols and conventions used see the end of the section, Some General Considerations.

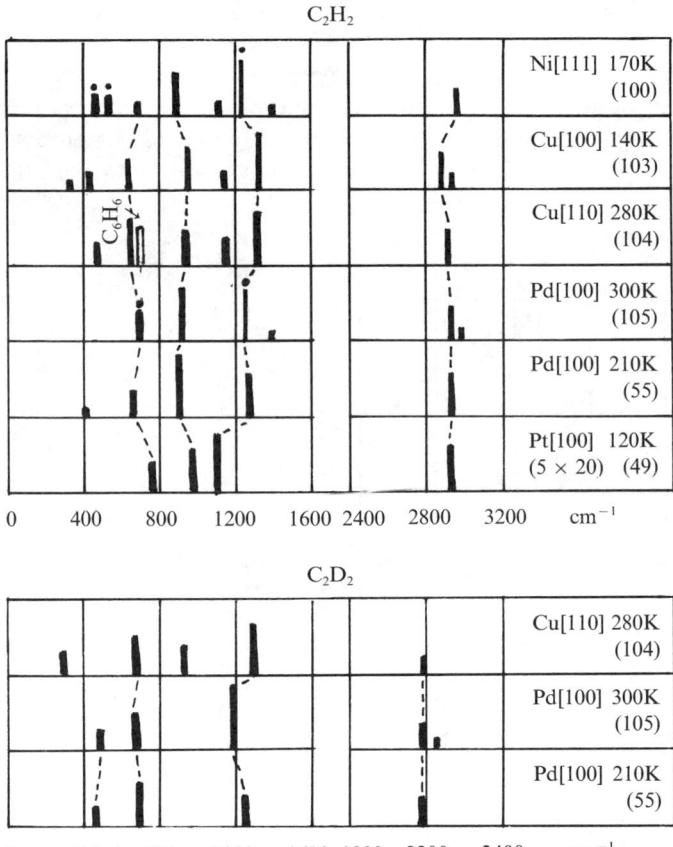

Figure 8 As for Figure 6, but Type A′ spectra from adsorbed acetylene.

recently by Hammer and colleagues (43, 100). A third spectrum from the same Ni[111] system by Bertolini et al (75) falls between the other two extremes. It is usually the case that Type A spectra have weak νCH losses near 3000 cm^{-1}, but Type A′ spectra have stronger ones. For reasons described above in the section on Electron Energy Loss Spectroscopy, the weak νCH spectra are obtained from EEL spectrometers with narrow-angle input, whereas the stronger ones involve a wider-angle input. The latter collect off-specular, impact-excited, bands more efficiently when the spectrometer is set at the mean specular angle. It seems that there are a number of such bands that are weak in the specular direction but much stronger off-specular. With this experimental consideration in mind, we shall here treat Type A and A' spectra as from the same general family of chemisorbed species.

In these spectra the νCC mode is identified by strong bands that shift little on deuterium substitution. However, on different surfaces they vary substantially in frequency between the extremes of 1320 cm^{-1} on Cu[100] and 1089 cm^{-1} on Pt[100] (5 × 20) (49). The latter is the reconstructed Pt[100] surface with a close-packed [111]-like arrangement of surface metal atoms. As has been noted above, these CC frequencies are very low in comparison with the value of 1974 cm^{-1} for gas-phase acetylene, and show that multiple bonding to the surface has caused a reduction of bond order from 3 to between 1.7 and 1.3.

This type of spectrum and the species associated with it seem not to be very sensitive to the arrangement of atoms in the surface. For example, all three [111] (95), [110] (104), and [100] (103) Cu planes give rise to the same general type of spectrum. It has been noted before in two cases, Cu[110] (104) and Pd[100] (105), that additional bands occur in the spectra; this could imply the presence of a proportion of coadsorbed benzene derived from the trimerization of acetylene. The bands in question are of moderate strength between 710 and 630 cm^{-1} from C_2H_2 adsorption and between 500 and 400 cm^{-1} from C_2D_2. It is worth noting, however, that Type B acetylene spectra (see below) also have analogous strong bands in these regions, although usually with slightly higher frequencies.

A particularly interesting spectrum was obtained from C_2H_2 on Ni[100] by Dinardo, Demuth & Avouris (102), in that a comparison was made between spectra obtained with the plane of incidence and collection of the electrons parallel with the sides of the square unit cell or parallel to its diagonal. With specular scattering the band at 944 cm^{-1} was present in the first case, but missing in the second. From the selection rules for impact scattering this shows that there is a plane of symmetry parallel to the diagonal to which the vibration in question is antisymmetrical. This led the authors to propose a structure with a CC bond along the diagonal that is approximately parallel, but tilted, with respect to the metal surface.

It is to be hoped that more experiments will be carried out in future in which changes in alignment of the incident plane are made with respect to a known reference direction in the metal surface. This would be particularly valuable for studies of [110] planes where the two principal directions at right angles are clearly distinguishable, see Figure 1. However, further experiments with [100] planes would also be valuable, although, as pointed out by Dinardo et al (102), the interpretation is then complicated by the likelihood of two mutually perpendicular domains of equivalent adsorbed species.

The completely symmetrical modes from Types A and A' spectra have the frequency range as follows:

νCH(νCD) 2970–2880 (2200–2140); νCC(νCC) 1320–1100 (1270–1060); δCH(δCD) 990–810 (700–630); γCH(γCD) 780–630 (550–450) cm^{-1}.

More recently, Chesters & McCash (106, 107) have obtained an excellent RAIR spectrum down to 800 cm^{-1} from acetylene on Cu[111] at 250K. This shows the (necessarily) completely symmetrical modes to have frequencies as follows:

νCH(νCD) 2913(2189); νCC(νCC) 1294(1276); δCH, 920 cm^{-1}.

These results are in very good general agreement with the earlier EELS results, but give much more accurate frequencies. The simplicity of the spectrum is consistent with a C_{2v} di-σ/di-π structure as in (VIII) for the adsorbed species.

THE TYPE B SPECTRA—PROBABLY A DI-σ/π ADSORBED SPECIES The updated set of these spectra on crystal faces Ni[110] (45, 95); Ru[001] (97, 108); Rh[111] (71); Pd[111] (58, 109, 110); Pd[110] (95); and Pt[111] (32) are summarized in Figure 9, together with the reference spectrum showing the completely symmetrical modes from the (CHCH) ligand in (C$_2$H$_2$)Os$_3$(CO)$_{10}$ that led to the original assignment of the di-σ/π structure (IV) to the corresponding adsorbed species (96). It is seen that such spectra occur on fcc [111] and [110] and hcp [001] faces. The di-σ/π structure seems natural for the close-packed fcc [111] or hcp [001] cases; on a [110] face it would seem to require interaction with metal atoms in both the first and second layers.

The completely symmetrical modes have frequency ranges as follows:

νCH(νCD) 3020–2930 (2240–2190); νCC(νCC) 1410–1260 (1360–1210); δCH 990–870 (730–620); γCH(γCD) 770–670 (570–500) cm^{-1}.

DISCUSSION OF TYPES A, A', AND B SPECTRA; ACETYLENE ADSORPTION ON Ag[110] Although the δCH and γCH modes do not differ much in frequency between Type A and Type B spectra, both the νCH and νCC modes, with mean values of 2925/1210 for Type A and 2975/1335 cm^{-1} for Type B, indicate that there is a lower C–C bond-order (presumably from more $d \to \pi^*$ back donation) for Type A adsorbates than for Type B. This would be consistent with the former (CHCH) group interacting with four metal atoms, and the latter with three, as has been concluded above in individual cases.

Three model cluster compounds with (CHCH) ligands have been cited in the search for analogies with surface-adsorbed acetylene. These are (CHCH)Co$_2$(CO)$_6$ (111), (CHCH)Os$_3$(CO)$_{10}$ (96) and (CHCH)Co$_4$(CO)$_{10}$

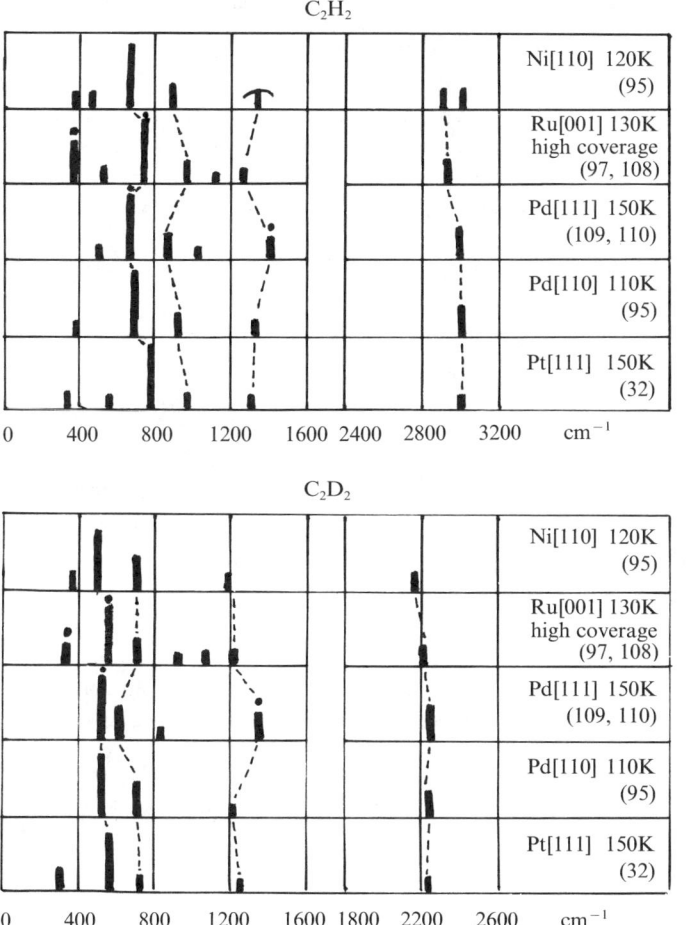

Figure 9 As for Figure 7, but Type B spectra from adsorbed acetylene. These have been assigned to a di-σ/π bonded C_2H_2/C_2D_2 surface species involving bonding to three metal atoms.

(112). The νCH and νCC modes for these compounds occur at *ca.* 3100/1403, 2996/1301, and 3010/1199 cm^{-1}, respectively. On the νCC criterion the model compounds in which (CHCH) interacts with three or four metal atoms do seem to provide better analogues to the adsorbed species. The Co_4 compound has the C–C group bonded diagonally across a nonplanar quadrilateral of metal atoms.

Acetylene adsorbed on Ag[110] at 100K shows an EEL spectrum indicating only slight perturbation from the gas-phase linear structure (113).

On Ag[110]O an AgC≡CH structure was proposed (113). In each case close to *sp* carbon hybridization is retained.

THE THERMAL EVOLUTION OF ACETYLENE SPECTRA   It has been shown in a number of cases that ethylidyne species can be readily detected on close-packed surfaces as the original acetylenic surface species is warmed up in the presence of small amounts of hydrogen. It can probably also be produced by disproportionation when a pair of C_2H_2 surface complexes reacts to give $C_2H + C_2H_3$.

Further thermal evolution has been studied in a considerable number of cases, usually toward CCH-type spectra at high temperatures. However, as we shall see, there is a tendency for the strongest absorption band from the out-of-plane CH deformation mode of this type of species to occur at lower frequencies than when it is derived from ethylene adsorption on the same surface.

A particularly instructive study by Gates & Kesmodel (58) showed that on Pd[111] a type-B acetylene spectrum obtained at 150K evolved into a clear-cut di-σ/π bonded vinylidene spectrum at 250K, and then into ethylidyne at 300K and into C_nH (3002, 762) at 500K. It is probable that the same sequence is followed on Pd[100] (105), with a Type A species up to 300K leading to C_nH at 450K (3000, 750). Tardy & Bertolini (55) show a somewhat different spectrum from a more unsaturated species at 300K but also a C_nH spectrum (3025, 725) at 500K. Piecing-together evidence from several cases the overall sequence of surface species seems to be

$$\underset{\text{Type A or B}}{(\text{CHCH})} \rightarrow CCH_3 \rightarrow \underset{\text{di-}\sigma/\pi}{C=CH_2} \rightarrow CCH \rightarrow C_nH(n>2) \rightarrow C(\text{graphitic})$$

with different metals giving different transition-temperatures from one species to the next.

Trisubstituted ethylenes have their out-of-plane γC=CH deformation vibrations with frequencies near 830 cm^{-1}, but this value will certainly be sensitive to carbon-to-metal substitutions. Isolated aromatic γCH modes have similar frequencies (9).

STRUCTURAL EVIDENCE FROM OTHER SINGLE-CRYSTAL TECHNIQUES Reference has already been made to a careful LEED and EELS study of acetylene on Ni[111] by Hammer et al (43, 100) in which adsorption of acetylene to adjacent bridged sites is proposed. NEXAFS studies have also been made of acetylene on Pt[111] (86) and Cu[100] (13). In agreement with the EELS results, the CC bonds are found to be parallel to the metal surface with bond lengths corresponding to a bond order of *ca.* 1.5.

Higher Alkynes

PROPYNE (METHYL ACETYLENE)   The EEL spectrum of propyne on Rh[111] has been studied by Bent et al (91) in conjunction with measurements by

LEED and TPD. Although the low temperature EEL spectrum at 240K was not analyzed in detail it was consistent with the retention of the methyl group and indicates that the molecule is bonded to the metal surface through its original C≡C bond. Furthermore, the absence of a vCH mode above 3000 cm^{-1} shows that, as with acetylene on most metals (see above), the carbon atoms have been substantially rehybridized to give a CC bond order of not greater than 1.6. Above 270K the original propyne molecule dissociates and the EEL spectrum at 310K shows clearly that a CC bond has broken to give ethylidyne together with probably a CH species. Isotopic studies with deuterium showed that it is the original C≡C bond that has been broken on the surface. Above 430K the ethylidyne further fragments in the usual manner (see sections on Further Thermal Evolution of the Spectra and the Thermal Evolution of Acetylene Spectra, above) to give C_nH.

Bent et al (91) have also shown that propadiene, isomeric with propyne, gives a somewhat similar low temperature spectra at 77 and 240K; it also decomposes to ethylidyne and CH by 310K.

Chesters & McCash (106, 107) have carried out a RAIRS study of the chemisorption of propyne on Cu[111] above 800 cm^{-1}. They conclude that at low temperature (150K) propyne is bonded to the surface intact via the original C≡C group, which lies parallel to the surface, to give a reduced bond-order of *ca.* 1.7. The frequencies and their assignments are as follows:

vCH$_3$as, 2923 (w); vCH$_3$s, 2883 (s); vCH, 2855 (ms); [vCD from CH$_3$CCD, 2157 (w)]; 2 × δCH$_3$as, 2828 (m); vCC, 1361 (ms).

No bands were observed for δCH$_3$as, δCH$_3$s, or CH$_3$ rock. The spectrum differs substantially from that obtained on Rh[111] by EELS at 240K (91).

BUT-2-YNE (DIMETHYL ACETYLENE) But-2-yne has also been studied on Cu[111] by Chesters & McCash with RAIRS (106, 107), and on Pt[111] by Avery & Sheppard with EELS (90). In both cases it is considered that at low temperature the molecule is chemisorbed intact via its C≡C group to once again give a much reduced bond order. On Pt[111] the spectrum is identical to those from the but-2-enes at room temperature and therefore the structure of the chemisorbed species is considered to be as in Formula VI given above in the section on the Higher Alkenes. This is a substituted form of the general type (VII) assigned to Type B spectra for unsubstituted acetylenes. A similar (CH$_3$CCCH$_3$) structure is deducted for but-2-yne on Cu[111] but with C_{2v} symmetry with the C_4 plane perpendicular to the metal surface, as for Type A spectra for unsubstituted acetylenes. The observed frequencies are as follows:

Pt[111], EELS (88, 90): vCH$_3$as, 2960 (m); δCH$_3$as 1435 (s); vCC ∼ 1390 (obscured); δCH_3, 1365 (s); CH_3 rock, 1045 (s); νCM, 410 (m). Cu[111], RAIRS down to 800 cm^{-1} (106, 107): νCH_3as, 2930 (w); νCH_3s, 2880 (s); $2 \times \delta CH_3$as, 2828 (w); νCC, 1392 (s); CH_3 rock, 1018 (w).

AROMATIC HYDROCARBONS

Benzene

The EEL spectra of benzene, C_6H_6 and of C_6D_6 on metal single-crystal faces have been the subject of intense study. Results to date are available on Ni[111] (114–117), Ni[100] (115), Ni[110] (118); Rh[111] (119, 120); Pd[111] (55, 121); Pd[100] (55, 122); Ag[111] (123); Pt[111] (116, 124); Pt[110] (125) and Re[001] (126). These spectra are summarized in Figures 10A and 10B.

Whereas with ethylene and acetylene the principal aim was to deduce the structures of the chemisorbed complexes from the observed EEL or RAIR spectra, with benzene there is general agreement that at low or normal temperatures benzene lies flat on the metal surfaces in an undissociated form. Sometimes, as with Ag[111] as studied by Avouris & Demuth (123) and with several crystal faces of Cu recently studied by N. R. Avery (personal communication), there are relatively slight changes in the frequencies of the adsorbed benzene relative to the gas-phase. However, in the majority of cases there are substantial frequency differences that imply notable CC bond-length changes. We shall discuss the latter spectra first.

If the benzene molecule were π-bonded to a smooth metal surface, its gas-phase symmetry of D_{6h} would be reduced to C_{6v}. The only completely symmetrical, and therefore dipole-active, modes would be of form $\nu CH(A_{1g} \to A_1)$, $\nu CC(A_{1g} \to A_1)$ and $\gamma CH(A_{2u} \to A_1)$ (128). Within the brackets the left-hand symmetry symbol relates to the gas-phase molecule, and the right-hand one to a C_{6v} adsorbed species. We shall use the Herzberg numbering of the modes of benzene (129), which has been favored by most EELS workers rather than that of Wilson (128). Under Herzberg's system, these modes are ν_1, ν_2, and ν_4, respectively.

It is clear that ν_1 and ν_4 can be assigned to the band-sequences shown in Figures 10A and 10B in the frequency ranges $\nu CH(\nu CD)$ 3040–2990 (2300–2240) and $\gamma CH(\gamma CD)$ 910–700 (660–510) cm^{-1}. The νCH modes are of moderate intensity, and the broad EELS bands (which have substantial impact character) may overlap with contributions from other less symmetrical νCH modes. As in the gas-phase spectra, the out-of-plane $\gamma CH(\gamma CD)$ modes are very intense. They also vary substantially in frequency, but give essentially the same frequency sequence for the different crystal planes as follows:

Ag[111] 675(−); Ni[110] 700(510); Pd[100] 710(520); Pd[111] 720(525); Re[001] 740(535); Ni[111] 745(545); Ni[100] 750(540); Rh[111] 805(565); Pt[111] 835(605); and Pt[110] 910(655) cm^{-1}.

There is some agreement that, except for the case of Ag[111] and the several planes of Cu, the νCC breathing mode ν_2 very probably falls in the frequency range 920–800 cm^{-1}. This is considerably lower than the gas-phase value of 992 cm^{-1}, and implies weakened CC force-constants and lengthened C–C bonds. Possible sequences of the ν_2 modes are shown in Figures 10A and 10B. However, it should be borne in mind that in the C_6H_6 case there is overlap with ν_{11} and in the C_6D_6 case there is overlap with ν_{10}, as discussed below. A ^{13}C isotopic-substitution experiment claimed to show that a band at 875 cm^{-1} for C_6H_6 on Ni[111] is of skeletal origin and therefore ν_2 (117). Also, analogous bands in the spectra on Pd[111] and Pd[100] at 810 and 870 cm^{-1}, respectively, show the expected dipole activity (121, 122). However, Waddill & Kesmodel (121, 122) have paired these two bands with others at 640 and 675 cm^{-1} in the corresponding C_6D_6 spectra, implying that they represent second modes of the γCH/γCD type. As they suggest, the best frequency-fit would then be to the ν_{11} mode (E_{1g}) of benzene.

It is probable, as Bertolini et al have suggested (117), that both ν_2 and ν_{11} overlap in the 900–800 cm^{-1} region of the C_6H_6 spectra. This ambiguity would best be resolved with the help of the higher resolution and stricter selection rules applicable to RAIRS. To date, one such measurement seems to have been made on Pt[111], when only the strongest γCH mode at 826 cm^{-1} was detected (130). In the case of the EEL spectrum on Pt[111] it has been suggested, because of coverage-variable intensities, that the pair of bands at 920 and 830 cm^{-1} may represent γCH modes of two different C_6H_6 surface species (116); others have not confirmed this variability (124). However, variable intensities have also been observed for bands at 819 and 776 cm^{-1} on Rh[111] (119, 120), and at 830 and 730 cm^{-1} on Ni[111] (116).

As the A_1 modes under C_{6v} symmetry would remain completely symmetrical under any possible symmetry for a benzene molecule parallel to the surface, it is only the presence (or absence) of other features in the spectra that can provide more detailed information about the probable site and symmetry of the adsorption complex in relation to the array of metal atoms on the surface.

It is at first sight a surprising feature of the C_6H_6/C_6D_6 spectra that a number of additional weaker features occur in all cases, regardless of the availability of different possible higher-symmetry sites provided by different crystal planes, such as on Ni[111], Ni[100], or Ni[110] faces. The

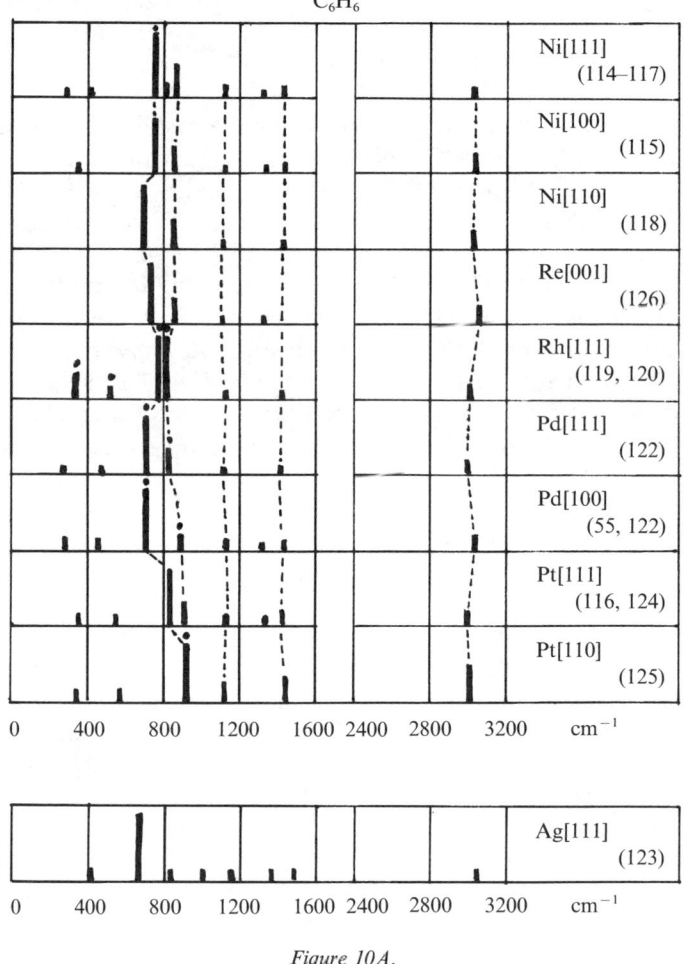

Figure 10A.

Figure 10 On-specular EEL spectra from benzene adsorbed on different metal surfaces that have been assigned to intact C_6H_6 molecules adsorbed parallel to the surfaces (A) for C_6H_6 and (B) for C_6D_6.

frequencies of these common additional bands are C_6H_6 1435–1405 and 1140–1100 cm^{-1}; C_6D_6 1390–1350 cm^{-1}. Some species from C_6H_6 (five out of ten) show an additional weak band at *ca.* 1325 cm^{-1}, and a corresponding number from C_6D_6 (five out of nine) show a weak band at *ca.* 1220 cm^{-1}.

Consideration of likely related gas-phase frequencies of C_6H_6 and C_6D_6

C_6D_6

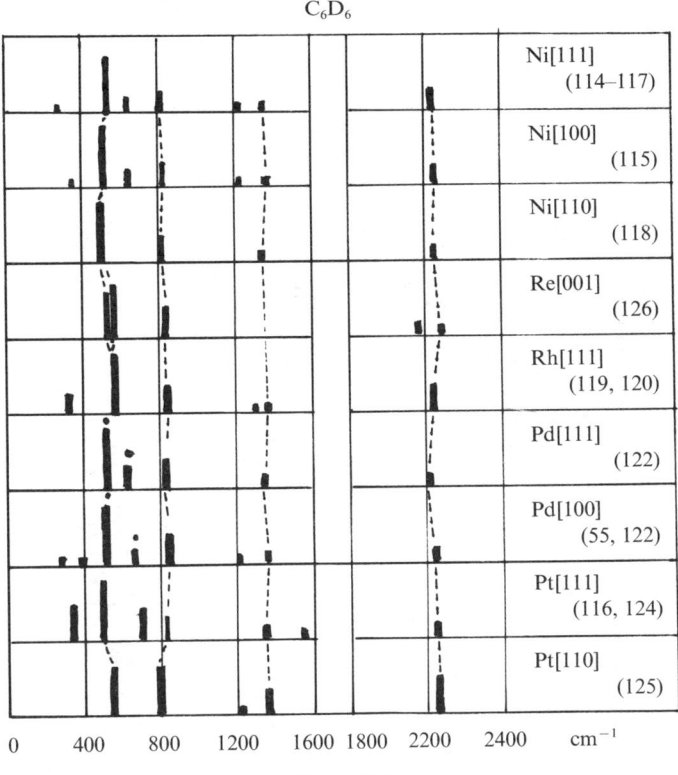

Figure 10B.

and expected isotopic ratios (131a,b) has led to fairly general agreement to the following mode assignments for these bands, using the mean values of the range cited above:

1420/1370, v_{13} (E_{1u}); 1325/1220, v_9 (B_{2u}); 1120/ca. 830, v_{10} (B_{2u})

with $v_{17}(E_{2g})$ as an alternative assignment for the latter bands. The symmetry symbols in brackets relate to the gas-phase fundamentals. In order to attempt to deduce site-symmetries from these data, it is necessary to consider the spectral activities of these various modes on sites of different symmetries. This is explored in Table 1 in relation to a range of site-symmetry groups considered as subgroups of the gas-phase D_{6h} symmetry group (128).

Completely symmetrical modes in each case are dipole or impact allowed on-specular, although usually the former mechanism contributes most

Table 1 The site-symmetry dependence of the symmetries of the fundamental modes of vibration of benzene C_6H_6

| | | C_{3v} | | | C_{2v} | | | C_s | | Modes | |
| | | | | | | | | | | Herzberg numbering (129) | Wilson numbering (128) |
D_{6h}	C_{6v}	σ_v	σ_d	C_3	σ_v	σ_d	C_2	σ_v	σ_d		
A_{1g}	A_1	A_1	A_1	A	A_1	A_1	A	A'	A'	ν_1, ν_2	ν_2, ν_1
A_{2g}	A_2	A_2	A_2	A	A_2	A_2	A	A''	A''	ν_3	ν_3
A_{2u}	A_1	A_1	A_1	A	A_1	A_1	A	A'	A'	ν_4	ν_{11}
B_{1u}	B_1^*	A_1	A_2	A	B_2^*	B_1^*	B^*	A'	A''	ν_5, ν_6	ν_{13}, ν_{12}
B_{2g}	B_1^*	A_1	A_2	A	B_2^*	B_1^*	B^*	A'	A''	ν_7, ν_8	ν_5, ν_4
B_{2u}	B_2^*	A_2	A_1	A	B_1^*	B_2^*	B^*	A''	A'	ν_9, ν_{10}	ν_{14}, ν_{15}
E_{1g}	E_1^*	E	E	E	$B_1^* + B_2^*$	$B_1^* + B_2^*$	$2B^*$	$A' + A''$	$A' + A''$	ν_{11}	ν_{10}
E_{1u}	E_1^*	E	E	E	$B_1^* + B_2^*$	$B_1^* + B_2^*$	$2B^*$	$A' + A''$	$A' + A''$	$\nu_{12}, \nu_{13}, \nu_{14}$	$\nu_{20}, \nu_{19}, \nu_{18}$
E_{2g}	E_2	E	E	E	$A_1 + A_2$	$A_1 + A_2$	$2A$	$A' + A''$	$A' + A''$	$\nu_{15}, \nu_{16}, \nu_{17}, \nu_{18}$	$\nu_7, \nu_8, \nu_9, \nu_6$
E_{2u}	E_2	E	E	E	$A_1 + A_2$	$A_1 + A_2$	$2A$	$A' + A''$	$A' + A''$	ν_{19}, ν_{20}	ν_{17}, ν_{16}

Notes: 1. The A_1, A, or A' modes are completely symmetrical and dipolar-active on-specular.
2. The *starred symmetry symbols* are antisymmetric with respect to the two-fold axis and therefore impact forbidden on-specular. They are also dipolar-forbidden. Other noncompletely symmetric modes may be impact-forbidden on-specular for certain relative orientations of symmetry planes and the plane of incidence of electrons (see text).
3. σ_v denotes symmetry planes perpendicular to the benzene ring that also include CH bonds; σ_d denotes such planes which bisect CC bonds.

to the intensity. Modes are impact forbidden on-specular, if they are antisymmetric with respect to a two-fold axis, or to a plane of symmetry either perpendicular or parallel to the plane of incidence of the electron beam (2, 26). The plane-of-symmetry criteria depend on particular orientations of the plane of incidence with respect to specific directions in the crystal plane and will often be more or less ineffective unless the experiment is carefully designed for this purpose. This will be even more so where the regular array of adsorbed molecules is of lower symmetry than that of the base crystal plane, so that different domains are present with different orientations. Hence the most effective symmetry element in this respect is likely to be the two-fold axis.

Considering the "extra" bands from v_9, v_{10}, or v_{17}, and v_{13} discussed above, as the C_{3v} and C_3 symmetry groups do not contain a two-fold axis, it is seen that all these modes are impact-allowed on-specular by this two-fold axis criterion, although the E modes might be weak. All but one, v_{13}, constitute the set of A_1 modes under $C_{3v}(\sigma_d)$ symmetry. Without reconstruction, however, such site symmetries are only possible on [111] planes. For [100] or [110] planes (and also for [111]) C_{3v} sites are possible but it is seen that under this site-symmetry only v_{17} is allowed in addition to v_1, v_2, and v_4. This is because B_1 and B_2 modes are antisymmetric with respect to the two-fold axis (62, 128). All the modes become allowed again under $C_s(\sigma_d)$ site-symmetries. We conclude therefore that, given the existence of regular arrays of adsorbed molecules, the observed spectra seem only to be consistent with substantially lowered site-symmetries for the adsorption complexes relative to C_{6v}. Of course incomplete arrays lead to reduced local asymmetries in the vicinity of edges, or at the sites of absent adsorbed molecules (19). But if that were the cause of the activities of some of the bands, such as v_9 or v_{10}, it would have been expected that occasionally the appropriate spectral feature would have been missing when, as has sometimes been the case, the EEL spectra were measured when the LEED diffractions patterns were particularly well defined.

Van Hove, Somorjai and their colleagues have carried out a number of full LEED studies for benzene adsorbed on Rh[111] and Pt[111] with or without coadsorbed CO, and have recently reviewed their results (132). The presence of adsorbed CO usually gave rise to better LEED diffraction patterns. For three Rh[111]/benzene regular arrays they have shown that the site symmetries appear to be $C_{3v}(\sigma_d)$, $C_s(\sigma_d)$, and C_2, respectively, the third one being for chemisorbed benzene alone. For a Pt[111] $C_6H_6/2CO$ lattice they deduced a site-symmetry of C_1. With the exception of the C_2 case, which was studied in less detail by LEED, the site symmetries are consistent with the EEL selection rules. In the first two cases on Rh[111] the LEED results showed the present of alternating CC bond lengths of

1.46/1.58 and 1.33/1.81, respectively, around the benzene ring; the short bonds were over Rh atoms and the longer ones were between pairs of Rh atoms (132). In the Pt[111] case the interaction was mainly a bridging one of the ring with two metal atoms; an opposite pair of bonds over Pt atoms had CC bond lengths of 1.65Å, and the other four bond lengths of 1.76Å (133). The mean CC bond lengths in these three well-analyzed cases are therefore 1.52, 1.57, and 1.72Å. Although the latter value seems very high, these distances reflect the presence of essentially single CC bonds. This is in good agreement with the conclusions of a normal-coordinate calculation based on the v_2 CC-breathing mode assigned at 875 cm^{-1} on Ni[111] (117).

C. E. Anson and D. B. Powell (51 and personal communication) have analyzed the spectra of some inorganic metal cluster complexes as models for adsorbed benzene. These can be μ_1, μ_2, or μ_3 in type, i.e. with the benzene ligand interacting with one, two, or three metal atoms. In the μ_3 case, for the complex $[\mu_3(C_6H_6)Ru_6C(CO)_{11}[C_6H_6]]$ (135) there is a measured alternation in CC bond lengths around the C_6 ring of 1.39/1.48 Å compared with the pairs of values of 1.48/1.58 and 1.35/1.81 for the monolayers on Rh[111] with C_3 or quasi-C_3 symmetries.

In spectroscopic terms, the mean frequencies for the modes of interest for the gas-phase molecules and the adsorbed molecules are compared with the mean values for the several types of model ligands below (C. E. Anson and D. B. Powell, personal communication) (Table 2).

It is seen that mean frequencies for the adsorbed species are (except for v_1 where the model compounds give consistently higher vCH frequencies) in somewhat better agreement with those for μ_2 or μ_3 ligands than for μ_1. For the latter the adsorption analogy would be a site with the center of the benzene ring over a metal atom. Unlike the conclusions from EELS and LEED, several angle-resolved UV photoelectron spectroscopic studies (ARUPS) on Rh (136, 137) and Pd[100] (138) have concluded that the site

Table 2 A comparison of the wavenumbers (cm^{-1}) of corresponding vibrations of gas-phase, cluster-coordinated and adsorbed benzene, C_6H_6

	v_1	v_{13}	v_9	(v_{10} or v_{17})		(v_2 or v_{11})		v_4
gas-phase	3059	1479	1309	1149	1178	992	849	670
μ_1^a	3116	1445	1315	1159	1151	971	909	803
μ_2^b	3075	1462	—	1127	1079	—	885	739
μ_3^c	3100	1399	1320	1163	1147	887	926	854
adsorbed	ca.3020	1420	1325	1120		ca.860		910–720

^{a} $(\mu_1\text{-}C_6H_6)Cr(CO)_3$ and $(\mu_1\text{-}C_6H_6)Ru_6C(CO)_{14}$.
^{b} $(C_5H_5)_2V_2(\mu\text{-}H_2)(\mu_2\text{-}C_6H_6)$.
^{c} $Os_3(CO)_9(\mu_3\text{-}C_6H_6)$, $HOs_3(CO)_9(\mu_3\text{-}C_6H_6)^+$ BF_4^- and $(\mu_3\text{-}C_6H_6)Ru_6C(CO)_{11}(\mu_1\text{-}C_6H_6)$.

symmetry could be as high as C_{6v}. It seems that the ARUPS method may be insufficiently sensitive to distortions from the latter symmetry. However, a NEXAFS study on Pt[111] (139) has also concluded that the absorbed benzene molecules lie flat on the surface with CC distances of 1.40 Å. These are little greater then for the gas-phase molecule, a finding that is difficult to reconcile with the EELS and LEED results, even allowing for the fact that the CC distances in the latter case do seem unexpectedly long.

It is mentioned above that the EEL spectra from adsorption of benzene on Ag[111] (123) and on [111], [100], and [110] faces of Cu, indicates much weaker perturbation of the benzene adsorbate. In each of these cases the observed frequencies (including the CC breathing frequency, v_2) are close to the gas-phase values. A comparison of the frequencies on Ag[111] with the gas phase values and symmetries is given below. Only the 675 cm^{-1} band is strong.

Ag[111]	3030	1590	1480	1360	1155	1000	820	675	410
gas-phase	3059	1596	1479	1346	1149	992	849	670	—
	v_1	v_{16}	v_{13}	v_3	v_{10}	v_2	v_{11}	v_4	vCMs
	A_{1g}	E_{2g}	E_{1u}	A_{2g}	B_{2u}	A_{1g}	E_{1g}	A_{2u}	

Considering the site-symmetries classified in Table 1, all the observed bands would be allowed for a $C_{3v}(\sigma_d)$ site, as deduced by the original authors (123), but not for a C_{6v}. In the latter case the B_2 and E_1 modes are antisymmetric with respect to the C_2 axis.

On the Cu faces the number of extra bands increases systematically with the facial sequence [111] → [100] → [110]. The results appear to be consistent with site-symmetries of C_{3v} (or even C_{6v}) on [111], C_{2v} on [100], and C_s on [110]. Once again the v_2 CC-breathing modes are within 12 cm^{-1} of the gas-phase value, showing that in these cases the adsorption has led to little change in the geometry of the adsorbed benzene.

The general conclusion from this work is that the Group IB metals Cu and Ag cause minor perturbations of the adsorbed benzene as shown by the EEL spectra, in forms that are consistent with expected symmetry sites on different surfaces. However, the Group VIII metals give rise to much greater structural perturbations on sites of low symmetry, resulting in substantial lengthening of the CC bonds. However, the similarity of the spectra on the presumed low symmetry sites and, in these circumstances, the relatively few additional fundamentals that give bands of appreciable intensity both remain surprising features of the spectra of benzene on the transition-metal surfaces.

In general, however, the spectroscopic and other evidence is consistent with the benzene ring essentially retaining its coplanarity and lying flat on the metal surface.

Other Aromatic Hydrocarbon Molecules

Only toluene seems to have been studied on Pt[111], by Bertolini et al (124). It was concluded that the benzene ring is once again essentially parallel to the metal surface, and bands consistent with the presence of methyl groups suggested that the adsorbate is nondissociated.

SATURATED HYDROCARBONS

There is a strong interest in the possibility of activating paraffinic CH bonds by metal-based catalysts. However, the interaction of saturated hydrocarbons with metal surfaces is expected to be weak, and rather few experimental studies have been reported on metal single-crystal surfaces.

At an early stage Horn & Pritchard (140) and Horn (141) reported reflection/absorption studies in the vCD region of ethane on Cu[110] at 77K and Pt[111] at 80K, respectively. It was concluded that this involved physisorption rather than chemisorption. More recently, Chesters' group has obtained very good RAIRS spectra from ethane on Cu[111] at 91K in both the vCH$_3$ and δCH$_3$ regions (142). These show that the molecules in the first monolayer are physisorbed with the C–C bond parallel to the surface, the observed fundamentals being 2961(s) vCH$_3$as; 2861(w), vCH$_3$s; and 1488, δCH$_3$as. In addition, two weak overtones of δCH$_3$as modes occurred at 2923 and 2903 cm^{-1}. These findings are in strict accord with expectations based on the metal-surface selection rule, the vCH$_3$s mode being completely symmetrical and weakly allowed through distortion of the molecule by interaction with the surface.

Most studies of adsorbed saturated hydrocarbons have been concerned with cycloalkanes. A particularly interesting initial EELS study was made of cyclohexane adsorbed on Ni[111] and Pt[111] at 140K by Demuth, Ibach & Lehwald (143). In each case they observed two bands in the vCH region, one at 2900 cm^{-1} in the normally expected vCH region for a saturated hydrocarbon, and a second strong and broad band at 2720 cm^{-1} on Ni[111] and 2590 cm^{-1} on Pt[111]. The remarkably low frequencies of the latter "soft" modes indicated substantial interaction of the CH bonds with the metal surface, leading to CH bond-weakening. The spectral characteristics of these bands (lower in frequency, strong and broad) were convincingly interpreted in terms of a hydrogen-bonding type of interaction of the CH bond with the metal surface. Unlike the usual type of hydrogen-bonding between an electron-deficient hydrogen atom as an acceptor and an electron-rich atom as a donor as in, for example, O–H $\cdots$ O, in this case the paraffinic hydrogen (which has the highest electron density among hydrocarbons) is the electron donor, and the electron-deficient metal is the acceptor. The dimensions of the cyclohexane molecule are such that, if the mean cyclohexane skeletal plan were parallel to the metal surface, three axial C–H bonds could make close contact with a triangle of directly bonded metal atoms on the surface, or alternatively to a triangle of three-fold hollow sites with the cyclohexane ring then centered over a metal atom.

Demuth et al (143) correlated the larger low frequency shift of the soft νCH mode on Pt[111] compared to Ni[111] with the fact that on Pt[111] cyclohexane dissociates to benzene at 200K, whereas on Ni[111] the molecule desorbs at 170K before decomposition. However, decomposition of cyclohexane to benzene has been observed at 225K on a stepped Ni(5[111] × [1$\bar{1}$0]) surface (56).

Recently, Chesters (142) has published a higher resolution RAIR spectrum of cyclohexane on Pt[111] in the νCH region that confirms the presence of the soft mode in the form of a very broad absorption at ca. 2600 cm^{-1}, with a half-width of 400 cm^{-1}. The normal νCH region was resolved into two bands at 2900 (νCH$_2$as) and 2850 (νCH$_2$s).

An EELS study of cyclohexane on Ru[001] by Weinberg et al (144, 145) gave a soft mode νCH value of 2580 cm^{-1}. They concluded once again that the cyclohexane molecule is adsorbed flat on the surface with C_{3v} symmetry and, based on an earlier electron-stimulated desorption-ion angular-distribution (146), that the three axial C–H bonds are directed to three-fold hollow sites. They pointed out (145) that electrical image-interactions with the metal surfaces could contribute to the lowering of the soft mode νCH, but concluded that chemical interactions are nevertheless responsible for the major part of the shift.

RAIRS studies have recently been made of cyclohexane on Cu[111] (147) and on Cu[100] (148) by Chesters, Parker & Raval. Also, an EEL spectrum was shown on Cu[111] (147). The RAIR spectrum led to the conclusion that at low coverages the cyclohexane ring lies parallel to the surface, giving rise once again to a soft mode. The bands observed on Cu[111] at 95K are as follows:

2905(s), νCH$_2$as; 2895(sh) νCH$_{eq}$; 2845(m) νCH$_2$s; 2770(s, v.bd) νCH$_{ax}$; 1444(w) δCH$_2$; and 1030 (w).

Here eq and ax refer to equatorial and axial CH bonds; the latter interact with the surface to give the 2770 cm^{-1} band. At higher coverages the soft νCH modes diminished in intensity, a finding that was taken to be evidence for a tilting of the cyclohexane molecules with respect to the surface in order to provide closer packing within the monolayer.

Hoffmann & Upton (149) have studied adsorption at 90K on Ru[001] for the series of cycloparaffins C_nH_{2n} where $n = 4, 5, 6,$ and 8 by EELS in the νCH region. They observed no separate lowered νCH frequency with cyclobutane, but substantial lowerings to 2610, 2516, and 2630 for $n = 5$, 6, and 8, respectively, for the others. They also concluded that the axial CH bonds interact with three-fold sites, as after preadsorption of oxygen in these sites the adsorbed cycloparaffins did not give soft-mode absorptions. They point out, however, as does Avery (150), that there is a smooth variation of monolayer desorption temperatures, and therefore adsorption energies, for all the cycloparaffins despite the fact that for $n = 3$ (see below) and $n = 4$ there are not soft-mode absorptions, whereas for $n = 5$ and 6 these are very prominent. They consider therefore that the soft-mode absorptions indicate C–H bond activation but not attractive hydrogen-bonding to the surface. However, the nature of the soft-mode spectroscopic phenomena is so similar to that characteristic of hydrogen-bonding that the reviewer prefers an alternative viewpoint whereby attractive hydrogen-bonding into three-fold sites is energetically offset by increased repulsions between the carbon skeleton and the underlying metal atom. Waddill & Kesmodel (151) have observed a soft mode at 2635 cm^{-1} for cyclohexane on Pd[111].

Avery (150) and Hoffmann et al (152) have carried out full EELS studies of cyclopentane on Pt[111] and Ru[001], showing soft modes at 2690 and 2610 cm^{-1}, respectively. Hoffmann et al (149, 152) suggest that the substantial frequency lowering exhibited by the soft-mode on Ru[001] occurs through puckering of the flexible C_5 ring, so that two C–H bonds interact with the same three-fold sites as with three in the case of cyclohexane. Avery (150) discusses the matter in terms of more uniform interactions of five C–H bonds with the surface. He also concludes that at higher temperatures the cyclopentane decomposes first to di-σ adsorbed C_5H_8 and then (as with 1:3-cyclopentadiene) (94) to a cyclopentadienyl surface species.

Weinberg's group (153a,b) have also made an EELS study of cyclopropane adsorbed on Ru[001]. They do not observe a soft νCH absorption, but conclude that, at 90K, the molecule is nondissociatively adsorbed and, through consideration of the strengths of the individual bands, that the C_3 plane of the molecule is inclined at a substantial angle to the metal surface.

INFRARED TRANSMISSION STUDIES ON FINELY DIVIDED METAL CATALYSTS

There is an extensive literature, published over the past 30 years, following the pioneering work of Pliskin & Eischens in 1956 on Ni/SiO$_2$ (5), on the infrared transmission spectra of hydrocarbons, mostly unsaturated, on finely divided oxide-supported metal catalysts. These spectra are complex because such catalysts can expose a range of different sites—for example, on crystalline catalysts associated with different facets such as [111] and [100] in type (63)—with the result that the observed spectra show overlapping contributions from several different adsorbed species. Also, the range of frequencies available on the oxide-supported catalysts are limited because of strong blackout absorption by the oxides themselves. For example, on SiO_2 or Al_2O_3 as supports the available transmission ranges are limited to above 1300 and above 1100 cm^{-1}, respectively. Furthermore, the early work did not take into account the effective operation of the metal-surface selection rule on metal particles exceeding in size a few nanometres (21, 23), with the result that the absence of certain expected group-characteristic absorption bands has been erroneously interpreted as implying the absence of the related species. The identification of the π-complex from ethylene on metal surfaces was delayed for this reason (21). Also, except for the smallest more volatile molecules such as ethane, ethylene, or acetylene, ready physical adsorption of hydrocarbons on the more extensive oxide surfaces can substantially limit studies at low temperatures, through their absorption bands overlying weaker absorption from chemisorbed species. These various factors have caused substantial difficulties in the identification of the structures of the adsorbed species. Some of the unassigned bands could not be related to their structural origins until EELS, and subsequently RAIRS, enabled results to be obtained, usually for one species at a time, over a wider range of frequencies, including the "fingerprint" region below 1400 cm^{-1}. An example here is the unexpected surface species ethylidyne, which was identified from its complete EEL spectra (18).

Despite these difficulties, the original motivation to interpret the spectra on finely divided metals remains very strong because these are much closer to working catalysts than the simplified single-crystal systems that have been considered in this review article. Furthermore, compared with EELS, infrared spectroscopy does enable spectra of adsorbed species to be obtained under pressure of gases similar to those used in real catalytic systems. The time is now ripe for a reinterpretation of the collected earlier spectra, by using the spectral results from the single-crystal systems as a valuable database, but this must await another review article.

This article is therefore restricted to a discussion of a "worked example" where the various procedures outlined above have been successfully used to interpret the more complex spectra. The example chosen is ethylene adsorbed on oxide-supported platinum (Pt/SiO_2 and Pt/Al_2O_3). Room temperature spectra in the νCH (3000 cm^{-1}) region for ethylene on Pt/SiO_2 were first published by Morrow & Sheppard (154). A weak band occurred near 3010 cm^{-1} that signified the presence of a C=CH grouping and was assigned to the dissociatively adsorbed species PtCH=CHPt. The stronger bands, clearly from saturated surface species (9), occurred at 2920(m), 2880(s), and 2795(w) cm^{-1} and were considered to arise from the expected di-σ species PtCH$_2$CH$_2$Pt. It was considered that there were probably two components to the 2920 cm^{-1} absorption, and that the second one could be assigned to the dissociatively adsorbed species M$_2$CHCHM$_2$. On hydrogenation at room temperature all the adsorbed species were removed. As only ethane was detected in the gas-phase, this showed that all the adsorbed species had two carbon atoms. Also the ca. weak 2795 cm^{-1} band was identified as arising from a PtCH$_3$ or PtCH$_2$ group, as such bands occur in the infrared spectra of metal-alkyl groups and arise from overtones of δCH$_3$ or δCH$_2$ angle-deformation vibrations near 1400 cm^{-1}.

The Pt/SiO$_2$ system was reinvestigated some years later by Prentice (155) with Fourier-transform and spectral-ratioing infrared techniques whereby it was possible to measure additional absorptions in the νC=C and angle-deformation regions. Such an absorption at 1500 cm^{-1} was immediately recognized as being typical of νC=C in the π-complex, and its intensity correlated with that at 3010 cm^{-1} (156, 157). This interpretation of the 3010 cm^{-1} band had earlier been rejected on the argument that a C=CH$_2$ group should have given a stronger additional band at ca. 3080 cm^{-1}. Coadsorption of ethylene with CO led to much weakening of the 2880 cm^{-1} absorption with no noticeable change near 2920 cm^{-1}. This implied that the former band, if assigned to the di-σ species, was the only fundamental observable from a species with as many as four C–H bonds. Both the problem of the missing 3080 cm^{-1} band from the π-complex, and the single observed νCH fundamental from a di-σ species, were readily explicable later when it was realized that the metal-surface selection rule was operative. This required that only one completely symmetrical νCH fundamental should be present in each case (21).

Soma (158) also observed absorptions from the π-complex on Pt/Al$_2$O$_3$, which were interpreted in the same way, and in addition reported bands in the angle-deformation region at 1420 (w, bd) and 1337 cm^{-1} (158, 159). It was shown that adsorption at the lower temperature of 195K gave only a weak 1337 cm^{-1} band but that this grew at the expense of the ca. 1495 cm^{-1} band from the π-complex on warming to room temperature. Following similar work on Rh/SiO$_2$ by Pearce (160), Prentice also observed the same three bands in the νCC/δCH region at room temperature on Pt/SiO$_2$ (155). As mentioned above, it was also shown that subsequent adsorption of CO led to a much reduced intensity of the strongest νCH band at 2880 cm^{-1}, whereas the ca. 1340 cm^{-1} band only shifted slightly upwards in frequency. The surprising implication seemed to be that the strongest bands in the νCH and δCH regions appeared to be associated with different adsorbed species. Neither Soma (159) nor the Norwich group could find a very satisfactory interpretation for the 1340 cm^{-1} band, although deuterium-substitution experiments showed that it was of the δCH/δCD type (155, 158).

At this stage the EEL spectrum on Pt[111] became available (32), which was in due course interpreted as arising from the unexpected ethylidyne species (8, 12, 18). This accounted for the 1340 cm^{-1} band (the more precise value subsequently obtained by RAIRS was 1339 cm^{-1}), which both in EELS (18) and RAIRS (33, 34) was accompanied by a companion band in the νCH region at 2884 cm^{-1} and the weak overtone band at 2795 cm^{-1}. All these bands in the transmission spectra were therefore assigned to ethylidyne independently by the Norwich group (37, 46a,b, 161) and by Beebe, Albert, and Yates (162). The remarkable weakening of the 2880 cm^{-1} absorption on CO coadsorption remains to be explained; it certainly delayed an overall understanding of the spectra on the finely divided Pt catalyst. This left two absorptions at 2920 cm^{-1} and ca. 1420 cm^{-1} in the room temperature spectrum to be assigned to a third surface species.

In the early study by Morrow & Sheppard (154), a low-temperature spectrum at 128K had given a single νCH absorption at 2907 cm^{-1}. This was converted to the usual spectrum on warming to room temperature. As it would normally be considered more probable that a nondissociatively adsorbed species, such as PtCH$_2$CH$_2$Pt, would occur at low temperatures, it was at that time puzzling that a single band was observed. It was therefore tentatively assigned, with some misgivings, to the dissociatively adsorbed species Pt$_2$CHCHPt$_2$, as the absorption fell in the region expected for sp^3 hydrized carbon. The theoretical difficulty disappeared once the effective operation of the metal surface selection on finely divided catalysts was realized (21).

De la Cruz & Sheppard (163) have recently reinvestigated the temperature evolution of the spectrum from ethylene on Pt/SiO$_2$ starting at 100K. On warming, the first bands to go are from multilayer physical adsorption. The remaining spectrum from the chemisorbed species showed at 189K the presence of the π-complex (νCH at 3022; νCC at 1500 cm^{-1}), plus two adjacent νCH bands at 2922 and 2906 cm^{-1} and a broad weak band at ca. 1419 cm^{-1}. No absorptions from the ethylidyne species were present at this temperature, but on warming to 294K the 2906 cm^{-1} band, and probably part of the ca. 1419 cm^{-1} band, were replaced by the 2881/1342 cm^{-1} bands of ethylidyne. The 2922 cm^{-1} νCH band and absorption (still weak and broad) at ca. 1426 cm^{-1} remained. Ibach et al (18, 32) had convincingly shown by EELS that a low-temperature spectrum from the di-σ adsorbed species is converted to ethylidyne before room temperature, and therefore the bands at 2906 and *ca.* 1419 cm^{-1} that behave likewise can be assigned to the di-σ adsorbed species on that crystal plane. The remaining companion bands at 2922 and 1426 cm^{-1} can therefore very probably be assigned to a di-σ species that remains stable up to room temperature on another crystal face. The comparable intensity of the 2922/2906 cm^{-1} bands suggests that at room temperature there are approximately equal numbers of di-σ and ethylidyne species. The bands from the π-species are weaker than the others, but, bearing in mind that νCH absorptions from ethylenic bonds are intrinsically weaker than from paraffinic ones (164a,b) it could be that there are also comparable amounts of the π-species present. We therefore finally have a convincing interpretation of the principal features in the room-temperature spectrum from ethylene on Pt/SiO$_2$ in terms of the coexistence of π-, di-σ, and ethylidyne adsorbed species.

Beebe & Yates (162), using Pt/Al$_2$O$_3$, which is transparent to lower frequencies, have also identified the νCC absorption of ethylidyne. They have also more recently proposed that the strength of the bands from the ethylidyne species, which normally require triangular metal sites, can be used in conjunction with quantitative CO adsorption measurements to evaluate the proportion of [111] faces in a metal catalyst (165). The authors also interpret bands at 2940 and 1411 cm^{-1} on Pt/Al$_2$O$_3$ as arising from a breakdown of the metal-surface selection rule so as to allow the νCH$_3$as and δCH$_3$as of the ethylidyne species to be observed (166). The interpretation is not necessarily in disagreement with that given above by de la Cruz & Sheppard, as the frequencies concerned are different and the mean particle size of the Pt/Al$_2$O$_3$ catalyst was very small at *ca.* 2 nm, so the metal surface selection rule may be less operative.

Sheppard et al some years ago (167) showed that essentially the same infrared spectrum is obtained under kinetic conditions of ethylene hydrogenation for ethylene-rich C$_2$H$_4$/H$_2$ mixtures. Beebe & Yates (168) have recently made similar measurements on Pt/Al$_2$O$_3$ and have concluded that the rate of ethylene hydrogenation seems to be independent of the amount of ethylidyne species on the surface. They therefore discount a suggestion from Zaera & Somorjai (169) that the hydrogenation process involves the interaction of gas-phase ethylene with a "carbonaceous layer" in the form of ethylidyne surface species (168). However, as Sheppard et al have noted (167), some care has to be taken in interpreting kinetic results for metal catalysts in pressed-disc form, because the rate of reaction can be dependent on processes involving diffusion of reagent or products into and out of the pores of the catalyst, rather than on the reaction at the metal surface.

Overall, however, the Pt/SiO$_2$ and Pt/Al$_2$O$_3$ work has now enabled a very complete interpretation of the infrared spectra to be achieved, making very good use of the single crystal results obtained by EELS and RAIRS. It is to be expected that other similar successful infrared studies of the species adsorbed on finely divided metal catalysts will be published in the future.

CONCLUSIONS AND FUTURE PROSPECTS
Summary of Results

Much of the wealth of information now available in the literature is seen to refer to results on the adsorption of the type-molecules ethylene, acetylene, benzene, and cyclohexane. In all cases the vibrational spectra have provided valuable, and often conclusive, evidence for the nature of the chemisorbed species. As might have been anticipated, the lowest temperature spectra are usually from nondissociatively adsorbed species, i.e. species in which the original hydrocarbon H/C ratio is retained, whereas progressing to higher temperatures leads to a gradual reduction in the H/C ratio and ultimately to graphitized carbon layers. However, different metals, even when they give rise to similar such sequences of adsorbed species, may cause structural transitions between adsorbed species at notably different temperatures.

Ethylene

For ethylene the low-temperature species are either π-complexes (Type II spectra) or di-σ complexes (Type I spectra). In some cases both types of species coexist; in others one type will occur on one crystal face of a given metal and the other on a different one. These observations, which are paralleled in the extensive results published for CO adsorption on metal surfaces (98), suggest that the two species—although chemically distinct and possibly significantly very different in their charge distributions and reactivities—are of relatively similar stability on the surfaces. Both types of species have CC bond orders between 2 and 1, with the π-complexes having values nearer 2 and the di-σ complexes nearer 1. On the well-established pattern of synergic bonding of unsaturated molecules on transition metals, this difference in bond order can be ascribed to a greater degree of d $\rightarrow \pi^*$ back bonding from the metal to the ethylene-derived ligand in the case of the di-σ complexes compared with the π-complexes. There seems to be an enthalpic compensation between the bonding within the hydrocarbon grouping (stronger for the π-complex with partial double-bond CC character) and the bonding to the surface (weaker C to metal bonding in the π-complex). It was earlier thought, on evidence for Type I' spectra, that there may be intermediate structures between the two types. However, it now seems more probable that these spectra represent the coexistence of the di-σ and π-complexes. There are substantial variations of frequencies of individual modes of vibration when different metals are involved in forming the same general type of complex. These differences may also be of significance in reactivity.

By room temperature, dissociatively adsorbed species normally occur; on fcc [111] and hcp [001] faces these are, with the single exception of Ni[111] (42), in the form of the ethylidyne (CH_3C) species, which is clearly stable on the three-fold triangular sites provided by these faces. On Ni[111] the species present at room temperature is an acetylenic C_2H_2 adsorbed species, and it remains to be seen whether the ethylidyne species has occurred as an intermediate at lower temperatures. This is probably the case, because ethylidyne-like spectra have been observed on Ni/SiO_2 at 195K (154). It is also probable that a di-σ adsorbed vinylidene ($C=CH_2$) species *plus adsorbed hydrogen* is an intermediate in the formation of ethylidyne. The same species in the absence of surface hydrogen is also probably the first ethylidyne dissociation species leading, at higher temperatures, to C_2H and then species with even lower H/C ratios. Another interesting regularity, which may have a bearing on catalytic processes involving ethylene in gas-streams with impurities, is that oxygen or carbon preadsorption seems invariably to lead to π-complexes.

Acetylene

The adsorption of acetylene at low temperatures also leads to two principal types of spectra, of which one (Type B spectra) seems reasonably well assigned to di-σ/π bonded species on three metal atoms. The structural nature of the other type of spectrum is less clear but, on grounds of lower average νCC and νCH frequencies, seems to involve a greater degree of d $\rightarrow \pi^*$ back donation. For the Cu[111] case, a di-σ/di-π type of interaction with four metal atoms is indicated (106). Once again, however, the two types of spectra can coexist, e.g. as a function of coverage, on the same metal surface as with Ru[001] (97). The bond-order of all the observed C_2H_2 species seems to fall between about 1.3 and 1.6, and is much reduced from the bond-order of 3 for the parent hydrocarbon. This occurs even for adsorption on Cu surfaces (95, 104). Hydrogen-deficient species similar to those observed with ethylene occur on heating to higher temperatures, and in the vicinity of room temperature the addition of hydrogen to acetylenic species can readily lead once again to the stable ethylidyne.

A Comparison of Nondissociated Adsorbed Species from Ethylene and Acetylene

Table 3 below summarizes and compares the results obtained to date for the nondissociatively adsorbed species from both ethylene and acetylene.

Table 3 A summary of the nondissociative surface species identified by vibrational spectroscopy from the chemisorption of ethylene and acetylene on metal single-crystal faces at low temperatures

Surface plane	[111]	[100]	[110]
Cu (fcc)	—/A	π/A	—/A
Ni (fcc)	di-σ/A	di-σ/A	di-σ/B
Pd (fcc)	π/B	π or (di-σ+π)/A	π/B
Pt (fcc)	di-σ/B	di-σ/A	
Fe (bcc)	di-σ+π/A	—	di-σ/A
Ru (hcp)	[001] di-σ+π/$A \to B$		

Notes: The ethylene species (π or di-σ) are given on the *left*; the acetylene species (Type A or Type B spectra, the latter probably corresponding to di-σ/π species) on the *right*. X+Y denotes the coexistence of such species; X or Y denotes the detection of different species in different experiments; X $\to$ Y denotes the detection of different species with increasing coverage.

In the former case di-σ or π denote the two types of species. In the latter case these are labeled according to the type of spectrum, A or B, because the structure or structures associated with Type A spectra are not well established. It is seen that the overall pattern is complex. For example, for a given type of crystal face, different metals give different species, e.g. [111] faces. Also for a given metal, the different faces can give different species, e.g. Ni or Pd. However, there is a reasonably good general relationship between the occurrence of ethylene π-complexes and Type B acetylenic spectra and particularly between ethylenic di-σ and Type A acetylenic spectra. The former pair relate to less, and the latter to more, d $\to \pi^*$ back bonding from the metal. However, there are some exceptions to this generalization as with Ni[110], Pt[111], and Cu[100]. In some cases pairs of species are found on certain metal faces, e.g. Fe[111], Ru[001], and Pd[100]. As is common in catalytic systems, there seems to be a degree of interplay between electronic and geometrical factors. It is noteworthy, however, that Pd in particular seems less inclined to provide electronic back-donation, and this may be related to the fact that the free atom has a completed d^{10} shell, which normally has a pronounced stability. There is some evidence for benzene formation from trimerization of acetylenic species.

Benzene

The adsorption of benzene leads to a picture simpler than that from the C_2 hydrocarbons because it seems to have been invariably concluded that a planar or near-planar cyclic C_6H_6 structure is retained that is adsorbed parallel to the metal surface. However, with the Group IB metals, Cu and Ag, only minor changes of geometry of the adsorbed molecule are indicated, as the vibration frequencies differ little from the gas-phase values. It may be that on these surfaces the benzene molecules are centered over a single metal atom and that the bonding to the surface is of the usual (σ-donation/d $\rightarrow \pi^*$ back-donation) type. For the transition-metals proper, however (Groups VIII and VIIA), the spectroscopic and LEED data show clearly that the CC bond length has on average become substantially closer to that of a single bond, with in several cases bond-variations, or regular bond-length alternation, around the ring. The interaction with the surface seems here to be through several metal atoms, e.g. in some cases with a triangle whereby short CC bonds lie over the metal atoms and longer bonds alternate with the short ones. However, there remains something of a mystery with regard to the uniformity of the additional allowed fundamentals that should be related to site-symmetries of less than C_{6v} on a variety of different planes.

Cyclohexane and Cyclopentane

For cyclohexane the most interesting finding is strong spectroscopic evidence for substantial CH bond-weakening, and presumably bond-activation. This is caused by hydrogen-bond-like interactions of particular CH bonds with the metal surface, possibly at three-fold hollow sites. The degree of the CH bond-weakening, as reflected in the lowering of the vCH frequencies, seems to correlate with the relative ease or difficulty of conversion of cyclohexane to benzene at higher temperatures on the different metal faces. These vCH "soft modes" are particularly pronounced over transition-metal surfaces such as Pt[111], Ni[111] and Ru[001], but they also occur on Cu[111] and Cu[100] faces (148, 149).

The flexible cyclopentane ring gives rise to similar soft vCH modes on Ru[001] (147), but such features do not arise through adsorption of either cyclobutane or cyclopropane.

Points for Further Spectroscopic Study

In general the application of EEL and RAIR vibrational spectroscopies has enabled the enormous steps forward made during the past 10 to 15 years in our knowledge of the structures of hydrocarbon-based adsorbed species on metal single-crystal surfaces. It has also opened up the prospect of obtaining a much better understanding of the species responsible for activity on finely-divided metal catalysts.

Clearly, much further work needs to be carried out on noncyclic paraffinic hydrocarbons at low and higher temperatures. For benzene itself the main need is for as accurate as possible LEED studies on non-[111] surfaces in order to attempt to explain the surprisingly uniform spectroscopic results. There is also good scope for studies of alkyl-substituted benzenes to see whether alkyl groups or the benzene-ring first dissociate on metal surfaces.

For ethylene and acetylene there are a number of gaps in the experimental data. More low temperature work on Rh and Ir would be particularly welcome. Further work on acetylenic adsorption on Ni[110] or Pt[111] and Cu[100] might throw light on the unusual juxtapositions of di-σ/B or π/A ethylene/acetylene spectra on these surfaces. Also, further work on unreconstructed Pt[110] and Pt[100] surfaces would be generally welcome. Some clarification is also needed for the circumstances under which more than one type of ethylene- (or acetylene-) derived species occur on particular faces, such as have been observed on Ru[001], or Pd[100] and Fe[111]. It is of considerable interest and surprise to have found that acetylene is very strongly bound to the several Cu faces. Similar explorations of Ag faces with acetylene and ethylene would be illuminating, as would ethylene studies on Cu[110] and Cu[111]. More work on the smaller alkyl-substituted ethylenes and acetylenes would also be welcome.

Methods and Techniques

The dipolar metal-surface selection rule affects in the same way on-specular intensities in EELS and intensities in RAIRS and the infrared spectra of finely-divided metals. Taking this into account, much structural information has been derived from a comparison of the spectra of adsorbed species and the bands from completely symmetrical modes in the infrared spectra of related hydrocarbon ligands on metal-cluster compounds. This is likely to remain a fruitful area of investigation.

The EEL and RAIR spectroscopic methods remain complementary. EELS provides additional information about non-completely symmetrical modes as observed best in off-specular experiments; RAIRS gives high resolution, unambiguous identification of completely symmetrical fundamentals or overtones/combinations, and the possibilities of making measurements under the higher pressures of gases normally employed in catalytic work. Whereas surface species involved in reactions of the Langmuir/Hinshelwood type between chemisorbed species might be detected by EELS, those of the Eley-Rideal type, involving reactions between chemisorbed species and physically adsorbed ones, can probably only be carried out at higher pressures by RAIRS.

Further very profitable EELS or RAIRS studies could be made on [100] and particularly [110] planes where the electron plane of incidence is carefully aligned along one or another well-defined direction on a surface. As for Ni[100] (102), this will certainly improve the quality of vibrational assignments and hence of structure determination using impact selection rules based on planes of symmetry. It goes without saying that spectroscopic results are likely to be particularly meaningful when taken under coverage conditions where there are well-defined and interpretable LEED patterns. Parallel TPD studies of overall surface compositions of hydrocarbon adsorbates will always be valuable. Also, any techniques that can minimize or preferably determine the number of defects on flat surfaces could be of help in explaining anomalous spectra at low coverages.

Kinetic Studies

It can be argued that a good proportion of the species so painstakingly identified under static conditions might persist on the surface, and play no useful role, during actual catalytic reactions. It is possible to investigate this question qualitatively without difficulty if a catalytic reaction involves two or more reacting species. If a partial coverage is achieved by one species of a pair, the introduction of the second reagent at catalytic reaction temperatures should lead to the disappearance of an active surface species with the production of the expected gas-phase product. Alternatively it should lead to the retention of an inactive species in the original or a modified form.

Reaction mechanisms, however, can only be properly investigated and evaluated by quantitative kinetic studies, whereby the rate of the overall reaction is related to surface concentrations of specific active species. Somorjai's group, for example, has done much meaningful catalytic work by carrying out a reaction in a gas-phase at higher pressures and then investigating the resulting surface species by EELS and other techniques after evacuation. However, in principle, by using infrared spectroscopy, it should be possible to carry out in situ kinetic studies of surface species under flowing conditions. Such work on usefully simplified single-crystal metal surfaces could be done using RAIRS, or on finely divided catalysts using infrared transmission spectroscopy. In the latter case, difficulties, due to diffusion-controlled (rather than surface-reaction controlled) kinetics in the narrow pores of catalyst pressed-discs, can now often be overcome by using diffuse reflectance or photoacoustic infrared methods on loose powdered catalysts. The generally low sensitivity of the latter methods has now been largely overcome by the use of Fourier-transform infrared techniques. Kinetic work involving the monitoring of adsorbed species, such as previously carried out by Zaera & Hall (45a,b) or Beebe & Yates (168) on hydrocarbon/metal systems, will surely form a main thrust of vibrational spectroscopic work during the next few decades. It will be much aided by the great progress that has already been achieved in identifying the static surface species by the vibrational spectroscopies.

Acknowledgments

The author would like to thank present and past colleagues, Dr. C. E. Anson, Dr. N. R. Avery, Dr. D. B. Powell, and particularly Dr. M. A. Chesters for agreeing to inclusion of discussions of some of their recent work, which is still in the process of publication, and also for more general discussions. He would also like to thank Dr. Anson and Dr. P. Gardner for substantial assistance with the bibliography on the sections on benzene and the saturated hydrocarbons, respectively. The U.K. Science and Engineering Council is to be thanked for much financial support of the work of the author's group on vibrational spectra of adsorbed molecules over the past 15 years.

Literature Cited

1. Yates, J. T. Jr., Madey, T. E. 1987. *Vibrational Spectroscopy of Molecules on Surfaces.* New York: Plenum
2. Ibach, H., Mills, D. L. 1982. *Electron Energy Loss Spectroscopy and Surface Vibrations.* New York: Academic
3. Chesters, M. A. 1986. *J. Electron Spectros. Relat. Phenom.* 38: 123
4. Eischens, R. P., Pliskin, W. A., Francis, S. A. 1954. *J. Chem. Phys.* 22: 1786
5. Pliskin, W. A., Eischens, R. P. 1956. *J. Chem. Phys.* 24: 482
6. Otto, A. 1984. In *Light Scattering in Solids,* ed. M. Cardona, G. Güntherodt, 4: 298–418. Berlin: Springer-Verlag
7. Nakamoto, K. 1986. *Infrared and Raman Spectra of Inorganic and Coordination Compounds.* New York: Wiley-Interscience. 4th ed.
8. Skinner, P., Howard, M. W., Oxton, I. A., Kettle, S. F. A., Powell, D. B., Sheppard, N. 1981. *J. Chem. Soc. Faraday Trans. 2* 77: 1203
9. Bellamy, L. J. 1975. *Infrared Spectra of Complex Molecules,* Vol. 1. London: Chapman & Hall. 3rd ed.
10. Roberts, M. W., McKee, C. S. 1978. *Chemistry of the Metal-Gas Interface.* Oxford: Oxford Univ. Press
11. Somorjai, G. A. 1981. *Chemistry in Two Dimensions: Surfaces,* Chap. 2 and 4. New York: Cornell Univ. Press
12. Kesmodel, L. L., Dubois, L. H., Somorjai, G. A. 1979. *J. Chem. Phys.* 70: 2180
13. Arvanitis, D., Dobler, U., Wenzel, L., Baberschke, K., Stöhr, J. 1986. *Surf. Sci.* 178: 686
14. Horn, K., Bradshaw, A. M., Jacobi, K. 1978. *Surf. Sci.* 72: 719
15. Wang, P. K., Slichter, C. P., Sinfelt, J. S. 1984. *Phys. Rev. Lett.* 53: 82
16. Salmeron, M., Somorjai, G. A. 1982. *J. Phys. Chem.* 86: 341
17. Demuth, J. E. 1979. *Surf. Sci.* 80: 367
18. Steininger, H., Ibach, H., Lehwald, S. 1982. *Surf. Sci.* 117: 685
19. Sheppard, N., Erkelens, J. 1984. *Appl. Spectros.* 38: 471
20. Sheppard, N., Richardson, N. V. 1987. Ref. 1, p. 1
21. Pearce, H. A., Sheppard, N. 1976. *Surf. Sci.* 59: 205
22. Greenler, R. G. 1966. *J. Chem. Phys.* 44: 310
23. Greenler, R. G., Snider, D. R., Witt, D., Sorbello, R. S. 1982. *Surf. Sci.* 118: 415
24. Ibach, H. 1977. *Surf. Sci.* 66: 56
25. Ho, W., Willis, R. F., Plummer, E. W. 1978. *Phys. Rev. Lett.* 40: 1463
26. Tong, S. Y., Li, C. H., Mills, D. L. 1980. *Phys. Rev. Lett.* 44: 407; 1981. *Phys. Rev. B* 24: 806
27. Cavanagh, R. R., Rush, J. J., Kelley, R. D. 1987. See Ref. 1, pp. 183–222
28. Moskovits, M. 1982. *J. Chem. Phys.* 77: 4408
29. Koel, B. E., Crowell, J. E., Bent, B. E., Mate, C. M., Somorjai, G. A. 1986. *J. Phys. Chem.* 90: 2949
30. Abon, M., Billy, J., Bertolini, J. C., Tardy, B. 1986. *Surf. Sci.* 167: L187
31. Marinova, T. S., Kostov, K. L. 1987. *Surf. Sci.* 181: 573
32. Ibach, H., Lehwald, S. 1978. *J. Vac. Sci. Technol.* 15: 407
33. Chesters, M. A., McCash, E. M. 1987. *Surf. Sci.* 187: L639
34. Malik, I. J., Brubaker, M. E., Mohsin, S. B., Trenary, M. 1987. *J. Chem. Phys.* 87: 5554

35. Nyberg, C., Tengstal, C. G., Andersson, S., Holmes, M. W. 1982. *Chem. Phys. Lett.* 87: 87
36. Sheppard, N. 1986. *J. Electron Spectro. Relat. Phenom.* 38: 175
37. Bandy, B. J., Chesters, M. A., James, D. I., McDougall, G. S., Pemble, M. E., Sheppard, N. 1986. *Philos. Trans. R. Soc. London Ser. A* 318: 141
38. Sheppard, N., de la Cruz, C. 1987. *React. Kinet. Catal. Lett.* 35: 21
39. Hamilton, J. C., Swanson, N., Waclawski, B. J., Celotta, R. J. 1981. *J. Chem. Phys.* 74: 4156
40. Backx, C., Willis, R. F. 1978. *Chem. Phys. Lett.* 53: 471
41. Backx, C., Willis, R. F., Feuerbacher, B., Fitton, B. 1977. *Surf. Sci.* 68: 516
42. Lehwald, S., Ibach, H. 1979. *Surf. Sci.* 89: 425
43. Hammer, L., Hertlein, T., Müller, K. 1986. *Surf. Sci.* 178: 693
44. Lehwald, S., Ibach, H., Steininger, H. 1982. *Surf. Sci.* 117: 342
45a. Zacra, F., Hall, R. B. 1987. *Surf. Sci.* 180: 1
45b. Zaera, F., Hall, R. B. 1987. *J. Phys. Chem.* 91: 4318
46a. Anson, C. E., Bandy, B. J., Chesters, M. A., Keiller, B., Oxton, I. A., Sheppard, N. 1983. *J. Electron Spectros. Relat. Phenom.* 29: 315
46b. McDougall, G. S. 1985. PhD thesis. Univ. East Anglia
47. Stroscio, J. A., Bare, S. R., Ho, W. 1984. *Surf. Sci.* 148: 499
48. Barteau, M. A., Broughton, J. Q., Menzel, D. 1984. *Appl. Surf. Sci.* 19: 92
49. Hatzikos, G. H., Masel, R. I. 1987. *Surf. Sci.* 185: 479
50. Erley, W., Baro, A. M., Ibach, H. 1982. *Surf. Sci.* 120: 273
51. Anson, C. E. 1988. PhD thesis. Univ. East Anglia
52. Stuve, E. M., Madix, R. J. 1985. *J. Phys. Chem,* 89: 105
53. Hills, M. M., Parmenter, J. E., Mullins, C. B., Weinberg, W. H. 1986. *J. Am. Chem. Soc.* 108: 3554
54. Seip, U., Tsai, M-C., Küppers, J., Ertl, G. 1984. *Surf. Sci.* 147: 65
55. Tardy, B., Bertolini, J. C. 1985. *J. Chim. Phys.* 82: 407
56. Lehwald, S., Ibach, H. 1979. *Surf. Sci.* 89: 425
57. Gates, J. A., Kesmodel, L. L. 1982. *Surf. Sci.* 120: L461
58. Gates, J. A., Kesmodel, L. L. 1982. *Surf. Sci.* 124: 68
59. Chesters, M. A., McDougall, G. S., Pemble, M. E., Sheppard, N. 1985. *Appl. Surf. Sci.* 22/23: 369
60. Grogan, M. J., Nakamoto, K. 1966. *J. Am. Chem. Soc.* 88: 5454
61. Powell, D. B., Scott, J. G. V., Sheppard, N. 1972. *Spectrochim. Acta* 28A: 327
62. Herzberg, G. 1945. *Infrared and Raman Spectra*. New York: Van Nostrand
63. Sheppard, N., Nguyen, T. T. 1978. *Adv. Infrared Raman Spectrosc.* 5: 67–148
64. Stuve, E. M., Madix, R. J. 1985. *Surf. Sci.* 160: 293
65. Backx, C., de Groot, C. P. M., Biloen, P. 1980. *Appl. Surf. Sci.* 6: 256
66. Backx, C., de Groot, C. P. M. 1982. *Surf. Sci.* 115: 382
67. Kostov, K. L., Marinova, T. S. 1987. *Surf. Sci.* 184: 359
68. Deleted in proof
69. Bond, G. C. 1986. *Philos. Trans. R. Soc. London Ser. A* 318: 160
70. Baro, A. M., Ibach, H. 1981. *J. Chem. Phys.* 74: 4194
71. Dubois, L. H., Castner, D. G., Somorjai, G. A. 1980. *J. Chem. Phys.* 72: 5234
72. Koel, B. E., Bent, B. E., Somorjai, G. A. 1984. *Surf. Sci.* 146: 211
73. Ibach, H. 1978. *Proc. Vibrations in Adsorbed Layers Conf.*, Jülich, Jülich GmbH, FDR: Kernforschungsanlage; and personal communication
74. Eady, C. R., Fernandez, J. M., Johnson, B. F. G., Lewis, J., Raithby, P. R., Sheldrick, G. M. 1978. *J. Chem. Soc. Chem. Commun.* 1978: 421
75. Bertolini, J. C., Rousseau, J. 1979. *Surf. Sci.* 83: 531
76. Hill, M. M., Parmenter, J. E., Weinberg, W. H. 1987. *J. Am. Chem. Soc.* 109: 597
77. Andrews, J., Kettle, S. F. A., Powell, D. B., Sheppard, N. 1982. *Inorg. Chem.* 21: 2874
78. Evans, J., McNulty, G. S. 1983. *J. Chem. Soc. Dalton Trans.* 1983: 639
79. Evans, J., McNulty, G. S. 1984. *J. Chem. Soc. Dalton Trans.* 1984: 79
80. Abon, M., Billy, J., Bertolini, J. C., Tardy, B. 1986. *Surf. Sci.* 167: 1
81. Koestner, R. J., Van Hove, M. A., Somorjai, G. A. 1983. *J. Phys. Chem.* 87: 203
82. Albert, M. R., Sneddon, L. G., Eberhardt, W., Greuter, F., Gustafsson, T., Plummer, E. W. 1982. *Surf. Sci.* 120: 19
83. Lloyd, D. R., Netzer, F. P. 1983. *Surf. Sci.* 129: L249
84. Kruger, B., Benndorf, C. 1986. *Surf. Sci.* 178: 704
85. Fischer, D. A., Döbler, U., Arvanitis, D., Wenzel, L., Baberschke, K., Stöhr, J. 1986. *Surf. Sci.* 177: 114

86. Stöhr, J., Sette, F., Johnson, A. L. 1984. *Phys. Rev. Lett.* 53: 1684
87. Koestner, R. J., Stöhr, J., Gland, J. L., Horsley, J. A. 1984. *Chem. Phys. Lett.* 105: 332
88. Avery, N. R., Sheppard, N. 1986. *Surf. Sci.* 169: L367
89. Avery, N. R., Sheppard, N. 1986. *Proc. R. Soc. London Ser. A* 405: 1
90. Avery, N. R., Sheppard, N. 1986. *Proc. R. Soc. London Ser. A* 405: 27
91. Bent, B. E., Mate, C. M., Crowell, J. E., Koel, B. E., Somorjai, G. A. 1987. *J. Phys. Chem.* 91: 1493
92. Deleted in proof
93. Deleted in proof
94. Avery, N. R. 1986. *J. Electron Spectrosc. Relat. Phenom.* 39: 1
95. Bandy, B. J., Chesters, M. A., Pemble, M. E., McDougall, G. S., Sheppard, N. 1984. *Surf. Sci.* 139: 87
96. Anson, C. E., Keiller, B. T., Oxton, I. A., Powell, D. B., Sheppard, N. 1983. *J. Chem. Soc. Chem. Commun.* 1983: 470
97. Jakob, P., Cassuto, A., Menzel, D. 1987. *Surf. Sci.* 187: 407
98. Bradshaw, A. M., Hoffmann, F. M. 1978. *Surf. Sci.* 72: 513
99. Ibach, H., Lehwald, S. 1981. *J. Vac. Sci. Technol.* 18: 625
100. Hammer, L. 1985. PhD thesis. Univ. of Erlangen-Nürnberg, FDR
101. Dinardo, N. J., Demuth, J. E., Avouris, P. 1983. *J. Vac. Sci. Tech. A* 1: 1244
102. Dinardo, N. J., Demuth, J. E., Avouris, P. 1983. *Phys. Rev. B* 27: 5832
103. Marinova, T. S., Stefanov, P. K. 1987. *Surf. Sci.* 191: 66
104. Avery, N. R. 1985. *J. Am. Chem. Soc.* 107: 6711
105. Kesmodel, L. L., Waddill, G. D., Gates, J. A. 1984. *Surf. Sci.* 138: 464
106. Chesters, M. A., McCash, E. M. 1987. *J. Electron Spectrosc. Relat. Phenom.* 44: 99
107. Chesters, M. A. 1988. *Proc. Eur. Molec. Spectrosc. Conf. 1987, Amsterdam.* To appear in *J. Mol. Spectrosc.*
108. Parmenter, J. E., Hills, M. M., Weinberg, W. H. 1986. *J. Am. Chem. Soc.* 108: 3563
109. Gates, J. A., Kesmodel, L. L. 1984. *Surf. Sci.* 138: 464
110. Gates, J. A., Kesmodel, L. L. 1982. *J. Chem. Phys.* 76: 4281
111. Iwashita, Y., Tamura, F., Nakamura, A. 1969. *Inorg. Chem.* 8: 1179
112. Gervasio, G., Rossetti, R., Stanghellini, P. L. 1985. *Organometallics* 4: 1612
113. Stuve, E. M., Madix, R. J., Sexton, B. A. 1982. *Surf. Sci.* 123: 491
114. Bertolini, J. C., Massardier, J., Dalmai-Imelik, G. 1978. *J. Chem. Soc. Faraday Trans. 1* 74: 1720
115. Bertolini, J. C., Rousseau, J. 1979. *Surf. Sci.* 89: 467
116. Lehwald, S., Ibach, H., Demuth, J. E. 1978. *Surf. Sci.* 78: 577
117. Jobic, H., Tardy, B., Bertolini, J. C. 1986. *J. Electron Spectrosc. Relat. Phenom.* 38: 55
118. Bertolini, J. C., Massardier, J., Tardy, B. 1981. *J. Chim. Phys.* 78: 939
119. Koel, B. E., Crowell, J. E., Mate, C. M., Somorjai, G. A. 1984. *J. Phys. Chem.* 88: 1988
120. Mate, C. M., Somorjai, G. A. 1985. *Surf. Sci.* 160: 542
121. Waddill, G. D., Kesmodel, L. L. 1985. *Phys. Rev. B* 32: 2107
122. Waddill, G. D., Kesmodel, L. L. 1985. *Phys. Rev. B* 31: 4940
123. Avouris, P., Demuth, J. E. 1981. *J. Chem. Phys.* 75: 4783
124. Abon, M., Bertolini, J. C., Billy, J., Massardier, J., Tardy, B. 1985. *Surf. Sci.* 162: 395
125. Surman, M., Bare, S. R., Hofmann, P., King, D. A. 1983. *Surf. Sci.* 126: 349
126. Tardy, B., Bertolini, J. C., Ducros, R. 1985. *Bull. Soc. Chim. Fr.* 3: 313
127. Deleted in proof
128. Wilson, E. B. Jr., Decius, J. C., Cross, P. C. 1955. *Molecular Vibrations.* New York: McGraw-Hill
129. Herzberg, G. 1945. See Ref. 62, p. 118
130. Schweitzer, E., Bradshaw, A. M. 1986. *J. Electron Spectrosc. Relat. Phenom.* 38: 141
131a. Mair, R. D., Hornig, D. F. 1949. *J. Chem. Phys.* 17: 1236
131b. Painter, P. C., Koenig, J. L. 1977. *Spectrochim. Acta A* 33: 1003
132. Lin, R. F., Blackman, G. S., Van Hove, M. A., Somorjai, G. A. 1987. *Acta Cryst. B* 43: 368
133. Ogletree, D. F., Van Hove, M. A., Somorjai, G. A. 1987. *Surf. Sci.* 183: 1
134. Deleted in proof
135. Gomez-Sal, M. P., Johnson, B. F. G., Lewis, J., Raithby, P. R., Wright, A. H. 1985. *J. Chem. Soc. Chem. Commun.* 1985: 1682
136. Bertel, E., Rosina, G., Netzer, F. P. 1986. *Surf. Sci.* 172: L515
137. Netzer, F. P., Rosina, G., Bertel, E., Saalfeld, H. 1987. *Surf. Sci.* 184: L397
138. Hoffmann, P., Horn, K., Bradshaw, A. M. 1981. *Surf. Sci.* 105: L260
139. Horsley, J. A., Stöhr, J., Hitchcock, A. P., Newbury, D. C., Sette, F. 1985. *J. Chem. Phys.* 83: 6099
140. Horn, K., Pritchard, J. 1975. *Surf. Sci.* 52: 437

141. Horn, K. 1978. See Ref. 73, p. 140
142. Chesters, M. A. 1988. In *Analytical Applications of Spectroscopy*, ed. C. S. Creaser, A. M. C. Davies, p. 201. London: Royal Soc. Chem.
143. Demuth, J. E., Ibach, H., Lehwald, S. 1978. *Phys. Rev. Lett.* 40: 1044
144. Hoffmann, F. M., Felter, T. E., Thiel, P. A., Weinberg, W. H. 1981. *J. Vac. Sci. Tech.* 18: 651
145. Hoffmann, F. M., Felter, T. E., Thiel, P. A., Weinberg, W. H. 1983. *Surf. Sci.* 130: 173
146. Madey, T. E., Yates, J. T. Jr. 1978. *Surf. Sci.* 76: 397
147. Chesters, M. A., Parker, S. F., Raval, R. 1986. *J. Electron Spectrosc. Relat. Phenom.* 39: 155
148. Chesters, M. A., Parker, S. F., Raval, R. 1986. *Surf. Sci.* 165: 179
149. Hoffmann, F. M., Upton, T. H. 1984. *J. Phys. Chem.* 88: 6209
150. Avery, N. R. 1985. *Surf. Sci.* 163: 357
151. Waddill, G. D., Kesmodel, L. L. 1986. *Chem. Phys. Lett.* 128: 208
152. Hoffmann, F. M., O'Brien, E. V., Hrbek, J., Paola, R. A. 1983. *J. Electron Spectrosc. Relat. Phenom.* 29: 301
153a. Hoffmann, F. M., Felter, T. E., Weinberg, W. H. 1982. *J. Chem. Phys.* 76: 3799
153b. Felter, T. E., Hoffmann, F. M., Thiel, P. A., Weinberg, W. H. 1983. *Surf. Sci.* 130: 163
154. Morrow, B. A., Sheppard, N. 1969. *Proc. R. Soc. London Ser. A* 311: 391
155. Prentice, J. D. 1977. PhD thesis. Univ. of East Anglia
156. Sheppard, N., Chenery, D. H., Lesiunas, A., Prentice, J. D., Pearce, H. A. 1976. *Proc. 12th Eur. Congr. on Mol. Spectrosc. Strasbourg*, ed. M. Grosmann, S. G. Elkomoss, J. Ringeissen, p. 345
157. Prentice, J. D., Lesiunas, A., Sheppard, N. 1976. *J. Chem. Soc. Chem. Commun.* 1976: 76
158. Soma, Y. 1976. *J. Chem. Soc. Chem. Commun.* 1976: 1004
159. Soma, Y. 1979. *J. Catal.* 59: 239
160. Pearce, H. A. 1974. PhD thesis. Univ. East Anglia
161. Sheppard, N., James, D. I., Lesiunas, A., Prentice, J. D. 1984. *Commun. Dept. Chem., Bulgarian Acad. Sci.* 17: 95
162. Beebe, T. P. Jr., Albert, M. R., Yates, J. T. Jr. 1985. *J. Catal.* 96: 1
163. de la Cruz, C., Sheppard, N. 1987. *J. Chem. Soc. Chem. Commun.* 1987: 1854
164a. Golike, R. C., Mills, I. M., Person, W. B., Crawford, B. Jr. 1956. *J. Chem. Phys.* 25: 1266
164b. Francis, S. A. 1950. *J. Chem. Phys.* 18: 861
165. Beebe, T. P. Jr., Yates, J. T. Jr. 1986. *Surf. Sci.* 173: L606
166. Beebe, T. P. Jr., Yates, J. T. Jr. 1987. *J. Phys. Chem.* 91: 254
167. Sheppard, N., Avery, N. R., Morrow, B. A., Young, R. P. 1970. *Chemisorption and Catalysis*, ed. P. Hepple, p. 135. London: Inst. Petroleum
168. Beebe, T. P. Jr., Yates, J. T. Jr. 1986. *J. Am. Chem. Soc.* 108: 663
169. Zaera, F., Somorjai, G. A. 1984. *J. Am. Chem. Soc.* 106: 2288

AUTHOR INDEX

A

Abbott, R. J., 470
Abe, H., 126, 128, 134-36, 140-43
Abe, N., 547, 566
Abon, M., 596, 608, 620, 628
Abraham, F. F., 150
Abrahams, E., 174
Abrahams, E. S., 283
Abram, I. I., 464
Abramson, E., 131
Abul-Haj, N. A., 345, 353
Abu-Salbi, N., 319, 325, 328
Achiba, Y., 65, 128, 135
Ackerman, C. C., 109
Adachi, H., 376-78, 383
Adam, A. G., 202
Adam, G., 156, 157
Adams, J. E., 319
Adams, J. T., 320
Adelman, S. A., 353, 354, 361
Aharonov, Y., 326
Ahlrichs, R., 186, 203, 206
Ahumada, J. J., 333
Akamatsu, R., 62
Akasaka, K., 523
Aker, P. M., 373
Akimoto, H., 370, 384
Albert, M. R., 609, 633, 634
Albrecht, A. C., 537, 545, 548, 551, 552, 575, 576
Alder, A. D., 525
Alder, B. J., 327, 360
Alekseyev, V. A., 78
Ali, D. P., 352, 354
Alla, M., 517
Allen, P. J., 527
Allin, E. J., 72, 88
Allison, J., 193
Almasy, F., 123
Almlöf, J., 185, 186, 206
Altmann, K., 79
Alvarado-Swaisgood, A.E., 193
Alwine, J. C., 292
Amaee, B., 322, 331
Amano, A., 227, 229
Amimoto, S. T., 373
Amirav, A., 127, 417, 419, 496
Amit, D. J., 166, 168
Anastasi, C., 375, 376, 383
Andersen, H. C., 150, 153, 161, 168
Anderson, A. B., 206, 489
Anderson, A. C., 117
Anderson, J. B., 318, 323
Anderson, J. G., 370, 371
Anderson, P. C., 375

Anderson, P. W., 111, 174, 283
Anderson, S., 598, 602
Andersson, S., 417
Ando, I., 524
Ando, S., 524
Andres, K., 109
Andresen, P., 48-50
Andrews, E. R., 517
Andrews, J., 607, 608
Angell, C. A., 149, 150, 154-57, 165
Anglister, J., 512
Anner, O., 141
Anselment, M., 138
Anson, C. E., 599, 612, 616, 626, 633
Antoniewicz, P. R., 409-11
Applebury, M. L., 321
Aquilanti, V., 325
Archer, B. J., 327, 329
Arepalli, S., 321, 331
Arguello, G., 381
Arimitsu, T., 470
Arison, B. H., 233
Arita, K., 134
Armentrout, P. B., 189, 193, 194
Armitage, I. M., 528
Arnold, W., 111
Arseneau, D. J., 327
Arvanitis, D., 591, 609, 618
ASHER, S. A., 537-88; 538, 543, 545, 551-53, 566, 574, 579-82
Ashfold, M. N. R., 40, 63, 64
Atabek, O., 45
Atkinson, G. H., 386, 388
Atkinson, R., 386
Audibert, M. M., 74
Aue, W. P., 518, 524
Auerbach, D. J., 55, 395, 417, 421, 485, 490, 494, 496-98, 500-2
Avery, N. R., 609, 613, 619, 630, 634, 636
Avouris, P., 395, 409, 412-14, 417, 612, 613, 615, 620, 622, 627, 639
Azuma, Y., 202

B

Baba, M., 141
Babamov, V. K., 320, 323, 325
Babernich, L., 491
Baberschke, K., 591, 609, 618

Bachovchin, W. W., 524, 529, 530
Bacis, R., 344
Back, M. H., 230
Backx, C., 421, 599, 604
Bado, P., 346
Baek, M., 575
Baer, M., 319, 325-28
Baer, Y., 406
Baerends, E. J., 206
Baggott, J. E., 40
Baghal-Voyjooee, M. H., 372
Bagus, P. S., 182, 184
Bahta, A., 378
Bajdor, K., 543, 575
Balalev, V. E., 328
Balasubramanian, K., 195
Baldwin, A. C., 375
Baldwin, R. L., 307
Balfour, W. J., 189, 194
Balint-Kurti, G. G., 41, 44, 49, 50
Balk, M. W., 353, 354, 361
Ball, R. C., 240, 252, 253, 255, 257, 265
Balooch, M., 480
Bambara, R., 299
Bamford, D. J., 55-57
Band, Y. B., 46-48, 55
Bandy, B. J., 598, 599, 602, 604, 611, 612, 615, 633, 636
Bang, J., 395
Bangcharoenpaurpong, O., 573
Bare, S. R., 599, 620, 622, 623
Barker, A. S. Jr., 469
Barker, J. A., 395, 417, 496, 502
Barker, J. R., 384
Barlow, A. J., 276
Barnett, R. N., 327, 412
Baro, A. M., 599, 605, 606, 613
Barrow, R. F., 199, 345
Barry, J. A., 202
Barteau, M. A., 599, 600, 602, 605, 606
Bartlett, P. D., 228
Barz, G. L. Jr., 252
Basco, N., 373, 374, 376-78, 383
Baseman, R. J., 376
Basilevsky, M. V., 224
Baskin, J. S., 134, 137
Bass, A. M., 372, 374
Batchelder, L. S., 531
Bates, R., 549, 570

645

AUTHOR INDEX

Batt, L., 384, 386
Baughcum, S. L., 374, 384, 385
Baulch, D. L., 330, 333, 370, 377, 381, 382, 390
Baumann, F. C., 102
BAUSCHLICHER, C. W. JR., 181-212; 182-86, 188, 189, 191, 192, 194, 196, 200-7, 209, 325
Bax, A., 511
Bayes, K. D., 370, 374, 375, 386, 388
Baykara, N. A., 204
Bearder, S. S., 347
Beasley, G. H., 216, 218, 220
Beauchamp, J. L., 189, 193, 194
Beck, D., 52
Becker, C. A., 494
Becker, K. H., 64
Beckerle, J. D., 480, 481, 485, 486, 488, 491, 492, 494, 495, 501, 502, 506
Beebe, T. P. Jr., 491, 492, 633, 634, 640
Beeken, P. B., 347
Bell, R. P., 320, 323
Bellamy, L. J., 591, 618, 632
Bellus, D., 222
Bendedouch, D., 442, 444, 445, 447
Bender, C. F., 389
Bendler, J. T., 269, 281, 282
Bengtzelius, U., 162, 164, 165, 167
Benndorf, C., 609
Benner, C. W., 217
Benoist d'Azy, O., 55
Ben-Shaul, A., 428, 437, 448
BENSON, S. W., 1-37; 321, 372, 382, 384
Bent, B. E., 596, 605, 606, 608-10, 618, 619
Benton, W. D., 292
Benzon, M. S., 217, 218
Berendsen, H. J. C., 439, 440
Berens, P. H., 345, 346
Berent, S. L., 292, 310
Berg, M., 347, 348, 356, 357, 477
Berger, A., 518
Berger, J. M., 269
Bergmann, K., 55, 432
Bergsma, J. P., 345, 346
Berkovitch-Yellin, Z., 139, 141
Berman, M., 413
Berman, R., 97-99
Bernal, J. D., 149
Bernardi, F., 224, 227, 233, 234
Bernasek, S. L., 496
Bernstein, E. R., 133, 139

Bernstein, R. B., 318, 320, 395, 398
Berquist, B. M., 372
Berry, R. S., 136
Bersohn, R., 40-42, 44, 46, 55, 58, 127, 409
Berson, J. A., 138, 221, 521
Bertel, E., 626
Bertolini, J. C., 596, 602, 603, 607, 608, 613, 614, 618, 620, 622, 623, 626, 628
Bertran, J., 224, 226
Beswick, J. A., 45-47, 55
Bethune, D. S., 497, 498, 501
Beushausen, V., 49
Beysens, D., 88
Bhattacharjee, S. M., 175
Biermasz, T., 98
Bilhorn, R. B., 544
Billy, J., 596, 608, 620, 628
Biloen, P., 421, 604
Binder, K., 84
Birch, F., 112
Bischel, W. K., 72, 73, 558, 566
Bishop, D. J., 164
Bittman, J. S., 334
Black, G., 558, 566
Blackman, G. S., 625, 626
Blais, N. C., 325, 327, 328
Blake, D. R., 367
Blazej, D. C., 575
Blocker, H., 307
Blomberg, M. R. A., 182, 195, 207-9, 327
Bloom, M., 517
Bloomfield, V. A., 303, 309
Blumen, A., 281, 284
Bodenhausen, G., 511, 521, 523
Bohm, D., 326
Bollen, A. P., 292, 310
Bonamy, J., 75, 77, 78
Bonamy, L., 78
Bond, G. C., 604
Bondi, D. K., 24, 324, 327, 331, 332
Bondybey, V. E., 347
Bonneau, R., 345
Boom, J. P., 159, 162
Booth, D., 341, 351
Bootsma, G. A., 491
Borcic, S., 216
BORDEN, W. T., 213-36; 216, 217, 219, 220, 222
Bordewijk, P., 276, 281
Borer, P. N., 307
Born, M., 95
Born, R., 434
Börnsen, K. D., 139
Bosanac, S., 52
Botet, R., 237, 238, 240, 252, 258, 263, 265

Böttger, H., 469
Bottoni, A., 224, 227
Boulanger, Y., 528
Bowers, M., 370
Bowman, J. M., 45, 51, 52, 55, 318-20, 324, 326, 328, 329, 331, 333, 334
Boyle, J. W., 374, 375, 378, 383
Bradbury, D., 517
Bradshaw, A. M., 592, 612, 621, 626, 635
Brady, R. M., 252, 253, 255, 257
Brafman, O., 575, 576
Bragg, J. K., 428
Bragg, P. W., 292, 310
Bragg, S. L., 72
Brahms, J., 575
Brako, R., 417
Brand, J. C. D., 345
Brandemark, U. B., 182
Brandes, R., 517
Brass, S. G., 486
Braun, W., 372
Brawer, S., 149, 150, 157, 159
Brawer, S. A., 170-72, 177
Brayman, H. C., 199, 200
Breen, P. J., 133
Breitenbach, L. P., 381, 382, 386, 389
Brenig, W., 409, 411
Breslauer, K. J., 307
Breslow, R., 445
Britt, D., 575
Britten, R. J., 292, 309
Brock, J. C. F., 99
Brode, S., 203, 206
Broida, H. P., 348
Brooks, C. L., 353, 354, 361
Broughton, J. Q., 599, 600, 602, 605, 606
Brout, R., 101
Brower, G. D., 118
Brown, A., 217
Brown, C. E., 525
Brown, C. M., 203
Brown, F. B., 325, 327
Brown, F. K., 224, 227, 228
BROWN, J. K., 341-66; 347, 355, 357, 360, 361
Brown, R. C., 137, 326
Brown, S. B., 230
Brown, W. D., 265
Broyer, M., 345
Brubaker, M. E., 598, 606, 633
Brucat, P. J., 183
Brueck, S. R. J., 80, 81, 85
Brühlmann, U., 55
Brumbaugh, D. V., 138-40
Brumer, P., 409
Bryant, J. L., 434
Bryant, R. G., 528, 529

AUTHOR INDEX 647

Buck, U., 51, 55-57, 61
Budd, D. L., 573
Buelow, S., 40, 324
Buenker, R. J., 56, 62
Buff, F. P., 346, 351, 355
Buldt, G., 440, 441
Bunker, D. L., 319, 346, 351, 354, 355
Buntin, S. A., 329, 330
Burak, I., 55
Burger, V. T., 525
Burke, F. P., 473
Burke, L. A., 224
Burns, A. R., 409, 411, 412
Burns, M. J., 464
Burrows, J. P., 380, 386
Burshtein, A. I., 78
Burum, D. P., 516
Buss, R. J., 373, 376
Busto-Latorre, P., 454
Butler, J. E., 384
Butz, K. W., 140, 141

C

Cabane, B., 442, 445-47
Cable, J. R., 575, 576
Cafiso, D. S., 443
CAHILL, D. G., 93-121; 111, 116, 118, 119
Cain, J. E., 453, 454
Calaway, W. F., 80, 85
Callear, A. B., 374
Callis, J. B., 455
Callis, P. R., 551
Calvert, J. G., 372, 374, 375, 378, 382-84, 386
Campbell, C. T., 494, 496
Campbell, I. M., 381
Cannell, D. S., 237
Cantor, C. R., 296, 303
Cantor, R. S., 440-42, 444, 445, 447
Cantrell, C. A., 372
Capelle, G. A., 348
Caracciolo, G., 334
Caramella, P., 226
Caravatti, P., 523
Cardillo, M. J., 395, 417, 480
Cardinal, J. R., 457
Carey, P. R., 570
Carrington, T., 55, 62, 319
Carstensen, J. T., 258
Casavecchia, P., 373
Casimir, H. B. G., 98
Cassuto, A., 612, 613, 616, 636
Castella, M., 141
Castleman, A. W. Jr., 128, 135, 142
Castner, D. G., 605, 606, 616
Caswell, D. S., 543, 570, 574
Cates, M. E., 238

Catlett, D. L., 140, 141
Cavanagh, R. R., 594
Cederbaum, L. S., 413
Celotta, R. J., 599
Ceperley, D. M., 327
CEYER, S. T., 479-510; 395, 417, 480, 481, 485, 486, 488, 491, 492, 494-96, 501, 502, 506
Chachaty, C., 442
Chadwick, R. R., 570
Chakravarty, S., 466, 468
Chamberlin, R. V., 269
Champion, P. M., 551, 573
Chan, C. K., 575, 576
Chandler, D. W., 55
Chandrasekhar, S., 351
Channin, D. J., 109
Charvolin, J., 443
Chatterjee, P. K., 296
Chaturvedi, B. K., 328
Chawla, G. K., 333
Chen, K. T., 140
Chen, P., 138
Chen, S.-H., 442, 444, 445, 447
Chen, Z., 244
Chenery, D. H., 632
Chesnoy, J., 84
Chesters, M. A., 589, 598, 599, 602-4, 606, 611, 612, 615, 616, 619, 620, 628, 629, 633, 636, 638
Cheung, L. M., 187
Cheung, W.-Y., 55
Chevaleyre, J., 345
Chewster, L. A., 136
Chiang, C.-M., 417
Chiba, N., 142
Child, M. S., 48
Chinsky, L., 537, 575
Chlenov, I. E., 224
Choi, B. H., 326, 328
Choi, S., 573
Chong, D. P., 183, 186, 189, 194
Chow, K. S., 118
Chowdhury, D., 149
Chraplyvy, A. R., 475, 476
Christoffel, K. M., 334
Chronister, E. L., 475, 476
Chu, S.-I., 327, 334
Chuang, T. J., 346, 348, 351
Chung, Y. C., 553, 583-85
Chupka, W. A., 138
Churassy, S., 344
Ciccotti, G., 353
Cicerone, R. J., 368, 372
Cimino, G. D., 292, 295-97, 301, 302, 304
Cini, M., 409
Citrin, P., 406
Clark, H., 112

Clark, J. H., 386, 388
Clary, D. C., 45, 321, 324, 327
Clements, W. R. L., 80, 85, 86
Cleveland, C. L., 412
Clinton, W. L., 409
CLOUTER, M. J., 69-91; 71, 74, 81-84, 87, 88
Clyne, M. A. A., 372
Coalson, R. D., 398, 548, 553, 554
Cobos, C. J., 374
Cohen, M. H., 284
Cohen, N., 330
Colnago, L. A., 531
Colson, S. D., 138
Colton, M. C., 322, 327, 329
Coltrin, M. E., 320, 323, 334
Colussi, A. J., 372
Combourieu, J., 372
Comes, F. J., 48, 55, 58-60, 64
Compton, R. N., 128
Comsa, G., 397, 422
Connor, J. N. L., 24, 318-20, 322-25, 327-29, 331, 332
Conrad, N. D., 223
Cool, T. A., 199, 200
Cooper, D. E., 128, 131, 135, 136
Copeland, R. A., 538, 574, 575
Copié, V., 527, 529
Cornelius, P. A., 470, 471, 474
Cosse, C., 206
Courtin, J. M. L., 520, 525, 527
Cowan, R. D., 184, 185
Cowin, J. P., 494, 500
Cox, R. A., 375-77, 380-84, 390
Crawford, B. Jr., 634
Craycraft, M. J., 183
Cross, P. C., 620, 623
Crothers, D. M., 303, 307, 309
Crowell, J. E., 596, 608-10, 618-20
Csizmadia, I. G., 568
Curtin, D. Y., 525
Cvetanovic, R. J., 370, 373
Cwikel, D., 575
Czarniecki, M. F., 445

D

Daccord, G., 252
Dagaut, P., 380
Dahlquist, F. W., 524
Dai, H. L., 131
Dalgarno, A., 334
Dalmai-Imelik, G., 620, 622, 623
Dalterio, R. A., 575
Dammel, R., 346, 351, 355
Damo, C. P., 138

AUTHOR INDEX

Dao, P. D., 128, 135, 142
Das, S. P., 160, 164, 177, 178
David, R., 397, 422
Davidson, E. R., 186, 219, 220, 222
Davidson, J. A., 372, 373
Davidson, N., 309, 341
Davis, D. D., 370-72, 381
Davis, J. H., 517
Davis, M. J., 326, 395, 409, 421
Davis, P. K., 360
Davis, R. W., 292
Dawes, J. M., 341, 352
Deacon, C. G., 83, 84, 88
Debarre, D., 57
de Boer, J., 395
DeBolt, M. A., 149, 157, 159
deBree, P., 470, 474
Debye, P., 95, 96, 101, 115, 269
Decius, J. C., 620, 623
D'Evelyn, M. P., 494, 495, 502
Degen, V., 72, 74, 75
deGennes, P.-G., 427, 428, 450
de Grip, W. J., 527
de Groot, C. P. M., 421, 604
De Groot, H. J. M., 527
Deguchi, K., 524
de Haas, W. J., 98
DeJong, A. F., 523
de la Cruz, C., 598, 633
Delalande, C., 360
Del Bene, J., 568
Delgado-Barrio, G., 45
Delley, B., 204
DeLouise, L. A., 497, 498, 500, 501
DeMeuse, M., 346, 351
Demmer, D. R., 143
DeMore, W. B., 372, 373, 381, 385, 390
Demuth, J., 395, 413, 414, 417
Demuth, J. E., 413, 592, 605, 612, 613, 615, 620, 622, 627-29, 639
Dengler, B., 307
Denton, M. B., 544
DePristo, A. E., 494
De Raedt, H., 164, 167, 171, 173
Derwent, R. G., 375, 384
Desiderio, R. A., 549, 570
Deutch, J. M., 353
Devito, V. L., 541, 570, 574, 579
Dewar, M. J. S., 213, 216-18, 222-24, 226, 227, 230-32, 234
De Yoreo, J. J., 105, 111, 117, 118
De Young, L., 454, 456

Diamond, J. M., 454
Diaz, L. E., 525
DiBartolo, B., 468
Dicker, A. I. M., 474
Diercksen, G. H. F., 56, 57
Diestler, D. J., 464, 470
Dietz, T. G., 128, 135, 137
Diffenderfer, R. N., 187
DiLella, D. P., 206
Dill, J. A., 442, 444, 445, 447
DILL, K. A., 425-61; 426, 428, 430, 435, 437, 438, 440-47, 451, 452, 454-58
DiMarzio, E. A., 157, 178, 427, 428, 432, 435, 437
DiMauro, L. F., 138
Dinardo, N. J., 612, 613, 615, 639
Dion, P., 72, 78
DiVerdi, J. A., 521
Dixon, R. N., 55, 58, 60, 62, 63
Dixon, W. T., 519
Dixon-Lewis, G., 370
Dlott, D. D., 475, 476
Doba, T., 321
Dobis, O., 372
Dobkowski, J., 474
Döbler, U., 591, 609, 618
Dobson, C. M., 524, 529, 531
Docken, K. K., 187
Docker, M. P., 55, 58
Doering, W. von E., 216-18, 220, 228
Dolan, A. K., 427
Doll, J. D., 326
Domcke, W., 413
Domelsmith, L. N., 226
Dominey, R. N., 443
Donn, B., 252, 260, 265
Donnell, J. T., 258
Doren, D. J., 499
Dorfman, L. M., 333
Dorsey, A. T., 466, 468
Dorsey, J. G., 456, 457
Doty, P., 292
Doyen, G., 494, 496
Drake, R. L., 263
Dreier, Th., 328
Drobits, J. C., 65
Drobny, G. P., 524
Drolshagen, G., 395, 398, 417, 421
Drysdale, D. D., 330, 333
Dube, L., 413
Dubois, L. H., 591, 592, 605, 606, 616, 633
Dubost, H., 464
Dubs, M., 55
Ducros, R., 620, 622, 623
Ducuing, J., 74

Dudik, J. M., 545, 552, 553, 566, 571, 575
Dugan, C. H., 55
Dulong, P. L., 93
Duncan, M. A., 128, 135, 137
Dunlap, B. I., 56
Dunn, T. M., 135
Dunne, L. J., 62
Dunning, T. H., 320, 322, 323, 325, 333
Duplessix, R., 446, 447
Dupuy, C. G., 346
Duquesne, M., 575
Dworkin, A., 155, 165
Dyer, M. J., 72, 73
Dymanus, A., 139
Dynes, R. C., 109

E

Eades, R. G., 517
Eady, C. R., 606
Earl, W. L., 516
EBATA, T., 123-47; 128, 130, 131, 133, 135, 136, 386, 388
Eberhardt, W., 406, 609
Ebno, Y., 376
Echargui, M., 78
Echargui, M. A., 86
Eckhardt, G., 545
Eden, M., 237
Edholm, O., 448
Edwards, J. O., 269
Edwards, S. F., 427
Egelhoff, W. F. Jr., 406
Egger, K. W., 230
Ehrhardt, C., 186
Ehrhardt, H., 55
Ehrlich, G., 486
Eigner, J., 292
Einstein, A., 94, 113, 116
Eischens, R. P., 590, 630
Eisele, F. L., 376, 377
Eisenthal, K. B., 346, 348, 351
Ekberg, S. A., 476
Elam, W. T., 252
Elber, R., 419, 496
Elbert, S. T., 187
Elchelberger, T. S., 130
Eley, D. D., 480
Elkind, J. L., 189, 193, 194
Elleman, D. D., 516
Ellena, J. F., 443
Ellet, J. D. Jr., 518
Ellis, D. E., 204
Ellis, P. D., 524, 528, 529
Ellis, R. L., 568
Ellis, T. H., 334
Elson, E. L., 307
Engel, V., 45, 46, 48-52, 54, 55, 58
Epperson, P. M., 544

AUTHOR INDEX 649

Erkelens, J., 592, 625
Erley, W., 599, 613
Ernst, M. H., 263
Ernst, R. R., 511, 522, 523
Ertl, G., 395, 417, 494, 496, 499, 500, 600, 602, 604, 613
Este, G. O., 334
Estrado, H., 413
Eucken, A., 94, 95, 111, 112
Evans, D. F., 345
Evans, G. T., 351
Evans, J., 607, 608
Even, U., 124, 127, 141
Evenson, K. M., 370, 389
Ewing, G. E., 80, 85, 133, 140, 141
Eyre, J. A., 333
Ezra, G. S., 46, 55

F

Fabre, F., 417, 421
Facelli, J. C., 523
Faegri, K., 185
Fairbank, H. A., 97, 98, 106
Faizulov, F., 78
Family, F., 238, 240, 252, 258, 263
Faraci, W. S., 529
Farr-Jones, S., 529
Fayer, M. D., 464, 477
Feder, J., 238, 252
Feibelman, P. J., 409
Feigerle, C. S., 189, 194, 195
Feller, D., 219, 220, 222
Felter, T. E., 629, 630
Femelat, B., 345
Feng, P. Y., 195
Feng, X., 485
Fenn, J. B., 494, 496
Ferguson, M., 296, 301, 302, 304
Fernandez, J. M., 606
Fernández-Alonso, J., 224, 226
Feuerbacher, B., 406, 599
Feulner, P., 409, 411
Field, L. R., 455
Field, M. J., 224, 227
Field, R. W., 131, 189, 191, 201, 344
Figuiere, P., 155, 165
Filatov, V. V., 328
Filipov, N. I., 321, 327
Filipov, V. V., 321, 327
Filseth, S. V., 55, 56
Finegold, L. X., 258
Finlayson, B. J., 375, 377
Fiquet-Fayard, F., 46
Firestone, R. A., 233
Fisanick, G. J., 130
Fischell, D. R., 199, 200
Fischer, D. A., 609

Fischer, E. W., 434
Fischer, G., 367
Fischer, H. E., 97, 98, 118, 119
Fischer, S., 370-72
Fisher, G. B., 492
Fisher, M. P. A., 466, 468
Fisher, W. H., 55
Fitton, B., 406, 599
Fiutak, J., 72, 75
Fixman, M., 351
Flaugh, P. L., 575
Fleer, G. J., 428, 437, 450, 452
Fleming, D. G., 327
Fleming, G. R., 343
Fletcher, C., 347
Fletcher, D. A., 200
Flory, P. J., 426, 428, 432-35, 437, 450, 452
Flouquet, F., 62
Flynn, G. W., 329, 333, 347
Fodor, S. P. A., 539, 575
Foltz, M. F., 55, 56
Fontijn, A., 330
Force, A. P., 373
Forch, B. E., 140
Ford, G. P., 217, 218, 222, 230
Forrest, S. R., 237
Forrest, T. M., 531
Foster, E. L., 97, 98
Foster, S. C., 138
Fouassier, M., 206
Fournier d'Albe, E. E., 242
Fournier de Violet, P., 345
Fox, J. R., 150, 153, 161
Francis, S. A., 590, 634
Franck, J., 341
Franck-Neumann, M., 228
Frank, F. C., 149
Frank, R., 307
Fredkin, D. R., 345
FREDRICKSON, G. H., 149-80; 168, 170-75, 177, 427
Freed, K. F., 45-48, 55, 409
Freeman, A., 204
Frennet, A., 491
Frey, H. M., 228
Frey, M. H., 520, 523, 524
Frey, T., 512
Friedlander, J., 527
Friedlander, S. K., 263
Friedman, J., 124
Fritz, B., 376, 377, 384
Frueholz, R. P., 128, 135, 136
Fry, H. A., 476
Frydman, B., 525
Frydman, L., 525
Frye, J. M., 138
Fuchs, A., 155, 165
Fueki, K., 327

Fujii, M., 126-28, 130, 131, 134-36
Fujito, T., 524
Fujiwara, I., 370
Fuke, K., 128, 135, 138, 142
Fukunaga, T., 226
Funabashi, K., 281
Fung, K. H., 141
Furue, H., 331
Fyfe, C. A., 455

G

GADZUK, J. W., 395-424; 395, 403, 406, 411-13, 415, 417-19, 421, 470, 495
Gaidai, N. A., 491
Gaines, G. L., 450
Gajewski, J. J., 216, 217, 223, 230
Gale, G. M., 360
Galica, G. E., 548, 553, 554
Gamper, H. B., 292, 295-97, 301, 302, 304
Ganapathy, S., 517, 523
Ganesh, K. N., 512
Gang, X., 473
Gangoda, M. E., 455
Garbow, J. R., 531
Garcia, L. A., 233
Garfunkel, E. L., 485
Garg, A., 466, 468
Garner, D. M., 327
Garrabos, Y., 88
Garrett, B. C., 317, 319, 320, 322-28, 331, 333-35
Gates, J. A., 602, 603, 605, 613, 616
Gaylord, R., 428, 433, 434
Gdowski, G. E., 494
Geiger, L. C., 333, 334
Geisel, T., 289
Gelbart, W. M., 46, 47, 55, 65, 428, 437, 448
Gentry, W. R., 329, 330, 404
George, T. F., 326, 352
Gerber, A., 413
Gerber, R. B., 395, 417, 419, 496
Gericke, K.-H., 48, 55, 58-60, 64
Gerrity, D. P., 563, 570
Gershenzon, Yu. M., 328
Gerstein, B. C., 516
Gervasio, G., 617
Gettins, P., 529
Geusic, M. E., 204
Ghormley, J. A., 374, 375, 378, 383
Gibbs, J. H., 156, 157, 178
Gibby, M. G., 516, 518
Gibson, L. L., 326
Gierasch, L. M., 524, 525

Giese, C. F., 329, 330
Gijzeman, P. L. J., 491
Gilbert, R. G., 352, 374
Gill, R. J., 386, 388
Gilman, N. W., 217
Gilpin, R. K., 455
Gingerich, K. A., 203
Ginter, M. L., 203
Gislason, E. A., 395
Gladstone, D. J., 480, 481
Gland, J. L., 609
Glarum, S. H., 276, 343
Glass, G. P., 328
Glauber, R. J., 169
Glogover, S. G., 548, 553, 554
Glotin, M., 432
Gochev, A. D., 321
Goddard, J. D., 56, 57, 233, 234
Goddard, W. A., 193, 204, 205, 389
Golden, D. M., 375, 381, 382, 384, 385, 390
Golden, D. W., 216
Goldfield, E. M., 46, 55
Goldstein, M., 149, 154, 156
Goldstein, M. J., 217, 218
Golike, R. C., 634
Gölzenleuchter, H., 64
Gomer, R., 321, 409
Gomez-Sal, M. P., 626
Gonohe, N., 128, 134-36, 141, 143
Goodgame, M. M., 204, 205
Goodman, D. W., 395, 491, 492
Gordon, A. S., 227, 229
Gordon, E. B., 328
Gordon, M. D., 226
Gordon, R. J., 321, 330, 331, 371
Gordon, S., 370, 372
Gortel, Z. W., 409, 411
Goto, A., 126, 128, 130, 131, 135, 136
Gotoh, K., 327
Götting, R., 329
Gottwald, E., 55
Götze, W., 162, 164, 165, 167
Goubau, W. M., 109
Goursaud, S., 46
Gouterman, M., 572
Graedel, T. E., 367
Gralla, J., 307
Grant, D. M., 516, 523, 525
Granville, M. F., 127
Grasiuk, A., 78
Graves, B., 295
Gray, C. G., 69, 72, 74, 75, 79
Gray, D. L., 488
Gray, J. A., 189, 191
Gray, S. K., 48
Green, S., 334

Greenburg, M. M., 521
Greenhill, P. G., 381
Greenler, R. G., 593, 631
Greenspan, C. M., 292
Greiner, N. R., 370
Grest, G. S., 284
Greuter, F., 609
Grev, R. S., 319
Grier, D., 252
Griffey, R. H., 511
Griffin, A. C., 224, 226
Griffin, B. G., 527
Griffin, D. C., 184, 185
GRIFFIN, R. G., 511-35; 512, 516-25, 527-29
Grinberg, H., 45
Grogan, M. J., 602
Groth, W., 64
Grotheer, H. H., 381
Gruen, D. W. R., 438, 445
Grump-Perkins, C. A., 386
Grunewald, A. U., 48, 55, 58-60
Grunstein, M., 292
Gubernatis, J. E., 335
Gubser, D. V., 252
Guczi, L., 491
Guest, M. F., 224, 227
Guillion, T., 522
Gulotty, R. G., 373
Gunnarsson, O., 485
Gunton, J. D., 84
Guo, H., 62
Gupta, P. K., 149, 157, 159
Gustafsson, T., 609
Gutman, D., 333, 334, 370, 377, 382, 384
Gutowsky, H. S., 529
Gutschick, V. P., 405
Guttman, C.-M., 427, 432, 437
Guyer, R. A., 106, 109

H

Haas, Y., 141
Haberkorn, R. A., 523
Haeberlen, U., 512, 516
Hagedorn, W. J., 432
Hagele, J., 381
Hager, J. W., 128, 130, 135, 136, 143
Hahn, K., 434
Haile, J. M., 442-44
Halberstadt, N., 140
Hale, M. O., 548, 553, 554
Hall, G. E., 55
Hall, R. B., 599, 608, 616, 640
Halle, L. F., 189, 193, 194
Haller, G. L., 494, 496
Halperin, B. I., 111, 160, 165, 168, 464, 470
Halpern, J. B., 55
Halsey, T. C., 257

Halstead, D., 499
Hamann, D., 397, 422
Hamill, W. H., 281
Hamilton, J. C., 599
Hammer, L., 599, 608, 613, 614, 618
Hammond, L., 454
Hampson, R. F., 377, 381, 382, 385, 390
Hamza, A. V., 485, 494, 497, 498, 501
Hanazaki, I., 141
Hancock, G., 40, 369
Handy, B. J., 381
Handy, N. C., 184, 319
Hansen, H.-J., 222
Hansen, J. P., 159, 162, 167
Hansen, S. G., 204
Hanson, E. A., 347
Hanson, R. C., 90
Hanst, P. L., 375
Haochang, P., 417, 421
Hara, K., 134
Harbison, G. S., 516, 520-23, 525
Harding, L. B., 333, 334
Haribson, G. S., 527
Harmon, P. A., 541, 543, 553, 558, 561, 562, 564, 565, 570, 574, 579, 581, 582
HARRIS, A. L., 341-66; 347, 348, 356, 357
HARRIS, C. B., 341-66; 347, 348, 354-57, 360, 361, 470, 471, 474
Harris, J., 417
Harris, S. J., 139
Harrison, J. F., 193, 196
Harrison, J. P., 104
Hartzell, C. J., 524
Hase, W. L., 317, 319
Hasegawa, T., 134
Hasselman, D., 228
Hassing, S., 545, 546
Hatano, M., 526
Hatzikos, G. H., 599, 600, 604, 613
Haug, K., 327, 329, 331
Häusler, D., 49, 50
Hawkins, C. M., 217
Hawkins, R. T., 132
Hawkins, W. G., 50
Hay, P. J., 46, 55, 56, 182, 184, 185, 191
Hayakawa, Y., 252
Hayashi, H., 526
Hayashi, M., 292
Hayden, C. C., 323, 325, 331, 334
Hayes, E. F., 325, 328
Haynam, C. A., 138-40
Hays, T. R., 575
Hayshi, M. N., 292

AUTHOR INDEX 651

He, G., 376
Healy, E. F., 223, 224
HEARST, J. E., 291-315; 292, 295-98, 301, 302, 304-6, 311, 312
Heather, R., 44, 335, 395, 398
Heaven, M., 138
Hecht, K. T., 59
Hefter, U., 55
Hehre, W. J., 220, 224, 225
Heiberg, A., 186
Heicklen, J., 373, 375, 377, 378, 383, 389
Heidner, R. F., 373
Helfand, E., 175, 427, 428, 430, 437, 450
Heller, E. J., 44, 46, 48, 54-56, 137, 345, 395, 399, 400, 402, 405, 409, 421, 548, 553
Hemley, R. J., 127
Hemminger, J. C., 139
Heneghan, S. P., 372
Hennig, S., 46, 54, 58
Henrichs, P. M., 523
Henriksen, N. E., 48
Henshaw, J. P., 45
Hepburn, J. W., 55, 56
Herrero, V., 329
Herrmann, H. J., 238
Hertlein, T., 599, 608, 613, 614, 618
Herzberg, G., 75, 79, 197, 200, 202, 374, 488, 603, 620, 624, 625
Herzenberg, A., 413
Herzfeld, J., 518, 520-25, 527
Herzfeld, K. F., 360
Hess, L. A., 475, 476
Hess, W., 549, 570
Hesselink, W. H., 464, 477
Hewitt, J. M., 523
Hexem, J. G., 520
Hiberty, P. C., 233
Higgs, T. P., 517
Hikida, T., 333
Hildebrand, P. G., 574
Hill, M. M., 607
Hill, R. K., 217
Hiller, C., 320, 323
Hillier, I. H., 224, 227
Hills, B. P., 82, 83
Hills, M. M., 600, 605, 606, 608, 616
Hines, M. A., 480, 485, 501
Hinze, J., 187
Hipes, P. G., 327, 329
Hippler, H., 341, 351
Hippler, H. H., 374
Hirakawa, A. Y., 537, 575
Hiraoka, S., 138
Hiroi, M., 132
Hiroike, E., 470

Hirota, E., 376
Hirschfelder, J. O., 324
Hitchcock, A. P., 627
Hiyama, Y., 531
Hizhnyakov, V., 464
Hjelmberg, H., 485
Ho, J., 203, 207
Ho, W., 594, 599
Hochanadel, C. J., 370, 374, 375, 378, 383
Hodge, I. M., 173
Hodgson, A., 55, 58, 499
Hoehener, A., 522
Hoeve, C. A. J., 427
Hoffman, G. W., 346, 348, 351
Hoffman, J. D., 427, 432, 437
Hoffman, R., 485
Hoffmann, F. M., 612, 629, 630, 635, 638
Hoffmann, H. M. R., 230
Hoffmann, R., 213, 217
Hofmann, P., 620, 622, 623, 626
Hogan, E. M., 106-8
Hogness, D. S., 292
Hohenberg, P. C., 160, 165, 168
Holloway, S., 403, 406, 417, 419, 485, 495, 499
Holmes, M. W., 598, 602
Holt, P. M., 375
Holtzclaw, K. W., 126
Holzer, W., 79
Homma, K., 139
Hong, K. M., 427
Honjo, H., 252
Honkonen, R. S., 524
Hopkins, J. B., 139
Hopkins, T. L., 531
Horii, F., 432
Horn, F. H., 118, 119
Horn, K., 406, 592, 626, 628
Horn, T. C. M., 417, 421
Hornburger, H., 137
Horne, D. G., 330, 333, 370
Horner, E. C. A., 389
Hornig, D. F., 623
Horowitz, A., 386
Horsley, J. A., 62, 609, 627
HOUK, K. N., 213-36; 224, 226-30, 233, 234
Houston, J. P., 55
Houston, P. L., 46, 50, 55, 333
Hovarth, C. S., 455
Howard, C. J., 369, 370, 373, 381, 385, 389, 390
Howard, J. A., 206
Howard, L. N., 434
Howard, M. W., 590, 592, 605, 633
Hoy, A. R., 345
Hrbek, J., 630
Hsu, D., 464, 468, 476

Hsu, Y. C., 138
Huang, G., 202
Huang, T.-H., 521, 529
Hubbell, W. L., 438
Huber, J. R., 55
Huber, K. P., 197, 200, 202
Huber, L. M., 516
Hucknall, D. J., 333
Hudgins, R. R., 508
Hudson, B. S., 538, 539, 541, 549, 554, 570, 575
Hughes, B. D., 274
Hulbert, H. M., 324
Hull, W. E., 524
Humski, K., 216
Hunklinger, S., 111
Hunt, R. H., 59
Hunziker, H. E., 376
Hurd, A. J., 252
Hurst, J. E., 500
Hurst, W. S., 71, 72, 74, 75
Husain, D., 370, 373
Husigen, R., 233
Hutley, M. C., 552
Huzinaga, S., 185
Hynes, J. T., 317, 319, 343, 346, 352, 355, 356, 360, 361

I

Ibach, H., 589, 592-94, 598-600, 602, 604, 607, 612, 613, 616, 617, 620, 625, 628, 629, 631, 633, 636
Ichijo, M., 140
Imajo, T., 128
Imre, D., 131, 409, 548, 553, 554
Inglefield, P. T., 282
Ingold, K. U., 321
Innes, K. K., 131
Inoue, G., 370, 384
Iosilevskii, Ya. A., 101
Ippen, E. P., 321
Ireton, R., 227, 229
Isaacs, S. T., 292, 296, 298, 301, 302, 304
Isaacson, A. D., 317, 319
Ishiwata, T., 370
Iskovskikh, A. S., 327
ITO, M., 123-47; 123, 126-28, 130-32, 134-36, 139, 140, 143
Itoh, M., 141
Ivanco, M., 128, 135, 136
Ivanov, A. V., 328
Ivanov, B. I., 328
Iwashita, Y., 616
Iyre, J. A., 333

AUTHOR INDEX

J

Jäckle, J., 149
Jackson, B., 395, 398, 417, 477
Jackson, H. E., 103, 109
Jackson, W. M., 40, 41, 55
Jacob, G. S., 531
Jacobi, K., 592
Jacobs, A., 55, 58, 60
Jacobsen, B. S., 346, 351, 354, 355
Jaffe, H. H., 568
Jain, R. K., 74, 82, 84
Jakob, P., 612, 613, 616, 636
Jakubetz, W., 319, 320, 325, 328, 329, 331
James, D. G. L., 374, 378, 383
James, D. I., 598, 599, 602, 604, 612, 633
James, F. C., 374
James, N., 523
Jameson, A. K., 140
Janda, K., 139
Jang, D. J., 345, 353
Jansen op de Haar, B. M. D. D., 327
Jay, E., 299
Jayanty, R., 373
Jeans, J., 260, 262
Jeffrey, K. P., 517
Jelinski, L. W., 523, 531
Jellinek, J., 318, 325, 328
Jennings, D. A., 373
Jennison, D. R., 409, 411, 412
Jensen, F., 229, 234
Jeong, K. M., 370-72
Jie, C., 218-23
Jobic, H., 620, 626
Joffrin, C., 74
Johnson, A. D., 485, 486, 488, 491, 492, 494, 501, 502, 506
Johnson, A. L., 609, 618
Johnson, B. F. G., 606, 626
Johnson, B. R., 331
Johnson, C. R., 538-42, 545, 552, 553, 566, 570, 571, 574, 575, 579, 580
Johnson, G. E., 173
Johnson, P. M., 137
Johnson, W. D., 386, 388
Johnson, W. L., 151
Jones, A. A., 282
Jones, C. M., 541, 551, 574, 575, 577, 579
Jones, I. S., 427
Jones, K. E., 464, 470, 471
Jones, L. H., 476
Jones, P. L., 55
Jonsson, B., 448
Jortner, J., 124, 127, 139, 141, 395, 417
Jøssang, T., 252
Joussot-Dubien, J., 345
Ju, G.-Z., 328
Jug, K., 224
Julien, L. M., 345
Jullien, R., 237-40, 252, 258, 263, 265
Jung, K., 55
Just, Th., 381
Jutila, R. E., 409

K

Kable, S. H., 132
Kagan, Yu., 101
Kagel, J. R., 230
Kajimoto, O., 134, 139, 140
Kakinuma, T., 128, 130, 131, 134-36
Kaliteevskii, M. Yu., 473
Kalkar, A. K., 345
Kallenbach, N. R., 303
Kamei, S., 126, 134
Kaminsky, M. E., 132
Kan, C. S., 378, 382, 383
Kang, K., 283
Kanne, D., 295
Kant, A., 183, 204
Kantor, Y., 544
Kapral, R., 351, 352
Kara, A., 494
Kardar, M., 240
Karlström, G., 182
Karner, H., 494, 497, 498, 501
Karplus, M., 319, 398
Kasemo, B., 395, 417
Kaski, K., 84
Katayama, D. H., 131
Kato, K., 541
Kato, S., 219, 220, 222
Katunin, A. Ya., 321, 327
Katz, J. J., 525
Katz, V., 454
Kaufman, F., 369-72, 376
Kauzmann, W., 157
Kawai, R., 409, 412
Kawasaki, K., 160, 162, 163, 165, 169
Kawashima, Y., 258
Kay, B. D., 491, 492
Kaya, K., 128, 131, 134, 135, 138, 142
Kaye, J. A., 325
Kaye, R. L., 228
Kearns, D. R., 517
Kearsey, S. V., 384
Keiffer, M., 374, 375
Keiller, B., 599, 633
Keiller, B. T., 612, 616
Keller, P., 443
Kelley, D. F., 345, 348, 353
Kelley, J. D., 72
Kelley, R. D., 594
Kelly, P. B., 549, 554, 563, 570
Kelusky, E. C., 455
Kemp, D. J., 292
Kendrick, R., 521
Keniry, M. A., 529
Kennedy, R. A., 138
Kenny, J. E., 140
Kentgens, A. P. M., 523, 527
Kerr, J. A., 375, 377, 381, 382, 390
Kertesz, J., 257
Kesmodel, L. L., 413, 591, 592, 602, 603, 605, 613, 616, 620, 630, 633
Kettle, S. F. A., 590, 592, 605, 607, 608, 633
Keyes, T., 160, 162, 163
Keyser, L. F., 372
Khachatryan, L. A., 377, 382
Khare, V., 328
Khodadoost, B., 575, 576
Khundkar, L. R., 137
Kiefte, H., 71, 74, 81-84, 87, 88
Kiehl, J. T., 368
Kikuchi, O., 224
Kim, H. L., 132
Kim, S.-H., 295
Kimman, J., 417, 421
Kimura, K., 65, 128, 135
King, D. A., 479, 480, 620, 622, 623
King, G. I., 453, 454
King, K. D., 216
Kinkead, S., 90
Kinsey, J. L., 131, 409, 548, 553, 554
Kirkpatrick, T. R., 164, 165
Kirschner, S., 224, 226
Kisliuk, P. J., 479, 501
Kistiakowsky, G. B., 229
Kitamura, R., 432
Kitsopoulos, T., 321, 331, 333
Kittrel, C., 131
Klafter, J., 273, 281, 283, 284, 286, 289
Klais, O., 375
Klapstein, D., 138
Klee, S., 55, 58, 60
Kleiman, G. G., 412
Kleiman, R. N., 164
Klein, J., 427
Klein, M. L., 449
Klein, R., 244
Klemens, P. G., 99
Klemm, R. B., 376
Klemperer, W., 139
Kley, D., 64
Kleyn, A. W., 55, 395, 417, 421
Klimcak, C. M., 128, 131, 135, 136

AUTHOR INDEX 653

Klitsner, T., 97, 98, 100
Klockner, H. W., 545-47
Knaak, W., 105, 111, 117, 118
Knee, J. L., 137
Kneppe, H., 376
Knight, A. E. W., 132
Knight, L. B., 205
Knotek, M. L., 409, 411
Knowles, P. J., 184
Kobayashi, T., 139, 140, 345, 575
Koda, S., 139, 376
Koehler, W. C., 445, 447
Koel, B. E., 596, 605, 606, 608-10, 618-20
Koenig, J. L., 623
Koestner, R. J., 609
Kohida, T., 142
Kohlrausch, F., 269
Kohn, W., 407
Kohne, D. E., 292, 309
Kolb, M., 237, 252, 257, 258, 263, 265
Kollman, P., 318
Kollmar, H. W., 216
Kolmogorov, A. N., 286
Kolodney, E., 419, 496
Komornicki, A., 217, 233, 234
Kono, H., 137
Kopelman, R., 269
Koppel, D. E., 442, 444, 445, 447
Korenowski, G. M., 537
Korotaev, O. N., 473
Korsch, H. J., 52, 55
Koshihara, S., 575
Kosloff, R., 398, 419, 421, 470
Kostov, K. L., 596, 604-6, 608
Koszykowski, M. L., 395
Kouri, D. J., 318, 319, 325, 327-29, 331, 335
Kovacs, A. J., 149, 157
Kowalski, F. V., 132
Kozlov, D. N., 86
Krajnovich, D., 140, 141
Kramer, K. J., 531
Kramers, H. A., 428, 435
Krauss, M., 185, 192, 202, 414
Krautwald, H. J., 63, 64
Kreevoy, M. M., 335
Krenos, J., 409, 548, 553, 554
Kreutter, N. M., 376, 377
Kreuzer, H. J., 409, 411
Kricheldorf, H. R., 524
Krischner, S., 216
Krivoglaz, M. A., 464, 470, 471
Kruger, B., 609
Krüger, H.-W., 224
Krumhansl, J. A., 109
Kubasek, W. L., 575
Kubiak, G. D., 421
Kubo, R., 353

Kucheryavii, S. I., 328
Kuchitsu, K., 139
Kuebler, N. A., 130
Kuehnlenz, G., 568
Kugler, S., 434
Kuhn, R., 138
Kuipers, H., 494
Kulander, K. C., 44, 45
Kumar, G. S., 97, 98
Kundla, E., 517
Kung, C. Y., 138
Kunimori, K., 494
Kuppermann, A., 318, 320, 322, 324-27, 329, 334
Küppers, J., 600, 602, 604, 613
Kuruoglu, Z. C., 325
Kurylo, M. J., 372, 375, 378, 380, 381, 383, 385, 390
Kuster, E. J., 412
Kutz, H., 55

L

Lacher, J. R., 229
Lacher, R. C., 434
Laemmel, H., 123
Lagana, A., 319, 320, 328, 329
Laguna, G. A., 374
Lahmani, F., 55
Laigle, A., 575
Lakshman, S. V. J., 202
La Mar, G. N., 573
Lamb, J., 276
Lampe, F. W., 341
Landau, D. P., 84, 238
Landman, U., 412
Landowne, S., 55
Lane, D., 292
Lang, N. D., 406-9, 412, 419, 485
Langford, A. O., 386, 388, 389
Langhoff, C. A., 346, 351
LANGHOFF, S. R., 181-212; 183, 184, 186, 189, 191, 192, 194, 196, 200, 201, 203, 206, 207
Langmuir, I., 479, 501
Langridge-Smith, P. R. R., 139
Langry, K. C., 573
Lardeux, C., 55
Laskowski, B. C., 206, 207
Laskowski, M., 574
Laufer, A. H., 374, 375
Lauterbur, P., 523
Law, K. S., 139
Lawrance, W. D., 132
Lawson, D. T., 97, 98
Lax, M., 273
Leach, G. W., 143
Leacock, R. A., 59
LeBras, G., 372
Leduff, Y., 79
Lee, D., 545

Lee, J. H., 369, 372, 494, 500
Lee, K.-P., 321, 327
Lee, K.-T., 328, 331, 334
Lee, L. C., 58
Lee, L. Y., 327
Lee, M. B., 480, 481, 485, 486, 491, 492, 494-96, 501, 502
Lee, S. B., 499, 500
Lee, S.-Y., 44, 48, 395, 399, 402, 421, 548, 553
Lee, Y. T., 323, 325, 331, 334, 373, 376
Leermakers, F. A. M., 428, 433, 434, 438, 448
Lefebvre, M., 57
Lefebvre, R., 396
Legay, F., 85
Legay-Sommaire, N., 85
Leggett, A. J., 466, 468
Le Grand, D. G., 269, 282, 450
Lehwald, S., 592, 598-600, 602, 604, 605, 607, 612, 613, 616, 617, 620, 628, 629, 631, 633, 636
Leibler, L., 427
LeMaster, D. M., 511
LeNeindre, B., 88
Lennard-Jones, J. E., 485
Lenormand, R., 252
Leone, S. R., 40, 41, 333
Leopold, D. G., 127, 203, 206, 207, 334
LeRoy, D. J., 321, 326, 327
Leroy, G., 224
LeRoy, R. J., 344
Lesclaux, R., 378, 380, 383, 386, 388
Lesiunas, A., 632, 633
Lester, M. I., 65, 138
Lester, W. A., 327
Leu, M. T., 372
Leung, K. M., 403
Leutheusser, E., 162, 171, 173
Leutwyler, S., 139, 141
Levesque, L., 80
Levi, G., 75, 77, 78
Levine, J. S., 367
Levine, M. D., 307
Levine, R. D., 318, 320, 324, 334, 395, 398
Levine, S., 428
Levit, S., 55
Levitt, M. H., 522, 523
Levy, B., 186
Levy, D. H., 123-25, 138-40
Levy, H., 368, 369, 382
Levy, P., 272
Lewis, A., 321
Lewis, J., 626
Leyvraz, F., 263
Li, C. H., 594, 625

Lian, L.-Y., 529, 531
Liang, R., 575, 576
Liao, M. Z., 195
Lienard, G., 491
Lifshitz, I. M., 99
Light, J. C., 44, 45, 324, 326, 327, 329
Lim, C., 417
Lim, E. C., 140
Lin, C. L., 372, 373
Lin, M. C., 395, 417, 494
Lin, M. Y., 244
Lin, R. F., 625, 626
Lin, S. H., 137
Lin, Y.-T., 224, 227, 228
Linder, M., 522, 523
Linderberg, J., 327
Lindgren, B., 189, 194
Lindsay, H. M., 244
Lineberger, W. C., 189, 194, 195, 203, 206, 207, 334
Lipkus, A. H., 346, 351, 355
Lippmaa, E., 517
Litovitz, T. A., 360
Liu, B., 204, 327
Liu, K., 183
Liu, W. K., 464
Liu, X., 138
Liverman, M. G., 137
Lloyd, D. R., 609
Lluch, J. M., 224
Logan, J. A., 368
Löhmannsröben, H. G., 137
Lomakin, L. A., 328
Lombardo, G., 104
LONCHARICH, R. J., 213-36; 230
Long, D. H., 538
Looi, E. C., 72, 73, 86
Lopez, V., 320, 323, 325
Lorenz, K., 376, 377, 381, 384
Los, J., 395, 419
Lowe, I. J., 517
Lowry, T. H., 350
Lubman, D. M., 135
Lucas, A. A., 403
Ludwig, M., 538, 541, 553, 574, 575, 579, 580
Luger, M., 494, 497, 498, 501
Lugtenburg, J., 520, 525, 527
Lugtenburg, J. A., 525
Lukashevich, I. I., 321, 327
Lülf, H., 49
Lundqvist, B. I., 395, 417, 485
Luntz, A. C., 55, 373, 395, 417, 470, 497, 498, 501
Lushnikov, A. A., 263
Luther, K., 137, 341, 351, 374
Lyklema, J., 438, 448
Lyne, M. P. J., 202

M

MacDonald, P. M., 517
Macedo, P. B., 149, 157, 159
Machado, G., 372
Macho, V., 521
Maciel, G. E., 455, 523
Madden, P. A., 82, 83
Madeleine, D., 252
Madey, T. E., 395, 409, 411, 412, 417, 589, 629
Madigosky, W. M., 360
Madix, R. J., 485, 494, 495, 497, 498, 500-2, 600, 608, 617, 618
Magde, D., 346
Magnuson, A. W., 319
Mahgerefteh, D., 55
Mahmoudi, M., 292, 310
Maier, J. P., 138
Maier, M., 80, 85
Mair, R. D., 623
Majors, T. J., 127
Maker, P. D., 381, 382, 386, 389
Makide, Y., 367
Makoshi, K., 417
Malcom, D. G., 370
Malik, I. J., 598, 606, 633
Malley, M. M., 346
Malojcic, R., 216
Maloney, J. J., 489
Måløy, K. J., 252
Mandelbrot, B. B., 237, 239, 248, 269, 273
Mandelkern, L., 432
Mandell, M. J., 150
Mandich, M. L., 189, 193, 194, 347
Mangle, E. A., 139
Manneback, C., 415
Manning, R. G., 372
Mansfield, M. L., 428, 433
Mantashyan, A. A., 377, 382
Mantell, D. A., 494
Manz, J., 320, 323, 331, 332
Marchetti, P. S., 524, 528, 529
Marcus, R. A., 319, 320, 323, 325, 334, 395
Margitan, J. J., 370, 371, 381, 385, 390
Mariani, C., 406-9
Maricq, M. M., 515, 521
Marinelli, W. J., 55
Marinova, T. S., 596, 604-6, 608, 613
Marks, S., 470, 471, 474
Markus, M. A., 321
Marky, L. A., 307
Marmur, J., 292
Maroncelli, M., 448, 449

MARQUSEE, J. A., 425-61; 428, 432, 437, 451, 457, 458
Marsault, J. P., 75, 78
Marsault-Herail, F., 75, 77, 78, 86
Marshall, P., 330
Marshall, R., 341
Martin, F., 344
Martin, J. E., 237
Martin, P. C., 166
Martin, R. L., 182, 184, 185, 191, 203, 206
Martin, S. W., 155, 165
Martinez, R. I., 386, 388
Masel, R. I., 599, 600, 604, 613
Massa, D. J., 523
Massardier, J., 620, 622, 623, 628
Massmann, H., 55
Mate, C. M., 596, 608-10, 618-20
Mathies, R. A., 395, 402, 421, 520, 525, 527, 551
Mathisen, K. B., 182
Matire, B., 352
Matsumi, Y., 138
Matsuoka, S., 172, 173
Matsushita, M., 252
Mattheus, A., 55
May, A. D., 72-75, 78, 79
May, W., 71
Mayama, S., 138
Mayer, E. W., 367
Mayne, C. L., 525
Mayne, H. R., 328
Mayne, L. C., 570
Mazenko, G. F., 160, 164, 177
Mazur, P., 353
McAdam, K., 378, 380, 383
McAfee, J. R., 373
McBane, G. C., 333
McCash, E. M., 598, 606, 612, 616, 619, 620, 633, 636
McClure, D. S., 473, 474
McConnell, H. M., 438, 512
McCumber, D. E., 464, 470, 471, 473
McCurdy, C. W., 413
McDonald, I. R., 159, 162, 167, 449
McDonald, J. D., 132
McDonald, J. R., 370, 384
McDonald, S. A., 131
McDouall, J. J. W., 233, 234
McDougall, G. S., 598, 599, 602-4, 611, 612, 615, 633, 636
McDowell, C. A., 523, 525
McElroy, M. B., 368
McGonigal, M., 480, 481

AUTHOR INDEX 655

McGrady, E. D., 258
McIver, J. W., 217
McIver, J. W. Jr., 224
McKay, R. A., 519
McKee, C. S., 591
McKee, M. L., 217, 218, 222
McLaren, I. A., 55
McLaughlin, D. F., 381
McLean, A. D., 203, 204
McLeod, D., 199
McMahon, T. J., 524
McMaster, B. N., 204
McMillan, G. R., 374, 384
McMillen, D. F., 382
McNelly, T. F., 97, 98, 103, 104, 109, 110
McNulty, G. S., 607, 608
McNutt, J. F., 325
McQuigg, R. D., 378, 383
McRae, J. E., 367
McTague, J. P., 150
Mead, C. A., 206, 326, 327
Meadows, L. F., 341, 351
MEAKIN, P., 237-67; 237, 240, 244, 252-54, 257, 258, 260, 263, 265
Meerts, W. L., 139
Mehring, M., 512, 513, 523
Meier, B. H., 516
Meier, U., 56, 58, 59, 381
Meissner, M., 105, 111, 117, 118
Mejean, T., 206
Melander, W. R., 455
Mely, B., 443
Menapace, J. A., 139
Menger, E. M., 520
Menzel, D., 409, 411, 599, 600, 602, 605, 606, 612, 613, 616, 636
Merer, A. J., 202
Merrill, R. A., 521
Metiu, H., 44, 335, 395, 398, 403, 418
Metz, R. B., 331, 333
Meyer, H. D., 55-57
Meyer, W., 187
Micha, D., 325
Michael, J. V., 333, 369, 372, 382, 386
Michalopoulos, D. L., 204
Michaud, M., 413
Michaud, P., 373
Michl, J., 523
Mies, F. H., 414
MIKAMI, N., 123-47; 123, 126, 128, 130-32, 134-36, 139-43
Mikula, R. J., 327
Mile, B., 206
Miles, T., 367
Millar, D. P., 341

Miller, D. R., 480
Miller, K. W., 454
Miller, T. A., 138
Miller, T. M., 334
Miller, W. H., 45, 55, 317, 319, 320, 323, 325-27, 335, 352, 354, 395
Mills, D. L., 589, 593, 594, 625
Mills, I. M., 634
Mitchell, D. N., 321
Mitchell, G. E., 492
Miura, H., 516
Miyazaki, T., 321, 327
Mizuno, H., 133, 135, 136
Moerner, W. E., 475, 476
Mohsin, S. B., 598, 606, 633
Molenkamp, L. W., 474
Molina, M. J., 381, 385, 390
Mollevanger, L. C. P. J., 527
Molnar, G., 368
Montague, D. C., 367
Montroll, E. W., 171, 269, 272, 273, 278, 280
Montrose, C. J., 149, 157, 159
Moore, B., 346, 351
Moore, C. B., 55-57, 386, 388, 389
Moore, C. E., 182
Moortgat, G. K., 380, 386
Morawitz, H., 417, 421
Morgan, S., 128, 135, 142
Mori, H., 353
Morin, F. G., 525
Morokuma, K., 219, 220, 222
Morris, E. D. Jr., 386
Morris, G. A., 511
Morrow, B. A., 631, 633, 634, 636
Morse, M. D., 46, 47, 55, 202, 206, 207
Mortenson, O. S., 545, 546
Morter, C., 139
Moskovits, M., 206, 595
Moss, D. B., 133
Mowrey, R. C., 335
Moynihan, C. T., 149, 157, 159
Mozurkewich, G., 269
Muckerman, J. T., 325
Mueller, P. H., 233
Mukamel, S., 83, 396, 464, 470, 545
Mukerjee, P., 457
Mulac, W. A., 370, 372
Mulcahy, M. F. R., 376
Mulder, P. P. J., 525
Müller, D., 524
Muller, J. E., 502
Müller, K., 599, 608, 613, 614, 618

Muller, N., 442, 446
Müller-Dethlefs, K., 136
Mulliken, R. S., 344, 345
Mullins, C. B., 600, 605, 606, 608
Munchausen, L. L., 233
Mündel, C., 413
Munowitz, M. G., 521, 522, 524
Murakami, J., 131, 134
Murray, K. K., 334
Murrell, J. N., 62, 346, 351, 355
Murtaugh, J., 539, 540, 542, 575
Murtaugh, J. L., 566, 568
Myers, A. B., 395, 402, 421, 551

N

Nachman, D. F., 200
Nadler, I., 55
Nadtochenko, V. A., 386, 388
NAGHIZADEH, J., 425-61; 428, 430, 437, 438, 443, 452
Naim, T. A., 575
Naito, A., 523
Nakamoto, K., 492, 590, 602
Nakamura, A., 616
Nakamura, H., 65
Nakamura, K., 327, 328
Nakanishi, H., 173
Nalbandyan, A. B., 374
Nangia, P. S., 382
Narasimhan, L. R., 477
Narayanamurti, V., 103, 105, 109
Naruge, Y., 370
Nava, D. F., 382, 386
Neftel, A., 367
Neiss, T. G., 516, 523
Nelin, C. J., 182, 184, 185, 204
Nelson, D. J., 230
Nelson, D. R., 149, 150
Nelson, H. H., 370
Nelson, W. H., 575
Nesbitt, D. J., 346, 355, 356, 360, 361
Nesbitt, F. L., 376
Netzer, F. P., 609, 626
Neuman, D. M., 380
Neumark, D. M., 323, 325, 331, 333, 334
Newbury, D. C., 627
Newns, D. M., 409, 412, 417, 419
Nguyen, T. T., 604, 631
Niazi, U., 233, 234
Niazyan, O. M., 377, 382

Nicovich, J. M., 372
Niederburger, W., 439, 440
Nieh, J.-C., 329, 330, 334
Niemeyer, L., 252
Nierwetberg, J., 289
Nieto, M. M., 405
Niki, H., 381, 382, 386, 389
Nikisha, L. V., 374
Nikitin, E. E., 398, 418
Nishimura, M., 526
Nishimura, Y., 537, 543, 575
Nittmann, J., 252, 257
Nitzan, A., 395, 398, 417, 418, 470
Noble, M., 40
Nobuyuki, M., 327, 328
Nocentini, S., 575
Noid, D. W., 395
Noolandi, J., 427
Norrish, R. G. W., 370, 373
Norton, P. R., 480
Nøskov, J. K., 417, 419, 485, 491
Novic, S. E., 139
Noyes, R. M., 341, 351
Nozawa, T., 526
Nyberg, C., 598, 602

O

Oas, T. G., 522, 524
Obi, K., 138, 370, 386, 388
O'Brien, E. V., 630
O'Brien, J. J. A., 373
O'Connell, J. P., 442-44
O'Connor, P., 269
O'Connor, S., 189, 194
O'Donnell, S., 538, 574
Oeschger, H., 367
O'Gara, J. F., 282
Ogletree, D. F., 626
O'Grady, B. V., 381
Ogren, P. J., 370, 374, 375, 378, 383
Ohanessian, G., 233
Ohmori, N., 126
Ohta, S., 252
Ohyanagi, Y., 140
Oikawa, A., 128, 134, 142
Okabe, A., 143
Okabe, H., 40, 41
Okajima, S., 140
Okamoto, M., 327, 328
Okazaki, M., 525
Okuda, M., 370, 384
Okuyama, K., 126, 133-36
Oldenborg, R. C., 384, 385
Oldfield, E., 517, 523, 529
Olejniczak, E. T., 519, 520, 528
Oliva, A., 224, 226
Olivella, S., 224, 226, 227, 231, 234

Oliver, D. W., 118, 119
Oliveria, M., 237
Olivieri, A. C., 525
Olszewski, W. V., 269, 282
Ondrechen, M. J., 139
Ondrey, G. S., 48, 58
O-Ohta, K., 62
Opella, S. J., 512, 517, 520, 521, 523-25, 531
Oppenheim, I., 353, 360
Orbach, R., 269
O'Reilly, J. M., 149, 154, 156
Orgel, L. E., 345
Orlandi, G., 134
Orlova, N. D., 86
Orozmamatov, S. T., 321, 327
Ortega, M., 224
Orwoll, R. A., 450
Osad'ko, I. S., 464
Osamura, Y., 219, 220, 222
Osborne, D. T., 333
Oschner, M., 138
Osgood, R. M. Jr., 80, 85
Osherov, V. I., 328
Osif, T. L., 389
Ostovic, D., 335
Ostrowsky, N., 238
Otis, C. E., 137
Otlewski, J., 574
Otsuka, H., 258
Otten, D., 61
Otto, A., 590
Otto, B., 341, 351
Ouellette, P. A., 378
Overend, R., 370, 381
Overton, W. C., 109
Owens, K. A., 221
Owensen, B., 448
Oxtoby, D. W., 69, 80, 83, 355, 356, 360, 470
Oxton, I. A., 590, 592, 599, 605, 612, 616, 633
Ozaki, T., 524
Ozawa, K., 138

P

Pacey, P. D., 331
Pack, D. W., 473, 474
Pack, R. T., 46, 55, 56, 327, 329
Padmanabhan, R., 299
Page, J. B., 575, 576
Paige, M. E., 348, 354, 356, 357, 360, 361
Painter, P. C., 623
Pake, G. E., 521
Pal, S., 409
Palings, I., 527
Palma, A., 48, 56
Palmer, R. G., 149, 153, 160, 161, 163, 174, 283
Pancir, J., 224

Paola, R. A., 630
Paraskevopoulos, G., 370, 373, 381
Pardoen, J. A., 520, 525, 527
Parisi, G., 240
Parker, G. A., 327, 329
Parker, S. F., 629, 638
Parkes, D. A., 374-76, 378, 382, 383
Parks, E. K., 183
Parmenter, C. S., 133, 140, 141
Parmenter, J. E., 600, 605-8, 616
Partridge, H., 183, 184, 189, 191, 192, 194, 196
Pasternack, L., 370
Pasternack, L. R., 384
Pate, C. T., 375, 377
Patil, A. O., 525
Patrick, K. G., 384
Pattengill, M. D., 46
Paul, D. M., 375, 383
Paul, I. C., 525
Pausak, S., 523
Pawliszyn, J., 139
Payne, W. A., 372, 382, 386
Pealat, M., 57
Pearce, H. A., 592, 593, 631-33
Pearson, G. S., 389
Peckler, S., 295
Pedersen, L., 199
Peierls, R. E., 96, 97
Pelissier, M., 186
Pelte, D., 55
Pemble, M. E., 598, 599, 602-4, 611, 612, 615, 633, 636
Pendley, R. D., 127
Peressini, P. P., 104
Perminov, A. P., 328
Perno, J. R., 575
Perrot, J. P., 345
Perry, M. B., 575, 576
Perry, R. A., 55
Person, W. B., 345, 634
Persson, B. N. J., 470
Persson, M., 417
Pesic, D. S., 203
Peters, C. W., 59
Peters, K. S., 321
Peterson, K. B., 230
Peticolas, W. L., 537, 543, 575
Petit, A. T., 93
Petsalakis, I. D., 56, 62
Pettersson, L. G. M., 183, 188, 189, 191, 192
Pettiette, C. L., 183
Peyser, P., 427
Pezolet, M., 537
Pfister, G., 269

AUTHOR INDEX 657

Pfnür, H. E., 485, 490, 494, 500
Phillips, J. C., 149, 150
Phillips, L. F., 345, 373
Phillips, M. C., 276
Phillips, R. B., 217, 218
Phillips, W. A., 111
Pierini, A. B., 224, 226
Pietronero, L., 238, 252, 255
Piette, J., 295
Pilling, M. J., 374, 375, 378
Pimentel, G. C., 386, 388
Pincus, P., 427
Pines, A., 516, 518, 523
Pinnick, D. A., 90
Pitts, J. N. Jr., 375, 377, 386
Pitzer, K. S., 195
Piuzzi, F., 141
Placzek, G., 545, 548
Plane, J. M., 370
Pliskin, W. A., 590, 630
Plonka, A., 269, 281
Plumb, I. C., 374-76
Plummer, E. W., 406, 594, 609
Pobo, L. G., 183
Poe, R. T., 326, 328
Pohl, F. M., 307
POHL, R. O., 93-121; 97-100, 102, 103, 105, 111, 112, 114, 116-19
Poll, J. D., 72-74
Pollak, E., 319, 320, 328
Polyak, S. S., 374
Pomeranchuk, I., 99
Pompi, R. L., 105
Ponomarev, A. V., 328
Pople, J. A., 220
Poppe, D., 61
Poppinger, D., 231
Porter, E.-G., 454
Porter, R. N., 319, 320
Post, D., 206
Pottinger, R., 228
Poulet, G., 372
Powell, D. B., 590, 592, 603, 605, 607, 608, 612, 616, 633
Powell, R. C., 473
Powers, D. E., 139, 204
Pozdnyakova, L. A., 86
Prasad, P. N., 475, 476
Prather, M. J., 368
Pratt, D. W., 125, 126
Pratt, G. L., 375
Pratt, L. R., 448
Prendergast, F. G., 574
Prentice, J. D., 632, 633
Presser, N., 321, 330, 331, 371
Preston, K. F., 206
Preston, R. K., 352, 398, 418
Prior, W. R. C., 88
Pritchard, J., 628
Prokhorov, A. M., 86

Protz, R., 85
Pullman, B., 395, 417
Pulu, A. C., 137
Purcell, F. J., 575
Pütz, O., 55
Pynn, R., 238
Pynn, R. A., 238

Q

Quarles, G. J., 473
Quinn, C. P., 375, 383

R

Rabinowitch, E., 341
Radford, H. E., 384, 389
Radhakrishnan, G., 40, 324
Rädle, M., 55
Radnoczi, G., 252
Radom, L., 220, 386
Raff, L. M., 320
Raffenetti, R., 185
Ragulsky, V., 78
Rahman, A., 150
Raithby, P. R., 626
Raleigh, D. P., 516, 519-21, 523, 527, 528
Ramaker, D. E., 395, 417
Ramanathan, V., 368
Ramanlal, P., 240, 253
Ramaswamy, S., 160, 164
Rammal, R., 161
Ramunni, G., 224, 225
Randeria, M., 118
Rao, T. V. R., 202
Raper, O. F., 373
Rapoport, H., 295, 298
Rapp, D., 404
Rasmussen, J. O., 55
Rastogi, A. K., 451
Ratajczak, E., 333, 334, 370
Ratner, M. A., 326
Rava, R. P., 539, 574, 575
Raval, R., 629, 638
RAVISHANKARA, A. R., 367-94; 370-74, 376, 377, 381, 385, 386, 390
Ray, S., 269
Raychaudhuri, A. K., 111, 112, 116
Razi Nagvi, K., 415
Redfield, A. G., 511
Redhead, P. E., 409
Redmon, M. J., 325
Redner, S., 283
Reed, D. A., 486
Rehfeld, S. J., 457
Reid, B. R., 511
Reid, I. D., 327
Reid, S., 132
Reilly, J. P., 386, 388
Reinsch, E.-A., 205

Reisler, H., 40, 55
Reisner, D. E., 131
Rendulic, K. D., 494, 497, 498, 501
Renner, G., 80, 85
Rentzepis, P. M., 321, 348
Rettner, C. T., 417, 421, 485, 490, 494, 497-501
Reynolds, P. J., 327
Rhasa, D., 381, 384
Rhim, W. K., 516
Rice, S. A., 140, 141, 419, 421, 470
Rice, S. F., 189, 191, 201
Richard, C., 230
Richards, F. M., 511
Richardson, D., 252
Richardson, K. S., 350
Richardson, L. F., 289
Richardson, N. V., 592
Richmond, P., 427
Richtsmeier, S. C., 183
Rideal, E. K., 480
Riesner, D., 309
Righini, R., 476
Riley, S. J., 183
Rinsland, C. P., 367
Riste, T., 238
Robb, M. A., 224, 227, 233, 234, 568
Robert, D., 75, 77, 78
Roberts, J. E., 516, 520-23
Roberts, M. W., 591
Robie, D. C., 321, 331
Robiette, A. G., 488
Robin, M. B., 40, 130, 565, 566
Robinson, G. N., 323, 325, 331, 384, 386
Robinson, K., 428
Robota, H. J., 494
Robson, R. C., 375, 383
Rockwell, A. L., 524
Roe, R. J., 427
Roebber, J. L., 127, 553, 554, 558, 583
Rogers, S. J., 109
Rohlfing, C. M., 185, 191
Rohlfing, E. A., 202, 206, 207, 554
Rollefson, R. J., 105, 109
Romanowski, H., 334
Römelt, J., 320, 323, 325, 331, 332
Romer, R., 309
Roos, B. O., 182, 184, 186, 195, 204
Rosasco, G. J., 71, 72, 74, 75
Rose, H. A., 166
Rosenberger, H., 516, 528
Rosina, G., 626
Rosman, R. L., 141
Ross, U., 52

AUTHOR INDEX

Rossetti, R., 617
Rossi, G., 257
Rossi, M., 216
Roth, W., 217
Rothe, E. W., 48
Rothgeb, T. M., 529
Rothschild, W. G., 69, 83
Rothwell, W. P., 521
Rousseau, D. L., 545
Rousseau, J., 607, 614, 620, 622, 623
Rowland, F. S., 367
Rowley, D., 227, 229
Roy, A. K., 282
Rozenshtein, V. B., 328
Ruben, D. J., 518
Rubin, R. J., 428, 435
Rudolph, P., 403
Ruedenberg, K., 187
Ruffolo, A., 334
Rumelfanger, R., 575, 576
Rush, J. J., 594
Russell, D. J., 348, 354, 356, 357, 360, 361
Russell, D. K., 384
Russell, J. A., 55
Rust, F., 372
Ryali, S. B., 494
Ryan, K. R., 374-76
Ryberg, R., 470
Ryckaert, J.-P., 353, 449
Rzepa, H. S., 217, 218, 222, 224, 226, 230, 231

S

Saalfeld, H., 626
Sachdev, S., 149, 160, 168
Sadowski, C. M., 55, 373
Saebo, S., 386
Saigusa, H., 140, 141
Saika, A., 516
Saito, H., 524
Sala, J. P., 75, 77
Salahub, D. R., 204
Salem, L., 224, 225
Salmeron, M., 592, 609, 610
Saltsburg, H., 395, 417
Samoson, A., 517
Sana, M., 224
Sancar, A., 295, 296
Sanche, L., 413
Sander, L. C., 455
Sander, L. M., 237, 238, 240, 250, 252-54, 257
Sander, M., 136
Sander, S. P., 376, 378, 383
Sanders, N., 384
Sanhueza, F., 383
Sano, M., 252
Sapozhnikov, M. N., 464
Sappey, A. D., 138
Sarkar, S. K., 531

Sarkisov, O. M., 386, 388
Sato, K., 65, 128, 135
Sato, M., 126
Sato, S., 325, 327, 328
Sato, T., 126
Sauer, J., 228, 229
Sauer, K., 303
Saunders, V. R., 184
Savage, C. M., 381, 382, 386, 389
Sawada, S.-I., 395, 398, 418
Sawada, Y., 252
Saxon, R. P., 326
Scacchi, G., 230
Sceats, M. G., 341, 346, 351, 352, 355
Schaefer, D. W., 237
Schaefer, H. F., 56, 57
Schaefer, H. F. III, 233, 234, 386
Schaefer, J., 517, 519, 522, 531
Schafer, T. P., 323, 325
Scharf, B., 138
Scharf, P., 186, 203, 206
SCHATZ, G. C., 317-40; 317, 320, 322, 326-29, 331, 333-35
Schauer, M., 139
Schawlow, A. L., 132
Scheffer, J. R., 525
Scheffler, I. E., 307
Scheiner, S., 139
Schell, M., 352
Schepper, W., 52
Scher, H., 257, 269, 273
Scherer, G. W., 149, 157, 159, 173
Scherer, N. F., 134, 137
Scheutjens, J. M. H. M., 428, 433, 434, 437, 438, 448, 450, 452
Schiferl, D., 90
Schiff, H. I., 373
Schiff, R., 370, 371
Schildkraut, C., 292
Schilling, J. B., 193
Schimmel, P. R., 303
SCHINKE, R., 39-68; 41, 44-46, 48-52, 54-58, 60-63
Schlaegel, J. E., 494, 500
Schlag, E. W., 136, 139, 141
Schlegel, H. B., 233, 234
Schlesinger, M., 238
SCHLESINGER, M. F., 269-90; 269, 281, 283, 289
Schleyer, P. von R., 220
Schmeisser, D., 413
Schmeltekopf, A. L., 373
Schmid, H., 222
Schmidt, A., 521
Schnabel, B., 516, 528
Schneider, W., 380
Schneidmessser, B., 97, 98

Schnepp, O., 575, 576
Schnieder, L., 63, 64
Schoeller, W., 217
Scholes, G., 334
Schomacker, K. T., 573
Schön, G., 403
Schosser, C. L., 475, 476
Schouten, F. C., 491
Schramm, S., 517
Schröder, J. O., 202
Schroeder, J., 341, 351
Schrotter, H. W., 545-47
Schubert, V., 341, 351
Schueller, K. E., 228
Schulberg, M. T., 480, 481
Schultz, W. R., 326, 327
Schwartz, R. N., 360, 405
Schwarz, H. A., 320
Schweitzer, E., 621
Schwenke, D. W., 325, 327-29, 331
Scott, J. G. V., 603
Scotto, M., 80, 85
Sears, T. J., 138
Seelig, A., 440, 441
Seelig, J., 439-41, 517
Seeman, P., 454
Sefcik, M. D., 519
Segal, G., 224, 225
Segev, E., 45
Segner, J., 494, 496
Seiler, W., 386
Seip, U., 600, 602, 604, 613
Seizle, H. L., 139, 141
Selman, J. I., 133
Selzer, E. A., 370, 374, 375
Semmes, D. H., 134, 137, 386
Senba, M., 327
Sension, R. J., 345, 570
Sentell, K. B., 456, 457
Sethna, J. P., 118
Sette, F., 609, 618, 627
Sexton, B. A., 617, 618
Shamov, A. G., 224
Shank, C. V., 321
Shapiro, M., 41, 42, 44-46, 48, 49, 55, 409
Sharma, R. D., 319
Shavitt, I., 184
Shaw, J. H., 382
Shea, K. J., 217, 218
Sheard, S., 486, 490
Shelby, R. M., 470, 471, 474
Sheldrick, G. M., 606
Shen, C. J., 298
Sheng, P., 244
Shepanski, J. F., 134, 137
SHEPPARD, N., 589-644; 590, 592, 593, 598, 599, 602, 604, 605, 607, 612, 615, 619, 625, 631, 633, 636
Shetter, R. E., 372
Shevtsov, V. A., 321, 327

Shi, Y., 295, 298
Shibata, F., 470
Shibuya, K., 386, 388
Shim, I., 203
Shima, Y., 328, 329
Shimada, K., 526
Shimizu, A., 128, 135
Shin, K. J., 351
Shlesinger, M. F., 171, 272-74, 281, 282, 284, 286, 288
Shobatake, K., 323, 325, 334
Shoemaker, C. L., 318, 323, 325
Shoji, A., 524
Shoosmith, J., 374
Shore, J. E., 281
Sibener, S. J., 373, 500
Sichina, W., 157
Siebrand, W., 321
Siegbahn, P. E. M., 182, 184, 186, 188, 195, 202, 205, 207-9
Sievers, A. J., 469, 475, 476
Siggia, E. D., 166
Silberberg, A., 427
Silbey, R., 273, 477
Silveston, P. L., 508
Simha, R., 149, 154, 156
Simmons, H. E., 226
Simmons, L. M., 405
Simon, S. A., 454
Simonaitis, R., 373, 375, 378, 383
Simons, J. P., 40, 55, 58, 62, 63
Simsohn, H., 442, 446
Sindorf, D. W., 455
Sinfelt, J. S., 592
Singh, H. B., 368
Sinha, A., 409, 548, 553, 554
Sitz, G. O., 421
Sivakumar, N., 55
Sizun, M., 46
Sjögren, L., 164, 167
Sjölander, A., 162, 164, 165, 167
Skene, J. M., 65
SKINNER, J. L., 463-78; 464, 468, 476
Skinner, P., 590, 592, 605, 633
Skjeltorp, A. J., 238
Sklyarevskii, V. V., 321, 327
Skodje, R. T., 320, 323, 325
Slack, G. A., 97, 98, 113, 115, 116, 118, 119
Slagle, I. R., 377, 382
Slater, N. K. H., 370
Slawsky, Z. I., 360
Slemr, F., 382
Slichter, C. P., 592
Sloan, J. J., 373
Small, G. J., 471, 473
Smalley, J. F., 376

Smalley, R. E., 123, 128, 135, 137, 139, 183, 204
Smilansky, U., 55
Smirnov, V. V., 86
Smith, A. J., 62, 63
Smith, D. E., 347, 348, 357, 360, 361
Smith, D. L., 156, 157
Smith, E., 360, 361
Smith, I. W. M., 375, 376, 383
Smith, K. M., 573
Smith, M. A., 128, 130, 135, 136
Smith, M. J. C., 374, 375, 378
Smith, R. H., 386
Smith, R. L., 517, 529
Smith, R. S., 138
SMITH, S. O., 511-35; 520, 521, 525, 527, 529
Smyth, K. C., 71, 72, 74, 75
Sneddon, L. G., 609
Snider, D. R., 593, 631
Snyder, R. G., 448, 449
Sobelman, I., 78
Soep, B., 140
Sokolova, N. A., 374
Solgadi, D., 55
Soma, Y., 632, 633
Somorjai, G. A., 395, 417, 491, 591, 592, 596, 605, 606, 608-10, 616, 618-20, 625, 626, 633, 634
Song, S., 568
Sonnenschein, M., 127
Sorbello, R. S., 593, 631
Southern, E. M., 292
Spangler, L. H., 125
Sparks, R. K., 323, 325, 334
Sparks, S. W., 511
Speirs, R. D., 531
Speis, H. J., 185
Spellmeyer, D. S., 229, 234
Spencer, R. B., 525
Sperry, J. F., 575
Spicer, C. W., 375, 377
Spiegelman, S., 292
Spiglanin, T. A., 55
Spiro, T. G., 538, 539, 543, 570, 573-75
Sprague, J., 252
Srdanov, V. I., 203
Sridharan, U. C., 376
Srivastava, R. P., 69, 72, 74
Stace, A. J., 346, 351, 355
Staemmler, V., 45, 48, 49, 56, 58-60
Stallard, B. R., 551
Stanghellini, P. L., 617
Staniaszek, P., 87, 88
Stanley, H. E., 238, 246, 252, 257
Stark, G. R., 292
Stark, R. E., 523

Stauffer, B., 367
Stechel, E. B., 324, 395, 405, 409, 411, 412, 421
Steckler, R., 325
Stefanov, P. K., 613-15
Stegeman, G. I., 71
Steimle, T. C., 200
Stein, D. L., 174, 283
Stein, H., 499
Steiner, H., 227, 229
Steinert, W., 370
Steinfeld, J. I., 348
Steininger, H., 592, 599, 600, 602, 604, 631, 633
Steinrück, H.-P., 485, 494, 495, 497, 498, 501, 502
Stejskal, E. O., 517, 519, 531
Stephenson, T. A., 140, 141
Steven, J. R., 376
Stevens, A. E., 189, 194, 195
Stevens, C. M., 372
Stevens, W. J., 185, 192, 202
Stewart, J. J. P., 224, 227, 234
Stewart, P. L., 512, 517, 531
Stickney, R. E., 480
Stief, L. J., 372, 382, 386
Stillinger, F. H., 150, 161, 174-77
Sting, 234
Stockbauer, R., 395, 417
Stöhr, J., 591, 609, 618, 627
Stoicheff, B. P., 72, 80, 85, 86
Stoltze, P., 491
Stone, B. M., 133
Stone, W. L., 454
St. Pierre, L. E., 451
Straub, K., 295
Strauss, B., 183, 204
Strauss, H. L., 83, 88, 345, 448, 449
Streit, G. E., 373
Strobl, G. R., 432
Strommen, D. P., 575
Strong, R. L., 345
Stroscio, J. A., 599
Struth, U., 434
Stryland, J. C., 72-75, 79, 86
Studencki, A. B., 310
Stuke, M., 137
Sturge, M. D., 464, 470, 471, 473
Stuve, E. M., 600, 608, 617, 618
Style, D. W. G., 389
Su, F., 386
Sugahara, Y., 128, 135, 139, 140
Sullivan, C. E., 531
Sullivan, M. J., 523
Summers, D., 389
Summers, D. W., 434
Sun, J. C., 326, 328
Sun, Y., 327

AUTHOR INDEX

Sun, Z., 448
Sunada, H., 258
Sundberg, R. L., 395, 402, 421, 548, 553, 554
Šunjić, M., 403, 406
Sunko, D. E., 216
Sur, A., 344, 345
Suraev, V. V., 321, 327
Surman, M., 620, 622, 623
Sustmann, R., 228, 229
Sutcliffe, R., 206
Sutherland, D. N., 237
Suto, M., 58
Suzuki, I., 143
Suzuki, M., 370
Suzuki, T., 126, 130, 132
Swanson, B. I., 476
Swanson, N., 599
Swartz, E. T., 118, 119
Sweedler, J. V., 544
Sworski, T., 370
Szczęsniak, M. M., 139
Szeverenyi, N. M., 523
Szleifer, I., 428, 437, 448
Szwarc, H., 155, 165

T

Tabeta, R., 524
Taborek, P., 164
Tagami, Y., 427
Takano, H., 173
Takayanagi, T., 325, 327, 328
Takeda, Y., 138
Takenaka, H., 258
Takeuchi, A., 327
Tal-Ezer, H., 398
Tamres, M., 345
Tamura, F., 616
Tanaka, I., 370, 386, 388
Tang, C., 257
Tang, K. T., 326, 328, 548, 552
Tang, S. L., 480, 481, 485, 486, 491, 492, 495, 501, 502
Tannor, D. J., 395, 402, 419, 421, 548, 553
Taran, J.-P. E., 57
Tardy, B., 596, 602, 603, 608, 613, 614, 618, 620, 622, 623, 626, 628
Tardy, D. C., 227, 229
Tave, J., 269
Taylor, J. B., 479, 501
Taylor, P. R., 184, 185, 186, 189, 203, 206, 207, 325
Teitel'boim, M. A., 377, 382
Teleman, O., 448
Tellinghuisen, J., 344, 345
Temkin, S. I., 78
Temps, F., 386, 388, 389
Tengstal, C. G., 598, 602

Terao, T., 516
Teraoka, J., 543, 553, 574, 579, 581, 582
Tessman, J. W., 296
Thacher, P. D., 99, 100
Thaddeus, P., 334
Theodorakopoulos, G., 56, 62
Theodorou, D. N., 428, 437, 438, 450, 451
Thiel, P. A., 629, 630
Thirumalai, D., 325
Thomlinson, M. M., 428
Thomlinson, W. C., 107
Thompson, B. R., 257
Thompson, R. L., 372
Thompson, T. C., 206, 325
Tikhomirov, V. A., 224
Timonen, R. S., 333, 334, 370
Tinoco, I., 303, 307, 309
Tirmizi, S. M. A., 97, 98
Titze, B., 48
Todhunter, I., 270
Toennies, J. P., 328, 329, 417
Tolk, N. H., 409, 411
Tomioka, T., 227, 229
Tomioka, Y., 142
Toner, J., 160, 164
Tong, S. Y., 594, 625
Tonks, D. L., 575, 576
Topp, M. R., 139
Torchia, D. A., 511, 512, 520, 523, 531
Torczynski, R. M., 292, 310
Torell, L. M., 150
Torrie, G. M., 352
Tosatti, E., 238
Toscano, V. G., 216, 218, 220
Toulouse, G., 161
Townshend, R. E., 224, 225
Tramer, A., 141
Tranquille, M., 206
Traum, M. M., 409, 411
Trembreull, R., 135
Trenary, M., 598, 606, 633
Trevor, P., 417
Troe, J., 341, 351, 374, 377, 381, 382, 390
Truhlar, D. G., 206, 317-20, 322-29, 331, 333, 334
Tsai, M-C., 600, 602, 604, 613
Tsang, P., 517
Tsang, W., 227, 229
Tsou, L., 496
Tsuboi, M., 537, 575
Tsuchiya, S., 131, 132, 138, 139, 376
Tsukada, M., 395, 417
Tu, A. T., 574
Tuckerman, R. T., 370
Tuerkes, P. R. H., 118, 119
Tufeu, R., 88
Tuherm, T., 517
Tully, F. P., 370-72

Tully, J. C., 352, 355, 360, 395, 398, 409, 411, 417, 418, 496, 499
Turkevich, L. A., 257
Turner, D. L., 511
Turner, J. L., 413
Turpin, P. Y., 537, 575
Tuszyński, J. A., 84
Tycko, R., 517
Tyler, S. C., 367, 372
Tyndall, G. S., 376, 380, 381, 383

U

Uchiyama, M., 227, 229
Udagawa, Y., 142
Ueba, H., 409
Uhlenbeck, O. C., 307
Untch, A., 46, 54, 58
Unterberg, U., 341, 351
Upton, T. H., 629, 630, 638
Ushakov, V. G., 328

V

Vaccaro, P. H., 131
Vaida, V., 127
Valbura, V., 334
Valentine, K. G., 512, 517, 524, 531
Valentini, J. J., 202, 206, 207, 329, 330, 334, 554
Valic, M. F., 517
Van Cleve, J. E., 97, 98
van den Berg, R., 527
Van den Bergh, H. E., 374
VanderHart, D. L., 531
van der Ploeg, P., 439, 440
Vandersande, J. W., 97, 98, 111, 112
van Dongen, P. G. J., 263
van Herpen, W. M., 139
Van Houten, B., 295, 296
Van Hove, M. A., 609, 625, 626
Van Kranendonk, J., 72, 75
van Lenthe, J. H., 184
Van Vechten, D., 252
van Veen, N., 58
van Willigen, H., 523
Van Willigen, W., 485
van Wunnik, J. N. M., 417, 419
Van Zee, R. J., 202, 205
Varandas, A. J. C., 327, 328
Vardeny, Z., 269
Varghese, G., 72, 73, 79
Varma, C. M., 111
Varotsis, C., 570
Vaughan, R. W., 516
Vedeneev, V. I., 377, 382, 386, 388

AUTHOR INDEX

Veeman, W. S., 520, 523, 527
Vega, S., 519-21, 528
Verges, J., 344
Vesely, F. J., 353
Veyret, B., 378, 380, 383, 386, 388
Vicsek, T., 240, 242, 252, 257, 258, 263
Vielhaber, W., 494
Vilesov, F. I., 63
Villa, A., 375, 377
Vinogradov, J. P., 63
Viovy, J. L., 428
Virasoro, M. A., 161
Visscher, W. M., 101
Viswanathan, K. S., 344, 345
Vitius, P., 230
Vitturi, A., 55
Vodegel, M., 329
Voges, H., 55
Vold, M. J., 237
Vold, R. L., 517
Vold, R. R., 517
Völker, S., 474
Vonk, C. G., 434
von Karman, Th., 95
Voss, R. F., 240

W

Wachters, A. J. H., 184
Waclawski, B. J., 599
Waddill, G. D., 613, 620, 630
Wade, L. E., 216-18, 222
Wadt, W. R., 185
Wagner, A. F., 319, 320, 322, 323, 325, 331, 333, 334, 370
Wagner, H. Gg., 386, 388, 389
Wagner, M., 99
Wagner, W. G., 545
Wahl, M., 55, 58, 60
Wahner, A., 370, 386
Wainwright, T. E., 360
Waite, B. A., 56
Wakayama, M., 547, 566
Walch, S. P., 182, 183, 185, 186, 189, 194, 202-7, 331
Waldenstrom, S., 415
Walker, C. T., 99, 103, 109, 575, 576
Walker, R. B., 46, 55, 56, 324, 325, 327-29
Walker, R. F., 372
Walker, R. W., 374, 384
Wallace, R. B., 310
Wallace, S. C., 128, 130, 135, 136, 143
Walling, J. C., 473
Wallington, T. J., 378, 380, 383
Wallis, R. F., 101
Walsh, C. A., 477, 529

Walsh, R., 230
Walsh, S. P., 184
Wang, H. Y., 333
Wang, J. C., 303
Wang, P. K., 592
Wang, S.-W., 195, 206
Wang, W. Y., 368
Wang, Y., 575
Wanna, J., 139
Wannier, G. H., 428, 435
Wantuck, P. J., 384, 385
Ward, H. R., 343
Warnatz, J., 330, 333
Warneck, P., 382, 386
Warren, J. A., 133
Warshel, A., 575, 576
Washida, N., 370, 374, 375, 386, 388
Wasilewski, J., 56, 58, 59
Wasserman, Z. R., 427
Watanabe, H., 138, 139
Watson, R., 372
Watson, R. T., 376-78, 381-83, 385, 390
Watson, S. K., 118, 119
Watts, D. C., 269, 276
Waugh, J. S., 515, 516, 518, 521, 523
Weaver, A., 331, 333
Webb, G. G., 522
Weber, H. F., 93, 94
Weber, T. A., 150, 174-77, 427, 437, 450
Webster, F., 327, 329
Wehrli, R., 222
Weide, K., 46, 54, 58, 62, 63
Weinberg, W. H., 486, 490, 600, 605-8, 616, 629, 630
Weis, J. J., 80
Weiss, G. H., 273, 278, 280
Weiss, S., 499, 500
Weisshaar, J. C., 56, 386
Weitz, D. A., 237, 244
Welge, K. H., 63, 64
Weller, R., 55, 58, 60
Wells, M. G., 480
Welsh, H. L., 69, 72-75, 79, 86
Welsshaar, J. C., 138
Weltner, W., 199, 205
Weltner, W. Jr., 202
Wemmer, D. E., 511
Wendt, H. R., 376
Wenzel, L., 591, 609, 618
Werner, H.-J., 187, 203, 205, 206
Wertheim, G. K., 406
Wertheimer, R. K., 470
Wessel, J. E., 128, 131, 135, 136
West, B. J., 284, 286, 288
Westberg, K. R., 330
Westenberg, A. A., 370

Weston, G. W., 333
Weston, R. E. Jr., 320, 329
Wetmur, J. G., 309
Wharton, L., 123, 494
Whitbeck, M. R., 378, 383
White, S. H., 453, 454
Whitehead, J. C., 331
Whitfield, M., 524
Whitmore, M. D., 427
Whyte, A. R., 345
Whytock, D. A., 372
Wiebe, H. A., 375, 377
Wiersma, D. A., 464, 470, 474, 477
Wiesenfeld, J. R., 373
Wiesmann, H. J., 252
Wight, C. A., 333
Wildman, T. A., 321
Williams, A. R., 406
Williams, C. J., 45
Williams, G., 269, 276
Williams, M. D., 497, 498, 501
Williams, P. F., 545
Williamson, A. D., 128
Willis, R. F., 406, 594, 599
Wilson, E. B. Jr., 620, 623
Wilson, G. E. Jr., 531
Wilson, K. R., 345, 346
Wilson, W. E., 370
Wiltzius, P., 237
Wilzen, L., 417
Wine, P. H., 372, 373, 376, 377, 386
Winkel, C., 527
Winkler, A., 494, 497, 498, 501
Winn, J. S., 139
Winn, K. R., 384, 385
Winter, N. W., 331, 389
Winters, H. F., 486, 490
Witkowicz, T., 72, 74, 78
Witt, D., 593, 631
Witt, J., 55
Witten, T. A., 237, 238, 244, 250, 252, 253, 257
Wittig, C., 40, 55, 324
Wodtke, A. M., 323, 325, 331
Wofsy, S. C., 368
Wokaun, A., 511
Wolf, S. A., 252
Wolfrum, J., 55, 58, 60, 328
Wong, Y.-F., 525
Wood, C. F., 333
Wood, K. A., 88
Wood, S. W., 375
Wood, W. C., 341
Woodcock, L. V., 150
Woods, M. C., 442-44
Woodward, R. B., 213, 217
Wright, A. H., 626
Wu, R., 299
Wu, S.-F., 324, 334
Wuthrich, K., 511

AUTHOR INDEX

Wyatt, R. E., 318-20, 323, 325, 326, 334
Wyttenbach, T., 138

X

Xi, L., 473
Xu, G. Q., 496

Y

Yabe, T., 142
Yamada, C., 376
Yamada, H., 131, 132
Yamada, Y., 258
Yamamoto, S , 135
Yamanobe, T., 524
Yamanouchi, K., 131, 132, 139
Yamasaki, K., 134
Yamashita, A. B., 345
Yamashita, I., 62
Yamazaki, H., 373
Yang, Q. Y., 480, 481, 485, 486, 488, 491, 492, 494, 496, 501, 502, 506
Yang, S., 183
Yannoni, C. S., 521
Yarkony, D. R., 187
Yates, J. T., 411, 412
Yates, J. T. Jr., 486, 490-92, 589, 629, 633, 634, 640
Yatsuda, N., 134
Yee, K. K., 345
Yip, S., 159, 162
Yoon, D. Y., 428, 432-34, 450
Yoshimori, A., 395, 417
Yoshiuchi, H., 128, 135, 142
Young, J. D., 118
Young, L., 139, 140
Young, P. E., 531
Young, R. P., 634
Yu, C. F., 500
Yu, M. L., 419
Yu, T. J., 537

Z

Zaccai, G., 440, 441
Zacherl, A., 289
Zaera, F., 599, 608, 616, 634, 640
Zahniser, M., 372
Zaidi, H. R., 69, 72, 74
Zaitsev, N. N., 464
Zalczer, G., 88
Zare, R. N., 421
Zegarski, B. R., 138
Zeller, R. C., 111, 112, 114, 116
Zellner, R., 370, 376, 377, 381, 384
Zemb, T., 442, 446, 447
Zerbetto, F., 134
Zewail, A. H., 134, 137, 464, 470, 471
Zhang, J. Z. H., 325, 327, 329, 331
Zhang, Y.-C., 240, 331
Zhang, Y. P., 553, 583, 585
Zheng, L. S., 183
Ziegler, G., 55
Ziegler, L. D., 537, 538, 541, 549, 553, 554, 558-60, 563, 570, 575, 583-85
Ziff, R. M., 258
Zilm, K. W., 516, 521, 522
Zimm, B. H., 303, 428
Zimmerman, H. E., 216
Zimmerman, J., 341
Zinck, J. J., 486, 490
Zoebisch, E. G., 224
Zulicke, L., 398, 418
Zumofen, G., 281, 284
Zwanzig, R. W., 281, 360
Zwerger, W., 466, 468

SUBJECT INDEX

A

ABCO
  ionic states of
    Rydberg series converging to, 136
  ionization threshold spectra of, 136
  two-color ionization dip spectroscopy and, 131
Absorption spectra
  photodissociation and, 39
Acetamide
  preresonance Raman studies of, 566-68
Acetone
  photochemical decomposition of, 10
  photolysis of, 19
  preresonance Raman studies of, 566
Acetophenone
  sensitized phosphorescence excitation spectroscopy and, 126
Acetylene
  adsorption on metal crystals, 607-8, 611-18, 636
  1,3-dipolar cycloaddition and, 231-34
  ion of
    laser-excited fluorescence and, 138
  stimulated emission spectroscopy and, 131
Acoustic phonons, 468-69
Acrolein
  pyrazoline ring cleavage and, 11
Acrylonitrile
  Diels-Alder reaction and, 228-30
Adam-Gibbs equation, 156
Adenine
  ultraviolet resonance Raman spectroscopy and, 575
Air/water interfaces
  amphiphiles and, 425
Alanine racemase
  nuclear magnetic resonance of, 528-29
Aldehydes
  bond strength in, 32
Alkaline-earth halides
  transition metal halides and, 195-200
Alkanes
  bond strength in, 32
  chain molecules of, 448-50

Alkenes
  adsorption on metal crystals, 598-611
Alkylbenzene
  triplet decay rates of, 137
Alkynes
  adsorption on metal crystals, 611-20
Amino acids
  nuclear magnetic resonance of, 523-24
  ultraviolet resonance Raman spectroscopy and, 574-75
2-Aminopyridine
  multiphoton ionization spectroscopy and, 128
2-Aminopyridine complexes
  ionization energies in, 143
Amphiphiles
  air/water interfaces and, 425
Aniline
  ionization threshold spectra of, 136
  Rydberg series of, 136
  stimulated emission spectroscopy and, 132
  two-color ionization dip spectroscopy and, 131
Aniline-rare gas complex
  electronic spectra of, 139
Anisotropy
  chemical shift
    magic angle spinning and, 518-19
  excited state potential energy surface and, 46
  photodissociation of water and, 48
Anthracene
  planar supersonic expansion technique and, 127
Aromatic amino acids
  ultraviolet resonance Raman spectroscopy and, 574-75
Aromatic hydrocarbons
  adsorption on metal crystals, 620-28
Atmosphere
  CH_2O reactions in, 386-87
  CH_3 reactions in, 374-75
  CH_3O reactions in, 384-86
  CH_3O_2 reactions in, 375-83
  HCO reactions in, 387-89
  methane oxidation in, 367-90
Auger electron spectroscopy, 479
1-Azacarbazole dimer
  tautomerism in, 142

7-Azaindole
  multiphoton ionization spectroscopy and, 128
7-Azaindole dimer
  tautomerism in, 142
Azulene
  planar supersonic expansion technique and, 127
  population labeling spectroscopy and, 132

B

Bacteriophage (fd)
  nuclear magnetic resonance of, 531
Bacteriorhodopsin
  nuclear magnetic resonance of, 525-26
  photoexcitation of, 321
Bell-Evans-Polanyi plots, 214
Benzaldehyde
  sensitized phosphorescence excitation spectroscopy and, 126
Benzene
  absorption spectra of, 123
  adsorption on metal crystals, 620-27, 637-38
  core/water interface and, 457
  excited states of
    ultraviolet Raman excitation and, 570
  ionization potential of, 136
  Raman spectra of, 558-63
  triplet decay rates of, 137
  two-color ionization dip spectroscopy and, 131
Benzene complexes
  vibrational predissociation in, 140
Benzoic acid
  sensitized phosphorescence excitation spectroscopy and, 126
Benzonitrile-Ar complex
  electronic spectra of, 139
Benzonitrile complexes
  vibrational predissociation in, 140
Benzophenone
  amplitude motions in, 134
  sensitized phosphorescence excitation spectroscopy and, 126
  Bernoulli scaling, 270-73
  fractal time and, 273-74
9,9'-Bianthryl
  amplitude motions in, 134

663

SUBJECT INDEX

Bilayer membranes
  chain configurations in, 438-40
  solute partitioning into, 453-54
Biphenyl
  absorption spectra of, 123
  amplitude motions in, 134
Bis(imidazole) Fe(III) protoporphyrin IX-dimethylester
  electronic transitions of, 570-71
Blacet-Hedman apparatus
  microanalysis of gases and, 15
Boltzmann distributions, 49
Boltzmann-Planck equation, 430
Born-Oppenheimer approximation, 184
Born-von Karman model
  lattice vibrations and, 95-96
Boron
  specific heat of
    temperature dependence of, 94
Boron hydrides
  stability of, 33
Boundary shift model
  liquid-glass transition and, 176
Brownian motion, 275
Butadiene
  Diels-Alder reaction and, 224-29
  excited states of
    ultraviolet Raman excitation and, 570
Butane
  dissociative chemisorption on nickel, 494
Butene-1
  cyclopentanone ring cleavage and, 10
But-2-yne
  adsorption on metal crystals, 619-20

C

Cantor set, 271
Carbon dioxide
  vibrational Raman spectra of, 89
Carbon monoxide
  cyclopentanone ring cleavage and, 10
  gas-phase
    vibrational Raman spectra of, 79
  liquid
    vibrational Raman spectra of, 85-86
Carbon tetrachloride
  geminate recombination dynamics and, 347-49
Carboxypeptidase A
  nuclear magnetic resonance of, 529
Chain molecules, 425-58
  alkane crystals and, 448-50
  bilayer membranes and, 438-40
  finite
    condensed at interfaces, 435-50
  long
    condensed at interfaces, 428-35
  micelles and, 440-48
  mixing properties of, 451-57
Chain reactions
  discovery of, 29
Chemical kinetics, 29
Chemical reactions
  quantum effects in, 317-35
Chemical shift anisotropy
  magic angle spinning and, 518-19
Chemisorption, 479-508
  collision-induced, 503-6
  experimental technique and, 481-84
  mechanisms for, 484-501
    direct, 484-96
    precursor, 496-501
  precursor molecules to, 501-2
Chlorophyll a
  nuclear magnetic resonance of, 525
Chromium
  orbital overlap in, 183
Clebsch-Gordon factors, 49
Cole-Cole plot, 276
Collagen
  nuclear magnetic resonance of, 531
Collinear resonances, 325
Colloidal aggregation, 237-65
  cluster-cluster, 257-63
  diffusion-limited, 250-57
  fractal geometry and, 241-50
Combustion
  methane oxidation in, 368
Computer simulation
  colloidal aggregation and, 237-38
Concanavilin A
  nuclear magnetic resonance of, 529
Condensed matter
  fractal time in, 269-89
Configurational entropy
  supercooled liquids and, 156
Configuration-interaction method, 184

Coordinates
  photodissociation processes and, 42
Cope rearrangement, 215-24
Copper
  surface-enhanced Raman spectroscopy and, 590
m-Cresol
  internal rotation in, 133
Crystals
  dephasing in, 463-77
    optical, 472-74
    vibrational, 471-72, 475-76
  lattice vibrations in, 93-119
    Born-von Karman model of, 95-96
    Debye model of, 95-96
    Einstein model of, 93-94, 111-19
    lattice defects and, 99-105
    phonon-phonon scattering and, 96-99
    Poiseuille heat flow and, 105-11
  specific heat of
    temperature dependence of, 94
Cycloaddition reactions, 213
  1,3-dipolar, 231-34
Cycloalkanes
  adsorption on metal crystals, 628
Cyclobutane
  cyclopentanone photolysis and, 16
  psoralen photoreactivity and, 293-95
Cyclobutanone
  carbon monoxide production from, 10
  photolysis of
    products of, 17
    preparation of, 11
Cyclobutene
  electrocyclic ring opening of, 213
Cycloheptanone
  carbon monoxide production from, 10
Cyclohexane
  adsorption on metal crystals, 628-29, 638
Cyclohexane-1,4-diyl
  bonds of, 220
  heat of formation of, 216
Cyclohexanone
  carbon monoxide production from, 10
  photolysis of
    products of, 16
Cyclohexene
  Diels-Alder reaction and, 227

SUBJECT INDEX 665

Cycloparaffins
  adsorption on metal crystals, 629-30
Cyclopentadiene
  adsorption on metal crystals, 611
Cyclopentane
  adsorption on metal crystals, 630, 638
  cyclohexanone photolysis and, 16
Cyclopentanone
  carbon monoxide production from, 10
  photochemical decomposition of, 10
  photolysis of
    products of, 16
  tetramethylene biradical of, 16
Cyclopropane
  adsorption on metal crystals, 630
  isomerization to propylene, 29
Cystine
  preresonance Raman studies of, 566
Cytosine
  ultraviolet resonance Raman spectroscopy and, 575

D

DABCO
  ionic states of
    Rydberg series converging to, 136
  ionization threshold spectra of, 136
  two-color ionization dip spectroscopy and, 131
Debye model
  lattice vibrations and, 95-96
Defect diffusion, 282
Deflection function
  elastic scattering and, 45
Desorption
  collision-induced, 506
Detergents
  polar headgroup of
    affinities of, 425
Deuterium
  gas-phase
    vibrational Raman spectra of, 74-75
  mixtures with rare gases
    vibrational Raman spectra of, 75-77
  ortho and *para* species in, 75
Diacetylene
  ion of
    laser-excited fluorescence and, 138

Diamond
  specific heat of
    temperature dependence of, 94
Diatomic molecules
  optical-optical double resonance spectroscopy and, 135
  population labeling spectroscopy and, 132
  vibrational autoionization of, 136
  vibrational Raman spectra of, 69-71
1,2-Diazacyclopetane
  See Pyrazoline
Diazomethane
  cyclobutanone production and, 11
Diborane
  bond strength in, 33
Dichlorethylene
  excited states of
    ultraviolet Raman excitation and, 570
cis-Dichloroammine platinum (II)
  ultraviolet resonance Raman spectroscopy and, 575
Dicke narrowing, 74
Dicopper, 202-4
1,1-Dicyanoethylene
  Diels-Alder reaction and, 228-29
1,4-Dicyano-1,5-hexadiene
  Cope rearrangement of, 222-23
2,5-Dicyano-1,5-hexadiene
  Cope rearrangement of, 222-23
Diels-Alder reaction, 224-30
Dienes
  Diels-Alder reaction of, 226
Diffusion
  photodissociation and, 350-51
p-Difluorobenzene
  ionic states of
    Rydberg series converging to, 136
  jet-cooled
    high excited Rydberg states of, 130
  stimulated emission spectroscopy and, 132
p-Difluorobenzene-Ar complex
  vibrational predissociation in, 141
Difluorobenzene complexes
  vibrational predissociation in, 140
Dimers
  transition metal, 202-6

Dimethyl acetylene
  adsorption on metal crystals, 619-20
Dimethylethylene
  excited states of
    ultraviolet Raman excitation and, 570
Diphenylacetylene
  torsional motions of, 133
2,4-Diphenyl-1,5-hexadiene
  Cope rearrangement of, 222
Dipolar-chemical shift spectroscopy, 521-22
Dipolar mechanism
  electron energy loss spectroscopy and, 593
Dipole effects
  chemical shift spectra and, 519-21
Disproportionation reactions, 32
DNA
  psoralen cycloadducts and, 295
  5'-TpA-3' sequences in
    photoreactivity of, 295
  x-ray studies of, 30
Dulong-Petit value, 93-94

E

Einstein model
  lattice vibrations and, 93-94, 111-19
Elastic scattering
  deflection function in, 45
Elastin
  nuclear magnetic resonance of, 531
Electromagnetic radiation
  intercalated psoralen and, 293
Electron energy loss spectroscopy
  chemisorbed hydrocarbon species and, 593-94
  chemisorption and, 482-84
  surface dynamics and, 413-17
Electronic motion
  Born-Oppenheimer separation of, 318
Electronic transitions
  surface dynamics and, 399-402
Ene reaction, 230-31
Energy sudden approximation, 44-45
Energy transfer
  collisional mechanisms for, 29
Enzymes
  nuclear magnetic resonance of, 528-31

SUBJECT INDEX

Ethane
  dissociative chemisorption on iridium, 494
  dissociative chemisorption on nickel, 494
  vibrational Raman spectra of, 89
Ethylene
  adsorption on metal crystals, 598-609, 635-36
  Diels-Alder reaction and, 224-29
  ene reaction and, 230-31
  excited states of
    ultraviolet Raman excitation and, 570
Ethylidyne
  adsorption on metal crystals, 604-6
Exciplex formation, 141
Excited states
  ultraviolet resonance Raman spectroscopy and, 564-74

F

Fabry-Perot interferometry, 71
Fatty acids
  polar headgroup of affinities of, 425
Fick's law, 350-51
Fictive temperature, 153
Flash photolysis, 29
Fluctuation-dissipation theorem, 165, 351
Fluids
  See Liquids
Fluorene
  planar supersonic expansion technique and, 127
Fluorescence excitation spectroscopy, 124-32
Fluorobenzene
  fluorescence excitation spectra of, 133
  multiphoton ionization spectroscopy and, 128
p-Fluorotoluene
  fluorescence excitation spectra of, 133
  stimulated emission spectroscopy and, 132
Formaldehyde
  ene reaction and, 231
  photodissociation of, 56-58
  stimulated emission spectroscopy and, 131
Fractal geometry
  colloidal aggregation and, 241-50
Fractal time, 269-89
  Bernoulli scaling and, 273-74
  mechanism generating, 275

Fractal time-space, 285-89
Franck-Condon limit
  photodissociation processes and, 46-47
Franck-Condon mapping
  photodissociation processes and, 47-48
Franck-Condon principle
  product state distributions and, 41
Franck-Condon transition, 345, 399
  photodissociation processes and, 45
Free radicals
  pyrolysis and, 10
Fulminic acid
  1,3-dipolar cycloaddition and, 231-34
Fumaronitrile
  Diels-Alder reaction and, 228-30
Furan
  ion of
    laser-excited fluorescence and, 138

G

Gases
  microanalysis of
    Blacet-Hedman apparatus and, 15
  vibrational Raman spectra of, 72-79
Gas phase kinetics, 29
Geminate recombination dynamics, 347-49
  curve crossing dynamics and, 350-55
Glasses
  lattice vibrations in, 93-119
    Born-von Karman model of, 95-96
    Debye model of, 95-96
    Einstein model of, 93-94, 111-19
    lattice defects and, 99-105
    phonon-phonon scattering and, 96-99
    Poiseuille heat flow and, 105-11
  methanol
    hydrogen abstraction in, 321
  relaxation in, 275-82
  spin, 149
    quenched disorder in, 161
    spin models of, 168-74
  structural, 149-50
  structural dynamics of, 159-62

Glass transition, 150-59
  ideal, 153
  laboratory, 153
Glass transition temperature, 152
Glycylglycine
  nuclear magnetic resonance of, 523-24
Glyoxal
  jet-cooled
    two-color fluorescence dip spectra of, 131
  sensitized phosphorescence excitation spectroscopy and, 126
  stimulated emission spectroscopy and, 131-32
Glyoxal complexes
  vibrational predissociation in, 140
Gold
  surface-enhanced Raman spectroscopy and, 590
Golden rule expression, 42-43
Graphite
  specific heat of
    temperature dependence of, 94
Group additivity, 32
Guanine
  ultraviolet resonance Raman spectroscopy and, 575
Gunpowder
  ingredients of, 2

H

Halides
  transition metal, 195-200
Halobenzene
  ion of
    laser-excited fluorescence and, 138
Hamiltonian
  alternative partitionings of, 319
  rigid-rotor, 42
  spin
    nuclear magnetic resonance and, 513-17
Harmonic oscillators
  surface dynamics and, 403-5
  two-level systems and, 465-67
Heavy water, 3
Heisenberg uncertainty principle, 2, 463
Helium
  solid
    Poiseuille heat flow in, 107
Heme proteins
  electronic structure of, 572
  nuclear magnetic resonance of, 528-29

SUBJECT INDEX 667

Hemoglobin
  nuclear magnetic resonance of, 529
Hexadiene
  planar supersonic expansion technique and, 127
1,5-Hexadiene
  Cope rearrangement of, 215-24
Hexafluorobenzene
  ground state ion of
    Jahn-Teller effect in, 138
Hybridization, 292-93
Hydrazine
  pyrazoline ring cleavage and, 11
Hydrides
  transition metal, 189-95
Hydrocarbons
  adsorption on metal crystals, 589-640
  aromatic
    adsorption on metal crystals, 620-28
  saturated
    adsorption on metal crystals, 628-30
Hydrogen
  gas-phase of
    vibrational Raman spectra of, 72-74
  liquid
    vibrational Raman spectra of, 87-89
  mixtures with rare gases
    vibrational Raman spectra of, 75-77
  ortho and *para* species in, 75
Hydrogen bonding
  laser spectroscopy and, 142-43
Hydrogen peroxide
  photodissociation of, 58-60
Hydrophobic/hydrophilic interfaces, 425
4'-Hydroxymethyl-4,5',8-trimethylpsoralen
  DNA photoreactivity and, 295
Hyper-Raman spectroscopy, 583-84

I

Ideal glass transition, 153
Imidazole
  preresonance Raman studies of, 566-68
Impact mechanism
  electron energy loss spectroscopy and, 594
Indole
  ionization threshold spectra of, 136
  multiphoton ionization spectroscopy and, 128
Inelastic neutron scattering
  chemisorbed hydrocarbon species and, 594
Infrared spectroscopy
  chemisorbed hydrocarbon species and, 592-93
Insect cuticle
  nuclear magnetic resonance of, 531
Interferometry
  Fabry-Perot, 71
Iodine
  photodissociation/recombination in liquids, 341-43
  V-T energy transfer in, 357-62
Iodine spectroscopy
  potential surfaces and, 343-46
Iodine X state potential
  vibrational relaxation of, 355-57
Iridium
  dissociative chemisorption of ethane on, 494
Isobutene
  adsorption on metal crystals, 610
Isoquinoline complexes
  vibrational predissociation in, 140

J

Jacobi-coordinates
  photodissociation processes and, 42, 58
Jahn-Teller effect
  hexafluorobenzene ground state ion and, 138
Josephson junctions, 289

K

Kauzmann paradox, 157
Ketenes
  cyclobutanone production and, 11
  reactivity of, 17
Ketones
  bond strength in, 32
  photochemical decomposition of, 10
Knotek-Feibelman model
  wave packets and, 409-11
Kohlrausch-Williams-Watts function, 154, 171, 281
Kramer-Heisenberg-Dirac formalism, 548

L

Laboratory glass transition, 153
Lagrange method, 431
Langevin dynamics
  photodissociation and, 351-54
Langmuir-Blodgett films, 425-26
Laser spectroscopy
  hydrogen bonding and, 142-43
  polyatomic molecules and, 123-44
  van der Waals complexes and, 138-42
Lattice vibrations, 93-119
  Born-von Karman model of, 95-96
  Debye model of, 95-96
  Einstein model of, 93-94, 111-19
  lattice defects and
    phonon scattering by, 99-105
  phonon-phonon scattering and, 96-99
  Poiseuille heat flow and, 105-11
  second sound and, 105-11
Lennard-Jones fluid
  viscosity of, 167
Lennard-Jones potentials, 356
Levy distributions, 272-73
Levy flight, 285-89
Levy walk, 285-89
Lipids
  polar headgroup of
    affinities of, 425
Liquid-glass transition, 149-79
  hydrodynamic models of, 162-68
  spin models of, 168-74
  square tiling model of, 174-76
Liquids
  photodissocation reactions in, 341-63
  supercooled
    configurational entropy of, 156
    hydrodynamic models of, 162-68
    linear response properties of, 154-57
    nonlinear response properties of, 157-59
    spin models of, 168-74
    square tiling model of, 174-76
    vibrational Raman spectra of, 79-89
Low-energy electron diffraction
  chemisorbed hydrocarbon species and, 591

SUBJECT INDEX

α-Lytic protease
    nuclear magnetic resonance of, 528-29

M

Magic angle spinning
    nuclear magnetic resonance and, 512, 517-21
Magnetization transfer
    nuclear magnetic resonance and, 522-23
Maleonitrile
    Diels-Alder reaction and, 228-29
Marcus-Coltrin theory, 334
Membrane proteins
    nuclear magnetic resonance of, 526-28
Menzel-Gomer-Redhead model
    wave packets and, 409-11
Metal catalysts
    infrared transmission studies on, 630-35
Metal crystals
    hydrocarbon chemisorption on, 589-640
Metalloproteins
    nuclear magnetic resonance of, 528-29
Metal surfaces
    adsorbate covered photoemission lineshapes of, 406-9
Metal surface selection rule, 593
Methane
    collision-induced desorption from nickel, 506
    collision-induced dissociation on nickel, 503-6
    dissociative chemisorption on nickel, 485-94
    liquid
        vibrational Raman spectra of, 86-87
Methane oxidation
    atmospheric, 367-90
    experimental methods and, 369-70
Methanol glasses
    hydrogen abstraction in, 321
N-Methylacetamide
    preresonance Raman studies of, 566-70
Methylacetylene
    adsorption on metal crystals, 618-19
Methyl β-cyanoacrylate
    Diels-Alder reaction and, 230
Micelles
    chain configurations in, 440-48

solute partitioning into, 456-57
Minimal aggregation model
    liquid-glass transition and, 175
Molecular dynamics, 318
    photodissociation and, 354-55
    of surfaces, 395-422
        applications of, 405-21
        methodology in, 397-405
        modeling of, 397-405
    ultraviolet resonance Raman spectroscopy and, 537-84
Molecular time scale generalized Langevin equation, 353-54
Monte Carlo simulation
    liquid-glass transition and, 171-75
Montroll-Weiss continuous-time random walk, 278-81
Morse-Clark formula, 203
Motional narrowing, 74-76
Multibond reactions
    Cope rearrangement and, 215-24
    Diels-Alder reaction and, 224-30
    1,3-dipolar cycloaddition and, 231-34
    ene reaction and, 230-31
    synchronicity in, 213-34
Multiphoton ionization spectroscopy, 124-32
    two-color, 127-28
Multiple collision effects
    photodissociation processes and, 60-64
Myoglobin
    nuclear magnetic resonance of, 529

N

Naphthalene
    ionization threshold spectra of, 136
    sensitized phosphorescence excitation spectroscopy and, 126
    ultraviolet resonance Raman spectroscopy and, 575-76
β-Naphthol
    m-substituted
        fluorescence excitation spectra of, 134
Navier-Stokes equation, 165
Near-edge x-ray absorption fine structure
    chemisorbed hydrocarbon species and, 591
Nickel
    atomic states of, 182-83

collision-induced desorption of methane from, 506
collision-induced dissociation of methane on, 503-6
dissociative chemisorption of methane on, 485-94
Nitrogen
    gas-phase
        vibrational Raman spectra of, 79
    liquid
        vibrational Raman spectra of, 80-84
    pyrazoline ring cleavage and, 11
Nitrogen pentoxide
    decomposition of, 29
Noble gases
    preparation in high fictive temperature glassy states, 153
Nuclear magnetic resonance
    chemisorbed hydrocarbon species and, 591
    enzymes and, 528-31
    magic angle spinning and, 512, 517-21
    membrane proteins and, 526-28
    polycarbonate defect and, 282
    solid-state
        methods of, 512-23
    proteins and, 523-32
    soluble proteins and, 528-31
    spin Hamiltonian and, 513-17
    structural proteins and, 531
Nuclear motion
    Born-Oppenheimer separation of, 318
Nuclear Overhauser effect, 522
Nucleic acid renaturation, 292
Nucleic acids
    ultraviolet resonance Raman spectroscopy and, 575

O

Octadine
    planar supersonic expansion technique and, 127
Oligonucleotide hybridization, 291-314
    experimental methods in, 296-98
    experimental results in, 299-303
    theory of, 303-13
Oligonucleotide probe
    photocrosslinkable
        kinetic advantages of, 308-10
    sensitivity of, 313

separation of, 298-99
synthesis of, 293-96
Optical dephasing
  crystals and, 472-74
Optical-optical double resonance
  spectroscopy, 135
Optical phonons, 469
Optics
  ultraviolet resonance Raman
    spectroscopy and, 542
Oxides
  transition metal, 200-2
Oxygen
  liquid
    vibrational Raman spectra
      of, 85-86
Ozone
  decomposition of, 31

P

Palladium
  atomic states of, 182-83
Parvalbumin
  nuclear magnetic resonance
    of, 529
Peptide folding
  ultraviolet resonance Raman
    spectroscopy and, 574
Peptidoglycan
  nuclear magnetic resonance
    of, 531
Pericyclic reactions, 213
Phenanthrene
  planar supersonic expansion
    technique and, 127
Phenol
  ionization threshold spectra
    of, 136
  m-substituted
    fluorescence excitation
      spectra of, 134
  multiphoton ionization spec-
    troscopy and, 128
Phenol-benzene complex
  predissociation of, 137
Phenol-dioxane complex
  ionization threshold of, 143
Phonons
  acoustic, 468-69
  optical, 469
  pseudolocal, 469-70
Phonon scattering
  lattice defects and, 99-105
  lattice vibrations and, 96-99
Phosphorescence excitation spectrum, 125
  sensitized, 126
Photodissociation, 341-63
  classical theory of, 45-46
  diffusion and, 350-51

dynamical mapping and, 51-
  56
extreme coupling and, 60-64
final state interaction in, 41
formaldehyde, 56-58
Franck-Condon limit and, 46-
  50
hydrogen peroxide, 58-60
iodine spectroscopy and, 343-
  46
Langevin dynamics and, 351-
  54
molecular dynamics and, 354-
  55
multiple collision effects and,
  60-64
picosecond spectroscopy and,
  346-47
quantum mechanical theory
  of, 42-45
rotational distributions in, 39-
  65
rotational reflection principle
  and, 51-60
steps in, 39
strong coupling and, 51-60
theories of, 41-46
ultraviolet resonance Raman
  spectroscopy and, 553-64
vibrational relaxation and,
  355-62
water in first absorption band,
  48-50
water in second absorption
  band, 61-64
weak coupling and, 46-50
Photolysis
  products of, 16-17
Picosecond spectroscopy
  photodissociation and, 346-47
Planar supersonic expansion
  technique, 127
Platinum
  atomic states of, 182-83
  ethylene adsorption on, 598-
    609
Poiseuille heat flow
  lattice vibrations and, 105-11
Poisson bracket relations, 165
Poisson law, 281
Polyatomic molecules
  amplitude motion of, 133-34
  laser spectroscopy of, 123-44
  photodissociation of, 39, 58,
    65
  rotational isomerism and,
    134-35
  vibrational autoionization of,
    136
  vibrational Raman spectra of,
    70
Polyatomic radicals, 138

Polycarbonate
  defect in
    mechanical relaxation and,
      282
Polymeric melts
  interfacial tension of, 450
Polypeptides
  nuclear magnetic resonance
    of, 524
Population labeling spectroscopy, 132
Porphyrins
  nuclear magnetic resonance
    of, 525
Potential surfaces
  iodine spectroscopy and, 343-
    46
Precursor geometry limited reactions, 324
Product state distributions
  Franck-Condon principle and,
    41
  photodissociation and, 39-40
Propane
  dissociative chemisorption on
    nickel, 494
Propene
  adsorption on metal crystals,
    609
  ene reaction and, 230-31
Propylene
  cyclopropane isomerization to,
    29
Propylidyne
  adsorption on metal crystals,
    610
Propyltoluene
  internal rotation in, 133
Propyne
  adsorption on metal crystals,
    618-19
Protein folding
  ultraviolet resonance Raman
    spectroscopy and, 574
Proteins
  adsorption of gases by, 26
  nuclear magnetic resonance
    of, 523-32
Pseudolocal phonons, 469-70
Psoralen
  photoreactivity of, 293-95
Pure vibrational spectra, 70-71
Pyrazine
  ionization threshold spectra
    of, 136
  jet-cooled
    two-color fluorescence dip
      spectra of, 131
  Rydberg series of, 136
  triplet decay rates of, 137
Pyrazoline
  ring cleavage of, 11

SUBJECT INDEX

Pyrene
  ultraviolet resonance Raman spectroscopy and, 575-76
Pyrimidine
  multiphoton ionization spectroscopy and, 128
  triplet decay rates of, 137
Pyrimidine-Ar complex
  electronic spectra of, 139
Pyrimidine complexes
  vibrational predissociation in, 140
Pyrolysis
  very low pressure, 33-34
Pyrolysis reactions
  free radicals and, 10
  unimolecular fission processes in, 29

Q

Quantitative analysis, 6-7
Quantum effects, 318-27
Quantum mechanical theory, 2
  time-independent photodissociation processes and, 42-45
Quasiclassical trajectory method, 319-20
Quinones
  nuclear magnetic resonance of, 525

R

Raman effect, 544-53
Raman excitation profile, 552
Raman-gain spectroscopy, 71
Raman spectroscopy
  chemisorbed hydrocarbon species and, 595
Random walk
  Montroll-Weiss continuous-time, 278-81
  Weierstrass, 273
Rare gases
  H_2-D_2 mixtures with vibrational Raman spectra of, 75-77
Renaturation, 292
Resonances, 323-25
  Cl + HCl and, 331-33
  H + CO and, 333-34
  H + H_2 and, 329-30
Retinal
  nuclear magnetic resonance of, 526-27
Reversed-phase liquid chromatography
  solute partitioning in, 455-56
Rhodopsin
  nuclear magnetic resonance of, 526-27

Ribonuclease A
  nuclear magnetic resonance of, 528-31
Rice-Ramsperger-Kassel theory, 28, 34
Rotational isomerism
  polyatomic molecules and, 134-35
Rotational rainbow, 51, 55, 57
Rotational reflection principle
  photodissociation processes and, 51-60
Rydberg series, 136
Rydberg states, 130
Rydberg transitions, 558

S

St. Petersburg paradox, 270-71
Sample handling
  ultraviolet resonance Raman spectroscopy and, 542
Saturated hydrocarbons
  adsorption on metal crystals, 628-30
Saturation Raman spectroscopy, 578-83
Scandium
  halides of, 195-200
Scanning tunneling microscopy, 479
Schrödinger equation, 42, 47, 184, 319
Silicon
  specific heat of
    temperature dependence of, 94
Silver
  surface-enhanced Raman spectroscopy and, 590
Small-angle neutron scattering
  micelles and, 445-48
Soluble proteins
  nuclear magnetic resonance of, 528-31
Solute partitioning
  bilayer membranes and, 453-54
  micelles and, 456-57
  reversed-phase liquid chromatography and, 455-56
Solvents
  geminate recombination dynamics and, 345-49
Spectrometer and detection systems
  ultraviolet resonance Raman spectroscopy and, 543-44
Spectroscopy
  Auger electron, 479
  dipolar-chemical shift, 521-22

electron energy loss
  chemisorbed hydrocarbon species and, 593-94
  chemisorption and, 482-84
  surface dynamics and, 413-17
fluorescence excitation, 124-32
hyper-Raman, 583-84
infrared
  chemisorbed hydrocarbon species and, 592-93
iodine
  potential surfaces and, 343-46
laser
  hydrogen bonding and, 142-43
  polyatomic molecules and, 123-44
  van der Waals complexes and, 138-42
multiphoton ionization, 124-32
  two-color, 127-28
optical-optical double resonance, 135
picosecond
  photodissocation and, 346-47
population labeling, 132
Raman
  chemisorbed hydrocarbon species and, 595
Raman-gain, 71
saturation Raman, 578-83
stimulated emission, 131-32
supersonic jet, 125-32
surface-enhanced Raman
  metal crystals and, 590
thermal desorption
  chemisorbed hydrocarbon species and, 591
transition metal hydrides and, 193-95
two-color ionization dip, 131
two-color phosphorescence dip, 131
ultraviolet photoelectron
  adsorbate covered metallic surfaces and, 406-9
  chemisorbed hydrocarbon species and, 591
ultraviolet resonance Raman, 537-84
  applications of, 553-78
  instrumentation in, 539-44
x-ray photoelectron
  adsorbate covered metallic surfaces and, 406-9
Spin glasses, 149
  quenched disorder in, 161

Spin Hamiltonian
  nuclear magnetic resonance and, 513-17
Square tiling model
  liquid-glass transition and, 174-76
Stilbene
  planar supersonic expansion technique and, 127
trans-Stilbene
  jet-cooled
    high excited Rydberg states of, 130
    stimulated emission spectroscopy and, 132
    torsional motions of, 133
Stimulated emission spectroscopy, 131-32
Stirling's approximation, 430
Stokes Raman scattering, 544
Stretched exponential law, 275-82
Structural glasses, 149-50
  spin models of, 168-74
Structural proteins
  nuclear magnetic resonance of, 531
Styrene
  excited states of
    ultraviolet Raman excitation and, 570
Supercooled liquids
  configurational entropy of, 156
  hydrodynamic models of, 162-68
  linear response properties of, 154-57
  nonlinear response properties of, 157-59
  spin models of, 168-74
  square tiling model of, 174-76
Supersonic jets
  polyatomic molecules in laser spectroscopy of, 123-44
Supersonic jet spectroscopy, 125-32
Surface-enhanced Raman spectroscopy
  metal crystals and, 590
Surfaces
  molecular dynamics of, 395-422
    applications of, 405-21
    methodology in, 397-405
    modeling of, 397-405
Surfactant monolayers, 425-26
Synchronicity, 213-34
  Cope rearrangment and, 215-24

  Diels-Alder reaction and, 224-30
  1,3-dipolar cycloaddition and, 231-34
  ene reaction and, 230-31

T

Tautomerism
  dimers and, 142
Temperature
  fictive, 153
  glass transition, 152
  specific heat of crystals and, 94
Tetracene-rare gas complex
  electronic spectra of, 139
Tetradeuteriobutadiene
  Diels-Alder reaction and, 228
1,2,4,5-Tetrafluorobenzene
  excited state of
    flexibility in, 134
Tetratomic molecules
  photodissociation of, 60
Tetrazine-HCL complex
  electronic spectra of, 139
Tetrazine-H_2O complex
  electronic spectra of, 139
Tetrazine-rare gas complex
  electronic spectra of, 139
Thermal conductivity
  model of minimum, 115-16
Thermal desorption spectroscopy
  chemisorbed hydrocarbon species and, 591
Thermoheological simplicity, 156
Threshold energies, 319
Thymine
  ultraviolet resonance Raman spectroscopy and, 575
Toluene
  adsorption on metal crystals, 628
  internal rotation in, 133
  triplet decay rates of, 137
  two-color ionization dip spectroscopy and, 131
o-Toluidine
  internal rotation in, 133
m-Toluidine
  internal rotation in, 133
Transition metals, 181-210
  atoms of, 187-89
  dimers of, 202-6
  features of, 181-87
  halides of, 195-200
  hydrides of, 189-95
  Ni + H_2 reaction and, 207-9
  oxides of, 200-2
  trimers of, 206-7
Transition state theory, 28
Trapping problem, 281

Trimers
  transition metal, 206-7
Triphenylene
  ultraviolet resonance Raman spectroscopy and, 575-76
Tropolone
  jet-cooled
    fluorescence excitation spectra of, 142
Tryptophan
  saturation Raman spectroscopy and, 581-82
Tungsten surfaces
  H atom diffusion on, 321
Tunneling, 320-23
  H + H_2 and, 327-29
  O + H_2 and, 330-31
Turbulent diffusion, 285-89
  Richardson's law of, 289
Two-color ionization dip spectroscopy, 131
Two-color multiphoton ionization spectroscopy, 127-28
Two-color phosphorescence dip spectroscopy, 131
Tyrosinate
  ultraviolet Raman spectra of, 579
Tyrosine
  saturation Raman spectroscopy and, 581-82

U

Ultraviolet photoelectron spectroscopy
  adsorbate covered metallic surfaces and, 406-9
  chemisorbed hydrocarbon species and, 591
Ultraviolet resonance Raman spectroscopy, 537-84
  applications of, 553-78
  instrumentation in, 539-44
Uncertainty principle, 2, 463
Unimolecular rate theory, 28-29
Uracil
  ultraviolet resonance Raman spectroscopy and, 575

V

Van der Waals complexes
  laser spectroscopy and, 138-42
Van der Waals resonances, 324
Very low pressure pyrolysis, 33-34
Vibrational dephasing, 75-76
  crystals and, 471-76
  indirect, 78
Vibrationally adiabatic ground state potential, 319

SUBJECT INDEX

Vibrational Raman spectra, 69-90
  CO and, 79
  D_2 and, 74-75
  D_2-rare gas mixtures and, 75-77
  gas-phase line widths and, 74-79
  gas-phase Raman shifts and, 72-74
  H_2 and, 74-75
  H_2-rare gas mixtures and, 75-77
  HD and HD-Ar mixtures and, 77-79
  liquid CF_4 and, 86-87
  liquid CH_4 and, 86-87
  liquid CO and, 85-86
  liquid H_2 and, 87-89
  liquid N_2 and, 80-84
  liquid O_2 and, 85-86
  liquids and, 79-89
  N_2 and, 79
Vibrational reflection principle, 51-56
Vibrational relaxation, 355-62
  iodine X state potential and, 355-57
Vibrational spectra
  pure, 70-71

Vinyl
  adsorption on metal crystals, 607-8
Vinylidene
  adsorption on metal crystals, 607-8
Vinylidene cyanide
  Diels-Alder reaction and, 230
Viral coat proteins
  nuclear magnetic resonance of, 531
Visual pigments
  nuclear magnetic resonance of, 525
Vogel law, 282
Vogel-Tamman-Fulcher equation, 156

W

Water
  photodissociation of
    first absorption band and, 48-50
    second absorption band and, 61-64
  preresonance Raman studies of, 566
Wave packet propagation, 398-402

Weierstrass function, 271
Weierstrass random walk, 273
Wigner distribution, 46
Wigner threshold rules, 328
Woodward-Hoffmann rules, 213

X

X-ray crystallography
  thymine furan-side monoadducts and, 295
X-ray photoelectron spectroscopy
  adsorbate covered metallic surfaces and, 406-9
Xylene
  internal rotation in, 133
p-Xylene
  multiphoton ionization spectroscopy and, 128

Y

Yttrium
  halides of, 195-200

Z

Zeeman Hamiltonian, 513
Zero point energy, 318-20

CUMULATIVE INDEXES

CONTRIBUTING AUTHORS, VOLUMES 35–39

A

Altkorn, R., 35:265–89
Anderson, J. G., 38:489–520
Asher, S. A., 39:537–88
Ausserré, D., 38:317–47
Avouris, P., 35:49–73

B

Baldridge, K. K., 38:211–52
Baumgärtner, A., 35:419–35
Bauschlicher, C. W. Jr., 39:181–212
Ben-Shaul, A., 36:179–211
Benson, S. W., 39:1–37
Berne, B. J., 37:401–24
Boatz, J. A., 38:211–52
Bondybey, V. E., 35:591–612
Borden, W. T., 39:212–36
Brown, J. K., 39:341–66
Budil, D. E., 38:561–83

C

Cahill, D. G., 39:93–121
Campion, A., 36:549–72
Castleman, A. W. Jr., 37:525–50
Ceyer, S. T., 39:479–510
Chang, C.-H., 38:561–83
Chen, S. H., 37:351–99
Christiansen, P. A., 36:407–32
Clouter, M. J., 39:69–91
Cohen, M. L., 35:537–62
Cole, R. G., 37:105–25
Cowley, J. M., 38:57–88
Crim, F. F., 35:657–91

D

Dalton, L. R., 37:459–91
Das, P., 35:507–36
Debacker, M. G., 38:271–301
de Leeuw, S. W., 37:245–70
Demuth, J., 35:49–73
Dill, K. A., 39:425–61
Dlott, D. D., 37:157–87
Doll, J. D., 38:413–31
Drobny, G. P., 36:451–89
Dye, J. L., 38:271–301

E

Ebata, T., 39:123–47
Elson, E. T., 36:379–406
Ermler, W. C., 36:407–32
Ertl, G., 37:587–615
Evans, G. T., 37:105–25
Ezra, G. S., 36:277–320

F

Fendler, J. H., 35:137–57
Field, R. W., 37:493–524
Fleming, G. R., 37:81–104
Flynn, G. W., 37:551–85
Fogarasi, G., 35:191–213
Fredrickson, G. H., 39:149–80
Frei, H., 36:491–524

G

Gadzuk, J. W., 39:395–424
Garrett, B. C., 35:159–89
Gast, P., 38:561–83
Gelbart, W. M., 36:179–211
Glaeser, R. M., 36:243–75
Goodman, D. W., 37:425–57
Gordon, M. S., 38:211–52
Grant, E. R., 36:277–320
Griffin, R. G., 39:511–35
Griffiths, J. F., 36:77–104
Gudeman, C. S., 35:387–418

H

Hamilton, C. E., 37:493–524
Harding, J. H., 37:53–80
Harris, A. L., 39:341–66
Harris, C. B., 39:341–66
Harris, S. J., 36:31–52
Haymet, A. D. J., 38:89–108
Hearst, J. E., 39:291–315
Heller, E. J., 35:563–89
Hervet, H., 38:317–47
Herzberg, G., 36:1–30; 38:27–55
Hirota, E., 36:53–76
Houk, K. N., 39:213–36
Huppert, D., 37:127–56
Hynes, J. T., 36:573–97

I

Ito, M., 39:123–47

J

Jasinski, J. M., 38:109–40
Johnston, H. S., 35:481–505
Jovin, T. M., 38:521–60

K

Kawaguchi, K., 36:53–76
Keesee, R. G., 37:525–50
Kinsey, J. S., 37:493–524
Klein, M. L., 36:525–48
Kollman, P., 38:303–16
Kommandeur, J., 38:433–62
Koseki, S., 38:211–52
Kosower, E. M., 37:127–56
Krauss, M., 35:357–85

L

Langhoff, S. R., 39:181–212
Leone, S. R., 35:109–35
Levelt Sengers, J. M. H., 37:189–222
Lin, M. C., 37:587–615
Loncharich, R. J., 39:212–36
Lowe, M. A., 36:213–41
Louie, S. G., 35:537–62

M

Madey, T. E., 35:215–40
Majewsky, W. A., 38:433–62
Mandel, M., 35:75–108
Marqusee, J. A., 39:425–61
McCaffery, A. J., 37:223–44
McIntosh, L. P., 38:521–60
Meakin, P., 39:237–67
Meerts, W. L., 38:433–62
Metiu, H., 35:507–36
Meyerson, B. S., 38:109–40
Mikami, N., 39:123–47
Milligan, R. F., 36:139–58
Munoz-Rojas, A., 38:191–210

N

Naghizadeh, J., 39:425–61
Newton, M. D., 35:437–80
Nibler, J. W., 38:349–81
Norris, J. W., 38:349–81

673

O

O'Brien, M. P. 38:383–411
Odijk, T., 35:75–108

P

Patterson, G. D., 38:191–210
Perram, J. W., 37:245–70
Peters, K., 38:253–70
Pimentel, G. C., 36:491–524
Pitzer, K. E., 36:407–32; 38:1–25
Pohl, R. O., 39:93–121
Pope, M., 35:613–55
Porschke, D., 36:159–78
Pratt, D. W., 38:433–62
Pratt, L. R., 36:433–49
Prestegard, J., 38:383–411
Proctor, M. J., 37:223–44
Pulay, P., 35:191–213

R

Ramaker, D. E., 35:215–40
Ravishankara, A. R., 39:367–94
Reid, B. R., 36:105–37
Reisler, H., 37:307–49
Rondelez, F., 38:317–47
Rossky, P. J., 36:321–46

S

Saykally, R. J., 35:387–418
Schatz, G. C., 39:317–40
Schiffer, M., 38:561–83
Schinke, R., 39:39–68
Schmidt, J., 38:141–61
Schmitz, K. S., 37:271–305
Schneider, F. W., 36:347–78
Schroeder, J., 38:163–90
Schurr, J. M., 37:271–305
Scott, B. A., 38:109–40
Sengers, J. V., 37:189–222
Sheppard, N., 39:589–644
Shlesinger, M. F., 39:269–90
Singel, D. J., 38:141–61
Skinner, J. L., 39:463–78
Slichter, C. P., 37:25–51
Smith, E. R., 37:245–70
Smith, S. O., 39:511–35
Spaepen, F., 35:241–63
Soumpasis, D. M., 38:521–60
Stechel, E. B., 35:563–89
Stephens, P. J., 36:213–41
Stevens, W. J., 35:357–85
Stockbauer, R., 35:215–40
Stockmayer, W. H., 35:1–21
Stoneham, E. M., 37:53–80
Stout, J. W., 37:1–23
Sturtevant, J. M., 38:463–88
Sutin, N., 35:437–80
Swenberg, C. E., 35:613–55

T

Thirumalai, D., 37:401–24
Thomas, G. A., 36:139–58
Tinoco, I. Jr., 35:329–55
Troe, J., 38:163–90
Truhlar, D. G., 35:150–89
Turnbull, D., 35:241–63

V

Van Zee, R. J., 35:291–327
Vistnes, A. I., 37:459–91
Voter, A. F., 38:413–31

W

Weiner, A. M., 36:31–52
Weltner, W. Jr., 35:291–327
Wemmer, D. E., 36:105–37
Weston, R. E. Jr., 37:551–85
Whetten, R. L., 36:277–320
Whitaker, B. J., 37:223–44
Whitney, D., 37:459–91
Williams, A. L., 35:329–55
Wittig, C., 37:307–49

Y

Yamakawa, H., 35:23–47
Yang, J. J., 38:349–81
Young, C. L., 37:459–91

Z

Zare, R. N., 35:265–89
Zimm, B. H., 35:1–21

CHAPTER TITLES, VOLUMES 35–39

BIOPHYSICAL CHEMISTRY

Interactions and Kinetics in Membrane Mimetic Systems	J. H. Fendler	35:137–57
Differential Absorption and Differential Scattering of Circularly Polarized Light: Applications to Biological Macromolecules	I. Tinoco, Jr., A. L. Williams, Jr.	35:329–55
High Resolution NMR Studies of Nucleic Acids and Proteins	D. E. Wemmer, B. R. Reid	36:105–37
Effects of Electric Fields on Biopolymers	D. Porschke	36:159–78
Electron Crystallography of Biological Macromolecules	R. M. Glaeser	36:243–75
Fluorescence Correlation and Photobleaching Recovery	E. L. Elson	36:379–406
Dynamic Light Scattering Studies of Biopolymers: Effects of Charge, Shape, and Flexibility	J. M. Schurr, K. S. Schmitz	37:189–222
Molecular Modeling	P. Kollman	38:306–16
Membrane and Vesicle Fusion	J. H. Prestegard, M. P. O'Brien	38:383–411
Chemical Applications of Differential Scanning Calorimetry	J. Sturtevant	38:463–88
The Transition Between B-DNA and Z-DNA	T. M. Jovin, D. M. Soumpasis, L. P. McIntosh	38:521–60
Three-Dimensional X-Ray Crystallography of Membrane Proteins: Insights into Electron Transfer	D. E. Budil, P. Gast, C.-H. Chang, M. Schiffer, J. R. Norris	38:561–83
A Photochemical Investigation of the Dynamics of Oligonucleotide Hybridization	J. E. Hearst	39:291–315
High-Resolution Solid-State NMR of Proteins	S. O. Smith, R. G. Griffin	39:511–35
UV Resonance Raman Studies of Molecular Structure and Dynamics: Applications in Physical and Biophysical Chemistry	S. A. Asher	39:537–88

CHEMICAL KINETICS—CONDENSED PHASE

Electron Transfer Reactions in Condensed Phases	M. D. Newton, N. Sutin	35:437–80
Periodic Perturbations of Chemical Oscillators: Experiments	F. W. Schneider	36:347–78
Chemical Reaction Dynamics in Solution	J. T. Hynes	36:573–97
Excited State Electron and Proton Transfers	E. M. Kosower, D. Huppert	37:127–56
Elementary Reactions in the Gas-Liquid Transition Range	J. Schroeder, J. Troe	38:163–90
Three-Dimensional X-Ray Crystallography of Membrane Proteins: Insights into Electron Transfer	D. E. Budil, P. Gast, C.-H. Chang, M. Schiffer, J. R. Norris	38:561–83
Synchronicity in Multibond Reactions	W. T. Borden, R. J. Loncharich, K. N. Houk	39:213–36
Fractal Time in Condensed Matter	M. F. Shlesinger	39:269–90
The Nature of Simple Photodissociation Reactions in Liquids on Ultrafast Time Scales	A. L. Harris, J. K. Brown, C. B. Harris	39:341–66

CHEMICAL KINETICS—GAS PHASE

Variational Transition State Theory	D. G. Truhlar	35:159–89
Selective Excitation Studies of Unimolecular Reaction Dynamics	F. F. Crim	35:657–91
Chemical Kinetics of Soot Particle Growth	S. J. Harris, A. M. Weiner	36:31–52
Thermokinetic Interactions in Simple Gaseous Reactions	J. F. Griffiths	36:77–104
Elementary Reactions in the Gas-Liquid Transition Range	J. Schroeder, J. Troe	38:163–90
Free Radicals in the Earth's Atmosphere: Their Measurement and Interpretation	J. G. Anderson	38:489–520
Kinetics of Radical Reactions in the Atmospheric Oxidation of CH_4	A. R. Ravishankara	39:367–94

CHEMICAL KINETICS—PHOTOCHEMISTRY AND RADIATION CHEMISTRY

Infrared Induced Photochemical Processes in Matrices	H. Frei, G. C. Pimentel	36:491–524
Hot Atoms Revisited: Laser Photolysis and Product Detection	G. W. Flynn, R. E. Weston, Jr.	37:551–85
Picosecond Organic Chemistry	K. Peters	38:253–70

CHEMICAL KINETICS—REACTION DYNAMICS

State-Resolved Molecular Reaction Dynamics	S. R. Leone	35:109–35
Mechanistic Studies of Chemical Vapor Deposition	J. M. Jasinski, B. S. Meyerson, B. A. Scott	38:109–40
Pyrazin: An "Exact" Solution to the Problem of Radiationless Transitions	J. Kommandeur, W. A. Majewski, W. L. Meerts, D. W. Pratt	38:433–62
Rotational Distributions in Direct Molecular Photodissociation	R. Schinke	39:39–68
Quantum Effects in Gas Phase Chemical Reactions	G. C. Schatz	39:317–40

COLLOIDS

Models for Colloidal Aggregation	P. Meakin	39:237–67

ELECTROCHEMISTRY

Physical and Chemical Properties of Alkalides and Electrides	J. L. Dye, M. G. DeBacker	38:271–301

GEOCHEMISTRY AND COSMOCHEMISTRY

Human Effects on the Global Atmosphere	H. S. Johnston	35:481–505
Free Radicals in the Earth's Atomosphere: Their Measurement and Interpretation	J. G. Anderson	38:489–520

LASER CHEMISTRY, ENERGY TRANSFER AND RELAXATION

Relaxation and Vibrational Energy Redistribution Processes in Polyatomic Molecules	V. E. Bondybey	35:591–612
Stimulated Emission Pumping: New Methods in Spectroscopy and Molecular Dynamics	C. E. Hamilton, J. S. Kinsey, R. W. Field	37:493–524
Rotational Energy Transfer: Polarization and Scaling	A. J. McAffery, M. J. Proctor, B. J. Whitaker	37:223–44
Photo-initiated Unimolecular Reactions	H. Reisler, C. Wittig	37:307–49
The Nature of Simple Photodissociation Reactions in Liquids on Ultrafast Time Scales	A. L. Harris, J. K. Brown, C. B. Harris	39:341–66

LIQUID STATE—SIMPLE FLUIDS

Dynamics of Polyatomic Fluids: A Kinetic Theory Approach	R. G. Cole, G. T. Evans	37:105–25
Vibrational Raman Spectra of Simple Fluids	M. J. Clouter	39:69–91

LIQUID STATE—SOLUTIONS OF ELECTROLYTES; FUSED SALTS

Dielectric Properties of Polyelectrolyte Solutions	M. Mandel, T. Odijk	35:75–108

LIQUID STATE—STRUCTURE

The Structure of Polar Molecular Liquids	P. J. Rossky	36:321–46
Theory of Hydrophobic Effects	L. R. Pratt	36:433–49
Computer Simulation of the Static Dielectric Constant of Systems with Permanent Electric Dipoles	S. W. de Leeuw, J. W. Perram, E. R. Smith	37:245–70

MAGNETIC RESONANCE (ELECTRON SPIN, NUCLEAR, QUADRUPOLE)

Multiple Quantum NMR: Studies of Molecules in Ordered Phases	G. P. Drobny	36:451–89
Fashioning Electron Spin Echoes into Spectroscopic Tools: A Study of Aza-aromatic Molecules in Metastable Triplet States	J. Schmidt, D. J. Singel	38:141–61
High-Resolution Solid-State NMR of Proteins	S. O. Smith, R. G. Griffin	39:511–35

MISCELLANEOUS

Metallic Glasses	F. Spaepen, D. Turnbull	35:241–63
Relativistic Effects in Chemical Systems	P. A. Christiansen, W. C. Ermler, K. E. Pitzer	36:407–32
High Resolution Electron Microscopy	J. M. Cowley	38:57–88

MOLECULAR STRUCTURE

Transition Metal Molecules	W. Weltner, Jr., R. J. Van Zee	35:291–327
Clusters: Properties and Formation	A. W. Castleman, Jr., R. G. Keesee	37:525–50
Molecular Modeling	P. Kollman	38:303–16

PHYSICAL ORGANIC

Picosecond Organic Photochemistry	K. Peters	38:253–70
Synchronicity in Multibond Reactions	W. T. Borden, R. J. Loncharich, K. N. Houk	39:213–36

PHYSICAL PHENOMENA—MISCELLANEOUS

Vibrational Circular Dichroism	P. J. Stephens, M. A. Lowe	36:213–41

POLYMERS AND MACROMOLECULES

Stiff-Chain Macromolecules	H. Yamakawa	35:23–47
Simulation of Polymer Motion	A. Baumgärtner	35:419–35
Theory of Chain Packing in Amphiphilic Aggregates	A. Ben-Shaul, W. M. Gelbart	36:179–211
ESR and ENDOR of Conducting Polymers	C. L. Young, D. Whitney, A. I. Vistnes, L. R. Dalton	37:459–91
Dynamic Light Scattering Near the Glass Transition	G. D. Patterson, A. Munoz-Rojas	38:191–210
Experimental Studies of Polymer Concentration Profiles at Solid-Liquid and Liquid-Gas Interfaces by Optical and X-Ray Evanescent Wave Techniques	F. Rondelez, D. Ausserré, H. Hervet	38:317–47
Chain Molecules at High Densities at Interfaces	K. A. Dill, J. Naghizadeh, J. A. Marqusee	39:425–61

CHAPTER TITLES

PREFATORY CHAPTERS

When Polymer Science Looked Easy	W. H. Stockmayer, B. H. Zimm	35:1–21
Molecular Spectroscopy: A Personal History	G. Herzberg	36:1–30
The Journal of Chemical Physics: The First 50 Years	J. W. Stout	37:1–23
Of Physical Chemistry and Other Activities	K. S. Pitzer	38:1–25
50 Years of Physical Chemistry, a Personal Account	S. W. Benson	39:1–37

QUANTUM CHEMISTRY

Ab Initio Vibrational Force Fields	G. Fogarasi, P. Pulay	35:191–213
Effective Potentials in Molecular Quantum Chemistry	M. Krauss, W. J. Stevens	35:357–85
Molecular Dynamics Beyond the Adiabatic Approximation: New Experiments and Theory	R. L. Whetten, G. S. Ezra, E. R. Grant	36:277–320
On the Simulation of Quantum Systems: Path Integral Methods	B. J. Berne, D. Thirumalai	37:401–24
Theoretical Studies of Silicon Chemistry	K. K. Baldridge, J. A. Boatz, S. Koseki, M. S. Gordon	38:211–52
Ab Initio Studies of Transition Metal Systems	S. R. Langhoff, C. W. Bauschlicher, Jr.	39:181–212

QUANTUM MECHANICS

Quantum Ergodicity and Spectral Chaos	E. B. Stechel, E. J. Heller	35:563–89

SCATTERING PHENOMENA—DYNAMICAL

Dynamic Light Scattering Near the Glass Transition	G. D. Patterson, A. Munoz-Rojas	38:191–210

SCATTERING PHENOMENA—STRUCTURAL

Small Angle Neutron Scattering Studies of the Structure and Interaction in Micellar and Microemulsion Systems	S. H. Chen	37:351–99
Experimental Studies of Polymer Concentration Profiles at Liquid-Solid and Liquid-Gas Interfaces by Optical and X-Ray Evanescent Wave Techniques	F. Rondelez, D. Ausserré, H. Hervet	38:317–47
Three-Dimensional X-Ray Crystallography of Membrane Proteins: Insights Into Electron Transfer	D. E. Budil, P. Gast, C.-H. Chang, M. Schiffer, J. R. Norris	38:561–83

SOLIDS AND ORDERED ARRAYS—STRUCTURE AND DYNAMICS

Electronic Processes in Organic Solids	M. Pope, C. E. Swenberg	35:613–55
The Metal-Insulator Transition	R. F. Milligan, G. A. Thomas	36:139–58
Computer Simulation Studies of Solids	M. L. Klein	36:525–48
Interatomic Potentials in Solid State Chemistry	A. M. Stoneham, J. H. Harding	37:53–80
Optical Phonon Dynamics in Molecular Crystals	D. D. Dlott	37:159–87
Theory of the Equilibrium Liquid-Solid Transition	A. D. J. Haymet	38:89–108
Physical and Chemical Properties of Alkalides and Electrides	J. L. Dye, M. G. DeBacker	38:271–301
Three-Dimensional X-Ray Crystallography of Membrane Proteins: Insights into Electron Transfer	D. E. Budil, P. Gast, C.-H. Chang, M. Schiffer, J. R. Norris	38:561–83
Lattice Vibrations and Heat Transport in Crystals and Glasses	D. G. Cahill, R. O. Pohl	39:93–121

Recent Developments in Dynamical Theories of the Liquid-Glass Transition	G. H. Fredrickson	39:149–80

SPECTROSCOPY—ELECTRONIC AND PHOTOELECTRONIC

Effects of Saturation on Laser-Induced Fluorescence Measurements of Population and Polarization	R. Altkorn, R. N. Zare	35:265–89
Subpicosecond Spectroscopy	G. R. Fleming	37:81–104
Rydberg Molecules	G. Herzberg	38:27–55
Pyrazine: An "Exact" Solution to the Problem of Radiationless Transitions	J. Kommandeur, W. A. Majewski, W. L. Meerts, D. W. Pratt	38:433–62
Laser Spectroscopy of Large Polyatomic Molecules in Supersonic Jets	M. Ito, T. Ebata, N. Mikami	39:123–47
Theory of Pure Dephasing in Crystals	J. L. Skinner	39:463–78

SPECTROSCOPY—VIBRATIONAL

Velocity Modulation Infrared Laser Spectroscopy of Molecular Ions	C. S. Gudeman, R. J. Saykally	35:387–418
The Electromagnetic Theory of Surface Enhanced Spectroscopy	H. Metiu, P. Das	35:507–36
High Resolution Infrared Studies of Molecular Dynamics	E. Hirota, K. Kawaguchi	36:53–76
Raman Spectroscopy of Molecules Adsorbed on Solid Surfaces	A. Campion	36:549–72
Nonlinear Raman Spectroscopy of Gases	J. W. Nibler, J. J. Yang	38:349–81
Vibrational Raman Spectra of Simple Fluids	M. J. Clouter	39:69–91
UV Resonance Raman Studies of Molecular Structure and Dynamics: Applications in Physical and Biophysical Chemistry	S. A. Asher	39:537–88
Vibrational Spectroscopic Studies of the Structure of Species Derived from the Chemisorption of Hydrocarbons on Metal Single-Crystal Surfaces	N. Sheppard	39:589–644

STATISTICAL MECHANICS

Thermodynamic Behavior of Fluids Near the Critical Point	J. V. Sengers, J. M. H. Levelt Sengers	37:189–222
Theory of the Equilibrium Liquid-Solid Transition	A. D. J. Haymet	38:89–108
Fractal Time in Condensed Matter	M. F. Shlesinger	39:269–90
Chain Molecules at High Densities at Interfaces	K. A. Dill, J. Naghizadeh, J. A. Marqusee	39:425–61
Theory of Pure Dephasing in Crystals	J. L. Skinner	39:463–78

SURFACES—ADSORPTION AND CATALYSIS

Catalytic Studies with Metal Single Crystals	D. W. Goodman	37:425–57
Mechanistic Studies of Chemical Vapor Depostion	J. M. Jasinski, B. S. Meyerson, B. A. Scott	38:109–40
Dissociative Chemisorption: Dynamics and Mechanisms	S. T. Ceyer	39:479–510

SURFACES—STRUCTURE AND DYNAMICS

Electron Energy Loss Spectroscopy in the Study of Surfaces	P. Avouris, J. Demuth	35:265–89
Characterization of Surfaces Through Electron and Photon Stimulated Desorption	T. E. Madey, D. E. Ramaker, R. Stockbauer	35:215–40
Electronic Properties of Surfaces	M. L. Cohen, S. G. Louie	35:537–62

Laser Probing of Molecules Desorbing and Scattering from Solid Surfaces	M. C. Lin, G. Ertl	37:587–615
Probing Phenomena at Metal Surfaces by NMR	C. P. Slichter	37:25–51
Recent Developments in the Theory of Surface Diffusion	J. D. Doll, A. F. Voter	38:413–31
The Semiclassical Way to Molecular Dynamics at Surfaces	J. W. Gadzuk	39:395–424
Vibrational Spectroscopic Studies of the Structure of Species Derived from the Chemisorption of Hydrocarbons on Metal Single-Crystal Surfaces	N. Sheppard	39:589–644

THERMOCHEMISTRY AND THERMODYNAMICS

Biochemical Applications of Differential Scanning Calorimetry	J. M. Sturtevant	38:463–88

Annual Reviews Inc.
A NONPROFIT SCIENTIFIC PUBLISHER

4139 El Camino Way
P.O. Box 10139
Palo Alto, CA 94303-0897 • USA

ORDER FORM
Now you can order
TOLL FREE
1-800-523-8635
(except California)

Annual Reviews Inc. publications may be ordered directly from our office by mail or use our Toll Free Telephone line (for orders paid by credit card or purchase order, and customer service calls only); through booksellers and subscription agents, worldwide; and through participating professional societies. Prices subject to change without notice.   ARI Federal I.D. #94-1156476

- **Individuals:** Prepayment required on new accounts by check or money order (in U.S. dollars, check drawn on U.S. bank) or charge to credit card — American Express, VISA, MasterCard.
- **Institutional buyers:** Please include purchase order number.
- **Students:** $10.00 discount from retail price, per volume. Prepayment required. Proof of student status must be provided (photocopy of student I.D. or signature of department secretary is acceptable. Students must send orders direct to Annual Reviews. Orders received through bookstores and institutions requesting student rates will be returned. You may order at the Student Rate for a maximum of 3 years.
- **Professional Society Members:** Members of professional societies that have a contractual arrangement with Annual Reviews may order books through their society at a reduced rate. Check with your society for information.
- **Toll Free Telephone orders:** Call 1-800-523-8635 (except from California) for orders paid by credit card or purchase order and customer service calls only. California customers and all other business calls use 415-493-4400 (not toll free). Hours: 8:00 AM to 4:00 PM, Monday-Friday, Pacific Time.

Regular orders: Please list the volumes you wish to order by volume number.
Standing orders: New volume in the series will be sent to you automatically each year upon publication. Cancellation may be made at any time. Please indicate volume number to begin standing order.
Prepublication orders: Volumes not yet published will be shipped in month and year indicated.
California orders: Add applicable sales tax.
Postage paid (4th class bookrate/surface mail) by **Annual Reviews Inc.** Airmail postage or UPS, extra.

ANNUAL REVIEWS SERIES		Prices Postpaid per volume USA & Canada/elsewhere	Regular Order Please send:	Standing Order Begin with:
			Vol. number	Vol. number
Annual Review of ANTHROPOLOGY				
Vols. 1-14	(1972-1985)	$27.00/$30.00		
Vols. 15-16	(1986-1987)	$31.00/$34.00		
Vol. 17	(avail. Oct. 1988)	$35.00/$39.00	Vol(s). _____	Vol. _____
Annual Review of ASTRONOMY AND ASTROPHYSICS				
Vols. 1-2, 4-20	(1963-1964; 1966-1982)	$27.00/$30.00		
Vols. 21-25	(1983-1987)	$44.00/$47.00		
Vol. 26	(avail. Sept. 1988)	$47.00/$51.00	Vol(s). _____	Vol. _____
Annual Review of BIOCHEMISTRY				
Vols. 30-34, 36-54	(1961-1965; 1967-1985)	$29.00/$32.00		
Vols. 55-56	(1986-1987)	$33.00/$36.00		
Vol. 57	(avail. July 1988)	$35.00/$39.00	Vol(s). _____	Vol. _____
Annual Review of BIOPHYSICS AND BIOPHYSICAL CHEMISTRY				
Vols. 1-11	(1972-1982)	$27.00/$30.00		
Vols. 12-16	(1983-1987)	$47.00/$50.00		
Vol. 17	(avail. June 1988)	$49.00/$53.00	Vol(s). _____	Vol. _____
Annual Review of CELL BIOLOGY				
Vol. 1	(1985)	$27.00/$30.00		
Vols. 2-3	(1986-1987)	$31.00/$34.00		
Vol. 4	(avail. Nov. 1988)	$35.00/$39.00	Vol(s). _____	Vol. _____

ANNUAL REVIEWS SERIES	Prices Postpaid per volume USA & Canada/elsewhere	Regular Order Please send:	Standing Order Begin with:
		Vol. number	Vol. number

Annual Review of COMPUTER SCIENCE
- Vols. 1-2   (1986-1987)................$39.00/$42.00
- Vol. 3   (avail. Nov. 1988).............$45.00/$49.00   Vol(s). _____   Vol. _____

Annual Review of EARTH AND PLANETARY SCIENCES
- Vols. 1-10   (1973-1982)................$27.00/$30.00
- Vols. 11-15   (1983-1987)................$44.00/$47.00
- Vol. 16   (avail. May 1988).............$49.00/$53.00   Vol(s). _____   Vol. _____

Annual Review of ECOLOGY AND SYSTEMATICS
- Vols. 2-16   (1971-1985)................$27.00/$30.00
- Vols. 17-18   (1986-1987)................$31.00/$34.00
- Vol. 19   (avail. Nov. 1988).............$34.00/$38.00   Vol(s). _____   Vol. _____

Annual Review of ENERGY
- Vols. 1-7   (1976-1982)................$27.00/$30.00
- Vols. 8-12   (1983-1987)................$56.00/$59.00
- Vol. 13   (avail. Oct. 1988).............$58.00/$62.00   Vol(s). _____   Vol. _____

Annual Review of ENTOMOLOGY
- Vols. 10-16, 18-30   (1965-1971; 1973-1985)........$27.00/$30.00
- Vols. 31-32   (1986-1987)................$31.00/$34.00
- Vol. 33   (avail. Jan. 1988).............$34.00/$38.00   Vol(s). _____   Vol. _____

Annual Review of FLUID MECHANICS
- Vols. 1-4, 7-17   (1969-1972, 1975-1985)........$28.00/$31.00
- Vols. 18-19   (1986-1987)................$32.00/$35.00
- Vol. 20   (avail. Jan. 1988).............$34.00/$38.00   Vol(s). _____   Vol. _____

Annual Review of GENETICS
- Vols. 1-19   (1967-1985)................$27.00/$30.00
- Vols. 20-21   (1986-1987)................$31.00/$34.00
- Vol. 22   (avail. Dec. 1988).............$34.00/$38.00   Vol(s). _____   Vol. _____

Annual Review of IMMUNOLOGY
- Vols. 1-3   (1983-1985)................$27.00/$30.00
- Vols. 4-5   (1986-1987)................$31.00/$34.00
- Vol. 6   (avail. April 1988).............$34.00/$38.00   Vol(s). _____   Vol. _____

Annual Review of MATERIALS SCIENCE
- Vols. 1, 3-12   (1971, 1973-1982)............$27.00/$30.00
- Vols. 13-17   (1983-1987)................$64.00/$67.00
- Vol. 18   (avail. August 1988)...........$66.00/$70.00   Vol(s). _____   Vol. _____

Annual Review of MEDICINE
- Vols. 1-3, 6, 8-9   (1950-1952, 1955, 1957-1958)
- 11-15, 17-36   (1960-1964, 1966-1985)........$27.00/$30.00
- Vols. 37-38   (1986-1987)................$31.00/$34.00
- Vol. 39   (avail. April 1988).............$34.00/$38.00   Vol(s). _____   Vol. _____

Annual Review of MICROBIOLOGY
- Vols. 18-39   (1964-1985)................$27.00/$30.00
- Vols. 40-41   (1986-1987)................$31.00/$34.00
- Vol. 42   (avail. Oct. 1988).............$34.00/$38.00   Vol(s). _____   Vol. _____